Formulas/Equations

Distance Formula

If $P_1 = (x_1, y_1)$ and $P_2 = (x_2, y_2)$, the distance from P_1 to P_2 is

$$d(P_1, P_2) = \sqrt{(x_2 - x_1)^2 + (y_2 - y_1)^2}$$

Equation of a Circle

The equation of a circle of radius r with center at (h, k) is

$$(x - h)^2 + (y - k)^2 = r^2$$

Slope Formula

The slope m of the line containing the points $P_1 = (x_1, y_1)$ and $P_2 = (x_2, y_2)$ is

$$m = \frac{y_2 - y_1}{x_2 - x_1} \qquad \text{if } x_1 \neq x_2$$

$$m \text{ is undefined} \qquad \text{if } x_1 = x_2$$

Point–Slope Equation of a Line

The equation of a line with slope m containing the point (x_1, y_1) is

$$y - y_1 = m(x - x_1)$$

Slope–Intercept Equation of a Line

The equation of a line with slope m and y-intercept b is

$$y = mx + b$$

Quadratic Formula

The solutions of the equation $ax^2 + bx + c = 0$, $a \neq 0$, are

$$x = \frac{-b \pm \sqrt{b^2 - 4ac}}{2a}$$

If $b^2 - 4ac > 0$, there are two real unequal solutions.

If $b^2 - 4ac = 0$, there is a repeated real solution.

Geometry Formulas

Circle

r = Radius, A = Area, C = Circumference

$$A = \pi r^2 \qquad C = 2\pi r$$

Triangle

b = Base, h = Altitude (Height), A = area

$$A = \tfrac{1}{2}bh$$

Rectangle

l = Length, w = Width, A = area, P = perimeter

$$A = lw \qquad P = 2l + 2w$$

Rectangular Box

l = Length, w = Width, h = Height, V = Volume

$$V = lwh$$

Sphere

r = Radius, V = Volume, S = Surface area

$$V = \tfrac{4}{3}\pi r^3 \qquad S = 4\pi r^2$$

College Algebra

Enhanced with Graphing Utilities

Michael Sullivan
Chicago State University

Michael Sullivan, III
South Suburban College

Prentice Hall, Upper Saddle River, NJ 07458

Acquisitions Editor: Sally Denlow
Editorial Assistant: Joanne Wendelken
Editor-in Chief of Development: Ray Mullaney
Director of Production and Manufacturing: David W. Riccardi
Managing Editor: Linda Behrens
Production Editor: Robert C. Walters
Marketing Manager: Jolene L. Howard
Marketing Assistant: William Paquin
Copy Editor: William Thomas
Interior Designer: Rosemarie Votta
Cover Designer: Rosemarie Votta
Creative Director: Paula Maylahn
Art Director: Amy Rosen
Assistant to Art Director: Rod Hernandez
Manufacturing Manager: Alan Fischer
Photo Researcher: Kathy Ringrose
Photo Editor: Lorinda Morris-Nantz
Supplements Editor: Audra J. Walsh

© 1996 by Prentice-Hall, Inc.
Simon & Schuster/A Viacom Company
Upper Saddle River, NJ 07458

Printed in the United States of America

10 9 8 7 6 5 4 3 2

ISBN 0-02-343751-0

PRENTICE-HALL INTERNATIONAL (UK) LIMITED, LONDON
PRENTICE-HALL OF AUSTRALIA PTY. LIMITED, SYDNEY
PRENTICE-HALL CANADA INC. TORONTO
PRENTICE-HALL HISPANOAMERICANA, S.A., MEXICO
PRENTICE-HALL OF INDIA PRIVATE LIMITED, NEW DELHI
PRENTICE-HALL OF JAPAN, INC., TOKYO
SIMON & SCHUSTER ASIA PTE. LTD., SINGAPORE
EDITORA PRENTICE-HALL DO BRASIL, LTDA., RIO DE JANEIRO

In Memory of Mary—
Wife and Mother

Contents

Preface to the Instructor   ix
Preface to the Student   xvii

CHAPTER 1   GRAPHS   1

1.1   Rectangular Coordinates; Graphing Utilities   2
1.2   Graphs of Equations   14
1.3   The Straight Line   35
1.4   Parallel and Perpendicular Lines; Circles   51
1.5   Linear Curve Fitting   64
1.6   Variation   71
        Chapter Review   78

CHAPTER 2   FUNCTIONS AND THEIR GRAPHS   85

2.1   Functions   86
2.2   More about Functions   102
2.3   Graphing Techniques   123
2.4   Operations on Functions; Composite Functions   135
2.5   One-to-One Functions; Inverse Functions   144
2.6   Mathematical Models: Constructing Functions   155
        Chapter Review   168

CHAPTER 3   EQUATIONS AND INEQUALITIES   173

3.1   Solving Equations Using a Graphing Utility   174
3.2   Linear and Quadratic Equations   179
3.3   Setting Up Equations: Applications   194
3.4   Complex Numbers; Quadratic Equations with a Negative
        Discriminant   206

3.5   Other Types of Equations   215
3.6   Inequalities   222
3.7   Other Inequalities   234
3.8   Equations and Inequalities Involving Absolute Value   243
        Chapter Review   250

CHAPTER 4   POLYNOMIAL AND RATIONAL FUNCTIONS   255

4.1   Quadratic Functions   256
4.2   Polynomial Functions   274
4.3   Rational Functions   285
4.4   The Zeros of a Polynomial Function   302
4.5   Complex Polynomials; Fundamental Theorem of Algebra   318
        Chapter Review   322

CHAPTER 5   EXPONENTIAL AND LOGARITHMIC FUNCTIONS   327

5.1   Exponential Functions   328
5.2   Logarithmic Functions   341
5.3   Properties of Logarithms   351
5.4   Logarithmic and Exponential Equations   358
5.5   Compound Interest   366
5.6   Growth and Decay   376
5.7   Nonlinear Curve Fitting   382
5.8   Logarithmic Scales   397
        Chapter Review   401

CHAPTER 6   THE CONICS; VECTORS   407

6.1   Preliminaries   408
6.2   The Parabola   409
6.3   The Ellipse   422
6.4   The Hyperbola   437
6.5   Vectors   454
        Chapter Review   465

CHAPTER 7   SYSTEMS OF EQUATIONS AND INEQUALITIES   469

7.1   Systems of Linear Equations: Substitution; Elimination   470
7.2   Systems of Linear Equations: Matrices   483
7.3   Systems of Linear Equations: Determinants   501
7.4   Matrix Algebra   513
7.5   Partial Fraction Decomposition   532
7.6   Systems of Nonlinear Equations   539
7.7   Systems of Inequalities   552
7.8   Linear Programming   562
        Chapter Review   569

CHAPTER 8 SEQUENCES; INDUCTION; COUNTING; PROBABILITY 577

8.1 Sequences 578
8.2 Arithmetic Sequences 588
8.3 Geometric Sequences; Geometric Series 593
8.4 Mathematical Induction 603
8.5 The Binomial Theorem 607
8.6 Sets and Counting 615
8.7 Permutations and Combinations 621
8.8 Probability 630
Chapter Review 640

APPENDIX REVIEW 647

1. Topics From Algebra and Geometry 647
2. Polynomial and Rational Expressions 659
3. Radicals; Rational Exponents 673
4. Solving Equations 679
5. Completing the Square 683
6. Synthetic Division 686

ANSWERS AN1

INDEX I1

Preface

To the Instructor

Times certainly are changing. This text represents two generations of mathematics learning and teaching in the Sullivan family.

As the father, I have been teaching at an urban public university for over 30 years. Many things have changed during those years. . . . Students today have many varied needs—some students have very little mathematical background and a fear of the subject, while others are extremely motivated and have a strong educational background. Many are trying to balance family, work and school. All of these added pressures enter into the classroom. To be able to reach these students and excite them about the subject I love is truly wonderful. Technology, too, is changing. For me, as with many of my colleagues, the move has been from slide rules to arithmetic/scientific calculators to graphing calculators and computer algebra systems. As a result, our careers have been exciting, challenging, and, most importantly, filled with a wide range of learning experiences.

As the son, I grew up with teaching and learning all around me. Having completed dual advanced degrees in economics and mathematics has impacted the way I think about college courses . . . applications and skills blend together. Unlike my father, technology was in every aspect of my college career. Now as a mathematics teacher at a community college, I have the opportunity to see the strengths of the traditional foundations of mathematics combined with the exploratory nature of technology and the power of real world applications. For many of my students, the visual nature of mathematics through technology compels them to become more active in the classroom. Being able to use real world data, do experiments, and work in groups has made learning interesting again to them.

Welcome to College Algebra: Enhanced with Graphing Utilities. We hope you enjoy using it as much as we do.

Michael Sullivan

Michael Sullivan, III

Our Philosophy on Technology

MAA, AMATYC, and NCTM all recommend the appropriate use of technology in the classroom and the mathematics laboratory. Appropriate is the difficult word to define. Our text assumes immediate access to graphing utilities by both students and instructors. We fully utilize the graphers ability to promote visualization, exploration, foreshadowing of concepts, data analysis, curve fitting and most of all, excitement in the classroom. We have not, however, thrown out all hand work. By blending together technology and traditional methods of problem solving, students are empowered with knowledge and ability that they will use throughout their lives.

How we integrate technology through the text is subtle—no big boxes, no icons. You will see both T1-82 screens and traditional line art where appropriate. Many examples are solved in two ways: with a graphing solution and with an algebraic solution. Many exercises are multi-tasked, beginning with questions that require the use of algebra and ending with questions that require not just a graph, but also an analysis of the graph. We encourage students to distinguish between solving a given problem using the full power of the technology vs. recognizing when the technology is limited or the simple nature of the problem makes a solution by hand the better choice.

One very effective use of a graphing utility is to develop the ability of students to recognize patterns. With a graphing utility, students can quickly and effectively start to recognize patterns; they can actually 'see' the mathematics, resulting in a better conceptual understanding of the concept at hand. Throughout the text, we have students graph related functions on the same screen, asking them leading questions to help them recognize patterns.

Another strong reason to use the technologies available is to move away from contrived and 'simple' numbers. Students can work with real information involving 'messy' numbers in a way that traditionally they could not do. Without taking away from the conceptual understanding of the objective at hand, students can analyze and interpret real data. Using this premise, we have included problems using real world data, as well as CBL exercises where students can actively collect their own data and manipulate it.

Technology can also be used to introduce ideas and solve problems that go beyond the traditional limitations of college algebra. For example, many concepts and problems typically covered in a calculus class can be investigated and solved using graphing utilities. Determining where a function is increasing and decreasing and where its local maxima and minima occur are topics that can now be introduced, discussed, and solved in College Algebra, thanks to the technology. Curve fitting is another topic that can be introduced, discussed, and solved, enabling students to relate algebra skills with data analysis.

These are a few of the uses made of technology in *College Algebra: Enhanced with Graphing Utilities*. A graphing utility is required for the student to use this text. If you or your school would like to use graphing utilities, but not require them, *College Algebra,* 4th Edition, by Michael Sullivan, which provides an optional use of graphing utilities, is also available. To receive an examination copy, contact your Prentice Hall representative.

About This Book

Content The content of this book is not different than that of a traditional College Algebra—but the emphasis and the approach is! For example, we approach the idea of Horizontal Shifts by first exploring the graphs of several functions using a graphing utility and then drawing a conclusion about what is happening. Also, many Examples and Exercises found in this text are ones that cannot be handled using only traditional algebra methods.

Organization With the use of fully integrated technology, the order of presentation of certain topics shifts.

To emphasize the focus of the book, Chapter 1 begins with graphing. In recognition of the fact that some students may not be fully prepared to start here, an Appendix has been carefully designed to provide necessary review. The sections of the Appendix that apply are referenced at the beginning of a Chapter and also within the body of the chapter.

Chapter 1 treats the graphs of certain key equations, including the line and the circle. A section on Linear Curve Fitting demonstrates the power of a graphing utility to solve applied problems in this area.

Chapter 2 develops the concept of a function, the graphs of functions, and properties of functions. Chapter 3 solves equations and inequalities utilizing both graphing and algebraic techniques.

Chapter 4 discusses Polynomial and Rational Functions, with an emphasis on analyzing their graphs. Chapter 5, Exponential and Logarithmic Functions, also places emphasis on the analysis of graphs, but, in addition, contains an entire section devoted to Nonlinear Curve Fitting.

Chapter 6, Conics, is mostly traditional, as is Chapter 8, Sequences; Induction; Counting; Probability. Chapter 7, Systems of Equations and Inequalities, discusses both graphing and algebraic approaches. Many examples and exercises found here can only be solved using technology.

Examples Recognizing that many students learn through working problems by hand as well as through the use of a graphing utility, we have provided both traditional and technological examples of solving problems. Examples are worked out in appropriate detail, starting with simple, reasonable problems and working gradually up to more challenging ones. There are no magic steps and we encourage and often show the final step of checking the result. Many of the examples involve applications that will be seen in calculus or in other disciplines. At the end of many examples there are "Now Work" suggestions which refer students to an odd numbered problem in section exercises which is similar to the worked out example shown. This allows students immediate feedback on whether they understand the concept clearly before moving on.

Exercises The exercises in the text are mostly of three types: visual-where students are asked to draw conclusions about a graph; technological—where the students will fully utilize the power of their graphing utility; and open-ended—where critical thinking, writing, research or collaborative effort is required. The text contains over 4500 tested and true exercises, over 800 are applied problems. Exercise sets begin with problems designed to build confidence, continue with problems which relate to worked out examples in the text and conclude with problems that are more challenging. Many of the problems, especially those at the beginning are visual in character—such as showing a graph and asking for conclusions.

Answers are given in the back of the text for all the odd-numbered problems. Fully worked out, step by step solutions for the odd-numbered problems are found in the **Student's Solutions Manual** while the even solutions are equally worked out in the **Instructor's Solutions Manual.**

Illustrations The design uses color effectively and functionally to help the student identify definitions, theorems, formulas and procedures. Included in the Preface To the Student is an overview of how these colorized elements can be helpful when studying. Many illustrations have been included to provide a dynamic realism to selected examples and exercises. All the graphing utility and line art has been computer generated for consistency and accuracy. For the purpose of clarity, all line art utilizes two colors. With over 1400 pieces of art throughout the text we aim to help the student to visualize mathematics and show how to use it to solve mathematical problems.

Applications Every opportunity has been taken to present understandable, realistic applications consistent with the abilities of the student, drawing from such sources as tax rate tables, the Guinness Book of World Records, Government publication and newspaper articles. For added interest, some of the applied exercises have been adapted from textbooks the students may be using in other courses (such as economics, chemistry, physics, etc.)

Communication and Problem Solving The recommendations of the NCTM, AMATYC and MAA support the inclusion of writing, verbalizing, research and critical thinking in mathematics. Throughout the text there are many ways we encourage these activities. In the exercises, these kind of problems are designated by a ![icon]. All of these problems not only ask the student to write, discuss or to do active problem solving, but they also drive forward the concepts of the section.

Collaborative Projects In each chapter, a full page "Mission Possible" has been devoted to collaborative learning. These multi-tasked projects, written by Hester Lewellen, one of the co-authors of the University of Chicago High School Mathematics Project, will help your students work together to solve some unusual problems. Some of the projects require utilizing the graphing utility but all require critical thinking and communication. Suggested solutions to the collaborative projects are found in the Instructor's Solutions Manual.

Calculus While many of your students may not be going on to calculus after this course, some may. To encourage students and to let them preview calculus, we have included many examples and exercises that foreshadow topics found in calculus.

CBL Projects In key places in the text, CBL (Calculator Based Laboratory) experiments are given in the exercises. Each demonstrates real world applications of the topics just covered. Students are asked to perform experiments and collect and analyze the data obtained.

Instructor's Supplementary Aids

Instructor's Solutions Manual Contains complete step-by-step worked out solutions to all the even numbered exercises in the textbook. Also included are strategies for using the collaborative learning projects found in each chapter. Transparency masters which duplicate important illustrations in the text can be found as well.

Video Review A new videotape series which has been created to accompany the Sullivan texts include a half hour review of the most important topics in each chapter. Each segment uses both traditional and technological ways of solving mathematical problems. Entertaining and educational, these videos provide an alternative process which can add to your students' success in this course. Also included is a graphing calculator video tutorial which walks one through the most common uses of the various calculators. Written by a mathematics teachers, these videos concentrate on those topics that 'get them every time'.

Written Test Item File Features six tests per chapter plus four forms of a final examination, prepared and ready to be photocopied. Of these tests, three are multiple choice and three are free response.

Prentice Hall Custom Test (Computerized Testing Generator - Mac and Windows) PH Custom Test is a fully networkable, easy to use test generator. Instructors may select questions by objective, section, chapter or use a ready-made test for each chapter. As the questions are algorithmically generated, an instructor can create up to 99 versions of each question while keeping the problem type and objective constant. PH Custom Test allows for on-line testing and offers a gradebook feature that not only organizes test grades but can be used for any other classroom grades or information. Instructors can download questions into word processing programs or create their own problems and insert them into PH Custom Test. Graphics and mathematical symbols are integrated into the program and other graphics from programs such as Mathematica, Maple, Derive or Matlab can be imported into the program.

Student's Supplementary Aids

Student's Solutions Manual Contains complete step-by-step worked out solutions to all the odd numbered exercises in the textbook. This is terrific for getting instant feedback for your students.

Prentice Hall Tutorial Program (Software tutorial both Mac and Windows) This computerized tutorial program is based on the highly successful Prentice Hall testing program. Utilizing the same algorithms programmed for the test generator, students are able to pretest their abilities, receive a diagnostic recommendation for further study, work through a step-by-step tutorial—complete with a graphing utility built into the program, and tutorial tests. The software is fully networkable and can be used with the gradebook feature in the Prentice Hall Custom Test.

Visual Precalculus A software package for IBM compatible computers which consists of two parts. Part One contains routines to graph and evaluate functions, graph conic sections, investigate series, carry out synthetic division and illuminate important concepts with animation. Part Two contains routines to solve triangles, graph systems of linear equations and inequalities, evaluate matrix expressions, apply Gaussian elimination to reduce or invert matrices and graphically solve linear programming problems. Those routines will provide additional insights to the material covered within the text.

X(PLORE) A powerful (yet inexpensive) fully programmable symbolic and numeric mathematical processor for IBM and Macintosh computers. This program will allow your students to evaluate expressions, graph curves, solve equations and use matrices. This software package may also be used for calculus or differential equations.

New York Times Supplement A free newspaper from Prentice Hall and the New York Times which includes interesting and current articles on mathematics in the world around us. Great for getting students to talk and write about mathematics. This supplement is created new each year.

For any of the above supplements, please contact your Prentice Hall representative.

Acknowledgments

Textbooks are written by an author, but evolve from an idea into final form through the efforts of many people. Special thanks to Don Dellen, who first suggested this book and the other books in this series.

There are many people we would like to thank for their input, encouragement, patience and support. They have our deepest thanks and appreciation. We apologize for any omissions . . .

James Africh, College of Du Page
Steve Agronsky, Cal Poly State University
Dave Anderson, South Suburban College
Joby Milo Anthony, University of Central Florida
James E. Arnold, University of Wisconsin, Milwaukee
Agnes Azzolino, Middlesex County College
Wilson P. Banks, Illinois State University
Dale R. Bedgood, East Texas State University
Beth Beno, South Suburban College
William H. Beyer, University of Akron
Richelle Blair, Lakeland Community College
Trudy Bratten, Grossmont College
William J. Cable, University of Wisconsin—Stevens Point
Lois Calamia, Brookdale Community College
Roger Carlsen, Moraine Valley Community College
John Collado, South Suburban College
Denise Corbett, East Carolina University
Theodore C. Coskey, South Seattle Community College
John Davenport, East Texas State University
Duane E. Deal, Ball State University
Vivian Dennis, Eastfield College
Karen R. Dougan, University of Florida
Louise Dyson, Clark College
Paul D. East, Lexington Community College
Don Edmondson, University of Texas, Austin
Christopher Ennis, University of Minnesota
Garret J. Etgen, University of Houston
W. A. Ferguson, University of Illinois, Urbana/Champaign
Iris B. Fetts, Clemson University

Mason Flake, student at Edison Community College
Merle Friel, Humboldt State University
Richard A. Fritz, Moraine Valley Community College
Carolyn Funk, South Suburban College
Dewey Furness, Ricke College
Wayne Gibson, Rancho Santiago College
Joan Goliday, Santa Fe Community College
Frederic Gooding, Goucher College
Ken Gurganus, University of North Carolina
James E. Hall, University of Wisconsin, Madison
Judy Hall, West Virginia University
Edward R. Hancock, DeVry Institute of Technology
Brother Herron, Brother Rice High School
Kim Hughes, California State College, San Bernardino
Ron Jamison, Brigham Young University
Richard A. Jensen, Manatee Community College
Sandra G. Johnson, St. Cloud State University
Moana H. Karsteter, Tallahassee Community College
Arthur Kaufman, College of Staten Island
Thomas Kearns, North Kentucky University
Keith Kuchar, Manatee Community College
Tor Kwembe, Chicago State University
Linda J. Kyle, Tarrant County Jr. College
H. E. Lacey, Texas A & M University
Christopher Lattin, Oakton Community College
Adele LeGere, Oakton Community College
Stanley Lukawecki, Clemson University
Virginia McCarthy, Iowa State University
James McCollow, DeVry Institute of Technology
Laurence Maher, North Texas State University
James Maxwell, Oklahoma State University, Stillwater
Carolyn Meitler, Concordia University
Eldon Miller, University of Mississippi
James Miller, West Virginia University
Michael Miller, Iowa State University
Jane Murphy, Middlesex Community College
Bill Naegele, South Suburban College
James Nymann, University of Texas, El Paso
Sharon O'Donnell, Chicago State University
Seth F. Oppenheimer, Mississippi State University
E. James Peake, Iowa State University
Thomas Radin, San Joaquin Delta College
Ken A. Rager, Metropolitan State College
Elsi Reinhardt, Truckee Meadows Community College
Jane Ringwald, Iowa State University
Stephen Rodi, Austin Community College
Howard L. Rolf, Baylor University
Edward Rozema, University of Tennessee at Chattanooga
Dennis C. Runde, Manatee Community College
John Sanders, Chicago State University
Susan Sandmeyer, Jamestown Community College
A.K. Shamma, University of West Florida
Martin Sherry, Lower Columbia College
Timothy Sipka, Alma College
John Spellman, Southwest Texas State University
Becky Stamper, Western Kentucky University
Neil Stephens, Hinsdale South High School
Diane Tesar, South Suburban College
Tommy Thompson, Brookhaven College
Richard J. Tondra, Iowa State University
Marvel Townsend, University of Florida
Jim Trudnowski, Carroll College
Richard G. Vinson, University of Southern Alabama
Darlene Whitkenack, Northern Illinois University
Chris Wilson, West Virginia University
Carlton Woods, Auburn University
George Zazi, Chicago State University

Recognition and thanks are due particularly to the following individuals for their valuable assistance in the preparation of this edition: Jerome Grant for his support and commitment; Sally Denlow, for her genuine interest and insightful direction; Bob Walters for his organizational skill as production supervisor; Jolene Howard and Evan Hanby for their innovative marketing efforts; Ray Mullaney for his specific editorial comments; the entire Prentice-Hall sales staff for their confidence; and to Katy Murphy for checking the answers to all the exercises.

Preface

As you begin your study of college algebra, you might feel overwhelmed by the number of theorems, definitions, procedures and equations that confront you. You may even wonder whether you can learn all this material in a single course. For many of you, this may be your last mathematics course, while for others, just the first in a series of many. Don't worry—either way, this text was written with you in mind.

This text was designed to help you—the student, master the terminology and basic concepts of college algebra. These aims have helped to shape every aspect of the book. Many learning aids are built into the format of the text to make your study of this material easier and more rewarding. This book is meant to be a "machine for learning," one that can help you to focus your efforts and get the most from the time and energy you invest.

This book requires that you have access to a graphing utility: a graphing calculator or a computer software package that has a graphing component. Be sure you have some familiarity with the device you are using before the course begins.

Here are some hints we give our students at the beginning of the course:

1. Take advantage of the feature PREPARING FOR THIS CHAPTER. At the beginning of each chapter, we have prepared a list of topics to review. Be sure to take the time to do this. It will help you proceed quicker and more confidently through the chapter.
2. Read the material in the book before the lecture. Knowing what to expect and what is in the book, you can take fewer notes and spend more time listening and understanding the lecture.
3. After each lecture, rewrite your notes as you re-read the book, jotting down any additional facts that seem helpful. Be sure to do the Now Work Problem x as you proceed through a section. After completing a section, be sure to do the assigned problems. Answers to the Odd ones are in the back of the book.
4. If you are confused about something, visit your instructor during office hours immediately, before you fall behind. Bring your attempted solutions to problems with you to show your instructor where you are having trouble.
5. To prepare for an exam, review your notes. Then proceed through the Chapter Review. It contains a capsule summary of all the important material of the chapter. If you are uncertain of any concept, go back into the chapter and study it further. Be sure to do the Review Exercises for practice.

Remember the two "golden rules" of college algebra:

1. DON'T GET BEHIND! The course moves too fast, and it's hard to catch up.
2. WORK LOTS OF PROBLEMS. Everyone needs to practice, and problems show where you need more work. If you can't solve the homework problems without help, you won't be able to do them on exams.

We encourage you to examine the following overview for some hints on how to use this text.

Best Wishes!

Michael Sullivan
Michael Sullivan, III

OVERVIEW

Chapter 2

PREPARING FOR THIS CHAPTER

Before getting started on this chapter, review the following concepts;
Topics from Algebra and Geometry (Appendix, Section 1)
Graphs of certain equations (Example 2, p. 15;
Example 4, p. 17; Example 5, p. 18; Example 16, p. 30)
Tests for symmetry of an equation (p. 28)
Procedure for finding intercepts of an equation (p. 22)

FUNCTIONS AND THEIR GRAPHS

2.1 Functions
2.2 More about Functions
2.3 Graphing Techniques
2.4 Operations
 on Functions;
 Composite Functions
2.5 One-to-One Functions;
 Inverse Functions
2.6 Mathematical Models:
 Constructing Functions
 Chapter Review

Preview Getting from an Island to Town

An island is 2 miles from the nearest point P on a straight shoreline. A town is 12 miles down the shore from P.
(a) If a person can row a boat at an average speed of 3 miles per hour and the same person can walk 5 miles per hour, express the time T it takes to go from the island to town as a function of the distance x from P to where the person lands the boat.
(b) What is the domain of T?
(c) How long will it take to travel from the island to town if the person lands the boat 4 miles from P?
(d) How long will it take if the person lands the boat 8 miles from P?
(e) Use a graphing utility to graph the function T = T(x).
(f) Use the TRACE function to see how the time T varies as x changes from 0 to 12.
(g) What value of x results in the least time?
[Example 9 in Section 2.1] ■

Each chapter begins with a list of concepts to review. Refresh your memory by turning to the pages listed.

A section by section outline is also provided. This is a good way to organize your notes for studying.

Highlighting an application from the chapter, you are given a "Preview" of a coming use of the mathematics introduced in this chapter.

Most chapters open with a brief historical discussion of "where this material came from". It is helpful to understand how others created and used these ideas to solve their everyday problems.

New terms appear in boldface type where they are defined.

he idea of using a system of rectangular coordinates dates back to ancient times, when such a system was used for surveying and city planning. Apollonius of Perga, in 200 BC, used a form of rectangular coordinates in his work on conics, although this use does not stand out as clearly as it does in modern treatments. Sporadic use of rectangular coordinates continued until the 1600's. By that time, algebra had developed sufficiently so that René Descartes (1596–1650) and Pierre de Fermat (1601–1665) could take the crucial step, which was the use of rectangular coordinates to translate geometry problems into algebra problems, and vice versa. This step was supremely important for two reasons. First, it allowed both geometers and algebraists to gain critical new insights into their subjects, which previously had been regarded as separate but now were seen to be connected in many important ways. Second, the insights gained made possible the development of calculus, which greatly enlarged the number of areas in which mathematics could be applied and made possible a much deeper understanding of these areas. With the advent of technology, in particular, graphing utilities, we are now able not only to visualize the dual roles of algebra and geometry, but we are also able to solve many problems that before this technology required advanced methods.

1.1

Rectangular Coordinates; Graphing Utilities

FIGURE 1

FIGURE 2

We locate a point on the real number line by assigning it a single real number, called the *coordinate of the point*. For work in a two-dimensional plane, we locate points by using two numbers.

We begin with two real number lines located in the same plane: one horizontal and the other vertical. We call the horizontal line the **x-axis,** the vertical line the **y-axis,** and the point of intersection the **origin** O. We assign coordinates to every point on these number lines as shown in Figure 1, using a convenient scale. In mathematics, we usually use the same scale on each axis; in applications, a different scale is often used on each axis.

The origin O has a value of 0 on both the x-axis and the y-axis. We follow the usual convention that points on the x-axis to the right of O are associated with positive real numbers, and those to the left of O are associated with negative real numbers. Those on the y-axis above O are associated with positive real numbers, and those below O are associated with negative real numbers. In Figure 1, the x-axis and y-axis are labeled as x and y, respectively, and we have used an arrow at the end of each axis to denote the positive direction.

The coordinate system described here is called a **rectangular,** or **Cartesian* coordinate system.** The plane formed by the x-axis and y-axis is sometimes called the **xy-plane,** and the x-axis and y-axis are referred to as the **coordinate axes.**

Any point P in the xy-plane can then be located by using an **ordered pair** (x, y) of real numbers. Let x denote the signed distance of P from the y-axis (*signed* in the sense that, if P is to the right of the y-axis, then $x > 0$, and if P is to the left of the y-axis, then $x < 0$); and let y denote the signed distance of P from the x-axis. The ordered pair (x, y), also called the **coordinates** of P, then gives us enough information to locate the point P in the plane.

For example, to locate the point whose coordinates are $(-3, 1)$, go 3 units along the x-axis to the left of O and then go straight up 1 unit. We **plot** this point by placing a dot at this location. See Figure 2, in which the points with coordinates $(-3, 1)$, $(-2, -3)$, $(3, -2)$, and $(3, 2)$ are plotted.

The origin has coordinates $(0, 0)$. Any point on the x-axis has coordinates of the form $(x, 0)$, and any point on the y-axis has coordinates of the form $(0, y)$.

If (x, y) are the coordinates of a point P, then x is called the **x-coordinate,** or **abscissa,** of P and y is the **y-coordinate,** or **ordinate,** of P. We identify the

*Named after René Descartes (1596–1650), a French mathematician, philosopher, and theologian.

2

Although each situation has its own unique features, we can provide an outline of the steps to follow in setting up applied problems.

Steps for Setting Up Applied Problems

STEP 1: Read the problem carefully, perhaps two or three times. Pay particular attention to the question being asked in order to identify what you are looking for. If you can, determine realistic possibilities for the answer.

STEP 2: Assign a letter (variable) to represent what you are looking for, and, if necessary, express any remaining unknown quantities in terms of this variable.

STEP 3: Make a list of all the known facts, and write down any relationships among them, especially any that involve the variable. These may take the form of an equation (or, later, an inequality) involving the variable. If possible, draw an appropriately labeled diagram to assist you. Sometimes, a table or chart helps.

STEP 4: Solve the equation for the variable, and then answer the question asked in the problem.

STEP 5: Check the answer with the facts in the problem. If it agrees, congratulations! If it does not agree, try again.

Let's look at an example.

EXAMPLE 2   *Determining an Hourly Wage*

Colleen grossed $435 one week by working 52 hours. Her employer pays time-and-a-half for all hours worked in excess of 40 hours. With this information, can you determine Colleen's regular hourly wage?

Solution   STEP 1: We are looking for an hourly wage. Our answer will be in dollars per hour.
STEP 2: Let x represent the regular hourly wage; x is measured in dollars per hour.
STEP 3: We set up a table:

	HOURS WORKED	HOURLY WAGE	SALARY
Regular	40	x	$40x$
Overtime	12	$1.5x$	$12(1.5x) = 18x$

The sum of regular salary plus overtime salary will equal $435. Thus, from the table, $40x + 18x = 435$.

STEP 4:
$$40x + 18x = 435$$
$$58x = 435$$
$$x = 7.50$$

STEP 5: Thus, Colleen's regular hourly wage is $7.50 per hour.
Forty hours yields a salary of $40(7.50) = 300, and 12 hours of overtime yields a salary of $12(1.5)(7.50) = 135, for a total of $435. ■

■ Now work Problem 15.

129. Explain what is wrong in the following steps:

$$x = 2 \qquad (1)$$
$$3x - 2x = 2 \qquad (2)$$
$$3x = 2x + 2 \qquad (3)$$
$$x^2 + 3x = x^2 + 2x + 2 \qquad (4)$$
$$x^2 + 3x - 10 = x^2 + 2x - 8 \qquad (5)$$
$$(x - 2)(x + 5) = (x - 2)(x + 4) \qquad (6)$$
$$x + 5 = x + 4 \qquad (7)$$
$$1 = 0 \qquad (8)$$

130. Which of the following pairs of equations are equivalent? Explain.
(a) $x^2 = 9; x = 3$ (b) $x = \sqrt{9}; x = 3$ (c) $(x - 1)(x - 2) = (x - 1)^2; x - 2 = x - 1$

131. The equation
$$\frac{5}{x + 3} + 3 = \frac{8 + x}{x + 3}$$
has no solution, yet when we go through the process of solving it we obtain $x = -3$. Write a brief paragraph to explain what causes this to happen.

132. Make up an equation that has no solution and give it to a fellow student to solve. Ask the fellow student to write a critique of your equation.

133. Describe three ways you might solve a quadratic equation. State your preferred method; explain why you chose it.

134. Explain the benefits of evaluating the discriminant of a quadratic equation before attempting to solve it.

135. Make up three quadratic equations: one having two distinct solutions, one having no real solution, and one having exactly one real solution.

136. The word *quadratic* seems to imply four (quad), yet a quadratic equation is an equation that involves a polynomial of degree 2. Investigate the origin of the term *quadratic* as it is used in the expression *quadratic equation*. Write a brief essay on your findings.

137. Write a program that will solve a quadratic equation:
{Enter the coefficient of x squared}   READ (a);
{Enter the coefficient of x}   READ (b);
{Enter the constant term}   READ (c);
IF $b^2 - 4ac < 0$
THEN {write no real solution}
ELSE IF $b^2 - 4ac = 0$
    THEN {write $-b/2a$ is a double root}
    ELSE {write $(-b + SQRT(b^2 - 4ac))/2a$
        or $(-b - SQRT(b^2 - 4ac))/2a$
        is a solution}

3.3
Setting Up Equations: Applications

The previous section provides the tools for solving equations. But, unfortunately, applied problems do not come in the form, "Solve the equation. . . ." Instead, they are narratives that supply information—hopefully, enough to answer the question that inevitably arises. Thus, to solve applied problems we must be able to translate the verbal description into the language of mathematics. We do this by using symbols (usually letters of the alphabet) to represent unknown quantities and then finding relationships (such as equations) that involve these symbols. The process of doing this is called **mathematical modeling**.

Solution   The voltage lies between 110 and 120, inclusive, so
$$110 \leq E \leq 120$$
$$110 \leq IR \leq 120 \qquad \text{Ohm's law. } E = IR$$
$$110 \leq I(10) \leq 120 \qquad R = 10$$
$$\frac{110}{10} \leq \frac{I(10)}{10} \leq \frac{120}{10} \qquad \text{Divide each part by 10.}$$
$$11 \leq I \leq 12 \qquad \text{Simplify.}$$
The air conditioner will draw between 11 and 12 amperes of current, inclusive. ■

HISTORICAL FEATURE   ■ Inequalities are a relatively new component of the algebra curriculum. They have been introduced in the last 25 years for two important reasons.
First, if approximations are made in a problem, inequalities allow calculation of how serious the error is likely to be. This use is important for practical applications of mathematics and is also critical for the understanding of calculus (in its modern version, due principally to Augustin Louis Cauchy, 1789–1857, and Karl Weierstrass, 1815–1897).
Second, linear inequalities are the basis for solving a kind of problem that involves finding a maximum or minimum of a quantity, depending on variables that are subjected to certain constraints. For example, we might wish to ship several products by truck from several different factories, each factory having a limited supply of each product, and to get enough of the products to a central point within 3 days using the minimum amount of gas possible. Such problems, called *linear programming problems*, are discussed later in this text.

3.6
Exercise 3.6

In Problems 1–8, fill in the blank with the correct inequality symbol.

1. If $x < 5$, then $x - 5$ _____ 0.
2. If $x < -4$, then $x + 4$ _____ 0.
3. If $x > -4$, then $x + 4$ _____ 0.
4. If $x > 6$, then $x - 6$ _____ 0.
5. If $x > -4$, then $3x$ _____ -12.
6. If $x < 3$, then $2x$ _____ 6.
7. If $x < 6$, then $-2x$ _____ -12.
8. If $x > -2$, then $-4x$ _____ 8.

In Problems 9–12, an inequality is given. Write the equivalent inequality obtained by:
(a) Adding -3 to each side of the given inequality.
(b) Subtracting 5 from each side of the given inequality.
(c) Multiplying each side of the given inequality by 3.
(d) Multiplying each side of the given inequality by -2.

9. $3 < 5$ 10. $2 > 1$ 11. $2x + 1 < 2$ 12. $1 - 2x > 5$

In Problems 13–48, solve each inequality (a) graphically and (b) algebraically. Graph the solution set.

13. $x + 1 < 5$ 14. $x - 6 < 1$ 15. $1 - 2x \leq 3$
16. $2 - 3x \leq 5$ 17. $3x - 7 > 2$ 18. $2x + 5 > 1$
19. $3x - 1 \geq 3 + x$ 20. $2x - 2 \geq 3 + x$ 21. $-2(x + 3) < 8$
22. $-3(1 - x) < 12$ 23. $4 - 3(1 - x) \leq 3$ 24. $8 - 4(2 - x) \leq -2x$

Annotations:

Important Procedures and STEPS are noted in the left column and are separated from the body of the text by two horizontal color rules. You will need to know these procedures and steps to do your homework problems and prepare for exams.

Examples are easy to locate and are titled to tell you what concept they highlight. Most examples work the solution first with the graphing utility and then algebraically.

The Now Work Problem xx feature asks you to do a particular problem before you go on in the section. This is to ensure that you have mastered the material just presented before going on. By following the practice of doing these Now Work Problems, you will gain confidence and save time.

The "pencil and book" icon is used to indicate open-ended questions for discussion, writing, group or research projects.

How to solve word problems is explained in a step-by-step manner using mathematical modeling.

Historical Features place the mathematics you are learning in a historical context. By learning how others have used similar concepts you will understand how they may be used in your own life.

Figure 82 shows the graphs.

FIGURE 82

■ Now work Problem 23.

Warning Be sure to use a square screen when you graph perpendicular lines. Otherwise, the angle between the two lines will appear distorted.

Circles

One advantage of a coordinate system is that it enables us to translate a geometric statement into an algebraic statement, and vice versa. Consider, for example, the following geometric statement that defines a circle.

Circle | A **circle** is a set of points in the xy-plane that are a fixed distance r from a fixed point (h, k). The fixed distance r is called the **radius**, and the fixed point (h, k) is called the **center** of the circle.

FIGURE 83

Figure 83 shows the graph of a circle. Is there an equation having this graph? If so, what is the equation? To find the equation, we let (x, y) represent the coordinates of any point on a circle with radius r and center (h, k). Then the distance between the points (x, y) and (h, k) must always equal r. That is, by the distance formula,

$$\sqrt{(x - h)^2 + (y - k)^2} = r$$

or, equivalently,

$$(x - h)^2 + (y - k)^2 = r^2$$

The **standard form of an equation of a circle** with radius r and center (h, k) is

Standard Form of an Equation of a Circle | $$(x - h)^2 + (y - k)^2 = r^2 \qquad (2)$$

"Seeing the Concept." Use your graphing utility to recognize patterns in the algebra.

Figure 58 shows the graph of the line $y = x$ on a square screen using the viewing rectangle given in Example 1(b). Notice that the line now bisects the first and third quadrants. Compare this illustration to Figure 57.

FIGURE 58

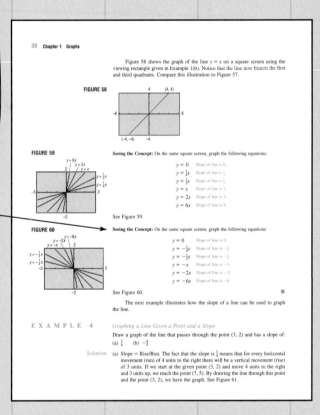

FIGURE 59

Seeing the Concept: On the same square screen, graph the following equations:

$y = 0$ Slope of line is 0.
$y = \frac{1}{4}x$ Slope of line is $\frac{1}{4}$.
$y = \frac{1}{2}x$ Slope of line is $\frac{1}{2}$.
$y = x$ Slope of line is 1.
$y = 2x$ Slope of line is 2.
$y = 6x$ Slope of line is 6.

See Figure 59.

FIGURE 60

Seeing the Concept: On the same square screen, graph the following equations:

$y = 0$ Slope of line is 0.
$y = -\frac{1}{4}x$ Slope of line is $-\frac{1}{4}$.
$y = -\frac{1}{2}x$ Slope of line is $-\frac{1}{2}$.
$y = -x$ Slope of line is -1.
$y = -2x$ Slope of line is -2.
$y = -6x$ Slope of line is -6.

See Figure 60.

The next example illustrates how the slope of a line can be used to graph the line.

EXAMPLE 4 | *Graphing a Line Given a Point and a Slope*

Draw a graph of the line that passes through the point $(3, 2)$ and has a slope of:

(a) $\frac{3}{4}$ (b) $-\frac{4}{5}$

Solution | (a) Slope = Rise/Run. The fact that the slope is $\frac{3}{4}$ means that for every horizontal movement (run) of 4 units to the right there will be a vertical movement (rise) of 3 units. If we start at the given point $(3, 2)$ and move 4 units to the right and 3 units up, we reach the point $(7, 5)$. By drawing the line through this point and the point $(3, 2)$, we have the graph. See Figure 61.

𝜋 ISSION POSSIBLE

Chapter 2

CONSULTING FOR THE SILVER SATELLITE & CABLE TV CO.

Your team works for the Silver Satellite & Cable TV Company in the Research & Development Department. You've been asked to come up with a formula to determine the cost of running cable from a connection box to a new cable household. The first example you are working with involves the Stevens family who own a rural home with a driveway two miles long extending to the house from a nearby highway. The nearest connection box is the highway but 5 miles from the driveway.

It costs the company $10 per mile to install cable along the highway and $14 per mile to install cable off the highway. Because the Steven house is surrounded by farmland which they own, it would be possible to run the cable overland to the house directly from the connection box or from any point between the connection box to the driveway.

1. Draw a sketch of this problem situation, assuming the highway is a straight road and the driveway is also a straight road perpendicular to the highway. Include two or more possible routes for the cable.
2. Suppose x represents the distance in miles the cable runs along the highway from the connection box before turning off toward the house. Express the total cost of installation as a function of x. (You may choose to answer #3 before #2 if you would like to examine concrete instances before creating the equation.)
3. Make a table of the possible integral values of x and the corresponding cost in each instance. Is there one choice which appears to cost the least?
4. If you charge the Stevens $80 for installation, would you be willing to let them choose which way the cable would go? Explain.
5. Using a graphing calculator, graph the function from #2 and determine if there is a non-integral choice for x that would make the installation cost even cheaper. ZOOM and TRACE until you have the lowest possible cost.
6. Before proceeding further with the installation, you check the local regulations for cable companies and find that there is a pending state legislation that says the cable cannot turn off the highway more than .5 miles from the Stevens' driveway. If this legislation passes, what will be the ultimate cost of installing the Stevens' cable?
7. If the cable company wishes to install cable in 5000 homes in this area, and assuming the figures for the Stevens installation are typical, how much will the new legislation cost the company over all if they cannot use the cheapest installation cost, but instead have to follow the new state regulations?

Hints or warnings are offered where appropriate. Sometimes there are short cuts or pitfalls that students should know about—we've included them.

Major definitions appear in large type enclosed within a color screen. These are important vocabulary items for you to know.

All important formulas are enclosed by a box and shown in color. This is done to alert you to important concepts.

Mission Possible: In the "real world" colleagues often collaborate to solve more difficult problems—or problems that may have more than one answer. Every chapter has a "Mission Possible" for you and your classmates to collaborate on. All of these projects will require you to verbally communicate or write up your answers. Good communication skills are very important to becoming successful—no matter what your future holds.

The Chapter Review is for your use in checking your understanding of the chapter materials.

"Things to Know" is the best place to start. Check your understanding of the concepts listed there.

"Important Functions" highlight the functions introduced and used throughout the chapter.

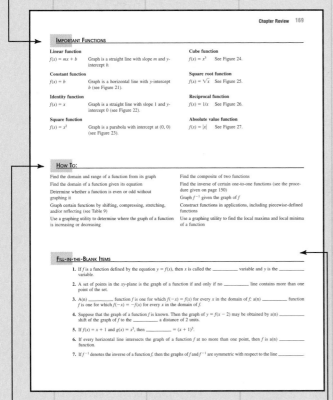

Demonstrate to yourself that you know "How to" deal with the concepts listed.

"Fill in the Blanks" will determine your comfort with vocabulary.

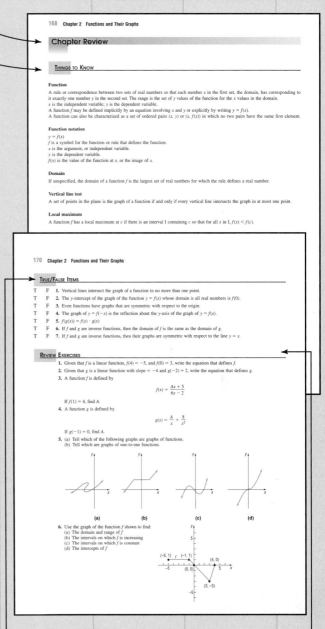

"True/False" is a stickler for knowing definitions!

The "Review Exercises" provide a comprehensive supply of exercises using all the concepts contained in the chapter.

Student's Supplementary Aids

You may find yourself seeking out extra help with this course. Many students have found the following items to be useful in becoming successful in college algebra. Your college bookstore should have these items available, but if not, they can order them for you.

Student's Solutions Manual Contains complete step-by-step worked out solutions to all the odd numbered exercises in the textbook. This is terrific for getting instant feedback on whether you are proceeding correctly while solving problems.

Visual Precalculus A software package for IBM compatible computers which consists of two parts. Part One contains routines to graph and evaluate functions, graph conic sections, investigate series, carry out synthetic division, and illuminate important concepts with animation. Part Two contains routines to solve triangles, graph systems of linear equations and inequalities, evaluate matrix expressions, apply Gaussian elimination to reduce or invert matrices and graphically solve linear programming problems. These routines will provide additional insights into the material covered within the text.

X (Plore) A powerful (yet inexpensive) fully programmable symbolic and numeric mathematical processor for IBM and Macintosh computers. This program will allow you to evaluate expressions, graph curves, solve equations and matrices. This software package may also be used for calculus or differential equations.

New York Times Supplement A free newspaper from Prentice Hall and the New York Times which includes interesting and current articles on mathematics in the world around us. Great for getting together to talk and write about mathematics! This supplement is created new each year.

Photo Credits

Chapter 1	NASA's Space Telescope	Photrl
Chapter 2	Port of Soller	Pedro Coll/The Stock Market
Chapter 3	P&O Line Cruise Ship Leaving San Francisco	Superstock
Chapter 4	Golden Gate Bridge	Deborah Davis/PhotoEdit
Chapter 5	Sobriety Testing	Bachmann/Photrl
Chapter 6	Satellite Station	Four by Five/Superstock
Chapter 7	Track Run	David Madison Photography
Chapter 8	Stained Glass Window	Wolfgang Koohler

Chapter 1

PREPARING FOR THIS CHAPTER

Before getting started on this chapter, review the following concepts:

Topics from Algebra and Geometry (Appendix, Section 1)
Solving equations (Appendix, Section 4)
Completing the square (Appendix, Section 5)

GRAPHS

1.1 Rectangular
    Coordinates;
    Graphing Utilities
1.2 Graphs of Equations
1.3 The Straight Line
1.4 Parallel and
    Perpendicular
    Lines; Circles
1.5 Linear Curve Fitting
1.6 Variation
    Chapter Review

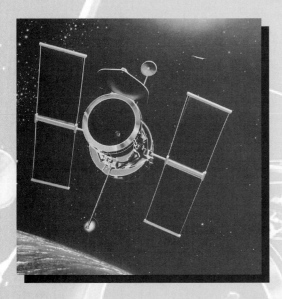

Preview   Communications Satellites

The speed v *required of a satellite to maintain a near-Earth circular orbit is directly proportional to the square root of the distance* r *of the satellite from the center of Earth. Some communications satellites remain stationary above a fixed point on the equator of Earth's surface. How high are such satellites and what is their common speed?*

[Problem 40 in Section 1.6] ■

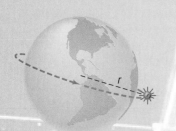

The idea of using a system of rectangular coordinates dates back to ancient times, when such a system was used for surveying and city planning. Apollonius of Perga, in 200 BC, used a form of rectangular coordinates in his work on conics, although this use does not stand out as clearly as it does in modern treatments. Sporadic use of rectangular coordinates continued until the 1600's. By that time, algebra had developed sufficiently so that René Descartes (1596–1650) and Pierre de Fermat (1601–1665) could take the crucial step, which was the use of rectangular coordinates to translate geometry problems into algebra problems, and vice versa. This step was supremely important for two reasons. First, it allowed both geometers and algebraists to gain critical new insights into their subjects, which previously had been regarded as separate but now were seen to be connected in many important ways. Second, the insights gained made possible the development of calculus, which greatly enlarged the number of areas in which mathematics could be applied and made possible a much deeper understanding of these areas. With the advent of technology, in particular, graphing utilities, we are now able not only to visualize the dual roles of algebra and geometry, but we are also able to solve many problems that before this technology required advanced methods.

Rectangular Coordinates; Graphing Utilities

FIGURE 1

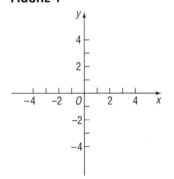

FIGURE 2

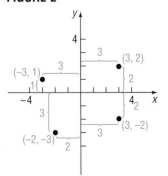

We locate a point on the real number line by assigning it a single real number, called the *coordinate of the point.* For work in a two-dimensional plane, we locate points by using two numbers.

We begin with two real number lines located in the same plane: one horizontal and the other vertical. We call the horizontal line the **x-axis,** the vertical line the **y-axis,** and the point of intersection the **origin O.** We assign coordinates to every point on these number lines as shown in Figure 1, using a convenient scale. In mathematics, we usually use the same scale on each axis; in applications, a different scale is often used on each axis.

The origin O has a value of 0 on both the x-axis and the y-axis. We follow the usual convention that points on the x-axis to the right of O are associated with positive real numbers, and those to the left of O are associated with negative real numbers. Those on the y-axis above O are associated with positive real numbers, and those below O are associated with negative real numbers. In Figure 1, the x-axis and y-axis are labeled as x and y, respectively, and we have used an arrow at the end of each axis to denote the positive direction.

The coordinate system described here is called a **rectangular,** or **Cartesian* coordinate system.** The plane formed by the x-axis and y-axis is sometimes called the **xy-plane,** and the x-axis and y-axis are referred to as the **coordinate axes.**

Any point P in the xy-plane can then be located by using an **ordered pair** (x, y) of real numbers. Let x denote the signed distance of P from the y-axis (*signed* in the sense that, if P is to the right of the y-axis, then $x > 0$, and if P is to the left of the y-axis, then $x < 0$); and let y denote the signed distance of P from the x-axis. The ordered pair (x, y), also called the **coordinates** of P, then gives us enough information to locate the point P in the plane.

For example, to locate the point whose coordinates are $(-3, 1)$, go 3 units along the x-axis to the left of O and then go straight up 1 unit. We **plot** this point by placing a dot at this location. See Figure 2, in which the points with coordinates $(-3, 1)$, $(-2, -3)$, $(3, -2)$, and $(3, 2)$ are plotted.

The origin has coordinates $(0, 0)$. Any point on the x-axis has coordinates of the form $(x, 0)$, and any point on the y-axis has coordinates of the form $(0, y)$.

If (x, y) are the coordinates of a point P, then x is called the **x-coordinate,** or **abscissa,** of P and y is the **y-coordinate,** or **ordinate,** of P. We identify the

*Named after René Descartes (1596–1650), a French mathematician, philosopher, and theologian.

FIGURE 3

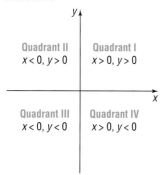

point P by its coordinates (x, y) by writing $P = (x, y)$. Usually, we will simply say "the point (x, y)" rather than "the point whose coordinates are (x, y)."

The coordinate axes divide the xy-plane into four sections, called **quadrants,** as shown in Figure 3. In quadrant I, both the x-coordinate and the y-coordinate of all points are positive; in quadrant II, x is negative and y is positive; in quadrant III, both x and y are negative; and in quadrant IV, x is positive and y is negative. Points on the coordinate axes belong to no quadrant.

Graphing Utilities

All graphing utilities, that is, all graphing calculators and all computer software graphing packages, graph equations by plotting points on a screen. The screen itself actually consists of small rectangles, called **pixels.** The more pixels the screen has, the better the resolution. Most graphing calculators have 48 pixels per square inch; most computer screens have 32 to 108 pixels per square inch. When a point to be plotted lies inside a pixel, the pixel is turned on (lights up). Thus, the graph of an equation is a collection of pixels. Figure 4 shows how the graph of $y = 2x$ looks on a T1-82 graphing calculator.

FIGURE 4
$y = 2x$

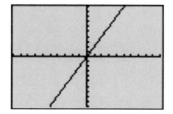

The screen of a graphing utility will display the coordinate axes of a rectangular coordinate system. However, you must set the scale on each axis. You must also include the smallest and largest values of x and y that you want included in the graph. This is called **setting the RANGE** and it gives the **viewing rectangle** or **window.**

Figure 5 illustrates a typical viewing rectangle.

FIGURE 5

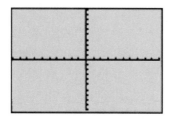

To select the viewing rectangle, we must give values to the following expressions:

Xmin: the smallest value of x

Xmax: the largest value of x

Xscl: the number of units per tick mark on the x-axis

Ymin: the smallest value of y

Ymax: the largest value of y

Yscl: the number of units per tick mark on the y-axis

Figure 6 illustrates these settings for a typical screen.

FIGURE 6

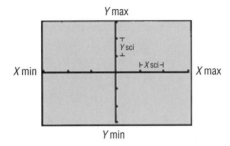

If the scale used on each axis is known, we can determine the minimum and maximum values of x and y shown on the screen by counting the tick marks. Look again at Figure 5. For a scale of 1 on each axis, the minimum and maximum values of x are -10 and 10, respectively; the minimum and maximum values of y are also -10 and 10. If the scale is 2 on each axis, then the minimum and maximum values of x are -20 and 20, respectively; the minimum and maximum value of y are -20 and 20, respectively.

Conversely, if we know the minimum and maximum values of x and y, we can determine the scales being used by counting the tick marks displayed. We shall follow the practice of showing the minimum and maximum values of x and y in our illustrations so that you will know how the RANGE was set. See Figure 7.

FIGURE 7

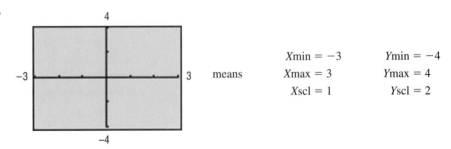

$$\text{means} \quad \begin{aligned} X\text{min} &= -3 & Y\text{min} &= -4 \\ X\text{max} &= 3 & Y\text{max} &= 4 \\ X\text{scl} &= 1 & Y\text{scl} &= 2 \end{aligned}$$

E X A M P L E   1

Finding the Coordinates of a Point Shown on a Graphing Utility Screen

Find the coordinates of the point shown in Figure 8.

FIGURE 8

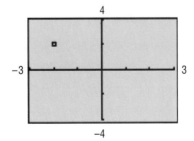

Solution   First we note that the range setting used in Figure 8 is

$$X\text{min} = -3 \qquad X\text{scl} = 1 \qquad Y\text{max} = 4$$
$$X\text{max} = 3 \qquad Y\text{min} = -4 \qquad Y\text{scl} = 2$$

The point shown is 2 tic units to the left on the horizontal axis (scale = 1) and 1 tic up on the vertical (scale = 2). Thus, the coordinates of the point shown are $(-2, 2)$. ■

■ Now work Problems 5 and 15.

Distance between Points

If the same units of measurement, such as inches, centimeters, and so on, are used for both the x-axis and the y-axis, then all distances in the xy-plane can be measured using this unit of measurement.

E X A M P L E   2

Finding the Distance between Two Points

Find the distance d between the points $(1, 3)$ and $(5, 6)$.

Solution

First we plot the points $(1, 3)$ and $(5, 6)$ as shown in Figure 9(a). Then we draw a horizontal line from $(1, 3)$ to $(5, 3)$ and a vertical line from $(5, 3)$ to $(5, 6)$, forming a right triangle, as in Figure 9(b). One leg of the triangle is of length 4 and the other is of length 3. By the Pythagorean Theorem (see the Appendix, Section 1), the square of the distance d we seek is

$$d^2 = 4^2 + 3^2 = 16 + 9 = 25$$
$$d = 5$$

FIGURE 9

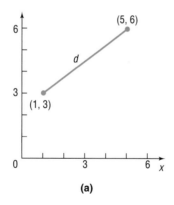

(a)

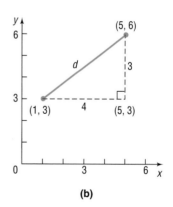

(b)

■

The **distance formula** provides a straightforward method for computing the distance between two points.

Theorem

The distance between two points $P_1 = (x_1, y_1)$ and $P_2 = (x_2, y_2)$, denoted by $d(P_1, P_2)$, is

Distance Formula

$$d(P_1, P_2) = \sqrt{(x_2 - x_1)^2 + (y_2 - y_1)^2} \qquad (1)$$

■

That is, to compute the distance between two points, find the difference of the x-coordinates, square it, and add this to the square of the difference of the y-coordinates. The square root of this sum is the distance. See Figure 10.

FIGURE 10

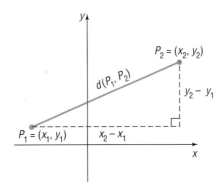

Proof of the Distance Formula Let (x_1, y_1) denote the coordinates of point P_1, and let (x_2, y_2) denote the coordinates of point P_2. Assume that the line joining P_1 and P_2 is neither horizontal nor vertical. Refer to Figure 11(a). The coordinates of P_3 are (x_2, y_1). The horizontal distance from P_1 to P_3 is the absolute value of the difference of the x-coordinates, $|x_2 - x_1|$. The vertical distance from P_3 to P_2 is the absolute value of the difference of the y-coordinates, $|y_2 - y_1|$. See Figure 11(b). The distance $d(P_1, P_2)$ that we seek is the length of the hypotenuse of the right triangle, so, by the Pythagorean Theorem, it follows that

$$[d(P_1, P_2)]^2 = |x_2 - x_1|^2 + |y_2 - y_1|^2$$
$$= (x_2 - x_1)^2 + (y_2 - y_1)^2$$
$$d(P_1, P_2) = \sqrt{(x_2 - x_1)^2 + (y_2 - y_1)^2}$$

FIGURE 11

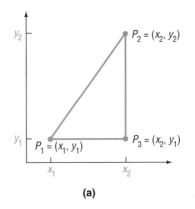

(a)

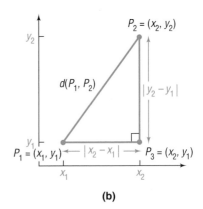

(b)

Now, if the line joining P_1 and P_2 is horizontal, then the y-coordinate of P_1 equals the y-coordinate of P_2; that is, $y_1 = y_2$. Refer to Figure 12(a). In this case, the distance formula (1) still works, because for $y_1 = y_2$, it reduces to

$$d(P_1, P_2) = \sqrt{(x_2 - x_1)^2 + 0^2} = \sqrt{(x_2 - x_1)^2} = |x_2 - x_1|$$

A similar argument holds if the line joining P_1 and P_2 is vertical. See Figure 12(b). Thus, the distance formula is valid in all cases.

FIGURE 12

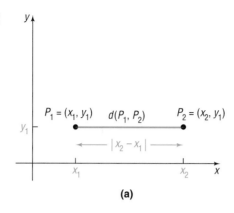
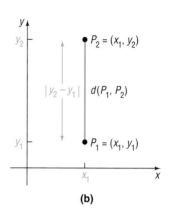

(a)   (b)

EXAMPLE   3   *Finding the Length of a Line Segment*

Find the length of the line segment shown in Figure 13.

FIGURE 13

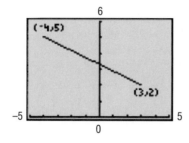

Solution   The length of the line segment is the distance between the points $(-4, 5)$ and $(3, 2)$. Using the distance formula (1), the length d is

$$d = \sqrt{[3-(-4)]^2 + (2-5)^2} = \sqrt{7^2 + (-3)^2}$$
$$= \sqrt{49 + 9} = \sqrt{58} \approx 7.62 \qquad \blacksquare$$

■   Now work Problem 25.

The distance between two points $P_1 = (x_1, y_1)$ and $P_2 = (x_2, y_2)$ is never a negative number. Furthermore, the distance between two points is 0 only when the points are identical, that is, when $x_1 = x_2$ and $y_1 = y_2$. Also, because $(x_2 - x_1)^2 = (x_1 - x_2)^2$ and $(y_2 - y_1)^2 = (y_1 - y_2)^2$, it makes no difference whether the distance is computed from P_1 to P_2 or from P_2 to P_1; that is, $d(P_1, P_2) = d(P_2, P_1)$.

The introduction to this chapter mentioned that rectangular coordinates enable us to translate geometry problems into algebra problems, and vice versa. The next example shows how algebra (the distance formula) can be used to solve geometry problems.

EXAMPLE   4   *Using Algebra to Solve Geometry Problems*

Consider the three points $A = (-2, 1)$, $B = (2, 3)$ and $C = (3, 1)$.

(a) Plot each point and form the triangle ABC.

(b) Find the length of each side of the triangle.

(c) Verify that the triangle is a right triangle.

(d) Find the area of the triangle.

Solution (a) Points A, B, C, and triangle ABC are plotted in Figure 14.

(b) $d(A, B) = \sqrt{[2 - (-2)]^2 + (3 - 1)^2} = \sqrt{16 + 4} = \sqrt{20} = 2\sqrt{5}$

$d(B, C) = \sqrt{(3 - 2)^2 + (1 - 3)^2} = \sqrt{1 + 4} = \sqrt{5}$

$d(A, C) = \sqrt{[3 - (-2)]^2 + (1 - 1)^2} = \sqrt{5^2 + 0^2} = 5$

(c) To show that the triangle is a right triangle, we need to show that the sum of the squares of the lengths of two of the sides equals the square of the length of the third side. (Why is this sufficient?) Looking at Figure 14, it seems reasonable to conjecture that the right angle is at vertex B. Thus, we shall check to see whether

$$[d(A, B)]^2 + [d(B, C)]^2 = [d(A, C)]^2$$

We find that

$$[d(A, B)]^2 + [d(B, C)]^2 = (2\sqrt{5})^2 + (\sqrt{5})^2$$
$$= 20 + 5 = 25 = [d(A, C)]^2$$

so it follows from the converse of the Pythagorean Theorem that triangle ABC is a right triangle.

(d) Because the right angle is at B, the sides AB and BC form the base and altitude of the triangle. Its area is therefore

$$\text{Area} = \frac{1}{2}(\text{Base})(\text{Altitude}) = \frac{1}{2}(2\sqrt{5})(\sqrt{5}) = 5 \text{ square units} \qquad \blacksquare$$

FIGURE 14

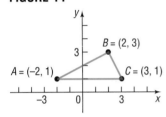

■ Now work Problem 43.

Midpoint Formula

We now derive a formula for the coordinates of the **midpoint of a line segment.** Let $P_1 = (x_1, y_1)$ and $P_2 = (x_2, y_2)$ be the endpoints of a line segment, and let $M = (x, y)$ be the point on the line segment that is the same distance from P_1 as it is from P_2. See Figure 15. The triangles P_1AM and MBP_2 are congruent.* [Do you see why? Angle AP_1M = Angle BMP_2,† Angle P_1MA = Angle MP_2B, and $d(P_1, M) = d(M, P_2)$ is given. Thus, we have Angle–Side–Angle.] Hence, corresponding sides are equal in length. That is,

$$x - x_1 = x_2 - x \quad \text{and} \quad y - y_1 = y_2 - y$$
$$2x = x_1 + x_2 \qquad\qquad 2y = y_1 + y_2$$
$$x = \frac{x_1 + x_2}{2} \qquad\qquad y = \frac{y_1 + y_2}{2}$$

FIGURE 15

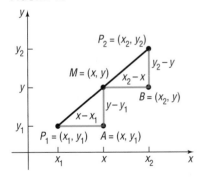

*The following statement is a postulate from geometry. Two triangles are congruent if their sides are the same length (SSS), or if two sides and the included angle are the same (SAS), or if two angles and the included side are the same (ASA).

†Another postulate from geometry states that the transversal $\overline{P_1P_2}$ forms equal corresponding angles with the parallel lines $\overline{P_1A}$ and $\overline{MB}$.

Theorem   The midpoint (x, y) of the line segment from $P_1 = (x_1, y_1)$ to $P_2 = (x_2, y_2)$ is

Midpoint Formula

$$(x, y) = \left(\frac{x_1 + x_2}{2}, \frac{y_1 + y_2}{2} \right) \qquad (2)$$

Thus, to find the midpoint of a line segment, we average the x-coordinates and the y-coordinates of the endpoints.

E X A M P L E   5   *Finding the Midpoint of a Line Segment*

Find the midpoint of a line segment from $P_1 = (-5, 3)$ to $P_2 = (3, 1)$. Plot the points P_1 and P_2 and their midpoint. Check your answer.

Solution   We apply the midpoint formula (2) using $x_1 = -5$, $x_2 = 3$, $y_1 = 3$, and $y_2 = 1$. Then the coordinates (x, y) of the midpoint M are

$$x = \frac{x_1 + x_2}{2} = \frac{-5 + 3}{2} = -1 \quad \text{and} \quad y = \frac{y_1 + y_2}{2} = \frac{3 + 1}{2} = 2$$

FIGURE 16

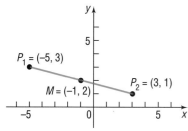

That is, $M = (-1, 2)$. See Figure 16.

Check: Because M is the midpoint, we check the answer by verifying that $d(P_1, M) = d(M, P_2)$:

$$d(P_1, M) = \sqrt{[-1 - (-5)]^2 + (2 - 3)^2} = \sqrt{16 + 1} = \sqrt{17}$$
$$d(M, P_2) = \sqrt{[3 - (-1)]^2 + (1 - 2)^2} = \sqrt{16 + 1} = \sqrt{17}$$

■   Now work Problem 53.

1.1

Exercise 1.1

In Problems 1 and 2, plot each point in the xy-plane. Tell in which quadrant or on what coordinate axis each point lies.

1. (a) $A = (-3, 2)$ (b) $B = (6, 0)$ (c) $C = (-2, -2)$
     (d) $D = (6, 5)$ (e) $E = (0, -3)$ (f) $F = (6, -3)$

2. (a) $A = (1, 4)$ (b) $B = (-3, -4)$ (c) $C = (-3, 4)$
     (d) $D = (4, 1)$ (e) $E = (0, 1)$ (f) $F = (-3, 0)$

3. Plot the points $(2, 0)$, $(2, -3)$, $(2, 4)$, $(2, 1)$, and $(2, -1)$. Describe the set of all points of the form $(2, y)$, where y is a real number.

4. Plot the points $(0, 3)$, $(1, 3)$, $(-2, 3)$, $(5, 3)$ and $(-4, 3)$. Describe the set of all points of the form $(x, 3)$, where x is a real number.

In Problems 5–8, determine the coordinates of the points shown. Tell in which quadrant each point lies.

5.

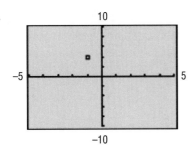

6.

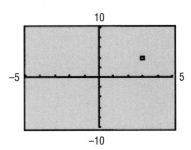

7.

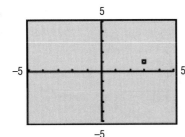

8.

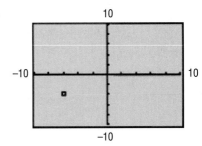

In Problems 9–14, select a RANGE setting so that each of the given points will lie within the viewing rectangle.

9. $(-10, 5)$, $(3, -2)$, $(4, -1)$

10. $(5, 0)$, $(6, 8)$, $(-2, -3)$

11. $(40, 20)$, $(-20, -80)$, $(10, 40)$

12. $(-80, 60)$, $(20, -30)$, $(-20, -40)$

13. $(0, 0)$, $(100, 5)$, $(5, 150)$

14. $(0, -1)$, $(100, 50)$, $(-10, 30)$

In Problems 15–24, determine the RANGE settings used for each viewing rectangle.

15.

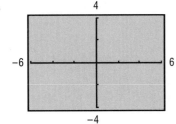

16.

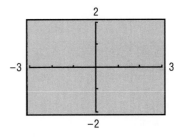

17.

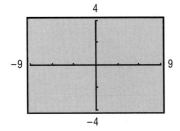

18.

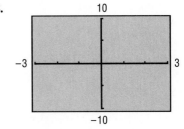

19.

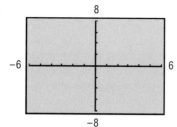

20.

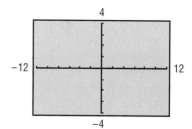

21.

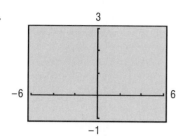

22.

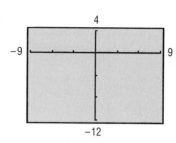

23.

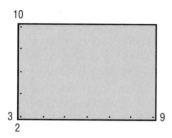

24.

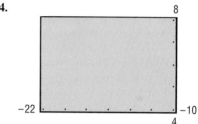

In Problems 25–38, find the distance $d(P_1, P_2)$ between the points P_1 and P_2.

25.

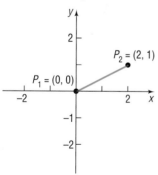

26.

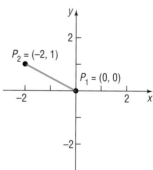

27.

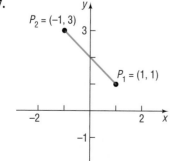

28.

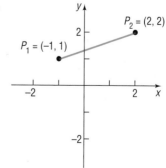

29. $P_1 = (3, -4);$ $P_2 = (5, 4)$

30. $P_1 = (-1, 0);$ $P_2 = (2, 4)$

31. $P_1 = (-3, 2);$ $P_2 = (6, 0)$

32. $P_1 = (2, -3);$ $P_2 = (4, 2)$

33. $P_1 = (4, -3);$ $P_2 = (6, 4)$

34. $P_1 = (-4, -3);$ $P_2 = (6, 2)$

35. $P_1 = (-0.2, 0.3);$ $P_2 = (2.3, 1.1)$

36. $P_1 = (1.2, 2.3);$ $P_2 = (-0.3, 1.1)$

37. $P_1 = (a, b);$ $P_2 = (0, 0)$

38. $P_1 = (a, a);$ $P_2 = (0, 0)$

In Problems 39–42, find the length of the line segment.

39.

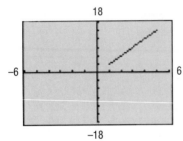

40.

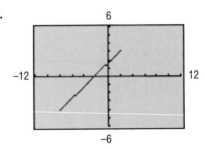

41.

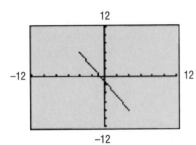

42.

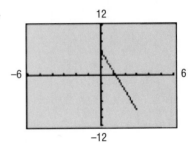

In Problems 43–48, plot each point and form the triangle ABC. Verify that the triangle is a right triangle. Find its area.

43. $A = (-2, 5);$ $B = (1, 3);$ $C = (-1, 0)$

44. $A = (-2, 5);$ $B = (12, 3);$ $C = (10, -11)$

45. $A = (-5, 3);$ $B = (6, 0);$ $C = (5, 5)$

46. $A = (-6, 3);$ $B = (3, -5);$ $C = (-1, 5)$

47. $A = (4, -3);$ $B = (0, -3);$ $C = (4, 2)$

48. $A = (4, -3);$ $B = (4, 1);$ $C = (2, 1)$

49.   Find all points having an x-coordinate of 2 whose distance from the point $(-2, -1)$ is 5.

50.   Find all points having a y-coordinate of -3 whose distance from the point $(1, 2)$ is 13.

51.   Find all points on the x-axis that are 5 units from the point $(4, -3)$.

52.   Find all points on the y-axis that are 5 units from the point $(4, 4)$.

In Problems 53–62, find the midpoint of the line segment joining the points P_1 and P_2.

53. $P_1 = (5, -4);$ $P_2 = (3, 2)$

54. $P_1 = (-1, 0);$ $P_2 = (2, 4)$

55. $P_1 = (-3, 2);$ $P_2 = (6, 0)$

56. $P_1 = (2, -3);$ $P_2 = (4, 2)$

57. $P_1 = (4, -3);$ $P_2 = (6, 1)$

58. $P_1 = (-4, -3);$ $P_2 = (2, 2)$

59. $P_1 = (-0.2, 0.3);$ $P_2 = (2.3, 1.1)$

60. $P_1 = (1.2, 2.3);$ $P_2 = (-0.3, 1.1)$

61. $P_1 = (a, b);$ $P_2 = (0, 0)$

62. $P_1 = (a, a);$ $P_2 = (0, 0)$

63. The **medians** of a triangle are the line segments from each vertex to the midpoint of the opposite side (see the figure in the margin). Find the lengths of the medians of the triangle with vertices at $A = (0, 0)$, $B = (0, 6)$, and $C = (8, 0)$.

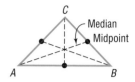

64. An **equilateral triangle** is one in which all three sides are of equal length. If two vertices of an equilateral triangle are $(0, 4)$ and $(0, 0)$, find the third vertex. How many of these triangles are possible?

*In Problems 65–68, find the length of each side of the triangle determined by the three points P_1, P_2, and P_3. State whether the triangle is an isosceles triangle, a right triangle, neither of these, or both. (An **isosceles triangle** is one in which at least two of the sides are of equal length.)*

65. $P_1 = (2, 1)$; $P_2 = (-4, 1)$; $P_3 = (-4, -3)$

66. $P_1 = (-1, 4)$; $P_2 = (6, 2)$; $P_3 = (4, -5)$

67. $P_1 = (-2, -1)$; $P_2 = (0, 7)$; $P_3 = (3, 2)$

68. $P_1 = (7, 2)$; $P_2 = (-4, 0)$; $P_3 = (4, 6)$

69. If r is a real number, prove that the point $P = (x, y)$ that divides the line segment from $P_1 = (x_1, y_1)$ to $P_2 = (x_2, y_2)$ in the ratio r, that is,

$$\frac{d(P_1, P)}{d(P_1, P_2)} = r$$

has coordinates

$$x = x_1 + r(x_2 - x_1) \quad \text{and} \quad y = y_1 + r(y_2 - y_1)$$

[*Hint:* Use similar triangles.*]

Problems 70–74 use the result of Problem 69.

70. Verify that the midpoint of a line segment divides the line segment from $P_1 = (x_1, y_1)$ to $P_2 = (x_2, y_2)$ in the ratio $r = \frac{1}{2}$.

71. What point P divides the line segment from P_1 to P_2 in the ratio $r = 1$?

72. What point P divides the line segment from P_1 to P_2 in the ratio $r = 0$?

73. Find the point P on the line joining $P_1 = (2, 4)$ and $P_2 = (5, 6)$ that is twice as far from P_1 as P_2 is from P_1 and that lies on the same side of P_1 as P_2 does.

74. Find the point P on the line joining $P_1 = (0, 4)$ and $P_2 = (-4, 1)$ that is three times as far from P_1 as P_2 is from P_1.

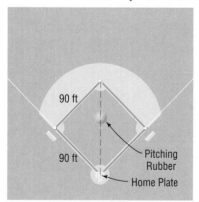

75. *Baseball*   A major league baseball "diamond" is actually a square, 90 feet on a side (see the figure). What is the distance directly from home plate to second base (the diagonal of the square)?

76. *Little League Baseball*   The layout of a Little League playing field is a square, 60 feet on a side.† How far is it directly from home plate to second base (the diagonal of the square)?

77. *Baseball*   Refer to Problem 75. Overlay a rectangular coordinate system on a major league baseball diamond so that the origin is at home plate, the positive x-axis lies in the direction from home plate to first base, and the positive y-axis lies in the direction from home plate to third base.
(a) What are the coordinates of first base, second base, and third base? Use feet as the unit of measurement.
(b) If the right fielder is located at $(310, 15)$, how far is it from there to second base?
(c) If the center fielder is located at $(300, 300)$, how far is it from there to third base?

78. *Little League Baseball*   Refer to Problem 76. Overlay a rectangular coordinate system on a Little League baseball diamond so that the origin is at home plate, the positive x-axis lies in the direction from home plate to first base, and the positive y-axis lies in the direction from home plate to third base.
(a) What are the coordinates of first base, second base, and third base? Use feet as the unit of measurement.
(b) If the right fielder is located at $(180, 20)$, how far is it from there to second base?
(c) If the center fielder is located at $(220, 220)$, how far is it from there to third base?

---

*Two triangles are **similar** if they have the same angles. In similar triangles, corresponding sides are in proportion.

†*Source: Little League Baseball, Official Regulations and Playing Rules, 1991.*

79. *Geometry* Find the midpoint of each diagonal of a square with side of length s. Draw the conclusion that the diagonals of a square intersect at their midpoints. [*Hint:* Use $(0, 0)$, $(0, s)$, $(s, 0)$, and (s, s) as the vertices of the square.]

80. *Geometry* Verify that the points $(0, 0)$, $(a, 0)$, and $(a/2, \sqrt{3}a/2)$ are the vertices of an equilateral triangle. Then show that the midpoints of the three sides are the vertices of a second equilateral triangle. (See Problem 64.)

81. An automobile and a truck leave an intersection at the same time. The automobile heads east at an average speed of 30 miles per hour, while the truck heads south at an average speed of 40 miles per hour. Find an expression for their distance apart d (in miles) at the end of t hours.

82. A hot air balloon, headed due east at an average speed of 15 miles per hour and at a constant altitude of 100 feet, passes over an intersection (see the figure). Find an expression for its distance d (measured in feet) from the intersection t seconds later.

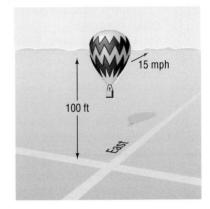

1.2

Graphs of Equations

Illustrations play an important role in helping us to visualize the relationships that exist between two variable quantities. For example, Figure 17 shows the variation in the price of oil from January to October 1991. Such illustrations are usually referred to as *graphs*. The **graph of an equation** in two variables x and y consists of the set of points in the xy-plane whose coordinates (x, y) satisfy the equation. Let's look at some examples.

FIGURE 17
Oil futures prices
Source: Dow Jones News Retrieval

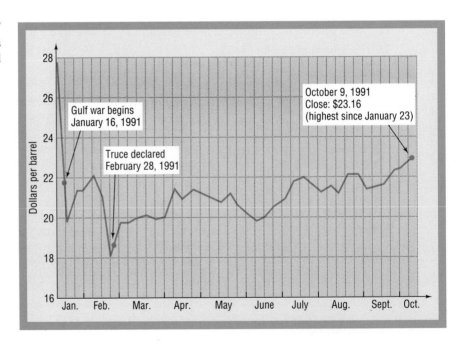

E X A M P L E   1   *Graphing an Equation by Hand*

Graph the equation: $y = 2x + 5$

Solution   We want to find all points (x, y) that satisfy the equation. To locate some of these points (and thus get an idea of the pattern of the graph), we assign some numbers to x and find corresponding values for y:

FIGURE 18
$y = 2x + 5$

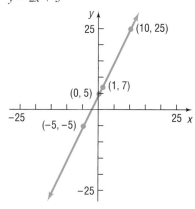

IF	THEN	POINT ON GRAPH
$x = 0$	$y = 2(0) + 5 = 5$	$(0, 5)$
$x = 1$	$y = 2(1) + 5 = 7$	$(1, 7)$
$x = -5$	$y = 2(-5) + 5 = -5$	$(-5, -5)$
$x = 10$	$y = 2(10) + 5 = 25$	$(10, 25)$

By plotting these points and then connecting them, we obtain the graph of the equation (a straight line), as shown in Figure 18.   ■

E X A M P L E   2

Graphing an Equation by Hand

Graph the equation: $y = x^2$

Solution   Table 1 provides several points on the graph. In Figure 19 we plot these points and connect them with a smooth curve to obtain the graph (a *parabola*).

TABLE 1

x	$y = x^2$	(x, y)
-4	16	$(-4, 16)$
-3	9	$(-3, 9)$
-2	4	$(-2, 4)$
-1	1	$(-1, 1)$
0	0	$(0, 0)$
1	1	$(1, 1)$
2	4	$(2, 4)$
3	9	$(3, 9)$
4	16	$(4, 16)$

FIGURE 19
$y = x^2$

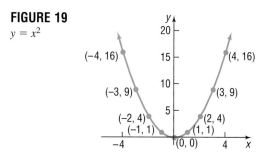

The graphs of the equations shown in Figures 18 and 19 do not show all points. For example, in Figure 18, the point $(20, 45)$ is a part of the graph of $y = 2x + 5$, but it is not shown. Since the graph of $y = 2x + 5$ could be extended out as far as we please, we use arrows to indicate that the pattern shown continues. Thus, it is important when illustrating a graph to present enough of the graph so that any viewer of the illustration will "see" the rest of it as an obvious continuation of what is actually there. This is referred to as a **complete graph.**

So, one way to obtain a complete graph of an equation is to plot a sufficient number of points on the graph until a pattern becomes evident. Then these points are connected with a smooth curve following the suggested pattern. But how many points are sufficient? Sometimes knowledge about the equation tells us. For example, we will learn in the next section that, if an equation is of the form $y = mx + b$, then its graph is a straight line. In this case, two points would suffice to obtain the graph.

One of the purposes of this book is to investigate the properties of equations in order to decide whether a graph is complete. Sometimes we shall graph equations by hand by plotting a sufficient number of points on the graph until a pattern becomes evident; then we connect these points with a smooth curve following the suggested pattern. Shortly, we shall investigate various techniques that will enable us to graph an equation by hand without plotting so many points. Other times, we shall graph equations using a graphing utility.

Using a Graphing Utility to Graph Equations

From Examples 1 and 2, we see that a graph can be obtained by plotting points in a rectangle coordinate system and connecting them. Graphing utilities perform these same steps when graphing an equation. For example, the T1-82 determines 95 evenly spaced input values*, uses the equation to determine the output values, plots these points on the screen and finally, (if in the connected mode), draws a line between consecutive points.

Most graphing utilities require the following steps in order to obtain the graph of an equation:

STEP 1: Solve the equation for y in terms of x.

STEP 2: Select the viewing rectangle.

STEP 3: Get into the graphing mode of your graphing utility. The screen will usually display $y =$, prompting you to enter the expression involving x that you found in STEP 1. (Consult your manual for the correct way to enter the expression; for example, $y = x^2$ might be entered as $x\text{^}2$ or as $x*x$ or as $x\ x^Y 2$).

STEP 4: Execute.

E X A M P L E   3   *Graphing an Equation on a Graphing Utility*

Graph the equation: $6x^2 + 3y = 24$

Solution   STEP 1: We solve for y in terms of x.

$$6x^2 + 3y = 24$$
$$3y = -6x^2 + 24$$
$$y = -2x^2 + 8$$

STEP 2: Select a viewing rectangle. We will use the one given next.

$$\text{Xmin} = -5$$
$$\text{Xmax} = 5$$
$$\text{Xscl} = 1$$
$$\text{Ymin} = -10$$
$$\text{Ymax} = 20$$
$$\text{Yscl} = 2$$

STEP 3: From the graphing mode, enter the expression $-2x^2 + 8$ after the prompt $y =$.

STEP 4: Execute

The screen should look like Figure 20.

*These input values depend on the values of Xmin and Xmax. For example, if Xmin $= -10$ and Xmax $= 10$, then the first input value will be -10 and the next input value will be $-10 + (10 - (-10))/94 = -9.7872$, and so on.

FIGURE 20

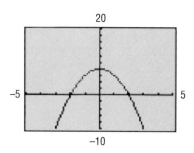

Now that we have seen the graph of $y = -2x^2 + 8$, we might want to graph it again using the settings given below:

$$\text{Xmin} = -5$$
$$\text{Xmax} = 5$$
$$\text{Xscl} = 1$$
$$\text{Ymin} = -10$$
$$\text{Ymax} = 10$$
$$\text{Yscl} = 1$$

Figure 21 shows the graph obtained. Notice the improvement over Figure 20.

FIGURE 21

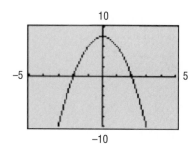

On some graphing utilities, you can scroll the graph up or down and left or right to see more of it, instead of setting a new viewing rectangle. With scrolling, the scale stays the same as it was originally; only the min/max values change.

■ Now work Problems 11(a) and 11(b).

E X A M P L E 4 *Graphing an Equation*

Graph the equation: $y = x^3$

Solution Using a graphing utility, we obtain Figure 22.

FIGURE 22

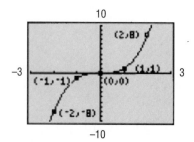

To graph by hand, we use the equation to obtain several points on the graph (Table 2) and then draw the graph on paper. See Figure 23.

TABLE 2

x	$y = x^3$	(x, y)
-3	-27	$(-3, -27)$
-2	-8	$(-2, -8)$
-1	-1	$(-1, -1)$
0	0	$(0, 0)$
1	1	$(1, 1)$
2	8	$(2, 8)$
3	27	$(3, 27)$

FIGURE 23

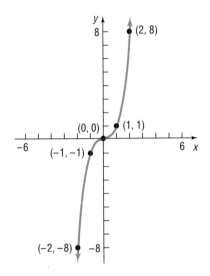

Table 2 can also be obtained using a graphing utility. (Check your manual to see how.) On a graphing utility, Table 2 takes the form shown in Table 3.

TABLE 3

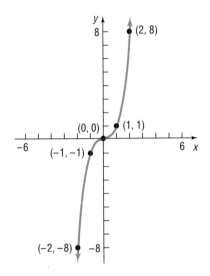

E X A M P L E   5   *Graphing an Equation*

Graph the equation: $x = y^2$

Solution   To graph an equation using a graphing utility, we must write the equation in the form $y = $ expression involving x. So to graph $x = y^2$, we must solve for y:

$$x = y^2$$
$$y^2 = x$$
$$y = \pm\sqrt{x}$$

Thus, to graph $x = y^2$, we need to graph both $y = \sqrt{x}$ and $y = -\sqrt{x}$ on the same screen. Figure 24 shows the result. Table 4 shows various values of y for a given value of x when $y_1 = \sqrt{x}$ and $y_2 = -\sqrt{x}$. Notice when $x < 0$, we get an error. Can you explain why?

TABLE 4

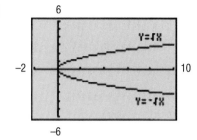

FIGURE 24

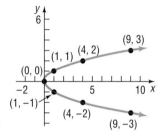

To graph by hand, we use the equation to obtain several points on the graph (Table 5) and then draw the graph on paper. See Figure 25.

TABLE 5

y	$x = y^2$	(x, y)
-3	9	$(9, -3)$
-2	4	$(4, -2)$
-1	1	$(1, -1)$
0	0	$(0, 0)$
1	1	$(1, 1)$
2	4	$(4, 2)$
3	9	$(9, 3)$
4	16	$(16, 4)$

FIGURE 25

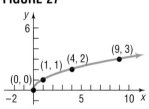

Look again at either Figure 24 or Figure 25. If we restrict y so that $y \geq 0$, the equation $x = y^2$, $y \geq 0$, may be written equivalently as $y = \sqrt{x}$. The portion of the graph of $x = y^2$ in quadrant I is therefore the graph of $y = \sqrt{x}$. See Figure 26 and Figure 27.

FIGURE 26

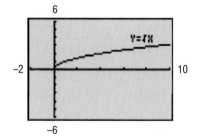

FIGURE 27

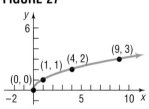

Now work Problem 47.

Obtaining a Complete Graph Using a Graphing Utility

Often the first choice for the RANGE setting does not show a complete graph. To obtain a complete graph requires adjusting the RANGE. One way to do this is to use ZOOM-OUT.

E X A M P L E   6   *Using ZOOM-OUT to Obtain a Complete Graph*

Graph the equation $y = x^3 - 11x^2 - 190x + 200$ using the following settings for the viewing rectangle

Xmin: -12
Xmax: 12
Xscl: 4
Ymin: -200
Ymax: 200
Yscl: 100

Solution   Figure 28 shows the graph.

FIGURE 28

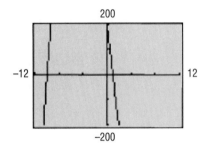

Notice how ragged the graph looks. The *y*-scale is clearly not adequate. The graph is not complete. We can use the ZOOM-OUT function to help obtain a complete graph.

After the first ZOOM-OUT, we obtain Figure 29.

FIGURE 29

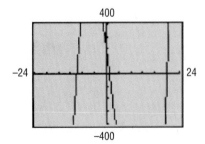

Notice that the min/max settings have been doubled, while the scales remained the same.* However, we still cannot see the top or bottom portion of the graph. ZOOM-OUT again to obtain the graph in Figure 30.

*On some graphing utilities, the default factor for the ZOOM-OUT function is 4, meaning that the original min/max setting will be multiplied by 4. Also, you can set the factor yourself, if you want; furthermore, the factor need not be the same for *x* and *y*.

FIGURE 30

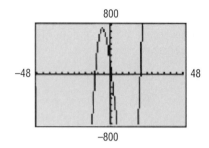

Notice that we can now see almost the entire graph with the exception of the bottom portion. Therefore ZOOMing-OUT further may not provide us with a complete graph. To get a complete graph, we instead choose to adjust the RANGE. After some experimentation, we will obtain Figure 31, a complete graph.

FIGURE 31
$y = x^3 - 11x^2 - 190x + 200$

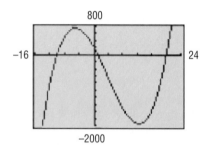

We said earlier that we would discuss techniques that reduce the number of points required to graph an equation by hand. Two such techniques involve finding *intercepts* and checking for *symmetry*.

Intercepts

The points, if any, at which a graph intersects the coordinate axes are called the **intercepts.** See Figure 32. The *x*-coordinate of a point at which the graph crosses or touches the *x*-axis is an ***x*-intercept,** and the *y*-coordinate of a point at which the graph crosses or touches the *y*-axis is a ***y*-intercept.**

FIGURE 32

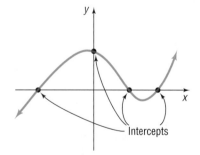

E X A M P L E   7   *Finding Intercepts on a Graph*

Find the intercepts of the graph in Figure 33. What are its x-intercepts? What are its y-intercepts?

FIGURE 33

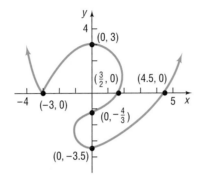

Solution   The intercepts of the graph are the points

$$(-3, 0), \quad (0, 3), \quad \left(\frac{3}{2}, 0\right), \quad \left(0, -\frac{4}{3}\right), \quad (0, -3.5), \quad (4.5, 0)$$

The x-intercepts are $-3, \frac{3}{2}, 4.5$; the y-intercepts are $-3.5, -\frac{4}{3}, 3$. ■

The intercepts of the graph of an equation can be found by using the fact that points on the x-axis have y-coordinates equal to 0, and points on the y-axis have x-coordinates equal to 0.

Procedure for Finding Intercepts

1. To find the x-intercept(s), if any, of the graph of an equation, let $y = 0$ in the equation and solve for x.
2. To find the y-intercept(s), if any, of the graph of an equation, let $x = 0$ in the equation and solve for y.

E X A M P L E   8   *Finding Intercepts from an Equation*

Find the x-intercept(s) and the y-intercept(s) of the graph of $y = x^2 - 4$.

Solution   To find the x-intercept(s), we let $y = 0$ and obtain the equation

$$x^2 - 4 = 0$$

The equation has two solutions, -2 and 2. Thus, the x-intercepts are -2 and 2.
To find the y-intercept(s), we let $x = 0$ and obtain the equation

$$y = -4$$

Thus, the y-intercept is -4. ■

FIGURE 34

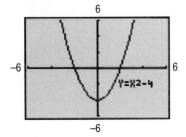

To graph $y = x^2 - 4$ using a graphing utility, we make the following observation.
Since $x^2 \geq 0$ for all x, we deduce from the equation $y = x^2 - 4$ that $y \geq -4$ for all x. This information, along with the intercepts, can now be used to set the viewing rectangle so a complete graph is obtained. See Figure 34.

To draw the graph by hand, we use the equation to obtain some additional points on the graph. See Table 6 and Figure 35.

TABLE 6

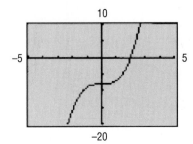

FIGURE 35

$y = x^2 - 4$

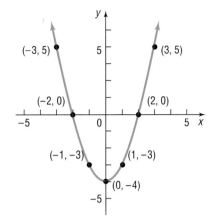

■   Now work Problem 57.

Using TRACE, ZOOM-IN, and BOX to Locate Intercepts

A graphing utility can be used to locate the intercepts of an equation. Three of the more effective ways to locate intercepts using a graphing utility are TRACE, ZOOM-IN, and BOX.

Most graphing utilities allow you to move from point to point along the graph, displaying on the screen the coordinates of each point. This feature is called the TRACE function.

E X A M P L E   9   *Using the TRACE Function to Locate Intercepts*

Graph the equation $y = x^3 - 8$ using the following viewing rectangle. Use the TRACE function to locate various points on the graph. In particular, locate the x-intercept.

RANGE

xMin = −5
xMax = 5
xScl = 1
yMin = −20
yMax = 10
yScl = 5

Solution   Figure 36 shows the graph of $y = x^3 - 8$.

FIGURE 36

Activate the TRACE function. As you move the cursor along the graph, you will see the coordinates of each point displayed. Just before you get to the *x*-axis, the display will look like the one in Figure 37. (Due to differences in graphing utilities, your display may be slightly different than the one shown here).

FIGURE 37

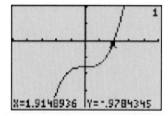

In Figure 37, the negative value of the *y*-coordinate indicates we are still below the *x*-axis. The next position of the cursor is shown in Figure 38.

FIGURE 38

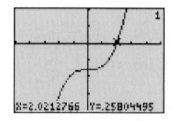

The positive value of the *y*-coordinate indicates we are now above the *x*-axis. This means that between these two points the *x*-axis was crossed. The *x*-intercept lies between 1.9148936 and 2.0212766.   ∎

Most graphing utilities also have a ZOOM-IN function that allows us to get better approximations.

E X A M P L E   1 0   *Using ZOOM-IN*

Graph the equation $y = x^3 - 8$ and use the ZOOM-IN function to improve on the approximation found in Example 9 for the *x*-intercept.

Solution   Using the viewing rectangle of Example 9, graph the equation and TRACE until the cursor is close to the *x*-intercept. See Figure 39.

FIGURE 39

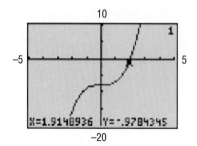

Activate the ZOOM-IN function. The result is shown in Figure 40.

FIGURE 40

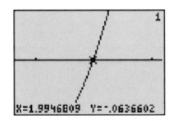

Move the cursor up. The result is shown in Figure 41.

FIGURE 41

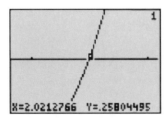

Notice the improvement. Now we know that the *x*-intercept lies between 1.9946809 and 2.0212766. ZOOM-IN again. The result is shown in Figure 42(a). Figure 42(b) is the result of moving the cursor down.

FIGURE 42

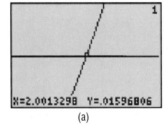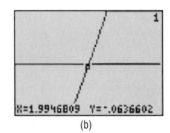

(a)   (b)

Now we know that the *x*-intercept lies between 1.9946809 and 2.0013298.   ■

Most graphing utilities have a BOX function that allows you to box in a specific part of the graph of an equation.

E X A M P L E   1 1   *Using the BOX Function*

Graph the equation $y = x^3 - 8$ and use the BOX function to improve on the approximation found in Example 9 for the *x*-intercept.

Solution   Using the viewing rectangle of Example 9, graph the equation. Activate the BOX function. (With some graphing utilities, this requires positioning the cursor at one corner of the box and then tracing out the sides of the box to the diagonal corner.) See Figure 43.

FIGURE 43

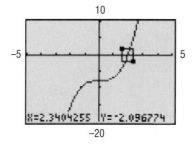

Once executed, the box becomes the viewing rectangle. The result is shown in Figure 44.

FIGURE 44

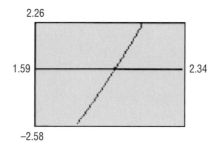

Now you can TRACE to get the approximations shown in Figure 45.

FIGURE 45

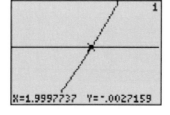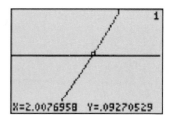

Now we know the x-intercept lies between 1.9997737 and 2.0076958.   ◼

Symmetry

We have just seen the role intercepts play in obtaining key points on the graph of an equation. Another helpful tool for graphing equations involves *symmetry*, particularly symmetry with respect to the x-axis, the y-axis, and the origin.

Symmetry with Respect
to the x-Axis

> A graph is said to be **symmetric with respect to the x-axis** if, for every point (x, y) on the graph, the point $(x, -y)$ is also on the graph.

Figure 46 illustrates the definition. Notice that when a graph is symmetric with respect to the *x*-axis the part of the graph above the *x*-axis is a reflection or mirror image of the part below it, and vice versa.

FIGURE 46

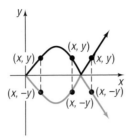

Symmetric with respect to
the *x*-axis

E X A M P L E   1 2   *Points Symmetric with Respect to the x-Axis*

If a graph is symmetric with respect to the *x*-axis and the point (3, 2) is on the graph, then the point (3, −2) is also on the graph. ∎

Symmetry with Respect
to the *y*-Axis

A graph is said to be **symmetric with respect to the y-axis** if, for every point (*x*, *y*) on the graph, the point (−*x*, *y*) is also on the graph.

Figure 47 illustrates the definition. Notice that when a graph is symmetric with respect to the *y*-axis the part of the graph to the right of the *y*-axis is a reflection of the part to the left of it, and vice versa.

FIGURE 47

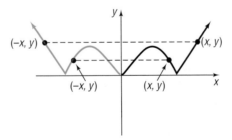

Symmetric with respect to the *y*-axis

E X A M P L E   1 3   *Points Symmetric with Respect to the y-Axis*

If a graph is symmetric with respect to the *y*-axis and the point (5, 8) is on the graph, then the point (−5, 8) is also on the graph. ∎

Symmetry with Respect
to the Origin

A graph is said to be **symmetric with respect to the origin** if, for every point (*x*, *y*) on the graph, the point (−*x*, −*y*) is also on the graph.

Figure 48 illustrates the definition. Notice that symmetry with respect to the origin may be viewed in two ways:

1. As a reflection about the *y*-axis, followed by a reflection about the *x*-axis.
2. As a projection along a line through the origin so that the distances from the origin are equal.

FIGURE 48

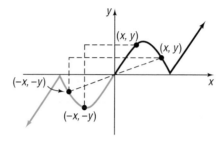

Symmetric with respect to
the origin

E X A M P L E   1 4   *Points Symmetric with Respect to the Origin*

If a graph is symmetric with respect to the origin and the point (4, 2) is on the graph, then the point $(-4, -2)$ is also on the graph.   ■

■   Now work Problems 1 and 31(b).

When the graph of an equation is symmetric with respect to a coordinate axis or the origin, the number of points that you need to plot in order to see the pattern is reduced. For example, if the graph of an equation is symmetric with respect to the *y*-axis, then, once points to the right of the *y*-axis are plotted, an equal number of points on the graph can be obtained by reflecting them about the *y*-axis. Thus, before we graph an equation, we first want to determine whether it has any symmetry. The following tests are used for that purpose.

Tests for Symmetry   To test the graph of an equation for symmetry with respect to the:

x-Axis   Replace *y* by $-y$ in the equation. If an equivalent equation results, the graph of the equation is symmetric with respect to the *x*-axis.

y-Axis   Replace *x* by $-x$ in the equation. If an equivalent equation results, the graph of the equation is symmetric with respect to the *y*-axis.

Origin   Replace *x* by $-x$ and *y* by $-y$ in the equation. If an equivalent equation results, the graph of the equation is symmetric with respect to the origin.

Let's look at an equation we have already graphed to see how these tests are used.

E X A M P L E   1 5   *Testing Equations for Symmetry*

(a) To test the graph of the equation $x = y^2$ for symmetry with respect to the *x*-axis, we replace *y* by $-y$ in the equation, as follows:

$$x = y^2 \qquad \text{Original equation}$$
$$x = (-y)^2 \qquad \text{Replace } y \text{ by } -y$$
$$x = y^2 \qquad \text{Simplify.}$$

When we replace y by $-y$, the result is the same equation. Thus, the graph is symmetric with respect to the x-axis.

(b) To test the graph of the equation $x = y^2$ for symmetry with respect to the y-axis, we replace x by $-x$ in the equation:

$$x = y^2 \quad \text{Original equation}$$
$$-x = y^2 \quad \text{Replace } x \text{ by } -x.$$

Because we arrive at the equation $-x = y^2$, which is not equivalent to the original equation, we conclude that the graph is not symmetric with respect to the y-axis.

(c) To test for symmetry with respect to the origin, we replace x by $-x$ and y by $-y$:

$$x = y^2 \quad \text{Original equation}$$
$$-x = (-y)^2 \quad \text{Replace } x \text{ by } -x \text{ and } y \text{ by } -y.$$
$$-x = y^2 \quad \text{Simplify.}$$

The resulting equation, $-x = y^2$, is not equivalent to the original equation. We conclude that the graph is not symmetric with respect to the origin. ■

Figure 49(a) illustrates the graph of $x = y^2$. In forming a table of points on the graph of $x = y^2$, we can restrict ourselves to points whose y-coordinates are positive. Once these are plotted and connected, a reflection about the x-axis (because of the symmetry) provides the rest of the graph.

Figures 49(b) and 49(c) illustrate two other equations we graphed earlier. Notice how the existence of symmetry reduces the number of points we need to plot.

FIGURE 49

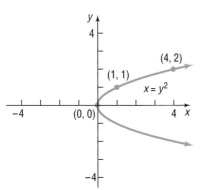

(a) Symmetry with respect to the x-axis

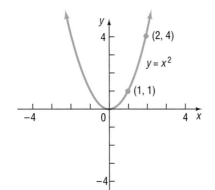

(b) Symmetry with respect to the y-axis

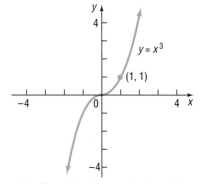

(c) Symmetry with respect to the origin

■   Now work Problem 89.

E X A M P L E   1 6   *Graphing the Equation y = 1/x*

Consider the equation $y = 1/x$

(a) Graph this equation using a graphing utility. Set the viewing rectangle as

RANGE

$\quad$ XMin = -3

$\quad$ XMax = 3

$\quad$ XScl = 1

$\quad$ YMin = -4

$\quad$ YMax = 4

$\quad$ YScl = 1

(b) Use algebra to find any intercepts and test for symmetry.

(c) Draw the graph by hand.

Solution   (a) Figure 50 illustrates the graph.

FIGURE 50

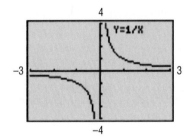

We infer from the graph that there are no intercepts; we may also infer that symmetry with respect to the origin is a possibility. The TRACE function on a graphing utility can provide further evidence of symmetry with respect to the origin. Using TRACE we observe that for any (x, y) ordered pair, the ordered pair $(-x, -y)$ is also a point on the graph. For example, the points $(0.95744681, 1.0444444)$ and $(-0.95744681, -1.0444444)$ both lie on the graph.

(b) We check for intercepts first. If we let $x = 0$, we obtain a 0 denominator, which is not allowed. Hence, there is no y-intercept. If we let $y = 0$, we get the equation $1/x = 0$, which has no solution. Hence, there is no x-intercept. Thus, the graph of $y = 1/x$ does not cross the coordinate axes.

$\qquad$ Next we check for symmetry:

x-Axis: Replacing y by $-y$ yields $-y = 1/x$, which is not equivalent to $y = 1/x$.

y-Axis: Replacing x by $-x$ yields $y = -1/x$, which is not equivalent to $y = 1/x$.

Origin: Replacing x by $-x$ and y by $-y$ yields $-y = -1/x$, which is equivalent to $y = 1/x$.

$\qquad$ The graph is symmetric with respect to the origin. This confirms the inferences drawn in part (a) of the solution.

(c) We can use the equation to obtain some points on the graph. Because of symmetry, we need only find points (x, y) for which x is positive. Also from the equation $y = 1/x$ we infer that if x is a large and positive number, then $y = 1/x$ is a positive number close to 0. We also infer that if x is a positive number close to 0, then $y = 1/x$ is a large and positive number. Armed with this information, we can graph the equation. Figure 51 illustrates some of these points and the graph of $y = 1/x$. Observe how the absence of intercepts and the existence of symmetry with respect to the origin were utilized.

TABLE 7

x	$y = 1/x$	(x, y)
$\frac{1}{10}$	10	$(\frac{1}{10}, 10)$
$\frac{1}{3}$	3	$(\frac{1}{3}, 3)$
$\frac{1}{2}$	2	$(\frac{1}{2}, 2)$
1	1	$(1, 1)$
2	$\frac{1}{2}$	$(2, \frac{1}{2})$
3	$\frac{1}{3}$	$(3, \frac{1}{3})$
10	$\frac{1}{10}$	$(10, \frac{1}{10})$

FIGURE 51

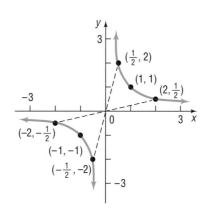

1.2

Exercise 1.2

In Problems 1–10, plot each point. Then plot the point that is symmetric to it with respect to:
(a) The x-axis   (b) The y-axis   (c) The origin

1. $(3, 4)$ **2.** $(5, 3)$ **3.** $(-2, 1)$ **4.** $(4, -2)$ **5.** $(1, 1)$

6. $(-1, -1)$ **7.** $(-3, -4)$ **8.** $(4, 0)$ **9.** $(0, -3)$ **10.** $(-3, 0)$

In Problems 11–30, graph each equation using the following RANGE settings:

(a) Xmin $= -5$	*(b)* Xmin $= -10$	*(c)* Xmin $= -10$	*(d)* Xmin $= -5$
Xmax $= 5$	Xmax $= 10$	Xmax $= 10$	Xmax $= 5$
Xscl $= 1$	Xscl $= 1$	Xscl $= 2$	Xscl $= 1$
Ymin $= -4$	Ymin $= -8$	Ymin $= -8$	Ymin $= -20$
Ymax $= 4$	Ymax $= 8$	Ymax $= 8$	Ymax $= 20$
Yscl $= 1$	Yscl $= 1$	Yscl $= 2$	Yscl $= 5$

11. $y = x + 2$ **12.** $y = x - 2$ **13.** $y = -x + 2$ **14.** $y = -x - 2$

15. $y = 2x + 2$ **16.** $y = 2x - 2$ **17.** $y = -2x + 2$ **18.** $y = -2x - 2$

19. $y = x^2 + 2$ **20.** $y = x^2 - 2$ **21.** $y = -x^2 + 2$ **22.** $y = -x^2 - 2$

23. $y = 2x^2 + 2$ **24.** $y = 2x^2 - 2$ **25.** $y = -2x^2 + 2$ **26.** $y = -2x^2 - 2$

27. $3x + 2y = 6$ **28.** $3x - 2y = 6$ **29.** $-3x + 2y = 6$ **30.** $-3x - 2y = 6$

In Problems 31–46, the graph of an equation is given.
(a) List the intercepts of the graph.
(b) Based on the graph, tell whether the graph is symmetric with respect to the x-axis, y-axis, and/or origin.

31.
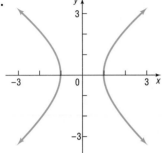

32.
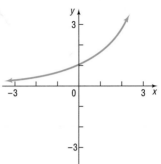

33.
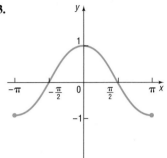

34.
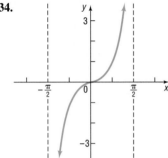

35.
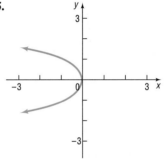

36.
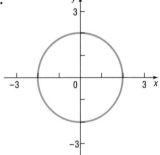

37.
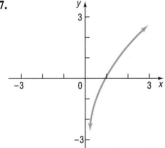

38.
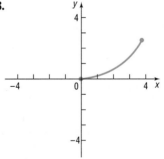

39.
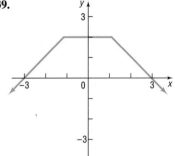

40.
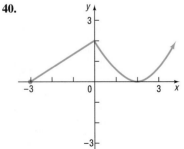

41.
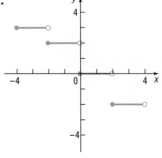

42.
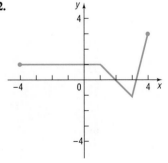

43.

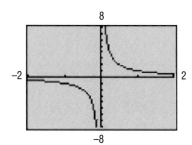

44.

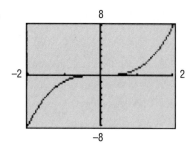

45.

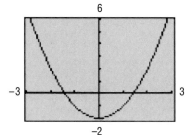

46.

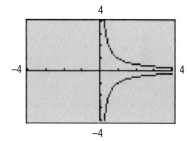

In Problems 47–52, tell whether the given points are on the graph of the equation.

47. Equation: $y = x^4 + \sqrt{x}$
Points: $(0, 0); (1, 1); (-1, 0)$

48. Equation: $y = x^3 - 2\sqrt{x}$
Points: $(0, 0); (1, 1); (1, -1)$

49. Equation: $y^2 = x^2 + 4$
Points: $(0, 2); (2, 0); (-2, 0)$

50. Equation: $y^3 = x + 1$
Points: $(1, 2); (0, 1); (-1, 0)$

51. Equation: $x^2 + y^2 = 4$
Points: $(0, 2); (-2, 2); (\sqrt{2}, \sqrt{2})$

52. Equation: $x^2 + 4y^2 = 4$
Points: $(0, 1); (2, 0); (2, \frac{1}{2})$

53. If $(a, 2)$ is a point on the graph of $y = 5x + 4$, what is a?

54. If $(2, b)$ is a point on the graph of $y = x^2 + 3x$, what is b?

55. If (a, b) is a point on the graph of $2x + 3y = 6$, write an equation that relates a to b.

56. If $(2, 0)$ and $(0, 5)$ are points on the graph of $y = mx + b$, what are m and b?

In Problems 57–76, use a graphing utility to graph each equation. State the viewing rectangle used and draw the graph by hand.

57. $3x + 5y = 75$ **58.** $3x - 5y = 75$ **59.** $3x + 5y = -75$ **60.** $3x - 5y = -75$

61. $y = (x - 10)^2$ **62.** $y = (x + 10)^2$ **63.** $y = x^2 - 100$ **64.** $y = x^2 + 100$

65. $x^2 + y^2 = 100$ **66.** $x^2 + y^2 = 164$ **67.** $3x^2 + y^2 = 900$ **68.** $4x^2 + y^2 = 1600$

69. $x^2 + 3y^2 = 900$ **70.** $x^2 + 4y^2 = 1600$ **71.** $y = x^2 - 10x$ **72.** $y = x^2 + 10x$

73. $y = x^2 - 18x$ **74.** $y = x^2 + 18x$ **75.** $y = x^2 - 36x$ **76.** $y = x^2 + 36x$

In Problems 77–80, use the graph on the right.

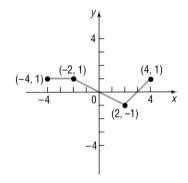

77. Extend the graph to make it symmetric with respect to the *x*-axis.

78. Extend the graph to make it symmetric with respect to the *y*-axis.

79. Extend the graph to make it symmetric with respect to the origin.

80. Extend the graph to make it symmetric with respect to the *x*-axis, *y*-axis, and origin.

In Problems 81–84, use the graph on the right.

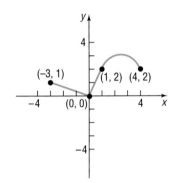

81. Extend the graph to make it symmetric with respect to the *x*-axis.

82. Extend the graph to make it symmetric with respect to the *y*-axis.

83. Extend the graph to make it symmetric with respect to the origin.

84. Extend the graph to make it symmetric with respect to the *x*-axis, *y*-axis, and origin.

In Problems 85–98, list the intercepts and test for symmetry. Graph each equation using a graphing utility.

85. $x^2 = y$

86. $y^2 = x$

87. $y = 3x$

88. $y = -5x$

89. $x^2 + y - 9 = 0$

90. $y^2 - x - 4 = 0$

91. $9x^2 + 4y^2 = 36$

92. $4x^2 + y^2 = 4$

93. $y = x^3 - 27$

94. $y = x^4 - 1$

95. $y = x^2 - 3x - 4$

96. $y = x^2 + 4$

97. $y = \dfrac{x}{x^2 + 9}$

98. $y = \dfrac{x^2 - 4}{x}$

In Problem 99, you may use a graphing utility, but it is not required.

99. (a) Graph $y = \sqrt{x^2}$, $y = x$, $y = |x|$, and $y = (\sqrt{x})^2$, noting which graphs are the same.
    (b) Explain why the graphs of $y = \sqrt{x^2}$ and $y = |x|$ are the same.
    (c) Explain why the graphs of $y = x$ and $y = (\sqrt{x})^2$ are not the same.
    (d) Explain why the graphs of $y = \sqrt{x^2}$ and $y = x$ are not the same.

100. Make up an equation with the intercepts (2, 0), (4, 0), and (0, 1). Compare your equation with a friend's equation. Comment on any similarities.

101. An equation is being tested for symmetry with respect to the *x*-axis, the *y*-axis, and the origin. Explain why, if two of these symmetries are present, then the remaining one must also be present.

102. Draw a graph that contains the points $(-2, -1)$, $(0, 1)$, $(1, 3)$, and $(3, 5)$. Compare your graph with those of other students. Are most of the graphs almost straight lines? How many are "curved"? Discuss the various ways these points might be connected.

1.3

The Straight Line

In this section we study a certain type of equation that contains two variables, called a *linear equation,* and its graph, a *straight line.*

Slope of a Line

FIGURE 52

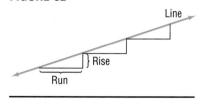

Consider the staircase illustrated in Figure 52. Each step contains exactly the same horizontal **run** and the same vertical **rise.** The ratio of the rise to the run, called the *slope,* is a numerical measure of the steepness of the staircase. For example, if the run is increased and the rise remains the same, the staircase becomes less steep. If the run is kept the same, but the rise is increased, the staircase becomes more steep. This important characteristic of a line is best defined using rectangular coordinates.

Slope of a Line

> Let $P = (x_1, y_1)$ and $Q = (x_2, y_2)$ be two distinct points with $x_1 \neq x_2$. The **slope m** of the nonvertical line L containing P and Q is defined by the formula
>
> $$m = \frac{y_2 - y_1}{x_2 - x_1} \qquad x_1 \neq x_2 \tag{1}$$
>
> If $x_1 = x_2$, L is a **vertical line** and the slope m of L is **undefined** (since this results in division by 0).

Figure 53(a) provides an illustration of the slope of a nonvertical line; Figure 53(b) illustrates a vertical line.

FIGURE 53

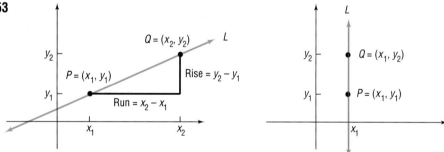

(a) Slope of L is $m = \dfrac{y_2 - y_1}{x_2 - x_1}$

(b) Slope is undefined; L is vertical

As Figure 53(a) illustrates, the slope m of a nonvertical line may be viewed as

$$m = \frac{y_2 - y_1}{x_2 - x_1} = \frac{\text{Rise}}{\text{Run}}$$

We can also express the slope m of a nonvertical line as

$$m = \frac{y_2 - y_1}{x_2 - x_1} = \frac{\text{Change in } y}{\text{Change in } x} = \frac{\Delta y}{\Delta x}$$

That is, the slope m of a nonvertical line L is the ratio of the change in the y-coordinates from P to Q, $\Delta y = y_2 - y_1$, to the change in the x-coordinates from P to Q, $\Delta x = x_2 - x_1$.

Two comments about computing the slope of a nonvertical line may prove helpful:

1. Any two distinct points on the line can be used to compute the slope of the line. (See Figure 54 for justification.)

FIGURE 54

Triangles *ABC* and *PQR* are similar (equal angles). Hence, ratios of corresponding sides are proportional. Thus:

Slope using P and $Q = \dfrac{y_2 - y_1}{x_2 - x_1}$

$\quad = $ Slope using A and $B = \dfrac{d(B, C)}{d(A, C)}$

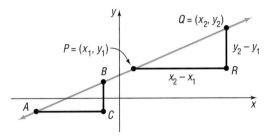

2. The slope of a line may be computed from $P = (x_1, y_1)$ to $Q = (x_2, y_2)$ or from Q to P, because

$$\frac{y_2 - y_1}{x_2 - x_1} = \frac{y_1 - y_2}{x_1 - x_2}$$

E X A M P L E   1   *Finding the Slope of a Line Joining Two Points*

The slope m of the line joining the points $(1, 2)$ and $(5, -3)$ may be computed as

$$m = \frac{-3 - 2}{5 - 1} = \frac{-5}{4} \quad \text{or as} \quad m = \frac{2 - (-3)}{1 - 5} = \frac{5}{-4} = \frac{-5}{4} \qquad \blacksquare$$

■ Now work Problem 7.

To get a better idea of the meaning of the slope m of a line L, consider the following example.

E X A M P L E   2   *Finding the Slopes of Various Lines Containing the Same Point (2, 3)*

Compute the slopes of the lines L_1, L_2, L_3, and L_4 containing the following pairs of points. Graph all four lines on the same set of coordinate axes.

$$L_1: \quad P = (2, 3) \qquad Q_1 = (-1, -2)$$
$$L_2: \quad P = (2, 3) \qquad Q_2 = (3, -1)$$
$$L_3: \quad P = (2, 3) \qquad Q_3 = (5, 3)$$
$$L_4: \quad P = (2, 3) \qquad Q_4 = (2, 5)$$

Solution   Let m_1, m_2, m_3, and m_4 denote the slopes of the lines L_1, L_2, L_3, and L_4, respectively. Then

$$m_1 = \frac{-2 - 3}{-1 - 2} = \frac{-5}{-3} = \frac{5}{3} \qquad \text{A rise of 5 divided by a run of 3}$$

$$m_2 = \frac{-1 - 3}{3 - 2} = \frac{-4}{1} = -4$$

$$m_3 = \frac{3 - 3}{5 - 2} = \frac{0}{3} = 0$$

m_4 is undefined

The graphs of these lines are given in Figure 55.

FIGURE 55

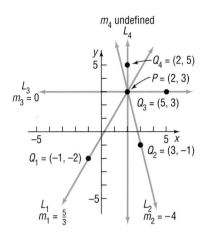

FIGURE 56

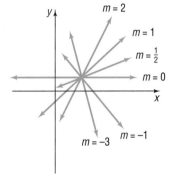

Figure 55 illustrates the following facts:

1. When the slope of a line is positive, the line slants upward from left to right (L_1).
2. When the slope of a line is negative, the line slants downward from left to right (L_2).
3. When the slope is 0, the line is horizontal (L_3).
4. When the slope is undefined, the line is vertical (L_4).

Figure 56 illustrates some additional facts about the slope of a line. Note that the closer the line is to the vertical position, the greater the magnitude of the slope.

FIGURE 57

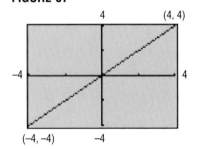

Square Screens

To get an undistorted view of slope, the same scale must be used on each axis. However, most graphing utilities have a rectangular screen. Because of this, using the same RANGE for both x and y will result in a distorted view. For example, Figure 57 shows the graph of the line $y = x$ connecting the points $(-4, -4)$ and $(4, 4)$.

We expect the line to bisect the first and third quadrants, but it doesn't. We need to adjust the selections for Xmin, Xmax, Ymin, and Ymax so that a **square screen** results. On most graphing utilities, this is accomplished by setting the ratio of x to y at $3:2$.* In other words,

$$2(\text{Xmax} - \text{Xmin}) = 3(\text{Ymax} - \text{Ymin})$$

E X A M P L E 3   *Examples of Viewing Rectangles that Result in Square Screens*

(a) Xmin = -3	(b) Xmin = -6	(c) Xmin = -6
Xmax = 3	Xmax = 6	Xmax = 6
Xscl = 1	Xscl = 2	Xscl = 2
Ymin = -2	Ymin = -4	Ymin = -4
Ymax = 2	Ymax = 4	Ymax = 4
Yscl = 1	Yscl = 2	Yscl = 1

*Some graphing utilities have a built-in function that automatically squares the screen. For example, the TI85 has a ZSQR function that does this. Some graphing utilities require a ratio other than $3:2$ to square the screen. For example, the HP48G requires the ratio of x to y to be $1:2$ for a square screen. Consult your manual.

Figure 58 shows the graph of the line $y = x$ on a square screen using the viewing rectangle given in Example 1(b). Notice that the line now bisects the first and third quadrants. Compare this illustration to Figure 57.

FIGURE 58

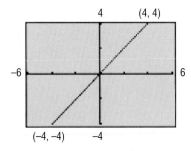

FIGURE 59

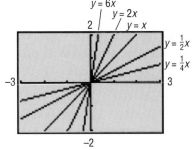

Seeing the Concept: On the same square screen, graph the following equations:

$y = 0$ Slope of line is 0.

$y = \frac{1}{4}x$ Slope of line is $\frac{1}{4}$.

$y = \frac{1}{2}x$ Slope of line is $\frac{1}{2}$.

$y = x$ Slope of line is 1.

$y = 2x$ Slope of line is 2.

$y = 6x$ Slope of line is 6.

See Figure 59.

FIGURE 60

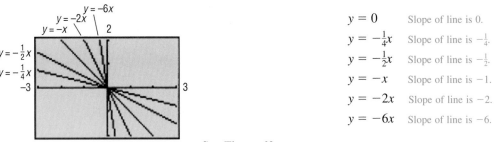

Seeing the Concept: On the same square screen, graph the following equations:

$y = 0$ Slope of line is 0.

$y = -\frac{1}{4}x$ Slope of line is $-\frac{1}{4}$.

$y = -\frac{1}{2}x$ Slope of line is $-\frac{1}{2}$.

$y = -x$ Slope of line is -1.

$y = -2x$ Slope of line is -2.

$y = -6x$ Slope of line is -6.

See Figure 60.

The next example illustrates how the slope of a line can be used to graph the line.

E X A M P L E   4   *Graphing a Line Given a Point and a Slope*

Draw a graph of the line that passes through the point (3, 2) and has a slope of:

(a) $\frac{3}{4}$ (b) $-\frac{4}{5}$

Solution   (a) Slope = Rise/Run. The fact that the slope is $\frac{3}{4}$ means that for every horizontal movement (run) of 4 units to the right there will be a vertical movement (rise) of 3 units. If we start at the given point (3, 2) and move 4 units to the right and 3 units up, we reach the point (7, 5). By drawing the line through this point and the point (3, 2), we have the graph. See Figure 61.

FIGURE 61
Slope $= \frac{3}{4}$

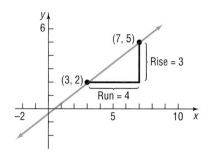

(b) The fact that the slope is

$$-\frac{4}{5} = \frac{-4}{5} = \frac{\text{Rise}}{\text{Run}}$$

means that for every horizontal movement of 5 units to the right there will be a corresponding vertical movement of -4 units (a downward movement). If we start at the given point $(3, 2)$ and move 5 units to the right and then 4 units down, we arrive at the point $(8, -2)$. By drawing the line through these points, we have the graph. See Figure 62.

FIGURE 62
Slope $= -\frac{4}{5}$

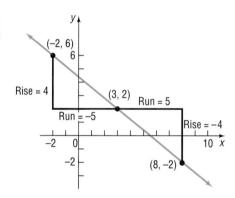

Alternatively, we can set

$$-\frac{4}{5} = \frac{4}{-5} = \frac{\text{Rise}}{\text{Run}}$$

so that for every horizontal movement of -5 units (a movement to the left) there will be a corresponding vertical movement of 4 units (upward). This approach brings us to the point $(-2, 6)$, which is also on the graph shown in Figure 62. ■

■   Now work Problem 17.

Equations of Lines

Now that we have discussed the slope of a line, we are ready to derive equations of lines. As we shall see, there are several forms of the equation of a line. Let's start with an example.

E X A M P L E   5   *Graphing a Line*

Graph the equation: $x = 3$

Solution   Using a graphing utility, we need to express the equation in the form $y =$ expression in x. But $x = 3$ cannot be put into this form. To overcome this problem, we must utilize a function in our graphing utility which allows vertical lines to be drawn. Consult your manual to determine the methodology required to draw vertical lines. Figure 63 shows the graph you should obtain.

FIGURE 63

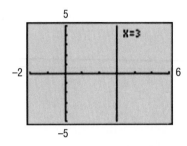

To graph $x = 3$ by hand, we recall that we are looking for all points (x, y) in the plane for which $x = 3$. Thus, no matter what y-coordinate is used, the corresponding x-coordinate always equals 3. Consequently, the graph of the equation $x = 3$ is a vertical line with x-intercept 3 and undefined slope. See Figure 64.

FIGURE 64
$x = 3$

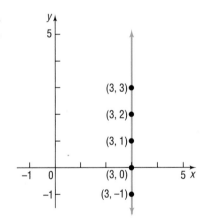

As suggested by Example 5, we have the following result:

Theorem   A vertical line is given by an equation of the form

Equation of a Vertical Line

$$x = a$$

where a is the x-intercept.   ∎

Now let L be a nonvertical line with slope m and containing the point (x_1, y_1). See Figure 65. For any other point (x, y) on L, we have

$$m = \frac{y - y_1}{x - x_1} \quad \text{or} \quad y - y_1 = m(x - x_1)$$

FIGURE 65

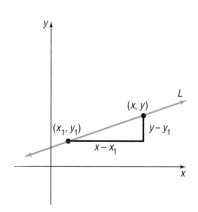

Theorem | An equation of a nonvertical line of slope m that passes through the point (x_1, y_1) is

Point-Slope Form
of an Equation of a Line

$$y - y_1 = m(x - x_1) \qquad (2)$$

E X A M P L E   6

Using the Point–Slope Form of a Line

An equation of the line with slope 4 and passing through the point $(1, 2)$ can be found by using the point–slope form with $m = 4$, $x_1 = 1$, and $y_1 = 2$:

$$y - y_1 = m(x - x_1)$$
$$y - 2 = 4(x - 1)$$
$$y = 4x - 2$$

See Figure 66.

FIGURE 66
$y = 4x - 2$

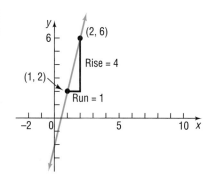

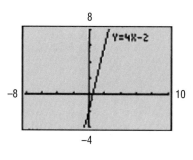

E X A M P L E   7

Finding the Equation of a Horizontal Line

Find an equation of the horizontal line passing through the point $(3, 2)$.

Solution | The slope of a horizontal line is 0. To get an equation, we use the point–slope form with $m = 0$, $x_1 = 3$, and $y_1 = 2$:

$$y - y_1 = m(x - x_1)$$
$$y - 2 = 0 \cdot (x - 3)$$
$$y - 2 = 0$$
$$y = 2$$

See Figure 67 for the graph.

FIGURE 67
$y = 2$

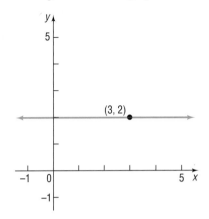
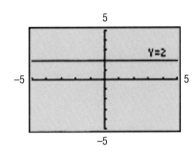

As suggested by Example 7, we have the following result:

Theorem   A horizontal line is given by an equation of the form

Equation of a Horizontal Line

$$y = b$$

where b is the y-intercept.

E X A M P L E   8   *Finding an Equation of a Line Given Two Points*

Find an equation of the line L passing through the points $(2, 3)$ and $(-4, 5)$. Graph the line L.

Solution   Since two points are given, we first compute the slope of the line:

$$m = \frac{5 - 3}{-4 - 2} = \frac{2}{-6} = \frac{-1}{3}$$

We use the point $(2, 3)$ and the fact that the slope $m = -\dfrac{1}{3}$ to get the point–slope form of the equation of the line:

$$y - 3 = -\frac{1}{3}(x - 2)$$

See Figure 68 for the graph.

FIGURE 68
$y - 3 = -\frac{1}{3}(x - 2)$

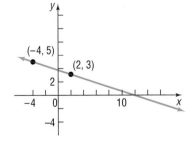
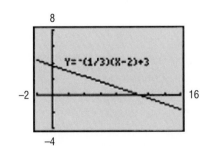

In the solution to Example 8, we could have used the other point, $(-4, 5)$, instead of the point $(2, 3)$. The equation that results, although it looks different, is equivalent to the equation we obtained in the example. (Try it for yourself.)

Another form of the equation of the line in Example 8 can be obtained by multiplying both sides of the point–slope equation by 3 and collecting terms:

$$y - 3 = -\frac{1}{3}(x - 2)$$

$$3(y - 3) = 3\left(-\frac{1}{3}\right)(x - 2) \quad \text{Multiply by 3.}$$

$$3y - 9 = -1(x - 2)$$

$$3y - 9 = -x + 2$$

$$x + 3y - 11 = 0$$

General Form of an Equation of a Line

> The equation of a line L is in **general form** when it is written as
>
> $$Ax + By + C = 0 \tag{3}$$
>
> where A, B, and C are three real numbers and A and B are not both 0.

■   Now work Problem 31.

Every line has an equation that is equivalent to an equation written in general form. For example, a vertical line whose equation is

$$x = a$$

can be written in the general form

$$1 \cdot x + 0 \cdot y - a = 0 \quad A = 1, B = 0, C = -a$$

A horizontal line whose equation is

$$y = b$$

can be written in the general form

$$0 \cdot x + 1 \cdot y - b = 0 \quad A = 0, B = 1, C = -b$$

Lines that are neither vertical nor horizontal have general equations of the form

$$Ax + By + C = 0 \quad A \neq 0 \text{ and } B \neq 0$$

Because the equation of every line can be written in general form, any equation equivalent to (3) is called a **linear equation.**

The next example illustrates one way of graphing a linear equation.

EXAMPLE 9 *Finding the Intercepts of a Line*

Find the intercepts of the line $2x + 3y - 6 = 0$. Graph this line.

Solution To find the point at which the graph crosses the x-axis, that is, to find the x-intercept, we need to find the number x for which $y = 0$. Thus, we let $y = 0$ to get

$$\begin{aligned} 2x + 3(0) - 6 &= 0 \\ 2x - 6 &= 0 \\ x &= 3 \end{aligned}$$

The x-intercept is 3. To find the y-intercept, we let $x = 0$ and solve for y:

$$\begin{aligned} 2(0) + 3y - 6 &= 0 \\ 3y - 6 &= 0 \\ y &= 2 \end{aligned}$$

The y-intercept is 2.

To graph the line by hand, we use the intercepts. Since the x-intercept is 3 and the y-intercept is 2, we know two points on the line: (3, 0) and (0, 2). Because two points determine a unique line, we do not need further information to graph the line. See Figure 69.

FIGURE 69 $2x + 3y - 6 = 0$

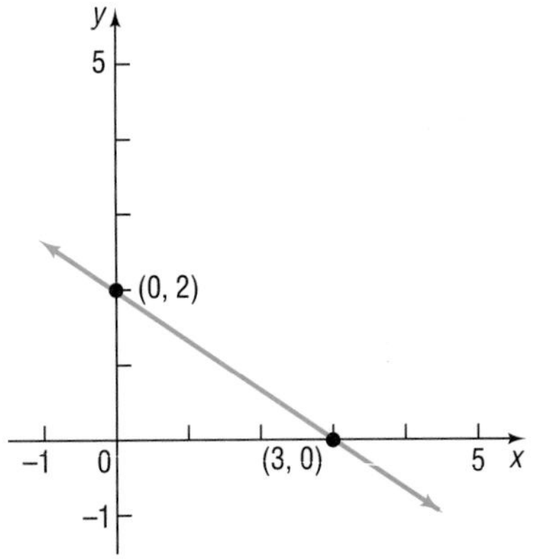

To graph the line using a graphing utility, we need to solve for y.

$$\begin{aligned} 2x + 3y - 6 &= 0 \\ 3y &= -2x + 6 \\ y &= -\tfrac{2}{3}x + 2 \end{aligned}$$

See Figure 70 for the graph.

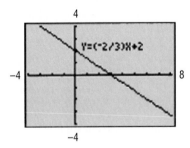

FIGURE 70

■

Another useful equation of a line is obtained when the slope m and y-intercept b are known. In this event, we know both the slope m of the line and a point $(0, b)$ on the line; thus, we may use the point–slope form, equation (2), to obtain the following equation:

$$y - b = m(x - 0) \quad \text{or} \quad y = mx + b$$

Theorem An equation of a line L with slope m and y-intercept b is

Slope–Intercept Form of an Equation of a Line

$$y = mx + b \tag{4}$$

■

Seeing the Concept: To see the role the slope m plays, graph the following lines on the same square screen

$$y = 2$$
$$y = x + 2$$
$$y = -x + 2$$
$$y = 3x + 2$$
$$y = -3x + 2$$

See Figure 71.

FIGURE 71
$y = mx + 2$

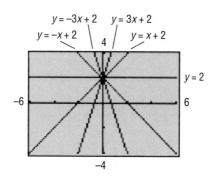

What do you conclude about the lines $y = mx + 2$?

Seeing the Concept: To see the role of the y-intercept b, graph the following lines on the same square screen.

$$y = 2x$$
$$y = 2x + 1$$
$$y = 2x - 1$$
$$y = 2x + 4$$
$$y = 2x - 4$$

FIGURE 72
$y = 2x + b$

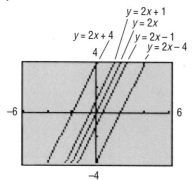

See Figure 72. What do you conclude about the lines $y = 2x + b$?

When the equation of a line is written in slope–intercept form, it is easy to find the slope m and y-intercept b of the line. For example, suppose the equation of a line is

$$y = -2x + 3$$

Compare it to $y = mx + b$:

$$y = -2x + 3$$
$$y = mx + b$$

The slope of this line is -2 and its y-intercept is 3.

Let's look at another example.

E X A M P L E   1 0   *Finding the Slope and y-Intercept of a Line*

Find the slope m and y-intercept b of the line $2x + 4y - 8 = 0$. Graph the line.

Solution   To obtain the slope and *y*-intercept, we transform the equation into its slope–intercept form. Thus, we need to solve for *y*:

$$2x + 4y - 8 = 0$$
$$4y = -2x + 8$$
$$y = -\tfrac{1}{2}x + 2$$

The coefficient of *x*, $-\tfrac{1}{2}$, is the slope, and the *y*-intercept is 2.
Figure 73 shows the graph using a graphing utility.

FIGURE 73

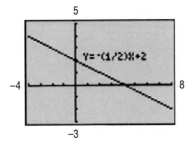

We can graph the line by hand in two ways:

1. Use the fact that the *y*-intercept is 2 and the slope is $-\tfrac{1}{2}$. Then, starting at the point (0, 2), go to the right 2 units and then down 1 unit to the point (2, 1). See Figure 74.

Or:

2. Locate the intercepts. Because the *y*-intercept is 2, we know one intercept is (0, 2). To obtain the *x*-intercept, let *y* = 0 and solve for *x*. When *y* = 0, we have

$$2x + 4 \cdot 0 - 8 = 0$$
$$2x - 8 = 0$$
$$x = 4$$

Thus, the intercepts are (4, 0) and (0, 2). See Figure 75.

FIGURE 74
$2x + 4y - 8 = 0$

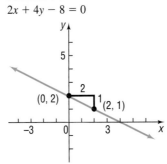

FIGURE 75
$2x + 4y - 8 = 0$

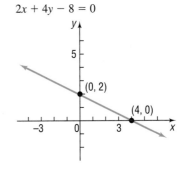

■   Now work Problem 47.

The next example illustrates a typical situation that requires the use of linear equations.

E X A M P L E   1 1   *Computing the Cost of Operating a Car*

The National Car Rental Company has determined that the cumulative cost of operating a vehicle is $0.41 per mile.

(a) Write an equation that relates the cumulative cost C, in dollars, of operating a car and the number x of miles it has been driven.

(b) What is the cost of operating a car with 1000 miles on it?

(c) What is the cost of operating a car with 2000 miles on it?

Solution   (a) If x is the number of miles the car has been driven, then the cumulative cost C, in dollars, is $0.41x$. Thus, an equation relating C and x is

$$C = 0.41x \qquad x \geq 0$$

The average cost per mile, $0.41, is the slope of the line $C = 0.41x$. In other words, the cost increases by $0.41 for each additional mile driven. See Figure 76.

(b) The cost of operating a car with 1000 miles on it is

$$C = 0.41x = 0.41(1000) = \$410$$

(c) The cost of operating a car with 2000 miles on it is

$$C = 0.41x = 0.41(2000) = \$820$$ ■

FIGURE 76
$C = 0.41x$

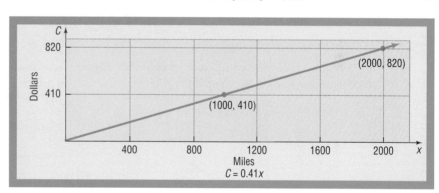

1.3

Exercise 1.3

In Problems 1–4, find the slope of the line.

1.

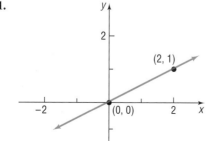

2.

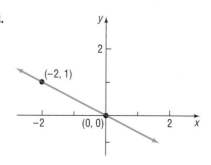

3.

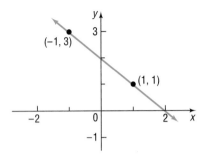

4.

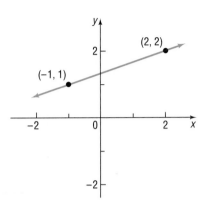

In Problems 5–14, plot each pair of points and determine the slope of the line containing them. By hand, graph the line.

5. (2, 3); (4, 0)

6. (4, 2); (3, 4)

7. (−2, 3); (2, 1)

8. (−1, 1); (2, 3)

9. (−3, −1); (2, −1)

10. (4, 2); (−5, 2)

11. (−1, 2); (−1, −2)

12. (2, 0); (2, 2)

13. ($\sqrt{2}$, 3); (1, $\sqrt{3}$)

14. (−2$\sqrt{2}$, 0); (4, $\sqrt{5}$)

In Problems 15–22, graph by hand the line passing through the point P and having slope m.

15. $P = (1, 2)$; $m = 3$

16. $P = (2, 1)$; $m = 4$

17. $P = (2, 4)$; $m = \frac{-3}{4}$

18. $P = (1, 3)$; $m = \frac{-2}{5}$

19. $P = (-1, 3)$; $m = 0$

20. $P = (2, -4)$; $m = 0$

21. $P = (0, 3)$; slope undefined

22. $P = (-2, 0)$; slope undefined

In Problems 23–26, find an equation of each line. Express your answer using either the general form or the slope–intercept form of the equation of a line, whichever you prefer.

23.

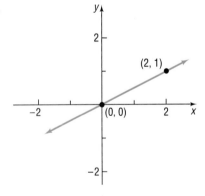

24.

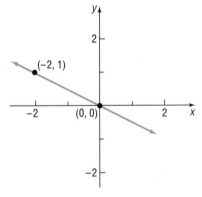

25.

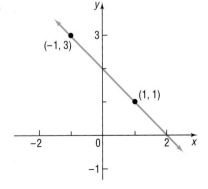

26.

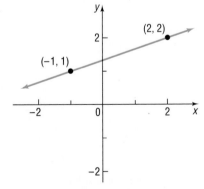

In Problems 27–38, find an equation for the line with the given properties. Express your answer using either the general form or the slope–intercept form of the equation of a line, whichever you prefer.

27. Slope $= 3$; passing through $(-2, 3)$

28. Slope $= 2$; passing through $(4, -3)$

29. Slope $= -\frac{2}{3}$; passing through $(1, -1)$

30. Slope $= \frac{1}{2}$; passing through $(3, 1)$

31. Passing through $(1, 3)$ and $(-1, 2)$

32. Passing through $(-3, 4)$ and $(2, 5)$

33. Slope $= -3$; y-intercept $= 3$

34. Slope $= -2$; y-intercept $= -2$

35. x-intercept $= 2$; y-intercept $= -1$

36. x-intercept $= -4$; y-intercept $= 4$

37. Slope undefined; passing through $(2, 4)$

38. Slope undefined; passing through $(3, 8)$

In Problems 39–58, find the slope and y-intercept of each line. By hand, graph the line. Check your graph using a graphing utility.

39. $y = 2x + 3$

40. $y = -3x + 4$

41. $\frac{1}{2}y = x - 1$

42. $\frac{1}{3}x + y = 2$

43. $y = \frac{1}{2}x + 2$

44. $y = 2x + \frac{1}{2}$

45. $x + 2y = 4$

46. $-x + 3y = 6$

47. $2x - 3y = 6$

48. $3x + 2y = 6$

49. $x + y = 1$

50. $x - y = 2$

51. $x = -4$

52. $y = -1$

53. $y = 5$

54. $x = 2$

55. $y - x = 0$

56. $x + y = 0$

57. $2y - 3x = 0$

58. $3x + 2y = 0$

59. Find an equation of the x-axis.

60. Find an equation of the y-axis.

61. *Measuring Temperature* The relationship between Celsius (°C) and Fahrenheit (°F) degrees for measuring temperature is linear. Find an equation relating °C and °F if 0°C corresponds to 32°F and 100°C corresponds to 212°F. Use the equation to find the Celsius measure of 70°F.

62. *Measuring Temperature* The Kelvin (K) scale for measuring temperature is obtained by adding 273 to the Celsius temperature.
(a) Write an equation relating K and °C.
(b) Write an equation relating K and °F (see Problem 61).

63. *Business: Computing Profit* Each Sunday, a newspaper agency sells x copies of a certain newspaper for $1.00 per copy. The cost to the agency of each newspaper is $0.50. The agency pays a fixed cost for storage, delivery, and so on, of $100 per Sunday.
(a) Write an equation that relates the profit P, in dollars, to the number x of copies sold. Graph this equation.
(b) What is the profit to the agency if 1000 copies are sold?
(c) What is the profit to the agency if 5000 copies are sold?

64. *Business: Computing Profit* Repeat Problem 63 if the cost to the agency is $0.45 per copy and the fixed cost is $125 per Sunday.

65. *Cost of Electricity* In 1991, Commonwealth Edison Company supplied electricity in the summer months to residential customers for a monthly customer charge of $9.06 plus 10.819¢ per kilowatt-hour supplied in the month.* Write an equation that relates the monthly charge C, in dollars, to the number x of kilowatt-hours in the month. Graph this equation. What is the monthly charge for using 300 kilowatt-hours? For using 900 kilowatt-hours?

66. Show that an equation for a line with nonzero x- and y-intercepts can be written as

$$\frac{x}{a} + \frac{y}{b} = 1$$

where a is the x-intercept and b is the y-intercept. This is called the **intercept form** of the equation of a line.

Source: Commonwealth Edison Co., Chicago, Illinois, 1991.

In Problems 67–70, match each graph with the correct equation:

(a) $y = x$ *(b)* $y = 2x$ *(c)* $y = x/2$ *(d)* $y = 4x$

67.
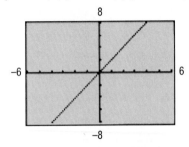

68.
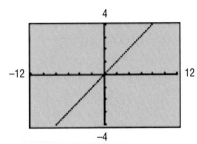

69.
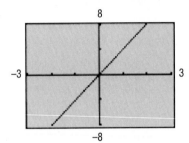

70.
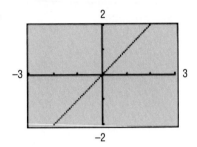

In Problems 71–76, write an equation of each line. Express your answer using either the general form or the slope–intercept form of the equation of a line, whichever you prefer.

71.
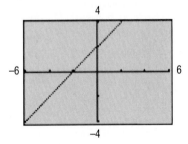

72.
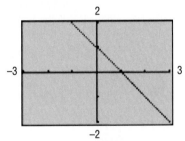

73.
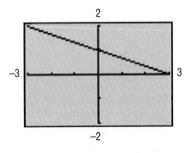

74.
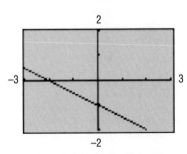

75.
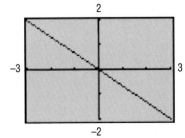

76.
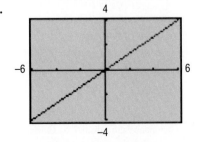

77. Which form of the equation of a line do you prefer to use? Justify your position with an example that shows that your choice is better than another. Have reasons.

78. Can every line be written in slope–intercept form? Explain.

79. Does every line have two distinct intercepts? Explain. Are there lines that have no intercepts? Explain.

80. What can you say about two lines that have equal slopes and equal y-intercepts?

81. What can you say about two lines with the same x-intercept and the same y-intercept? Assume that the x-intercept is not 0.

82. If two lines have the same slope, but different x-intercepts, can they have the same y-intercept?

83. If two lines have the same y-intercept, but different slopes, can they have the same x-intercept? What is the only way this can happen?

84. The accepted symbol used to denote the slope of a line is the letter *m*. Investigate the origin of this symbolism. Begin by consulting a French dictionary and looking up the French word *monter*. Write a brief essay on your findings.

85. The term *grade* is used to describe the inclination of a road. How does this term relate to the notion of slope of a line? Is a 4% grade very steep? Investigate the grades of some mountainous roads and determine their slopes. Write a brief essay on your findings.

86. *Carpentry* Carpenters use the term pitch to describe the steepness of staircases and roofs. How does pitch relate to slope? Investigate typical pitches used for stairs and for roofs. Write a brief essay on your findings.

1.4

Parallel and Perpendicular Lines; Circles

Parallel and Perpendicular Lines

When two lines (in the plane) have no points in common, they are said to be **parallel.** Look at Figure 77. There we have drawn two lines and have constructed two right triangles by drawing sides parallel to the coordinate axes. These lines are par-

FIGURE 77
The lines are parallel if and only if their slopes are equal.

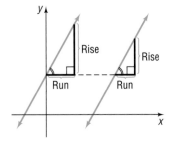

allel if and only if the right triangles are similar. (Do you see why? Two angles are equal.) But the triangles are similar if and only if the ratios of corresponding sides are equal.

This suggests the following result:

Theorem   Two distinct nonvertical lines are parallel if and only if their slopes are equal. ■

The use of the words "if and only if" in the preceding theorem means that actually two statements are being made, one the converse of the other.

If two distinct nonvertical lines are parallel, then their slopes are equal.

If two distinct nonvertical lines have equal slopes, then they are parallel.

■   Now work Problem 1(a).

E X A M P L E   1   *Showing That Two Lines Are Parallel*

Show that the lines given by the following equations are parallel:

$$L:\quad 2x + 3y - 6 = 0 \qquad M:\quad 4x + 6y = 0$$

Solution   To determine whether these lines have equal slopes, we write each equation in slope–intercept form:

$$
\begin{array}{ll}
L:\quad 2x + 3y - 6 = 0 & \qquad M:\quad 4x + 6y = 0 \\
\qquad\quad 3y = -2x + 6 & \qquad\qquad\quad 6y = -4x \\
\qquad\quad\; y = -\tfrac{2}{3}x + 2 & \qquad\qquad\quad\; y = -\tfrac{2}{3}x \\
\qquad\quad \text{Slope} = -\tfrac{2}{3} & \qquad\qquad\quad \text{Slope} = -\tfrac{2}{3}
\end{array}
$$

Because these lines have the same slope, $-\tfrac{2}{3}$, but different y-intercepts, the lines are parallel. See Figure 78.

FIGURE 78
Parallel lines

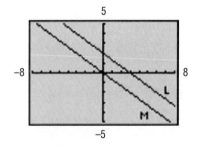 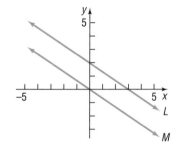

E X A M P L E   2   *Finding a Line That Is Parallel to Some Other Line*

Find an equation for the line that contains the point $(2, -3)$ and is parallel to the line $2x + y - 6 = 0$.

Solution   The slope of the line we seek equals the slope of the line $2x + y - 6 = 0$, since the two lines are to be parallel. Thus, we begin by writing the equation of the line $2x + y - 6 = 0$ in slope–intercept form:

$$2x + y - 6 = 0$$
$$y = -2x + 6$$

The slope is -2. Since the line we seek contains the point $(2, -3)$, we use the point–slope form to obtain

$$y + 3 = -2(x - 2)$$
$$2x + y - 1 = 0 \qquad \text{\small General form}$$
$$y = -2x + 1 \qquad \text{\small Slope–intercept form}$$

This line is parallel to the line $2x + y - 6 = 0$ and contains the point $(2, -3)$. See Figure 79.

FIGURE 79

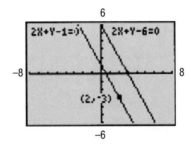
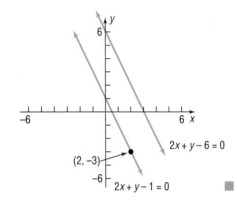

FIGURE 80

Perpendicular lines

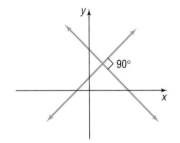

■ Now work Problem 15.

When two lines intersect at a right angle ($90°$), they are said to be **perpendicular.** See Figure 80.

The following result gives a condition, in terms of their slopes, for two lines to be perpendicular:

Theorem Two nonvertical lines are perpendicular if and only if the product of their slopes is -1.   ■

Here, we shall prove the "only if" part of the statement:

If two nonvertical lines are perpendicular, then the product of their slopes is -1.

In Problem 75, you are asked to prove the "if" part of the theorem; that is:

If two nonvertical lines have slopes whose product is -1, then the lines are perpendicular.

Proof Let m_1 and m_2 denote the slopes of the two lines. There is no loss in generality (that is, neither the angle nor the slopes are affected) if we situate the lines so that they meet at the origin. See Figure 81. The point $A = (1, m_2)$ is on the line having slope m_2, and the point $B = (1, m_1)$ is on the line having slope m_1. (Do you see why this must be true?)

FIGURE 81

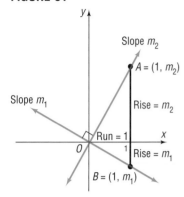

Suppose that the lines are perpendicular. Then triangle OAB is a right triangle. As a result of the Pythagorean Theorem, it follows that

$$[d(O, A)]^2 + [d(O, B)]^2 = [d(A, B)]^2 \qquad (1)$$

By the distance formula, we can write each of these distances as

$$[d(O, A)]^2 = (1 - 0)^2 + (m_2 - 0)^2 = 1 + m_2^2$$

$$[d(O, B)]^2 = (1 - 0)^2 + (m_1 - 0)^2 = 1 + m_1^2$$

$$[d(A, B)]^2 = (1 - 1)^2 + (m_2 - m_1)^2 = m_2^2 - 2m_1 m_2 + m_1^2$$

Using these facts in equation (1), we get

$$(1 + m_2^2) + (1 + m_1^2) = m_2^2 - 2m_1 m_2 + m_1^2$$

which, upon simplification, can be written as

$$m_1 m_2 = -1$$

Thus, if the lines are perpendicular, the product of their slopes is -1. ■

You may find it easier to remember the condition for two nonvertical lines to be perpendicular by observing that the equality $m_1 m_2 = -1$ means that m_1 and m_2 are negative reciprocals of each other; that is, either $m_1 = -1/m_2$ or $m_2 = -1/m_1$.

E X A M P L E   3   *Finding the Slope of a Line Perpendicular to Another Line*

If a line has slope $\frac{3}{2}$, any line having slope $-\frac{2}{3}$ is perpendicular to it. ■

E X A M P L E   4   *Finding the Equation of a Line Perpendicular to Another Line*

Find an equation of the line passing through the point $(1, -2)$ and perpendicular to the line $x + 3y - 6 = 0$. Graph the two lines.

Solution   We first write the equation of the given line in slope–intercept form to find its slope:

$$x + 3y - 6 = 0$$
$$3y = -x + 6$$
$$y = -\tfrac{1}{3}x + 2$$

The given line has slope $-\frac{1}{3}$. Any line perpendicular to this line will have slope 3. Because we require the point $(1, -2)$ to be on this line with slope 3, we use the point–slope form of the equation of a line:

$$y - (-2) = 3(x - 1)$$
$$y + 2 = 3(x - 1)$$

This equation is equivalent to the forms

$$3x - y - 5 = 0 \qquad \text{General form}$$
$$y = 3x - 5 \qquad \text{Slope–intercept form}$$

Figure 82 shows the graphs.

FIGURE 82

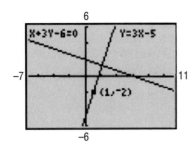

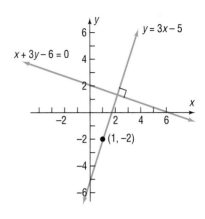

■   Now work Problem 23.

Warning Be sure to use a square screen when you graph perpendicular lines. Otherwise, the angle between the two lines will appear distorted.

Circles

One advantage of a coordinate system is that it enables us to translate a geometric statement into an algebraic statement, and vice versa. Consider, for example, the following geometric statement that defines a circle.

Circle

> A **circle** is a set of points in the xy-plane that are a fixed distance r from a fixed point (h, k). The fixed distance r is called the **radius,** and the fixed point (h, k) is called the **center** of the circle.

FIGURE 83

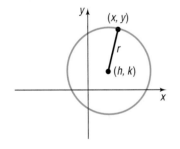

Figure 83 shows the graph of a circle. Is there an equation having this graph? If so, what is the equation? To find the equation, we let (x, y) represent the coordinates of any point on a circle with radius r and center (h, k). Then the distance between the points (x, y) and (h, k) must always equal r. That is, by the distance formula,

$$\sqrt{(x - h)^2 + (y - k)^2} = r$$

or, equivalently,

$$(x - h)^2 + (y - k)^2 = r^2$$

The **standard form of an equation of a circle** with radius r and center (h, k) is

Standard Form
of an Equation of a Circle

$$(x - h)^2 + (y - k)^2 = r^2 \qquad (2)$$

Conversely, by reversing the steps, we conclude: The graph of any equation of the form of equation (2) is that of a circle with radius r and center (h, k).

E X A M P L E   5   *Graphing a Circle*

Graph the equation: $(x + 3)^2 + (y - 2)^2 = 16$

Solution   The graph of the equation is a circle. To graph a circle on a graphing utility we must write the equation in the form $y =$ expression involving x*. Thus, we must solve for y in the equation

$$(x + 3)^2 + (y - 2)^2 = 16$$
$$(y - 2)^2 = 16 - (x + 3)^2$$
$$y - 2 = \pm\sqrt{16 - (x + 3)^2}$$
$$y = 2 \pm \sqrt{16 - (x + 3)^2}$$

To graph the circle, we first graph the top half

$$y = 2 + \sqrt{16 - (x + 3)^2}$$

and then graph the bottom half

$$y = 2 - \sqrt{16 - (x + 3)^2}$$

Also, be sure to use a square screen. Otherwise the circle will appear distorted. Figure 84 shows the graph.

FIGURE 84

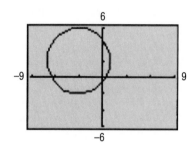

To graph the equation by hand, we first compare the given equation to the standard form of the equation of a circle. The comparison yields information about the circle:

$$(x + 3)^2 + (y - 2)^2 = 16$$
$$(x - (-3))^2 + (y - 2)^2 = 4^2$$
$$(x - h)^2 + (y - k)^2 = r^2$$

FIGURE 85

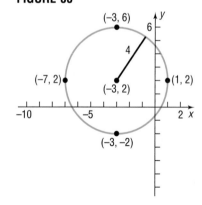

We see that $h = -3$, $k = 2$, and $r = 4$. Hence, the circle has center $(-3, 2)$ and a radius of 4 units. To graph this circle, we first plot the center $(-3, 2)$. Since the radius is 4, we can locate four points on the circle by going out 4 units to the left, to the right, up, and down from the center. These four points can then be used as guides to obtain the graph. See Figure 85.   ∎

■   Now work Problem 43.

*Some graphing utilities (e.g. TI-82, TI-85) have a CIRCLE function which allows the user to enter only the coordinates of the center of the circle and its radius to graph the circle.

ISSION POSSIBLE

Chapter 1

PREDICTING THE FUTURE OF OLYMPIC TRACK EVENTS

Some people think that women athletes are beginning to "catch up" to men in the Olympic Track Events. In fact, researchers at UCLA published data in 1992 that seemed to indicate that within 65 years, the top men and women runners would be able to compete on an equal basis. Other researchers disagreed. Here is some data on the 200-meter run with winning times for the Olympics in the year given.

	MEN'S TIME IN SECONDS	WOMEN'S TIME IN SECONDS
1948	21.1	24.4
1952	20.7	23.7
1956	20.6	23.4
1960	20.5	24.0
1964	20.3	23.0
1968	19.83	22.5
1972	20.00	22.40
1976	20.23	22.37
1980	20.19	22.03
1984	19.80	21.81
1988	19.75	21.34
1992	19.73	21.72

(You may notice that the introduction of better timing devices meant more accurate measures starting in 1968 for the men and 1972 for the women.)

1. To begin your investigation of the UCLA conjecture, make a graph of these data. To get the kind of accuracy you need, you should use graph paper. You will need to use only the first quadrant. On your x-axis place the Olympic years from 1948 to 2048, counting by 4's. On your y-axis place the numbers from 15 to 25 which represent the seconds; if you're using graph paper, allow about four squares per one second. Use X's to represent the men's times on the graph and O's to represent the women's times. These should form what we call a "scatter plot," not a neat line or curve.

2. Using a ruler or straightedge, draw a line that represents roughly the slope and direction indicated by the men's scores. Do the same for the women's scores. Write a sentence or two explaining why you think your line is a good representation of the scatter plot.

3. Next find the equation for each line, using your graph to estimate the slope and y-intercept for each. Try to be as accurate as possible, remembering your units and using the points where the line crosses intersections of the grid. The slope will be in seconds per year.

4. Do your two lines appear to cross? In what year do they cross? If you solve the two equations algebraically, do you get the same answer?

5. Some graphing calculators enable you do this problem in a statistics mode. They will plot the individual point that you type in and find a line that represents the data. Use your graphing calculator to find out whether its equation matches yours.

6. Make a group decision about whether or not you think that women's times will "catch up" to men's times in the future. Write out 2-3 sentences explaining why you believe they will or will not.

E X A M P L E   6   *Writing the Standard Form of the Equation of a Circle*

Write the standard form of the equation of the circle with radius 3 and center $(1, -2)$.

Solution   Using the form of equation (2) and substituting the values $r = 3$, $h = 1$, and $k = -2$, we have

$$(x - h)^2 + (y - k)^2 = r^2$$
$$(x - 1)^2 + (y + 2)^2 = 9$$ ∎

■ Now work Problem 27.

The standard form of an equation of a circle of radius r with center at the origin $(0, 0)$ is

$$x^2 + y^2 = r^2$$

If the radius $r = 1$, the circle whose center is at the origin is called the **unit circle** and has the equation

$$x^2 + y^2 = 1$$

See Figure 86.

FIGURE 86
Unit circle $x^2 + y^2 = 1$

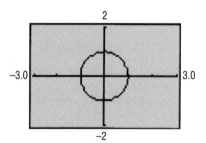
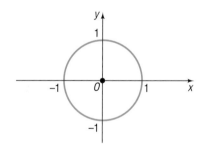

If we eliminate the parentheses from the standard form of the equation of the circle obtained in Example 6, we get

$$(x - 1)^2 + (y + 2)^2 = 9$$
$$x^2 - 2x + 1 + y^2 + 4y + 4 = 9$$

which we find, upon simplifying, is equivalent to

$$x^2 + y^2 - 2x + 4y - 4 = 0$$

By completing the squares on both the x- and y-terms, it can be shown that any equation of the form

$$x^2 + y^2 + ax + by + c = 0$$

has a graph that is a circle, or a point, or has no graph at all. For example, the graph of the equation $x^2 + y^2 = 0$ is the single point $(0, 0)$. The equation $x^2 + y^2 + 5 = 0$, or $x^2 + y^2 = -5$, has no graph, because sums of squares of real numbers are never negative. When its graph is a circle, the equation

**General Form
of the Equation of a Circle**

$$x^2 + y^2 + ax + by + c = 0$$

is referred to as the **general form of the equation of a circle.**

The next example shows how to transform an equation in the general form to an equivalent equation in standard form. As we said earlier, the idea is to use the method of completing the square on both the x- and y-terms. See the Appendix, Section 5, for a discussion of completing the square.

E X A M P L E   7   *Graphing a Circle Whose Equation Is in General Form*

Graph the equation: $x^2 + y^2 + 4x - 6y + 12 = 0$

Solution   We rearrange the equation as follows:

$$(x^2 + 4x) + (y^2 - 6y) = -12$$

Next, we complete the square of each expression in parentheses. Remember that any number added on the left also must be added on the right:

$$(x^2 + 4x + 4) + (y^2 - 6y + 9) = -12 + 4 + 9$$
$$(x + 2)^2 + (y - 3)^2 = 1$$

We recognize this equation as the standard form of the equation of a circle with radius 1 and center $(-2, 3)$.

To graph the equation using a graphing utility, we need to solve for y:

$$(y - 3)^2 = 1 - (x + 2)^2$$
$$y - 3 = \pm\sqrt{1 - (x + 2)^2}$$
$$y = 3 \pm \sqrt{1 - (x + 2)^2}$$

To graph the equation by hand, we use the center $(-2, 3)$ and the radius of 1. Figure 87 illustrates the graph.

FIGURE 87
$x^2 + y^2 + 4x - 6y + 12 = 0$

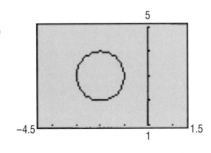
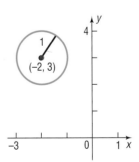

Now work Problem 45.

EXAMPLE 8   *Finding the General Equation of a Circle*

Find the general equation of the circle whose center is $(1, -2)$ and whose graph contains the point $(4, 2)$.

Solution   To find the equation of a circle, we need to know its center and its radius. Here, we know that the center is $(1, -2)$. Since the point $(4, 2)$ is on the graph, the radius r will equal the distance from $(4, 2)$ to the center $(1, -2)$. See Figure 88. Thus,

$$r = \sqrt{(4 - 1)^2 + [2 - (-2)]^2}$$
$$= \sqrt{9 + 16} = 5$$

The standard form of the equation of the circle is

$$(x - 1)^2 + (y + 2)^2 = 25$$

Eliminating the parentheses and rearranging terms, we get the general equation

$$x^2 + y^2 - 2x + 4y - 20 = 0 \qquad \blacksquare$$

FIGURE 88
$x^2 + y^2 - 2x + 4y - 20 = 0$

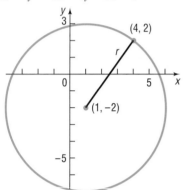

Overview

The preceding discussion about lines and circles dealt with two main types of problems that can be generalized as follows:

1. Given an equation, classify it and graph it.
2. Given a graph, or information about a graph, find its equation.

This text deals mainly with the first type of problem. We shall study various equations, classify them, and graph them. The second type of problem is usually more difficult to solve than the first. However, we shall tackle such problems when it is practical to do so.

1.4

Exercise 1.4

In Problems 1–10, the equation of a line L is given. Find the slope of a line that is: (a) Parallel to L
(b) Perpendicular to L

1. $y = 4x$ **2.** $y = -5x$ **3.** $y = -\frac{1}{2}x + 2$ **4.** $y = \frac{2}{3}x - 1$

5. $2x - 4y + 5 = 0$ **6.** $3x + y = 4$ **7.** $3x + 5y - 10 = 0$ **8.** $4x - 3y + 7 = 0$

9. $x = 4$ **10.** $y = 5$

In Problems 11–14, find an equation for the line L. Express your answer using either the general form or the slope–intercept form of the equation of a line, whichever you prefer.

11.

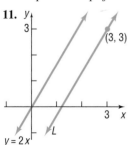

L is parallel to $y = 2x$

12.

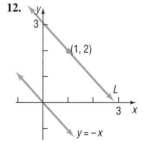

L is parallel to $y = -x$

13.

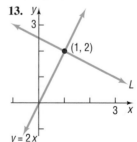

L is perpendicular to $y = 2x$

14.

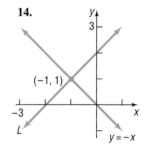

L is perpendicular to $y = -x$

In Problems 15–26, find an equation for the line with the given properties. Express your answer using either the general form or the slope–intercept form of the equation of a line, whichever you prefer.

15. Parallel to the line $y = 2x$; passing through $(-1, 2)$

16. Parallel to the line $y = -3x$; passing through $(-1, 2)$

17. Parallel to the line $2x - y + 2 = 0$; passing through $(0, 0)$

18. Parallel to the line $x - 2y + 5 = 0$; passing through $(0, 0)$

19. Parallel to the line $x = 5$; passing through $(4, 2)$

20. Parallel to the line $y = 5$; passing through $(4, 2)$

21. Perpendicular to the line $y = \frac{1}{2}x + 4$; passing through $(1, -2)$

22. Perpendicular to the line $y = 2x - 3$; passing through $(1, -2)$

23. Perpendicular to the line $2x + y - 2 = 0$; passing through $(-3, 0)$

24. Perpendicular to the line $x - 2y + 5 = 0$; passing through $(0, 4)$

25. Perpendicular to the line $x = 8$; passing through $(3, 4)$

26. Perpendicular to the line $y = 8$; passing through $(3, 4)$

In Problems 27–30, find the center and radius of each circle. Write the standard form of the equation.

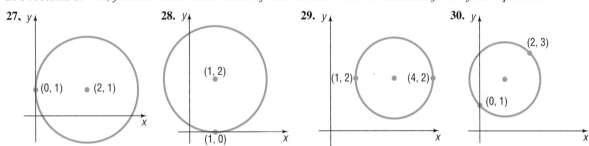

27. with $(0, 1)$ and $(2, 1)$

28. with $(1, 2)$ and $(1, 0)$

29. with $(1, 2)$ and $(4, 2)$

30. with $(2, 3)$ and $(0, 1)$

In Problems 31–40, write the standard form of the equation and the general form of the equation of each circle of radius r and center (h, k). By hand, graph each circle.

31. $r = 1$; $(h, k) = (1, -1)$

32. $r = 2$; $(h, k) = (-2, 1)$

33. $r = 2$; $(h, k) = (0, 2)$

34. $r = 3$; $(h, k) = (1, 0)$

35. $r = 5$; $(h, k) = (4, -3)$

36. $r = 4$; $(h, k) = (2, -3)$

37. $r = 2$; $(h, k) = (0, 0)$

38. $r = 3$; $(h, k) = (0, 0)$

39. $r = \frac{1}{2}$; $(h, k) = \left(\frac{1}{2}, 0\right)$

40. $r = \frac{1}{2}$; $(h, k) = \left(0, -\frac{1}{2}\right)$

In Problems 41–50, find the center (h, k) and radius r of each circle. By hand, graph each circle.

41. $x^2 + y^2 = 4$

42. $x^2 + (y - 1)^2 = 1$

43. $(x - 3)^2 + y^2 = 4$

44. $(x + 1)^2 + (y - 1)^2 = 2$

45. $x^2 + y^2 + 4x - 4y - 1 = 0$

46. $x^2 + y^2 - 6x + 2y + 9 = 0$

47. $x^2 + y^2 - x + 2y + 1 = 0$

48. $x^2 + y^2 + x + y - \frac{1}{2} = 0$

49. $2x^2 + 2y^2 - 12x + 8y - 24 = 0$

50. $2x^2 + 2y^2 + 8x + 7 = 0$

In Problems 51–56, find the general form of the equation of each circle.

51. Center at the origin and containing the point $(-2, 3)$

52. Center $(1, 0)$ and containing the point $(-3, 2)$

53. Center $(2, 3)$ and tangent to the x-axis

54. Center $(-3, 1)$ and tangent to the y-axis

55. With endpoints of a diameter at $(1, 4)$ and $(-3, 2)$

56. With endpoints of a diameter at $(4, 3)$ and $(0, 1)$

57. *Geometry* Use slopes to show that the triangle whose vertices are $(-2, 5)$, $(1, 3)$, and $(-1, 0)$ is a right triangle.

58. *Geometry* Use slopes to show that the quadrilateral whose vertices are $(1, -1)$, $(4, 1)$, $(2, 2)$, and $(5, 4)$ is a parallelogram.

59. *Geometry* Use slopes to show that the quadrilateral whose vertices are $(-1, 0)$, $(2, 3)$, $(1, -2)$, and $(4, 1)$ is a rectangle.

60. *Geometry* Use slopes and the distance formula to show that the quadrilateral whose vertices are $(0, 0)$, $(1, 3)$, $(4, 2)$, and $(3, -1)$ is a square.

In Problems 61–64, match each graph with the correct equation.
(a) $(x - 3)^2 + (y + 3)^2 = 9$ (c) $(x - 1)^2 + (y + 2)^2 = 4$
(b) $(x + 1)^2 + (y - 2)^2 = 4$ (d) $(x + 3)^2 + (y - 3)^2 = 9$

61.

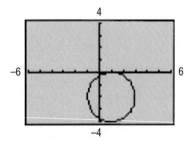

62.

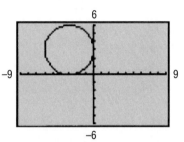

63.

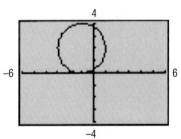

64.

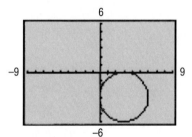

In Problems 65–68, find the standard form of the equation of each circle.

65.

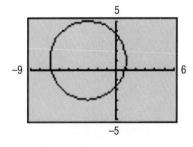

66.

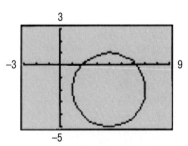

67.

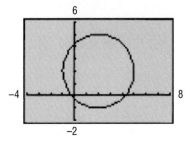

68.

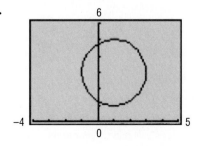

In Problems 69–74, find the area of the parallelogram with the given vertices. [Hint: The area of a parallelogram is the product of the base times the altitude to that base. See the figure.]

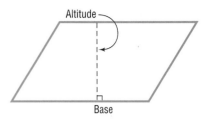

69. (0, 0); (4, 0); (1, 3); (5, 3)   **70.** (0, 0); (0, 2); (3, 1); (3, −1)

71. (2, 1); (4, 2); (2, 3); (4, 4)   **72.** (1, 2); (2, 0); (3, 0); (2, 2)

73. (1, 1); (3, 2); (2, −1); (4, 0)   **74.** (1, 3); (0, 1); (3, 2); (2, 0)

75. Prove that if two nonvertical lines have slopes whose product is −1, then the lines are perpendicular. [*Hint:* Refer to Figure 81, and use the converse of the Pythagorean Theorem.]

76. *Weather Satellites* Earth is represented on a map of a portion of the solar system so that its surface is the circle with equation $x^2 + y^2 + 2x + 4y - 4091 = 0$. A weather satellite circles 0.6 unit above Earth with the center of its circular orbit at the center of Earth. Find the equation for the orbit of the satellite on this map.

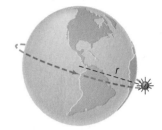

77. The **tangent line** to a circle may be defined as the line that intersects the circle in a single point, called the **point of tangency** (see the figure). If the equation of the circle is $x^2 + y^2 = r^2$ and the equation of the tangent line is $y = mx + b$, show that:

(a) $r^2(1 + m^2) = b^2$ [*Hint:* the quadratic equation $x^2 + (mx + b)^2 = r^2$ has exactly one solution.]

(b) The point of tangency is $(-r^2 m/b, r^2/b)$.

(c) The tangent line is perpendicular to the line containing the center of the circle and the point of tangency.

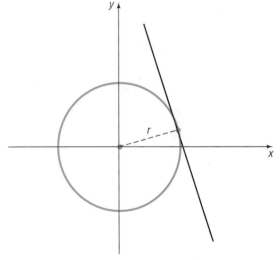

78. The Greek method for finding the equation of the tangent line to a circle used the fact that at any point on a circle the line containing the radius and the tangent line are perpendicular (see Problem 77). Use this method to find an equation of the tangent line to the circle $x^2 + y^2 = 9$ at the point $(1, 2\sqrt{2})$.

79. Use the Greek method described in Problem 78 to find an equation of the tangent line to the circle $x^2 + y^2 - 4x + 6y + 4 = 0$ at the point $(3, 2\sqrt{2} - 3)$.

80. Refer to Problem 77. The line $x - 2y + 4 = 0$ is tangent to a circle at $(0, 2)$. The line $y = 2x - 7$ is tangent to the same circle at $(3, -1)$. Find the center of the circle.

81. Find an equation of the line containing the centers of the two circles

$$x^2 + y^2 - 4x + 6y + 4 = 0 \quad \text{and} \quad x^2 + y^2 + 6x + 4y + 9 = 0$$

82. Show that the line containing the points (a, b) and (b, a) is perpendicular to the line $y = x$. Also show that the midpoint of (a, b) and (b, a) lies on the line $y = x$.

83. The equation $2x - y + C = 0$ defines a **family of lines,** one line for each value of C. On one set of coordinate axes, graph the members of the family when $C = -4$, $C = 0$, and $C = 2$. Can you draw a conclusion from the graph about each member of the family?

84. Rework Problem 83 for the family of lines $Cx + y + 4 = 0$.

85. If a circle of radius 2 is made to roll along the x-axis, what is an equation for the path of the center of the circle?

1.5

Linear Curve Fitting

Curve fitting is an area of statistics in which a relation between two or more variables is explained through an equation. For example, the equation Sales = $100000 + 12Advertising implies that if advertising expenditures were $0, sales would be $100,000 + 12(0) = $100,000 and if advertising expenditures were $10,000, sales would be $100,000 + 12($10,000) = $220,000. In this model, the variable advertising is called the predictor (independent) variable and sales is called the response (dependent) variable because if the level of advertising is known, it can be used to predict sales. Curve fitting is used to find an equation that relates two or more variables, using observed or experimental data.

There are three steps to follow to determine whether a relation exists between two variables.

STEP 1: Ask whether the variables are logically related to each other.
STEP 2: Obtain data and verify a relation exists. Then plot the points. The graph obtained is called a **scatter diagram.**
STEP 3: Find an equation which describes this relation.

Scatter Diagrams

Scatter diagrams are used to help us determine whether a relation exists between two variables. They are useful because of the ease with which they are constructed and also because they allow us to see the type of relation that might exist between the two variables.

To construct a scatter diagram simply plot the data with the independent variable as the x-coordinate and the corresponding dependent variable as the y-coordinate.

E X A M P L E   1   *Drawing a Scatter Diagram*

Use the following data to construct a scatter diagram. Comment on the type of relation you think exists between the two variables.

X(INDEPENDENT)	Y(DEPENDENT)
3	6
5	10
5	8
8	14
9	15
9	18
10	20
11	23

Solution   To draw a scatter diagram by hand, we plot the points (3, 6), (5, 10), (5, 8), and so on, to obtain a scatter diagram. See Figure 89.

FIGURE 89

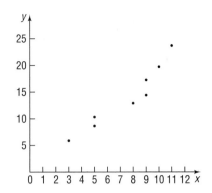

Scatter diagrams can also be constructed using a graphing utility (consult your owners manual). We enter the data into our utility and obtain the scatter diagram shown in Figure 90.

FIGURE 90

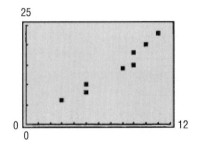

The scatter diagram appears to follow a line with positive slope.   ■

You may observe scatter plots which show other patterns. For example, the data in Figure 91 are not linear, but a pattern is evident. Fitting an equation to this type of data will be discussed in subsequent chapters.

FIGURE 91

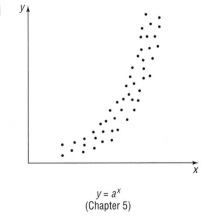

$y = a^x$
(Chapter 5)

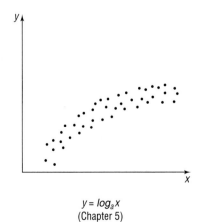

$y = \log_a x$
(Chapter 5)

Lines of Best Fit

E X A M P L E   2   *Finding the Equation of a Line From a Scatter Diagram*

A farmer collected the following data, which shows crop yields for various amounts of fertilizer used.

(a) Use a graphing utility to draw a scatter diagram.

(b) Use a graphing utility to fit a straight line to the data.

PLOT	1	2	3	4	5	5	6	7	8	9	10	11
FERTILIZER, X (POUNDS/100 FT²)	0	0	5	5	10	10	15	15	20	20	25	25
YIELD, Y (POUNDS)	4	6	10	7	12	10	15	17	18	21	23	22

Solution   (a) The data collected indicates a relation exists between the amount of fertilizer used and crop yield. To draw a scatter diagram, we plot points, using fertilizer as the x-coordinate and yield as the y-coordinate. See Figure 92. From the scatter diagram, it appears a linear relation exists between the amount of fertilizer used and yield.

FIGURE 92

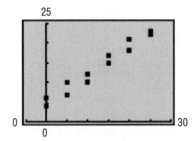

(b) Graphing utilities contain built-in programs that find the linear equation of "best fit" for a collection of points in a scatter diagram.* (Look in your owner's manual under Linear Regression or Line of Best Fit for details on how to execute the program.) Upon executing the LINear REGression program, we obtain the results shown in Figure 93. The output the utility provides shows us the equation, $y = ax + b$, where a is the slope of the line and b is the y-intercept. The line of best fit which relates fertilizer and yield is:

$$\text{Yield} = 0.7171428571(\text{fertilizer}) + 4.785714286$$

FIGURE 93

```
LinReg
y=ax+b
a=.7171428571
b=4.785714286
r=.9803266536
```

*We shall not discuss in this book the underlying mathematics of lines of best fit. Most books in statistics and many in linear algebra discuss this topic.

FIGURE 94

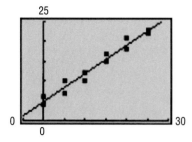

Figure 94 shows the graph of this line, along with the scatter diagram.   ■

   Does the line of best fit appear to be a good "fit"? In other words, does the line appear to accurately describe the relation between yield and fertilizer?

   And just how 'good' is this line of 'best fit'? How accurately does this line describe the relationship between yield and fertilizer? The answers are given by what is called the *correlation coefficient*.

■   Now work Problem 7.

Correlation Coefficients

Look again at Figure 93. The last line of output is 0.98. This number, called the **correlation coefficient, *r*,** $0 \leq |r| \leq 1$, is a measure of the strength of the linear relation that exists between two variables. The closer $|r|$ is to 1, the more perfect the linear relationship is. If r is close to 0, there is little or no linear relationship between the variables. A negative value of r, $r < 0$, indicates that as x increases, then y decreases; a positive value of r, $r > 0$, indicates that as x increases, then y does also. Thus, the data given in Example 2, having a correlation coefficient of 0.98, are strongly indicative of a linear relationship.

   Figure 95 illustrates a variety of scatter diagrams and the relations they suggest.

FIGURE 95

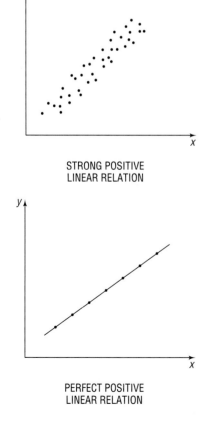

STRONG POSITIVE
LINEAR RELATION

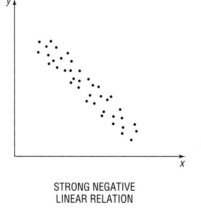

STRONG NEGATIVE
LINEAR RELATION

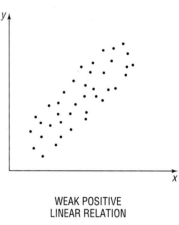

WEAK POSITIVE
LINEAR RELATION

PERFECT POSITIVE
LINEAR RELATION

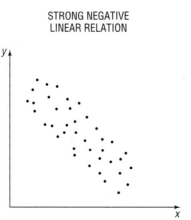

WEAK NEGATIVE
LINEAR RELATION

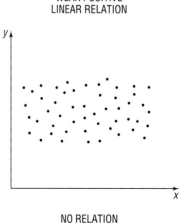

NO RELATION

Prediction and Accuracy

Once an equation with a correlation coefficient close to one has been obtained, then the equation can be used to predict values for the dependent variable for a given independent variable.

E X A M P L E   3   *Prediction*

Use the results of Example 2 to estimate the yield if the farmer uses 17 pounds of fertilizer/100 square feet.

Solution   To determine the yield, we substitute the value $x = 17$ pounds into the equation found in Example 2.

$$(\text{Yield}) = 0.7171428571(17) + 4.785714286 \approx 17 \text{ pounds}$$

So, 17 pounds of fertilizer per 100 square feet will, on average, produce a yield of 17 pounds. Notice that we rounded our prediction to the nearest whole number. This is because our data is measured to the nearest whole number as well. Our predictions cannot be more precise than our data.   ∎

Warning: It is important not to round the estimates of the slope or the y-intercept of the line of best fit, since this will decrease the accuracy of predictions.

E X A M P L E   4   *Interpreting the Slope*

Use the equation of the line found in Example 2 and interpret the slope.

Solution   The slope of the line is 0.7171428571. This can be interpreted as follows: For every 1 pound per square foot increase in fertilizer, the yield is expected to increase by 0.7171428571 pounds.

It is important that predictions are made within the scope of the model. That is, we can only make predictions regarding this model for $0 \leq \text{fertilizer} \leq 25$, since this is the range for which we have observable data. Unless we collect additional data for $x > 25$, we can't make predictions for $x > 25$. The reason should be clear. It is not clear whether adding more fertilizer will continue to increase crop yield. In fact, it is conceivable that adding more fertilizer may actually reduce crop yield.   ∎

1.5

Exercise 1.5

In Problems 1–6, examine the scatter diagram and determine if there is a strong or weak linear relation, a nonlinear relation, or no relation at all.

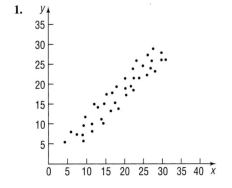

1.

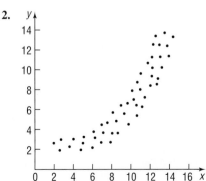

2.

3.

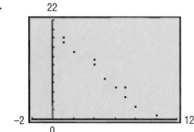

4.

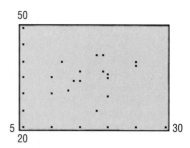

5.

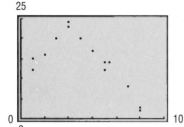

6.

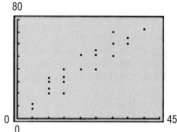

In Problems 7–14, (a) use a graphing utility to draw a scatter diagram; (b) use a graphing utility to fit a straight line to the data.

7.

x	3	4	5	6	7	8	9
y	4	6	7	10	12	14	16

8.

x	3	5	7	9	11	13
y	0	2	3	6	9	11

9.

x	-2	-1	0	1	2
y	-4	0	1	4	5

10.

x	-2	-1	0	1	2
y	7	6	3	2	0

11.

x	20	30	40	50	60
y	100	95	91	83	70

12.

x	5	10	15	20	25
y	2	4	7	11	18

13.

x	-20	-17	-15	-14	-10
y	100	120	118	130	140

14.

x	-30	-27	-25	-20	-14
y	10	12	13	13	18

15. *Consumption and Disposable Income* An economist wishes to estimate a line which relates personal consumption expenditures (C) and disposable income (I). Both C and I are in thousands of dollars. She interviews 8 heads of households for families of size 4 and obtains the following data:

C (000)	I (000)
16	20
18	20
13	18
21	27
27	36
26	37
36	45
39	50

Let I represent the independent variable and C the dependent variable.
(a) Use a graphing utility to draw a scatter diagram.
(b) Use a graphing utility to fit a straight line to the data.
(c) Interpret the slope. The slope of this line is called the **marginal propensity to consume.**
(d) Predict the consumption of a family whose disposable income is $42,000.

16. *Marginal Propensity to Save* The same economist as the one in Problem 15 wants to estimate a line which relates savings (*S*) and disposable income (*I*). Let $S = I - C$ be the dependent variable and *Y* the independent variable. The slope of this line is called the **marginal propensity to save.**
(a) Use a graphing utility to draw a scatter diagram.
(b) Use a graphing utility to fit a straight line to the data.
(c) Interpret the slope.
(d) Predict the savings of a family whose income is $42,000.

17. *Average Speed of a Car* An individual wanted to determine the relation that might exist between speed and miles per gallon of an automobile. Let *X* be the average speed of a car on the highway measured in miles per hour and let *Y* represent the miles per gallon of the automobile. The following data is collected:

X	50	55	55	60	60	62	65	65
Y	28	26	25	22	20	20	17	15

(a) Use a graphing utility to draw a scatter diagram.
(b) Use a graphing utility to fit a straight line to the data.
(c) Interpret the slope.
(d) Predict the miles per gallon of a car traveling 61 miles per hour.

18. *Height versus Weight* A doctor wished to determine whether a relation exists between the height of a female and her weight. She obtained the heights and weights of 10 females aged 18–24. Let height be the independent variable, *X*, measured in inches, and weight be the dependent variable, *Y*; measured in pounds.

X (height)	60	61	62	62	64	65	65	67	68	68
Y (weight)	105	110	115	120	120	125	130	135	135	145

(a) Use a graphing utility to draw a scatter diagram.
(b) Use a graphing utility to fit a straight line to the data.
(c) Interpret the slope.
(d) Predict the weight of a female aged 18–24 whose height is 66 inches.

19. *Sales Data versus Income* The following data represent sales and net income before taxes (both are in billions of dollars) for all manufacturing firms within the United States for 1980–1989. Treat sales as the independent variable and net income before taxes as the dependent variable.

YEAR	SALES	NET INCOME BEFORE TAXES
1980	1912.8	92.6
1981	2144.7	101.3
1982	2039.4	70.9
1983	2114.3	85.8
1984	2335.0	107.6
1985	2331.4	87.6
1986	2220.9	83.1
1987	2378.2	115.6
1988	2596.2	154.6
1989	2745.1	136.3

(a) Use a graphing utility to draw a scatter diagram.
(b) Use a graphing utility to fit a straight line to the data.
(c) Interpret the slope.
(d) Predict the net income before taxes of manufacturing firms in 1990, if sales are $2456.4 billion.

Source: Economic Report of the President, February, 1995.

20. *Employment and the Labor Force* The following data represent the civilian labor force (people aged 16 years and older, excluding those serving in the military) and the number of employed people in the United States for the years 1981–1991. Treat the size of the labor force as the independent variable and the number employed as the dependent variable. Both the size of the labor force and the number employed are measured in thousands of people.

YEAR	CIVILIAN LABOR FORCE	NUMBER EMPLOYED
1981	108,670	100,397
1982	110,204	99,526
1983	111,550	100,834
1984	113,544	105,005
1985	115,461	107,150
1986	117,834	109,597
1987	119,865	112,440
1988	121,669	114,968
1989	123,869	117,342
1990	124,787	117,914
1991	125,303	116,877

(a) Use a graphing utility to draw a scatter diagram.
(b) Use a graphing utility to fit a straight line to the data.
(c) Interpret the slope.
(d) Predict the number of employed if the civilian labor force is 122,340,000 people.

Source: Business Statistics, 1963–1991, U.S. Department of Commerce, Economics and Statistics Administration, Bureau of Economic Analysis, June 1992.

1.6

Variation

When a mathematical model is developed for a real world problem, it often involves relationships between quantities that are expressed in terms of proportionality:

Force is proportional to acceleration.

For an ideal gas held at a constant temperature, pressure and volume are inversely proportional.

The force of attraction between two heavenly bodies is inversely proportional to the square of the distance between them.

Revenue is directly proportional to sales.

Each of these statements illustrates the idea of **variation,** or how one quantity varies in relation to another quantity. Quantities may vary *directly, inversely,* or *jointly.*

Direct Variation

Let x and y denote two quantities. Then y **varies directly** with x, or y is **directly proportional to** x, if there is a nonzero number k such that

$$y = kx$$

The number k is called the **constant of proportionality.**

The graph in Figure 96 illustrates the relationship between y and x if y varies directly with x and $k > 0$, $x \geq 0$. Note that the constant of proportionality is, in fact, the slope of the line.

FIGURE 96
$y = kx, k > 0, x \geq 0$

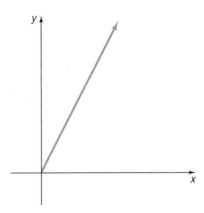

If we know that two quantities vary directly, then knowing the value of each quantity in one instance enables us to write a formula that is true in all cases.

E X A M P L E   1

Chemistry: Gas Law

For a certain gas enclosed in a container of fixed volume, the pressure P (in newtons per square meter) varies directly with temperature T (in kelvins). If the pressure is found to be 100 newtons per square meter at a temperature of 300 K, find a formula that relates pressure P to temperature T. Then find the pressure P when $T = 360$ K.

Solution   Because P varies directly with T, we know that

$$P = kT$$

for some constant k. Because $P = 100$ when $T = 300$,

$$100 = k(300)$$
$$k = \tfrac{1}{3}$$

Thus, in all cases,

$$P = \tfrac{1}{3}T$$

In particular, when $T = 360\ K$, we find

$$P = \tfrac{1}{3}(360) = 120 \text{ newtons per square meter}$$

Figure 97 illustrates the relationship between the pressure P and the temperature T.

FIGURE 97

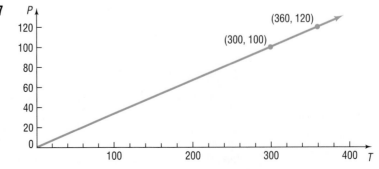

FIGURE 98

$y = \dfrac{k}{x}$; $k > 0$, $x > 0$

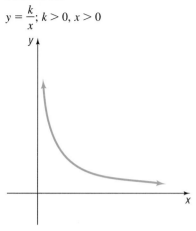

■ Now work Problem 1.

Inverse Variation

Let x and y denote two quantities. Then y **varies inversely** with x, or y is **inversely proportional to** x, if there is a nonzero constant k such that

$$y = \frac{k}{x}$$

The graph in Figure 98 illustrates the relationship between y and x if y varies inversely with x and $k > 0$, $x > 0$.

E X A M P L E   2   *Safe Weight That Can Be Supported by a Piece of Pine*

The weight W that can be safely supported by a 2 inch by 4 inch piece of lumber varies inversely with its length l. See Figure 99. Experiments indicate that the maximum weight a 10 foot pine 2 by 4 can support is 500 pounds. Write a general formula relating the safe weight W (in pounds) to length l (in feet). Find the maximum weight W that can be safely supported by a length of 25 feet.

FIGURE 99

Solution   Because W varies inversely with l, we know that

$$W = \frac{k}{l}$$

for some constant k. Because $W = 500$ when $l = 10$, we have

$$500 = \frac{k}{10}$$

$$k = 5000$$

Thus, in all cases,

$$W = \frac{5000}{l}$$

In particular, the maximum weight W that can be safely supported by a piece of pine 25 feet in length is

$$W = \frac{5000}{25} = 200 \text{ pounds}$$

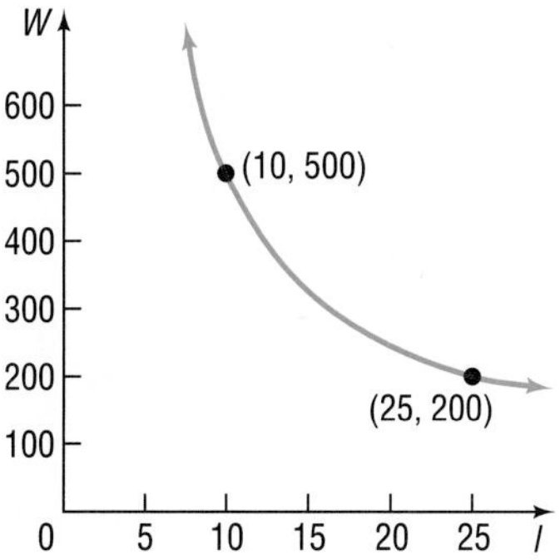

Figure 100 illustrates the relationship between the weight W and the length l. ■

FIGURE 100

In direct or inverse variation, the quantities that vary may be raised to powers. For example, in the early seventeenth century, Johannes Kepler (1571–1630) discovered that the square of the period T of a planet varies directly with the cube of its mean distance a from the Sun. That is, $T^2 = ka^3$, where k is the constant of proportionality.

Joint Variation and Combined Variation

When a variable quantity Q is proportional to the product of two or more other variables, we say that Q **varies jointly** with these quantities. Finally, combinations of direct and/or inverse variation may occur. This is usually referred to as **combined variation.**

Let's look at an example.

EXAMPLE 3   *Loss of Heat through a Wall*

The loss of heat through a wall varies jointly with the area of the wall and the difference between the inside and outside temperatures, and varies inversely with the thickness of the wall. Write an equation that relates these quantities.

Solution   We begin by assigning symbols to represent the quantities:

$$L = \text{Heat loss} \qquad T = \text{Temperature difference}$$
$$A = \text{Area of wall} \qquad d = \text{Thickness of wall}$$

Then

$$L = k\frac{AT}{d}$$

where k is the constant of proportionality.   ■

EXAMPLE 4   *Force of the Wind on a Window*

The force F of the wind on a flat surface positioned at a right angle to the direction of the wind varies jointly with the area A of the surface and the square of the speed v of the wind. A wind of 30 miles per hour blowing on a window measuring 4 feet by 5 feet has a force of 150 pounds. (See Figure 101.) What is the force on a window measuring 3 feet by 4 feet caused by a wind of 50 miles per hour?

FIGURE 101

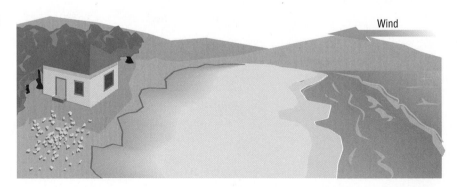

Wind

Solution   Since F varies jointly with A and v^2, we have

$$F = kAv^2$$

where k is the constant of proportionality. We are told that $F = 150$ when $v = 30$ and $A = 4 \cdot 5 = 20$. Thus,

$$150 = k(20)(900)$$

$$k = \frac{1}{120}$$

The general formula is therefore

$$F = \frac{1}{120}Av^2$$

For a wind of 50 miles per hour blowing on a window whose area is $A = 3 \cdot 4 = 12$ square feet, the force F is

$$F = \frac{1}{120}(12)(2500) = 250 \text{ pounds} \qquad \blacksquare$$

1.6

Exercise 1.6

In Problems 1–12, write a general formula to describe each variation.

1. y varies directly with x; $y = 2$ when $x = 10$

2. v varies directly with t; $v = 16$ when $t = 2$

3. A varies directly with x^2; $A = 4\pi$ when $x = 2$

4. V varies directly with x^3; $V = 36\pi$ when $x = 3$

5. F varies inversely with d^2; $F = 10$ when $d = 5$

6. y varies inversely with $\sqrt{x}$; $y = 4$ when $x = 9$

7. z varies directly with the sum of the squares of x and y; $z = 5$ when $x = 3$ and $y = 4$

8. T varies jointly with the cube root of x and the square of d; $T = 18$ when $x = 8$ and $d = 3$

9. M varies directly with the square of d and inversely with the square root of x; $M = 24$ when $x = 9$ and $d = 4$

10. z varies directly with the sum of the cube of x and the square of y; $z = 1$ when $x = 2$ and $y = 3$

11. The square of T varies directly with the cube of a and inversely with the square of d; $T = 2$ when $a = 2$ and $d = 4$

12. The cube of z varies directly with the sum of the squares of x and y; $z = 2$ when $x = 9$ and $y = 4$

In Problems 13–20, write an equation that relates the quantities.

13. *Geometry* The volume V of a sphere varies directly with the cube of its radius r. The constant of proportionality is $4\pi/3$.

14. *Geometry* The square of the hypotenuse c of a right triangle varies directly with the sum of the squares of the legs a and b. The constant of proportionality is 1.

15. *Geometry* The area A of a triangle varies jointly with the lengths of the base b and the height h. The constant of proportionality is $\frac{1}{2}$.

16. *Geometry* The perimeter p of a rectangle varies directly with the sum of the lengths of its sides l and w. The constant of proportionality is 2.

17. *Geometry* The volume V of a right circular cylinder varies jointly with the square of its radius r and its height h. The constant of proportionality is π. (See the figure).

18. *Geometry* The volume V of a right circular cone varies jointly with the square of its radius r and its height h. The constant of proportionality is $\pi/3$. (See the figure.)

19. *Physics: Newton's Law* The force F (in newtons) of attraction between two bodies varies jointly with their masses m and M (in kilograms) and inversely with the square of the distance d (in meters) between them. The constant of proportionality is $G = 6.67 \times 10^{-11}$.

20. *Physics: Simple Pendulum* The *period* of a pendulum is the time required for one oscillation; the pendulum is usually referred to as *simple* when the angle made to the vertical is less than 5°. The period T of a simple pendulum (in seconds) varies directly with the square root of its length l (in feet). The constant of proportionality is $2\pi/\sqrt{32}$.

21. *Physics: Falling Objects* The distance s an object falls is directly proportional to the square of the time t of the fall. If an object falls 16 feet in 1 second, how far will it fall in 3 seconds? How long will it take an object to fall 64 feet?

22. *Physics: Falling Objects* The velocity v of a falling object is directly proportional to the time t of the fall. If, after 2 seconds, the velocity of the object is 64 feet per second, what will its velocity be after 3 seconds?

23. *Physics: Stretching a Spring* The elongation E of a spring balance varies directly with the applied weight W (see the figure). If $E = 3$ when $W = 20$, find E when $W = 15$.

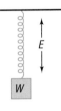

24. *Physics: Vibrating String* The rate of vibration of a string under constant tension varies inversely with the length of the string. If a string is 48 inches long and vibrates 256 times per second, what is the length of a string that vibrates 576 times per second?

25. *Weight of a Body* The weight of a body above the surface of Earth varies inversely with the square of the distance from the center of Earth. If a certain body weighs 55 pounds when it is 4×10^3 miles from the center of Earth, how much will it weigh when it is 4.4×10^3 miles from the center?

26. *Force of the Wind on a Window* The force exerted by the wind on a plane surface varies jointly with the area of the surface and the square of the velocity of the wind. If the force on an area of 20 square feet is 11 pounds when the wind velocity is 22 miles per hour, find the force on a surface area of 47.125 square feet when the wind velocity is 36.5 miles per hour.

27. *Horsepower* The horsepower that a shaft can safely transmit varies jointly with its speed (in revolutions per minute, rpm) and the cube of its diameter. If a shaft of a certain material 2 inches in diameter can transmit 36 horsepower at 75 rpm, what diameter must the shaft have in order to transmit 45 horsepower at 125 rpm?

28. *Weight of a Body* The weight of a body varies inversely with the square of its distance from the center of Earth. Assuming that the radius of Earth is 4000 miles, how much would a man weigh at an altitude of 1 mile above Earth's surface if he weighs 200 pounds on Earth's surface?

29. *Physics: Kinetic Energy* The kinetic energy K of a moving object varies jointly with its mass m and the square of its velocity v. If an object weighing 25 pounds and moving with a velocity of 100 feet per second has a kinetic energy of 400 foot-pounds, find its kinetic energy when the velocity is 150 feet per second.

30. *Electrical Resistance of a Wire* The electrical resistance of a wire varies directly with the length of the wire and inversely with the square of the diameter of the wire. If a wire 432 feet long and 4 millimeters in diameter has a resistance of 1.24 ohms, find the length of a wire of the same material whose resistance is 1.44 ohms and whose diameter is 3 millimeters.

31. *Measuring the Stress of Materials* The stress in the material of a pipe subject to internal pressure varies jointly with the internal pressure and the internal diameter of the pipe and inversely with the thickness of the pipe. The stress is 100 pounds per square inch when the diameter is 5 inches, the thickness is 0.75 inch, and the internal pressure is 25 pounds per square inch. Find the stress when the internal pressure is 40 pounds per square inch, if the diameter is 8 inches and the thickness is 0.50 inch.

32. *Safe Load for a Beam* The maximum safe load for a horizontal rectangular beam varies jointly with the width of the beam and the square of the thickness of the beam and inversely with its length. If an 8 foot beam will support up to 750 pounds when the beam is 4 inches wide and 2 inches thick, what is the maximum safe load in a similar beam 10 feet long, 6 inches wide, and 2 inches thick?

33. *Resistance due to a Conductor* The resistance (in ohms) of a circular conductor varies directly with the length of the conductor and inversely with the square of the radius of the conductor. If 50 feet of a wire with a radius of 6×10^{-3} inch has a resistance of 10 ohms, what would be the resistance of 100 feet of the same wire if the radius is increased to 7×10^{-3} inch?

34. *Chemistry: Gas Laws* The volume V of an ideal gas varies directly with the temperature T and inversely with the pressure P. Write an equation relating V, T, and P using k as the constant of proportionality. If a cylinder contains oxygen at a temperature of 300 K and a pressure of 15 atmospheres in a volume of 100 liters, what is the constant of proportionality k? If a piston is lowered into the cylinder, decreasing the volume occupied by the gas to 80 liters and raising the temperature to 310 K, what is the gas pressure?

Problems 36–40 use the result obtained in Problem 35.

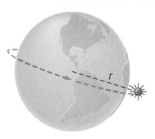

35. *Satellites in Orbit* The speed v required of a satellite to maintain a near-Earth circular orbit is directly proportional to the square root of the distance r of the satellite from the center of Earth.* The constant of proportionality is $\sqrt{g}$, where g is the acceleration of gravity for Earth. Write an equation that shows the relationship between v and r. (The radius of Earth is approximately 3960 miles.)

36. *Speed Required to Stay in Orbit* What speed is required to maintain a communications satellite in a circular orbit 500 miles above Earth's surface?

37. *Speed Required to Stay in Orbit* Find the speed of a satellite that moves in a circular orbit 100 miles above Earth's surface.

38. *Distance of a Satellite from Earth* Find the distance of a satellite from the surface of Earth as it moves around Earth in a circular orbit at a constant speed of 18,630 miles per hour.

39. *Distance of a Satellite from Earth* A weather satellite orbits Earth in a circle every 1.5 hours. How high is it above Earth?

40. *Distance and Speed of Stationary Satellites* Some communications satellites remain stationary above a fixed point on the equator of Earth's surface. How high are such satellites, and what is their common speed? (Assume that Earth turns once every 24 hours.)

Problems 42–46 use the result obtained in Problem 41.

41. *Physical Forces* The force F (in newtons) required to maintain an object in a circular path varies jointly with the mass m (in kilograms) of the object and the square of its speed v (in meters per second) and inversely with the radius r (in meters) of the circular path. The constant of proportionality is 1. Write an equation relating F, m, v, and r.

42. *Physical Forces* A motorcycle with mass 150 kilograms is driven at a constant speed of 120 kilometers per hour on a circular track with a radius of 100 meters. To keep the motorcycle from skidding, what frictional force must be exerted by the tires on the track?

43. *Physical Forces* If the speed of the motorcycle described in Problem 42 is increased by 10%, by how much is the frictional force of the tires increased?

*Near-Earth orbits are at least 100 miles above Earth's surface (out of Earth's atmosphere) and up to an altitude of approximately 15,000 miles. The effect of the gravitational attraction of other bodies is ignored. Although the acceleration of gravity at such altitudes is somewhat less than $g \approx 32$ feet per second per second $\approx 79{,}036$ miles per hour per hour, we shall ignore this discrepancy in our calculations.

44. *Physical Forces* If the radius of the track described in Problem 42 is cut in half, how much slower should the motorcycle be driven to maintain the same frictional force?

45. *Physical Forces* A woman is spinning a bucket of water in a horizontal plane at the end of a rope of length L (see the figure). If she triples the speed of the bucket, how many times as hard must she pull on the rope?

46. *Physical Forces* If the woman in Problem 45 doubles the length of the rope and maintains the same speed for the bucket, will she have to pull on the rope more or less? How much?

 47. The formula on page 74 attributed to Johannes Kepler is one of the famous three Keplerian Laws of Planetary Motion. Go to the library and research these laws. Write a brief paper about these laws and Kepler's place in history.

Chapter Review

THINGS TO KNOW

Formulas

Distance formula	$d = \sqrt{(x_2 - x_1)^2 + (y_2 - y_1)^2}$
Midpoint formula	$(x, y) = \left(\dfrac{x_1 + x_2}{2}, \dfrac{y_1 + y_2}{2} \right)$
Slope	$m = \dfrac{y_2 - y_1}{x_2 - x_1}$, if $x_1 \neq x_2$; undefined if $x_1 = x_2$
Parallel lines	Equal slopes ($m_1 = m_2$)
Perpendicular lines	Product of slopes is -1 ($m_1 \cdot m_2 = -1$)
Direct variation	$y = kx$
Inverse variation	$y = \dfrac{k}{x}$

Equations

Vertical line	$x = a$
Horizontal line	$y = b$
Point–slope form of the equation of a line	$y - y_1 = m(x - x_1)$; m is the slope of the line, (x_1, y_1) is a point on the line
General form of the equation of a line	$Ax + By + C = 0$, A, B not both 0
Slope–intercept form of the equation of a line	$y = mx + b$; m is the slope of the line, b is the y-intercept
Standard form of the equation of a circle	$(x - h)^2 + (y - k)^2 = r^2$; r is the radius of the circle, (h, k) is the center of the circle
General form of the equation of a circle	$x^2 + y^2 + ax + by + c = 0$
Equation of the unit circle	$x^2 + y^2 = 1$

How To:

Use the distance formula

Graph equations by plotting points

Find the intercepts of a graph

Test an equation for symmetry

Find the slope and intercepts of a line, given the equation

Graph lines by hand

Graph equations using a graphing utility

Obtain the equation of a line

Obtain the equation of a circle

Find the center and radius of a circle, given the equation

Graph circles by hand

Solve variation problems

Find the equation of a line using data

Fill-In-The-Blank Items

1. If (x, y) are the coordinates of a point P in the xy-plane, then x is called the _____ of P and y is the _____ of P.

2. If three points P, Q, and R all lie on a line and if $d(P, Q) = d(Q, R)$, then Q is called the _____ of the line segment from P to R.

3. If for every point (x, y) on a graph, the point $(-x, y)$ is also on the graph, then the graph is symmetric with respect to the _____.

4. The set of points in the xy-plane that are a fixed distance from a fixed point is called a(n) _____. The fixed distance is called the _____; the fixed point is called the _____.

5. The slope of a vertical line is _____; the slope of a horizontal line is _____.

6. Two nonvertical lines have slopes m_1 and m_2, respectively. The lines are parallel if _____; the lines are perpendicular if _____.

7. If z varies jointly as x^2 and y^3 and inversely as $\sqrt{t}$, then $z =$ _____, where k is the constant of proportionality.

True/False Items

T   F   1. The distance between two points is sometimes a negative number.

T   F   2. The graph of the equation $y = x^4 + x^2 + 1$ is symmetric with respect to the y-axis.

T   F   3. Vertical lines have undefined slope.

T   F   4. The slope of the line $2y = 3x + 5$ is 3.

T   F   5. Perpendicular lines have slopes that are reciprocals of one another.

T   F   6. The radius of the circle $x^2 + y^2 = 9$ is 3.

T   F   7. If y varies inversely with x, then as x increases in value, y will decrease in value.

Review Exercises

In Problems 1–10, find an equation of the line having the given characteristics. Express your answer using either the general form or the slope–intercept form of the equation of a line, whichever you prefer.

1. Slope $= -2$; passing through $(3, -1)$

2. Slope $= 0$; passing through $(-5, 4)$

3. Slope undefined; passing through $(-3, 4)$

4. x-intercept $= 2$; passing through $(4, -5)$

5. y-intercept $= -2$; passing through $(5, -3)$

6. Passing through $(3, -4)$ and $(2, 1)$

7. Parallel to the line $2x - 3y + 4 = 0$; passing through $(-5, 3)$

8. Parallel to the line $x + y - 2 = 0$; passing through $(1, -3)$

9. Perpendicular to the line $x + y - 2 = 0$; passing through $(4, -3)$

10. Perpendicular to the line $3x - y + 4 = 0$; passing through $(-2, 4)$

In Problems 11–16, graph each line using a graphing utility. Use BOX and TRACE to find the x-intercept and y-intercept. Also draw each graph by hand, labeling any intercepts.

11. $4x - 5y + 20 = 0$

12. $3x + 4y - 12 = 0$

13. $\dfrac{1}{2}x - \dfrac{1}{3}y + \dfrac{1}{6} = 0$

14. $-\dfrac{3}{4}x + \dfrac{1}{2}y = 0$

15. $\sqrt{2}x + \sqrt{3}y = \sqrt{6}$

16. $\dfrac{x}{3} + \dfrac{y}{4} = 1$

In Problems 17–20, find the center and radius of each circle. Graph each circle using a graphing utility.

17. $x^2 + y^2 - 2x + 4y - 4 = 0$

18. $x^2 + y^2 + 4x - 4y - 1 = 0$

19. $3x^2 + 3y^2 - 6x + 12y = 0$

20. $2x^2 + 2y^2 - 4x = 0$

21. Find the slope of the line containing the points $(7, 4)$ and $(-3, 2)$. What is the distance between these points? What is their midpoint?

22. Find the slope of the line containing the points $(2, 5)$ and $(6, -3)$. What is the distance between these points? What is their midpoint?

23. The figure on the right shows the graph of two parallel lines. Which of the following pairs of equations might have such a graph?

(a) $x - 2y = 3$
      $x + 2y = 7$

(b) $x + y = 2$
      $x + y = -1$

(c) $x - y = -2$
      $x - y = 1$

(d) $x - y = -2$
      $2x - 2y = -4$

(e) $x + 2y = 2$
      $x + 2y = -1$

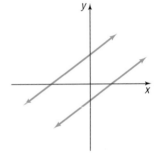

24. The figure on the right shows the graph of two perpendicular lines. Which of the following pairs of equations might have such a graph?

(a) $y - 2x = 2$
      $y + 2x = -1$

(b) $y - 2x = 0$
      $2y + x = 0$

(c) $2y - x = 2$
      $2y + x = -2$

(d) $y - 2x = 2$
      $x + 2y = -1$

(e) $2x + y = -2$
      $2y + x = -2$

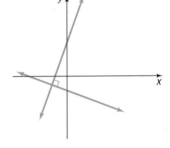

In Problems 25–32, list the intercepts and test for symmetry.

25. $2x = 3y^2$

26. $y = 5x$

27. $x^2 + 4y^2 = 16$

28. $9x^2 - y^2 = 9$

29. $y = x^4 + 2x^2 + 1$

30. $y = x^3 - x$

31. $x^2 + x + y^2 + 2y = 0$

32. $x^2 + 4x + y^2 - 2y = 0$

For Problems 33–36, (a) use a graphing utility to draw a scatter diagram; (b) use a graphing utility to fit a straight line to the data; and (c) interpret the slope.

33.

x	3	4	5	6	7	8	9
y	3	5	6	8	10	11	13

34.

x	10	12	13	15	16	18	20
y	34	27	26	23	20	18	17

35.

x	100	110	125	130	140	145	150	160	170	175
y	300	340	365	380	400	410	425	430	450	460

36.

x	200	220	230	235	245	250	265	275	280	300
y	1000	990	975	960	955	940	935	920	910	895

37. *Relating Algebra and Calculus Scores* The following data represent scores in an Algebra achievement test and Calculus achievement test for the same student. Treat the Algebra test score as the independent variable and the calculus test score as the dependent variable.

ALGEBRA SCORE	CALCULUS SCORE
17	73
21	66
11	64
16	61
15	70
11	71
24	90
27	68
19	84
8	52

(a) Use a graphing utility to draw a scatter diagram.
(b) Use a graphing utility to fit a straight line to the data.
(c) Interpret the slope.
(d) Predict the score a student would receive on the Calculus achievement test if she scored a 20 on the Algebra achievement test.

Source: "Factors Affecting Achievement in the First Course in Calculus," by Edge and Friedberg, *Journal of Experimental Education,* Vol. 52, No. 3.

38. *Relating Emission Levels of Hydrocarbons and Carbon Monoxide* The following data represent emissions levels for different vehicles. Measurements are given in grams per meter. Treat hydrocarbons (HC) as the independent variable and carbon monoxide (CO) as the dependent variable.

HC	CO
0.65	14.7
0.55	12.3
0.72	14.6
0.83	15.1
0.57	5.0
0.51	4.1
0.43	3.8
0.37	4.1

(a) Use a graphing utility to draw a scatter diagram.
(b) Use a graphing utility to fit a straight line to the data.
(c) Interpret the slope.
(d) Predict the level of CO in a vehicle's exhaust if the level of HC is 0.67.

Source: "Determining Statistical Characteristics of a Vehicle Emissions Audit Procedure," by Lorenzen, *Technometrics,* Vol. 22, No. 4.

39. *Geometry* The area of an equilateral triangle varies directly with the square of the length of a side. If the area of the equilateral triangle whose sides are of length 1 centimeter is $\sqrt{3}/4$, find the length s of each side of an equilateral triangle whose area A is 16 square centimeters.

40. *Vibrating Strings* In a vibrating string, the pitch varies directly with the square root of the tension of the string. If a certain string vibrates 300 times per second under a tension of 9 pounds, find the tension required to cause the string to vibrate 400 times per second.

41. *Kepler's Third Law of Planetary Motion* Kepler's third law of planetary motion states that the square of the period T of revolution of a planet is proportional to the cube of its mean distance from the Sun. If the mean distance of Earth from the Sun is 93 million miles, what is the mean distance a of the planet Mercury from the Sun, given that Mercury has a "year" of 88 days?

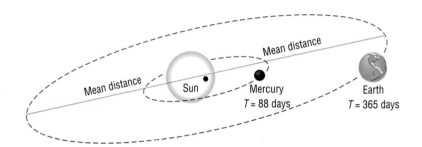

42. Use Problem 41 to find the mean distance of the planet Jupiter from the Sun, given that Jupiter circles the Sun every $5\sqrt{5}$ years.

43. Show that the midpoint of the hypotenuse of a right triangle is the same distance from each of the three vertices [*Hint:* Use the figure below.]

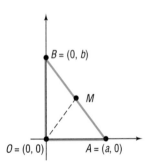

44. Show that the line joining the midpoints of two sides of a triangle is parallel to the third side.

45. Show that the points $A = (3, 4)$, $B = (1, 1)$, and $C = (-2, 3)$ are the vertices of an isosceles triangle.

46. Show that the points $A = (-2, 0)$, $B = (-4, 4)$, and $C = (8, 5)$ are the vertices of a right triangle in two ways:
(a) By using the converse of the Pythagorean Theorem
(b) By using the slopes of the lines joining the vertices

47. Show that the points $A = (2, 5)$, $B = (6, 1)$, and $C = (8, -1)$ lie on a straight line by using slopes.

48. Show that the points $A = (1, 5)$, $B = (2, 4)$, and $C = (-3, 5)$ lie on a circle with center $(-1, 2)$. What is the radius of this circle?

49. The endpoints of the diameter of a circle are $(-3, 2)$ and $(5, -6)$. Find the center and radius of the circle. Write the general equation of this circle.

50. Find two numbers y such that the distance from $(-3, 2)$ to $(5, y)$ is 10.

51. Make up four problems you might be asked to do given the two points $(-3, 4)$ and $(6, 1)$. Each problem should involve a different concept. Be sure your directions are clearly stated.

52. Describe each of the following graphs. Give justification.
(a) $x = 0$ (b) $y = 0$ (c) $x + y = 0$ (d) $xy = 0$ (e) $x^2 + y^2 = 0$

53. Suppose that you have a rectangular field that requires watering. Your watering system consists of an arm of variable length that rotates so that the watering pattern is a circle. Decide where to position the arm and what length it should be so that the entire field is watered most efficiently. When does it become desirable to use more than one arm? *Hint:* Use a rectangular coordinate system positioned so that the axes bisect the rectangle (see the figure). Write equations for the circle(s) swept out by the watering arm(s).

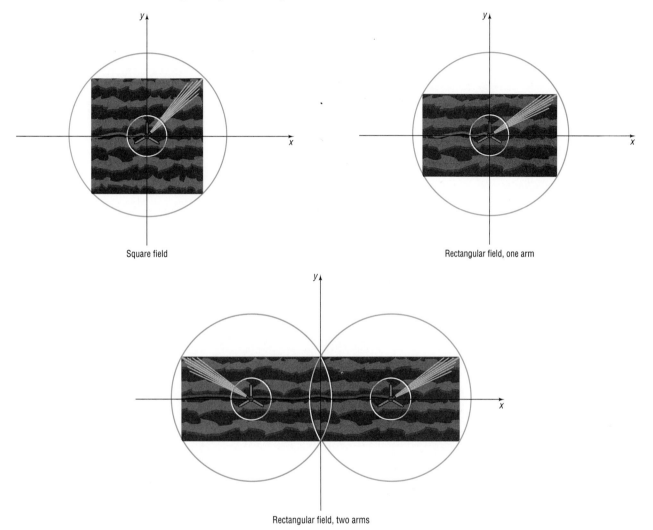

Square field

Rectangular field, one arm

Rectangular field, two arms

54. Graph $y = x$ and $y = 0.99x + 0.01$ on the same screen. It appears the graphs are identical. Explain why this happens. Provide a way to correct this situation.

55. Why does the graph of $y = x/6$, a straight line, appear to consist of a collection of tiny horizontal line segments?

PREPARING FOR THIS CHAPTER

Before getting started on this chapter, review the following concepts;

Topics from Algebra and Geometry (Appendix, Section 1)
Graphs of certain equations (Example 2, p. 15;
Example 4, p. 17; Example 5, p. 18; Example 16, p. 30)
Tests for symmetry of an equation (p. 28)
Procedure for finding intercepts of an equation (p. 22)

FUNCTIONS AND THEIR GRAPHS

2.1 Functions
2.2 More about Functions
2.3 Graphing Techniques
2.4 Operations
     on Functions;
     Composite Functions
2.5 One-to-One Functions;
     Inverse Functions
2.6 Mathematical Models:
     Constructing Functions
     Chapter Review

Preview   Getting from an Island to Town

An island is 2 miles from the nearest point P on a straight shoreline. A town is 12 miles down the shore from P.

(a) If a person can row a boat at an average speed of 3 miles per hour and the same person can walk 5 miles per hour, express the time T it takes to go from the island to town as a function of the distance x from P to where the person lands the boat.

(b) What is the domain of T?

(c) How long will it take to travel from the island to town if the person lands the boat 4 miles from P?

(d) How long will it take if the person lands the boat 8 miles from P?

(e) Use a graphing utility to graph the function T = T(x).

(f) Use the TRACE function to see how the time T varies as x changes from 0 to 12.

(g) What value of x results in the least time?
[Example 9 in Section 2.1] ■

erhaps the most central idea in mathematics is the notion of a *function*. This important chapter deals with what a function is, how to graph functions, how to perform operations on functions, and how functions are used in applications.

The word *function* apparently was introduced by René Descartes in 1637. For him, a function simply meant any positive integral power of a variable *x*. Gottfried Wilhelm von Leibniz (1646–1716), who always emphasized the geometric side of mathematics, used the word function to denote any quantity associated with a curve, such as the coordinates of a point on the curve. Leonhard Euler (1707–1783) employed the word to mean any equation or formula involving variables and constants. His idea of a function is similar to the one most often used today in courses that precede calculus. Later, the use of functions in investigating heat flow equations led to a very broad definition, due to Lejeune Dirichlet (1805–1859), which describes a function as a rule or correspondence between two sets. It is his definition that we use here.

2.1

Functions

In many applications, a correspondence often exists between two sets of numbers. For example, the revenue R resulting from the sale of x items selling for \$10 each is $R = 10x$ dollars. If we know how many items have been sold, then we can calculate the revenue by using the rule $R = 10x$. This rule is an example of a *function*.

As another example, if an object is dropped from a height of 64 feet above the ground, the distance s (in feet) of the object from the ground after t seconds is given (approximately) by the formula $s = 64 - 16t^2$. When $t = 0$ seconds, the object is $s = 64$ feet above the ground. After 1 second, the object is $s = 64 - 16(1)^2 = 48$ feet above the ground. After 2 seconds, the object strikes the ground. The formula $s = 64 - 16t^2$ provides a way of finding the distance s when the time t ($0 \le t \le 2$) is prescribed. There is a correspondence between each time t in the interval $0 \le t \le 2$ and the distance s. We say that the distance s is a *function* of the time t because:

1. There is a correspondence between the set of times and the set of distances.
2. There is exactly one distance s obtained for a prescribed time t in the interval $0 \le t \le 2$.

Let's now look at the definition of a function.

Definition of Function

Function

Let X and Y be two nonempty sets of real numbers.* A **function** from X into Y is a rule or a correspondence that associates with each element of X a unique element of Y. The set X is called the **domain** of the function. For each element x in X, the corresponding element y in Y is called the **value** of the function at x, or the **image** of x. The set of all images of the elements of the domain is called the **range** of the function.

When we select a viewing rectangle to graph a function, the values of Xmin, Xmax give the domain we wish to view, while Ymin, Ymax give the range we wish to view. These settings usually do not represent the actual domain and range of the function.

*The two sets X and Y can also be sets of complex numbers, and then we have defined a complex function. In the broad definition (due to Lejeune Dirichlet), X and Y can be any two sets.

FIGURE 1

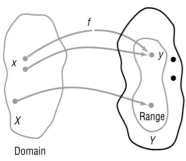

Domain

Warning: Do not confuse the two meanings given for the word *range*. When used in connection with a function, it means the set of all images of the elements of the domain of the function. When the word RANGE is used in connection with a graphing utility, it means the settings used for the viewing rectangle.

Refer to Figure 1. Since there may be some elements in Y that are not the image of some x in X, it follows that the range of a function may be a subset of Y.

The rule (or correspondence) referred to in the definition of a function is most often given as an equation in two variables, usually denoted x and y.

E X A M P L E   1

Example of a Function

Consider the function defined by the equation

$$y = 2x - 5 \qquad 1 \le x \le 6$$

The domain $1 \le x \le 6$ specifies that the number x is restricted to the real numbers from 1 to 6, inclusive. The rule $y = 2x - 5$ specifies that the number x is to be multiplied by 2 and then 5 is to be subtracted from the result to get y. For example, the value of the function at $x = \frac{3}{2}$ (that is, the image of $x = \frac{3}{2}$) is $y = 2 \cdot \frac{3}{2} - 5 = -2$. ∎

Functions are often denoted by letters such as f, F, g, G, and so on. If f is a function, then for each number x in its domain the corresponding image in the range is designated by the symbol $f(x)$, read as "f of x" or as "f at x." We refer to $f(x)$ as the **value of f at the number x.** Thus, $f(x)$ is the number that results when x is given and the rule for f is applied; $f(x)$ does *not* mean "f times x." For example, the function given in Example 1 may be written as $f(x) = 2x - 5$, $1 \le x \le 6$.

Figure 2 illustrates some other functions. Note that for each of the functions illustrated, to each x in the domain, there is one value in the range.

FIGURE 2

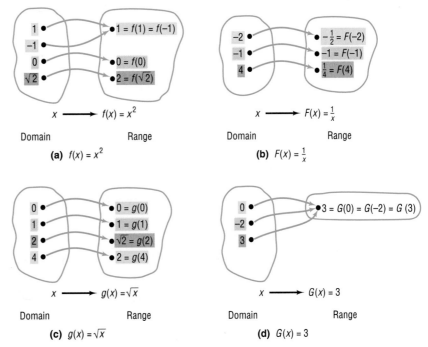

(a) $f(x) = x^2$

(b) $F(x) = \frac{1}{x}$

(c) $g(x) = \sqrt{x}$

(d) $G(x) = 3$

E X A M P L E   2   *Finding Values of a Function*

For the function

$$f(x) = x^2 + 3x - 4 \qquad -5 \le x \le 5$$

find the value of f at:

(a) $x = 0$ (b) $x = 1$ (c) $x = -4$ (d) $x = 5$

Solution   (a) The value of f at $x = 0$ is found by replacing x by 0 in the stated rule. Thus,

$$f(0) = 0^2 + 3(0) - 4 = -4$$

(b) $f(1) = 1^2 + 3(1) - 4 = 1 + 3 - 4 = 0$
(c) $f(-4) = (-4)^2 + 3(-4) - 4 = 16 - 12 - 4 = 0$
(d) $f(5) = 5^2 + 3(5) - 4 = 25 + 15 - 4 = 36$ ∎

■   Now work Problem 3.

In general, when the rule that defines a function f is given by an equation in x and y, we say that the function f is given **implicitly.** If it is possible to solve the equation for y in terms of x, then we write $y = f(x)$ and say that the function is given **explicitly.** In fact, we usually write "the function $y = f(x)$" when we mean "the function f defined by the equation $y = f(x)$." Although this usage is not entirely correct, it is rather common and should not cause any confusion. For example:

IMPLICIT FORM	EXPLICIT FORM
$3x + y = 5$	$y = f(x) = -3x + 5$
$x^2 - y = 6$	$y = f(x) = x^2 - 6$
$xy = 4$	$y = f(x) = 4/x$

Not all equations in x and y define a function $y = f(x)$. If an equation is solved for y and two or more values of y can be obtained for a given x, then the equation does not define a function $y = f(x)$. For example, consider the equation $x^2 + y^2 = 1$, which defines a circle. If we solve for y, we obtain $y = \pm\sqrt{1 - x^2}$ so that two values of y will result for numbers x between -1 and 1. Thus, $x^2 + y^2 = 1$ does not define a function.

The explicit form of a function is the form required by a graphing calculator. Now do you see why it is necessary to graph a circle in two "pieces"?

We list below a summary of some important facts to remember about a function f.

Summary of Important Facts
about Functions

1. $f(x)$ is the image of x, or the value of f at x, when the rule f is applied to an x in the domain.
2. To each x in the domain of f, there is one and only one image $f(x)$ in the range.
3. f is the symbol we use to denote the function. It is symbolic of the domain and the rule we use to get from an x in the domain to $f(x)$ in the range.

Function Keys

Most graphing utilities have special keys that enable you to find the value of certain commonly used functions. For example, you should be able to find the square function, $f(x) = x^2$; the square root function, $f(x) = \sqrt{x}$; the reciprocal function, $f(x) = 1/x = x^{-1}$; and many others that will be discussed later in this book (such as $\ln x$, $\log x$, and so on). Verify the results of Example 3 on your graphing utility.

E X A M P L E   3   *Finding Values of a Function on a Calculator*

(a) $f(x) = x^2$; $f(1.234) = 1.522756$

(b) $F(x) = 1/x$; $F(1.234) = 0.8103727715$

(c) $g(x) = \sqrt{x}$; $g(1.234) = 1.110855526$ ■

Domain of a Function

Often, the domain of a function f is not specified; instead, only a rule or equation defining the function is given. In such cases, we agree that the domain of f is the largest set of real numbers for which the rule makes sense or, more precisely, for which the value $f(x)$ is a real number. Thus, the domain of f is the same as the domain of the variable x in the expression $f(x)$.

E X A M P L E   4   *Finding the Domain of a Function*

Find the domain of each of the following functions:

(a) $f(x) = \dfrac{3x}{x^2 - 4}$ (b) $g(x) = \sqrt{4 - 3x}$

Solution   (a) The rule f tells us to divide $3x$ by $x^2 - 4$. Since division by 0 is not allowed, the denominator $x^2 - 4$ can never be 0. Thus, x can never equal 2 or -2. The domain of the function f is $\{x \mid x \neq -2, x \neq 2\}$.

(b) The rule g tells us to take the square root of $4 - 3x$. But only nonnegative numbers have real square roots. Hence, we require that

$$4 - 3x \geq 0$$
$$-3x \geq -4$$
$$x \leq \tfrac{4}{3}$$

The domain of g is $\{x \mid -\infty < x \leq \tfrac{4}{3}\}$ or the interval $(-\infty, \tfrac{4}{3}]$.   ■

We can use a graphing utility to estimate the domain of a function. For example, Figure 3 shows the graph of $y = \sqrt{4 - 3x}$. We can approximate the

FIGURE 3

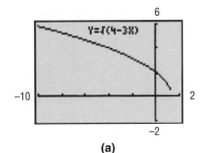

(a)

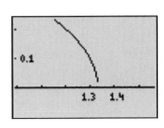

(b)

domain by noting that the domain consists of all numbers x less than or equal to the number where the graph begins. By utilizing the TRACE and ZOOM functions, experiment to see how close you can come to $x = \frac{4}{3}$, the largest value of x in the domain.

■ Now work Problem 41.

If x is in the domain of a function f, we shall say that f **is defined at x,** or $f(x)$ **exists.** If x is not in the domain of f, we say that f **is not defined at x,** or $f(x)$ **does not exist.** For example, if $f(x) = x/(x^2 - 1)$, then $f(0)$ exists, but $f(1)$ and $f(-1)$ do not exist. (Do you see why?)

We have not said much about finding the range of a function. The reason is that when a function is defined by an equation, it is often difficult to find the range. Therefore, we shall usually be content to find just the domain of a function when only the rule for the function is given. We shall express the domain of a function using interval notation, set notation, or words, whichever is most convenient.

When we use functions in applications, the domain may be restricted by physical or geometric considerations. For example, the domain of the function f defined by $f(x) = x^2$ is the set of all real numbers. However, if f is used as the rule for obtaining the area of a square when the length x of a side is known, then we must restrict the domain of f to the positive real numbers, since the length of a side can never be 0 or negative.

Independent Variable; Dependent Variable

Consider a function $y = f(x)$. The variable x is called the **independent variable,** because it can be assigned any of the permissible numbers from the domain. The variable y is called the **dependent variable,** because its value depends on x.

Any symbol can be used to represent the independent and dependent variables. For example, if f is the *cube function,* then f can be defined by $f(x) = x^3$ or $f(t) = t^3$ or $f(z) = z^3$. All three rules are identical: each tells us to cube the independent variable. In practice, the symbols used for the independent and dependent variables are based on common usage.

E X A M P L E   5   *Construction Cost*

The cost per square foot to build a house is $110. Express the cost C as a function of x, the number of square feet. What is the cost to build a 2000 square foot house?

Solution   The cost C of building a house containing x square feet is $110x$ dollars. A function expressing this relationship is

$$C(x) = 110x$$

where x is the independent variable and C is the dependent variable. In this setting, the domain is $\{x \mid x > 0\}$ since a house cannot have 0 or negative square feet. The cost to build a 2000 square foot house is

$$C(2000) = 110(2000) = \$220{,}000 \qquad ■$$

It is worth observing that in the solution to Example 5 we used the symbol C in two ways; it is used to name the function, and it is used to symbolize the dependent variable. This double use is common in applications and should not cause any difficulty.

E X A M P L E   6   *Area of a Circle*

Express the area of a circle as a function of its radius.

Solution   We know that the formula for the area A of a circle of radius r is $A = \pi r^2$. If we use r to represent the independent variable and A to represent the dependent variable, the function expressing this relationship is

$$A(r) = \pi r^2$$

In this setting, the domain is $\{r \mid r > 0\}$. (Do you see why?)   ■

■ Now work Problem 61.

The Graph of a Function

In applications, a graph often demonstrates more clearly the relationship between two variables than, say, an equation or table would. For example, Figure 4 shows the price per share (vertical axis) of McDonald's Corp. stock at the end of each week from Nov. 4, 1994 to Jan. 27, 1995 (horizontal axis). We can see from the graph that the price of the stock was falling over the few days preceding Nov. 25 and was rising over the days from Jan. 20 through Jan. 27. The graph also shows that the lowest price during this period occurred on Dec. 9 while the highest occurred on Jan. 27. Equations and tables, on the other hand, usually require some calculations and interpretation before this kind of information can be "seen."

FIGURE 4
Weekly closing prices of McDonald's Corp. stock

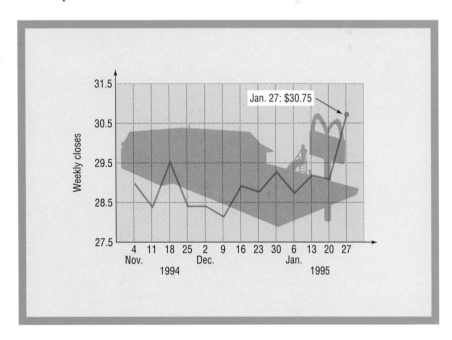

Look again at Figure 4. The graph shows that, for each time on the horizontal axis, there is only one price on the vertical axis. Thus, the graph represents a function, although the exact rule for getting from time to price is not given.

When the rule that defines a function f is given by an equation in x and y, the **graph of f** is the graph of the equation, that is, the set of points (x, y) in the xy-plane that satisfies the equation.

Not every collection of points in the xy-plane represents the graph of a function. Remember, for a function f, each number x in the domain of f has one and only one image $f(x)$. Thus, the graph of a function f cannot contain two points with the same x-coordinate and different y-coordinates. Therefore, the graph of a function must satisfy the following **vertical-line test:**

Theorem   A set of points in the xy-plane is the graph of a function if and only if a vertical
Vertical-Line Test   line intersects the graph in at most one point.   ■

It follows that, if any vertical line intersects a graph at more than one point, the graph is not the graph of a function.

E X A M P L E   7   *Identifying the Graph of a Function*

Which of the graphs in Figure 5 are graphs of functions?

FIGURE 5

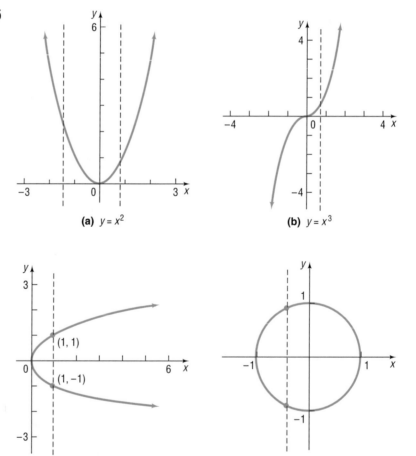

(a) $y = x^2$ **(b)** $y = x^3$

(c) $x = y^2$ **(d)** $x^2 + y^2 = 1$

Solution   The graphs in Figures 5(a) and 5(b) are graphs of functions, because a vertical line intersects each graph in at most one point. The graphs in Figures 5(c) and 5(d) are not graphs of functions, because some vertical line intersects each graph in more than one point.   ■

Ordered Pairs

The preceding discussion provides an alternative way to think of a function. We may consider a function f as a set of **ordered pairs** (x, y) or $(x, f(x))$, in which no two pairs have the same first element. The set of all first elements is the domain of the function, and the set of all second elements is its range. Thus, there is associated with each element x in the domain a unique element y in the range. An example is the set of all ordered pairs (x, y) such that $y = x^2$. Some of the pairs in this set are

$$(2, 2^2) = (2, 4) \qquad (0, 0^2) = (0, 0)$$

$$(-2, (-2)^2) = (-2, 4) \qquad \left(\frac{1}{2}, \left(\frac{1}{2}\right)^2\right) = \left(\frac{1}{2}, \frac{1}{4}\right)$$

In this set, no two pairs have the same *first* element (although there are pairs that have the same *second* element). This set is the *square function,* which associates with each real number x the number x^2. Look again at Figure 5(a).

On the other hand, the ordered pairs (x, y) for which $y^2 = x$ do not represent a function, because there are ordered pairs with the same first element but different second elements. For example, $(1, 1)$ and $(1, -1)$ are ordered pairs obeying the relationship $y^2 = x$ with the same first element but different second elements. Look again at Figure 5(c).

The next example illustrates how to determine the domain and range of a function if its graph is given.

E X A M P L E   8   *Obtaining Information from the Graph of a Function*

Let f be the function whose graph is given in Figure 6. Some points on the graph are labeled.

(a) What is the value of the function when $x = -6$, $x = -4$, $x = 0$, and $x = 6$?

(b) What is the domain of f?

(c) What is the range of f?

(d) List the intercepts. (Recall that these are the points, if any, where the graph crosses the coordinate axes.)

FIGURE 6

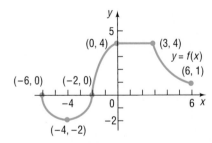

Solution   (a) Since $(-6, 0)$ is on the graph of f, the y-coordinate 0 must be the value of f at the x-coordinate -6; that is, $f(-6) = 0$. In a similar way, we find that when $x = -4$ then $y = -2$, or $f(-4) = -2$; when $x = 0$, then $y = 4$, or $f(0) = 4$; and when $x = 6$, then $y = 1$, or $f(6) = 1$.

(b) To determine the domain of f, we notice that the points on the graph of f all have x-coordinates between -6 and 6, inclusive; and, for each number x between -6 and 6, there is a point $(x, f(x))$ on the graph. Thus, the domain of f is $\{x | -6 \leq x \leq 6\}$, or the interval $[-6, 6]$.

(c) The points on the graph all have y-coordinates between -2 and 4, inclusive; and, for each such number y, there is at least one number x in the domain. Hence, the range of f is $\{y| -2 \le y \le 4\}$, or the interval $[-2, 4]$.

(d) The intercepts are $(-6, 0)$, $(-2, 0)$, and $(0, 4)$.   ■

    When the graph of a function is given, its domain may be viewed as the shadow created by the graph on the x-axis by vertical beams of light. Its range can be viewed as the shadow created by the graph on the y-axis by horizontal beams of light. Try this technique with the graph given in Figure 6.

■   Now work Problems 25 and 27.

EXAMPLE   9   *Getting from an Island to Town*

An island is 2 miles from the nearest point P on a straight shoreline. A town is 12 miles down the shore from P.

(a) If a person can row a boat at an average speed of 3 miles per hour and the same person can walk 5 miles per hour, express the time T it takes to go from the island to town as a function of the distance x from P to where the person lands the boat. See Figure 7.

(b) What is the domain of T?

(c) How long will it take to travel from the island to town if the person lands the boat 4 miles from P?

(d) How long will it take if the person lands the boat 8 miles from P?

(e) Use a graphing utility to graph the function $T = T(x)$.

(f) Use the TRACE function to see how the time T varies as x changes from 0 to 12.

(g) What value of x results in the least time?

Solution   (a) Figure 7 illustrates the situation. The distance d_1 from the island to the landing point satisfies the equation

$$d_1^2 = 4 + x^2$$

FIGURE 7

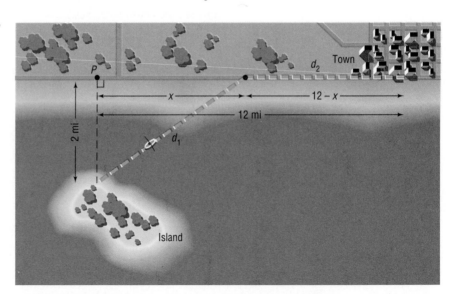

Since the average speed of the boat is 3 miles per hour, the time t_1 it takes to cover the distance d_1 is

$$d_1 = 3t_1$$

Thus,

$$t_1 = \frac{d_1}{3} = \frac{\sqrt{4 + x^2}}{3}$$

The distance d_2 from the landing point to town is $12 - x$, and the time t_2 it takes to cover this distance at an average walking speed of 5 miles per hour obey the equation

$$d_2 = 5t_2$$

Thus,

$$t_2 = \frac{d_2}{5} = \frac{12 - x}{5}$$

The total time T of the trip is $t_1 + t_2$. Thus,

$$T(x) = \frac{\sqrt{4 + x^2}}{3} + \frac{12 - x}{5}$$

(b) Since x equals the distance from P to where the boat lands, it follows that the domain of T is $0 \le x \le 12$.

FIGURE 8

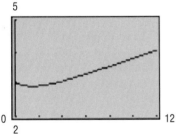

(c) If the boat is landed 4 miles from P, then $x = 4$. The time T the trip takes is

$$T(4) = \frac{\sqrt{20}}{3} + \frac{8}{5} \approx 3.09 \text{ hours}$$

(d) If the boat is landed 8 miles from P, then $x = 8$. The time T the trip takes is

$$T(8) = \frac{\sqrt{68}}{3} + \frac{4}{5} \approx 3.55 \text{ hours}$$

(e) See Figure 8.

(f) As x varies from 0 to 12, the time T varies from about 2.93 to about 4.05.

(g) Using the TRACE function, for x approximately 1.53 miles, the time T is least, about 2.93 hours. See Figure 9(a).

NOTE: Most graphing utilities have a function minimum command that determines the minimum value of a function for a specified domain. Consult your manual. Using this command, we find that x is approximately 1.50 miles with the time T about 2.93 hours. See Figure 9(b).

FIGURE 9

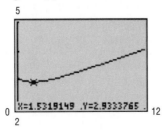

(a) Using TRACE

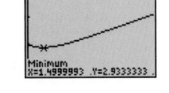

(b) Using minimum on a TI-82

■ Now work Problem 63.

Summary

We list here some of the important vocabulary introduced in this section, with a brief description of each term.

Function	A rule or correspondence between two sets of real numbers so that each number x in the first set, the domain, has corresponding to it exactly one number y in the second set.
	A set of ordered pairs (x, y) or $(x, f(x))$ in which no two pairs have the same first element.
	The range is the set of y values of the function for the x values in the domain.
	A function f may be defined implicitly by an equation involving x and y or explicitly by writing $y = f(x)$.
Unspecified domain	If a function f is defined by an equation and no domain is specified, then the domain will be taken to be the largest set of real numbers for which the rule defines a real number.
Function notation	$y = f(x)$
	f is a symbol for the rule that defines the function.
	x is the independent variable.
	y is the dependent variable.
	$f(x)$ is the value of the function at x, or the image of x.
Graph of a function	The collection of points (x, y) that satisfies the equation $y = f(x)$.
	A collection of points is the graph of a function provided vertical lines intersect the graph in at most one point (vertical-line test).

2.1

Exercise 2.1

In Problems 1–8, find the following values for each function:

(a) f(0)   (b) f(1)   (c) f(−1)   (d) f(3)

1. $f(x) = -3x^2 + 2x - 4$

2. $f(x) = 2x^2 + x - 1$

3. $f(x) = \dfrac{x}{x^2 + 1}$

4. $f(x) = \dfrac{x^2 - 1}{x + 4}$

5. $f(x) = |x| + 4$

6. $f(x) = \sqrt{x^2 + x}$

7. $f(x) = \dfrac{2x + 1}{3x - 5}$

8. $f(x) = 1 - \dfrac{1}{(x + 2)^2}$

In Problems 9–20, use the graph of the function f *given in the figure.*

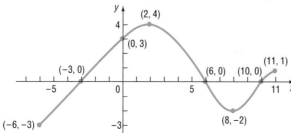

9. Find $f(0)$ and $f(-6)$.

10. Find $f(6)$ and $f(11)$.

11. Is $f(2)$ positive or negative?

12. Is $f(8)$ positive or negative?

13. For what numbers x is $f(x) = 0$?

14. For what numbers x is $f(x) > 0$?

15. What is the domain of f?

16. What is the range of f?

17. What are the x-intercepts?

18. What are the y-intercepts?

19. How often does the line $y = \frac{1}{2}$ intersect the graph?

20. How often does the line $y = 3$ intersect the graph?

In Problems 21–24, answer the questions about the given function.

21. $f(x) = \dfrac{x + 2}{x - 6}$

    (a) Is the point (3, 14) on the graph of f?

    (b) If $x = 4$, what is $f(x)$?

    (c) If $f(x) = 2$, what is x?

    (d) What is the domain of f?

22. $f(x) = \dfrac{x^2 + 2}{x + 4}$

    (a) Is the point $(1, \frac{3}{5})$ on the graph of f?

    (b) If $x = 0$, what is $f(x)$?

    (c) If $f(x) = \frac{1}{2}$, what is x?

    (d) What is the domain of f?

23. $f(x) = \dfrac{2x^2}{x^4 + 1}$

    (a) Is the point $(-1, 1)$ on the graph of f?

    (b) If $x = 2$, what is $f(x)$?

    (c) If $f(x) = 1$, what is x?

    (d) What is the domain of f?

24. $f(x) = \dfrac{2x}{x - 2}$

    (a) Is the point $(\frac{1}{2}, -\frac{2}{3})$ on the graph of f?

    (b) If $x = 4$, what is $f(x)$?

    (c) If $f(x) = 1$, what is x?

    (d) What is the domain of f?

In Problems 25–36, determine whether the graph is that of a function by using the vertical-line test. If it is, use the graph to find:

(a) Its domain and range

(b) The intercepts, if any

(c) Any symmetry with respect to the x-axis, y-axis, or origin

25.

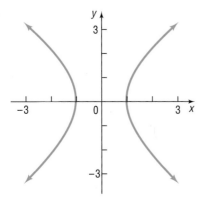

26.

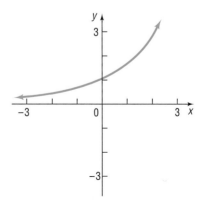

27.

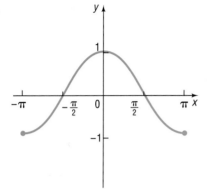

28.

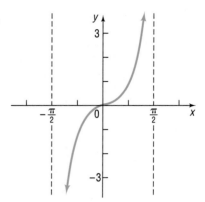

29.

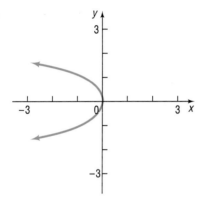

30.

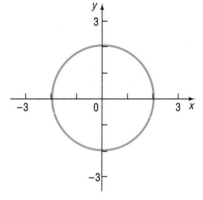

31.

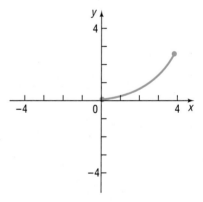

32.

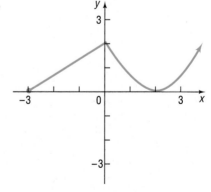

33.

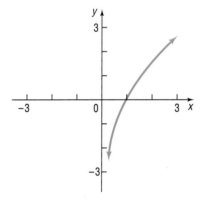

34.

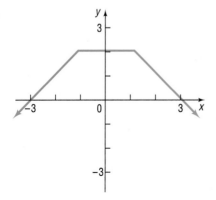

35.

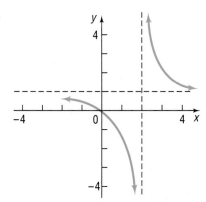

36.

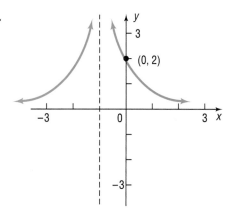

In Problems 37–50, find the domain of each function. Verify your results by graphing each function using a graphing utility.

37. $f(x) = 3x + 4$

38. $f(x) = 5x^2 + 2$

39. $f(x) = \dfrac{x}{x^2 + 1}$

40. $f(x) = \dfrac{x^2}{x^2 + 1}$

41. $g(x) = \dfrac{x}{x^2 - 1}$

42. $h(x) = \dfrac{x}{x - 1}$

43. $F(x) = \dfrac{x - 2}{x^3 + x}$

44. $G(x) = \dfrac{x + 4}{x^3 - 4x}$

45. $h(x) = \sqrt{3x - 12}$

46. $G(x) = \sqrt{1 - x}$

47. $f(x) = \sqrt{x^2 - 9}$

48. $f(x) = \dfrac{1}{\sqrt{x^2 - 4}}$

49. $p(x) = \sqrt{\dfrac{x - 2}{x - 1}}$

50. $q(x) = \sqrt{x^2 - x - 2}$

51. If $f(x) = 2x^3 + Ax^2 + 4x - 5$ and $f(2) = 5$, what is the value of A?

52. If $f(x) = 3x^2 - Bx + 4$ and $f(-1) = 12$, what is the value of B?

53. If $f(x) = (3x + 8)/(2x - A)$ and $f(0) = 2$, what is the value of A?

54. If $f(x) = (2x - B)/(3x + 4)$ and $f(2) = \frac{1}{2}$, what is the value of B?

55. If $f(x) = (2x - A)/(x - 3)$ and $f(4) = 0$, what is the value of A? Where is f not defined?

56. If $f(x) = (x - B)/(x - A)$, $f(2) = 0$, and $f(1)$ is undefined, what are the values of A and B?

57. *Effect of Gravity on Earth* If a rock falls from a height of 20 meters on Earth, the height H (in meters) after x seconds is approximately

$$H(x) = 20 - 4.9x^2$$

(a) Use a graphing utility to graph the function H.
(b) What is the height of the rock when $x = 1$ second? $x = 1.1$ seconds? $x = 1.2$ seconds? $x = 1.3$ seconds?
(c) When is the height of the rock 15 meters? When is it 10 meters? When is it 5 meters?
(d) When does the rock strike the ground?

58. *Effect of Gravity on Jupiter* If a rock falls from a height of 20 meters on the planet Jupiter, its height H (in meters) after x seconds is approximately

$$H(x) = 20 - 13x^2$$

(a) Use a graphing utility to graph the function H.
(b) What is the height of the rock when $x = 1$ second? $x = 1.1$ seconds? $x = 1.2$ seconds?
(c) When is the height of the rock 15 meters? When is it 10 meters? When is it 5 meters?
(d) When does the rock strike the ground?

59. *Geometry* Express the area A of a rectangle as a function of the length x if the length is twice the width of the rectangle.

60. *Geometry* Express the area A of an isosceles right triangle as a function of the length x of one of the two equal sides.

61. Express the gross salary G of a person who earns $5 per hour as a function of the number x of hours worked.

62. A commissioned salesperson earns $100 base pay plus $10 per item sold. Express the gross salary G as a function of the number x of items sold.

63. A cable TV company is asked to provide service to a customer whose house is located 2 miles from the road along which the cable is buried. The nearest connection box for the cable is located 5 miles down the road (see the figure).

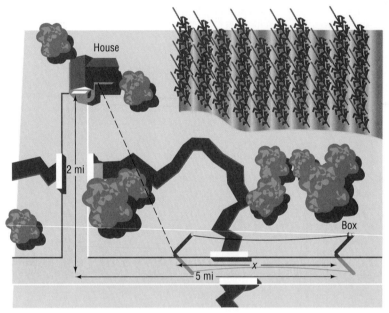

(a) If the installation cost is $10 per mile along the road and $14 per mile off the road, express the total cost C of installation as a function of the distance x (in miles) from the connection box to the point where the cable installation turns off the road. Give the domain.

(b) Compute the cost if $x = 1$ mile.

(c) Compute the cost if $x = 3$ miles.

(d) Graph the function $C = C(x)$. Use TRACE to see how the cost C varies as x changes from 0 to 5.

(e) What value of x results in the least cost?

64. An island is 3 miles from the nearest point P on a straight shoreline. A town is located 20 miles down the shore from P. (Refer to Figure 7 for a similar situation.)

(a) If a person has a boat that averages 12 miles per hour and the same person can run 5 miles per hour, express the time T it takes to go from the island to town as a function of x, where x is the distance from P to where the person lands the boat. Give the domain.

(b) How long will it take to travel from the island to town if you land the boat 8 miles from P?

(c) How long will it take if you land the boat 12 miles from P?

(d) Graph the function $T = T(x)$. Use TRACE to see how the time T varies as x changes from 0 to 20.

(e) What value of x results in the least time?

65. A page with dimensions of $8\frac{1}{2}$ inches by 11 inches has a border of uniform width x surrounding the printed matter of the page, as shown in the figure.

(a) Write a formula for the area A of the printed part of the page as a function of the width x of the border.

(b) Give the domain and range of A.

(c) Find the area of the printed page for borders of widths 1 inch, 1.2 inches, and 1.5 inches.

(d) Graph the function $A = A(x)$.

(e) Use TRACE to determine what margin should be used to obtain an area of 70 square inches and of 50 square inches.

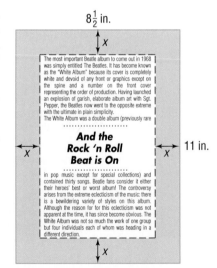

66. *Cost of Trans-Atlantic Travel* An airplane crosses the Atlantic Ocean (3000 miles) with an airspeed of 500 miles per hour. The cost C (in dollars) per passenger is given by

$$C(x) = 100 + \frac{x}{10} + \frac{36{,}000}{x}$$

where x is the ground speed (airspeed $\pm$ wind).
(a) What is the cost per passenger for quiescent (no wind) conditions?
(b) What is the cost per passenger with a head wind of 50 miles per hour?
(c) What is the cost per passenger with a tail wind of 100 miles per hour?
(d) What is the cost per passenger with a head wind of 100 miles per hour?
(e) Graph the function $C = C(x)$.
(f) As x varies from 400 to 600 miles per hour, how does the cost vary?

67. *Period of a Pendulum* The period T (in seconds) of a simple pendulum is a function of its length l (in feet) defined by the equation

$$T(l) = 2\pi \sqrt{\frac{l}{g}}$$

where $g \approx 32.2$ feet per second per second is the acceleration due to gravity.
(a) Use a graphing utility to graph the function $T = T(l)$.
(b) Use the TRACE function to see how the period T varies as 1 changes from 1 to 10.
(c) What length should be used if a period of 10 seconds is required?

68. *Effect of Elevation on Weight* If an object weighs m pounds at sea level, then its weight W (in pounds) at a height of h miles above sea level is given approximately by

$$W(h) = m\left(\frac{4000}{4000 + h}\right)^2$$

(a) If a woman weighs 120 pounds at sea level, how much will she weight on Pike's Peak, which is 14,110 feet above sea level?
(b) Use a graphing utility to graph the function $W = W(h)$. Use $m = 120$ pounds.
(c) Use the TRACE function to see how weight W varies as h changes from 0 to 5 miles.
(d) At what height will the 120 pound woman weigh 121 pounds?
(e) Does your answer to part (d) seem reasonable?

In Problems 69–76, tell whether the set of ordered pairs (x, y) defined by each equation is a function.

69. $y = x^2 + 2x$ **70.** $y = x^3 - 3x$ **71.** $y = \dfrac{2}{x}$ **72.** $y = \dfrac{3}{x} - 3$

73. $y^2 = 1 - x^2$ **74.** $y = \pm\sqrt{1 - 2x}$ **75.** $x^2 + y = 1$ **76.** $x + 2y^2 = 1$

77. Some functions f have the property that $f(a + b) = f(a) + f(b)$ for all real numbers a and b. Which of the following functions have this property?
(a) $h(x) = 2x$ (b) $g(x) = x^2$ (c) $F(x) = 5x - 2$ (d) $G(x) = 1/x$

78. Draw the graph of a function whose domain is $\{x \mid -3 \le x \le 8, x \ne 5\}$ and whose range is $\{y \mid -1 \le y \le 2, y \ne 0\}$. What point(s) in the rectangle $-3 \le x \le 8$, $-1 \le y \le 2$ cannot be on the graph? Compare your graph with those of other students. What differences do you see?

79. Are the functions $f(x) = x - 1$ and $g(x) = (x^2 - 1)/(x + 1)$ the same? Explain.

80. Describe how you would proceed to find the domain and range of a function if you were given its graph. How would your strategy change if, instead, you were given the equation defining the function?

81. How many x-intercepts can the graph of a function have? How many y-intercepts can it have?

82. Is a graph that consists of a single point the graph of a function? Can you write the equation of such a function?

83. Is there a function whose graph is symmetric with respect to the x-axis?

84. Investigate when, historically, the use of function notation $y = f(x)$ first appeared. Write a brief essay on the early uses of this notation.

2.2

More about Functions

Function Notation

The independent variable of a function is sometimes called the **argument** of the function. Thinking of the independent variable as an argument sometimes can make it easier to apply the rule of the function. For example, if f is the function defined by $f(x) = x^3$, then f is the rule that tells us to cube the argument. Thus, $f(2)$ means to cube 2, $f(a)$ means to cube the number a, and $f(x + h)$ means to cube the quantity $x + h$.

EXAMPLE 1

Finding Values of a Function

For the function G defined by $G(x) = 2x^2 - 3x$, evaluate:

(a) $G(3)$ (b) $G(x) + G(3)$ (c) $G(-x)$

(d) $-G(x)$ (e) $G(x + 3)$

Solution   (a) We replace x by 3 in the rule for G to get

$$G(3) = 2(3)^2 - 3(3) = 18 - 9 = 9$$

(b) $G(x) + G(3) = (2x^2 - 3x) + (9) = 2x^2 - 3x + 9$

(c) We replace x by $-x$ in the rule for G:

$$G(-x) = 2(-x)^2 - 3(-x) = 2x^2 + 3x$$

(d) $-G(x) = -(2x^2 - 3x) = -2x^2 + 3x$

(e) $G(x + 3) = 2(x + 3)^2 - 3(x + 3)$ Notice the use of parentheses here.

$$= 2(x^2 + 6x + 9) - 3x - 9$$
$$= 2x^2 + 12x + 18 - 3x - 9$$
$$= 2x^2 + 9x + 9$$

Notice in this example that $G(x + 3) \neq G(x) + G(3)$.

■   Now work Problem 31.

Example 1 illustrates certain uses of **function notation.** Let's look at another use.

EXAMPLE 2

Using Function Notation

For the function $f(x) = x^2 + 1$, find: $\dfrac{f(x) - f(1)}{x - 1}, x \neq 1.$

Solution   First, we find $f(1)$:

$$f(1) = (1)^2 + 1 = 2$$

Then

$$\frac{f(x) - f(1)}{x - 1} = \frac{(x^2 + 1) - (2)}{x - 1} = \frac{x^2 - 1}{x - 1} = \frac{(x + 1)(x - 1)}{x - 1} = x + 1 \quad ■$$

Expressions like the one we worked with in Example 2 occur frequently in calculus.

Difference Quotient

For a number c in the domain of a function f, the expression

$$\frac{f(x) - f(c)}{x - c} \qquad x \neq c \qquad (1)$$

is called the **difference quotient** of f at c.

The difference quotient of a function has an important geometric interpretation. Look at the graph of $y = f(x)$ in Figure 10. We have labeled two points on the graph: $(c, f(c))$ and $(x, f(x))$. The slope of the line containing these two points is

$$\frac{f(x) - f(c)}{x - c}$$

This line is called a **secant line.** Thus, the difference quotient of a function equals the slope of a secant line containing two points on its graph.

FIGURE 10

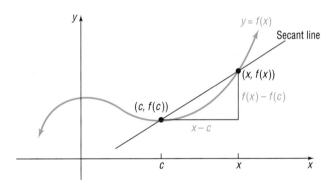

EXAMPLE 3

Finding a Difference Quotient

Find the difference quotient of $f(x) = 2x^2 - x + 1$ at $c = 2$.

Solution From expression (1), we seek

$$\frac{f(x) - f(2)}{x - 2} \qquad x \neq 2$$

We begin by finding $f(2)$:

$$f(2) = 2(2)^2 - (2) + 1 = 8 - 2 + 1 = 7$$

Then the difference quotient of f at 2 is

$$\frac{f(x) - f(2)}{x - 2} = \frac{2x^2 - x + 1 - 7}{x - 2} = \frac{2x^2 - x - 6}{x - 2} = \frac{(2x + 3)(x - 2)}{x - 2} = 2x + 3$$

■

■ Now work Problem 43.

Increasing and Decreasing Functions

Consider the graph given in Figure 11. If you look from left to right along the graph of this function, you will notice that parts of the graph are rising, parts are falling, and parts are horizontal. In such cases, the function is described as *increasing, decreasing,* and *constant,* respectively.

FIGURE 11

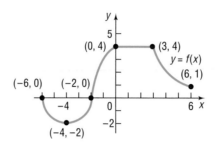

E X A M P L E 4   *Determining Where a Function Is Increasing, Decreasing, or Constant*

Where is the function in Figure 11 increasing? Where is it decreasing? Where is it constant?

Solution

To answer the question of where a function is increasing, where it is decreasing, and where it is constant, we use inequalities involving the independent variable x or we use intervals of x-coordinates. The graph in Figure 11 is rising (increasing) from the point $(-4, -2)$ to the point $(0, 4)$, so we conclude that it is increasing on the interval $[-4, 0]$ (or for $-4 \leq x \leq 0$). The graph is falling (decreasing) from the point $(-6, 0)$ to the point $(-4, -2)$ and from the point $(3, 4)$ to the point $(6, 1)$. We conclude that the graph is decreasing on the intervals $[-6, -4]$ and $[3, 6]$ (or for $-6 \leq x \leq -4$ and $3 \leq x \leq 6$). The graph is constant on the interval $[0, 3]$ (or for $0 \leq x \leq 3$). ■

More precise definitions follow.

Increasing Function

A function f is **increasing** on an interval I if, for any choice of x_1 and x_2 in I, with $x_1 < x_2$, we have $f(x_1) < f(x_2)$.

Decreasing Function

A function f is **decreasing** on an interval I if, for any choice of x_1 and x_2 in I, with $x_1 < x_2$, we have $f(x_1) > f(x_2)$.

Constant Function

A function f is **constant** on an interval I if, for all choices of x in I, the values $f(x)$ are equal.

Thus, the graph of an increasing function goes up from left to right, the graph of a decreasing function goes down from left to right, and the graph of a constant function remains at a fixed height. Figure 12 illustrates the definitions.

FIGURE 12

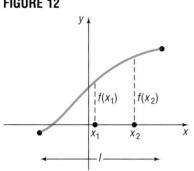

(a) For $x_1 < x_2$ in I,
$f(x_1) < f(x_2)$;
f is increasing

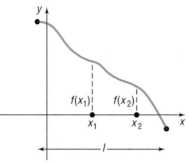

(b) For $x_1 < x_2$ in I,
$f(x_1) > f(x_2)$;
f is decreasing

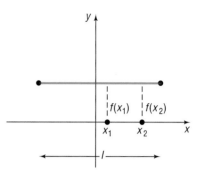

(c) Values of f are equal;
f is constant

E X A M P L E   5   *Using a Graphing Utility to Determine Where A Function Is Increasing and Decreasing*

Use a graphing utility to graph the function $f(x) = x^3 - 2x^2 + 3$ for $-1 \le x \le 2$. Determine where f is increasing and where it is decreasing.

Solution   Figure 13 shows the graph of f. Use the TRACE function, beginning at the point $(-1, 0)$. As we proceed along the graph, the values of y increase until we cross the y-axis (where $y = 3$). We infer that f is increasing on the interval $[-1, 0]$.

As we continue along the graph to the right of the y-axis, the values of f decrease until we reach $x \approx 1.33$ (where $y \approx 1.815$). We infer that f is decreasing on the interval $[0, 1.33]$.

As we continue, the values of y increase from $x = 1.33$ to $x = 2$. We infer that f is increasing on the interval $[1.33, 2]$.

FIGURE 13

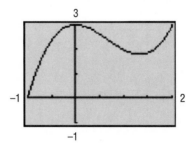

Local Maximum; Local Minimum

When the graph of a function is increasing to the left of $x = c$ and decreasing to the right of $x = c$, then at c the value of f is largest. This value is called a *local maximum* of f.

When the graph of a function is decreasing to the left of $x = c$ and is increasing to the right of $x = c$, then at c the value of f is the smallest. This value is called a *local minimum* of f.

Thus, if f has a local maximum at c, then the value of f at c is greater than or equal to the values of f near c. If f has a local minimum at c, then the value of

f at c is less than or equal to the values of f near c. The word *local* is used to suggest that it is only near c that the value $f(c)$ is largest or smallest. See Figure 14.

FIGURE 14

f has local maximum at x_1 and x_3;
f has a local minimum at x_2.

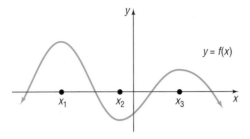

A function f has a **local maximum at c** if there is an interval I containing c so that, for all x in I, $f(x) < f(c)$. We call $f(c)$ a **local maximum of f.**

A function f has a **local minimum at c** if there is an interval I containing c so that, for all x in I, $f(x) > f(c)$. We call $f(c)$ a **local minimum of f.**

For example, in Figure 13, f has a local maximum at 0 and a local minimum at 1.33. The local maximum is $f(0) = 3$; the local minimum is $f(1.33) = 1.815$.

To locate the exact value at which a function f has a local maximum or a local minimum usually requires calculus. However, a graphing utility may be used to approximate these values.

E X A M P L E   6

Using a Graphing Utility to Locate Local Maxima and Minima

Use a graphing utility to graph $f(x) = 6x^3 - 12x + 5$ for $-2 \leq x \leq 2$. Determine where f has a local maximum and where f has a local minimum.

Solution   Figure 15 shows the graph of f on the interval $[-2, 2]$.

Using the TRACE function, we find the local maximum to occur at approximately -0.80 and the local minimum to occur at 0.80, correct to two decimal places. The local maximum is 11.53 and the local minimum is -1.53, correct to two decimal places. See Figures 15(a) and (b).

FIGURE 15

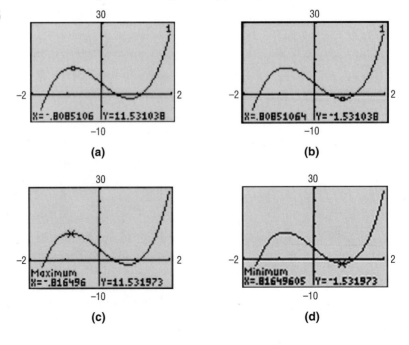

(a)   (b)

(c)   (d)

Most graphing utilities have a function that finds the maximum or minimum point of a graph within a given domain. Using this function, we find the maximum occurs at $(-0.81, 11.53)$ and the minimum occurs at $(0.81, -1.53)$, correct to two decimal places, for $-2 \le x \le 2$, as seen in Figures 15(c) and (d). ■

■ Now work Problem 103.

Even and Odd Functions

Even Function

A function f is **even** if for every number x in its domain the number $-x$ is also in the domain and

$$f(-x) = f(x)$$

Odd Function

A function f is **odd** if for every number x in its domain the number $-x$ is also in the domain and

$$f(-x) = -f(x)$$

Refer to Section 1.2, where the tests for symmetry are listed. The following results are then evident:

Theorem

A function is even if and only if its graph is symmetric with respect to the y-axis. A function is odd if and only if its graph is symmetric with respect to the origin. ■

E X A M P L E   7

Determining Even and Odd Functions from the Graph

Determine whether each graph given in Figure 16 is the graph of an even function, an odd function, or a function that is neither even nor odd.

FIGURE 16

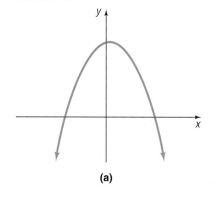

(a)

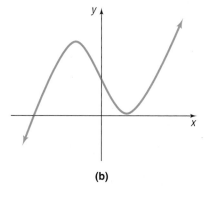

(b)

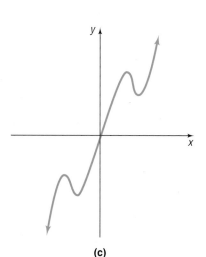

(c)

Solution   The graph in Figure 16(a) is that of an even function, because the graph is symmetric with respect to the y-axis. The function whose graph is given in Figure 16(b) is neither even nor odd, because the graph is neither symmetric with respect to the y-axis nor symmetric with respect to the origin. The function whose graph is given in Figure 16(c) is odd, because its graph is symmetric with respect to the origin.   ■

■ Now work Problem 9.

A graphing utility can be used to conjecture whether a function is even, odd, or neither. As stated, a function is even if $f(-x) = f(x)$. This condition implies that when an even function contains the point (x, y) it must also contain the point $(-x, y)$. Therefore, if TRACE indicates that both the point (x, y) and the point $(-x, y)$ are on the graph for every x, then we would conjecture that the function is even.*

In addition, a function is odd if $f(-x) = -f(x)$. This condition implies that an odd function contains the points $(-x, -y)$ and (x, y). TRACE could be used in the same way to conjecture that the function is odd.*

In the next example, we show how to verify whether a function is even, odd, or neither.

EXAMPLE 8   *Identifying Even and Odd Functions*

Determine whether each of the following functions is even, odd, or neither. Then determine whether the graph is symmetric with respect to the y-axis or with respect to the origin.

(a) $f(x) = x^2 - 5$ (b) $g(x) = x^3 - 1$

(c) $h(x) = 5x^3 - x$ (d) $F(x) = |x|$

(a) Graphing Solution   Graph the function. Use TRACE to determine different pairs of points (x, y) and $(-x, y)$. For example, the point $(1.9574468, -1.168402)$ is on the graph. See Figure 17(a). Now move the cursor to -1.9574468 and determine the corresponding y-coordinate. See Figure 17(b). Since $(1.9574468, -1.168402)$ and $(-1.9574468, -1.168402)$ both lie on the graph, we have evidence that the function is even. Repeating this procedure yields similar results. Therefore, we conjecture that the function is even.

FIGURE 17

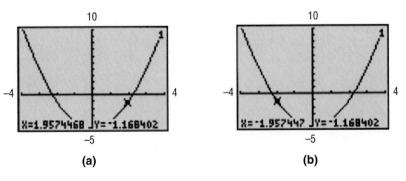

(a)   (b)

*−Xmin and Xmax must be equal for this to work.

Algebraic Solution   We replace x by $-x$ in $f(x) = x^2 - 5$. Then

$$f(-x) = (-x)^2 - 5 = x^2 - 5$$

Since $f(-x) = f(x)$, we conclude that f is an even function, and the graph is symmetric with respect to the y-axis.

(b) Graphing Solution   Graph the function. Using TRACE, the point $(1.787234, 4.7087929)$ is on the graph. See Figure 18(a). Now move the cursor to -1.787234 and determine the corresponding y-coordinate, -6.708793. See Figure 18(b). We conjecture that the function is neither even nor odd since (x, y) does not equal $(-x, y)$ (so it is not even) and (x, y) also does not equal $(-x, -y)$ (so it is not odd).

FIGURE 18

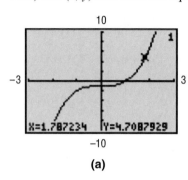   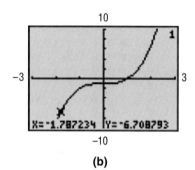

(a)   (b)

Algebraic Solution   We replace x by $-x$. Then

$$g(-x) = (-x)^3 - 1 = -x^3 - 1$$

Since $g(-x) \neq g(x)$ and $g(-x) \neq -g(x)$, we conclude that g is neither even nor odd. The graph is not symmetric with respect to the y-axis nor with respect to the origin.

(c) Graphing Solution   Graph the function. Using TRACE, the point $(1.0212766, 4.3047109)$ is on the graph. See Figure 19(a). We move the cursor to -1.021277 and determine the corresponding y-coordinate, -4.304711. See Figure 19(b). Since (x, y) equals $(-x, -y)$, correct to five decimal places we have evidence that the function is odd. An examination of other pairs of points leads to the conjecture that the function is odd.

FIGURE 19

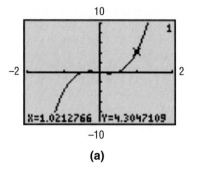   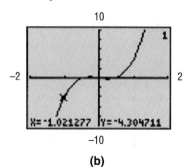

(a)   (b)

Algebraic Solution   We replace x by $-x$ in $h(x) = 5x^3 - x$. Then

$$h(-x) = 5(-x)^3 - (-x) = -5x^3 + x$$

Since $h(-x) = -h(x)$, h is an odd function, and the graph of h is symmetric with respect to the origin.

(d) Graphing Solution

Graph the function. Using TRACE, the point $(-5.744681, 5.7446809)$ is on the graph. See Figure 20(a). We move the cursor to 5.7446809 and determine the corresponding y-coordinate, 5.7446809. See Figure 20(b). Since (x, y) equals $(-x, y)$, we have evidence that the function is even. Repeating this procedure yields similar results. We conjecture that the function is even.

FIGURE 20

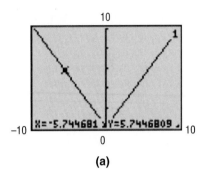
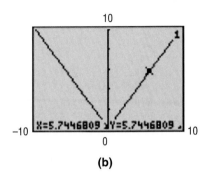

(a)   (b)

Algebraic Solution

We replace x by $-x$ in $F(x) = |x|$. Then

$$F(-x) = |-x| = |x|$$

Since $F(-x) = F(x)$, F is an even function, and the graph of F is symmetric with respect to the y-axis. ∎

■ Now work Problem 53.

Important Functions

We now give names to some of the functions we have encountered. In going through this list, pay special attention to the characteristics of each function, particularly to the shape of each graph.

Linear Function

$$f(x) = mx + b \qquad m \text{ and } b \text{ are real numbers}$$

The domain of the **linear function** f consists of all real numbers. The graph of this function is a nonvertical straight line with slope m and y-intercept b. A linear function is increasing if $m > 0$, decreasing if $m < 0$, and constant if $m = 0$.

Constant Function

$$f(x) = b \qquad b \text{ a real number}$$

See Figure 21.

FIGURE 21

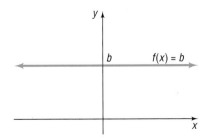

A **constant function** is a special linear function ($m = 0$). Its domain is the set of all real numbers; its range is the set consisting of a single number b. Its graph is a horizontal line whose y-intercept is b. The constant function is an even function whose graph is constant over its domain.

Identity Function

$$f(x) = x$$

See Figure 22.

FIGURE 22

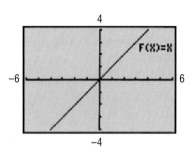 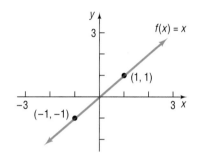

The **identity function** is also a special linear function. Its domain and its range are the set of all real numbers. Its graph is a line whose slope is $m = 1$ and whose y-intercept is 0. The line consists of all points for which the x-coordinate equals the y-coordinate. The identity function is an odd function that is increasing over its domain. Note that the graph bisects quadrants I and III.

Square Function

$$f(x) = x^2$$

See Figure 23.

FIGURE 23

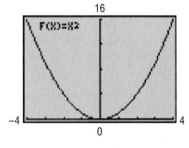 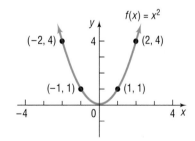

The domain of the **square function** f is the set of all real numbers; its range is the set of nonnegative real numbers. The graph of this function is a parabola, whose intercept is at $(0, 0)$. The square function is an even function that is decreasing on the interval $(-\infty, 0]$ and increasing on the interval $[0, \infty)$.

Cube Function

$$f(x) = x^3$$

See Figure 24.

FIGURE 24

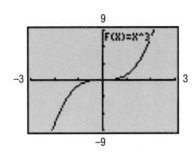

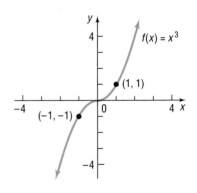

The domain and range of the **cube function** are the set of all real numbers. The intercept of the graph is at (0, 0). The cube function is odd and is increasing on the interval $(-\infty, \infty)$.

Square Root Function

$$f(x) = \sqrt{x}$$

See Figure 25.

FIGURE 25

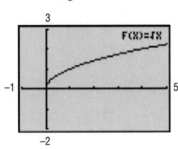

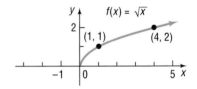

The domain and range of the **square root function** are the set of nonnegative real numbers. The intercept of the graph is at (0, 0). The square root function is neither even nor odd and is increasing on the interval $[0, \infty)$.

Reciprocal Function

$$f(x) = \frac{1}{x}$$

Refer to Example 16, p. 30, for a discussion of the equation $y = 1/x$. See Figure 26.

FIGURE 26

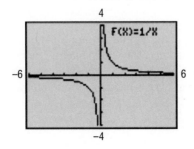

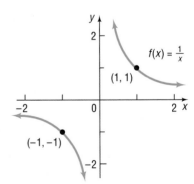

The domain and range of the **reciprocal function** are the set of all nonzero real numbers. The graph has no intercepts. The reciprocal function is decreasing on the intervals $(-\infty, 0)$ and $(0, \infty)$ and is an odd function.

Absolute Value Function	$f(x) = |x|$

See Figure 27.

FIGURE 27

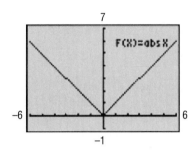 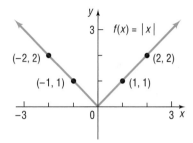

The domain of the **absolute value function** is the set of all real numbers; its range is the set of nonnegative real numbers. The intercept of the graph is at $(0, 0)$. If $x \geq 0$, then $f(x) = x$ and the graph of f is part of the line $y = x$; if $x < 0$, then $f(x) = -x$ and the graph of f is part of the line $y = -x$. The absolute value function is an even function; it is decreasing on the interval $(-\infty, 0]$ and increasing on the interval $[0, \infty)$.

Comment: If your utility has no built-in absolute value function, you can still graph $f(x) = |x|$ by using the fact that $|x| = \sqrt{(x^2)}$.

The symbol $[[x]]$, read as **"bracket x,"** stands for the largest integer less than or equal to x. For example,

$$[[1]] = 1 \qquad [[2.5]] = 2 \qquad [[\tfrac{1}{2}]] = 0 \qquad [[-\tfrac{3}{4}]] = -1 \qquad [[\pi]] = 3$$

This type of correspondence occurs frequently enough in mathematics that we give it a name.

Greatest-integer Function	$f(x) = [[x]] = $ Greatest integer less than or equal to x

We obtain the graph of $f(x) = [[x]]$ by plotting several points. See Table 1. For values of x, $-1 \leq x < 0$, the value of $f(x) = [[x]]$ is -1; for values of x, $0 \leq x < 1$, the value of f is 0. See Figure 28 for the graph.

TABLE 1

X	Y1
-1	-1
-.75	-1
-.5	-1
-.25	-1
0	0
.25	0
.5	0

Y1⊟int X

FIGURE 28

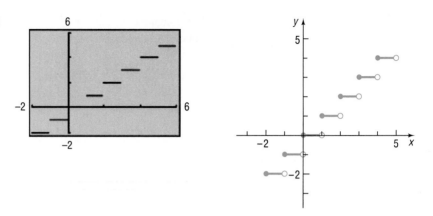

The domain of the **greatest integer function** is the set of all real numbers; its range is the set of integers. The y-intercept of the graph is at 0. The x-intercepts lie in the interval $[0, 1)$. The greatest-integer function is neither even nor odd. It is constant on every interval of the form $[k, k + 1)$, for k an integer. In Figure 28, we use a solid dot to indicate, for example, that at $x = 1$ the value of f is $f(1) = 1$; we use an open circle to illustrate that the function does not assume the value of 0 at $x = 1$.

From the graph of the greatest-integer function, we can see why it is also called a **step function**. At $x = 0$, $x = \pm 1$, $x = \pm 2$, and so on, this function exhibits what is called a *discontinuity;* that is, at integer values, the graph suddenly "steps" from one value to another without taking on any of the intermediate values. For example, to the immediate left of $x = 3$, the y-coordinates are 2, and to the immediate right of $x = 3$, the y-coordinates are 3.

Comment: When graphing a function, you can choose either the **connected mode,** in which points plotted on the screen are connected, making the graph appear without any breaks, or the **dot mode,** in which only the points plotted appear. When graphing the greatest integer function with a graphing utility, it is necessary to be in the **dot mode.** This is to prevent the utility from "connecting the dots" when $f(x)$ changes from one integer value to the next. See Figure 29.

FIGURE 29

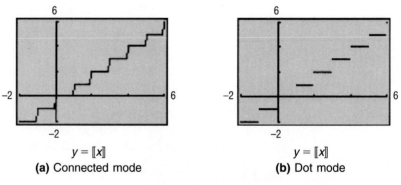

$y = [\![x]\!]$
(a) Connected mode

$y = [\![x]\!]$
(b) Dot mode

The functions we have discussed so far are basic. Whenever you encounter one of them, you should see a mental picture of its graph. For example, if you encounter the function $f(x) = x^2$, you should see in your mind's eye a picture like Figure 23.

■   Now work Problem 69.

Piecewise Defined Functions

Sometimes, a function is defined by a rule consisting of two or more equations. The choice of which equation to use depends on the value of the independent variable x. For example, the absolute value function $f(x) = |x|$ is actually defined by two equations: $f(x) = x$ if $x \geq 0$ and $f(x) = -x$ if $x < 0$. For convenience, we generally combine these equations into one expression as

$$f(x) = |x| = \begin{cases} x & \text{if } x \geq 0 \\ -x & \text{if } x < 0 \end{cases}$$

When functions are defined by more than one equation, they are called **piecewise defined** functions.

Let's look at another example of a piecewise defined function.

E X A M P L E   9   *Analyzing a Piecewise Defined Function*

For the following function f,

$$f(x) = \begin{cases} -x + 1 & \text{if } -1 \leq x < 1 \\ 2 & \text{if } x = 1 \\ x^2 & \text{if } x > 1 \end{cases}$$

(a) Find $f(0)$, $f(1)$, and $f(2)$.   (b) Determine the domain of f.

(c) Graph f.   (d) Use the graph to find the range of f.

Solution (a) To find $f(0)$, we observe that when $x = 0$ the equation for f is given by $f(x) = -x + 1$. So we have

$$f(0) = -0 + 1 = 1$$

When $x = 1$, the equation for f is $f(x) = 2$. Thus

$$f(1) = 2$$

When $x = 2$, the equation for f is $f(x) = x^2$. So

$$f(2) = 2^2 = 4$$

(b) To find the domain of f, we look at its definition. We conclude that the domain of f is $\{x | x \geq -1\}$, or $[-1, \infty)$.

(c) On a graphing utility, the procedure for graphing a piecewise defined function varies depending on the particular utility. In general, you need to enter each piece as a function with a restricted domain.

To graph the middle piece [the point $(1, 2)$], use the $\boxed{\begin{smallmatrix}\text{STAT}\\\text{PLOT}\end{smallmatrix}}$ function. See Figure 30(a).

In graphing piecewise defined functions, it is usually better to be in the dot mode, since such functions may have breaks.

To graph f on paper, we graph "each piece." Thus, we first graph the line $y = -x + 1$ and keep only the part for which $-1 \leq x < 1$. Then we plot the point $(1, 2)$, because when $x = 1$, $f(x) = 2$. Finally, we graph the parabola $y = x^2$ and keep only the part for which $x > 1$. See Figure 30(b).

FIGURE 30

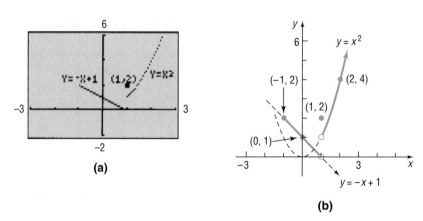

(a)

(b)

(d) From the graph, we conclude that the range of f is $\{y|y > 0\}$, or $(0, \infty)$. ∎

■ Now work Problem 83.

E X A M P L E   1 0   *Cost of Electricity*

In the winter, Commonwealth Edison Company supplies electricity to residences for a monthly customer charge of $9.06 plus 10.819¢ per kilowatt-hour (kWhr) for the first 400 kWhr supplied in the month, and 7.093¢ per kWhr for all usage over 400 kWhr in the month.*

(a) What is the charge for using 300 kWhr in a month?

(b) What is the charge for using 700 kWhr in a month?

(c) If C is the monthly charge for x kWhr, express C as a function of x.

Solution   (a) For 300 kWhr, the charge is $9.06 plus 10.819¢ = $0.10819 per kWhr. Thus,

$$\text{Charge} = \$9.06 + \$0.10819(300) = \$41.52$$

(b) For 700 kWhr, the charge is $9.06 plus 10.819¢ for the first 400 kWhr plus 7.093¢ for the 300 kWhr in excess of 400. Thus,

$$\text{Charge} = \$9.06 + \$0.10819(400) + \$0.07093(300) = \$73.62$$

(c) If $0 \le x \le 400$, the monthly charge C (in dollars) can be found by multiplying x times $0.10819 and adding the monthly customer charge of $9.06. Thus, if $0 \le x \le 400$, then $C(x) = 0.10819x + 9.06$. For $x > 400$, the charge is $0.10819(400) + 9.06 + 0.07093(x - 400)$, since $x - 400$ equals the usage in excess of 400 kWhr, which costs $0.07093 per kWhr. Thus, if $x > 400$, then

$$C(x) = 0.10819(400) + 9.06 + 0.07093(x - 400)$$
$$= 52.336 + 0.07093(x - 400)$$
$$= 0.07093x + 23.964$$

The rule for computing C follows two rules:

$$C(x) = \begin{cases} 0.10819x + 9.06 & \text{if } 0 \le x \le 400 \\ 0.07093x + 23.964 & \text{if } x > 400. \end{cases}$$

*Source: Commonwealth Edison Co., Chicago, Illinois, 1991.

See Figure 31 for the graph.

FIGURE 31

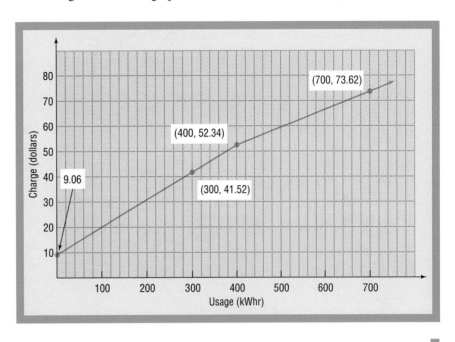

2.2

Exercise 2.2

In Problems 1–8, match each graph to the function listed whose graph most resembles the one given.

A. *Constant function*   B. *Linear function*
C. *Square function*   D. *Cube function*
E. *Square root function*   F. *Reciprocal function*
G. *Absolute value function*   H. *Greatest integer function*

1. 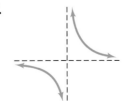   **2.**   **3.**   **4.**

5.   **6.**   **7.**    **8.**

In Problems 9–24, the graph of a function is given. Use the graph to find:

(a) *Its domain and range*

(b) *The intervals on which it is increasing, decreasing, or constant*

(c) *Whether it is even, odd, or neither*

(d) *The intercepts, if any*

9.

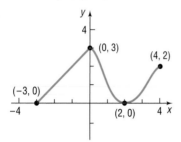

10.

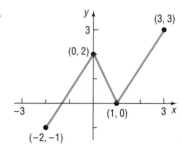

11.

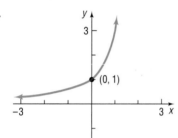

12.

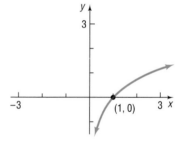

13.

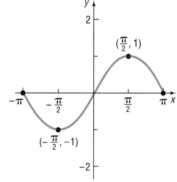

14.

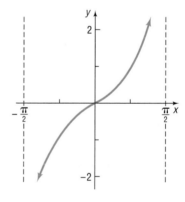

15.

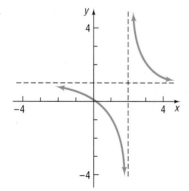

16.

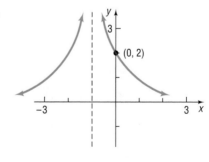

17.

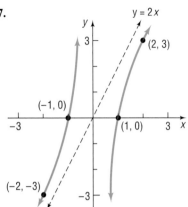

18.

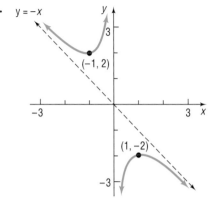

19.

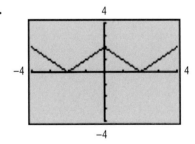

20.

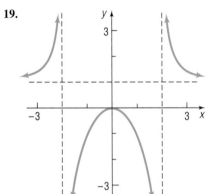

For Problems 21–24, assume that the graph shown is complete.

21.

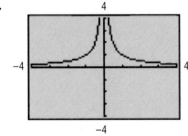

22.

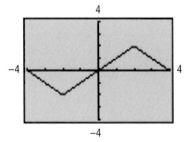

23.

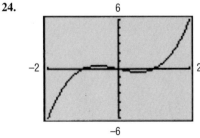

24.

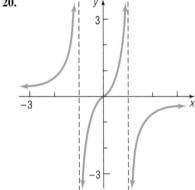

25. If $f(x) = [[2x]]$, find: (a) $f(1.2)$ (b) $f(1.6)$ (c) $f(-1.8)$
26. If $f(x) = [[x/2]]$, find: (a) $f(1.2)$ (b) $f(1.6)$ (c) $f(-1.8)$
27. If
$$f(x) = \begin{cases} x^2 & \text{if } x < 0 \\ 2 & \text{if } x = 0 \\ 2x + 1 & \text{if } x > 0 \end{cases}$$

find: (a) $f(-2)$ (b) $f(0)$ (c) $f(2)$

28. If
$$f(x) = \begin{cases} x^3 & \text{if } x < 0 \\ 3x + 2 & \text{if } x \geq 0 \end{cases}$$

find: (a) $f(-1)$ (b) $f(0)$ (c) $f(1)$

In Problems 29–40, find the following for each function:

(a) $f(-x)$ (b) $-f(x)$ (c) $f(2x)$ (d) $f(x - 3)$ (e) $f(1/x)$ (f) $1/f(x)$

29. $f(x) = 2x + 5$ **30.** $f(x) = 3 - x$ **31.** $f(x) = 2x^2 - 4$ **32.** $f(x) = x^3 + 1$

33. $f(x) = x^3 - 3x$ **34.** $f(x) = x^2 + x$ **35.** $f(x) = \dfrac{x}{x^2 + 1}$ **36.** $f(x) = \dfrac{x^2}{x^2 + 1}$

37. $f(x) = |x|$ **38.** $f(x) = \dfrac{1}{x}$ **39.** $f(x) = 1 + \dfrac{1}{x}$ **40.** $f(x) = 4 + \dfrac{2}{x}$

In Problems 41–52, find the difference quotient,

$$\frac{f(x) - f(1)}{x - 1} \qquad x \neq 1$$

for each function. Be sure to simplify.

41. $f(x) = 3x$ **42.** $f(x) = -2x$ **43.** $f(x) = 1 - 3x$ **44.** $f(x) = x^2 + 1$
45. $f(x) = 3x^2 - 2x$ **46.** $f(x) = 4x - 2x^2$ **47.** $f(x) = x^3 - x$ **48.** $f(x) = x^3 + x$
49. $f(x) = \dfrac{2}{x + 1}$ **50.** $f(x) = \dfrac{1}{x^2}$ **51.** $f(x) = \sqrt{x}$ **52.** $f(x) = \sqrt{x + 3}$

In Problems 53–64, tell whether each function is even, odd, or neither without drawing a graph. Graph each function and use TRACE to verify your results.

53. $f(x) = 4x^3$ **54.** $f(x) = 2x^4 - x^2$ **55.** $g(x) = 2x^2 - 5$ **56.** $h(x) = 3x^3 + 2$
57. $F(x) = \sqrt[3]{x}$ **58.** $G(x) = \sqrt{x}$ **59.** $f(x) = x + |x|$ **60.** $f(x) = \sqrt[3]{2x^2 + 1}$
61. $g(x) = \dfrac{1}{x^2}$ **62.** $h(x) = \dfrac{x}{x^2 - 1}$ **63.** $h(x) = \dfrac{x^3}{3x^2 - 9}$ **64.** $F(x) = \dfrac{x}{|x|}$

65. How many x-intercepts can a function defined on an interval have if it is increasing on that interval? Explain.
66. How many y-intercepts can a function have? Explain.

In Problems 67–92:

(a) *Find the domain of each function.*   (b) *Locate any intercepts.*
(c) *Graph each function by hand.*   (d) *Based on the graph, find the range.*
(e) *Verify your results using a graphing utility.*

67. $f(x) = 3x - 3$ **68.** $f(x) = 4 - 2x$ **69.** $g(x) = x^2 - 4$
70. $g(x) = x^2 + 4$ **71.** $h(x) = -x^2$ **72.** $F(x) = 2x^2$

73. $f(x) = \sqrt{x - 2}$ **74.** $g(x) = \sqrt{x} + 2$ **75.** $h(x) = \sqrt{2 - x}$

76. $F(x) = -\sqrt{x}$ **77.** $f(x) = |x| + 3$ **78.** $g(x) = |x + 3|$

79. $h(x) = -|x|$ **80.** $F(x) = |3 - x|$

81. $f(x) = \begin{cases} 2x & \text{if } x \neq 0 \\ 0 & \text{if } x = 0 \end{cases}$ **82.** $f(x) = \begin{cases} 3x & \text{if } x \neq 0 \\ 4 & \text{if } x = 0 \end{cases}$

83. $f(x) = \begin{cases} 1 + x & \text{if } x < 0 \\ x^2 & \text{if } x \geq 0 \end{cases}$ **84.** $f(x) = \begin{cases} 1/x & \text{if } x < 0 \\ \sqrt{x} & \text{if } x \geq 0 \end{cases}$

85. $f(x) = \begin{cases} |x| & \text{if } -2 \leq x < 0 \\ 1 & \text{if } x = 0 \\ x^3 & \text{if } x > 0 \end{cases}$ **86.** $f(x) = \begin{cases} 3 + x & \text{if } -3 \leq x < 0 \\ 3 & \text{if } x = 0 \\ \sqrt{x} & \text{if } x > 0 \end{cases}$

87. $g(x) = \begin{cases} 1 & \text{if } x \text{ is an integer} \\ -1 & \text{if } x \text{ is not an integer} \end{cases}$ **88.** $g(x) = \begin{cases} x & \text{if } x \geq 1 \\ 1 & \text{if } x < 1 \end{cases}$

89. $h(x) = 2[[x]]$ **90.** $f(x) = [[2x]]$

91. $F(x) = \begin{cases} 4 - x^2 & \text{if } |x| \leq 2 \\ x^2 - 4 & \text{if } |x| > 2 \end{cases}$ **92.** $G(x) = |x^2 - 4|$

Problems 93–96, require the following definition. Secant Line *The slope of the secant line containing the two points $(x, f(x))$ and $(x + h, f(x + h))$ on the graph of a function $y = f(x)$ may be given as*

$$\frac{f(x + h) - f(x)}{h}$$

In Problems 93–96, express the slope of the secant line of each function in terms of x and h. Be sure to simplify your answer.

93. $f(x) = 2x + 5$ **94.** $f(x) = -3x + 2$

95. $f(x) = x^2 + 2x$ **96.** $f(x) = 1/x$

In Problems 97–100, the graph of a piecewise-defined function is given. Write a definition for each function.

97.

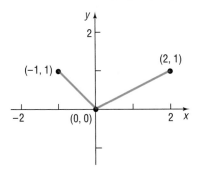

98.

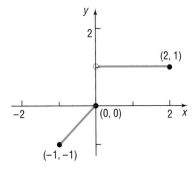

99.

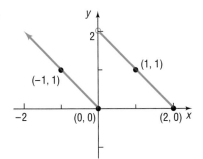

100.

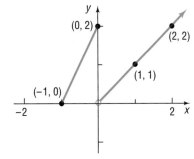

In Problems 101 and 102, decide whether each function is even. Give a reason.

101. $f(x) = \begin{cases} x^2 + 4 & \text{if } x \neq 2 \\ 6 & \text{if } x = 2 \end{cases}$

102. $f(x) = \begin{cases} x^2 + 4 & \text{if } x \neq 2 \\ 5 & \text{if } x = 2 \end{cases}$

In Problems 103–106, use a graphing utility to graph each function over the indicated interval. Determine where the function is increasing and where it is decreasing. Approximate any local maxima and local minima.

103. $f(x) = x^3 - 3x + 2$ $[-2, 2]$

104. $f(x) = x^3 - 3x^2 + 5$ $[-1, 3]$

105. $f(x) = x^5 - x^3$ $[-2, 2]$

106. $f(x) = x^4 - x^2$ $[-2, 2]$

107. Graph $y = x^2$. Then on the same screen graph $y = x^2 + 2$, followed by $y = x^2 + 4$, followed by $y = x^2 - 2$. What pattern do you observe? Can you predict the graph of $y = x^2 - 4$? Of $y = x^2 + 5$?

108. Graph $y = x^2$. Then on the same screen graph $y = (x - 2)^2$, followed by $y = (x - 4)^2$, followed by $y = (x + 2)^2$. What pattern do you observe? Can you predict the graph of $y = (x + 4)^2$? Of $y = (x - 5)^2$?

109. Graph $y = |x|$. Then on the same screen graph $y = 2|x|$, followed by $y = 4|x|$, followed by $y = \frac{1}{2}|x|$. What pattern do you observe? Can you predict the graph of $y = \frac{1}{4}|x|$? Of $y = 5|x|$?

110. Graph $y = x^2$. Then on the same screen graph $y = -x^2$. What pattern do you observe? Now try $y = |x|$ and $y = -|x|$. What do you conclude?

111. Graph $y = \sqrt{x}$. Then on the same screen graph $y = \sqrt{-x}$. What pattern do you observe? Now try $y = 2x + 1$ and $y = 2(-x) + 1$. What do you conclude?

112. Graph $y = x^3$. Then on the same screen graph $y = (x - 1)^3 + 2$. Could you have predicted the result?

113. Graph $y = x^2$, $y = x^4$, and $y = x^6$ on the same screen. What do you notice is the same about each graph? What do you notice that is different?

114. Graph $y = x^3$, $y = x^5$, and $y = x^7$ on the same screen. What do you notice is the same about each graph? What do you notice that is different?

115. *Cost of Natural Gas* On November 12, 1991, the Peoples Gas Light and Coke Company had the following rate schedule* for natural gas usage in single-family residences:

Monthly service charge	$7.00
Per therm service charge, 1st 90 therms	$0.21054/therm
Over 90 therms	$0.11242/therm
Gas charge	$0.26341/therm

(a) What is the charge for using 50 therms in a month?
(b) What is the charge for using 500 therms in a month?
(c) Construct a function that relates the monthly charge C for x therms of gas.
(d) Graph this function.

116. *Cost of Natural Gas* On November 19, 1991, Northern Illinois Gas Company had the following rate schedule† for natural gas usage in single-family residences:

Monthly customer charge	$4.00
Distribution charge, 1st 50 therms	$0.1402/therm
Over 50 therms	$0.0547/therm
Gas supply charge	$0.2406/therm

(a) What is the charge for using 40 therms in a month?
(b) What is the charge for using 202 therms in a month?
(c) Construct a function that gives the monthly charge C for x therms of gas.
(d) Graph this function.

*Source: The Peoples Gas Light and Coke Company, Chicago, Illinois.
†Source: Northern Illinois Gas Company, Naperville, Illinois.

117. Let f denote any function with the property that, whenever x is in its domain, then so is $-x$. Define the functions $E(x)$ and $O(x)$ to be

$$E(x) = \frac{1}{2}[f(x) + f(-x)] \qquad O(x) = \frac{1}{2}[f(x) - f(-x)]$$

(a) Show that $E(x)$ is an even function.   (b) Show that $O(x)$ is an odd function.
(c) Show that $f(x) = E(x) + O(x)$.
(d) Draw the conclusion that any such function f can be written as the sum of an even function and an odd function.

118. Let f and g be two functions defined on the same interval $[a, b]$. Suppose that we define two functions min (f, g) and max (f, g) as follows:

$$\min(f, g)(x) = \begin{cases} f(x) & \text{if } f(x) \le g(x) \\ g(x) & \text{if } f(x) > g(x) \end{cases} \qquad \max(f, g)(x) = \begin{cases} g(x) & \text{if } f(x) \le g(x) \\ f(x) & \text{if } f(x) > g(x) \end{cases}$$

Show that

$$\min(f, g)(x) = \frac{f(x) + g(x)}{2} - \frac{|f(x) - g(x)|}{2}$$

Develop a similar formula for max (f, g).

119. Consider the equation

$$y = \begin{cases} 1 & \text{if } x \text{ is rational} \\ 0 & \text{if } x \text{ is irrational} \end{cases}$$

Is this a function? What is its domain? What is its range? What is its y-intercept, if any? What are its x-intercepts, if any? Is it even, odd, or neither? How would you describe its graph?

120. Define some functions that pass through $(0, 0)$ and $(1, 1)$ and are increasing for $x \ge 0$. Begin your list with $y = \sqrt{x}$, $y = x$, and $y = x^2$. Can you propose a general result about such functions?

121. Can you think of a function that is both even and odd?

2.3

Graphing Techniques

At this stage, if you were asked to graph any of the functions defined by $y = x^2$, $y = x^3$, $y = x$, $y = \sqrt{x}$, $y = |x|$, or $y = 1/x$, your response should be, "Yes, I recognize these functions and know the general shapes of their graphs." (If this is not your answer, review the previous section and Figures 21 through 29.)

Sometimes, we are asked to graph a function that is "almost" like one we already know how to graph. In this section, we look at some of these functions and develop techniques for graphing them.

Vertical Shifts

EXAMPLE 1   *Vertical Shifts*

On the same screen, graph each of the following functions:

$$f(x) = x^2$$
$$f(x) = x^2 + 1$$
$$f(x) = x^2 + 2$$
$$f(x) = x^2 - 1$$
$$f(x) = x^2 - 2$$

Solution   Figure 32 illustrates the graphs. You should have observed a general pattern. With $y = x^2$ on the screen, the graph of $y = x^2 + 1$ is identical to that of $y = x^2$, except that it is shifted vertically up 1 unit. Similarly, $y = x^2 + 2$ is identical to that of $y = x^2$, except that it is shifted vertically up 2 units. The graph of $y = x^2 - 1$ is identical to that of $y = x^2$, except that it is shifted vertically down 1 unit.

FIGURE 32

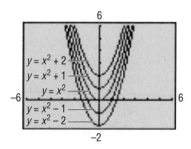

We are led to the following conclusion: If a real number c is added to the right side of a function $y = f(x)$, the graph of the new function $y = f(x) + c$ is the graph of f **shifted vertically** up (if $c > 0$) or down (if $c < 0$). Let's look at another example.

E X A M P L E   2   *Vertical Shift Down*

Use the graph of $f(x) = x^2$ to obtain the graph of $h(x) = x^2 - 4$.

Solution   Table 2 lists some points on the graphs of $f = Y_1$ and $h = Y_2$. The graph of h is identical to that of f, except that it is shifted down 4 units. See Figure 33.

TABLE 2

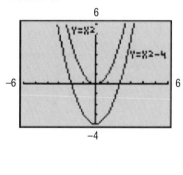

FIGURE 33

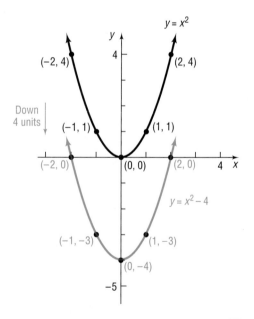

Horizontal Shifts

E X A M P L E 3

On the same screen, graph each of the following functions:

$$Y = x^2$$
$$Y = (x - 1)^2$$
$$Y = (x - 3)^2$$
$$Y = (x + 2)^2$$

Solution Figure 34 illustrates the graphs.

FIGURE 34

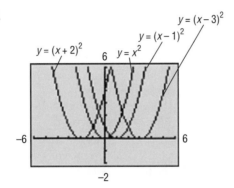

You should have observed the following pattern. With the graph of $y = x^2$ on the screen, the graph of $y = (x - 1)^2$ is identical to that of $y = x^2$, except it is shifted horizontally to the right 1 unit. Similarly, the graph of $y = (x - 3)^2$ is identical to that of $y = x^2$, except it is shifted horizontally to the right 3 units. Finally, the graph of $y = (x + 2)^2$ is identical to that of $y = x^2$, except it is shifted horizontally to the left 2 units. ∎

We are led to the following conclusion.

If a real number c is added to the argument x of a function f, the graph of the new function $g(x) = f(x + c)$ is the graph of f shifted horizontally left (if $c > 0$) or right (if $c < 0$).

■ Now work Problem 31.

Vertical and horizontal shifts are sometimes combined.

E X A M P L E 4

Combining Vertical and Horizontal Shifts

Graph the function: $f(x) = (x - 1)^3 + 3$

Solution We graph f in steps. First, we note that the rule for f is basically a cube function. Thus, we begin with the graph of $y = x^3$. See Figure 35(a). Next, to get the graph of $y = (x - 1)^3$, we shift the graph of $y = x^3$ horizontally 1 unit to the right. See Figure 35(b). Finally, to get the graph of $y = (x - 1)^3 + 3$, we shift the graph of $y = (x - 1)^3$ vertically up 3 units. See Figure 35(c). Note the three points that have been plotted on each graph. Using key points such as these can be helpful in keeping track of just what is taking place.

FIGURE 35

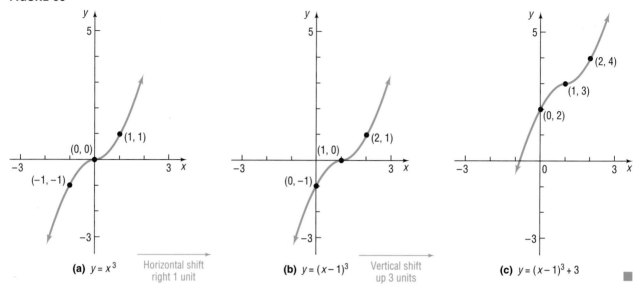

(a) $y = x^3$ → Horizontal shift right 1 unit   **(b)** $y = (x-1)^3$ → Vertical shift up 3 units   **(c)** $y = (x-1)^3 + 3$ ■

In Example 4, if the vertical shift had been done first, followed by the horizontal shift, the result would have been the same. (Try it for yourself.)

■ Now work Problem 43.

Compressions and Stretches

E X A M P L E   5   On the same screen, graph each of the following functions:

$$f(x) = |x|$$
$$f(x) = 2|x|$$
$$f(x) = 3|x|$$
$$f(x) = \tfrac{1}{2}|x|$$

Solution   Figure 36 illustrates the graphs. You should have observed the following pattern. The graphs of $y = 2|x|$ and $y = 3|x|$ can be obtained from the graph of $y = |x|$ by a vertical *stretch* with factors of 2 and 3, respectively. The graph of $y = \tfrac{1}{2}|x|$ can be obtained from the graph of $y = |x|$ by a vertical *compression* with a factor of $\tfrac{1}{2}$.

FIGURE 36

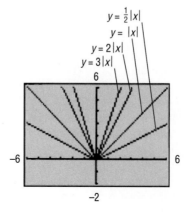

Look at Tables 3 and 4, where $Y_1 = |x|$, $Y_2 = 3|x|$, and $Y_3 = \frac{1}{2}|x|$. Notice that the values for Y_2 in Table 4 are three times the values of Y_1. Therefore, the graph of Y_2 will be vertically *stretched* by a factor of 3. Likewise, the values of Y_3 in Table 5 are half the values of Y_1. Therefore, the graph of Y_3 will be vertically *compressed* by a factor of $\frac{1}{2}$.

TABLE 3

X	Y₁	Y₂
-2	2	6
-1	1	3
0	0	0
1	1	3
2	2	6
3	3	9
4	4	12

Y₂ ■ 3abs X

TABLE 4

X	Y₁	Y₃
-2	2	1
-1	1	.5
0	0	0
1	1	.5
2	2	1
3	3	1.5
4	4	2

Y₃ ■ .5abs X

■

When the right side of a function $y = f(x)$ is multiplied by a positive number k, the graph of the new function $y = kf(x)$ is a vertically *compressed* (if $0 < k < 1$) or *stretched* (if $k > 1$) version of the graph of $y = f(x)$.

■ Now work Problem 33.

If the argument x of a function $y = f(x)$ is multiplied by a positive number k, the graph of the new function $y = f(kx)$ is also a compressed or stretched version of the graph of $y = f(x)$, but, in this case, it occurs horizontally rather than vertically. To see why, we look at the following example.

E X A M P L E   6

On the same screen, graph each of the following functions:

$$Y_1 = f(x) = x^2 + x$$
$$Y_2 = f(2x) = (2x)^2 + (2x) = 4x^2 + 2x$$
$$Y_3 = f\left(\frac{1}{2}x\right) = \left(\frac{1}{2}x\right)^2 + \left(\frac{1}{2}x\right) = \frac{1}{4}x^2 + \frac{1}{2}x$$

Solution   See Figure 37. The effect of multiplying the argument x by a factor results in a horizontal stretch or compression of the original graph.

FIGURE 37

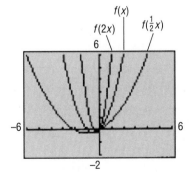

Look at Tables 5 and 6, where

$$Y_1 = f(x) = x^2 + x, \qquad Y_2 = f(2x) = 4x^2 + 2x, \qquad Y_3 = f\left(\frac{1}{2}x\right) = \frac{1}{4}x^2 + \frac{1}{2}x$$

Notice that the values for Y_2 in Table 5 are larger than the values of Y_1 for every x, except $x = 0$. Therefore, the graph of Y_2 is horizontally *compressed*. So, $f(kx)$ will have a graph that is horizontally *compressed* [that is, its graph will be steeper than $f(x)$] when $k > 1$. Likewise, the values of Y_3 in Table 6 are smaller than the values of Y_1 for every x, except $x = 0$. Therefore, the graph of Y_3 will be horizontally *stretched*. So, $f(kx)$ will have a graph that is horizontally *stretched* [that is, its graph will be flatter than $f(x)$] when $0 < k < 1$.

TABLE 5

X	Y₁	Y2
-3	6	30
-2	2	12
-1	0	2
0	0	0
1	2	6
2	6	20
3	12	42

Y₂⬛4X²+2X

TABLE 6

X	Y₁	Y3
-3	6	.75
-2	2	0
-1	0	-.25
0	0	0
1	2	.75
2	6	2
3	12	3.75

Y₃⬛.25X²+.5X

Reflections about the *x*-Axis and the *y*-Axis

E X A M P L E   7   *Reflection about the x-Axis*

(a) Graph $y = x^2$ followed by $y = -x^2$.

(b) Graph $y = |x|$ followed by $y = -|x|$.

(c) Graph $y = x^2 - 4$ followed by $y = -(x^2 - 4) = -x^2 + 4$.

Solution   See Tables 7(a), (b), and (c) and Figures 38(a), (b), and (c). In each instance, the second graph is the reflection about the *x*-axis of the first graph.

TABLE 7

X	Y₁	Y2
-3	9	-9
-2	4	-4
-1	1	-1
0	0	0
1	1	-1
2	4	-4
3	9	-9

Y₂⬛-X²

(a)

X	Y₁	Y2
-3	3	-3
-2	2	-2
-1	1	-1
0	0	0
1	1	-1
2	2	-2
3	3	-3

Y₂⬛-abs X

(b)

X	Y₁	Y2
-3	5	-5
-2	0	0
-1	-3	3
0	-4	4
1	-3	3
2	0	0
3	5	-5

Y₂⬛-X²+4

(c)

FIGURE 38

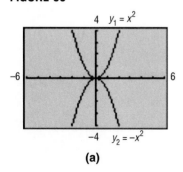

(a)

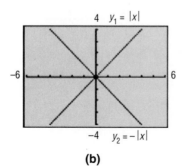

(b)

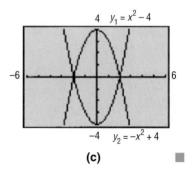

(c)

When the right side of the equation $y = f(x)$ is multiplied by -1, the graph of the new function $y = -f(x)$ is the **reflection about the *x*-axis** of the graph of the function $y = f(x)$.

■ Now work Problem 39.

E X A M P L E   8

(a) Graph $y = \sqrt{x}$ followed by $y = \sqrt{-x}$.

(b) Graph $y = x + 1$ followed by $y = -x + 1$.

(c) Graph $y = x^4 + x$ followed by $y = (-x)^4 + (-x) = x^4 - x$.

Solution

See Tables 8(a), (b), and (c) and Figures 39(a), (b), and (c). In each instance, the second graph is the reflection about the y-axis of the first graph.

TABLE 8

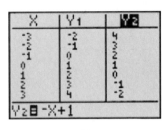

(a)

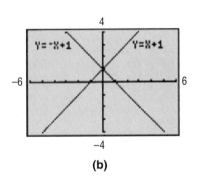

(b)

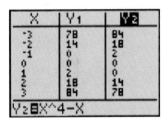

(c)

FIGURE 39

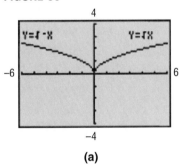

(a)

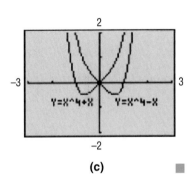

(b)

(c) ■

When the graph of the function $y = f(x)$ is known, the graph of the new function $y = f(-x)$ is the **reflection about the y-axis** of the graph of the function $y = f(x)$.

Summary of Graphing Techniques

Table 9 summarizes the graphing procedures we have just discussed.

TABLE 9

To Graph:	Draw the Graph of f and:
Vertical shifts	
$y = f(x) + c, \quad c > 0$	Raise the graph of f by c units.
$y = f(x) - c, \quad c > 0$	Lower the graph of f by c units.
Horizontal shifts	
$y = f(x + c), \quad c > 0$	Shift the graph of f to the left c units.
$y = f(x - c), \quad c > 0$	Shift the graph of f to the right c units.
Compressing or stretching	
$y = kf(x), \quad k > 0$	Compress or stretch the graph of f by a factor of k.
$y = f(kx), \quad k > 0$	
Reflection about the x-axis	
$y = -f(x)$	Reflect the graph of f about the x-axis.
Reflection about the y-axis	
$y = f(-x)$	Reflect the graph of f about the y-axis.

The examples that follow combine some of the procedures outlined in this section to get the required graph.

E X A M P L E   9   *Combining Graphing Procedures*

Graph the function: $f(x) = \dfrac{3}{x-2} + 1$

Solution   We use the following steps to obtain the graph of *f*:

STEP 1: $y = \dfrac{1}{x}$ Reciprocal function

STEP 2: $y = \dfrac{3}{x}$ Vertical stretch of the graph of $y = \dfrac{1}{x}$ by a factor of 3

STEP 3: $y = \dfrac{3}{x-2}$ Horizontal shift to the right 2 units; replace *x* by $x - 2$

STEP 4: $y = \dfrac{3}{x-2} + 1$ Vertical shift up 1 unit; add 1

See Figure 40.

FIGURE 40

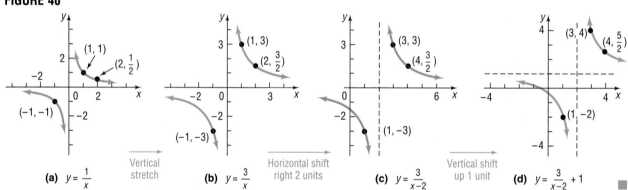

(a) $y = \dfrac{1}{x}$ Vertical stretch   (b) $y = \dfrac{3}{x}$ Horizontal shift right 2 units   (c) $y = \dfrac{3}{x-2}$ Vertical shift up 1 unit   (d) $y = \dfrac{3}{x-2} + 1$

There are other orderings of the steps shown in Example 9 that would also result in the graph of *f*. For example, try this one;

STEP 1: $y = \dfrac{1}{x}$ Reciprocal function

STEP 2: $y = \dfrac{1}{x-2}$ Horizontal shift to the right 2 units; replace *x* by $x - 2$

STEP 3: $y = \dfrac{3}{x-2}$ Vertical stretch of the graph of $y = \dfrac{1}{x-2}$ by factor of 3

STEP 4: $y = \dfrac{3}{x-2} + 1$ Vertical shift up 1 unit; add 1

■ Now work Problem 45.

E X A M P L E   1 0   *Combining Graphing Procedures*

Graph the function: $f(x) = \sqrt{1-x} + 2$

Solution We use the following steps to get the graph of $y = \sqrt{1 - x} + 2$:

STEP 1: $y = \sqrt{x}$ Square root function

STEP 2: $y = \sqrt{x + 1}$ Replace x by $x + 1$; horizontal shift left 1 unit

STEP 3: $y = \sqrt{-x + 1} = \sqrt{1 - x}$ Replace x by $-x$; reflect about y-axis

STEP 4: $y = \sqrt{1 - x} + 2$ Vertical shift up 2 units

See Figure 41.

FIGURE 41

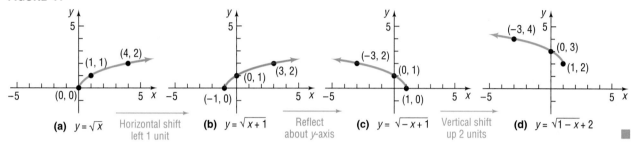

(a) $y = \sqrt{x}$ Horizontal shift left 1 unit   **(b)** $y = \sqrt{x + 1}$ Reflect about y-axis   **(c)** $y = \sqrt{-x + 1}$ Vertical shift up 2 units   **(d)** $y = \sqrt{1 - x} + 2$

2.3

Exercise 2.3

In Problems 1–12, match each graph to one of the following functions:

A. $y = x^2 + 2$ B. $y = -x^2 + 2$ C. $y = |x| + 2$ D. $y = -|x| + 2$

E. $y = (x - 2)^2$ F. $y = -(x + 2)^2$ G. $y = |x - 2|$ H. $y = -|x + 2|$

I. $y = 2x^2$ J. $y = -2x^2$ K. $y = 2|x|$ L. $y = -2|x|$

1.
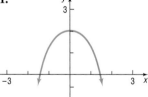

2.
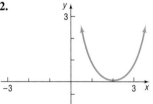

3.
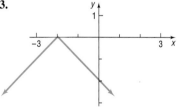

4.
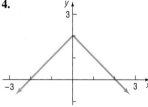

5.
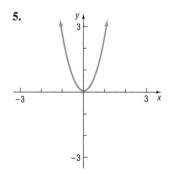

6.
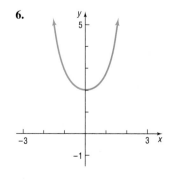

7.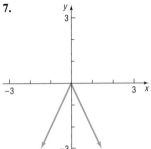

8.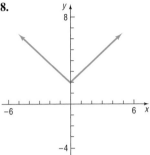

9.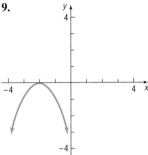

10.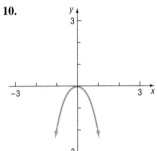

11.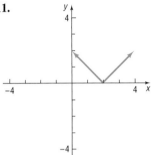

12.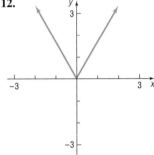

In Problems 13–16, match each graph to one of the following functions:

A. $y = 2x^3$ *B.* $y = (x + 2)^3$ *C.* $y = -2x^3$ *D.* $y = x^3 + 2$

13.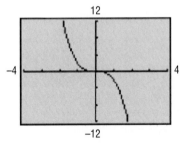

14.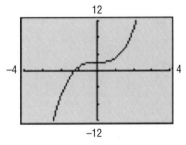

15.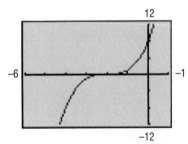

16.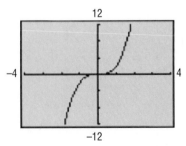

In Problems 17–24, use the function $f(x) = x^3$. Write the function whose graph is the graph of $y = x^3$, but is:

17. Shifted to the right 4 units

18. Shifted to the left 4 units

19. Shifted up 4 units

20. Shifted down 4 units

21. Reflected about the y-axis

22. Reflected about the x-axis

23. Vertically stretched by a factor of 4

24. Horizontally stretched by a factor of 4

In Problems 25–54, graph each function by hand using the techniques of shifting, compressing, stretching, and/or reflecting. Start with the graph of the basic function (for example, $y = x^2$) and show all stages. Verify your answer by using a graphing utility.

25. $f(x) = x^2 - 1$

26. $f(x) = x^2 + 4$

27. $g(x) = x^3 + 1$

28. $g(x) = x^3 - 1$

29. $h(x) = \sqrt{x - 2}$

30. $h(x) = \sqrt{x + 1}$

31. $f(x) = (x - 1)^3$

32. $f(x) = (x + 2)^3$

33. $g(x) = 4\sqrt{x}$

34. $g(x) = \frac{1}{2}\sqrt{x}$

35. $h(x) = \dfrac{1}{2x}$

36. $h(x) = \dfrac{4}{x}$

37. $f(x) = -|x|$

38. $f(x) = -\sqrt{x}$

39. $g(x) = -\dfrac{1}{x}$

40. $g(x) = -x^3$

41. $h(x) = [[-x]]$

42. $h(x) = \dfrac{1}{-x}$

43. $f(x) = (x + 1)^2 - 3$

44. $f(x) = (x - 2)^2 + 1$

45. $g(x) = \sqrt{x - 2} + 1$

46. $g(x) = |x + 1| - 3$

47. $h(x) = \sqrt{-x} - 2$

48. $h(x) = \dfrac{4}{x} + 2$

49. $f(x) = (x + 1)^3 - 1$

50. $f(x) = 4\sqrt{x - 1}$

51. $g(x) = 2|1 - x|$

52. $g(x) = 4\sqrt{2 - x}$

53. $h(x) = 2[[x - 1]]$

54. $h(x) = -x^3 + 2$

In Problems 55–60, the graph of a function f is illustrated. Use the graph of f as the first step toward graphing each of the following functions:

(a) $F(x) = f(x) + 3$

(b) $G(x) = f(x + 2)$

(c) $P(x) = -f(x)$

(d) $Q(x) = \frac{1}{2}f(x)$

(e) $g(x) = f(-x)$

(f) $h(x) = 3f(x)$

55.
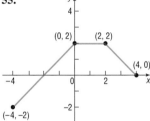

56.
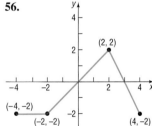

57.
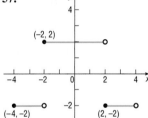

58.
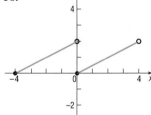

59.
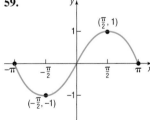

60.
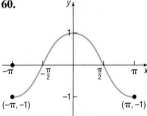

61. (a) Use a graphing utility to graph $y = x + 1$ and $y = |x + 1|$.
  (b) Graph $y = 4 - x^2$ and $y = |4 - x^2|$.
  (c) Graph $y = x^3 + x$ and $y = |x^3 + x|$.
  (d) What do you conclude about the relationship between the graphs of $y = f(x)$ and $y = |f(x)|$?

62. (a) Use a graphing utility to graph $y = x + 1$ and $y = |x| + 1$.
  (b) Graph $y = 4 - x^2$ and $y = 4 - |x|^2$.
  (c) Graph $y = x^3 + x$ and $y = |x|^3 + |x|$.
  (d) What do you conclude about the relationship between the graphs of $y = f(x)$ and $y = f(|x|)$?

63. The graph of a function f is illustrated in the figure.

(a) Draw the entire graph of $y = |f(x)|$.

(b) Draw the entire graph of $y = f(|x|)$.

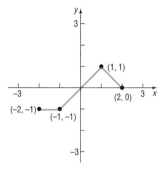

64. Repeat Problem 63 for the graph shown.

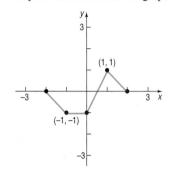

In Problems 65–70, complete the square of each quadratic expression. Then graph each function by hand using the technique of shifting. Verify your results using a graphing utility.

65. $f(x) = x^2 + 2x$ **66.** $f(x) = x^2 - 6x$ **67.** $f(x) = x^2 - 8x + 1$

68. $f(x) = x^2 + 4x + 2$ **69.** $f(x) = x^2 + x + 1$ **70.** $f(x) = x^2 - x + 1$

71. The equation $y = (x - c)^2$ defines a *family of parabolas,* one parabola for each value of c. On one set of coordinate axes, graph the members of the family for $c = 0$, $c = 3$, $c = -2$.

72. Repeat Problem 71 for the family of parabolas $y = x^2 + c$.

73. *Temperature Measurements* The relationship between the Celsius (°C) and Fahrenheit (°F) scales for measuring temperature is given by the equation

$$F = \frac{9}{5}C + 32$$

The relationship between the Celsius (°C) and Kelvin (K) scales is $K = C + 273$. Graph the equation $F = \frac{9}{5}C + 32$ using degrees Fahrenheit on the y-axis and degrees Celsius on the x-axis. Use the techniques introduced in this section to obtain the graph showing the relationship between Kelvin and Fahrenheit temperatures.

74. *Period of a Pendulum* The period T (in seconds) of a simple pendulum is a function of its length l (in feet) defined by the equation

$$T = 2\pi\sqrt{\frac{l}{g}}$$

where $g \approx 32.2$ feet per second per second is the acceleration of gravity.

(a) Use a graphing utility to graph the function $T = T(l)$.

(b) Now graph the functions $T = T(l + 1)$, $T = T(l + 2)$, and $T = T(l + 3)$.

(c) Discuss how adding to the length l changes the period T.

(d) Now graph the functions $T = T(2l)$, $T = T(3l)$, and $T = T(4l)$.

(e) Discuss how multiplying the length l by factors of 2, 3, and 4 changes the period T.

2.4

Operations on Functions; Composite Functions

In this section, we introduce some operations on functions. We shall see that functions, like numbers, can be added, subtracted, multiplied, and divided. For example, if $f(x) = x^2 + 9$ and $g(x) = 3x + 5$, then

$$f(x) + g(x) = (x^2 + 9) + (3x + 5) = x^2 + 3x + 14$$

The new function $y = x^2 + 3x + 14$ is called the *sum function $f + g$*. Similarly,

$$f(x) \cdot g(x) = (x^2 + 9)(3x + 5) = 3x^3 + 5x^2 + 27x + 45$$

The new function $y = 3x^3 + 5x^2 + 27x + 45$ is called the *product function $f \cdot g$*. The general definitions are given next.

Sum Function

If f and g are functions:
Their **sum $f + g$** is the function defined by

$$(f + g)(x) = f(x) + g(x)$$

Difference Function

Their **difference $f - g$** is the function defined by

$$(f - g)(x) = f(x) - g(x)$$

Product Function

Their **product $f \cdot g$** is the function defined by

$$(f \cdot g)(x) = f(x) \cdot g(x)$$

Quotient Function

Their **quotient f/g** is the function defined by

$$\left(\frac{f}{g}\right)(x) = \frac{f(x)}{g(x)}, \quad g(x) \neq 0$$

In each case, the domain of the resulting function consists of the numbers x that are common to the domains of f and g, but the numbers x for which $g(x) = 0$ must be excluded from the domain of the quotient f/g.

Thus, the sum function, $f + g$, is defined as the sum of the values of the functions f and g, and so on.

EXAMPLE 1

Operations on Functions

Let f and g be two functions defined as

$$f(x) = \sqrt{x + 2} \quad \text{and} \quad g(x) = \sqrt{x - 3}$$

Find the following, and determine the domain in each case:

(a) $(f + g)(x)$ (b) $(f - g)(x)$ (c) $(f \cdot g)(x)$ (d) $(f/g)(x)$

Solution   (a) $(f + g)(x) = f(x) + g(x) = \sqrt{x + 2} + \sqrt{x - 3}$

(b) $(f - g)(x) = f(x) - g(x) = \sqrt{x + 2} - \sqrt{x - 3}$

(c) $(f \cdot g)(x) = f(x) \cdot g(x) = (\sqrt{x + 2})(\sqrt{x - 3}) = \sqrt{(x + 2)(x - 3)}$

(d) $\left(\dfrac{f}{g}\right)(x) = \dfrac{f(x)}{g(x)} = \dfrac{\sqrt{x + 2}}{\sqrt{x - 3}} = \sqrt{\dfrac{x + 2}{x - 3}}$

The domain of f consists of all numbers x for which $x \geq -2$; the domain of g consists of all numbers x for which $x \geq 3$. The numbers x common to both these domains are those for which $x \geq 3$. As a result, the numbers x for which $x \geq 3$ comprise the domain of the sum function $f + g$, the difference function $f - g$, and the product function $f \cdot g$. For the quotient function f/g, we must exclude from this set the number 3, because the denominator, g, has the value 0 when $x = 3$. Thus, the domain of f/g consists of all x for which $x > 3$. ∎

■   Now work Problem 1.

In calculus, it is sometimes helpful to view a complicated function as the sum, difference, product, or quotient of simpler functions. For example,

$F(x) = x^2 + \sqrt{x}$ is the sum of $f(x) = x^2$ and $g(x) = \sqrt{x}$.

$H(x) = (x^2 - 1)/(x^2 + 1)$ is the quotient of $f(x) = x^2 - 1$ and $g(x) = x^2 + 1$.

One use of this view of functions is to obtain a graph. The next example illustrates this graphing technique when the function to be graphed is the sum of two simpler functions. In this instance, the method used is called **adding y-coordinates.**

E X A M P L E   2   *Graphing by Adding y-Coordinates*

Graph the function: $F(x) = x + \sqrt{x}$

Solution   First, we notice that the domain of F is $x \geq 0$. Next, we graph the two functions $f(x) = x$ and $g(x) = \sqrt{x}$ for $x \geq 0$. See Figures 42(a) and 42(b). To plot a point $(x, F(x))$ on the graph of F, we select a nonnegative number x and add the y-coordinates $f(x)$ and $g(x)$ to get the y-coordinate $F(x) = f(x) + g(x)$. For example, when $x = 1$, then $f(1) = 1$, $g(1) = 1$, and $F(1) = f(1) + g(1) = 1 + 1 = 2$. When $x = 4$, then $f(4) = 4$, $g(4) = 2$, and $F(4) = f(4) + g(4) = 4 + 2 = 6$, and so on. See Table 10. Figure 42(c) illustrates the graph of F. Figure 42(d) illustrates the graph of F on a graphing utility.

FIGURE 42

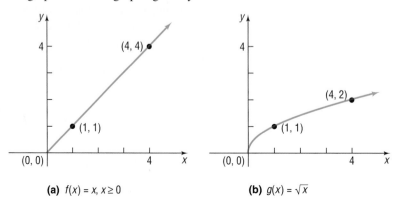

(a) $f(x) = x,\ x \geq 0$ **(b)** $g(x) = \sqrt{x}$

TABLE 10

x	$f(x) = x$	$g(x) = \sqrt{x}$	$F(x) = x + \sqrt{x}$
0	0	0	0
1	1	1	2
4	4	2	6

FIGURE 42 (continued)

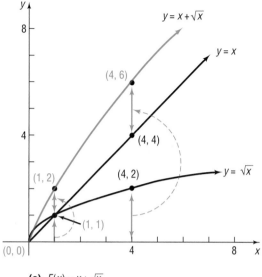

(c) $F(x) = x + \sqrt{x}$

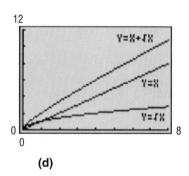

(d)

Now work Problem 13.

Composite Functions

Consider the function $y = (2x + 3)^2$. If we write $y = f(u) = u^2$ and $u = g(x) = 2x + 3$, then, by a substitution process, we can obtain the original function: $y = f(u) = f(g(x)) = (2x + 3)^2$. This process is called **composition.** In general, suppose that f and g are two functions, and suppose that x is a number in the domain of g. By evaluating g at x, we get $g(x)$. If $g(x)$ is in the domain of f, then we may evaluate f at $g(x)$ and thereby obtain the expression $f(g(x))$. If we do this for all x such that x is in the domain of g and $g(x)$ is in the domain of f, the resulting correspondence from x to $f(g(x))$ is called a *composite function $f \circ g$.* See Figure 43.

FIGURE 43

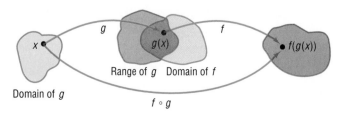

Composite Function

> Given the two functions f and g, the **composite function,** denoted by $f \circ g$ (read as "f composed with g"), is defined by
>
> $$(f \circ g)(x) = f(g(x))$$
>
> where the domain of $f \circ g$ is the set of all numbers x in the domain of g such that $g(x)$ is in the domain of f.

Figure 44 provides a second illustration of the definition. Notice that the "inside" function g in $f(g(x))$ is done first.

FIGURE 44

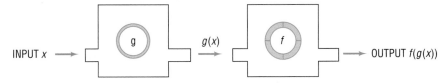

Let's look at some examples.

E X A M P L E 3

Evaluating a Composite Function

Suppose that $f(x) = 2x^2 - 3$ and $g(x) = 4x$. Find:

(a) $(f \circ g)(1)$ (b) $(g \circ f)(1)$ (c) $(f \circ f)(-2)$ (d) $(g \circ g)(-1)$

Solution (a) $(f \circ g)(1) = f(g(1)) = f(4) = 2 \cdot 16 - 3 = 29$

$g(x) = 4x \quad f(x) = 2x^2 - 3$
$g(1) = 4$

(b) $(g \circ f)(1) = g(f(1)) = g(-1) = 4 \cdot (-1) = -4$

$f(x) = 2x^2 - 3 \quad g(x) = 4x$
$f(1) = -1$

(c) $(f \circ f)(-2) = f(f(-2)) = f(5) = 2 \cdot 25 - 3 = 47$

$f(-2) = 5$

(d) $(g \circ g)(-1) = g(g(-1)) = g(-4) = 4 \cdot (-4) = -16$

$g(-1) = -4$

■ Now work Problem 25.

E X A M P L E 4

Finding a Composite Function

Suppose that $f(x) = \sqrt{x}$ and $g(x) = x^3 - 1$. Find the following composite functions, and then find the domain of each composite function:

(a) $f \circ g$ (b) $g \circ f$ (c) $f \circ f$ (d) $g \circ g$

Solution (a) $(f \circ g)(x) = f(g(x)) = f(x^3 - 1) = \sqrt{x^3 - 1}$

The domain of $f \circ g$ is the interval $[1, \infty)$, which is found by determining those x in the domain of g for which $x^3 - 1 \geq 0$.

(b) $(g \circ f)(x) = g(f(x)) = g(\sqrt{x}) = (\sqrt{x})^3 - 1 = x^{3/2} - 1$

The domain of $g \circ f$ is $[0, \infty)$.

𝒫ISSION POSSIBLE

Chapter 2

CONSULTING FOR THE SILVER SATELLITE & CABLE TV CO.

Your team works for the Silver Satellite & Cable TV Company in the Research & Development Department. You've been asked to come up with a formula to determine the cost of running cable from a connection box to a new cable household. The first example you are working with involves the Stevens family who own a rural home with a driveway two miles long extending to the house from a nearby highway. The nearest connection box is along the highway but 5 miles from the driveway.

It costs the company $10 per mile to install cable along the highway and $14 per mile to install cable off the highway. Because the Steven house is surrounded by farmland which they own, it would be possible to run the cable overland to the house directly from the connection box or from any point between the connection box to the driveway.

1. Draw a sketch of this problem situation, assuming the highway is a straight road and the driveway is also a straight road perpendicular to the highway. Include two or more possible routes for the cable.
2. Suppose x represents the distance in miles the cable runs along the highway from the connection box before turning off toward the house. Express the total cost of installation as a function of x. (You may choose to answer #3 before #2 if you would like to examine concrete instances before creating the equation.)
3. Make a table of the possible integral values of x and the corresponding cost in each instance. Is there one choice which appears to cost the least?
4. If you charge the Stevens $80 for installation, would you be willing to let them choose which way the cable would go? Explain.
5. Using a graphing calculator, graph the function from #2 and determine if there is a non-integral choice for x that would make the installation cost even cheaper. ZOOM and TRACE until you have the lowest possible cost.
6. Before proceeding further with the installation, you check the local regulations for cable companies and find that there is a pending state legislation that says the cable cannot turn off the highway more than .5 miles from the Stevens' driveway. If this legislation passes, what will be the ultimate cost of installing the Stevens' cable?
7. If the cable company wishes to install cable in 5000 homes in this area, and assuming the figures for the Stevens installation are typical, how much will the new legislation cost the company over all if they cannot use the cheapest installation cost, but instead have to follow the new state regulations?

(c) $(f \circ f)(x) = f(f(x)) = f(\sqrt{x}) = \sqrt{\sqrt{x}} = \sqrt[4]{x}$

The domain of $f \circ f$ is $[0, \infty)$.

(d) $(g \circ g)(x) = g(g(x)) = g(x^3 - 1) = (x^3 - 1)^3 - 1$

The domain of $g \circ g$ is the set of all real numbers.   ■

■   Now work Problem 35.

Examples 4(a) and 4(b) illustrate that, in general, $f \circ g \neq g \circ f$. However, sometimes $f \circ g$ does equal $g \circ f$, as shown in the next example.

E X A M P L E   5   *Showing Two Composite Functions Equal*

If $f(x) = 3x - 4$ and $g(x) = \frac{1}{3}(x + 4)$, show that $(f \circ g)(x) = (g \circ f)(x) = x$ for every x.

Solution $(f \circ g)(x) = f(g(x))$

$\qquad = f\left(\dfrac{x + 4}{3}\right)$ $g(x) = \frac{1}{3}(x + 4) = \dfrac{x + 4}{3}$

$\qquad = 3\left(\dfrac{x + 4}{3}\right) - 4$ Substitute $g(x)$ into the rule for f, $f(x) = 3x - 4$.

$\qquad = x + 4 - 4 = x$

$(g \circ f)(x) = g(f(x))$

$\qquad = g(3x - 4)$ $f(x) = 3x - 4$

$\qquad = \frac{1}{3}[(3x - 4) + 4)]$ Substitute $f(x)$ into the rule for g, $g(x) = \frac{1}{3}(x + 4)$.

$\qquad = \frac{1}{3}(3x) = x$

Thus, $(f \circ g)(x) = (g \circ f)(x) = x$.   ■

In the next section, we shall see that there is an important relationship between functions f and g for which $(f \circ g)(x) = (g \circ f)(x) = x$.

■   Now work Problem 49.

Calculus Application

Some techniques in calculus require that we be able to determine the components of a composite function. For example, the function $H(x) = \sqrt{x + 1}$ is the composition of the functions f and g, where $f(x) = \sqrt{x}$ and $g(x) = x + 1$, because $H(x) = (f \circ g)(x) = f(g(x)) = f(x + 1) = \sqrt{x + 1}$.

E X A M P L E   6   *Finding the Components of a Composite Function*

Find functions f and g such that $f \circ g = H$ if $H(x) = (x^2 + 1)^{50}$.

Solution   The function H takes $x^2 + 1$ and raises it to the power 50. A natural way to decompose H is to raise the function $g(x) = x^2 + 1$ to the power 50. Thus, if we let $f(x) = x^{50}$ and $g(x) = x^2 + 1$, then

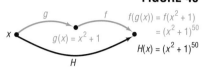

FIGURE 45

$$(f \circ g)(x) = f(g(x))$$
$$= f(x^2 + 1)$$
$$= (x^2 + 1)^{50} = H(x)$$

See Figure 45. ■

Other functions f and g may be found for which $f \circ g = H$ in Example 6. For example, if $f(x) = x^2$ and $g(x) = (x^2 + 1)^{25}$, then

$$(f \circ g)(x) = f(g(x)) = f((x^2 + 1)^{25}) = [(x^2 + 1)^{25}]^2 = (x^2 + 1)^{50}$$

Thus, although the functions f and g found as a solution to Example 6 are not unique, there is usually a "natural" selection for f and g that comes to mind first.

E X A M P L E   7   *Finding the Components of a Composite Function*

Find functions f and g such that $f \circ g = H$ if $H(x) = 1/(x + 1)$.

Solution   Here, H is the reciprocal of $g(x) = x + 1$. Thus, if we let $f(x) = 1/x$ and $g(x) = x + 1$, we find that

$$(f \circ g)(x) = f(g(x)) = f(x + 1) = \frac{1}{x + 1} = H(x)$$ ■

2.4

Exercise 2.4

In Problems 1–10, for the given functions f and g, find the following functions and state the domain of each:

(a) $f + g$ (b) $f - g$ (c) $f \cdot g$ (d) f/g

1. $f(x) = 3x + 4;$ $g(x) = 2x - 3$

2. $f(x) = 2x + 1;$ $g(x) = 3x - 2$

3. $f(x) = x - 1;$ $g(x) = 2x^2$

4. $f(x) = 2x^2 + 3;$ $g(x) = 4x^3 + 1$

5. $f(x) = \sqrt{x};$ $g(x) = 3x - 5$

6. $f(x) = |x|;$ $g(x) = x$

7. $f(x) = 1 + \dfrac{1}{x};$ $g(x) = \dfrac{1}{x}$

8. $f(x) = 2x^2 - x;$ $g(x) = 2x^2 + x$

9. $f(x) = \dfrac{2x + 3}{3x - 2};$ $g(x) = \dfrac{4x}{3x - 2}$

10. $f(x) = \sqrt{x + 1};$ $g(x) = \dfrac{2}{x}$

11.   Given $f(x) = 3x + 1$ and $(f + g)(x) = 6 - \frac{1}{2}x$, find the function g.

12.   Given $f(x) = 1/x$ and $(f/g)(x) = (x + 1) / (x^2 - x)$, find the function g.

In Problems 13–16, find the equation that defines the function $f + g$. Use a graphing utility to graph the functions f, g, and $f + g$ on the same screen.

13. $f(x) = x, g(x) = \dfrac{1}{x}$

14. $f(x) = x, g(x) = \dfrac{1}{x^2}$

15. $f(x) = x^2, g(x) = \dfrac{1}{x}$

16. $f(x) = x^2, g(x) = \dfrac{1}{x^2}$

In Problems 17–20, find the equation that defines the function f · g. Use a graphing utility to graph the functions f, g, and f · g on the same screen.

17. $f(x) = x,\ g(x) = \dfrac{1}{x^2 + 1}$

18. $f(x) = x,\ g(x) = \dfrac{1}{x^2 - 1}$

19. $f(x) = x^2,\ g(x) = \dfrac{1}{x^2 + 1}$

20. $f(x) = x^2,\ g(x) = \dfrac{1}{x^2 - 1}$

In Problems 21–24, use the method of adding y-coordinates to graph each function on the interval [0, 2]. Verify your result with a graphing utility.

21. $f(x) = |x| + x^2$ **22.** $f(x) = |x| + \sqrt{x}$ **23.** $f(x) = x^3 + x$ **24.** $f(x) = x^3 + x^2$

In Problems 25–34, for the given functions f and g, find:

(a) $(f \circ g)(4)$ *(b) $(g \circ f)(2)$*   *(c) $(f \circ f)(1)$*   *(d) $(g \circ g)(0)$*

25. $f(x) = 2x;\ \ g(x) = 3x^2 + 1$

26. $f(x) = 3x + 2;\ \ g(x) = 2x^2 - 1$

27. $f(x) = 4x^2 - 3;\ \ g(x) = 3 - \frac{1}{2}x^2$

28. $f(x) = 2x^2;\ \ g(x) = 1 - 3x^2$

29. $f(x) = \sqrt{x};\ \ g(x) = 2x$

30. $f(x) = \sqrt{x + 1};\ \ g(x) = 3x$

31. $f(x) = |x|;\ \ g(x) = \dfrac{1}{x^2 + 1}$

32. $f(x) = |x - 2|;\ \ g(x) = \dfrac{3}{x^2 + 2}$

33. $f(x) = \dfrac{3}{x^2 + 1};\ \ g(x) = \sqrt{x}$

34. $f(x) = x^3;\ \ g(x) = \dfrac{2}{x^2 + 1}$

In Problems 35–48, for the given functions f and g, find;

(a) $f \circ g$ *(b) $g \circ f$*   *(c) $f \circ f$*   *(d) $g \circ g$*

35. $f(x) = 2x + 3;\ \ g(x) = 3x$

36. $f(x) = -x;\ \ g(x) = 2x - 4$

37. $f(x) = 3x + 1;\ \ g(x) = x^2$

38. $f(x) = \sqrt{x + 1};\ \ g(x) = x + 4$

39. $f(x) = \sqrt{x};\ \ g(x) = x^2 - 1$

40. $f(x) = \sqrt{x + 1};\ \ g(x) = \dfrac{1}{x^2}$

41. $f(x) = \dfrac{x - 1}{x + 1};\ \ g(x) = \dfrac{1}{x}$

42. $f(x) = x + \dfrac{1}{x};\ \ g(x) = x^2$

43. $f(x) = x^2;\ \ g(x) = \sqrt{x}$

44. $f(x) = 2x + 4;\ \ g(x) = \frac{1}{2}x - 2$

45. $f(x) = \dfrac{1}{2x + 3};\ \ g(x) = 2x + 3$

46. $f(x) = \dfrac{x + 1}{x - 1};\ \ g(x) = \dfrac{x - 1}{x + 1}$

47. $f(x) = ax + b;\ \ g(x) = cx + d$

48. $f(x) = \dfrac{ax + b}{cx + d};\ \ g(x) = mx$

In Problems 49–56, show that $(f \circ g)(x) = (g \circ f)(x) = x$.

49. $f(x) = 2x;\ \ g(x) = \frac{1}{2}x$

50. $f(x) = 4x;\ \ g(x) = \frac{1}{4}x$

51. $f(x) = x^3;\ \ g(x) = \sqrt[3]{x}$

52. $f(x) = x + 5;\ \ g(x) = x - 5$

53. $f(x) = 2x - 6;\ \ g(x) = \frac{1}{2}(x + 6)$

54. $f(x) = 4 - 3x;\ \ g(x) = \frac{1}{3}(4 - x)$

55. $f(x) = ax + b;\ \ g(x) = \dfrac{1}{a}(x - b),\ a \neq 0$

56. $f(x) = \dfrac{1}{x};\ \ g(x) = \dfrac{1}{x}$

57. If $f(x) = 2x^3 - 3x^2 + 4x - 1$ and $g(x) = 2$, find $(f \circ g)(x)$ and $(g \circ f)(x)$.

58. If $f(x) = x/(x - 1)$, find $(f \circ f)(x)$.

In Problems 59–62, use $f(x) = x^2$, $g(x) = \sqrt{x} + 2$, and $h(x) = 1 - 3x$ to find the indicated composite function.

59. $f \circ (g \circ h)$ **60.** $(f \circ g) \circ h$ **61.** $(f + g) \circ h$ **62.** $(f \circ h) + (g \circ h)$

In Problems 63–70, let $f(x) = x^2$, $g(x) = 3x$, and $h(x) = \sqrt{x} + 1$. Express each function as a composite of f, g, and/or h.

63. $F(x) = 9x^2$ **64.** $G(x) = 3x^2$ **65.** $H(x) = |x| + 1$

66. $p(x) = 3\sqrt{x} + 3$ **67.** $q(x) = x + 2\sqrt{x} + 1$ **68.** $R(x) = 9x$

69. $P(x) = x^4$ **70.** $Q(x) = \sqrt{\sqrt{x} + 1} + 1$

In Problems 71–78, find functions f and g so that $f \circ g = H$.

71. $H(x) = (2x + 3)^4$ **72.** $H(x) = (1 + x^2)^{3/2}$ **73.** $H(x) = \sqrt{x^2 + x + 1}$

74. $H(x) = \dfrac{1}{1 + x^2}$ **75.** $H(x) = \left(1 - \dfrac{1}{x^2}\right)^2$ **76.** $H(x) = |2x^2 + 3|$

77. $H(x) = [[x^2 + 1]]$ **78.** $H(x) = (4 - x^2)^{-4}$

79. If $f(x) = 2x^2 + 5$ and $g(x) = 3x + a$, find a so that the graph of $f \circ g$ crosses the y-axis at 23.

80. If $f(x) = 3x^2 - 7$ and $g(x) = 2x + a$, find a so that the graph of $f \circ g$ crosses the y-axis at 68.

81. The surface area S (in square meters) of a hot air balloon is given by

$$S(r) = 4\pi r^2$$

where r is the radius of the balloon (in meters). If the radius r is increasing with time t (in seconds) according to the formula $r(t) = \frac{2}{3}t^3$, $t \geq 0$, find the surface area S of the balloon as a function of the time t.

82. The volume V (in cubic meters) of the hot air balloon described in Problem 81 is given by $V(r) = \frac{4}{3}\pi r^3$. If the radius r is the same function of t as in Problem 81, find the volume V as a function of the time t.

83. *Automobile Production* The number N of cars produced at a certain factory in 1 day after t hours of operation is given by $N(t) = 100t - 5t^2$, $0 \leq t \leq 10$. If the cost C (in dollars) of producing x cars is $C(x) = 15,000 + 8000x$, find the cost C as a function of the time t of operation of the factory.

84. *Environmental Concerns* The spread of oil leaking from a tanker is in the shape of a circle. If the radius r (in feet) of the spread after t hours is $r(t) = 200\sqrt{t}$, find the area A of the oil slick as a function of the time t.

85. *Production Cost* The price p of a certain product and the quantity x sold obey the demand equation

$$p = -\tfrac{1}{4}x + 100 \qquad 0 \leq x \leq 400$$

Suppose that the cost C of producing x units is

$$C = \frac{\sqrt{x}}{25} + 600$$

Assuming that all items produced are sold, find the cost C as a function of the price p. [*Hint:* Solve for x in the demand equation, and then form the composite.]

86. *Cost of a Commodity* The price p of a certain commodity and the quantity x sold obey the demand equation

$$p = -\tfrac{1}{5}x + 200 \qquad 0 \leq x \leq 1000$$

Suppose that the cost C of producing x units is

$$C = \frac{\sqrt{x}}{10} + 400$$

Assuming that all items produced are sold, find the cost C as a function of the price p.

87. If f and g are odd functions, show that the composite function $f \circ g$ is also odd.

88. If f is an odd function and g is an even function, show that the composite functions $f \circ g$ and $g \circ f$ are also even.

2.5

One-to-One Functions; Inverse Functions

Suppose that (x_1, y_1) and (x_2, y_2) are any two *distinct* points on the graph of a function $y = f(x)$. Then it follows that $x_1 \neq x_2$. For some functions, it also happens that the y-coordinates of distinct points are always unequal. Such functions are called *one-to-one* functions. See Figure 46.

FIGURE 46

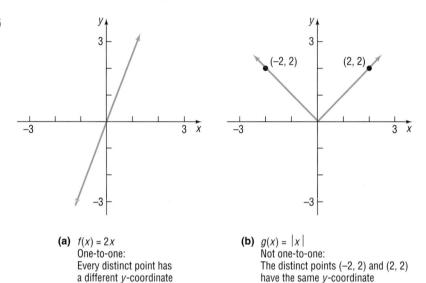

(a) $f(x) = 2x$
One-to-one:
Every distinct point has
a different y-coordinate

(b) $g(x) = |x|$
Not one-to-one:
The distinct points $(-2, 2)$ and $(2, 2)$
have the same y-coordinate

One-to-One Function

> A function f is said to be **one-to-one** if, for any choice of numbers x_1 and x_2, $x_1 \neq x_2$, in the domain of f, then $f(x_1) \neq f(x_2)$.

In other words, if f is a one-to-one function, then for each x in the domain of f, there is exactly one y in the range, and no y in the range is the image of more than one x in the domain. See Figure 47.

FIGURE 47

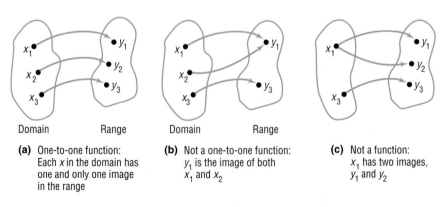

Domain   Range   Domain   Range

(a) One-to-one function:
Each x in the domain has
one and only one image
in the range

(b) Not a one-to-one function:
y_1 is the image of both
x_1 and x_2

(c) Not a function:
x_1 has two images,
y_1 and y_2

As Figure 48 illustrates, if the graph of a function f is known, there is a simple test, called the **horizontal-line test,** to determine whether f is one-to-one.

Theorem
Horizontal Line Test

If horizontal lines intersect the graph of a function f in at most one point, then f is one-to-one. ■

The reason this test works can be seen in Figure 48, where the horizontal line $y = h$ intersects the graph at two distinct points, (x_1, h) and (x_2, h), with the same second element. Thus, f is not one-to-one.

FIGURE 48
$f(x_1) = f(x_2) = h$, but $x_1 \neq x_2$;
f is not a one-to-one function

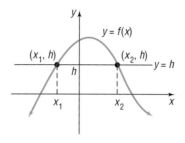

E X A M P L E   1

Using the Horizontal Line Test

For each given function, use the graph to determine whether the function is one-to-one.

(a) $f(x) = x^2$

(b) $g(x) = x^3$

Solution
(a) Figure 49(a) illustrates the horizontal line test for $f(x) = x^2$. The horizontal line $y = 1$ meets the graph of f twice, at $(1, 1)$ and at $(-1, -1)$, so f is not one-to-one.

(b) Figure 49(b) illustrates the horizontal line test for $g(x) = x^3$. Because each horizontal line will intersect the graph of g exactly once, it follows that g is one-to-one.

FIGURE 49

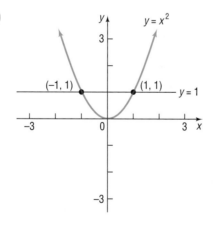

(a) A horizontal line intersects the graph twice; thus, f is not one-to-one

(b) Horizontal lines intersect the graph exactly once; thus, g is one-to-one ∎

∎ Now work Problem 1.

Let's look more closely at the one-to-one function $g(x) = x^3$. This function is an increasing function. Because an increasing (or decreasing) function will always have different y values for unequal x values, it follows that a function that is increasing (or decreasing) on its domain is also a one-to-one function.

Theorem
An increasing (decreasing) function is a one-to-one function. ∎

Inverse of a Function

We mentioned earlier that a function $y = f(x)$ can be thought of as a rule that tells us to do something to the argument x. For example, the function $f(x) = 2x$ multiplies the argument by 2. An *inverse function* of f undoes whatever f does. For example, the function $g(x) = \frac{1}{2}x$, which divides the argument by 2, is an inverse of $f(x) = 2x$. See Figure 50.

For a function $y = f(x)$ to have an inverse function, f must be one-to-one. Then for each x in its domain there is exactly one y in its range; furthermore, to each y in the range, there corresponds exactly one x in the domain. The correspondence from the range of f onto the domain of f is, therefore, also a function. It is this function that is the *inverse of f*. A definition is given next.

FIGURE 50

$f(x) = 2x;\ g(x) = \frac{1}{2}x$

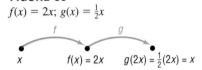

Inverse of f

Let f denote a one-to-one function $y = f(x)$. The **inverse of f,** denoted by f^{-1}, is a function such that $f^{-1}(f(x)) = x$ for every x in the domain of f and $f(f^{-1}(x)) = x$ for every x in the domain of f^{-1}.

Warning: Be careful! The -1 used in f^{-1} is not an exponent. Thus, f^{-1} does *not* mean the reciprocal of f; it means the inverse of f.

FIGURE 51

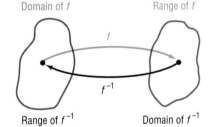

Domain of f Range of f

f

f^{-1}

Range of f^{-1} Domain of f^{-1}

Figure 51 illustrates the definition.

Two facts are now apparent about a function f and its inverse f^{-1}.

Domain of f = Range of f^{-1} Range of f = Domain of f^{-1}

Look again at Figure 51 to visualize the relationship. If we start with x, apply f, and then apply f^{-1}, we get x back again. If we start with x, apply f^{-1}, and then apply f, we get the number x back again. To put it simply, what f does, f^{-1} undoes, and vice versa:

| Input x | $\xrightarrow{\text{Apply } f}$ | $f(x)$ | $\xrightarrow{\text{Apply } f^{-1}}$ | $f^{-1}(f(x)) = x$ |

| Input x | $\xrightarrow{\text{Apply } f^{-1}}$ | $f^{-1}(x)$ | $\xrightarrow{\text{Apply } f}$ | $f(f^{-1}(x)) = x$ |

In other words,

$$f^{-1}(f(x)) = x \quad \text{and} \quad f(f^{-1}(x)) = x$$

The preceding conditions can be used to verify that a function is, in fact, the inverse of f, as Example 2 demonstrates.

EXAMPLE 2

Verifying Inverse Functions

(a) We verify that the inverse of $g(x) = x^3$ is $g^{-1}(x) = \sqrt[3]{x}$ by showing that

$$g^{-1}(g(x)) = g^{-1}(x^3) = \sqrt[3]{x^3} = x$$

and

$$g(g^{-1}(x)) = g(\sqrt[3]{x}) = (\sqrt[3]{x})^3 = x$$

(b) We verify that the inverse of $h(x) = 3x$ is $h^{-1}(x) = \frac{1}{3}x$ by showing that

$$h^{-1}(h(x)) = h^{-1}(3x) = \frac{1}{3}(3x) = x$$

and

$$h(h^{-1}(x)) = h(\frac{1}{3}x) = 3(\frac{1}{3}x) = x$$

(c) We verify that the inverse of $f(x) = 2x + 3$ is $f^{-1}(x) = \frac{1}{2}(x - 3)$ by showing that

$$f^{-1}(f(x)) = f^{-1}(2x + 3) = \frac{1}{2}[(2x + 3) - 3] = \frac{1}{2}(2x) = x$$

and

$$f(f^{-1}(x)) = f(\frac{1}{2}(x - 3)) = 2[\frac{1}{2}(x - 3)]+3 = (x - 3) + 3 = x. \qquad \blacksquare$$

■   Now work Problem 13.

Exploration: Graph $y = x$ on a square screen, using the viewing rectangle $-3 \le x \le 7$, $-2 \le y \le 2$. Then graph $y = x^3$ followed by its inverse, $y = \sqrt[3]{x}$. What do you observe about the graphs of $y = x^3$, its inverse $y = \sqrt[3]{x}$, and the line $y = x$?
    Do you see the symmetry of the graph of f and its inverse with respect to the line $y = x$? ■

    For the functions in Example 2(c), we list points on the graph of $f = Y_1$ and on the graph of $f^{-1} = Y_2$ in Table 11.

TABLE 11

X	Y1
-5	-7
-4	-5
-3	-3
-2	-1
-1	1
0	3
1	5

Y₁■2X+3

X	Y2
-7	-5
-5	-4
-3	-3
-1	-2
1	-1
3	0
5	1

Y₂■.5(X-3)

We notice that whenever (a, b) is on the graph of f then (b, a) is on the graph of f^{-1}. Figure 52 shows these points plotted. Also shown there is the graph of $y = x$, which you should observe is a line of symmetry of the points.

FIGURE 52

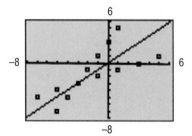

Geometric Interpretation

Suppose that (a, b) is a point on the graph of the one-to-one function f defined by $y = f(x)$. Then $b = f(a)$. This means that $a = f^{-1}(b)$, so (b, a) is a point on the graph of the inverse function f^{-1}. The relationship between the point (a, b) on f

and the point (b, a) on f^{-1} is shown in Figure 53. The line joining (a, b) and (b, a) is perpendicular to the line $y = x$ and is bisected by the line $y = x$. (Do you see why?) It follows that the point (b, a) on f^{-1} is the reflection about the line $y = x$ of the point (a, b) on f.

FIGURE 53

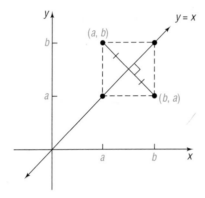

Theorem   The graph of a function f and the graph of its inverse f^{-1} are symmetric with respect to the line $y = x$.   ◼

Figure 54 illustrates this result. Notice that, once the graph of f is known, the graph of f^{-1} may be obtained by folding the paper along the line $y = x$.

FIGURE 54

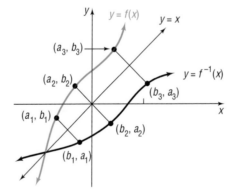

E X A M P L E   3   *Graphing the Inverse Function*

The graph in Figure 55(a) is that of a one-to-one function $y = f(x)$. Draw the graph of its inverse.

Solution   We begin by adding the graph of $y = x$ to Figure 55(a). Since the points $(-2, -1)$, $(-1, 0)$, and $(2, 1)$ are on the graph of f, we know that the points $(-1, -2)$, $(0, -1)$, and $(1, 2)$ must be on the graph of f^{-1}. Keeping in mind that the graph of f^{-1} is the reflection about the line $y = x$ of the graph of f, we can draw f^{-1}. See Figure 55(b).

FIGURE 55

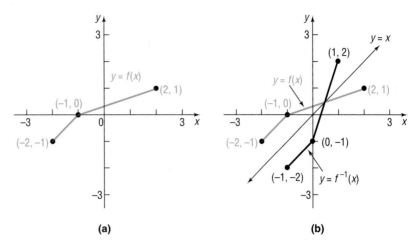

(a)   (b)

■   Now work Problem 7.

Finding the Inverse Function

The fact that the graph of a one-to-one function f and its inverse are symmetric with respect to the line $y = x$ tells us more. It says that we can obtain f^{-1} by interchanging the roles of x and y in f. Look again at Figure 54. That is, if f is defined by the equation

$$y = f(x)$$

then f^{-1} is defined by the equation

$$x = f(y)$$

Be careful! The equation $x = f(y)$ defines f^{-1} implicitly. If we can solve this equation for y, we will have the explicit form of f^{-1}, that is,

$$y = f^{-1}(x)$$

Let's use this procedure to find the inverse of $f(x) = 2x + 3$. (Since f is a linear function and is increasing, we know that f is one-to-one.)

E X A M P L E   4

Finding the Inverse Function

Find the inverse of $f(x) = 2x + 3$. Also find the domain and range of f and f^{-1}. Graph f and f^{-1} on the same coordinate axes.

Solution   In the equation $y = 2x + 3$, interchange the variables x and y. The result,

$$x = 2y + 3$$

is an equation that defines the inverse f^{-1} implicitly. Solving for y, we obtain

$$2y + 3 = x$$
$$2y = x - 3$$
$$y = \tfrac{1}{2}(x - 3)$$

The explicit form of the inverse f^{-1} is therefore

$$f^{-1}(x) = \tfrac{1}{2}(x - 3)$$

which we verified in Example 2(c).

Then we find

$$\text{Domain } f = \text{Range } f^{-1} = (-\infty, \infty)$$
$$\text{Range } f = \text{Domain } f^{-1} = (-\infty, \infty)$$

The graphs of $y_1 = f(x) = 2x + 3$ and its inverse $y_2 = f^{-1}(x) = \frac{1}{2}(x - 3)$ are shown in Figure 56. Note the symmetry of the graphs with respect to the line $y = x$.

FIGURE 56

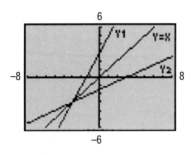
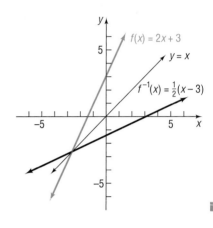

We outline next the steps to follow for finding the inverse of a one-to-one function.

Procedure for Finding the Inverse of a One-to-One Function

STEP 1: In $y = f(x)$, interchange the variables x and y to obtain

$$x = f(y)$$

This equation defines the inverse function f^{-1} implicitly.

STEP 2: If possible, solve the implicit equation for y in terms of x to obtain the explicit form of f^{-1}:

$$y = f^{-1}(x)$$

STEP 3: Check the result by showing that

$$f^{-1}(f(x)) = x \quad \text{and} \quad f(f^{-1}(x)) = x$$

E X A M P L E   5

Finding the Inverse Function

The function

$$f(x) = \frac{2x + 1}{x - 1}, x \neq 1$$

is one-to-one. Find its inverse and check the result.

Solution   **STEP 1:** Interchange the variables x and y in

$$y = \frac{2x + 1}{x - 1}$$

to obtain

$$x = \frac{2y + 1}{y - 1}$$

STEP 2:   Solve for y:

$$x = \frac{2y + 1}{y - 1}$$

$$x(y - 1) = 2y + 1$$

$$xy - x = 2y + 1$$

$$xy - 2y = x + 1$$

$$(x - 2)y = x + 1$$

$$y = \frac{x + 1}{x - 2}$$

The inverse is

$$f^{-1}(x) = \frac{x + 1}{x - 2}, \ x \neq 2$$

STEP 3:   *Check:*

$$f^{-1}(f(x)) = f^{-1}\left(\frac{2x + 1}{x - 1}\right) = \frac{\dfrac{2x + 1}{x - 1} + 1}{\dfrac{2x + 1}{x - 1} - 2} = \frac{2x + 1 + x - 1}{2x + 1 - 2(x - 1)} = \frac{3x}{3} = x$$

$$f(f^{-1}(x)) = f\left(\frac{x + 1}{x - 2}\right) = \frac{2\left(\dfrac{x + 1}{x - 2}\right) + 1}{\dfrac{x + 1}{x - 2} - 1} = \frac{2(x + 1) + x - 2}{x + 1 - (x - 2)} = \frac{3x}{3} = x$$

Exploration: We found that, if $f(x) = (2x + 1)/(x - 1)$, then $f^{-1}(x) = (x + 1)/(x - 2)$. Graph $y = f(f^{-1}(x))$ on a square screen. What do you see? Are you surprised?   ■

■   Now work Problem 25.

If a function is not one-to-one, then it will have no inverse. Sometimes, though, an appropriate restriction on the domain of such a function will yield a new function that is one-to-one. Let's look at an example of this common practice.

E X A M P L E   6   *Finding the Inverse Function*

Find the inverse of $y = f(x) = x^2$ if $x \geq 0$.

Solution   The function $f(x) = x^2$ is not one-to-one. [Refer to Example 1(a).] However, if we restrict f to only that part of its domain for which $x \geq 0$, as indicated, we have a new function that is increasing and therefore is one-to-one. As a result, the function defined by $y = x^2$, $x \geq 0$, has an inverse, f^{-1}.

We follow the steps given previously to find f^{-1}:

STEP 1:   In the equation $y = x^2$, $x \geq 0$, interchange the variables x and y. The result is

$$x = y^2 \qquad y \geq 0$$

This equation defines (implicitly) the inverse function.

STEP 2:   We solve for y to get the explicit form of the inverse. Since $y \geq 0$, only one solution for y is obtained:

$$y = \sqrt{x}$$

so that $f^{-1}(x) = \sqrt{x}$.

STEP 3:   *Check:* $f^{-1}(f(x)) = f^{-1}(x^2) = \sqrt{x^2} = |x| = x,$ since $x \geq 0$

$$f(f^{-1})(x)) = f(\sqrt{x}) = (\sqrt{x})^2 = x$$

Figure 57 illustrates the graphs of $f(x) = x^2$, $x \geq 0$, and $f^{-1}(x) = \sqrt{x}$.

FIGURE 57

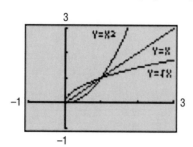

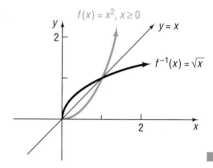

Summary

1. If a function f is one-to-one, then it has an inverse f^{-1}.
2. Domain f = Range f^{-1}; Range f = Domain f^{-1}.
3. To verify that f^{-1} is the inverse of f, show that $f^{-1}(f(x)) = x$ and $f(f^{-1}(x)) = x$.
4. The graphs of f and f^{-1} are symmetric with respect to the line $y = x$.

2.5

Exercise 2.5

In Problems 1–6, the graph of a function f is given. Use the horizontal line test to determine whether f is one-to-one.

1.

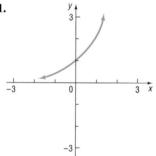

2.

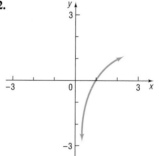

3.

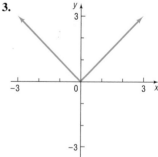

4.

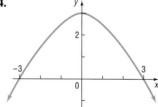

5.

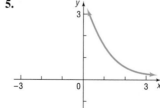

6.

In Problems 7–12, the graph of a one-to-one function f *is given. Find the graph of the inverse function* f^{-1}. *For convenience (and as a hint), the graph of* y = x *is also given.*

7.

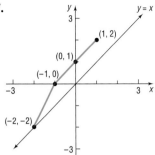

8.

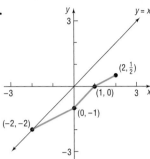

9.

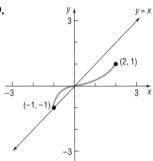

10.

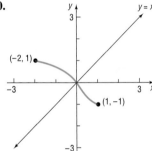

11.

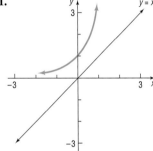

12.

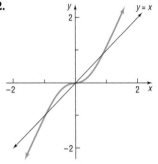

In Problems 13–22, verify that the functions f *and* g *are inverses of each other by showing that* f(g(x)) = x *and* g(f(x)) = x. *Using a graphing utility, graph* f, g, *and* y = x *on the same square screen.*

13. $f(x) = 3x + 4; \quad g(x) = \frac{1}{3}(x - 4)$

14. $f(x) = 3 - 2x; \quad g(x) = -\frac{1}{2}(x - 3)$

15. $f(x) = 4x - 8; \quad g(x) = \frac{x}{4} + 2$

16. $f(x) = 2x + 6; \quad g(x) = \frac{1}{2}x - 3$

17. $f(x) = x^3 - 8; \quad g(x) = \sqrt[3]{x + 8}$

18. $f(x) = (x - 2)^2, \quad x \geq 2; g(x) = \sqrt{x} + 2, \quad x \geq 0$

19. $f(x) = \frac{1}{x}; \quad g(x) = \frac{1}{x}$

20. $f(x) = x; \quad g(x) = x$

21. $f(x) = \frac{2x + 3}{x + 4}; \quad g(x) = \frac{4x - 3}{2 - x}$

22. $f(x) = \frac{x - 5}{2x + 3}; \quad g(x) = \frac{3x + 5}{1 - 2x}$

In Problems 23–34, the function f *is one-to-one. Find its inverse and check your answer. State the domain and range of* f *and* f^{-1}. *Graph* f, f^{-1}, *and* y = x *on the same coordinate axes. Check your results using a graphing utility.*

23. $f(x) = 3x$

24. $f(x) = -4x$

25. $f(x) = 4x + 2$

26. $f(x) = 1 - 3x$

27. $f(x) = x^3 - 1$

28. $f(x) = x^3 + 1$

29. $f(x) = x^2 + 4, \quad x \geq 0$

30. $f(x) = x^2 + 9, \quad x \geq 0$

31. $f(x) = \frac{4}{x}$

32. $f(x) = -\frac{3}{x}$

33. $f(x) = \frac{1}{x - 2}$

34. $f(x) = \frac{4}{x + 2}$

In Problems 35–46, the function f *is one-to-one. Find its inverse and check your answer. State the domain and range of* f *and* f^{-1}. *Using a graphing utility, graph* f, f^{-1}, *and* y = x *on the same square screen.*

35. $f(x) = \dfrac{2}{3 + x}$

36. $f(x) = \dfrac{4}{2 - x}$

37. $f(x) = (x + 2)^2, \quad x \geq -2$

38. $f(x) = (x - 1)^2, \quad x \geq 1$

39. $f(x) = \dfrac{2x}{x - 1}$

40. $f(x) = \dfrac{3x + 1}{x}$

41. $f(x) = \dfrac{3x + 4}{2x - 3}$

42. $f(x) = \dfrac{2x - 3}{x + 4}$

43. $f(x) = \dfrac{2x + 3}{x + 2}$

44. $f(x) = \dfrac{-3x - 4}{x - 2}$

45. $f(x) = 2\sqrt[3]{x}$

46. $f(x) = \dfrac{4}{\sqrt{x}}$

47. Find the inverse of the linear function $f(x) = mx + b, m \neq 0$.

48. Find the inverse of the function $f(x) = \sqrt{r^2 - x^2}, 0 \leq x \leq r$.

49. Can an even function be one-to-one? Explain.

50. Is every odd function one-to-one? Explain.

51. A function f has an inverse. If the graph of f lies in quadrant I, in which quadrant does the graph of f^{-1} lie?

52. A function f has an inverse. If the graph of f lies in quadrant II, in which quadrant does the graph of f^{-1} lie?

53. The function $f(x) = |x|$ is not one-to-one. Find a suitable restriction on the domain of f so that the new function that results is one-to-one. Then find the inverse of f.

54. The function $f(x) = x^4$ is not one-to-one. Find a suitable restriction on the domain of f so that the new function that results is one-to-one. Then find the inverse of f.

55. *Temperature Conversion* To convert from x degrees Celsius to y degrees Fahrenheit, we use the formula $y = f(x) = \frac{9}{5}x + 32$. To convert from x degrees Fahrenheit to y degrees Celsius, we use the formula $y = g(x) = \frac{5}{9}(x - 32)$. Show that f and g are inverse functions.

56. *Demand for Corn* The demand for corn obeys the equation $p(x) = 300 - 50x$, where p is the price per bushel (in dollars) and x is the number of bushels produced, in millions. Express the production amount x as a function of the price p.

57. *Period of a Pendulum* The period T (in seconds) of a simple pendulum is a function of its length l (in feet), given by $T(l) = 2\pi\sqrt{l/g}$, where $g \approx 32.2$ feet per second per second is the acceleration of gravity. Express the length l as a function of the period T.

58. Give an example of a function whose domain is the set of real numbers and that is neither increasing nor decreasing on its domain, but is one-to-one. [*Hint:* Use a piecewise defined function.]

59. Given
$$f(x) = \frac{ax + b}{cx + d}$$
find $f^{-1}(x)$. If $c \neq 0$, under what conditions on a, b, c, and d is $f = f^{-1}$?

60. We said earlier that finding the range of a function f is not easy. However, if f is one-to-one, we can find its range by finding the domain of the inverse function f^{-1}. Use this technique to find the range of each of the following one-to-one functions:

(a) $f(x) = \dfrac{2x + 5}{x - 3}$

(b) $g(x) = 4 - \dfrac{2}{x}$

(c) $F(x) = \dfrac{3}{4 - x}$

For Problems 61–66, write a program that will graph the inverse of a function y = f(x). *Then graph the function* f *and its inverse on the same screen. Compare your answers with those of Problems 23–28.*

61. $f(x) = 3x$

62. $f(x) = -4x$

63. $f(x) = 4x + 2$

64. $f(x) = 1 - 3x$

65. $f(x) = x^3 - 1$

66. $f(x) = x^3 + 1$

67. If the graph of a function and its inverse intersect, where must this necessarily occur? Can they intersect anywhere else? Must they intersect?

68. Can a one-to-one function and its inverse be equal? What must be true about the graph of *f* for this to happen? Give some examples to support your conclusion.

69. Draw the graph of a one-to-one function that contains the points $(-2, -3)$, $(0, 0)$, and $(1, 5)$. Now draw the graph of its inverse. Compare your graph to those of other students. Discuss any similarities. What differences do you see?

2.6

Mathematical Models: Constructing Functions

Real world problems often result in mathematical models that involve functions. These functions need to be constructed or built based on the information given. In constructing functions, we must be able to translate the verbal description into the language of mathematics. We do this by assigning symbols to represent the independent and dependent variables and then finding the function or rule that relates these variables.

EXAMPLE 1

Area of a Rectangle with Fixed Perimeter

The perimeter of a rectangle is 50 feet. Express its area A as a function of the length x of a side.

Solution

Consult Figure 58. If the length of the rectangle is x and if w is its width, then the sum of the lengths of the sides is the perimeter, 50.

FIGURE 58

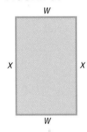

$$x + w + x + w = 50$$
$$2x + 2w = 50$$
$$x + w = 25$$
$$w = 25 - x$$

The area A is length times width, so

$$A = xw = x(25 - x)$$

The area A as a function of x is

$$A(x) = x(25 - x)$$ ■

Note that we use the symbol A as the dependent variable and also as the name of the function that relates the length x to the area A. As we mentioned earlier, this double usage is common in applications and should cause no difficulties.

EXAMPLE 2

Economics: Demand Equations

In economics, revenue R is defined as the amount of money derived from the sale of a product and is equal to the unit selling price p of the product times the number x of units actually sold. That is,

$$R = xp$$

Usually, p and x are related: as one increases, the other decreases. Suppose that p and x are related by the following **demand equation:**

$$p = -\tfrac{1}{10}x + 20 \qquad 0 \le x \le 200$$

Express the revenue R as a function of the number x of units sold.

Solution   Since $R = xp$ and $p = -\frac{1}{10}x + 20$, it follows that

$$R(x) = xp = x\left(-\frac{1}{10}x + 20\right) = -\frac{1}{10}x^2 + 20x$$ ◼

◼   Now work Problem 3.

E X A M P L E   3   *Finding the Distance from the Origin to a Point on a Graph*

Let $P = (x, y)$ be a point on the graph of $y = x^2 - 1$.

(a) Express the distance d from P to the origin O as a function of x.
(b) What is d if $x = 0$?   (c) What is d if $x = 1$?
(d) What is d if $x = \sqrt{2}/2$?
(e) Use a graphing utility to graph the function $d = d(x)$. Correct to two decimal places, find the value of x at which d has a local minimum. [This gives the point(s) on the graph of $y = x^2 - 1$ closest to the origin.]

Solution   (a) Figure 59 illustrates the graph. The distance d from P to O is

FIGURE 59
$y = x^2 - 1$

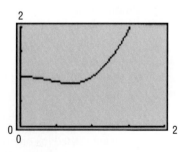

$$d = \sqrt{x^2 + y^2}$$

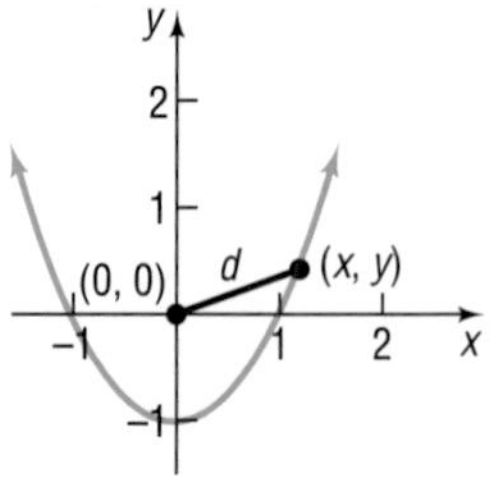

Since P is a point on the graph of $y = x^2 - 1$, we have

$$d(x) = \sqrt{x^2 + (x^2 - 1)^2} = \sqrt{x^4 - x^2 + 1}$$

Thus, we have expressed the distance d as a function of x.

(b) If $x = 0$, the distance d is

$$d(0) = \sqrt{1} = 1$$

(c) If $x = 1$, the distance d is

$$d(1) = \sqrt{1 - 1 + 1} = 1$$

(d) If $x = \sqrt{2}/2$, the distance d is

$$d\left(\frac{\sqrt{2}}{2}\right) = \sqrt{\left(\frac{\sqrt{2}}{2}\right)^4 - \left(\frac{\sqrt{2}}{2}\right)^2 + 1} = \sqrt{\frac{1}{4} - \frac{1}{2} + 1} = \frac{\sqrt{3}}{2}$$

(e) Figure 60 shows the graph of $y = \sqrt{x^4 - x^2 + 1}$.

FIGURE 60

2

0 ⌞_____⌟ 2
0

Using the minimum function on a graphing utility, we find that, when $x \approx 0.70$ the value of d is smallest ($d \approx 0.86$) correct to two decimal places.   ◼

◼   Now work Problem 13.

E X A M P L E   4

Filling a Swimming Pool

A rectangular swimming pool 20 meters long and 10 meters wide is 4 meters deep at one end and 1 meter deep at the other. Figure 61 illustrates a cross-sectional view of the pool. Water is being pumped into the pool at the deep end.

(a) Find a function that expresses the volume V of water in the pool as a function of the height x of the water at the deep end.

(b) Find the volume when the height is 1 meter.

(c) Find the volume when the height is 2 meters.

(d) Use a graphing utility to graph the function $V = V(x)$. At what height is the volume 20 cubic meters? 100 cubic meters?

FIGURE 61

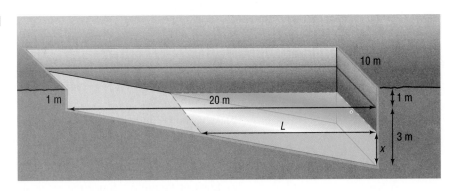

Solution   (a) Let L denote the distance (in meters) measured at water level from the deep end to the short end. Notice that L and x form the sides of a triangle that is similar to the triangle whose sides are 20 meters by 3 meters. Thus, L and x are related by the equation

$$\frac{L}{x} = \frac{20}{3} \quad \text{or} \quad L = \frac{20x}{3} \quad 0 \le x \le 3$$

The volume V of water in the pool at any time is

$$V = \left(\begin{array}{c} \text{Cross-sectional} \\ \text{triangular area} \end{array} \right)(\text{Width}) = \left(\tfrac{1}{2}Lx\right)(10) \text{ cubic meters}$$

Since $L = 20x/3$, we have

$$V(x) = \left(\frac{1}{2} \cdot \frac{20x}{3} \cdot x \right)(10) = \frac{100}{3}x^2 \text{ cubic meters}$$

(b) When the height x of the water is 1 meter, the volume $V = V(x)$ is

$$V(1) = \frac{100}{3} \cdot 1^2 = 33.3 \text{ cubic meters}$$

(c) When the height x of the water is 2 meters, the volume $V = V(x)$ is

$$V(2) = \frac{100}{3} \cdot 2^2 = \frac{400}{3} = 133.3 \text{ cubic meters}$$

(d) See Figure 62.

FIGURE 62

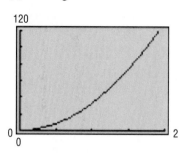

When $x \approx 0.77$ meters, the volume is 20 cubic meters. When $x \approx 1.73$ meters, the volume is 100 cubic meters.   ∎

E X A M P L E   5

Area of an Isosceles Triangle

Consider an isosceles triangle of fixed perimeter p.

(a) If x equals the length of one of the two equal sides, express the area A as a function of x.

(b) What is the domain of A?

(c) Use a graphing utility to graph $A = A(x)$ for $p = 4$.

(d) For what value of x is the area largest?

Solution

(a) Look at Figure 63. Since the equal sides are of length x, the third side must be of length $p - 2x$. (Do you see why?) We know that the area A is

$$A = \tfrac{1}{2}(\text{Base})(\text{Height})$$

FIGURE 63

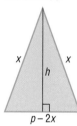

To find the height h, we drop the perpendicular to the base of length $p - 2x$ and use the fact that the perpendicular bisects the base. Then, by the Pythagorean Theorem, we have

$$h^2 = x^2 - \left(\frac{p - 2x}{2}\right)^2 = x^2 - \frac{1}{4}(p^2 - 4px + 4x^2)$$

$$= px - \frac{1}{4}p^2 = \frac{4px - p^2}{4}$$

$$h = \sqrt{\frac{4px - p^2}{4}} = \frac{\sqrt{p}}{2}\sqrt{4x - p}$$

The area A is given by

$$A = \frac{1}{2} \cdot (p - 2x)\frac{\sqrt{p}}{2}\sqrt{4x - p} = \frac{\sqrt{p}}{4}(p - 2x)\sqrt{4x - p}$$

(b) The domain of A is found as follows. Because of the expression $\sqrt{4x - p}$, we require that

$$4x - p > 0$$

$$x > \frac{p}{4}$$

Since $p - 2x$ is a side of the triangle, we also require that

$$p - 2x > 0$$
$$-2x > -p$$
$$x < \frac{p}{2}$$

Thus, the domain of A is $p/4 < x < p/2$, or $(p/4, p/2)$, and we state the functions as

$$A(x) = \frac{\sqrt{p}}{4}(p - 2x)\sqrt{4x - p} \qquad \frac{p}{4} < x < \frac{p}{2}$$

(c) See Figure 64 for the graph of $A(x) = 2(2 - x)\sqrt{x - 1}$.

FIGURE 64

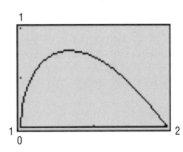

(d) The area is largest (approximately 0.77) when $x \approx 1.33$.   ■

■   Now work Problem 7.

E X A M P L E   6   *Minimum Payments and Interest Charged for Credit Cards*

(a) Holders of credit cards issued by banks, department stores, oil companies, and so on, receive bills each month that state minimum amounts that must be paid by a certain due date. The minimum due depends on the total amount owed. One such credit card company uses the following rules: For a bill of less than $10, the entire amount is due. For a bill of at least $10 but less than $500, the minimum due is $10. There is a minimum of $30 due on a bill of at least $500 but less than $1000, a minimum of $50 due on a bill of at least $1000 but less than $1500, and a minimum of $70 due on bills of $1500 or more. Find the function f that describes the minimum payment due on a bill of x dollars. Graph f by hand.

(b) The card holder may pay any amount between the minimum due and the total owed. The organization issuing the card charges the card holder interest of 1.5% per month for the first $1000 owed and 1% per month on any unpaid balance over $1000. Find the function g that gives the amount of interest charged per month on a balance of x dollars. Graph g by hand.

Solution   (a) The function f that describes the minimum payment due on a bill of x dollars is

$$f(x) = \begin{cases} x & \text{if} & 0 \le x < 10 \\ 10 & \text{if} & 10 \le x < 500 \\ 30 & \text{if} & 500 \le x < 1000 \\ 50 & \text{if} & 1000 \le x < 1500 \\ 70 & \text{if} & 1500 \le x \end{cases}$$

To graph this function f, we proceed as follows: For $0 \leq x < 10$, draw the graph of $y = x$; for $10 \leq x < 500$, draw the graph of the constant function $y = 10$; for $500 \leq x < 1000$, draw the graph of the constant function $y = 30$; and so on. The graph of f is given in Figure 65.

FIGURE 65

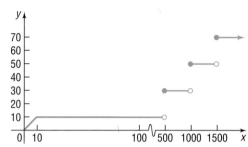

FIGURE 66

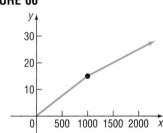

(b) If $g(x)$ is the amount of interest charged per month on a balance of x, then $g(x) = 0.015x$ for $0 \leq x \leq 1000$. The amount of the unpaid balance above $\$1000$ is $x - 1000$. If the balance due is $x > 1000$, then the interest is $0.015(1000) + 0.01(x - 1000) = 15 + 0.01x - 10 = 5 + 0.01x$, so

$$g(x) = \begin{cases} 0.015x & \text{if } 0 \leq x \leq 1000 \\ 5 + 0.01x & \text{if } x > 1000 \end{cases}$$

See Figure 66. ■

E X A M P L E   7   *Finding the Cost of a Can*

A company that manufactures aluminum cans requires a cylindrical container with a capacity of 500 cubic centimeters $\left(\frac{1}{2} \text{ liter}\right)$. The top and bottom of the can will be made of a special aluminum alloy that costs 0.05¢ per square centimeter. The sides of the can are to be made of material that costs 0.02¢ per square centimeter.

(a) Express the cost of material for the can as a function of the radius r of the can.

(b) Use a graphing utility to graph the function $C = C(r)$.

(c) What value of r will result in the least cost? What is this least cost?

Solution   (a) Figure 67 illustrates the situation. Notice that the material required to produce a cylindrical can of height h and radius r consists of a rectangle of area $2\pi rh$ and two circles, each of area πr^2. The total cost C (in cents) of manufacturing the can is therefore

FIGURE 67

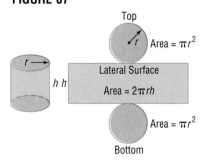

$$C = \text{Cost of top and bottom} + \text{Cost of side}$$
$$= \underbrace{2(\pi r^2)}_{\substack{\text{Total area} \\ \text{of top and} \\ \text{bottom}}} \underbrace{(0.05)}_{\substack{\text{Cost/unit} \\ \text{area}}} + \underbrace{(2\pi rh)}_{\substack{\text{Total} \\ \text{area of} \\ \text{side}}} \underbrace{(0.02)}_{\substack{\text{Cost/unit} \\ \text{area}}}$$
$$= 0.10\pi r^2 + 0.04\pi rh$$

But we have the additional restriction that the height h and radius r must be chosen so that the volume V of the can is 500 cubic centimeters. Since $V = \pi r^2 h$, we have

$$500 = \pi r^2 h \quad \text{or} \quad h = \frac{500}{\pi r^2}$$

Thus, the cost C, in cents, as a function of the radius r is

$$C(r) = 0.10\pi r^2 + 0.04\pi r\left(\frac{500}{\pi r^2}\right) = 0.10\pi r^2 + \frac{20}{r}$$

(b) See Figure 68.

FIGURE 68

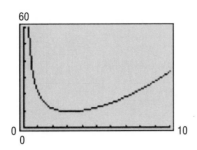

(c) The least cost is 9.46 cents for a radius of about 3.16 centimeters. ▪

E X A M P L E   8   *Making a Playpen*

A manufacturer of children's playpens makes a square model that can be opened at one corner and attached at right angles to a wall, or, perhaps, the side of a house. If each side is 3 feet in length, the open configuration doubles the available area in which the child can play from 9 square feet to 18 square feet. See Figure 69.

FIGURE 69

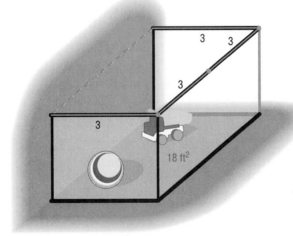

Now, suppose we place hinges at the outer corners to allow for a configuration like the one shown in Figure 70.

(a) Express the area A of this configuration as a function of the distance x between the two parallel sides.

(b) Find the domain of A.   (c) Find A if $x = 5$.

(d) Graph $A = A(x)$. For what value of x is the area largest? What is the maximum area?*

---

*Adapted from *Proceedings, Summer Conference for College Teachers on Applied Mathematics* (University of Missouri, Rolla), 1971.

FIGURE 70

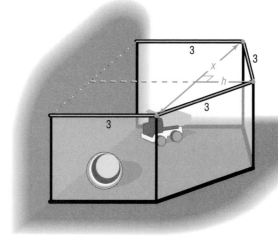

Solution   (a) Refer to Figure 70. The area A that we seek consists of the area of a rectangle (with width 3 and length x) and the area of an isosceles triangle (with base x and two equal sides of length 3). The height h of the triangle may be found using the Pythagorean Theorem:

$$h^2 = 3^2 - \left(\frac{x}{2}\right)^2 = 9 - \frac{x^2}{4} = \frac{36 - x^2}{4}$$

$$h = \tfrac{1}{2}\sqrt{36 - x^2}$$

The area A enclosed by the playpen is

$$A = \text{Area of rectangle} + \text{Area of triangle} = 3x + \tfrac{1}{2}x\left(\tfrac{1}{2}\sqrt{36 - x^2}\right)$$

$$A(x) = 3x + \frac{x\sqrt{36 - x^2}}{4}$$

Thus, we have expressed the area A as a function of x.

(b) To find the domain of A, we note first that $x > 0$, since x is a length. Also, the expression under the radical must be positive, so

$$36 - x^2 > 0$$
$$x^2 < 36$$
$$-6 < x < 6$$

Combining these restrictions, we find that the domain of A is $0 < x < 6$, or $(0, 6)$.

(c) If $x = 5$, the area is

$$A(5) = 3(5) + \frac{5}{4}\sqrt{36 - (5)^2} \approx 19.15 \text{ square feet}$$

Thus, if the width of the playpen is 5 feet, its area is 19.15 square feet.

(d) The maximum area is about 19.81 square feet, obtained when x is about 5.58 feet. See Figure 71.

FIGURE 71

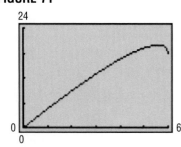

2.6

Exercise 2.6

1. *Volume of a Cylinder* The volume V of a right circular cylinder of height h and radius r is $V = \pi r^2 h$. If the height is twice the radius, express the volume V as a function of r.

2. *Volume of a Cone* The volume V of a right circular cone is $V = \frac{1}{3}\pi r^2 h$. If the height is twice the radius, express the volume V as a function of r.

3. *Demand Equation* The price p and the quantity x sold of a certain product obey the demand equation

$$p = -\tfrac{1}{6}x + 100 \qquad 0 \le x \le 600$$

   (a) Express the revenue R as a function of x. (Remember, $R = xp$.)
   (b) Graph the revenue function using a graphing utility.
   (c) What quantity x maximizes revenue? What is the maximum revenue?
   (d) What price should the company charge to maximize revenue?

4. *Demand Equation* The price p and the quantity x sold of a certain product obey the demand equation

$$p = -\tfrac{1}{3}x + 100 \qquad 0 \le x \le 300$$

   (a) Express the revenue R as a function of x.
   (b) Graph the revenue function using a graphing utility.
   (c) What quantity x maximizes revenue? What is the maximum revenue?
   (d) What price should the company charge to maximize revenue?

5. *Demand Equation* The price p and the quantity x sold of a certain product obey the demand equation

$$x = -5p + 100 \qquad 0 \le p \le 20$$

   (a) Express the revenue R as a function of x.
   (b) Graph the revenue function using a graphing utility.
   (c) What quantity x maximizes revenue? What is the maximum revenue?
   (d) What price should the company charge to maximize revenue?

6. *Demand Equation* The price p and the quantity x sold of a certain product obey the demand equation

$$x = -20p + 500 \qquad 0 \le p \le 25$$

   (a) Express the revenue R as a function of x.
   (b) Graph the revenue function using a graphing utility.
   (c) What quantity x maximizes revenue? What is the maximum revenue?
   (d) What price should the company charge to maximize revenue?

7. *Enclosing a Rectangular Field* A farmer has available 400 yards of fencing and wishes to enclose a rectangular area.
   (a) Express the area A of the rectangle as a function of the width x of the rectangle.
   (b) What is the domain of A?
   (c) Graph $A = A(x)$ using a graphing utility. For what value of x is the area largest?

8. *Enclosing a Rectangular Field along a River* A farmer has 3000 feet of fencing available to enclose a rectangular field. One side of the field lies along a river, so only three sides require fencing.
   (a) Express the area A of the rectangle as a function of x, where x is the length of the side parallel to the river.
   (b) Graph $A = A(x)$ using a graphing utility. For what value of x is the area largest?

9. A wire of length x is bent into the shape of a circle.
   (a) Express the circumference of the circle as a function of x.
   (b) Express the area of the circle as a function of x.

10. A wire of length x is bent into the shape of a square.
   (a) Express the perimeter of the square as a function of x.
   (b) Express the area of the square as a function of x.

11. A right triangle has one vertex on the graph of $y = x^3$, $x > 0$, at (x, y), another at the origin, and the third on the positive y-axis at $(0, y)$, as shown in the figure. Express the area A of the triangle as a function of x.

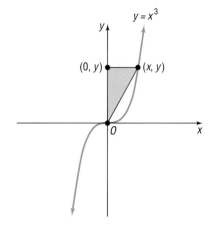

12. A right triangle has one vertex on the graph of $y = 9 - x^2$, $x > 0$, at (x, y), another at the origin, and the third on the positive x-axis at $(x, 0)$. Express the area A of the triangle as a function of x.

13. Let $P = (x, y)$ be a point on the graph of $y = x^2 - 8$.
(a) Express the distance d from P to the origin as a function of x.
(b) What is d if $x = 0$?
(c) What is d if $x = 1$?
(d) Use a graphing utility to graph $d = d(x)$.
(e) For what values of x is d smallest?

14. Let $P = (x, y)$ be a point on the graph of $y = x^2 - 8$.
(a) Express the distance d from P to the point $(0, -1)$ as a function of x.
(b) What is d if $x = 0$?
(c) What is d if $x = -1$?
(d) Use a graphing utility to graph $d = d(x)$.
(e) For what values of x is d smallest?

15. Let $P = (x, y)$ be a point on the graph of $y = \sqrt{x}$.
(a) Express the distance d from P to the point $(1, 0)$ as a function of x.
(b) Use a graphing utility to graph $d = d(x)$.
(c) For what values of x is d smallest?

16. Let $P = (x, y)$ be a point on the graph of $y = 1/x$.
(a) Express the distance d from P to the origin as a function of x.
(b) Use a graphing utility to graph $d = d(x)$.
(c) For what values of x is d smallest?

17. Two cars leave an intersection at the same time. One is headed south at a constant speed of 30 miles per hour; the other is headed west at a constant speed of 40 miles per hour (see the figure). Express the distance d between the cars as a function of the time t. [*Hint:*   At $t = 0$, the cars leave the intersection.]

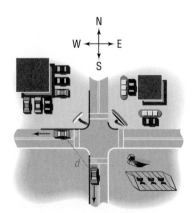

18. Two cars are approaching an intersection. One is 2 miles south of the intersection and is moving at a constant speed of 30 miles per hour. At the same time, the other car is 3 miles east of the intersection and is moving at a constant speed of 40 miles per hour.
  (a) Express the distance d between the cars as a function of time t. [*Hint:* At $t = 0$, the cars are 2 miles south and 3 miles east of the intersection, respectively.]
  (b) Use a graphing utility to graph $d = d(t)$. For what value of t is d smallest?

19. *Constructing an Open Box* An open box with a square base is to be made from a square piece of cardboard 24 inches on a side by cutting out a square from each corner and turning up the sides (see the figure).
  (a) Express the volume V of the box as a function of the length x of the side of the square cut from each corner.
  (b) Graph $V = V(x)$. For what value of x is V largest?

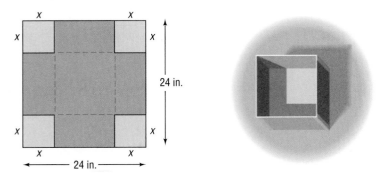

20. *Constructing an Open Box* An open box with a square base is required to have a volume of 10 cubic feet.
  (a) Express the amount A of material used to make such a box as a function of the length x of a side of the square base.
  (b) Graph $A = A(x)$. For what value of x is A smallest?

21. *Constructing a Closed Box* A closed box with a square base is required to have a volume of 10 cubic feet.
  (a) Express the amount A of material used to make such a box as a function of the length x of a side of the square base.
  (b) Graph $A = A(x)$. For what value of x is A smallest?

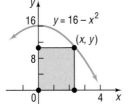

22. *Spheres* The volume V of a sphere of radius r is $V = \frac{4}{3}\pi r^3$; the surface area S of this sphere is $S = 4\pi r^2$. Express the volume V as a function of the surface area S. If the surface area doubles, how does the volume change?

23. A rectangle has one corner on the graph of $y = 16 - x^2$, another at the origin, a third on the positive y-axis, and the fourth on the positive x-axis (see the figure).
  (a) Express the area A of the rectangle as a function of x.
  (b) What is the domain of A?
  (c) Graph $A = A(x)$. For what value of x is A largest?

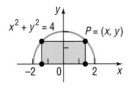

24. A rectangle is inscribed in a semicircle of radius 2 (see the figure). Let $P = (x, y)$ be the point in quadrant I that is a vertex of the rectangle and is on the circle.
  (a) Express the area A of the rectangle as a function of x.
  (b) Express the perimeter p of the rectangle as a function of x.
  (c) Graph $A = A(x)$. For what value of x is A largest?
  (d) Graph $p = p(x)$. For what value of x is p largest?

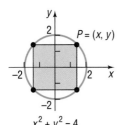

25. A rectangle is inscribed in a circle of radius 2 (see the figure). Let $P = (x, y)$ be the point in quadrant I that is a vertex of the rectangle and is on the circle.
  (a) Express the area A of the rectangle as a function of x.
  (b) Express the perimeter p of the rectangle as a function of x.
  (c) Graph $A = A(x)$. For which value of x is A largest?
  (d) Graph $p = p(x)$. For what value of x is p largest?

26. A circle of radius *r* is inscribed in a square (see the figure).
   (a) Express the area *A* of the square as a function of the radius *r* of the circle.
   (b) Express the perimeter *p* of the square as a function of *r*.

27. *Cost of a Can* A can in the shape of a right circular cylinder is required to have a volume of 500 cubic centimeters. The top and bottom are made of material that costs 6¢ per square centimeter, while the sides are made of material that costs 4¢ per square centimeter.
   (a) Express the total cost *C* of the material as a function of the radius *r* of the cylinder. (Refer to Figure 67.)
   (b) Graph $C = C(r)$. For what value of *r* is the cost *C* least?

28. *Material Needed to Make a Drum* A steel drum in the shape of a right circular cylinder is required to have a volume of 100 cubic feet.
   (a) Express the amount *A* of material required to make the drum as a function of the radius *r* of the cylinder.
   (b) How much material is required if the drum is of radius 3 feet?
   (c) Of radius 4 feet?
   (d) Of radius 5 feet?
   (e) Graph $A = A(r)$. For what value of *r* is *A* smallest?

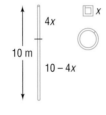

29. A wire 10 meters long is to be cut into two pieces. One piece will be shaped as a square, and the other piece will be shaped as a circle (see the figure).
   (a) Express the total area *A* enclosed by the pieces of wire as a function of the length *x* of a side of the square.
   (b) What is the domain of *A*?
   (c) Graph $A = A(x)$. For what value of *x* is *A* smallest?

30. A wire 10 meters long is to be cut into two pieces. One piece will be shaped as an equilateral triangle, and the other piece will be shaped as a circle.
   (a) Express the total area *A* enclosed by the pieces of wire as a function of the length *x* of a side of the equilateral triangle.
   (b) What is the domain of *A*?
   (c) Graph $A = A(x)$. For what value of *x* is *A* smallest?

31. A semicircle of radius *r* is inscribed in a rectangle so that the diameter of the semicircle is the length of the rectangle (see the figure).
   (a) Express the area *A* of the rectangle as a function of the radius *r* of the semicircle.
   (b) Express the perimeter *p* of the rectangle as a function of *r*.

32. An equilateral triangle is inscribed in a circle of radius *r*. See the figure. Express the circumference *C* of the circle as a function of the length *x* of a side of the triangle. [*Hint:* First show that $r^2 = x^2/3$.]

33. An equilateral triangle is inscribed in a circle of radius *r*. See the figure. Express the area *A* within the circle, but outside the triangle, as a function of the length *x* of a side of the triangle.

34. *Cost of Transporting Goods* A trucking company transports goods between Chicago and New York, a distance of 960 miles. The company's policy is to charge, for each pound, $0.50 per mile for the first 100 miles, $0.40 per mile for the next 300 miles, $0.25 per mile for the next 400 miles, and no charge for the remaining 160 miles.
   (a) Graph the relationship between the cost of transportation in dollars and mileage over the entire 960 mile route.
   (b) Find the cost as a function of mileage for hauls between 100 and 400 miles from Chicago.
   (c) Find the cost as a function of mileage for hauls between 400 and 800 miles from Chicago.

35. *Car Rental Costs* An economy car rented in Florida from National Car Rental® on a weekly basis costs $95 per week.* Extra days cost $24 per day until the daily rate exceeds the weekly rate, in which case the weekly rate applies. Find the cost *C* of renting an economy car as a piecewise-defined function of the number *x* of days used, where $7 \le x \le 14$. Graph this function. [*Note:* Any part of a day counts as a full day.]

36. Rework Problem 35 for a luxury car, which costs $219 on a weekly basis with extra days at $45 per day.

Source: National Car Rental®, 1995.

37. Water is poured into a container in the shape of a right circular cone with radius 4 feet and height 16 feet (see the figure). Express the volume V of water in the cone as a function of the height h of the water. [*Hint:* The volume V of a cone of radius r and height h is $V = \frac{1}{3}\pi r^2 h$.]

38. *Federal Income Tax* Two 1994 Tax Rate Schedules are given in the accompanying table. If x equals the amount on Form 1040, line 37, and y equals the tax due, construct a function f for each schedule.

1994 TAX RATE SCHEDULES

SCHEDULE X—USE IF YOUR FILING STATUS IS SINGLE				SCHEDULE Y-1—USE IF YOUR FILING STATUS IS MARRIED FILING JOINTLY OR QUALIFYING WIDOW(ER)			
If the amount on Form 1040, line 37, is: Over—	But not over—	Enter on Form 1040, line 38	of the amount over—	If the amount on Form 1040, line 37, is: Over—	But not over—	Enter on Form 1040, line 38	of the amount over—
$0	$22,750	- - - - 15%	$0	$0	$38,000	- - - - 15%	$0
22,750	55,100	$3,412.50 + 28%	22,750	38,000	91,850	$5,700.00 + 28%	38,000
55,100	115,000	12,470.50 + 31%	55,100	91,850	140,000	20,778.00 + 31%	91,850
115,000	250,000	31,039.50 + 36%	115,000	140,000	250,000	35,704.50 + 36%	140,000
250,000	- - - - -	79,639.50 + 39.6%	250,000	250,000	- - - - -	75,304.50 + 39.6%	250,000

39. *Linear Curve Fitting* The manager of a clothing store collected the following prices and quantity demanded for shirts.

PRICE ($/SHIRT)	QUANTITY PURCHASED
15	1000
17	920
18	890
19	870
21	800
23	720
24	670
25	620
26	580

(a) Use your graphing utility to find the line of best fit that relates price and quantity purchased. Treat price as the independent variable and quantity as the dependent variable.
(b) Determine the revenue function $R(p)$, where p is the price.
(c) Graph $R(p)$ using a graphing utility.
(d) What is the revenue maximizing price?
(e) What quantity will maximize revenue? What is the maximum revenue?

40. *Linear Curve Fitting* The manager of a clothing store collected the following prices and quantity demanded for pants.

PRICE ($/PAIR)	QUANTITY PURCHASED
29	1000
31	985
32	960
34	920
36	875
37	845
39	790
40	765
41	740
43	695

(a) Use your graphing utility to find the line of best fit that relates price and quantity purchased. Treat price as the independent variable and quantity as the dependent variable.
(b) Determine the revenue function $R(p)$, where p is the price.
(c) Graph $R(p)$ using a graphing utility.
(d) What is the revenue maximizing price?
(e) What quantity will maximize revenue? What is the maximum revenue?

Chapter Review

Function

A rule or correspondence between two sets of real numbers so that each number x in the first set, the domain, has corresponding to it exactly one number y in the second set. The range is the set of y values of the function for the x values in the domain.
x is the independent variable; y is the dependent variable.
A function f may be defined implicitly by an equation involving x and y or explicitly by writing $y = f(x)$.
A function can also be characterized as a set of ordered pairs (x, y) or $(x, f(x))$ in which no two pairs have the same first element.

Function notation

$y = f(x)$
f is a symbol for the function or rule that defines the function.
x is the argument, or independent variable.
y is the dependent variable.
$f(x)$ is the value of the function at x, or the image of x.

Domain

If unspecified, the domain of a function f is the largest set of real numbers for which the rule defines a real number.

Vertical line test

A set of points in the plane is the graph of a function if and only if every vertical line intersects the graph in at most one point.

Local maximum

A function f has a local maximum at c if there is an interval I containing c so that for all x in I, $f(x) < f(c)$.

Local minimum

A function f has a local minimum at c if there is an interval I containing c so that for all x in I, $f(x) > f(c)$.

Even function f

$f(-x) = f(x)$ for every x in the domain ($-x$ must also be in the domain).

Odd function f

$f(-x) = -f(x)$ for every x in the domain ($-x$ must also be in the domain).

One-to-one function f

If $x_1 \neq x_2$, then $f(x_1) \neq f(x_2)$ for any choice of x_1 and x_2 in the domain.

Horizontal line test

If horizontal lines intersect the graph of a function f in at most one point, then f is one-to-one.

Inverse function f^{-1} of f

Domain of f = Range of f^{-1}; Range of f = Domain of f^{-1}
$f^{-1}(f(x)) = x$ and $f(f^{-1}(x)) = x$.
Graphs of f and f^{-1} are symmetric with respect to the line $y = x$.

IMPORTANT FUNCTIONS

Linear function

$f(x) = mx + b$ Graph is a straight line with slope m and y-intercept b.

Constant function

$f(x) = b$ Graph is a horizontal line with y-intercept b (see Figure 21).

Identity function

$f(x) = x$ Graph is a straight line with slope 1 and y-intercept 0 (see Figure 22).

Square function

$f(x) = x^2$ Graph is a parabola with intercept at $(0, 0)$ (see Figure 23).

Cube function

$f(x) = x^3$ See Figure 24.

Square root function

$f(x) = \sqrt{x}$ See Figure 25.

Reciprocal function

$f(x) = 1/x$ See Figure 26.

Absolute value function

$f(x) = |x|$ See Figure 27.

HOW TO:

Find the domain and range of a function from its graph

Find the domain of a function given its equation

Determine whether a function is even or odd without graphing it

Graph certain functions by shifting, compressing, stretching, and/or reflecting (see Table 9)

Use a graphing utility to determine where the graph of a function is increasing or decreasing

Find the composite of two functions

Find the inverse of certain one-to-one functions (see the procedure given on page 150)

Graph f^{-1} given the graph of f

Construct functions in applications, including piecewise-defined functions

Use a graphing utility to find the local maxima and local minima of a function

FILL-IN-THE-BLANK ITEMS

1. If f is a function defined by the equation $y = f(x)$, then x is called the _____ variable and y is the _____ variable.

2. A set of points in the xy-plane is the graph of a function if and only if no _____ line contains more than one point of the set.

3. A(n) _____ function f is one for which $f(-x) = f(x)$ for every x in the domain of f; a(n) _____ function f is one for which $f(-x) = -f(x)$ for every x in the domain of f.

4. Suppose that the graph of a function f is known. Then the graph of $y = f(x - 2)$ may be obtained by a(n) _____ shift of the graph of f to the _____ a distance of 2 units.

5. If $f(x) = x + 1$ and $g(x) = x^3$, then _____ $= (x + 1)^3$.

6. If every horizontal line intersects the graph of a function f at no more than one point, then f is a(n) _____ function.

7. If f^{-1} denotes the inverse of a function f, then the graphs of f and f^{-1} are symmetric with respect to the line _____ .

TRUE/FALSE ITEMS

T   F   **1.** Vertical lines intersect the graph of a function in no more than one point.

T   F   **2.** The y-intercept of the graph of the function $y = f(x)$ whose domain is all real numbers is $f(0)$.

T   F   **3.** Even functions have graphs that are symmetric with respect to the origin.

T   F   **4.** The graph of $y = f(-x)$ is the reflection about the y-axis of the graph of $y = f(x)$.

T   F   **5.** $f(g(x)) = f(x) \cdot g(x)$

T   F   **6.** If f and g are inverse functions, then the domain of f is the same as the domain of g.

T   F   **7.** If f and g are inverse functions, then their graphs are symmetric with respect to the line $y = x$.

REVIEW EXERCISES

1. Given that f is a linear function, $f(4) = -5$, and $f(0) = 3$, write the equation that defines f.

2. Given that g is a linear function with slope $= -4$ and $g(-2) = 2$, write the equation that defines g.

3. A function f is defined by

$$f(x) = \frac{Ax + 5}{6x - 2}$$

If $f(1) = 4$, find A.

4. A function g is defined by

$$g(x) = \frac{A}{x} + \frac{8}{x^2}$$

If $g(-1) = 0$, find A.

5. (a) Tell which of the following graphs are graphs of functions.
   (b) Tell which are graphs of one-to-one functions.

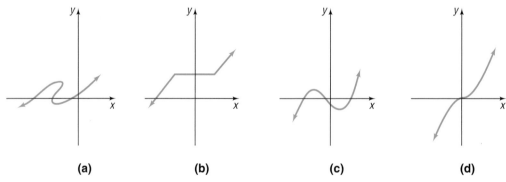

(a)   (b)   (c)   (d)

6. Use the graph of the function f shown to find:
   (a) The domain and range of f
   (b) The intervals on which f is increasing
   (c) The intervals on which f is constant
   (d) The intercepts of f

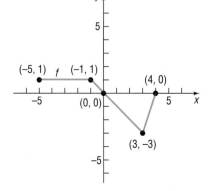

In Problems 7–12, find the following for each function:
(a) $f(-x)$ (b) $-f(x)$ (c) $f(x + 2)$ (d) $f(x - 2)$

7. $f(x) = \dfrac{3x}{x^2 - 4}$

8. $f(x) = \dfrac{x^2}{x + 2}$

9. $f(x) = \sqrt{x^2 - 4}$

10. $f(x) = |x^2 - 4|$

11. $f(x) = \dfrac{x^2 - 4}{x^2}$

12. $f(x) = \dfrac{x^3}{x^2 - 4}$

In Problems 13–18, determine whether the given function is even, odd, or neither without drawing a graph.

13. $f(x) = x^3 - 4x$

14. $g(x) = \dfrac{4 + x^2}{1 + x^4}$

15. $h(x) = \dfrac{1}{x^4} + \dfrac{1}{x^2} + 1$

16. $F(x) = \sqrt{1 - x^3}$

17. $G(x) = 1 - x + x^3$

18. $H(x) = 1 + x + x^2$

In Problems 19–30, find the domain of each function.

19. $f(x) = \dfrac{x}{x^2 - 9}$

20. $f(x) = \dfrac{3x^2}{x - 2}$

21. $f(x) = \sqrt{2 - x}$

22. $f(x) = \sqrt{x + 2}$

23. $h(x) = \dfrac{\sqrt{x}}{|x|}$

24. $g(x) = \dfrac{|x|}{x}$

25. $f(x) = \dfrac{x}{x^2 + 2x - 3}$

26. $F(x) = \dfrac{1}{x^2 - 3x - 4}$

27. $G(x) = \begin{cases} |x| & \text{if } -1 \le x \le 1 \\ 1/x & \text{if } x > 1 \end{cases}$

28. $H(x) = \begin{cases} 1/x & \text{if } 0 < x < 4 \\ x - 4 & \text{if } 4 \le x \le 8 \end{cases}$

29. $f(x) = \begin{cases} 1/(x - 2) & \text{if } x > 2 \\ 0 & \text{if } x = 2 \\ 3x & \text{if } 0 \le x < 2 \end{cases}$

30. $g(x) = \begin{cases} |1 - x| & \text{if } x < 1 \\ 3 & \text{if } x = 1 \\ x + 1 & \text{if } 1 < x \le 3 \end{cases}$

In Problems 31–50:
(a) Find the domain of each function. (b) Locate any intercepts.
(c) Graph each function. (d) Based on the graph, find the range.
(e) Use a graphing utility to verify your answers.

31. $F(x) = |x| - 4$

32. $f(x) = |x| + 4$

33. $g(x) = -|x|$

34. $g(x) = \frac{1}{2}|x|$

35. $h(x) = \sqrt{x - 1}$

36. $h(x) = \sqrt{x} - 1$

37. $f(x) = \sqrt{1 - x}$

38. $f(x) = -\sqrt{x}$

39. $F(x) = \begin{cases} x^2 + 4 & \text{if } x < 0 \\ 4 - x^2 & \text{if } x \ge 0 \end{cases}$

40. $H(x) = \begin{cases} |1 - x| & \text{if } 0 \le x \le 2 \\ |x - 1| & \text{if } x > 2 \end{cases}$

41. $h(x) = (x - 1)^2 + 2$

42. $h(x) = (x + 2)^2 - 3$

43. $g(x) = (x - 1)^3 + 1$

44. $g(x) = (x + 2)^3 - 8$

45. $f(x) = \begin{cases} 2\sqrt{x} & \text{if } x \ge 4 \\ x & \text{if } 0 < x < 4 \end{cases}$

46. $f(x) = \begin{cases} 3|x| & \text{if } x < 0 \\ \sqrt{1 - x} & \text{if } 0 \le x \le 1 \end{cases}$

47. $g(x) = \dfrac{1}{x - 1} + 1$

48. $g(x) = \dfrac{1}{x + 2} - 2$

49. $h(x) = [\![-x]\!]$

50. $h(x) = -[\![x]\!]$

In Problems 51–56, the function f is one-to-one. Find the inverse of each function and check your answer. Find the domain and range of f and f^{-1}. Use a graphing utility to graph f, f^{-1}, and $y = x$ on the same square screen.

51. $f(x) = \dfrac{2x + 3}{5x - 2}$

52. $f(x) = \dfrac{2 - x}{3 + x}$

53. $f(x) = \dfrac{1}{x - 1}$

54. $f(x) = \sqrt{x - 2}$

55. $f(x) = \dfrac{3}{x^{1/3}}$

56. $f(x) = x^{1/3} + 1$

In Problems 57–62, for the given functions f and g, find:

(a) $(f \circ g)(2)$ (b) $(g \circ f)(-2)$ (c) $(f \circ f)(4)$ (d) $(g \circ g)(-1)$

57. $f(x) = 3x - 5$; $g(x) = 1 - 2x^2$ **58.** $f(x) = 4 - x$; $g(x) = 1 + x^2$

59. $f(x) = \sqrt{x + 2}$; $g(x) = 2x^2 + 1$ **60.** $f(x) = 1 - 3x^2$; $g(x) = \sqrt{4 - x}$

61. $f(x) = \dfrac{1}{x^2 + 4}$; $g(x) = 3x - 2$ **62.** $f(x) = \dfrac{2}{1 + 2x^2}$; $g(x) = 3x$

In Problems 63–68, find $f \circ g$, $g \circ f$, $f \circ f$, and $g \circ g$ for each pair of functions.

63. $f(x) = \dfrac{2 - x}{x}$; $g(x) = 3x + 1$ **64.** $f(x) = \dfrac{2x}{x + 1}$; $g(x) = \dfrac{2x}{x - 1}$

65. $f(x) = 3x^2 + x + 1$; $g(x) = |3x|$ **66.** $f(x) = \sqrt{3x}$; $g(x) = 1 + x + x^2$

67. $f(x) = \dfrac{x + 1}{x - 1}$; $g(x) = \dfrac{1}{x}$ **68.** $f(x) = \sqrt{x^2 - 3}$; $g(x) = \sqrt{3 - x^2}$

69. For the graph of the function f shown:
   (a) Draw the graph of $y = f(-x)$.
   (b) Draw the graph of $y = -f(x)$.
   (c) Draw the graph of $y = f(x + 2)$.
   (d) Draw the graph of $y = f(x) + 2$.
   (e) Draw the graph of $y = f(2 - x)$.
   (f) Draw the graph of f^{-1}.

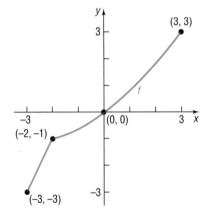

70. Repeat Problem 69 for the graph of the function g shown here.

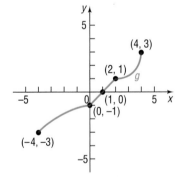

71. *Temperature Conversion* The temperature T of the air is approximately a linear function of the altitude h for altitudes within 10,000 meters of the surface of Earth. If the surface temperature is 30°C and the temperature at 10,000 meters is 5°C, find the function $T = T(h)$.

72. *Speed as a Function of Time* The speed v (in feet per second) of a car is a linear function of the time t (in seconds) for $10 \leq t \leq 30$. If after each second the speed of the car has increased by 5 feet per second, and if after 20 seconds the speed is 80 feet per second, how fast is the car going after 30 seconds? Find the function $v = v(t)$.

73. *Strength of a Beam* The strength of a rectangular wooden beam is proportional to the product of the width and the cube of its depth (see the figure). If the beam is to be cut from a log in the shape of a cylinder of radius 3 feet, express the strength S of the beam as a function of the width x. What is the domain of S?

Chapter 3

PREPARING FOR THIS CHAPTER

Before getting started on this chapter, review the following concepts:

Solving Equations (Appendix, Section 4)
Completing the Square (Appendix, Section 5)
Factoring Polynomials (Appendix, Section 2)
Inequalities; Intervals (Appendix, Section 1)
Absolute Value (Appendix, Section 1)

EQUATIONS AND INEQUALITIES

3.1 Solving Equations Using a Graphing Utility
3.2 Linear and Quadratic Equations
3.3 Setting Up Equations: Applications
3.4 Complex Numbers; Quadratic Equations with a Negative Discriminant
3.5 Other Types of Equations
3.6 Inequalities
3.7 Other Inequalities
3.8 Equations and Inequalities Involving Absolute Value
    Chapter Review

Preview   Rescue at Sea

A ship that is in danger of sinking radios the Coast Guard for assistance. When the rescue craft leaves the Coast Guard station, the ship is 60 miles away and heading directly toward the station. If the average speed of the ship is 10 miles per hour and the average speed of the rescue craft is 20 miles per hour, how long will it take for the rescue craft to reach the ship? [Problem 55 in Exercise 3.3] ■

The investigation of equations and their solutions has played a central role in algebra for many hundreds of years. In fact, as early as 200 BC, the Babylonians had a well-developed algebra that included a solution for quadratic equations (see the Historical Feature in Section 3.2). Of late, the study of inequalities (Sections 3.6–3.8) has become equally important. For example, the problem of optimizing airline scheduling leads to linear inequalities. As you go through this chapter, you will see evidence of the importance of the topics presented here for solving problems that occur in business, engineering, and the social and natural sciences.

We shall limit our discussion in this chapter to equations and inequalities containing a single variable. Later, we will study equations and inequalities containing more than one variable.

3.1

Solving Equations Using a Graphing Utility

You may wish to read *Solving Equations* in the Appendix, Section 4, before going on.

In solving the equation $2x = 6$, we are able to obtain an exact solution, $x = 3$. To solve the equation $x^3 = 25x$ requires some additional work:

$$x^3 = 25x$$
$$x^3 - 25x = 0 \qquad \text{Rewrite with 0 on the right side.}$$
$$x(x^2 - 25) = 0 \qquad \text{Factor}$$
$$x(x - 5)(x + 5) = 0 \qquad \text{Factor}$$
$$x = 0 \qquad x - 5 = 0 \qquad x + 5 = 0 \qquad \text{Set each factor equal to 0.}$$
$$x = 0 \qquad x = 5 \qquad x = -5 \qquad \text{Solve}$$

Again, exact solutions are obtained.

We shall see as we proceed through this book that some equations can be solved using algebraic techniques that result in exact solutions being obtained. Whenever algebraic techniques exist, we shall discuss them and use them to obtain exact solutions.

For many equations, though, there are no algebraic techniques that lead to a solution. For such equations, a graphing utility can often be used to investigate possible solutions. When a graphing utility is used to solve an equation, usually *approximate* solutions are obtained. Unless otherwise stated, we shall follow the practice of giving approximate solutions as decimals *correct to two decimal places.*

Let's look at an example that explains this terminology.

E X A M P L E   1   *Writing a Number Correct to n Decimal Places*

WRITING THE NUMBER	1/7	$\sqrt{2}$	π
Correct to 1 decimal place	0.1	1.4	3.1
Correct to 2 decimal places	0.14	1.41	3.14
Correct to 3 decimal places	0.142	1.414	3.141
Correct to 4 decimal places	0.1428	1.4142	3.1415
Correct to 5 decimal places	0.14285	1.41421	3.14159
Correct to 6 decimal places	0.142857	1.414213	3.141592 ■

As the example illustrates, **correct to *n* decimal places** means the decimal that results from truncation after the *n*th decimal.

■ Now work Problem 1.

We learned in Chapter 2 that the x-intercepts, if any, of a function $y = f(x)$ are found by letting $y = 0$ and solving the equation $f(x) = 0$. Thus, solving the equation $f(x) = 0$ is equivalent to finding the x-intercepts of the graph of the function $y = f(x)$. It is this relationship between the solutions of an equation and the graph of a function that enables us to use a graphing utility to solve equations.

We can be sure the graph of a function has an x-intercept by using the Intermediate Value Theorem.

Intermediate Value Theorem

The Intermediate Value Theorem requires that the function be *continuous*. Although it requires calculus to explain precisely the meaning, the *idea* of a continuous function is easy to understand. Very basically, a function f is continuous when its graph can be drawn without lifting pencil from paper, that is, when the graph contains no "holes" or "jumps" or "gaps".

Intermediate Value Theorem   Let f denote a continuous function. If $a < b$ and if $f(a)$ and $f(b)$ are of opposite sign, then the graph of f has at least one x-intercept between a and b.   ■

Although the proof of this result requires advanced methods in calculus, it is easy to "see" why the result is true. Look at Figure 1.

FIGURE 1

If $f(a) < 0$ and $f(b) > 0$, there is an x-intercept between a and b.

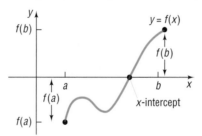

Since many of the functions we deal with are continuous, the Intermediate Value Theorem together with a graphing utility provide a basis for finding x-intercepts and solving equations.

E X A M P L E   2   *Using the Intermediate Value Theorem and a Graphing Utility to Locate x-Intercepts*

Show that the graph of $f(x) = x^5 - x^3 - 1$ has an x-intercept between 1 and 2.

Solution   Using a graphing utility, graph f. TRACE along the function. As the cursor moves from $x = 1$ to $x = 2$, notice the value of y changes from a negative value to a positive value (Figure 2(a) and (b)). According to the Intermediate Value Theorem, there is an x-intercept between $x = 1$ and $x = 2$. In fact, there is an x-intercept between 1.212766 and 1.2553191.

FIGURE 2

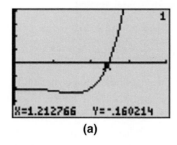

(a)

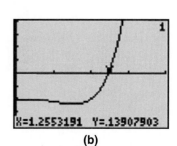

(b)

■

■   Now work Problem 7.

Solving Equations

E X A M P L E   3   *Using a Graphing Utility to Approximate Solutions of an Equation*

Find the smaller of the two solutions of the equation $x^2 - 6x + 7 = 0$. Express the answer correct to two decimal places.

Solution   The solutions of the equation $x^2 - 6x + 7 = 0$ are the same as the x-intercepts of the function $f(x) = x^2 - 6x + 7$. We begin by graphing the function f using a scale of 1 on each axis.

Figure 3 shows the graph.

FIGURE 3

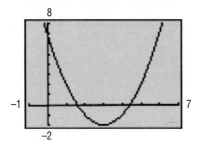

The smaller of the two x-intercepts (solutions of the equation) lies between 1 and 2. (Remember that Xscl = 1.) Thus, correct to 0 decimal places, the smaller solution of the equation is $x = 1$.

Next, we BOX the graph from approximately $x = 1$ to $x = 2$ and from $y = -1$ to $y = 1$. See Figure 4.

FIGURE 4

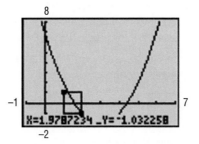

Adjust the RANGE setting so that Xscl = 0.1 and Yscl = 0.1. Graph f again. See Figure 5.

FIGURE 5

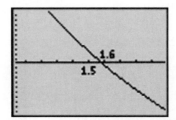

The x-intercept lies between 1.5 and 1.6. (Remember that the BOX was from $x = 1$ to $x = 2$ and Xscl = 0.1.) Thus, correct to one decimal place, the smaller solution of the equation is $x = 1.5$.

Next, BOX again from $x = 1.5$ to $x = 1.6$ and from $y = -0.1$ to $y = 0.1$. See Figure 6. Adjust the RANGE so that Xscl = 0.01 and Yscl = 0.01 and graph f. See Figure 7.

FIGURE 6

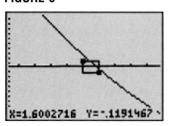

FIGURE 7

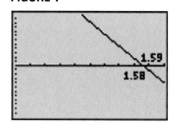

The x-intercept lies between 1.58 and 1.59. Remember that the BOX was from $x = 1.5$ to $x = 1.6$ with Xscl = 0.01. Thus, correct to two decimal places, the smaller of the two solutions is $x = 1.58$. ∎

The steps to follow for approximating solutions of equations correct to any desired number of decimals are given next.

Steps for Approximating Solutions of Equations	
STEP 1:	Write the equation in the form $f(x) = 0$.
STEP 2:	Graph the function f, using Xscl = 1.
STEP 3:	BOX the graph so that the x-intercept is within the BOX. Adjust Xscl and Yscl to one-tenth of their current value and graph f again.
STEP 4:	If additional accuracy is desired, repeat Step 3.

In these steps, each repetition of Step 3 gives one more decimal place of accuracy. Thus, when Xscl = 0.01, the x-intercept will be known correct to two decimal places. Let's look at another example.

E X A M P L E 4   *Use a Graphing Utility to Approximate the Solution(s) of the Equation*

$$x^3 - 1.5x^2 + x - 1.2 = 0$$

Solution   **STEP 1:** The equation is already in the form $f(x) = 0$.
STEP 2: Graph the function

$$f(x) = x^3 - 1.5x^2 + x - 1.2$$

Figure 8 shows the graph of f. Notice that the scale is 1 on each axis. Also notice that the x-intercept lies between 1 and 2.

FIGURE 8

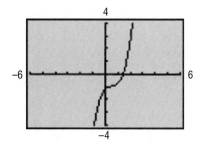

STEP 3: Figure 9 shows the box chosen. Notice that its width is about 1 unit and its height is about 2 units. Change Xscl = 0.1, Yscl = 0.1 and graph.

Figure 10 shows the result. Xscl and Yscl have both been changed to 0.1. From Figure 10 we see the x-intercept is near 1.4.

FIGURE 9

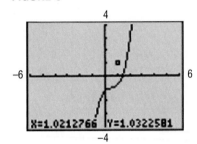

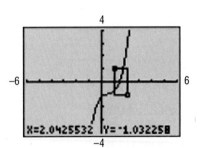

FIGURE 10

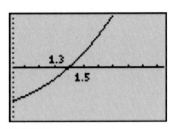

STEP 4: We repeat Step 3, using a BOX from approximately $x = 1.39$ to approximately $x = 1.41$ and changing Xscl = 0.01 and Yscl = 0.01. This gives us Figure 11. The x-intercept is 1.39, correct to two decimal places.

FIGURE 11

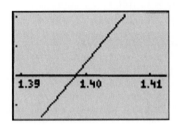

NOTE: If the x-intercept you are trying to approximate appears to fall on a tick mark, then you should evaluate the function at that tick mark. If the value is zero, then there is an exact solution. If the value is not zero, then you know whether the graph is above or below the x-axis at the tick mark by noting the sign at the tick mark.

3.1

Exercise 3.1

In Problems 1–6, write each expression as a decimal correct to two decimal places.

1. $\frac{3}{7}$ **2.** $\frac{2}{11}$ **3.** $\sqrt{5}$ **4.** $\sqrt{6}$ **5.** $\sqrt[3]{2}$ **6.** $\sqrt[3]{3}$

In Problems 7–12, use the Intermediate Value Theorem to show that the graph of each function has an x-intercept in the given interval. Approximate the x-intercept correct to two decimal places.

7. $f(x) = 8x^4 - 2x^2 + 5x - 1$;   [0, 1]   **8.** $f(x) = x^4 + 8x^3 - x^2 + 2$;   [-1, 0]

9. $f(x) = 2x^3 + 6x^2 - 8x + 2$;   [-5, -4]   **10.** $f(x) = 3x^3 - 10x + 9$;   [-3, -2]

11. $f(x) = x^5 - x^4 + 7x^3 - 7x^2 - 18x + 18$;   [1.4, 1.5]   **12.** $f(x) = x^5 - 3x^4 - 2x^3 + 6x^2 + x + 2$;   [1.7, 1.8]

In Problems 13–18, use a graphing utility to approximate the smaller of the two solutions of each equation. Express the answer correct to two decimal places.

13. $x^2 + 4x + 2 = 0$ **14.** $x^2 + 4x - 3 = 0$ **15.** $2x^2 + 4x + 1 = 0$

16. $3x^2 + 5x + 1 = 0$ **17.** $2x^2 - 3x - 1 = 0$ **18.** $2x^2 - 4x - 1 = 0$

*In Problems 19–26, use a graphing utility to approximate the **positive** solutions for each equation. Express the answer correct to two decimal places.*

19. $x^3 + 3.2x^2 - 16.83x - 5.31 = 0$

20. $x^3 + 3.2x^2 - 7.25x - 6.3 = 0$

21. $x^4 - 1.4x^3 - 33.71x^2 + 23.94x + 292.41 = 0$

22. $x^4 + 1.2x^3 - 7.46x^2 - 4.692x + 15.2881 = 0$

23. $\pi x^3 - (8.88\pi + 1)x^2 - (42.066\pi - 8.88)x + 42.066 = 0$

24. $\pi x^3 - (5.63\pi + 2)x^2 - (108.392\pi - 11.26)x + 216.784 = 0$

25. $x^3 + 19.5x^2 - 1021x + 1000.5 = 0$

26. $x^3 + 14.2x^2 - 4.8x - 12.4 = 0$

3.2

Linear and Quadratic Equations

In this section we discuss two types of equations that can be solved algebraically to obtain exact solutions.

Linear Equations

Linear equations are equations such as

$$3x + 12 = 0 \qquad -2x + 5 = 0 \qquad 4x - 3 = 0$$

A general definition is given next.

Linear Equation in One Variable	A **linear equation in one variable** is equivalent to an equation of the form

$$ax + b = 0$$

where a and b are real numbers and $a \neq 0$.

Sometimes, a linear equation is called a **first-degree equation,** because the left side is a polynomial in x of degree 1.

It is relatively easy to solve a linear equation:

$$ax + b = 0$$

$$ax = -b \qquad \text{Subtract } b \text{ from both sides.}$$

$$x = \frac{-b}{a} \qquad \text{Divide both sides by } a, a \neq 0.$$

The linear equation $ax + b = 0$ has the single solution given by the formula $x = -b/a$.

Whenever it is possible to solve a linear equation in your head, do so. For example,

The solution of $2x = 8$ is $x = 4$.

The solution of $3x - 15 = 0$ is $x = 5$.

Often, though, some rearrangement of the terms is necessary.

E X A M P L E 1   *Solving a Linear Equation*

Solve the equation: $\frac{1}{2}(p + 5) - 3 = \frac{1}{3}(2p - 1)$

Solution   To clear the equation of fractions, we multiply both sides by 6, the least common multiple of the denominators of the fractions $\frac{1}{2}$ and $\frac{1}{3}$:

$$\frac{1}{2}(p + 5) - 3 = \frac{1}{3}(2p - 1)$$

$$6\left[\frac{1}{2}(p + 5) - 3\right] = 6\left[\frac{1}{3}(2p - 1)\right] \quad \text{Multiply both sides by 6, the LCM of 2 and 3.}$$

$$3(p + 5) - 18 = 2(2p - 1) \quad \begin{array}{l}\text{Use the distributive property on the left and} \\ \text{the associative property on the right.}\end{array}$$

$$3p + 15 - 18 = 4p - 2 \quad \text{Use the distributive property.}$$

$$3p - 3 = 4p - 2 \quad \text{Combine like terms.}$$

$$3p = 4p + 1 \quad \text{Add 3 to each side.}$$

$$-p = 1 \quad \text{Subtract } 4p \text{ from each side.}$$

$$p = -1 \quad \text{Multiply both sides by } -1.$$

Check: $\dfrac{1}{2}(p + 5) - 3 = \dfrac{1}{2}(-1 + 5) - 3 = \dfrac{1}{2}(4) - 3 = 2 - 3 = -1$

$$\frac{1}{3}(2p - 1) = \frac{1}{3}(-2 - 1) = \frac{1}{3}(-3) = -1$$

Since the two expressions are equal, the solution $p = -1$ checks.   ■

E X A M P L E   2   *Solving a Linear Equation Using a Calculator*

Solve the equation: $2.78x + \dfrac{2}{17.931} = 54.06$

Express the answer correct to two decimal places.

Solution   To avoid rounding errors, we solve for x before using the calculator. We proceed as in Example 1.

$$2.78x + \frac{2}{17.931} = 54.06$$

$$2.78x = 54.06 - \frac{2}{17.931} \quad \begin{array}{l}\text{Subtract 2/17.931 from} \\ \text{each side.}\end{array}$$

$$x = \frac{54.06 - (2/17.931)}{2.78} \quad \text{Divide each side by 2.78.}$$

Now use your calculator. See Figure 12.

FIGURE 12

```
(54.06-2/17.931)
/2.78
          19.40592134
```

The solution, correct to two decimal places, is 19.40.

Check: We store the solution in memory and proceed to check.

$$(2.78)(19.40592134) + (2/17.931) = 54.06$$ ■

■ Now work Problems 19 and 67.

Equations That Lead to Linear Equations

The next two examples illustrate the solution of equations that are not linear but lead to linear equations upon simplification.

E X A M P L E   3   *Solving Equations*

Solve the equation: $(2y + 1)(y - 1) = (y + 5)(2y - 5)$

Solution

$$(2y + 1)(y - 1) = (y + 5)(2y - 5)$$

$$2y^2 - y - 1 = 2y^2 + 5y - 25 \qquad \text{Multiply, and combine like terms.}$$

$$-y - 1 = 5y - 25 \qquad \text{Subtract } 2y^2 \text{ from each side.}$$

$$-y = 5y - 24 \qquad \text{Add 1 to each side.}$$

$$-6y = -24 \qquad \text{Subtract } 5y \text{ from each side.}$$

$$y = 4 \qquad \text{Divide both sides by } -6.$$

Check: $(2y + 1)(y - 1) = (8 + 1)(3) = (9)(3) = 27$
$(y + 5)(2y - 5) = (4 + 5)(8 - 5) = (9)(3) = 27$

Since the two expressions are equal, the solution $y = 4$ checks. ■

E X A M P L E   4   *Solving Equations*

Solve the equation: $\dfrac{3}{x - 2} = \dfrac{1}{x - 1} + \dfrac{7}{(x - 1)(x - 2)}$

Solution   First, we note that the domain of the variable is $\{x \mid x \neq 1, x \neq 2\}$. As we did in Example 3, we clear the equation of fractions by multiplying both sides by the least common multiple of the denominators of the three fractions $(x - 1)(x - 2)$:

$$\frac{3}{x - 2} = \frac{1}{x - 1} + \frac{7}{(x - 1)(x - 2)}$$

$$(x - 1)(x - 2)\frac{3}{x - 2} = (x - 1)(x - 2)\left[\frac{1}{x - 1} + \frac{7}{(x - 1)(x - 2)}\right] \qquad \begin{array}{l}\text{Multiply both sides by } (x - 1)(x - 2).\\ \text{Cancel on the left.}\end{array}$$

$$3x - 3 = (x - 1)(x - 2)\frac{1}{x - 1} + (x - 1)(x - 2)\frac{7}{(x - 1)(x - 2)} \qquad \begin{array}{l}\text{Use the distributive property on each side;}\\ \text{cancel on the right.}\end{array}$$

$$3x - 3 = (x - 2) + 7$$

$$3x - 3 = x + 5 \qquad \text{Combine like terms.}$$

$$2x = 8 \qquad \text{Add 3 to each side. Subtract } x \text{ from each side.}$$

$$x = 4 \qquad \text{Divide by 2.}$$

Check: $\dfrac{3}{x - 2} = \dfrac{3}{4 - 2} = \dfrac{3}{2}$

$$\frac{1}{x - 1} + \frac{7}{(x - 1)(x - 2)} = \frac{1}{3} + \frac{7}{(3)(2)} = \frac{1}{3} + \frac{7}{6} = \frac{9}{6} = \frac{3}{2}$$

Since the two expressions are equal, the solution $x = 4$ checks. ■

■ Now work Problem 33.

Summary

Steps for Solving a Linear Equation Algebraically

To solve a linear equation, follow these steps:

STEP 1: If necessary, clear the equation of fractions by multiplying both sides by the least common multiple (LCM) of the denominators of all the fractions.

STEP 2: Remove all parentheses and simplify.

STEP 3: Collect all terms containing the variable on one side and all other terms on the other side.

STEP 4: Check your solution(s).

Quadratic Equations

Quadratic equations are equations such as

$$2x^2 + x + 8 = 0 \qquad 3x^2 - 5x + 6 = 0 \qquad x^2 - 9 = 0$$

A general definition is given next.

Quadratic Equation

A **quadratic equation** is an equation equivalent to one of the form

$$ax^2 + bx + c = 0 \tag{1}$$

where a, b, and c are real numbers and $a \neq 0$.

A quadratic equation written in the form $ax^2 + bx + c = 0$ is said to be in **standard form.**

Sometimes, a quadratic equation is called a **second-degree equation,** because the left side is a polynomial of degree 2. We shall discuss two ways of solving quadratic equations algebraically: by factoring and by use of the quadratic formula. We shall also solve quadratic equations using a graphing utility.

Factoring

When a quadratic equation is written in standard form, $ax^2 + bx + c = 0$, it may be possible to factor the expression on the left side as the product of two first-degree polynomials. Then, by setting each factor equal to 0 and solving the resulting linear equations, we obtain the solutions of the quadratic equation.

Let's look at an example.

E X A M P L E   5

Solving a Quadratic Equation by Graphing and by Factoring

Solve the equation: $x^2 - 5x + 6 = 0$

Graphing Solution

Figure 13 shows the graph of $f(x) = x^2 - 5x + 6$. We see that there are two *x*-intercepts: one near 2, the other near 3. The equation has two solutions. After BOXing, we conjecture that the two *x*-intercepts are 2 and 3, since the graphs of the function appears to cross the *x*-intercept at the tick marks. To verify that 2 and 3 are *x*-intercepts (and hence solutions to the equation), we evaluate $f(2)$ and $f(3)$.

$$f(2) = 2^2 - 5(2) + 6 = 0$$
$$f(3) = 3^2 - 5(3) + 6 = 0$$

So both $x = 2$ and $x = 3$ are solutions. The solution set is $\{2, 3\}$.

FIGURE 13

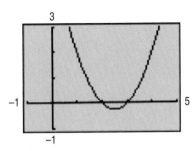

Algebraic Solution   The equation is in the standard form specified in equation (1). The left side may be factored as

$$x^2 - 5x + 6 = 0$$
$$(x - 2)(x - 3) = 0$$

We set each factor equal to 0 and solve the resulting first-degree equations.

$$x - 2 = 0 \quad \text{or} \quad x - 3 = 0$$
$$x = 2 \qquad\qquad x = 3$$

The solution set is $\{2, 3\}$.

E X A M P L E   6   *Solving a Quadratic Equation by Graphing and by Factoring*
Solve the equation: $x^2 = 12 - x$

Graphing Solution   We put the equation in standard form by adding $x - 12$ to each side:

$$x^2 = 12 - x$$
$$x^2 + x - 12 = 0$$

Figure 14 shows the graph of the function

$$f(x) = x^2 + x - 12$$

From the graph it appears there are two x-intercepts, one near -4, the other near 3. We conjecture, after using the BOX function, that the function has two x-intercepts: -4 and 3. This is verified by evaluating $f(-4)$ and $f(3)$. Since $f(-4) = f(3) = 0$, the solution set is $\{-4, 3\}$.

FIGURE 14

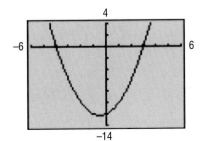

Algebraic Solution   We put the equation in standard form by adding $x - 12$ to each side:

$$x^2 = 12 - x$$
$$x^2 + x - 12 = 0$$

The left side may now be factored as

$$(x + 4)(x - 3) = 0$$

so that

$$x + 4 = 0 \quad \text{or} \quad x - 3 = 0$$
$$x = -4 \qquad\qquad x = 3$$

The solution set is $\{-4, 3\}$. ∎

When the left side factors into two linear equations with the same solution, the quadratic equation is said to have a **repeated solution.** We also call this solution a **root of multiplicity 2,** or a **double root.**

E X A M P L E   7   *Solving a Quadratic Equation by Graphing and by Factoring*

Solve the equation: $x^2 - 6x + 9 = 0$

Graphing Solution   Figure 15 shows the graph of the function

$$f(x) = x^2 - 6x + 9$$

From the graph it appears there is one x-intercept, 3. We conjecture, after using the BOX function, that the function has one x-intercept: 3. Since $f(3) = 0$, the equation has only the repeated solution 3.

FIGURE 15

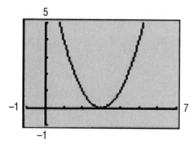

Algebraic Solution   This equation is already in standard form, and the left side can be factored:

$$x^2 - 6x + 9 = 0$$
$$(x - 3)(x - 3) = 0$$

so

$$x = 3 \quad \text{or} \quad x = 3$$

This equation has only the repeated solution 3. ∎

■   Now work Problems 37 and 45.

The Quadratic Formula

We begin with a preliminary result. Suppose we wish to solve the quadratic equation

$$x^2 = p \tag{2}$$

where $p \geq 0$ is a nonnegative number. We proceed as in the earlier examples:

$$x^2 - p = 0 \qquad \text{Put in standard form.}$$
$$(x - \sqrt{p})(x + \sqrt{p}) = 0 \qquad \text{Factor (over the real numbers).}$$
$$x = \sqrt{p} \quad \text{or} \quad x = -\sqrt{p} \quad \text{Solve.}$$

Thus, we have the following result:

$$\text{If } x^2 = p \text{ and } p \geq 0, \text{ then } x = \sqrt{p} \text{ or } x = -\sqrt{p}. \tag{3}$$

Note that if $p > 0$ the equation $x^2 = p$ has two solutions, $x = \sqrt{p}$ and $x = -\sqrt{p}$. We usually abbreviate these solutions as $x = \pm \sqrt{p}$, read as "x equals plus or minus the square root of p." For example, the two solutions of the equation

$$x^2 = 4$$

are

$$x = \pm \sqrt{4}$$

and since $\sqrt{4} = 2$, we have

$$x = \pm 2$$

The solution set is $\{-2, 2\}$.

EXAMPLE 8   *Solving Quadratic Equations Using Equation (3)*

Solve each equation.

(a) $x^2 = 5$ (b) $(x - 2)^2 = 16$

Solution   (a) We use the result in equation (3) to get

$$x^2 = 5$$
$$x = \pm \sqrt{5}$$
$$x = \sqrt{5} \quad \text{or} \quad x = -\sqrt{5}$$

The solution set is $\{-\sqrt{5}, \sqrt{5}\}$.

(b) We use the result in equation (3) to get

$$(x - 2)^2 = 16$$
$$x - 2 = \pm \sqrt{16}$$
$$x - 2 = \sqrt{16} \quad \text{or} \quad x - 2 = -\sqrt{16}$$
$$x - 2 = 4 \qquad\qquad x - 2 = -4$$
$$x = 6 \qquad\qquad\quad x = -2$$

The solution set is $\{-2, 6\}$.

We can use the method of completing the square* to obtain a general formula for solving the quadratic equation

$$ax^2 + bx + c = 0 \qquad a \neq 0$$

We begin by rearranging the terms as

$$ax^2 + bx = -c$$

Since $a \neq 0$, we can divide both sides by a to get

$$x^2 + \frac{b}{a}x = -\frac{c}{a}$$

Now the coefficient of x^2 is 1. To complete the square on the left side, add the square of $\frac{1}{2}$ the coefficient of x; that is, add

$$\left(\frac{1}{2} \cdot \frac{b}{a}\right)^2 = \frac{b^2}{4a^2}$$

to each side. Then,

$$x^2 + \frac{b}{a}x + \frac{b^2}{4a^2} = \frac{b^2}{4a^2} - \frac{c}{a}$$

$$\left(x + \frac{b}{2a}\right)^2 = \frac{b^2 - 4ac}{4a^2} \tag{4}$$

Provided $b^2 - 4ac \geq 0$, we now can apply the result in equation (3) to get

$$x + \frac{b}{2a} = \pm\sqrt{\frac{b^2 - 4ac}{4a^2}}$$

$$x = -\frac{b}{2a} \pm \frac{\sqrt{b^2 - 4ac}}{2a} = \frac{-b \pm \sqrt{b^2 - 4ac}}{2a}$$

What if $b^2 - 4ac$ is negative? Then equation (4) states that the left expression (a real number squared) equals the right expression (a negative number). Since this occurrence is impossible for real numbers, we conclude that if $b^2 - 4ac < 0$ the quadratic equation has no *real* solution. (We discuss quadratic equations for which the quantity $b^2 - 4ac < 0$ in detail in Section 3.4.)

We now state the *quadratic formula.*

Theorem   Consider the quadratic equation

$$ax^2 + bx + c = 0 \qquad a \neq 0$$

If $b^2 - 4ac < 0$, this equation has no real solution.
If $b^2 - 4ac \geq 0$, the real solution(s) of this equation is (are) given by the **quadratic formula:**

Quadratic Formula

$$x = \frac{-b \pm \sqrt{b^2 - 4ac}}{2a} \tag{5}$$

*Refer to the Appendix, Section 5.

The quantity $b^2 - 4ac$ is called the **discriminant** of the quadratic equation, because its value tells us whether the equation has real solutions. In fact, it also tells us how many solutions to expect.

Discriminant of a Quadratic Equation

For a quadratic equation $ax^2 + bx + c = 0$:

1. If $b^2 - 4ac > 0$, there are two unequal real solutions.
2. If $b^2 - 4ac = 0$, there is a repeated real solution, a root of multiplicity 2.
3. If $b^2 - 4ac < 0$, there is no real solution.

Thus, when asked to find the real solutions, if any, of a quadratic equation, always evaluate the discriminant first to see how many real solutions there are.

E X A M P L E   9   *Solving a Quadratic Equation by Graphing and by Using the Quadratic Formula*

Find the real solutions, if any, of the equation: $3x^2 - 5x + 1 = 0$

Solution   The equation is in standard form, so we compare it to $ax^2 + bx + c = 0$ to find a, b, and c:

$$3x^2 - 5x + 1 = 0$$
$$ax^2 + bx + c = 0$$

With $a = 3$, $b = -5$, and $c = 1$, we evaluate the discriminant $b^2 - 4ac$:

$$b^2 - 4ac = (-5)^2 - 4(3)(1) = 25 - 12 = 13$$

Since $b^2 - 4ac > 0$, there are two real solutions.

Graphing Solution   Figure 16 shows the graph of the function

$$f(x) = 3x^2 - 5x + 1$$

As expected, we see that there are two x-intercepts: one between 0 and 1, the other between 1 and 2. After using the BOX function, the solutions to the equation are 0.23 and 1.43 correct to two decimal places.

FIGURE 16

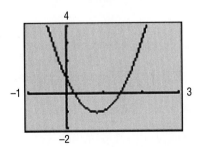

Algebraic Solution   We use the quadratic formula with $a = 3$, $b = -5$, $c = 1$, and $b^2 - 4ac = 13$.

$$x = \frac{-b \pm \sqrt{b^2 - 4ac}}{2a} = \frac{5 \pm \sqrt{13}}{6}$$

The solution set is $\{(5 - \sqrt{13})/6, (5 + \sqrt{13})/6\}$. These solutions are exact. ■

■ Now work Problem 97.

E X A M P L E   1 0   *Solving Quadratic Equations by Graphing and by Using the Quadratic Formula*

Find the real solutions, if any, of the equation:

$$\tfrac{25}{2}x^2 - 30x + 18 = 0$$

Solution   The equation is given in standard form. However, to simplify the arithmetic, we clear the fractions by multiplying both sides of the equation by two:

$$\tfrac{25}{2}x^2 - 30x + 18 = 0$$

$$25x^2 - 60x + 36 = 0 \quad \text{Clear fractions.}$$

$$ax^2 + bx + c = 0 \quad \text{Compare to standard form.}$$

With $a = 25$, $b = -60$, and $c = 36$, we evaluate the discriminant:

$$b^2 - 4ac = (-60)^2 - 4(25)(36) = 3600 - 3600 = 0$$

The equation has a repeated solution.

Graphing Solution   Figure 17 shows the graph of the function

$$f(x) = \frac{25}{2}x^2 - 30x + 18$$

From the graph, we see that the one x-intercept, is between 1 and 2. After using the BOX function, the solution to the equation is 1.20 correct to two decimal places.

FIGURE 17

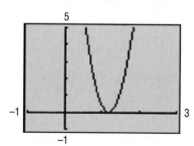

Algebraic Solution   We use $a = 25$, $b = -60$, $c = 36$, and $b^2 - 4ac = 0$ to solve

$$25x^2 - 60x + 36 = 0$$

Using the quadratic formula, we find

$$x = \frac{-b \pm \sqrt{b^2 - 4ac}}{2a} = \frac{60 \pm \sqrt{0}}{50} = \frac{60}{50} = \frac{6}{5}$$

The repeated solution is $\tfrac{6}{5}$, which is exact.   ■

E X A M P L E   1 1   *Solving Quadratic Equations by Graphing and by Using the Quadratic Formula*

Find the real solutions, if any, of the equation:

$$3x^2 + 2 = 4x$$

Solution   The equation, as given, is not in standard form.

$$3x^2 + 2 = 4x$$

$$3x^2 - 4x + 2 = 0 \quad \text{Put in standard form.}$$

$$ax^2 + bx + c = 0 \quad \text{Compare to standard form.}$$

With $a = 3$, $b = -4$, and $c = 2$, we find

$$b^2 - 4ac = 16 - 24 = -8$$

Since $b^2 - 4ac < 0$, the equation has no real solution.

Graphing Solution   We use the standard form of the equation and graph the function

$$f(x) = 3x^2 - 4x + 2$$

See Figure 18. We see that there are no x-intercepts, so the equation has no real solution, as expected.

FIGURE 18

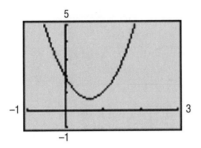

■   Now work Problem 83.

Sometimes a given equation can be transformed into a quadratic equation so that it can be solved using the quadratic formula.

E X A M P L E   1 2   *Solving Quadratic Equations by Graphing and by Using the Quadratic Formula*

Find the real solutions, if any, of the equation: $9 + \dfrac{3}{x} - \dfrac{2}{x^2} = 0,\ x \neq 0$

Graphing Solution   Figure 19 shows the graph of the function

$$f(x) = 9 + \frac{3}{x} - \frac{2}{x^2}$$

From the graph, we conjecture that there are two x-intercepts, one between -1 and 0, the other between 0 and 1. After using the BOX function, the solutions to the equation are -0.66 and 0.33 correct to two decimal places.

FIGURE 19

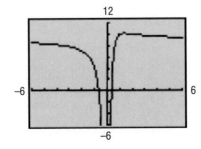

Comment: The graph of the preceding function on most graphing utilities will appear ragged near $x = 0$. This is caused because the function is not defined at $x = 0$. As a result, the graphing utility is unable to plot points when x gets close to zero. This problem can be easily overcome. Simply multiply both sides of the equation by x^2 and obtain $g(x) = 9x^2 + 3x - 2$.

The resulting equation will have the same x-intercepts as the original equation. We can use the BOX function to find the x-intercepts of $g(x)$ and these will be the same as the x-intercepts of $f(x)$. See Figure 20.

FIGURE 20

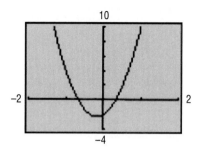

Algebraic Solution   In its present form, the equation

$$9 + \frac{3}{x} - \frac{2}{x^2} = 0$$

is not a quadratic equation. However, it can be transformed into one by multiplying each side by x^2. The result is

$$9x^2 + 3x - 2 = 0$$

Although we multiplied each side by x^2, we know that $x^2 \neq 0$ (do you see why?), so this quadratic equation is equivalent to the original equation.

Using $a = 9$, $b = 3$, and $c = -2$, the discriminant is

$$b^2 - 4ac = 9 + 72 = 81$$

Since $b^2 - 4ac > 0$, the new equation has two real solutions:

$$x = \frac{-b \pm \sqrt{b^2 - 4ac}}{2a} = \frac{-3 \pm \sqrt{81}}{18} = \frac{-3 \pm 9}{18}$$

$$x = \frac{-3 + 9}{18} = \frac{6}{18} = \frac{1}{3} \quad \text{or} \quad x = \frac{-3 - 9}{18} = \frac{-12}{18} = \frac{-2}{3}$$

The solution set is $\{-\frac{2}{3}, \frac{1}{3}\}$. These are the exact solutions. ■

Summary

Procedure for Solving
a Quadratic Equation
Algebraically

To solve a quadratic equation, first put it in standard form:

$$ax^2 + bx + c = 0$$

Then:

STEP 1:   Identify a, b, and c.
STEP 2:   Evaluate the discriminant, $b^2 - 4ac$.
STEP 3:   (a) If the discriminant is negative, the equation has no real solution.
          (b) If the discriminant is nonnegative, determine whether the left side can be factored. If you can easily spot factors, use the factoring method to solve the equation. Otherwise, use the quadratic formula.

HISTORICAL FEATURE   ■ The solution of equations is among the oldest of mathematical activities, and efforts to systematize this activity determined much of the shape of modern mathematics.

Consider the following problem, and its solution, using only words: Solve the problem of how many apples Jim has, given that

"Bob's five apples and Jim's apples together make twelve apples" by thinking,

"Jim's apples are all twelve apples less Bob's five apples" and then concluding,

"Jim has seven apples."

The mental steps translated into algebra are

$$5 + x = 12$$
$$x = 12 - 5$$
$$x = 7$$

The solution of this problem using only words is the earliest form of algebra. Such problems were solved exactly this way in Babylonia in 1800 BC. We know almost nothing of mathematical work before this date, although most authorities believe the sophistication of the earliest known texts indicates that a long period of previous development must have occurred. The method of writing out equations in words persisted for thousands of years, and although it now seems extremely cumbersome, it was used very effectively by many generations of mathematicians. The Arabs developed a good deal of the theory of cubic equations while writing out all the equations in words. About AD 1500, the tendency to abbreviate words in the written equations began to lead in the direction of modern notation; for example, the Latin word *et* (meaning *and*) developed into the plus sign, +. Although the occasional use of letters to represent variables dates back to AD 1200, the practice did not become common until about AD 1600. Development thereafter was rapid, and by 1635, algebraic notation did not differ essentially from what we use now.

Problems using quadratic equations are found in the oldest known mathematical literature. Babylonians and Egyptians were solving such problems before 1800 BC. Euclid solved quadratic equations geometrically in his *Data* (300 BC), and the Hindus and Arabs gave rules for solving any quadratic equation with real roots. Because negative numbers were not freely used before AD 1500, there were several different types of quadratic equations, each with its own rule. Thomas Harriot (1560–1621) introduced the method of factoring to obtain solutions, and François Viète (1540–1603) introduced a method that is essentially completing the square.

Until modern times it was usual to neglect the negative roots (if there were any), and equations involving square roots of negative quantities were regarded as unsolvable until the 1500's. ■

HISTORICAL PROBLEMS

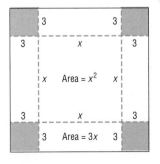

1. *One of al-Khowârizmî's solutions*   We solve $x^2 + 12x = 85$ by drawing the square shown. The area of the unshaded part is $x^2 + 12x$. We then set this expression equal to 85 to get the equation $x^2 + 12x = 85$. If we add the four shaded squares, we will have a larger square of known area. Complete the solution.

2. *Viète's method*   We solve $x^2 + 12x - 85 = 0$ by letting $x = u + z$. Then

$$(u + z)^2 + 12(u + z) - 85 = 0$$
$$u^2 + (2z + 12)u + (z^2 + 12z - 85) = 0$$

Now select z so that $2z + 12 = 0$ and finish the solution.

3. *Another method to get the quadratic formula*   Look at equation (4), page 186. Rewrite the right side as $(\sqrt{b^2 - 4ac}/2a)^2$ and then subtract it from each side. The right side is now 0 and the left side is a difference of two squares. If you factor this difference of two squares, you will easily be able to get the quadratic formula, and, moreover, the quadratic expression is factored, which is sometimes useful. ■

3.2

Exercise 3.2

In Problems 1–8, mentally solve each equation.

1. $3x = 21$
2. $3x = -24$
3. $5x + 15 = 0$
4. $3x + 18 = 0$
5. $2x - 3 = 5$
6. $3x + 4 = -8$
7. $\frac{1}{3}x = \frac{5}{12}$
8. $\frac{2}{3}x = \frac{9}{2}$

In Problems 9–66, solve each equation algebraically.

9. $3x + 2 = x + 6$
10. $2x + 7 = 3x + 5$
11. $2t - 6 = 3 - t$

12. $5y + 6 = -18 - y$
13. $6 - x = 2x + 9$
14. $3 - 2x = 2 - x$

15. $3 + 2n = 5n + 7$
16. $3 - 2m = 3m + 1$
17. $2(3 + 2x) = 3(x - 4)$

18. $3(2 - x) = 2x - 1$
19. $8x - (2x + 1) = 3x - 10$
20. $5 - (2x - 1) = 10$

21. $\frac{3}{2}x + 2 = \frac{1}{2} - \frac{1}{2}x$
22. $\frac{1}{3}x = 2 - \frac{2}{3}x$
23. $\frac{1}{2}x - 5 = \frac{3}{4}x$

24. $1 - \frac{1}{2}x = 6$
25. $\frac{2}{3}p = \frac{1}{2}p + \frac{1}{3}$
26. $\frac{1}{2} - \frac{1}{3}p = \frac{4}{3}$

27. $0.9t = 0.4 + 0.1t$
28. $0.9t = 1 + t$
29. $\dfrac{x + 1}{3} + \dfrac{x + 2}{7} = 5$

30. $\dfrac{2x + 1}{3} + 16 = 3x$
31. $\dfrac{2}{y} + \dfrac{4}{y} = 3$
32. $\dfrac{4}{y} - 5 = \dfrac{5}{2y}$

33. $\dfrac{1}{2} + \dfrac{2}{x} = \dfrac{3}{5}$
34. $\dfrac{3}{x} - \dfrac{1}{3} = \dfrac{1}{4}$
35. $x^2 = 9x$

36. $x^2 = -4x$
37. $x^2 - 25 = 0$
38. $x^2 - 9 = 0$

39. $z^2 + z - 12 = 0$
40. $v^2 + 7v + 12 = 0$
41. $2x^2 - 5x - 3 = 0$

42. $3x^2 + 5x + 2 = 0$
43. $3t^2 - 48 = 0$
44. $2y^2 - 50 = 0$

45. $x(x - 7) + 12 = 0$
46. $x(x + 1) = 12$
47. $4x^2 + 9 = 12x$

48. $25x^2 + 16 = 40x$
49. $6(p^2 - 1) = 5p$
50. $2(2u^2 - 4u) + 3 = 0$

51. $6x - 5 = \dfrac{6}{x}$
52. $x + \dfrac{12}{x} = 7$
53. $\dfrac{4(x - 2)}{x - 3} + \dfrac{3}{x} = \dfrac{-3}{x(x - 3)}$

54. $\dfrac{5}{x + 4} = 4 + \dfrac{3}{x - 2}$
55. $(x + 7)(x - 1) = (x + 1)^2$
56. $(x + 2)(x - 3) = (x - 3)^2$

57. $x(2x - 3) = (2x + 1)(x - 4)$
58. $x(1 + 2x) = (2x - 1)(x - 2)$
59. $z(z^2 + 1) = 3 + z^3$

60. $w(4 - w^2) = 8 - w^3$
61. $\dfrac{x}{x - 3} + 3 = \dfrac{3}{x - 3}$
62. $\dfrac{3x}{x + 2} = \dfrac{-6}{x + 2} - 2$

63. $x^2 = 4x$
64. $x^3 = x^2$
65. $t^3 - 9t^2 = 0$

66. $4z^3 - 8z^2 = 0$

In Problems 67–70, use a calculator to solve each equation. Express the solution correct to two decimal places.

67. $3.2x + \dfrac{21.3}{65.871} = 19.23$

68. $6.2x - \dfrac{19.1}{83.72} = 0.195$

69. $14.72 - 21.58x = \dfrac{18}{2.11}x + 2.4$

70. $18.63x - \dfrac{21.2}{2.6} = \dfrac{14x}{2.32} - 20$

In Problems 71–76, solve each equation. The letters a, b, and c are constants.

71. $ax - b = c, \quad a \neq 0$
72. $1 - ax = b, \quad a \neq 0$
73. $\dfrac{x}{a} + \dfrac{x}{b} = c, \quad a \neq 0, b \neq 0, a \neq -b$

74. $\dfrac{a}{x} + \dfrac{b}{x} = c, \quad c \neq 0$
75. $\dfrac{1}{x - a} + \dfrac{1}{x + a} = \dfrac{2}{x - 1}$
76. $\dfrac{b + c}{x + a} = \dfrac{b - c}{x - a}, \quad c \neq 0, a \neq 0$

In Problems 77–96, find the real solutions, if any, of each equation. Use the quadratic formula.

77. $x^2 - 4x + 2 = 0$

78. $x^2 + 4x + 2 = 0$

79. $x^2 - 4x - 1 = 0$

80. $x^2 + 6x + 1 = 0$

81. $2x^2 - 5x + 3 = 0$

82. $2x^2 + 5x + 3 = 0$

83. $4y^2 - y + 2 = 0$

84. $4t^2 + t + 1 = 0$

85. $4x^2 = 1 - 2x$

86. $2x^2 = 1 - 2x$

87. $4x^2 = 9x$

88. $5x = 4x^2$

89. $9t^2 - 6t + 1 = 0$

90. $4u^2 - 6u + 9 = 0$

91. $3x^2 - 2x - 2 = 0$

92. $2x^2 - 3x - 1 = 0$

93. $4 - \dfrac{1}{x} - \dfrac{2}{x^2} = 0$

94. $4 + \dfrac{1}{x} - \dfrac{1}{x^2} = 0$

95. $3x = 1 - \dfrac{1}{x}$

96. $x = 1 - \dfrac{4}{x}$

In Problems 97–104, use a graphing utility to approximate the real solutions, if any, of each equation. Then use the quadratic formula to obtain exact solutions. Compare the two results. Experiment further to see how close you can make the graphing utility solution to the exact solution.

97. $x^2 - 4x + 2 = 0$

98. $x^2 + 4x + 2 = 0$

99. $x^2 + \sqrt{3}x - 3 = 0$

100. $x^2 + \sqrt{2}x - 2 = 0$

101. $\pi x^2 - x - \pi = 0$

102. $\pi x^2 + \pi x - 2 = 0$

103. $3x^2 + 8\pi x + \sqrt{29} = 0$

104. $\pi x^2 - 15\sqrt{2}x + 20 = 0$

In Problems 105–116, use a graphing utility to approximate the real solutions, if any, of each equation. Then use any algebraic method that you wish to obtain exact solutions. Compare the two results. Experiment further to see how close you can make the graphing utility solution to the exact solution.

105. $x^2 - 7 = 0$

106. $x^2 - 8 = 0$

107. $16x^2 - 8x + 1 = 0$

108. $9x^2 - 6x + 1 = 0$

109. $10x^2 - 19x - 15 = 0$

110. $6x^2 + 7x - 20 = 0$

111. $2 + z = 6z^2$

112. $2 = y + 6y^2$

113. $x^2 + \sqrt{2}x = \frac{1}{2}$

114. $\frac{1}{2}x^2 = \sqrt{2}x + 1$

115. $x^2 + x = 4$

116. $x^2 + x = 1$

In Problems 117–122, use the discriminant to determine whether each quadratic equation has two unequal real solutions, a repeated real solution, or no real solution, without solving the equation. Use a graphing utility to verify your result.

117. $2x^2 - 6x + 7 = 0$

118. $x^2 + 4x + 7 = 0$

119. $9x^2 - 30x + 25 = 0$

120. $25x^2 - 20x + 4 = 0$

121. $3x^2 + 5x - 2 = 0$

122. $2x^2 - 3x - 4 = 0$

Problems 123–128 list some formulas that occur in applications. Solve each formula for the indicated variable.

123. *Electricity* $\dfrac{1}{R} = \dfrac{1}{R_1} + \dfrac{1}{R_2}$ for R

124. *Finance* $A = P(1 + rt)$ for r

125. *Mechanics* $F = \dfrac{mv^2}{R}$ for R

126. *Chemistry* $PV = nRT$ for T

127. *Mathematics* $S = \dfrac{a}{1 - r}$ for r

128. *Mechanics* $v = -gt + v_0$ for t

129. Explain what is wrong in the following steps:

$$x = 2 \tag{1}$$
$$3x - 2x = 2 \tag{2}$$
$$3x = 2x + 2 \tag{3}$$
$$x^2 + 3x = x^2 + 2x + 2 \tag{4}$$
$$x^2 + 3x - 10 = x^2 + 2x - 8 \tag{5}$$
$$(x - 2)(x + 5) = (x - 2)(x + 4) \tag{6}$$
$$x + 5 = x + 4 \tag{7}$$
$$1 = 0 \tag{8}$$

130. Which of the following pairs of equations are equivalent? Explain.

(a) $x^2 = 9; \; x = 3$ (b) $x = \sqrt{9}; \; x = 3$ (c) $(x - 1)(x - 2) = (x - 1)^2; \; x - 2 = x - 1$

131. The equation

$$\frac{5}{x + 3} + 3 = \frac{8 + x}{x + 3}$$

has no solution, yet when we go through the process of solving it we obtain $x = -3$. Write a brief paragraph to explain what causes this to happen.

132. Make up an equation that has no solution and give it to a fellow student to solve. Ask the fellow student to write a critique of your equation.

133. Describe three ways you might solve a quadratic equation. State your preferred method; explain why you chose it.

134. Explain the benefits of evaluating the discriminant of a quadratic equation before attempting to solve it.

135. Make up three quadratic equations: one having two distinct solutions, one having no real solution, and one having exactly one real solution.

136. The word *quadratic* seems to imply four (*quad*), yet a quadratic equation is an equation that involves a polynomial of degree 2. Investigate the origin of the term *quadratic* as it is used in the expression *quadratic equation*. Write a brief essay on your findings.

137. Write a program that will solve a quadratic equation:

```
{Enter the coefficient of x squared}      READ (a);
{Enter the coefficient of x}              READ (b);
{Enter the constant term}                 READ (c);
IF b² − 4ac < 0
THEN {write no real solution}
ELSE IF b² − 4ac = 0
      THEN {write −b/2a is a double root}
      ELSE {write (−b + SQRT(b² − 4ac))/2a
         or (−b − SQRT(b² − 4ac))/2a
         is a solution}
```

3.3

Setting Up Equations: Applications

The previous section provides the tools for solving equations. But, unfortunately, applied problems do not come in the form, "Solve the equation. . . ." Instead, they are narratives that supply information—hopefully, enough to answer the question that inevitably arises. Thus, to solve applied problems we must be able to translate the verbal description into the language of mathematics. We do this by using symbols (usually letters of the alphabet) to represent unknown quantities and then finding relationships (such as equations) that involve these symbols. The process of doing this is called **mathematical modeling.**

Any solution to the mathematical problem must be checked against the mathematical problem, the verbal description, and the real problem. See Figure 21 for an illustration of the modeling process.

FIGURE 21

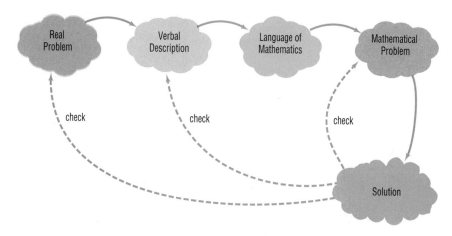

Let's look at a few examples that will help you to translate certain words into mathematical symbols.

E X A M P L E   1   *Translating Verbal Descriptions into Mathematical Expressions*

(a) The area of a rectangle is the product of its length times its width.

Translation: If A is used to represent the area, l the length, and w the width, then $A = lw$.

(b) For uniform motion, the velocity of an object equals the distance traveled divided by the time required.

Translation: If v is the velocity, s the distance, and t the time, then $v = s/t$.

(c) A total of \$5000 is invested, some in stocks and some in bonds. If the amount invested in stocks is x, express the amount invested in bonds in terms of x.

Translation: If y is the amount invested in bonds, then $x + y = 5000$. Thus, if x is the amount invested in stocks, then the amount invested in bonds is $y = 5000 - x$.

(d) Let x denote a number.

The number 5 times as large as x is $5x$.

The number 3 less than x is $x - 3$.

The number that exceeds x by 4 is $x + 4$.

The number that, when added to x, gives 5 is $5 - x$.   ∎

■ Now work Problem 1.

Mathematical equations that represent real situations should be consistent in terms of the units used. In Example 1(a), if l is measured in feet, then w also must be expressed in feet, and A will be expressed in square feet. In Example 1(b), if v is measured in miles per hour, then the distance s must be expressed in miles and the time t must be expressed in hours. It is a good practice to check units to be sure that they are consistent and make sense.

Although each situation has its own unique features, we can provide an outline of the steps to follow in setting up applied problems.

Steps for Setting Up Applied Problems

STEP 1: Read the problem carefully, perhaps two or three times. Pay particular attention to the question being asked in order to identify what you are looking for. If you can, determine realistic possibilities for the answer.

STEP 2: Assign a letter (variable) to represent what you are looking for, and, if necessary, express any remaining unknown quantities in terms of this variable.

STEP 3: Make a list of all the known facts, and write down any relationships among them, especially any that involve the variable. These may take the form of an equation (or, later, an inequality) involving the variable. If possible, draw an appropriately labeled diagram to assist you. Sometimes, a table or chart helps.

STEP 4: Solve the equation for the variable, and then answer the question asked in the problem.

STEP 5: Check the answer with the facts in the problem. If it agrees, congratulations! If it does not agree, try again.

Let's look at an example.

E X A M P L E   2

Determining an Hourly Wage

Colleen grossed $435 one week by working 52 hours. Her employer pays time-and-a-half for all hours worked in excess of 40 hours. With this information, can you determine Colleen's regular hourly wage?

Solution   **STEP 1:** We are looking for an hourly wage. Our answer will be in dollars per hour.

STEP 2: Let x represent the regular hourly wage; x is measured in dollars per hour.

STEP 3: We set up a table:

	HOURS WORKED	HOURLY WAGE	SALARY
Regular	40	x	$40x$
Overtime	12	$1.5x$	$12(1.5x) = 18x$

The sum of regular salary plus overtime salary will equal $435. Thus, from the table, $40x + 18x = 435$.

STEP 4:
$$40x + 18x = 435$$
$$58x = 435$$
$$x = 7.50$$

Thus, Colleen's regular hourly wage is $7.50 per hour.

STEP 5: Forty hours yields a salary of $40(7.50) = \$300$, and 12 hours of overtime yields a salary of $12(1.5)(7.50) = \$135$, for a total of $435. ∎

■   Now work Problem 15.

Interest

The next example involves **interest.** Interest is money paid for the use of money. The total amount borrowed (whether by an individual from a bank in the form of a loan or by a bank from an individual in the form of a savings account) is called the **principal.** The **rate of interest,** expressed as a percent, is the amount charged for the use of the principal for a given period of time, usually on a yearly (that is, per annum) basis.

If a principal of P dollars is borrowed for a period of t years at a per annum interest rate r, expressed as a decimal, the interest I charged is

Simple Interest Formula

$$I = Prt \qquad (1)$$

Interest charged according to formula (1) is called **simple interest.**

E X A M P L E   3

Financial Planning

An investor with $70,000 decides to place part of her money in corporate bonds paying 12% per year and the rest in a Certificate of Deposit paying 8% per year. If she wishes to obtain an overall return of 9% per year, how much should she place in each investment?

Solution

STEP 1:   The question is asking for two dollar amounts: the principal to invest in the corporate bonds and the principal to invest in the Certificate of Deposit.
STEP 2:   We let x represent the amount (in dollars) to be invested in the bonds. Then $70,000 - x$ is the amount that will be invested in the certificate. (Do you see why?)
STEP 3:   We set up a table:

	PRINCIPAL $	RATE	TIME yr	INTEREST $
Bonds	x	12% = 0.12	1	0.12x
Certificate	70,000 − x	8% = 0.08	1	0.08(70,000 − x)
Total	70,000	9% = 0.09	1	0.09(70,000) = 6300

Since the total interest from the investments is to equal 0.09(70,000) = 6300, we must have the equation

$$0.12x + 0.08(70,000 - x) = 6300$$

(Note that the units are consistent: the unit is dollars on each side.)

STEP 4:
$$
\begin{aligned}
0.12x + 5600 - 0.08x &= 6300 \\
0.04x &= 700 \\
x &= 17{,}500
\end{aligned}
$$

Thus, the investor should place $17,500 in the bonds and $70,000 − $17,500 = $52,500 in the certificate.
STEP 5:   The interest on the bonds after 1 year is 0.12($17,500) = $2100; the interest on the certificate after 1 year is 0.08($52,500) = $4200. The total annual interest is $6300, the required amount.   ■

■   Now work Problem 23.

Mixture Problems

The next example is a type usually referred to as a **mixture problem.**

E X A M P L E   4

Chemistry: Mixing Acids

In a chemistry laboratory the concentration of one solution is 10% hydrochloric acid (HCl) and that of a second solution is 60% HCl. How many milliliters (mL) of each should be mixed to obtain 50 mL of a 30% HCl solution?

Solution Let x represent the number of milliliters of the 10% solution. Then $50 - x$ equals the number of milliliters of the 60% solution. See Figure 22.

FIGURE 22

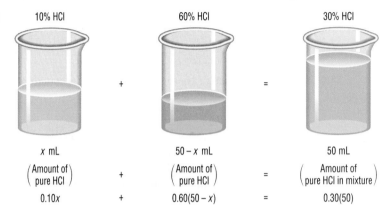

10% HCl   60% HCl   30% HCl

x mL $50 - x$ mL   50 mL

$\begin{pmatrix} \text{Amount of} \\ \text{pure HCl} \end{pmatrix}$ + $\begin{pmatrix} \text{Amount of} \\ \text{pure HCl} \end{pmatrix}$ = $\begin{pmatrix} \text{Amount of} \\ \text{pure HCl in mixture} \end{pmatrix}$

$0.10x$ + $0.60(50 - x)$ = $0.30(50)$

Based on the figure, we form a table:

	AMOUNT mL	CONCENTRATION OF HCl	AMOUNT OF PURE ACID mL
10% HCl	x	10% = 0.10	$0.10x$
60% HCl	$50 - x$	60% = 0.60	$0.60(50 - x)$
30% HCl	50	30% = 0.30	$0.30(50) = 15$

The amount of HCl in the 30% solution (15 milliliters) must equal the sum of the amounts of HCl found in the 10% solution and the 60% solution. Thus, we have the equation

$$0.10x + 0.60(50 - x) = 15$$
$$0.10x + 30 - 0.60x = 15$$
$$-0.50x = -15$$
$$x = 30 \text{ milliliters}$$

Thus, 30 milliliters of the 10% acid solution, when mixed with 20 milliliters of the 60% acid solution, yields 50 milliliters of a 30% acid solution.

Check: To check this answer, we note that there are $0.10(30) = 3$ milliliters of acid in the 10% solution and $0.60(20) = 12$ milliliters of acid in the 60% solution. The 50 milliliter mixture therefore contains 15 milliliters of acid, for an acid concentration of $\frac{15}{50} = 0.30 = 30\%$ acid solution. ∎

■ Now work Problem 58.

Uniform Motion

The next example deals with moving objects.

If an object moves at an average velocity v, the distance s covered in time t is given by the formula

$$s = vt \qquad (2)$$

That is, Distance = Velocity · Time. Objects that are moving in accordance with formula (2) are said to be in **uniform motion.**

EXAMPLE 5

Physics: Uniform Motion

A friend of yours, who is a long-distance runner, runs at an average velocity of 8 miles per hour. Two hours after your friend leaves your house, you leave in your car and follow the same route as your friend. If your average velocity is 40 miles per hour, how long will it be before you reach your friend? How far will each of you be from your house?

Solution

Refer to Figure 23. We use t to represent the time (in hours) that it takes the car to catch up with the runner. When this occurs, the total time elapsed for the runner is $t + 2$ hours.

FIGURE 23

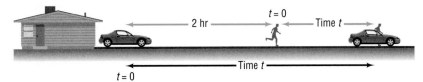

Set up the following table:

	VELOCITY mi/hr	TIME hr	DISTANCE mi
Runner	8	$t + 2$	$8(t + 2)$
Car	40	t	$40t$

Since the distance traveled is the same, we are led to the following equation:

$$8(t + 2) = 40t$$
$$8t + 16 = 40t$$
$$32t = 16$$
$$t = \frac{1}{2} \text{ hour}$$

It will take you $\frac{1}{2}$ hour to reach your friend. Each of you will have gone 20 miles.

Check: In 2.5 hours, the runner travels a distance of $(2.5)(8) = 20$ miles. In $\frac{1}{2}$ hour, the car travels a distance of $\left(\frac{1}{2}\right)(40) = 20$ miles.   ∎

■   Now work Problem 51.

EXAMPLE 6

Physics: Uniform Motion

A motorboat heads upstream a distance of 24 miles on a river whose current is running at 3 miles per hour. The trip up and back takes 6 hours. Assuming that the motorboat maintained a constant speed relative to the water, what was its speed?

Solution

See Figure 24. We use v to represent the constant speed of the motorboat relative to the water. Then the true speed going upstream is $v - 3$ miles per hour, and the true speed going downstream is $v + 3$ miles per hour. Since Distance = Velocity × Time, then Time = Distance/Velocity. We set up a table.

FIGURE 24

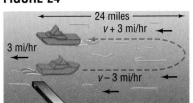

	VELOCITY mi/hr	DISTANCE mi	TIME = DISTANCE/VELOCITY hr
Upstream	$v - 3$	24	$\dfrac{24}{v - 3}$
Downstream	$v + 3$	24	$\dfrac{24}{v + 3}$

Since the total time up and back is 6 hours, we have

$$\frac{24}{v-3} + \frac{24}{v+3} = 6$$

$$\frac{24(v+3) + 24(v-3)}{(v-3)(v+3)} = 6$$

$$\frac{48v}{v^2 - 9} = 6$$

$$48v = 6(v^2 - 9)$$

$$6v^2 - 48v - 54 = 0$$

$$v^2 - 8v - 9 = 0$$

$$(v-9)(v+1) = 0$$

$$v = 9 \quad \text{or} \quad v = -1$$

We discard the solution $v = -1$ mile per hour, so the speed of the motorboat relative to the water is 9 miles per hour. ∎

Other Applied Problems

The next two examples illustrate problems that you will probably see again in a slightly different form if you study calculus.

E X A M P L E   7   *Preview of a Calculus Problem*

From each corner of a square piece of sheet metal, remove a square of side 9 centimeters. Turn up the edges to form an open box. If the box is to hold 144 cubic centimeters, what should be the dimensions of the piece of sheet metal?

Solution   We use Figure 25 as a guide. We have labeled by x the length of a side of the square piece of sheet metal. The box will be of height 9 centimeters and its square base will have $x - 18$ as the length of a side. The volume (Length × Width × Height) of the box is therefore

$$9(x - 18)(x - 18) = 9(x - 18)^2$$

FIGURE 25

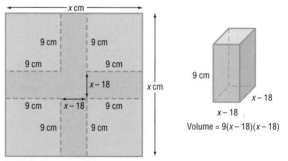

Since the volume of the box is to be 144 cubic centimeters, we have

$$9(x - 18)^2 = 144$$

$$(x - 18)^2 = 16$$

$$x - 18 = \pm 4$$

$$x = 18 \pm 4$$

$$x = 22 \quad \text{or} \quad x = 14$$

We discard the solution $x = 14$ (do you see why?) and conclude that the sheet metal should be 22 centimeters by 22 centimeters.

Check: If we begin with a piece of sheet metal 22 centimeters by 22 centimeters, cut out a 9 centimeter square from each corner, and fold up the edges, we get a box whose dimensions are 9 by 4 by 4, with volume $9 \times 4 \times 4 = 144$ cubic centimeters, as required. ■

■ Now work Problem 31.

E X A M P L E 8   *Preview of a Calculus Problem*

A piece of wire 8 feet in length is to be cut into two pieces. Each piece will then be bent into a square. Where should the cut in the wire be made if the sum of the areas of these squares is to be 2 square feet?

Solution   We use Figure 26 as a guide. We have labeled by x the length of one of the pieces of wire after it has been cut. The remaining piece will be of length $8 - x$. If each length is bent into a square, then one of the squares has a side of length $x/4$ and the other a side of length $(8 - x)/4$. Since the sum of the areas of these two squares is 2, we have the equation

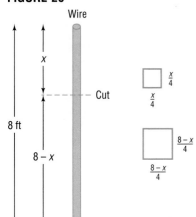

FIGURE 26

$$\left(\frac{x}{4}\right)^2 + \left(\frac{8 - x}{4}\right)^2 = 2$$

$$\frac{x^2}{16} + \frac{64 - 16x + x^2}{16} = 2$$

$$2x^2 - 16x + 64 = 32$$

$$2x^2 - 16x + 32 = 0 \quad \text{Put in standard form.}$$

$$x^2 - 8x + 16 = 0 \quad \text{Divide by 2.}$$

$$(x - 4)^2 = 0 \quad \text{Factor}$$

$$x = 4$$

Since $x = 4$, $8 - x = 4$, and the original piece of wire should be cut into two pieces, each of length 4 feet.

Check: If the length of each piece of wire is 4 feet, then each piece can be formed into a square whose side is 1 foot. The area of each square is then 1 square foot, so the sum of the areas is 2 square feet, as required. ■

3.3

Exercise 3.3

In Problems 1–10, translate each sentence into a mathematical equation. Be sure to identify the meaning of all symbols.

1. *Geometry* The area of a circle is the product of the number π times the square of the radius.
2. *Geometry* The circumference of a circle is the product of the number π times twice the radius.
3. *Geometry* The area of a square is the square of the length of a side.
4. *Geometry* The perimeter of a square is four times the length of a side.
5. *Physics* Force equals the product of mass times acceleration.

6. *Physics* Pressure is force per unit area.

7. *Physics* Work equals force times distance.

8. *Physics* Kinetic energy is one-half the product of the mass times the square of the velocity.

9. *Business* The total variable cost of manufacturing x dishwashers is $150 per dishwasher times the number of dishwashers manufactured.

10. *Business* The total revenue derived from selling x dishwashers is $250 per dishwasher times the number of dishwashers sold.

11. *Finance* A total of $20,000 is to be invested, some in bonds and some in Certificates of Deposit (CD's). If the amount invested in bonds is to exceed that in CD's by $2000, how much will be invested in each type of instrument?

12. *Finance* A total of $10,000 is to be divided up between Katy and Mike, with Mike to receive $2000 less than Katy. How much will each receive?

13. *Finance* An inheritance of $900,000 is to be divided among Katy, Mike, and Dan in the following manner: Mike is to receive $\frac{3}{4}$ of what Katy gets, while Dan gets $\frac{1}{2}$ of what Katy gets. How much does each receive?

14. *Sharing the Cost of a Pizza* Mike and Colleen agree to share the cost of an $18 pizza based on how much each ate. If Colleen ate $\frac{2}{3}$ the amount Mike ate, how much should each pay?

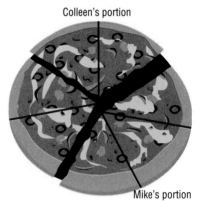
Colleen's portion

Mike's portion

15. *Computing Hourly Wages* A worker who is paid time-and-a-half for hours worked in excess of 40 hours had gross weekly wages of $442 for 48 hours worked. What is the regular hourly rate?

16. *Computing Hourly Wages* Colleen is paid time-and-a-half for hours worked in excess of 40 hours and double-time for hours worked on Sunday. If Colleen had gross weekly wages of $342 for working 50 hours, 4 of which were on Sunday, what is her regular hourly rate?

17. *Football* In an NFL football game, one team scored a total of 41 points, including one safety (2 points) and two field goals (3 points each). After scoring a touchdown (6 points), a team is given the chance to score 1 extra point. The team missed 2 extra points after scoring touchdowns. How many touchdowns did they get?

18. *Basketball* In a basketball game, one team scored a total of 70 points and made three times as many field goals (2 points each) as free throws (1 point each). How many field goals did they have?

19. *Geometry* The perimeter of a rectangle is 60 feet. Find its length and width if the length is 8 feet longer than the width.

20. *Geometry* The perimeter of a rectangle is 42 meters. Find its length and width if the length is twice the width.

21. *Enclosing a Garden* A gardener has 46 feet of fencing to be used to enclose a rectangular garden that has a border 2 feet wide surrounding it (see the figure).
    (a) If the length of the garden is to be twice its width, what will be the dimensions of the garden?
    (b) What is the area of the garden?
    (c) If the length and width of the garden were to be the same, what would be the dimensions of the garden?
    (d) What would be the area of the square garden?

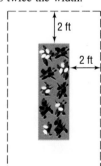
2 ft

2 ft

22. *Chemistry: Sugar Molecules* A sugar molecule has twice as many atoms of hydrogen as it does oxygen and one more atom of carbon than oxygen. If a sugar molecule has a total of 45 atoms, how many are oxygen? How many are hydrogen?

23. *Financial Planning* A recent retiree requires $6000 per year in extra income. She has $50,000 to invest and can invest in B-rated bonds paying 15% per year or in a Certificate of Deposit (CD) paying 7% per year. How much money should be invested in each to realize exactly $6000 in interest per year?

24. *Financial Planning* After 2 years, the retiree referred to in Problem 23 finds she now will require $7000 per year. Assuming that the remaining information is the same, how should the money be reinvested?

25. *Banking* A bank loaned out $12,000, part of it at the rate of 8% per year and the rest at the rate of 18% per year. If the interest received totaled $1000, how much was loaned at 8%?

26. *Banking* A loan officer at a bank has $1,000,000 to lend and is required to obtain an average return of 18% per year. If she can lend at the rate of 19% or the rate of 16%, how much can she lend at the 16% rate and still meet her requirement?

27. *Dimensions of a Window* The area of the opening of a rectangular window is to be 143 square feet. If the length is to be 2 feet more than the width, what are the dimensions?

28. *Dimensions of a Window* The area of a rectangular opening is to be 306 square centimeters. If the length exceeds the width by 1 centimeter, what are the dimensions?

29. *Geometry* Find the dimensions of a rectangle whose perimeter is 26 meters and whose area is 40 square meters.

30. *Watering a Field* An adjustable water sprinkler that sprays water in a circular pattern is placed at the center of a square field whose area is 1250 square feet (see the figure). What is the shortest radius setting that can be used if the field is to be completely enclosed within the circle?

31. *Constructing a Box* An open box is to be constructed from a square piece of sheet metal by removing a square of side 1 foot from each corner and turning up the edges. If the box is to hold 4 cubic feet, what should be the dimensions of the sheet metal?

32. *Constructing a Box* Rework Problem 31 if the piece of sheet metal is a rectangle whose length is twice its width.

33. *Physics* A ball is thrown vertically upward from the top of a building 96 feet tall with an initial velocity of 80 feet per second. The distance s (in feet) of the ball from the ground after t seconds is $s = 96 + 80t - 16t^2$.
   (a) After how many seconds does the ball strike the ground?
   (b) After how many seconds will the ball pass the top of the building on its way down?

34. *Constructing a Coffee Can* A 39 ounce can of Hills Bros.® coffee requires 188.5 square inches of aluminum. If its height is 7 inches, what is its radius? (The surface area A of a right circular cylinder is $A = 2\pi r^2 + 2\pi rh$, where r is the radius and h is the height.)

35. *Business: Discount Pricing* A builder of tract homes reduced the price of a model by 15%. If the new price is $125,000, what was its original price? How much can be saved by purchasing the model?

36. *Business: Discount Pricing* A car dealer, at a year-end clearance, reduces the list price of last year's models by 15%. If a certain four-door model has a discounted price of $8000, what was its list price? How much can be saved by purchasing last year's model?

37. *Business: Marking Up the Price of Books* A college book store marks up the price it pays the publisher for a book by 25%. If the selling price of a book is $56.00, how much did the book store pay for the book?

38. *Personal Finance: Cost of a Car* The suggested list price of a new car is $12,000. The dealer's cost is 85% of list. How much will you pay if the dealer is willing to accept $100 over cost for the car?

39. *Working Together on a Job* Mike can deliver his newspapers in 30 minutes. It takes Danny 20 minutes to do the same route. How long would it take them to deliver the newspapers if they work together?

40. *Working Together on a Job* A painter by himself can paint four rooms in 10 hours. If he hires a helper, they can do the same job together in 6 hours. If he lets the helper work alone, how long will it take for the helper to paint four rooms?

41. *Computing Grades* Going into the final exam, which will count as two tests, Colleen has test scores of 80, 83, 71, 61, and 95. What score does Colleen need on the final in order to have an average score of 80?

42. *Computing Grades* Going into the final exam, which will count as two-thirds of the final grade, Dan has test scores of 86, 80, 84, and 90. What score does Dan need on the final in order to earn a B, which requires an average score of 80? What does he need to earn an A, which requires an average of 90?

43. *Football* A tight end can run the 100 yard dash in 12 seconds. A defensive back can do it in 10 seconds. The tight end catches a pass at his own 20 yard line with the defensive back at the 15 yard line. (See the figure.) If no other players are nearby, at what yard line will the defensive back catch up to the tight end?

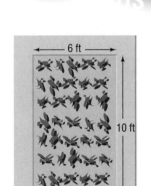

44. *Computing Business Expenses* Debbie, an outside saleswoman, uses her car for both business and pleasure. Last year, she traveled 30,000 miles, using 900 gallons of gasoline. Her car gets 40 miles per gallon on the highway and 25 in the city. She can deduct all highway travel, but no city travel, on her taxes. How many miles should Debbie be allowed as a business expense?

45. *Constructing a Border Around a Garden* A landscaper, who just completed a rectangular flower garden measuring 6 feet by 10 feet, orders 1 cubic yard of premixed cement, all of which is to be used to create a border of uniform width around the garden. If the border is to have a depth of 3 inches, how wide will the border be? (1 cubic yard = 27 cubic feet)

46. *Physics* An object is propelled vertically upward with an initial velocity of 20 meters per second. The distance s (in meters) of the object from the ground after t seconds is $s = -4.9t^2 + 20t$.
   (a) When will the object be 15 meters above the ground?
   (b) When will it strike the ground?
   (c) Will the object reach a height of 100 meters?
   (d) What is the maximum height?

47. *Reducing the Size of a Candy Bar* A jumbo chocolate bar with a rectangular shape measures 12 centimeters in length, 7 centimeters in width, and 3 centimeters in thickness. Due to escalating costs of cocoa, management decides to reduce the volume of the bar by 10%. To accomplish this reduction, management decides the new bar should have the same 3 centimeter thickness, but the length and width each should be reduced an equal number of centimeters. What should be the dimensions of the new candy bar?

48. *Reducing the Size of a Candy Bar* Rework Problem 47 if the reduction is to be 20%.

49. *Constructing a Border Around a Pool* A pool in the shape of a circle measures 10 feet across. One cubic yard of concrete is to be used to create a circular border of uniform width around the pool. If the border is to have a depth of 3 inches, how wide will the border be? (1 cubic yard = 27 cubic feet)

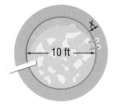

50. *Constructing a Border Around a Pool* Rework Problem 49 if the depth of the border is 4 inches.

51. *Physics: Uniform Motion* A motorboat can maintain a constant speed of 16 miles per hour relative to the water. The boat makes a trip upstream to a certain point in 20 minutes; the return trip takes 15 minutes. What is the speed of the current? (See the figure.)

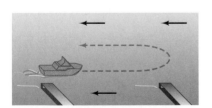

52. *Purity of Gold* The purity of gold is measured in karats, with pure gold being 24 karats. Other purities of gold are expressed as proportional parts of pure gold. Thus, 18 karat gold is $\frac{18}{24}$, or 75% pure gold; 12 karat gold is $\frac{12}{24}$, or 50%, pure gold; and so on. How much 12 karat gold should be mixed with pure gold to obtain 60 grams of 16 karat gold?

53. *Running a Race* Mike can run the mile in 6 minutes, and Dan can run the mile in 9 minutes. If Mike gives Dan a head start of 1 minute, how far from the start will Mike pass Dan? (See the figure.) How long does it take?

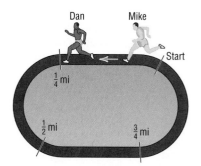

54. *Physics: Uniform Motion* A motorboat heads upstream on a river that has a current of 3 miles per hour. The trip upstream takes 5 hours, while the return trip takes 2.5 hours. What is the speed of the motorboat? (Assume that the motorboat maintains a constant speed relative to the water.)

55. *Rescue at Sea* A ship that is in danger of sinking radios the Coast Guard for assistance. When the rescue craft leaves the Coast Guard station, the ship is 60 miles away and heading directly toward the station. If the average speed of the ship is 10 miles per hour and the average speed of the rescue craft is 20 miles per hour, how long will it take for the rescue craft to reach the ship? (See the figure).

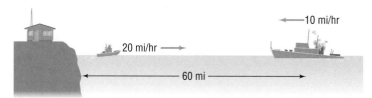

56. *Physics: Uniform Motion* Two cars enter the Florida Turnpike at Commercial Boulevard at 8:00 AM, each heading for Wildwood. One car's average speed is 10 miles per hour more than the other's. The faster car arrives at Wildwood at 11:00 AM, $\frac{1}{2}$ hour before the other car. What is the average speed of each car? How far did each travel?

57. *Emptying Oil Tankers* An oil tanker can be emptied by the main pump in 4 hours. An auxiliary pump can empty the tanker in 9 hours. If the main pump is started at 9 AM, when should the auxiliary pump be started so that the tanker is emptied by noon?

58. *Cement Mix* A 20 pound bag of Economy brand cement mix contains 25% cement and 75% sand. How much pure cement must be added to produce a cement mix that is 40% cement?

59. *Emptying a Tub* A bathroom tub will fill in 15 minutes with both faucets open and the stopper in place. With both faucets closed and the stopper removed, the tub will empty in 20 minutes. How long will it take for the tub to fill if both faucets are open and the stopper is removed?

60. *Range of an Airplane* An air rescue plane averages 300 miles per hour in still air. It carries enough fuel for 5 hours of flying time. If, upon takeoff, it encounters a wind of 30 miles per hour, how far can it fly and return safely? (Assume that the wind remains constant.)

61. *Home Equity Loans* Suppose you obtain a home equity loan of $100,000 that requires only a monthly interest payment at 10% per annum, with the principal due after 5 years. You decide to invest part of the loan in a 5 year CD that pays 9% compounded and paid monthly and part in a B+ rated bond due in 5 years that pays 12% compounded and paid monthly. What is the most you can invest in the CD to ensure that the monthly home equity loan payment is made?

62. *Comparing Olympic Heroes* In the 1984 Olympics, Carl Lewis of the United States won the gold medal in the 100 meter race with a time of 9.99 seconds. In the 1896 Olympics, Thomas Burke, also of the United States, won the gold medal in the 100 meter race in 12.0 seconds. If they ran in the same race repeating their respective times, by how many meters would Lewis beat Burke?

63. *Computing Average Speed* In going from Chicago to Atlanta, a car averages 45 miles per hour, and in going from Atlanta to Miami, it averages 55 miles per hour. If Atlanta is halfway between Chicago and Miami, what is the average speed from Chicago to Miami? Discuss an intuitive solution. Write a paragraph defending your intuitive solution. Then solve the problem algebraically. Is your intuitive solution the same as the algebraic one? If not, find the flaw.

64. *Speed of a Plane* On a recent flight from Phoenix to Kansas City, a distance of 919 nautical miles, the plane arrived 20 minutes early. On leaving the aircraft, I asked the captain, "What was our tail wind?" He replied, "I don't know, but our ground speed was 550 knots." How can you determine if enough information is provided to find the tail wind? If possible, find the tail wind. (1 knot = 1 nautical mile per hour)

65. *Critical Thinking* You are the manager of a clothing store and have just purchased 100 dress shirts for $20.00 each. After 1 month of selling the shirts at the regular price, you plan to have a sale giving 40% off the original selling price. However, you still want to make a profit of $4 on each shirt at the sale price. What should you price the shirts at initially to ensure this? If, instead of 40% off at the sale, you give 50% off, by how much is your profit reduced?

66. *Critical Thinking* Make up a word problem that requires solving a linear equation as part of its solution. Exchange problems with a friend. Write a critique of your friend's problem.

67. *Critical Thinking* Without solving, explain what is wrong with the following mixture problem: How many liters of 25% ethanol should be added to 20 liters of 48% ethanol to obtain a solution of 58% ethanol? Now go through an algebraic solution. What happens?

3.4

Complex Numbers; Quadratic Equations with a Negative Discriminant

One property of a real number is that its square is nonnegative. For example, there is no real number x for which

$$x^2 = -1$$

To remedy this situation, we introduce a number called the **imaginary unit,** which we denote by i and whose square is -1. Thus,

$$i^2 = -1$$

This should not surprise you. If our universe were to consist only of integers, there would be no number x for which $2x = 1$. This unfortunate circumstance was remedied by introducing numbers such as $\frac{1}{2}$ and $\frac{2}{3}$, the *rational numbers.* If our universe were to consist only of rational numbers, there would be no number x whose square equals 2. That is, there would be no number x for which $x^2 = 2$. To remedy this, we introduced numbers such as $\sqrt{2}$ and $\sqrt[3]{5}$, the *irrational numbers.* The *real numbers,* you will recall, consist of the rational numbers and the irrational numbers. Now, if our universe were to consist only of real numbers, then there would be no number x whose square is -1. To remedy this, we introduce a number i, whose square is -1.

In the progression outlined, each time we encountered a situation that was unsuitable, we introduced a new number system to remedy this situation. And each new number system contained the earlier number system as a subset. The number system that results from introducing the number i is called the **complex number system.**

Complex Numbers

Complex numbers are numbers of the form $a + bi$, where a and b are real numbers. The real number a is called the **real part** of the number $a + bi$; the real number b is called the **imaginary part** of $a + bi$.

For example, the complex number $-5 + 6i$ has the real part -5 and the imaginary part 6.

When a complex number is written in the form $a + bi$, where a and b are real numbers, we say it is in **standard form.** However, if the imaginary part of a complex number is negative, such as in the complex number $3 + (-2)i$, we agree to write it instead in the form $3 - 2i$.

Also, the complex number $a + 0i$ is usually written merely as a. This serves to remind us that the real numbers are a subset of the complex numbers. The complex number $0 + bi$ is usually written as bi. Sometimes the complex number bi is called a **pure imaginary number.**

Equality, addition, subtraction, and multiplication of complex numbers are defined so as to preserve the familiar rules of algebra for real numbers. Thus, two complex numbers are equal if and only if their real parts are equal and their imaginary parts are equal. That is,

Equality of Complex Numbers

$$a + bi = c + di \quad \text{if and only if} \quad a = c \text{ and } b = d \qquad (1)$$

Two complex numbers are added by forming the complex number whose real part is the sum of the real parts and whose imaginary part is the sum of the imaginary parts. That is,

Sum of Complex Numbers

$$(a + bi) + (c + di) = (a + c) + (b + d)i \qquad (2)$$

To subtract two complex numbers, we follow the rule

Difference of Complex Numbers

$$(a + bi) - (c + di) = (a - c) + (b - d)i \qquad (3)$$

E X A M P L E   1   *Adding and Subtracting Complex Numbers*

(a) $(3 + 5i) + (-2 + 3i) = [3 + (-2)] + (5 + 3)i = 1 + 8i$

(b) $(6 + 4i) - (3 + 6i) = (6 - 3) + (4 - 6)i = 3 + (-2)i = 3 - 2i$ ∎

∎   Now work Problem 5.

Products of complex numbers are calculated as illustrated in Example 2.

E X A M P L E   2   *Multiplying Complex Numbers*

$$(5 + 3i) \cdot (2 + 7i) = 5 \cdot (2 + 7i) + 3i(2 + 7i) = 10 + 35i + 6i + 21i^2$$

   ↑ Distributive property   ↑ Distributive property

$$= 10 + 41i + 21(-1)$$

   ↑ $i^2 = -1$

$$= -11 + 41i$$ ∎

Based on the procedure of Example 2, we define the **product** of two complex numbers by the formula

Product of Complex Numbers

$$(a + bi) \cdot (c + di) = (ac - bd) + (ad + bc)i \qquad (4)$$

Do not bother to memorize formula (4). Instead, whenever it is necessary to multiply two complex numbers, follow the usual rules for multiplying two binomials, as in Example 2, remembering that $i^2 = -1$. For example,

$$(2i)(2i) = 4i^2 = -4$$

$$(2 + i)(1 - i) = 2 - 2i + i - i^2 = 3 - i$$

■   Now work Problem 11.

Algebraic properties for addition and multiplication, such as the commutative, associative, and distributive properties, hold for complex numbers. Of these, the property that every nonzero complex number has a multiplicative inverse, or reciprocal, requires a closer look.

Conjugates

Conjugate

If $z = a + bi$ is a complex number, then its **conjugate,** denoted by $\bar{z}$, is defined as

$$\bar{z} = \overline{a + bi} = a - bi$$

For example, $\overline{2 + 3i} = 2 - 3i$ and $\overline{-6 - 2i} = -6 + 2i$.

E X A M P L E   3   *Multiplying a Complex Number by Its Conjugate*

Find the product of the complex number $z = 3 + 4i$ and its conjugate $\bar{z}$.

Solution   Since $\bar{z} = 3 - 4i$, we have

$$z\bar{z} = (3 + 4i)(3 - 4i) = 9 + 12i - 12i - 16i^2 = 9 + 16 = 25 \qquad ■$$

The result obtained in Example 3 has an important generalization:

Theorem   The product of a complex number and its conjugate is a nonnegative real number. Thus, if $z = a + bi$, then

$$z\bar{z} = a^2 + b^2 \qquad (5)$$

Proof: If $z = a + bi$, then

$$z\bar{z} = (a + bi)(a - bi) = a^2 - (bi)^2 = a^2 - b^2i^2 = a^2 + b^2 \qquad ■$$

To express the reciprocal of a nonzero complex number z in standard form, multiply the numerator and denominator by its conjugate $\bar{z}$. Thus, if $z = a + bi$ is a nonzero complex number, then

$$\frac{1}{a+bi} = \frac{1}{z} = \frac{1}{z} \cdot \frac{\bar{z}}{\bar{z}} = \frac{\bar{z}}{z\bar{z}} = \frac{a-bi}{(a+bi)(a-bi)}$$

$$= \frac{a-bi}{a^2+b^2}$$

$\uparrow$
Use (5).

$$= \frac{a}{a^2+b^2} - \frac{b}{a^2+b^2}i$$

E X A M P L E   4   *Writing the Reciprocal of a Complex Number in Standard Form*

Write $\dfrac{1}{3+4i}$ in standard form $a+bi$; that is, find the reciprocal of $3+4i$.

Solution   The idea is to multiply the numerator and denominator by the conjugate of $3+4i$, that is, the complex number $3-4i$. The result is

$$\frac{1}{3+4i} = \frac{1}{3+4i} \cdot \frac{3-4i}{3-4i} = \frac{3-4i}{9+16} = \frac{3}{25} - \frac{4}{25}i \qquad \blacksquare$$

To express the quotient of two complex numbers in standard form, we multiply the numerator and denominator of the quotient by the conjugate of the denominator.

E X A M P L E   5   *Writing the Quotient of Complex Numbers in Standard Form*

Write each of the following in standard form:

(a) $\dfrac{1+4i}{5-12i}$ (b) $\dfrac{2-3i}{4-3i}$

Solution   (a) $\dfrac{1+4i}{5-12i} = \dfrac{1+4i}{5-12i} \cdot \dfrac{5+12i}{5+12i} = \dfrac{5+20i+12i+48i^2}{25+144}$

$$= \frac{-43+32i}{169} = \frac{-43}{169} + \frac{32}{169}i$$

(b) $\dfrac{2-3i}{4-3i} = \dfrac{2-3i}{4-3i} \cdot \dfrac{4+3i}{4+3i} = \dfrac{8-12i+6i-9i^2}{16+9} = \dfrac{17-6i}{25} = \dfrac{17}{25} - \dfrac{6}{25}i$ $\blacksquare$

■ Now work Problem 19.

E X A M P L E   6   *Writing Other Expressions in Standard Form*

If $z = 2 - 3i$ and $w = 5 + 2i$, write each of the following expressions in standard form:

(a) $\dfrac{z}{w}$ (b) $\overline{z+w}$ (c) $z+\bar{z}$

Solution   (a) $\dfrac{z}{w} = \dfrac{z \cdot \overline{w}}{w \cdot \overline{w}} = \dfrac{(2 - 3i)(5 - 2i)}{(5 + 2i)(5 - 2i)} = \dfrac{10 - 15i - 4i + 6i^2}{25 + 4}$

$= \dfrac{4 - 19i}{29} = \dfrac{4}{29} - \dfrac{19}{29}i$

(b) $\overline{z + w} = \overline{(2 - 3i) + (5 + 2i)} = \overline{7 - i} = 7 + i$

(c) $z + \overline{z} = (2 - 3i) + (2 + 3i) = 4$ ■

The conjugate of a complex number has certain general properties that we shall find useful later.

For a real number $a = a + 0i$, the conjugate is $\overline{a} = \overline{a + 0i} = a - 0i = a$. That is,

Theorem   The conjugate of a real number is the real number itself.   ■

Other properties of the conjugate that are direct consequences of the definition are given next. In each statement, z and w represent complex numbers.

Theorem   The conjugate of the conjugate of a complex number is the complex number itself:

$$\overline{(\overline{z})} = z \qquad (6)$$

The conjugate of the sum of two complex numbers equals the sum of their conjugates:

$$\overline{z + w} = \overline{z} + \overline{w} \qquad (7)$$

The conjugate of the product of two complex numbers equals the product of their conjugates:

$$\overline{z \cdot w} = \overline{z} \cdot \overline{w} \qquad (8)$$

■

We leave the proofs of equations (6), (7), and (8) as exercises.

Powers of i

The **powers of i** follow a pattern that is useful to know:

$$
\begin{array}{ll}
i^1 = i & i^5 = i^4 \cdot i = 1 \cdot i = i \\
i^2 = -1 & i^6 = i^4 \cdot i^2 = -1 \\
i^3 = i^2 \cdot i = -i & i^7 = i^4 \cdot i^3 = -i \\
i^4 = i^2 \cdot i^2 = (-1)(-1) = 1 & i^8 = i^4 \cdot i^4 = 1
\end{array}
$$

And so on. Thus, the powers of i repeat with every fourth power.

EXAMPLE 7   *Evaluating Powers of i*

(a) $i^{27} = i^{24} \cdot i^3 = (i^4)^6 \cdot i^3 = 1^6 \cdot i^3 = -i$

(b) $i^{101} = i^{100} \cdot i^1 = (i^4)^{25} \cdot i = 1^{25} \cdot i = i$ ∎

EXAMPLE 8   *Writing the Power of a Complex Number in Standard Form*

Write $(2 + i)^3$ in standard form.

Solution   We use the special product formula for $(x + a)^3$:

$$(x + a)^3 = x^3 + 3ax^2 + 3a^2x + a^3$$

Thus,

$$(2 + i)^3 = 2^3 + 3 \cdot i \cdot 2^2 + 3 \cdot i^2 \cdot 2 + i^3$$
$$= 8 + 12i + 6(-1) + (-i)$$
$$= 2 + 11i$$ ∎

∎   Now work Problem 33.

Quadratic Equations with a Negative Discriminant

Quadratic equations with a negative discriminant have no real number solution. However, if we extend our number system to allow complex numbers, quadratic equations will always have a solution. Since the solution to a quadratic equation involves the square root of the discriminant, we begin with a discussion of square roots of negative numbers.

Principal Square Root of $-N$

If N is a positive real number, we define the **principal square root of $-N$,** denoted by $\sqrt{-N}$, as

$$\sqrt{-N} = \sqrt{N}\,i$$

where i is the imaginary unit and $i^2 = -1$.

EXAMPLE 9   *Evaluating the Square Root of Negative Numbers*

(a) $\sqrt{-1} = \sqrt{1}\,i = i$ (b) $\sqrt{-4} = \sqrt{4}\,i = 2i$

(c) $\sqrt{-8} = \sqrt{8}\,i = 2\sqrt{2}\,i$ ∎

EXAMPLE 10   *Solving Equations*

Solve each equation in the complex number system.

(a) $x^2 = 4$ (b) $x^2 = -9$

Solution   (a) $x^2 = 4$

$x = \pm\sqrt{4} = \pm 2$

The equation has two solutions, -2 and 2.

(b) $x^2 = -9$

$$x = \pm \sqrt{-9} = \pm \sqrt{9}i = \pm 3i$$

The equation has two solutions, $-3i$ and $3i$. ∎

■ Now work Problem 45.

Warning: When working with square roots of negative numbers, do not set the square root of a product equal to the product of the square roots (which can be done with positive numbers). To see why, look at this calculation: We know that $\sqrt{100} = 10$. However, it is also true that $100 = (-25)(-4)$, so

$$10 = \sqrt{100} = \sqrt{(-25)(-4)} \underset{\uparrow}{=} \sqrt{-25}\sqrt{-4}$$

$$\text{Here is the error.}$$

$$= (\sqrt{25}i)(\sqrt{4}i) = (5i)(2i) = 10i^2 = -10$$

Because we have defined the square root of a negative number, we now can restate the quadratic formula without restriction.

Theorem In the complex number system, the solutions of the quadratic equation $ax^2 + bx + c = 0$, where a, b, and c are real numbers and $a \neq 0$, are given by the formula

Quadratic Formula

$$x = \frac{-b \pm \sqrt{b^2 - 4ac}}{2a} \tag{9}$$

∎

E X A M P L E   1 1 *Solving Quadratic Equations in the Complex Number System*
Solve the equation $x^2 - 4x + 8 = 0$ in the complex number system.

Solution Here $a = 1$, $b = -4$, $c = 8$, and $b^2 - 4ac = 16 - 4(8) = -16$. Using equation (9), we find

$$x = \frac{4 \pm \sqrt{-16}}{2} = \frac{4 \pm \sqrt{16}i}{2} = \frac{4 \pm 4i}{2} = 2 \pm 2i$$

The equation has the solution set $\{2 - 2i, 2 + 2i\}$.

Check:

$$2 + 2i: \quad (2 + 2i)^2 - 4(2 + 2i) + 8 = 4 + 8i + 4i^2 - 8 - 8i + 8$$
$$= 4 - 4 = 0$$
$$2 - 2i: \quad (2 - 2i)^2 - 4(2 - 2i) + 8 = 4 - 8i + 4i^2 - 8 + 8i + 8$$
$$= 4 - 4 = 0$$
∎

■ Now work Problem 51.

The discriminant, $b^2 - 4ac$, of a quadratic equation still serves as a way to determine the character of the solutions.

Discriminant of a Quadratic
Equation

In the complex number system, consider a quadratic equation $ax^2 + bx + c = 0$ with real coefficients.

1. If $b^2 - 4ac > 0$, the equation has two unequal real solutions.
2. If $b^2 - 4ac = 0$, the equation has a repeated real solution—a double root.
3. If $b^2 - 4ac < 0$, the equation has two complex solutions that are not real. The solutions are conjugates of each other.

The third conclusion in the display is a consequence of the fact that if $b^2 - 4ac = -N < 0$ then, by the quadratic formula, the solutions are

$$x = \frac{-b + \sqrt{b^2 - 4ac}}{2a} = \frac{-b + \sqrt{-N}}{2a} = \frac{-b + \sqrt{N}\,i}{2a} = \frac{-b}{2a} + \frac{\sqrt{N}}{2a}i$$

and

$$x = \frac{-b - \sqrt{b^2 - 4ac}}{2a} = \frac{-b - \sqrt{-N}}{2a} = \frac{-b - \sqrt{N}\,i}{2a} = \frac{-b}{2a} - \frac{\sqrt{N}}{2a}i$$

which are conjugates of each other.

E X A M P L E   1 2

Determining the Character of the Solution of a Quadratic Equation

Without solving, determine the character of the solution of each equation in the complex number system.

(a) $3x^2 + 4x + 5 = 0$

(b) $2x^2 + 4x + 1 = 0$

(c) $9x^2 - 6x + 1 = 0$

Solution

(a) Here, $a = 3$, $b = 4$, and $c = 5$, so $b^2 - 4ac = 16 - 4(3)(5) = -44$. The solutions are complex numbers that are not real and are conjugates of each other.

(b) Here, $a = 2$, $b = 4$, and $c = 1$, so $b^2 - 4ac = 16 - 8 = 8$. The solutions are two unequal real numbers.

(c) Here, $a = 9$, $b = -6$, and $c = 1$, so $b^2 - 4ac = 36 - 4(9)(1) = 0$. The solution is a repeated real number, that is, a double root.

Graphing Solution

(a) Figure 27(a) shows the graph of $f(x) = 3x^2 + 4x + 5$. Since the graph never touches or crosses the x-axis, there are no real values of x such that $f(x) = 0$. The equation has no real solution, as expected.

(b) Figure 27(b) shows the graph of $f(x) = 2x^2 + 4x + 1$. As expected, we see there are two real solutions since the graph crosses the x-axis at two points.

(c) Figure 27(c) shows the graph of $f(x) = 9x^2 - 6x + 1$. As expected, there is one real solution between 0 and 1, where the graph touches the x-axis.

FIGURE 27

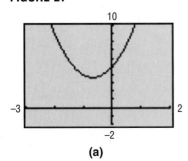

(a)

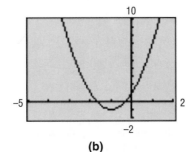

(b)

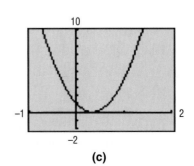

(c)

3.4

Exercise 3.4

In Problems 1–38, write each expression in the standard form $a + bi$.

1. $(2 - 3i) + (6 + 8i)$
2. $(4 + 5i) + (-8 + 2i)$
3. $(-3 + 2i) - (4 - 4i)$
4. $(3 - 4i) - (-3 - 4i)$
5. $(2 - 5i) - (8 + 6i)$
6. $(-8 + 4i) - (2 - 2i)$
7. $3(2 - 6i)$
8. $-4(2 + 8i)$
9. $2i(2 - 3i)$
10. $3i(-3 + 4i)$
11. $(3 - 4i)(2 + i)$
12. $(5 + 3i)(2 - i)$
13. $(-6 + i)(-6 - i)$
14. $(-3 + i)(3 + i)$
15. $\dfrac{10}{}$
16. $\dfrac{13}{}$
17. $\dfrac{2 + i}{i}$
18. $\dfrac{2 - i}{-2i}$
19. $\dfrac{6 - i}{1 + i}$
20. $\dfrac{2 + 3i}{1 - i}$
21. $\left(\dfrac{1}{2} + \dfrac{\sqrt{3}}{2}i\right)^2$
22. $\left(\dfrac{\sqrt{3}}{2} - \dfrac{1}{2}i\right)^2$
23. $(1 + i)^2$
24. $(1 - i)^2$
25. i^{23}
26. i^{14}
27. i^{-15}
28. i^{-23}
29. $i^6 - 5$
30. $4 + i^3$
31. $6i^3 - 4i^5$
32. $4i^3 - 2i^2 + 1$
33. $(1 + i)^3$
34. $(3i)^4 + 1$
35. $i^7(1 + i^2)$
36. $2i^4(1 + i^2)$
37. $i^6 + i^4 + i^2 + 1$
38. $i^7 + i^5 + i^3 + i$

In Problems 39–44, perform the indicated operations and express your answer in the form $a + bi$.

39. $\sqrt{-4}$
40. $\sqrt{-9}$
41. $\sqrt{-25}$
42. $\sqrt{-64}$
43. $\sqrt{(3 + 4i)(4i - 3)}$
44. $\sqrt{(4 + 3i)(3i - 4)}$

In Problems 45–64, solve each equation in the complex number system.

45. $x^2 + 4 = 0$
46. $x^2 - 4 = 0$
47. $x^2 - 16 = 0$
48. $x^2 + 25 = 0$
49. $x^2 - 6x + 13 = 0$
50. $x^2 + 4x + 8 = 0$
51. $x^2 - 6x + 10 = 0$
52. $x^2 - 2x + 5 = 0$
53. $8x^2 - 4x + 1 = 0$
54. $10x^2 + 6x + 1 = 0$
55. $5x^2 + 2x + 1 = 0$
56. $13x^2 + 6x + 1 = 0$
57. $x^2 + x + 1 = 0$
58. $x^2 - x + 1 = 0$
59. $x^3 - 8 = 0$
60. $x^3 + 27 = 0$
61. $x^4 - 16 = 0$
62. $x^4 - 1 = 0$
63. $x^4 + 13x^2 + 36 = 0$
64. $x^4 + 3x^2 - 4 = 0$

In Problems 65–70, without solving, determine the character of the solutions of each equation in the complex number system. Verify your answer using a graphing utility.

65. $3x^2 - 3x + 4 = 0$
66. $2x^2 - 4x + 1 = 0$
67. $2x^2 + 3x - 4 = 0$
68. $x^2 + 2x + 6 = 0$
69. $9x^2 - 12x + 4 = 0$
70. $4x^2 + 12x + 9 = 0$

71. $2 + 3i$ is a solution of a quadratic equation with real coefficients. Find the other solution.

72. $4 - i$ is a solution of a quadratic equation with real coefficients. Find the other solution.

In Problems 73–76, $z = 3 - 4i$ and $w = 8 + 3i$. Write each expression in the standard form $a + bi$.

73. $z + \bar{z}$
74. $w - \bar{w}$
75. $z\bar{z}$
76. $\overline{z - w}$

77. Use $z = a + bi$ to show that $z + \bar{z} = 2a$ and that $z - \bar{z} = 2bi$.

78. Use $z = a + bi$ to show that $\overline{(\bar{z})} = z$.

79. Use $z = a + bi$ and $w = c + di$ to show that $\overline{z + w} = \bar{z} + \bar{w}$.

80. Use $z = a + bi$ and $w = c + di$ to show that $\overline{z \cdot w} = \bar{z} \cdot \bar{w}$.

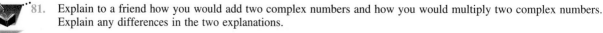

81. Explain to a friend how you would add two complex numbers and how you would multiply two complex numbers. Explain any differences in the two explanations.

82. Write a brief paragraph that compares the method used to rationalize the denominator of a rational expression and the method used to write a complex number in standard form.

3.5

Other Types of Equations

In this section we look at other types of equations, most of which can be solved using variations of the techniques already discussed. For the rest of this chapter, we shall be working in the real number system.

Equations Containing Radicals

When the variable in an equation occurs in a square root, cube root, and so on, that is, when it occurs in a radical, the equation is called a **radical equation.** Sometimes, a suitable operation will change a radical equation to one that is linear or quadratic. A commonly used procedure is to isolate the most complicated radical on one side of the equation and then eliminate it by raising each side to a power equal to the index of the radical. Care must be taken, however, because apparent solutions that are not, in fact, solutions of the original equation may result. These are called **extraneous solutions.** Thus, we need to check all answers when working with radical equations.

E X A M P L E   1   *Solving a Radical Equation*

Find the real solutions of the equation: $\sqrt[3]{2x - 4} - 2 = 0$

Graphing Solution   Figure 28(a) shows the graph of the function

$$f(x) = \sqrt[3]{2x - 4} - 2$$

From the graph, we see one x-intercept near 6. After using the BOX function, we conjecture that the x-intercept is 6. Remember, if upon repeated ZOOMing it appears that the graph crosses the x-axis at a tick mark, we conjecture that the tick mark is the x-intercept. See Figure 28(b). We verify this by checking if $f(6) = 0$.

$$f(6) = \sqrt[3]{2(6) - 4} - 2 = \sqrt[3]{8} - 2 = 0$$

Since $f(6) = 0$, 6 is the x-intercept. Thus, the only solution is $x = 6$.

FIGURE 28

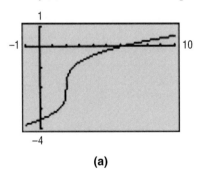

(a)

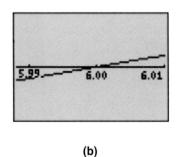

(b)

Algebraic Solution   The equation contains a radical whose index is 3. We isolate it on the left side:

$$\sqrt[3]{2x - 4} - 2 = 0$$
$$\sqrt[3]{2x - 4} = 2$$

Now raise each side to the third power (the index of the radical is 3) and solve:

$$(\sqrt[3]{2x - 4})^3 = 2^3$$
$$2x - 4 = 8$$
$$2x = 12$$
$$x = 6$$

Check: $\sqrt[3]{2(6) - 4} - 2 = \sqrt[3]{12 - 4} - 2 = \sqrt[3]{8} - 2 = 2 - 2 = 0$

The solution is $x = 6$. ∎

Sometimes, we need to raise each side to a power more than once in order to solve a radical equation algebraically.

E X A M P L E   2   *Solving a Radical Equation*

Find the real solutions of the equation: $\sqrt{2x + 3} - \sqrt{x + 2} = 2$

Graphing Solution   First, we must put the equation into the form $f(x) = 0$. Figure 29 shows the graph of the function

$$f(x) = \sqrt{2x + 3} - \sqrt{x + 2} - 2$$

From the graph, we see one x-intercept near 23. After using the BOX function, we conjecture that the x-intercept is 23. We verify this by checking if $f(23) = 0$.

$$f(23) = \sqrt{2(23) + 3} - \sqrt{23 + 2} - 2 = \sqrt{49} - \sqrt{25} - 2 = 0$$

The solution is 23.

FIGURE 29

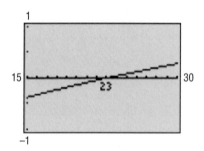

Algebraic Solution   First, we choose to isolate the more complicated radical expression (in this case, $\sqrt{2x + 3}$) on the left side:

$$\sqrt{2x + 3} = \sqrt{x + 2} + 2$$

Now square both sides (the index of the radical is 2):

$$(\sqrt{2x + 3})^2 = (\sqrt{x + 2} + 2)^2$$
$$2x + 3 = (\sqrt{x + 2})^2 + 4\sqrt{x + 2} + 4$$
$$2x + 3 = x + 2 + 4\sqrt{x + 2} + 4$$

Because the equation still contains a radical, we combine like terms, isolate the remaining radical on the right side, and again square both sides:

$$x - 3 = 4\sqrt{x + 2}$$
$$(x - 3)^2 = 16(x + 2)$$
$$x^2 - 6x + 9 = 16x + 32$$
$$x^2 - 22x - 23 = 0$$
$$(x - 23)(x + 1) = 0$$
$$x = 23 \quad \text{or} \quad x = -1$$

The original equation appears to have the solution set $\{-1, 23\}$. However, we have not yet checked.

Check: $\sqrt{2(23) + 3} - \sqrt{23 + 2} = \sqrt{49} - \sqrt{25} = 7 - 5 = 2$
$\sqrt{2(-1) + 3} - \sqrt{-1 + 2} = \sqrt{1} - \sqrt{1} = 1 - 1 = 0$

Thus, the equation has only one real solution, 23; the solution -1 is extraneous.

∎

Now work Problem 7.

Equations Quadratic in Form

The equation $x^4 + x^2 - 12 = 0$ is not quadratic in x, but it is quadratic in x^2. That is, if we let $u = x^2$, we get $u^2 + u - 12 = 0$, a quadratic equation. This equation can be solved for u and, in turn, by using $u = x^2$, we can find the solutions x of the original equation.

In general, if an appropriate substitution u transforms an equation into one of the form

$$au^2 + bu + c = 0 \qquad a \neq 0$$

then the original equation is called an **equation of the quadratic type,** or an **equation quadratic in form.**

The difficulty of solving such an equation algebraically lies in the determination that the equation is, in fact, quadratic in form. After you are told an equation is quadratic in form, it is easy enough to see it, but some practice is needed to enable you to recognize them on your own.

E X A M P L E   3

Solving Equations That Are Quadratic in Form

Find the real solutions of the equation: $(x^2 - 1)^2 + (x^2 - 1) - 12 = 0$

Graphing Solution   Figure 30 shows the graph of the function

$$f(x) = (x^2 - 1)^2 + (x^2 - 1) - 12$$

From the graph, we see two x-intercepts: one near -2; the other near 2. After using the BOX function, we conjecture that the x-intercepts are -2 and 2. We verify this by checking if $f(-2) = f(2) = 0$.

$$f(-2) = [(-2)^2 - 1]^2 + [(-2)^2 - 1] - 12 = 9 + 3 - 12 = 0$$
$$f(2) = (2^2 - 1)^2 + (2^2 - 1) - 12 = 9 + 3 - 12 = 0$$

The solution set is $\{-2, 2\}$.

FIGURE 30

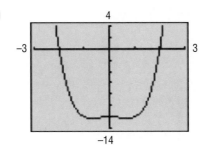

Algebraic Solution   For the equation $(x^2 - 1)^2 + (x^2 - 1) - 12 = 0$, we let $u = x^2 - 1$ so that $u^2 = (x^2 - 1)^2$. Then the original equation

$$(x^2 - 1)^2 + (x^2 - 1) - 12 = 0$$

becomes

$$u^2 + u - 12 = 0 \quad {\scriptstyle u = x^2 - 1}$$
$$(u + 4)(u - 3) = 0 \quad {\scriptstyle \text{Factor.}}$$
$$u = -4 \quad \text{or} \quad u = 3 \quad {\scriptstyle \text{Solve.}}$$

But remember that we want to solve for x. Because $u = x^2 - 1$, we have

$$x^2 - 1 = -4 \quad \text{or} \quad x^2 - 1 = 3$$
$$x^2 = -3 \qquad\qquad x^2 = 4$$

The first of these has no real solution; the second has the solution set $\{-2, 2\}$.

Check: $x = -2$: $(4 - 1)^2 + (4 - 1) - 12 = 9 + 3 - 12 = 0$
$x = 2$: $(4 - 1)^2 + (4 - 1) - 12 = 9 + 3 - 12 = 0$

Thus, $\{-2, 2\}$ is the solution set of the original equation.   ∎

The function f used in the graphing solution of Example 3 is even $[f(-x) = f(x)]$, so its graph is symmetric with respect to the y-axis. As a result, if a is an x-intercept, so is $-a$. Similarly, if a function f is odd $[f(-x) = -f(x)]$, then its graph is symmetric with respect to the origin. As a result, if a is an x-intercept, so is $-a$. When using a graphing utility to solve an equation $f(x) = 0$, check to see whether f is even or odd. This will save time in listing the solutions of the equation.

E X A M P L E   4   *Solving Equations That Are Quadratic in Form*
Find the real solutions of the equation: $x + 2\sqrt{x} - 3 = 0$

Graphing Solution   Figure 31 shows the graph of the function

$$f(x) = x + 2\sqrt{x} - 3$$

From the graph, we see one x-intercept near 1. After using the BOX function, we conjecture that the x-intercept is 1. This is verified since $f(1) = 0$. The only solution is $x = 1$.

FIGURE 31

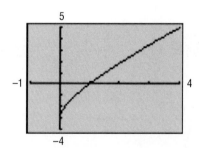

Algebraic Solution   For the equation $x + 2\sqrt{x} - 3 = 0$, let $u = \sqrt{x}$. Then $u^2 = x$, and the original equation,

$$x + 2\sqrt{x} - 3 = 0$$

becomes

$$u^2 + 2u - 3 = 0 \quad {\scriptstyle u = \sqrt{x}}$$
$$(u + 3)(u - 1) = 0 \quad {\scriptstyle \text{Factor.}}$$
$$u = -3 \quad \text{or} \quad u = 1 \quad {\scriptstyle \text{Solve.}}$$

Since $u = \sqrt{x}$, we have $\sqrt{x} = -3$ or $\sqrt{x} = 1$. The first of these, $\sqrt{x} = -3$, has no real solution, since the square root of a real number is never negative. The second one, $\sqrt{x} = 1$, has the solution $x = 1$.

Check: $1 + 2\sqrt{1} - 3 = 1 + 2 - 3 = 0$

Thus, $x = 1$ is the only solution of the original equation. ■

■ Another algebraic method for solving Example 4 would be to treat it as a radical equation. Solve it this way for practice.

The idea should now be clear. If an equation contains an expression and that same expression squared, make a substitution for the expression. You may get a quadratic equation.

■ Now work Problems 31 and 35.

Factorable Equations

We have already used factoring as a means of algebraically solving certain quadratic equations. This method can also be used to solve *any* equation that can be factored with 0 on one side of the equation. The solutions are then found by setting each factor equal to 0.

E X A M P L E   5   *Solving Equations by Factoring*

Find the real solutions of the equation: $x^3 - x^2 - 4x + 4 = 0$

Graphing Solution   Figure 32 shows the graph of the function

$$f(x) = x^3 - x^2 - 4x + 4$$

From the graph we see three x-intercepts, one near -2, one near 1, the other near 2. After using the BOX function, we conjecture that the x-intercepts are -2, 1, and 2. This is verified since $f(-2) = f(1) = f(2) = 0$.

FIGURE 32

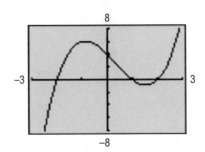

Algebraic Solution Do you recall the method of factoring by grouping? We group the terms of $x^3 - x^2 - 4x + 4 = 0$ as follows:

$$(x^3 - x^2) - (4x - 4) = 0$$

Factor out x^2 from the first grouping and 4 from the second:

$$x^2(x - 1) - 4(x - 1) = 0$$

This reveals the common factor $(x - 1)$, so we have

$$(x^2 - 4)(x - 1) = 0$$

$$(x - 2)(x + 2)(x - 1) = 0 \qquad \text{Factor again.}$$

$$x - 2 = 0 \quad \text{or} \quad x + 2 = 0 \quad \text{or} \quad x - 1 = 0 \qquad \text{Set each factor equal to 0.}$$

$$x = 2 \qquad\qquad x = -2 \qquad\qquad x = 1 \qquad \text{Solve.}$$

The solution set is $\{-2, 1, 2\}$. ◼

◼ Now work Problem 67.

3.5

Exercise 3.5

In Problems 1–26, find the real solutions of each equation (a) graphically and (b) algebraically.

1. $\sqrt{2t - 1} = 1$
2. $\sqrt{3t + 4} = 2$
3. $\sqrt{3t + 1} = -4$
4. $\sqrt{5t + 4} = -3$
5. $\sqrt[3]{1 - 2x} - 3 = 0$
6. $\sqrt[3]{1 - 2x} - 1 = 0$
7. $\sqrt{15 - 2x} = x$
8. $\sqrt{12 - x} = x$
9. $x = 2\sqrt{x - 1}$
10. $x = 2\sqrt{-x - 1}$
11. $\sqrt{x^2 - x - 4} = x + 2$
12. $\sqrt{3 - x + x^2} = x - 2$
13. $3 + \sqrt{3x + 1} = x$
14. $2 + \sqrt{12 - 2x} = x$
15. $\sqrt{2x + 3} - \sqrt{x + 1} = 1$
16. $\sqrt{3x + 7} + \sqrt{x + 2} = 1$
17. $\sqrt{3x + 1} - \sqrt{x - 1} = 2$
18. $\sqrt{3x - 5} - \sqrt{x + 7} = 2$
19. $\sqrt{3 - 2\sqrt{x}} = \sqrt{x}$
20. $\sqrt{10 + 3\sqrt{x}} = \sqrt{x}$
21. $(3x + 1)^{1/2} = 4$
22. $(3x - 5)^{1/2} = 2$
23. $(x - 1)^{1/3} = 2$
24. $(2x + 1)^{1/3} = -2$
25. $(x^2 + 9)^{1/2} = 5$
26. $(x^2 - 16)^{1/2} = 9$

In Problems 27–52, find the real solutions of each equation (a) graphically and (b) algebraically.

27. $(x + 1)^2 + 7(x + 1) + 12 = 0$
28. $(2x + 3)^2 - (2x + 3) - 6 = 0$
29. $(3x + 4)^2 - 6(3x + 4) + 9 = 0$
30. $(2 - x)^2 + (2 - x) - 20 = 0$
31. $2(s + 1)^2 - 5(s + 1) = 3$
32. $3(1 - y)^2 + 5(1 - y) + 2 = 0$
33. $x - 4\sqrt{x} = 0$
34. $x + 8\sqrt{x} = 0$
35. $x + \sqrt{x} = 20$
36. $x + \sqrt{x} - 6 = 0$
37. $t^{1/2} - 2t^{1/4} + 1 = 0$
38. $z^{1/2} - 2z^{1/4} + 1 = 0$
39. $4x^{1/2} - 9x^{1/4} + 4 = 0$
40. $x^{1/2} - 3x^{1/4} + 2 = 0$
41. $\sqrt[4]{5x^2 - 6} = x$
42. $\sqrt[4]{4 - 5x^2} = x$
43. $x^2 + 3x + \sqrt{x^2 + 3x} = 6$
44. $x^2 - 3x - \sqrt{x^2 - 3x} = 2$
45. $\dfrac{1}{(x + 1)^2} = \dfrac{1}{x + 1} + 2$
46. $\dfrac{1}{(x - 1)^2} + \dfrac{1}{x - 1} = 12$
47. $3x^{-2} - 7x^{-1} - 6 = 0$
48. $2x^{-2} - 3x^{-1} - 4 = 0$
49. $2x^{2/3} - 5x^{1/3} - 3 = 0$
50. $3x^{4/3} + 5x^{2/3} - 2 = 0$
51. $\left(\dfrac{v}{v + 1}\right)^2 + \dfrac{2v}{v + 1} = 8$
52. $\left(\dfrac{y}{(y - 1)}\right)^2 = 6\left(\dfrac{y}{y - 1}\right) + 7$

In Problems 53–72, find the real solutions of each equation by factoring. Verify any solution(s) by solving the equation graphically.

53. $x^3 = x$

54. $x^4 = 4x^3$

55. $x^4 - 5x^2 + 4 = 0$

56. $x^4 - 10x^2 + 25 = 0$

57. $3x^4 - 2x^2 - 1 = 0$

58. $2x^4 - 5x^2 - 12 = 0$

59. $x^6 + 7x^3 - 8 = 0$

60. $x^6 - 7x^3 - 8 = 0$

61. $x = 6\sqrt{x}$

62. $x = 4\sqrt{x}$

63. $x^{3/2} - 2x^{1/2} = 0$

64. $x^{3/4} - 4x^{1/4} = 0$

65. $x^3 + x^2 - 20x = 0$

66. $x^3 + 6x^2 - 7x = 0$

67. $x^3 + x^2 + x + 1 = 0$

68. $x^3 + x^2 - x - 1 = 0$

69. $x^3 - 3x^2 - 4x + 12 = 0$

70. $x^3 - 3x^2 - x + 3 = 0$

71. $t^6 - t^4 - t^2 + 1 = 0$

72. $y^6 - 4y^4 - y^2 + 4 = 0$

In Problems 73–78, find the real solutions of each equation. Use a calculator to express solutions correct to two decimal places.

73. $x - 4x^{1/2} + 2 = 0$

74. $x^{2/3} + 4x^{1/3} + 2 = 0$

75. $x^4 + \sqrt{3}x^2 - 3 = 0$

76. $x^4 + \sqrt{2}x^2 - 2 = 0$

77. $\pi(1 + t)^2 = \pi + 1 + t$

78. $\pi(1 + r)^2 = 2 + \pi(1 + r)$

79. If $k = \dfrac{x + 3}{x - 3}$ and $k^2 - k = 12$, find x.

80. If $k = \dfrac{x + 3}{x - 4}$ and $k^2 - 3k = 28$, find x.

81. *Physics: Using Sound to Measure Distance* The depth of a well can sometimes be found by dropping an object into the well and measuring the time elapsed until a sound is heard. If t_1 is the time (measured in seconds) it takes for the object to strike the bottom of the well, then t_1 will obey the equation $s = 16t_1^2$, where s is the distance (measured in feet). It follows that $t_1 = \sqrt{s}/4$. Suppose that t_2 is the time it takes for the sound of the impact to reach your ears. Because sound waves are known to travel at a speed of approximately 1100 feet per second, the time t_2 to travel the distance s will be $t_2 = s/1100$. Now $t_1 + t_2$ is the total time that elapses from the moment the object is dropped to the moment a sound is heard. Thus, we have the equation

$$\text{Total time elapsed} = \frac{\sqrt{s}}{4} + \frac{s}{1100}$$

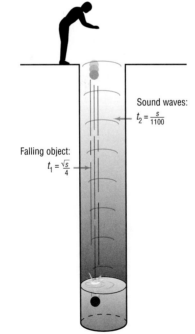

Sound waves:
$t_2 = \dfrac{s}{1100}$

Falling object:
$t_1 = \dfrac{\sqrt{s}}{4}$

(a) Find the depth of a well if the total time elapsed from dropping a rock to hearing it hit bottom is 4 seconds.

(b) Using a graphing utility, graph the function

$$Y_1 = \frac{\sqrt{x}}{4} + \frac{x}{1100}$$

for $0 \le x \le 300$ and $0 \le Y_1 \le 5$.

(c) Set up a table with ΔTbl $= 5$ and TblMin $= 0$. From the table, we can approximate the slope of the graph found in part (b) by calculating $\dfrac{\Delta Y_1}{\Delta x}$. Approximate the slope as x increases from 0 to 5, 40 to 45, and 100 to 105. What is happening to the slope as x (the depth of the well) increases?

(d) What conclusions can you make about the relationship between the depth of a well and the time that elapses before a sound is heard based on your answers to parts (b) and (c)?

82. Make up a radical equation that has no solution.

83. Make up a radical equation that has an extraneous solution.

84. Discuss what there is in the solving process for radical equations that leads to the possibility of extraneous solutions. Why is there no such possibility for linear and quadratic equations?

3.6

Inequalities

Properties of Inequalities

In working with inequalities, we will need to know certain properties that they obey.

We begin with the **trichotomy property,** which states that either two numbers are equal or one of them is less than the other.

For any pair of numbers a and b,

Trichotomy Property

$$a < b \quad \text{or} \quad a = b \quad \text{or} \quad b < a$$

If $b = 0$, the trichotomy property states that, for any real number a,

$$a < 0 \quad \text{or} \quad a = 0 \quad \text{or} \quad a > 0$$

That is, any real number is negative or 0 or positive, a fact we have already noted.

The product of two positive real numbers is positive, the product of two negative real numbers is positive, and the product of 0 and 0 is 0. Thus, for any real number a, the value of a^2 is 0 or positive; that is, a^2 is nonnegative. This is called the **nonnegative property.**

For any real number a, we have

Nonnegative Property

$$a^2 \geq 0$$

FIGURE 33

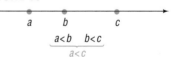

In Figure 33, we can see that, if a lies to the left of b and b lies to the left of c, then it follows that a must also lie to the left of c. This is called the **transitive property** for the inequality $<$. There is also a corresponding property for $>$.

Transitive Property
of Inequalities

If $a < b$ and $b < c$, then $a < c$.   (1a)
If $a > b$ and $b > c$, then $a > c$.   (1b)

■ Draw an illustration similar to Figure 33 that depicts the transitive property (1b) for $>$.

If we add the same number to both sides of an inequality, we obtain an equivalent inequality. For example, since $3 < 5$, then $3 + 4 < 5 + 4$ or $7 < 9$. This is called the **addition property** of inequalities.

Addition Property
of Inequalities

If $a < b$, then $a + c < b + c$.   (2a)
If $a > b$, then $a + c > b + c$.   (2b)

ISSION POSSIBLE

Chapter 3

"SAVING THE ECONOMIC FUTURE OF KRISPY KRUNCHY KANDY BAR CO."

The Krispy Krunchy Kandy Bar Company is facing a financial crisis because the cost of shipping cocoa beans from Ghana has increased. The CEO has decided that the way to stay afloat would be to reduce the size of their GIANT KRISPY KRUNCHY BAR by 10% but keep the price the same. He doesn't want to lose any customers, however. Therefore, he wants the change in size to be as unobtrusive as possible. The present dimensions of the GIANT KRISPY KRUNCHY BAR are 12 cm in length, 7 cm in width, and 3 cm in thickness. The CEO has asked your consulting firm to come up with the best way to shrink the candy bar. Because millions of dollars are riding on this decision, you will need to find all answers in centimeters correct to three decimal places and all percents correct to two decimal places.

1. Make a sketch of the candy bar, roughly to scale, and label it.
2. What is the present volume of the candy bar?
3. What would be the new volume after a 10% reduction?
4. What would be the new volume if each dimension were reduced by 10%. Is this the same as your answer to #3? (It shouldn't be.) Explain the difference.
5. Consider reducing only one of the dimension. There are three possibilities. What would the new dimensions be in each case?
6. Consider reducing two (but not three) of the dimensions by the same amount. What would the new dimensions be in each case? (There are three possibilities; in each case you want the volume to be 10% less than the original volume.)
7. Consider reducing all three of the dimensions by the same amount. What would the new dimensions be?
8. Within your group make a decision about which of the seven possibilities you found would be the best one to recommend to the CEO of Krispy Krunchy. Write out two or three sentences to justify your choice.
9. Would you mind if you discovered that your favorite candy bar had been reduced in size while the price stayed the same? Do you think a candy company might actually do this to improve their financial standing? What is your protection as a consumer from being fooled?

The addition property states that the sense, or direction, of an inequality remains unchanged if the same number is added to each side. Figure 34 illustrates the addition property (2a). In Figure 34(a), we see that a lies to the left of b. If c is positive, then $a + c$ and $b + c$ each lie c units to the right of a and b, respectively. Consequently, $a + c$ must lie to the left of $b + c$; that is, $a + c < b + c$. Figure 34(b) illustrates the situation if c is negative.

FIGURE 34

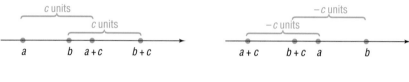

(a) If $a < b$ and $c > 0$, then $a + c < b + c$.

(b) If $a < b$ and $c < 0$, then $a + c < b + c$.

■ Draw an illustration similar to Figure 34 that illustrates the addition property (2b).

E X A M P L E   1   *Addition Property of Inequalities*

(a) If $x < -5$, then $x + 5 < -5 + 5$ or $x + 5 < 0$.

(b) If $x > 2$, then $x + (-2) > 2 + (-2)$ or $x - 2 > 0$.   ■

■ Now work Problem 3.

We'll use two examples to arrive at our next property.

E X A M P L E   2   *Multiplying an Inequality by a Positive Number*

Express as an inequality the result of multiplying each side of the inequality $3 < 7$ by 2.

Solution   We begin with

$$3 < 7$$

Multiplying each side by 2 yields the numbers 6 and 14, so we have

$$6 < 14$$ ■

E X A M P L E   3   *Multiplying an Inequality by a Negative Number*

Express as an inequality the result of multiplying each side of the inequality $9 > 2$ by -4.

Solution   We begin with

$$9 > 2$$

Multiplying each side by -4 yields the numbers -36 and -8, so we have

$$-36 < -8$$ ■

Note that the effect of multiplying both sides of $9 > 2$ by the negative number -4 is that the direction of the inequality symbol is reversed.

Examples 2 and 3 illustrate the following general **multiplication properties** for inequalities:

Multiplication Properties for Inequalities

> If $a < b$ and if $c > 0$, then $ac < bc$.
> If $a < b$ and if $c < 0$, then $ac > bc$.   (3a)
>
> If $a > b$ and if $c > 0$, then $ac > bc$.
> If $a > b$ and if $c < 0$, then $ac < bc$.   (3b)

The multiplication properties state that the sense, or direction, of an inequality *remains the same* if each side is multiplied by a *positive* real number, while the direction is *reversed* if each side is multiplied by a *negative* real number.

E X A M P L E 4   *Multiplication Property of Inequalities*

(a) If $2x < 6$, then $\frac{1}{2}(2x) < \frac{1}{2}(6)$ or $x < 3$.

(b) If $\dfrac{x}{-3} > 12$, then $-3\left(\dfrac{x}{-3}\right) < -3(12)$ or $x < -36$.

(c) If $-4x > -8$, then $\dfrac{-4x}{-4} < \dfrac{-8}{-4}$ or $x < 2$.

(d) If $-x < 8$, then $(-1)(-x) > (-1)(8)$ or $x > -8$. ■

■   Now work Problem 7.

The **reciprocal property** states that the reciprocal of a positive real number is positive and that the reciprocal of a negative real number is negative.

Reciprocal Property for Inequalities

> If $a > 0$, then $\dfrac{1}{a} > 0$.   (4a)
>
> If $a < 0$, then $\dfrac{1}{a} < 0$.   (4b)

Solving Inequalities

An **inequality in one variable** is a statement involving two expressions, at least one containing the variable, separated by one of the inequality symbols, $<$, $\le$, $>$, or $\ge$. To **solve an inequality** means to find all values of the variable for which the statement is true. These values are called **solutions** of the inequality.

For example, the following are all inequalities involving one variable, x:

$$x + 5 < 8 \qquad 2x - 3 \ge 4 \qquad x^2 - 1 \le 3 \qquad \frac{x + 1}{x - 2} > 0$$

Two inequalities having exactly the same solution set are called **equivalent inequalities.**

As with equations, one method for solving an inequality algebraically is to replace it by a series of equivalent inequalities, until an inequality with an obvious solution, such as $x < 3$, is obtained. We obtain equivalent inequalities by applying some of the same operations as those used to find equivalent equations. The addition property and the multiplication properties form the basis for the following procedures.

Procedures That Leave the Inequality Symbol Unchanged

1. Simplify both sides of the inequality by combining like terms and eliminating parentheses:

$$\text{Replace} \quad (x + 2) + 6 > 2x + (x + 1)$$
$$\text{by} \quad x + 8 > 3x + 1$$

2. Add or subtract the same expression on both sides of the inequality:

$$\text{Replace} \quad 3x - 5 < 4$$
$$\text{by} \quad (3x - 5) + 5 < 4 + 5$$

3. Multiply or divide both sides of the inequality by the same *positive* expression:

$$\text{Replace} \quad 4x > 16 \quad \text{by} \quad \frac{4x}{4} > \frac{16}{4}$$

Procedures That Reverse the Sense or Direction of the Inequality Symbol

1. Interchange the two sides of the inequality:

$$\text{Replace} \quad 3 < x \quad \text{by} \quad x > 3$$

2. Multiply or divide both sides of the inequality by the same *negative* expression:

$$\text{Replace} \quad -2x > 6 \quad \text{by} \quad \frac{-2x}{-2} < \frac{6}{-2}$$

To solve an inequality using a graphing utility, we follow these steps:

Steps for Solving Inequalities Graphically

STEP 1: Write the inequality in one of the following forms:

$$f(x) < 0 \qquad f(x) > 0 \qquad f(x) \le 0 \qquad f(x) \ge 0$$

STEP 2: Graph: $y = f(x)$

If the inequality is of the form $f(x) < 0$, determine on what interval the graph is below the x-axis.

If the inequality is of the form $f(x) > 0$, determine on what interval the graph is above the x-axis.

If the inequality is not strict, include the x-intercepts in the solution.

Linear Inequalities

A **linear inequality in one variable** is an inequality equivalent to one of the forms

$$ax + b < 0 \qquad ax + b > 0$$
$$ax + b \le 0 \qquad ax + b \ge 0$$

where a and b are real numbers and $a \ne 0$.

The remainder of this section deals with solving linear inequalities. In the next section, we discuss the solution of other types of inequalities. As the examples that follow illustrate, we solve linear inequalities algebraically using many of the same steps we would use to solve a linear equation. In writing the solution of an inequality, we may use either set notation or interval notation, whichever is more convenient.

E X A M P L E   5   *Solving Linear Inequalities*

Solve the inequality $3 - 2x < 5$, and draw a graph to illustrate the solution.

Graphing Solution   We write the inequality in the form $f(x) < 0$:

$$3 - 2x < 5$$

$$-2 - 2x < 0 \quad \text{Subtract 5 from both sides.}$$

We graph $f(x) = -2 - 2x$. See Figure 35. After using the BOX function, we decide that the x-intercept is at $x = -1$. Since the graph of f is below the x-axis for $x > -1$, the solution set of the inequality is $\{x | x > -1\}$ or, using interval notation, all numbers in the interval $[-1, \infty]$.

FIGURE 35

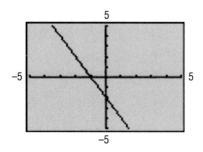

Algebraic Solution

$$3 - 2x < 5$$

$$3 - 2x - 3 < 5 - 3 \quad \text{Subtract 3 from both sides.}$$

$$-2x < 2 \qquad \text{Simplify.}$$

$$\frac{-2x}{-2} > \frac{2}{-2} \qquad \text{Divide both sides by } -2. \text{ (The sense of the inequality symbol is reversed.)}$$

$$x > -1 \qquad \text{Simplify.}$$

The solution set is $\{x | -1 < x < \infty\}$ or, using interval notation, all numbers in the interval $(-1, \infty)$. See Figure 36 for the graph of the solution set.

FIGURE 36
$-1 < x < \infty$ or $(-1, \infty)$

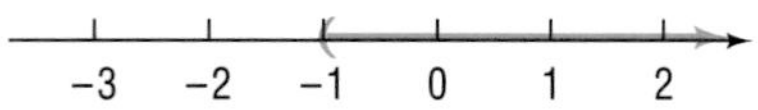

$$\begin{array}{ccccccc} & \mid & \mid & (& \mid & \mid & \mid \rightarrow \\ -3 & -2 & -1 & 0 & 1 & 2 \end{array}$$

∎

E X A M P L E   6   *Solving Linear Inequalities*

Solve the inequality $4x + 7 \geq 2x - 3$, and draw a graph to illustrate the solution.

Graphing Solution   We write the inequality in the form $f(x) \geq 0$:

$$4x + 7 \geq 2x - 3$$

$$4x + 7 + 3 - 2x \geq 2x - 3 + 3 - 2x \quad \text{Add } 3 - 2x \text{ to both sides.}$$

$$2x + 10 \geq 0 \qquad\qquad \text{Simplify.}$$

We graph $f(x) = 2x + 10$. See Figure 37. The graph of f is above the x-axis for $x > -5$. Since the inequality is not strict, we include the x-intercept, -5, in the solution. The solution set is $\{x | x \geq -5\}$ or, using interval notation, $[-5, \infty)$.

FIGURE 37

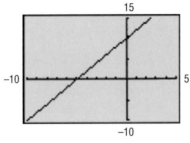

Algebraic Solution

$$4x + 7 \geq 2x - 3$$

$$4x + 7 - 7 \geq 2x - 3 - 7 \qquad \text{Subtract 7 from both sides.}$$

$$4x \geq 2x - 10 \qquad \text{Simplify.}$$

$$4x - 2x \geq 2x - 10 - 2x \qquad \text{Subtract } 2x \text{ from both sides.}$$

$$2x \geq -10 \qquad \text{Simplify.}$$

$$\frac{2x}{2} \geq \frac{-10}{2} \qquad \text{Divide both sides by 2. (The sense of the inequality symbol is unchanged.)}$$

$$x \geq -5 \qquad \text{Simplify.}$$

FIGURE 38

$-5 \leq x < \infty$ or $[-5, \infty)$

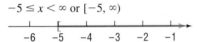

The solution set is $\{x \mid -5 \leq x < \infty\}$ or, using interval notation, all numbers in the interval $[-5, \infty)$.

See Figure 38 for the graph. ∎

■ Now work Problem 19.

E X A M P L E 7

Solving Combined Inequalities

Solve the inequality $-5 < 3x - 2 < 1$ and draw a graph to illustrate the solution.

Graphing Solution

To solve a combined inequality, we graph each part: $Y_1 = -5$, $Y_2 = 3x - 2$, and $Y_3 = 1$. We seek the values of x for which the graph of Y_2 is between the graphs of Y_1 and Y_3. See Figure 39. The point of intersection of Y_1 and Y_2 is $(-1, 5)$, and the point of intersection of Y_2 and Y_3 is $(1, 1)$. The inequality is true for all values of x between these two intersection points. Therefore, the solution set is $\{x \mid -1 < x < 1\}$ or, using interval notation, $(-1, 1)$.

FIGURE 39

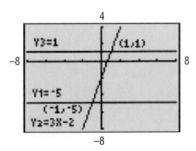

Algebraic Solution

Recall that the inequality

$$-5 < 3x - 2 < 1$$

is equivalent to the two inequalities

$$-5 < 3x - 2 \quad \text{and} \quad 3x - 2 < 1$$

We will solve each of these inequalities separately. For the first inequality,

$$-5 < 3x - 2$$
$$-5 + 2 < 3x - 2 + 2 \quad \text{Add 2 to both sides.}$$
$$-3 < 3x \quad \text{Simplify.}$$
$$\frac{-3}{3} < \frac{3x}{3} \quad \text{Divide both sides by 3.}$$
$$-1 < x \quad \text{Simplify.}$$

The second inequality is solved as follows:

$$3x - 2 < 1$$
$$3x - 2 + 2 < 1 + 2 \quad \text{Add 2 to both sides.}$$
$$3x < 3 \quad \text{Simplify.}$$
$$\frac{3x}{3} < \frac{3}{3} \quad \text{Divide both sides by 3.}$$
$$x < 1 \quad \text{Simplify.}$$

The solution set of the original pair of inequalities consists of all x for which

$$-1 < x \quad \text{and} \quad x < 1$$

FIGURE 40
$-1 < x < 1$ or $(-1, 1)$

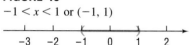

This may be written more compactly as $\{x \mid -1 < x < 1\}$. In interval notation, the solution is $(-1, 1)$. See Figure 40 for the graph.   ■

We observe in the preceding process that the two inequalities we solved required exactly the same steps. A shortcut to solving the original inequality algebraically is to deal with the two inequalities at the same time, as follows:

$$-5 < \quad 3x - 2 \quad < 1$$
$$-5 + 2 < 3x - 2 + 2 < 1 + 2 \quad \text{Add 2 to each part}$$
$$-3 < \quad 3x \quad < 3 \quad \text{Simplify.}$$
$$\frac{-3}{3} < \quad \frac{3x}{3} \quad < \frac{3}{3} \quad \text{Divide each part by 3.}$$
$$-1 < \quad x \quad < 1 \quad \text{Simplify.}$$

We use this shortcut in the algebraic solution of the next example.

E X A M P L E   8

Solving Combined Inequalities

Solve the inequality:

$$-1 \le \frac{3 - 5x}{2} \le 9$$

Graph the solution set.

Graphing Solution

Figure 41 shows the graphs of each part: $Y_1 = -1$, $Y_2 = \dfrac{3 - 5x}{2}$, and $Y_3 = 9$. The two points of intersection are $(-3, 9)$ and $(1, -1)$. The inequality is true for all values of x between these two intersection points. Therefore, the solution set is $\{x \mid -3 \le x \le 1\}$ or, using interval notation, $[-3, 1]$.

FIGURE 41

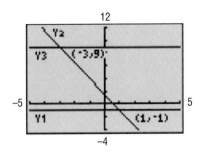

Algebraic Solution

$$-1 \le \frac{3 - 5x}{2} \le 9$$

$$2(-1) \le 2\left(\frac{3 - 5x}{2}\right) \le 2(9) \qquad \text{Multiply each part by 2 to remove the denominator.}$$

$$-2 \le \quad 3 - 5x \quad \le 18 \qquad \text{Simplify.}$$

$$-2 - 3 \le 3 - 5x - 3 \le 18 - 3 \qquad \text{Subtract 3 from each part to isolate the term containing } x.$$

$$-5 \le \quad -5x \quad \le 15 \qquad \text{Simplify.}$$

$$\frac{-5}{-5} \ge \quad \frac{-5x}{-5} \quad \ge \frac{15}{-5} \qquad \text{Divide each part by } -5 \text{ (change the sense of each inequality symbol).}$$

$$1 \ge \quad x \quad \ge -3 \qquad \text{Simplify.}$$

$$-3 \le \quad x \quad \le 1 \qquad \text{Reverse the order so that the numbers get larger as you read from left to right.}$$

FIGURE 42

$-3 \le x \le 1$ or $[-3, 1]$

```
 ├──┼──[──┼──┼──┼──]──┼──┤
-4  -3  -2  -1   0   1   2
```

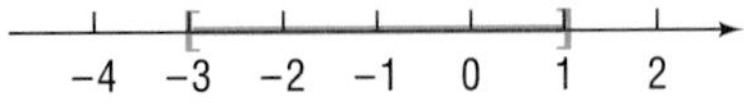

The solution set is $\{x \mid -3 \le x \le 1\}$, that is, all x in $[-3, 1]$. Figure 42 illustrates the graph. ∎

■ Now work Problem 33.

E X A M P L E   9

Creating Equivalent Inequalities

If $-1 < x < 4$, find a and b so that $a < 2x + 1 < b$.

Solution

$$-1 < \quad x \quad < 4$$
$$-2 < \quad 2x \quad < 8 \qquad \text{Multiply each part by 2.}$$
$$-1 < 2x + 1 < 9 \qquad \text{Add 1 to each part.}$$

Thus, $a = -1$ and $b = 9$. ∎

■ Now work Problem 51.

Let's look at an applied problem involving linear inequalities.

E X A M P L E   1 0

Physics: Ohm's Law

In electricity, Ohm's law states that $E = IR$, where E is the voltage (in volts), I is the current (in amperes), and R is the resistance (in ohms). An air-conditioning unit is rated at a resistance of 10 ohms. If the voltage varies from 110 to 120 volts, inclusive, what corresponding range of current will the air conditioner draw?

Solution   The voltage lies between 110 and 120, inclusive, so

$$110 \leq \quad E \quad \leq 120$$

$$110 \leq \quad IR \quad \leq 120 \qquad \text{Ohm's law, } E = IR$$

$$110 \leq I(10) \leq 120 \qquad R = 10$$

$$\frac{110}{10} \leq \frac{I(10)}{10} \leq \frac{120}{10} \qquad \text{Divide each part by 10.}$$

$$11 \leq \quad I \quad \leq 12 \qquad \text{Simplify.}$$

The air conditioner will draw between 11 and 12 amperes of current, inclusive. ■

HISTORICAL FEATURE   ■ Inequalities are a relatively new component of the algebra curriculum. They have been introduced in the last 25 years for two important reasons.

First, if approximations are made in a problem, inequalities allow calculation of how serious the error is likely to be. This use is important for practical applications of mathematics and is also critical for the understanding of calculus (in its modern version, due principally to Augustin Louis Cauchy, 1789–1857, and Karl Weierstrass, 1815–1897).

Second, linear inequalities are the basis for solving a kind of problem that involves finding a maximum or minimum of a quantity, depending on variables that are subjected to certain constraints. For example, we might wish to ship several products by truck from several different factories, each factory having a limited supply of each product, and to get enough of the products to a central point within 3 days using the minimum amount of gas possible. Such problems, called *linear programming problems,* are discussed later in this text. ■

3.6

Exercise 3.6

In Problems 1–8, fill in the blank with the correct inequality symbol.

1. If $x < 5$, then $x - 5$ _____ 0.

2. If $x < -4$, then $x + 4$ _____ 0.

3. If $x > -4$, then $x + 4$ _____ 0.

4. If $x > 6$, then $x - 6$ _____ 0.

5. If $x > -4$, then $3x$ _____ -12.

6. If $x < 3$, then $2x$ _____ 6.

7. If $x < 6$, then $-2x$ _____ -12.

8. If $x > -2$, then $-4x$ _____ 8.

In Problems 9–12, an inequality is given. Write the equivalent inequality obtained by:
(a) Adding -3 to each side of the given inequality.
(b) Subtracting 5 from each side of the given inequality.
(c) Multiplying each side of the given inequality by 3.
(d) Multiplying each side of the given inequality by -2.

9. $3 < 5$ **10.** $2 > 1$ **11.** $2x + 1 < 2$ **12.** $1 - 2x > 5$

In Problems 13–48, solve each inequality (a) graphically and (b) algebraically. Graph the solution set.

13. $x + 1 < 5$ **14.** $x - 6 < 1$ **15.** $1 - 2x \leq 3$

16. $2 - 3x \leq 5$ **17.** $3x - 7 > 2$ **18.** $2x + 5 > 1$

19. $3x - 1 \geq 3 + x$ **20.** $2x - 2 \geq 3 + x$ **21.** $-2(x + 3) < 8$

22. $-3(1 - x) < 12$ **23.** $4 - 3(1 - x) \leq 3$ **24.** $8 - 4(2 - x) \leq -2x$

25. $\frac{1}{2}(x - 4) > x + 8$

26. $3x + 4 > \frac{1}{3}(x - 2)$

27. $\frac{x}{2} \geq 1 - \frac{x}{4}$

28. $\frac{x}{3} \geq 2 + \frac{x}{6}$

29. $0 \leq 2x - 6 \leq 4$

30. $4 \leq 2x + 2 \leq 10$

31. $-5 \leq 4 - 3x \leq 2$

32. $-3 \leq 3 - 2x \leq 9$

33. $-3 < \frac{2x - 1}{4} < 0$

34. $0 < \frac{3x + 2}{2} < 4$

35. $1 < 1 - \frac{1}{2}x < 4$

36. $0 < 1 - \frac{1}{3}x < 1$

37. $(x + 2)(x - 3) > (x - 1)(x + 1)$

38. $(x - 1)(x + 1) > (x - 3)(x + 4)$

39. $x(4x + 3) \leq (2x + 1)^2$

40. $x(9x - 5) \leq (3x - 1)^2$

41. $\frac{1}{2} \leq \frac{x + 1}{3} < \frac{3}{4}$

42. $\frac{1}{3} < \frac{x + 1}{2} \leq \frac{2}{3}$

43. $(4x + 2)^{-1} < 0$

44. $(2x - 1)^{-1} > 0$

45. $0 < \frac{1}{x} < \frac{1}{5}$

46. $0 < \frac{1}{x} < \frac{1}{3}$

47. $0 < (2x - 4)^{-1} < \frac{1}{2}$

48. $0 < (3x + 6)^{-1} < \frac{1}{3}$

In Problems 49–58, find a *and* b.

49. If $-1 < x < 1$, then $a < x + 4 < b$.

50. If $-3 < x < 2$, then $a < x - 6 < b$.

51. If $2 < x < 3$, then $a < -4x < b$.

52. If $-4 < x < 0$, then $a < \frac{1}{2}x < b$.

53. If $0 < x < 4$, then $a < 2x + 3 < b$.

54. If $-3 < x < 3$, then $a < 1 - 2x < b$.

55. If $-3 < x < 0$, then $a < \frac{1}{x + 4} < b$.

56. If $2 < x < 4$, then $a < \frac{1}{x - 6} < b$.

57. If $6 < 3x < 12$, then $a < x^2 < b$.

58. If $0 < 2x < 6$, then $a < x^2 < b$.

59. If $-10 < x < 5$, solve: $\frac{2}{x + 10} > \frac{-1}{x - 5}$.

60. If $x > 3$, solve: $\frac{x}{x - 3} > 2$.

61. What is the domain of the variable in the expression $\sqrt{3x + 6}$? Verify your result using a graphing utility.

62. What is the domain of the variable in the expression $\sqrt{8 + 2x}$? Verify your result using a graphing utility.

63. If $a \leq b$ and $c > 0$, show that $ac \leq bc$. [*Hint:* Since $a \leq b$, it follows that $a - b \leq 0$. Now multiply each side by c.]

64. If $a \leq b$ and $c < 0$, show that $ac \geq bc$.

65. *Arithmetic Mean* If $a < b$, show that $a < (a + b)/2 < b$. The number $(a + b)/2$ is called the **arithmetic mean** of a and b.

66. Refer to Problem 65. Show that the arithmetic mean of a and b is equidistant from a and b.

67. *Geometric Mean* If $0 < a < b$, show that $a < \sqrt{ab} < b$. The number $\sqrt{ab}$ is called the **geometric mean** of a and b.

68. Refer to Problems 65 and 67. Show that the geometric mean of a and b is less than the arithmetic mean of a and b.

69. *Harmonic Mean* For $0 < a < b$, let h be defined by

$$\frac{1}{h} = \frac{1}{2}\left(\frac{1}{a} + \frac{1}{b}\right)$$

Show that $a < h < b$. The number h is called the **harmonic mean** of a and b.

70. Refer to Problems 65, 67, and 69. Show that the harmonic mean of a and b equals the geometric mean squared divided by the arithmetic mean.

71. A young adult may be defined as someone older than 21, but less than 30 years of age. Express this statement using inequalities.

72. Middle-aged may be defined as being 40 or more and less than 60. Express this statement using inequalities.

73. *Life Expectancy* Metropolitan Life Insurance Co. reported that an average 25-year-old male in 1996 could expect to live at least 48.4 more years, and an average 25-year-old female in 1996 could expect to live at least 54.7 more years.

(a) To what age can an average 25-year-old male expect to live? Express your answer as an inequality.

(b) To what age can an average 25-year-old female expect to live? Express your answer as an inequality.

(c) Who can expect to live longer, a male or a female? By how many years?

74. *General Chemistry* For a certain ideal gas, the volume V (in cubic centimeters) equals 20 times the temperature T (in degrees Celsius). If the temperature varies from 80° to 120°C, inclusive, what is the corresponding range of the volume of the gas?

75. *Real Estate* A real estate agent agrees to sell a large apartment complex according to the following commission schedule: $45,000 plus 25% of the selling price in excess of $900,000. Assuming the complex will sell at some price between $900,000 and $1,100,000, inclusive, over what range does the agent's commission vary? How does the commission vary as a percent of selling price?

76. *Sales Commission* A used car salesperson is paid a commission of $25 plus 40% of the selling price in excess of owner's cost. The owner claims that used cars typically sell for at least owner's cost plus $70 and at most owner's cost plus $300. For each sale made, over what range can the salesperson expect the commission to vary?

77. *Federal Tax Withholding* The percentage method of withholding for federal income tax (1995)* states that a single person whose weekly wages, after subtracting withholding allowances, are over $476, but not over $999, shall have $63.90 plus 28% of the excess over $476 withheld. Over what range does the amount withheld vary if the weekly wages vary from $500 to $550, inclusive?

78. *Federal Tax Withholding* Rework Problem 77 if the weekly wages vary from $600 to $700, inclusive.

79. *Electricity Rates* Commonwealth Edison Company's summer charge for electricity is 10.819¢ per kilowatt-hour.† In addition, each monthly bill contains a customer charge of $9.06. If last summer's bills ranged from a low of $82.14 to a high of $279.63, over what range did usage vary (in kilowatt-hours)?

80. *Water Bills* The Village of Oak Lawn charges homeowners $21.60 per quarter year plus $1.70 per 1000 gallons for water usage in excess of 12,000 gallons.‡ In 1995, one homeowner's quarterly bill ranged from a high of $65.75 to a low of $28.40. Over what range did water usage vary?

81. *Markup of a New Car* The markup over dealer's cost of a new car ranges from 12% to 18%. If the sticker price is $8800, over what range will the dealer's cost vary?

82. *IQ Tests* A standard intelligence test has an average score of 100. According to statistical theory, of the people who take the test, the 2.5% with the highest scores will have scores of more than 1.95σ above the average, where σ (sigma, a number called the *standard deviation*) depends on the nature of the test. If $\sigma = 12$ for this test and there is (in principle) no upper limit to the score possible on the test, write the interval of possible test scores of the people in the top 2.5%.

83. In your Economics 101 class, you have scores of 68, 82, 87, and 89 on the first four of five tests. To get a grade of B, the average of the five test scores must be greater than or equal to 80 and less than 90. Solve an inequality to find the range of the score you need on the last test to get a B.

84. Repeat Problem 83 if the fifth test counts double.

85. A car that averages 25 miles per gallon has a tank that holds 20 gallons of gasoline. After a trip that covered at least 300 miles, the car ran out of gasoline. What is the range of the amount of gasoline (in gallons) that was in the tank at the start of the trip?

86. Repeat Problem 85 if the same car runs out of gasoline after a trip of no more than 250 miles.

87. Make up a linear inequality that has $-3 < x < 4$ as a solution.

88. Do you prefer to use inequality notation or interval notation to express the solution to an inequality? Give your reasons. Are there particular circumstances when you prefer one to the other? Cite examples.

89. How would you explain to a fellow student the underlying reason for the multiplication property for inequalities (page 225); that is, the sense or direction of an inequality remains the same if each side is multiplied by a positive real number, while the direction is reversed if each side is multiplied by a negative real number?

*Source: Employer's Tax Guide, Department of the Treasury, Internal Revenue Service, 1995.
†Source: Commonwealth Edison Co., Chicago, Illinois, 1991.
‡Source: Village of Oak Lawn, Illinois, 1995.

3.7

Other Inequalities

In this section, we solve inequalities that contain polynomials of degree 2 and higher, as well as some that contain rational expressions. Before we can solve such inequalities, we rearrange them so that the polynomial or rational expression is on the left side and 0 is on the right side.

We will solve these inequalities using both a graphing and an algebraic approach. To utilize the algebraic approach, we will need to be able to factor the left side of the inequality.

E X A M P L E   1   *Solving Quadratic Inequalities*

Solve the inequality $x^2 + x - 12 > 0$, and graph the solution set.

Graphing Solution   Figure 43 shows the graph of the function $Y_1 = x^2 + x - 12$. The x-intercepts of the function are -4 and 3. From the graph we see that $Y_1 > 0$ when the graph is above the x-axis. Therefore, the solution set is $\{x | -\infty < x < -4 \text{ or } 3 < x < \infty\}$ or, using interval notation, $(-\infty, -4)$ or $(3, \infty)$.

FIGURE 43

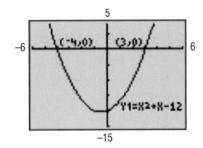

Algebraic Solution   We factor the left side, obtaining

$$x^2 + x - 12 > 0$$
$$(x + 4)(x - 3) > 0$$

The product of two real numbers is positive either when both factors are positive or when both factors are negative.

BOTH NEGATIVE	**OR**	**BOTH POSITIVE**

BOTH NEGATIVE

$x + 4 < 0$ and $x - 3 < 0$
  $x < -4$ and $x < 3$

The numbers x that are less than -4 and at the same time less than 3 are simply

$$x < -4$$

OR

BOTH POSITIVE

$x + 4 > 0$ and $x - 3 > 0$
  $x > -4$ and $x > 3$

The numbers x that are greater than -4 and at the same time greater than 3 are simply

$$x > 3$$

or

FIGURE 44

$-\infty < x < -4$ or $3 < x < \infty$;
$(-\infty, -4)$ or $(3, \infty)$

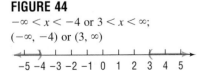

The solution set is $\{x | -\infty < x < -4 \text{ or } 3 < x < \infty\}$. In interval notation, we write the solution as $(-\infty, -4)$ or $(3, \infty)$. See Figure 44.   ∎

We can also obtain the algebraic solution to the inequality of Example 1 by another method. The left-hand side of the inequality is factored so that it becomes $(x + 4)(x - 3) > 0$, as before. We then use the real number line to construct a graph that uses the solutions to the equation

$$x^2 + x - 12 = (x + 4)(x - 3) = 0$$

which are $x = -4$ and $x = 3$. These numbers separate the real number line into three intervals: $-\infty < x < -4$, $-4 < x < 3$, and $3 < x < \infty$. See Figure 45(a).

Now, if $x < -4$, then $x + 4 < 0$. We indicate this fact about the expression $x + 4$ by placing minus signs $(- - -)$ to the left of -4. See Figure 45(b). If $x > -4$, then $x + 4 > 0$. We indicate this fact about $x + 4$ by placing plus signs $(+ + +)$ to the right of -4.

Similarly, if $x < 3$, then $x - 3 < 0$. We indicate this fact about $x - 3$ by placing minus signs to the left of 3. If $x > 3$, then $x - 3 > 0$. We indicate this fact about $x - 3$ by placing plus signs to the right of 3. See Figure 45(b).

Next we prepare Figure 45(c) as follows: Since we know that the expressions $x + 4$ and $x - 3$ are both negative for $x < -4$, it follows that their product will be positive for $x < -4$. Since we know that $x - 3$ is negative and $x + 4$ is positive for $-4 < x < 3$, it follows that their product is negative for $-4 < x < 3$. Finally, since both expressions are positive for $x > 3$, their product is positive for $x > 3$. We place plus and minus signs as shown in Figure 45(c) to indicate these facts.

Finally, in Figure 45(d) we show on the number line where the expression $(x + 4)(x - 3)$ is positive. The solution to the original inequality $(x + 4)(x - 3) = x^2 + x - 12 > 0$ is $\{x \mid -\infty < x < -4 \text{ or } 3 < x < \infty\}$, as before.

FIGURE 45

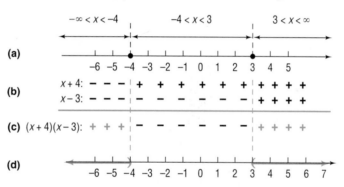

The preceding discussion demonstrates that the sign of each factor of an expression is the same on each interval that the real number line was divided into. Consequently, an alternative, and simpler, approach to obtaining Figure 45(c) would be to select a **test number** in each interval and use it to evaluate each factor to see if it is positive or negative. You may choose any number in the interval as a test number. See Figure 46.

FIGURE 46

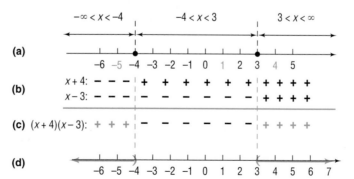

In Figure 46(a), the test numbers, -5, 1, 4 that we selected are in brown. For $x + 4$, we insert $- - -$ under $-\infty < x < -4$ because, for the test number -5, the value of $x + 4$ is $-5 + 4 = -1$, a negative number. Continuing across, we insert $+ + +$ under $-4 < x < 3$ because, for the test number 1, the value of $x + 4$ is $1 + 4 = 5$, a positive number. This process is continued for each factor and for each interval to obtain Figure 46(b). The rest of Figure 46 is obtained as before.

We shall employ the method of using a test number to solve inequalities algebraically. Here is another example showing all the details.

E X A M P L E   2   *Solving Quadratic Inequalities*

Solve the inequality $x^2 \leq 4x + 12$, and graph the solution set.

Graphing Solution   First, we rearrange the inequality so that 0 is on the right side. Thus, we subtract $4x + 12$ from both sides of the inequality and obtain

$$x^2 - 4x - 12 \leq 0$$

Figure 47 shows the graph of $Y_1 = x^2 - 4x - 12$. The x-intercepts are -2 and 6. From the graph we see that $Y_1 \leq 0$ when the graph is on or below the x-axis. Therefore, the solution set is $\{x | -2 \leq x \leq 6\}$ or, using interval notation, $[-2, 6]$.

FIGURE 47

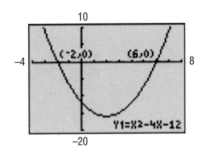

Algebraic Solution   First, we rearrange the inequality so that 0 is on the right side:

$$x^2 \leq 4x + 12$$
$$x^2 - 4x - 12 \leq 0$$
$$(x + 2)(x - 6) \leq 0 \qquad \text{Factor.}$$

Next, we set the left side equal to 0 and solve the resulting equation:

$$(x + 2)(x - 6) = 0$$

The solutions of the equation are -2 and 6, and they separate the real number line into three intervals:

$$-\infty < x < -2 \qquad -2 < x < 6 \qquad 6 < x < \infty$$

See Figure 48(a).

Now, for the test number -3, we find that $x + 2 = -3 + 2 = -1$, a negative number, so we place minus signs under the interval $-\infty < x < -2$ across from $x + 2$. For the test number 1, we find $x + 2 = 1 + 2 = 3$, a positive number, so we place plus signs under the interval $-2 < x < 6$ across from $x + 2$. Continuing in this fashion, we obtain Figure 48(b).

Next we enter the signs of the product $(x + 2)(x - 6)$ in Figure 48(c). Since we want to know where the product $(x + 2)(x - 6)$ is negative, we conclude that the solutions are numbers x for which $-2 < x < 6$. However, because the origi-

FIGURE 48

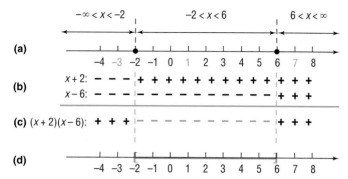

nal inequality is nonstrict, numbers x that satisfy the equation $x^2 = 4x + 12$ are also solutions of the inequality $x^2 \le 4x + 12$. Thus, we include -2 and 6, and the solution set of the given inequality is $\{x \mid -2 \le x \le 6\}$; that is, all x in $[-2, 6]$. See Figure 48(d). ∎

■ Now work Problems 3 and 7.

We have been solving inequalities by rearranging the inequality so that 0 is on the right side, setting the left side equal to 0, and solving the resulting equation. The solutions are then used to separate the real number line into intervals. But what if the resulting equation has no real solution? In this case, we rely on the following result.

Theorem   If a polynomial equation has no real solutions, the polynomial is either always positive or always negative. ∎

For example, the equation

$$x^2 + 5x + 8 = 0$$

has no real solutions. (Do you see why? Its discriminant, $b^2 - 4ac = 25 - 32 = -7$, is negative.) The value of $x^2 + 5x + 8$ is therefore always positive or always negative. To see which is true, we test its value at some number (0 is the easiest). Because $0^2 + 5(0) + 8 = 8$ is positive, we conclude that $x^2 + 5x + 8 > 0$ for all x.

This result can also be verified using a graphing utility. Figure 49 shows the graph of $Y_1 = x^2 + 5x + 8$. We see from the graph that the equation $x^2 + 5x + 8 = 0$ has no real solutions, and $Y_1 > 0$ for all values of x.

FIGURE 49

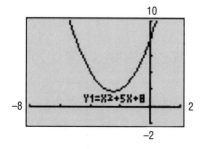

E X A M P L E   3   *Solving a Polynomial Inequality*

Solve the inequality $x^4 \le x$, and graph the solution set.

Graphing Solution First, we rearrange the inequality so that 0 is on the right side. Thus, we subtract x from both sides of the inequality and obtain

$$x^4 - x \le 0$$

Figure 50 shows the graph of $Y_1 = x^4 - x$. The x-intercepts of the function are 0 and 1. From the graph we see that $Y_1 \le 0$ when the graph is on or below the x-axis. Therefore, the solution set is $\{x \mid 0 \le x \le 1\}$ or, using interval notation, [0, 1].

FIGURE 50

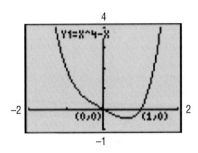

Algebraic Solution We rewrite the inequality so that 0 is on the right side:

$$x^4 \le x$$
$$x^4 - x \le 0$$

Then we proceed to factor the left side:

$$x^4 - x \le 0$$
$$x(x^3 - 1) \le 0$$
$$x(x - 1)(x^2 + x + 1) \le 0$$

The solutions of the equation

$$x(x - 1)(x^2 + x + 1) = 0$$

are just 0 and 1, since the equation $x^2 + x + 1 = 0$ has no real solutions. We note that the expression $x^2 + x + 1$ is always positive. (Do you see why?) Next, we use 0 and 1 to separate the real number line into three intervals:

$$-\infty < x < 0 \qquad 0 < x < 1 \qquad 1 < x < \infty$$

Then we construct Figure 51, showing the signs of x, $x - 1$, and $x^2 + x + 1$. Note that, since $x^2 + x + 1 > 0$ for all x, we enter plus signs next to it.

FIGURE 51

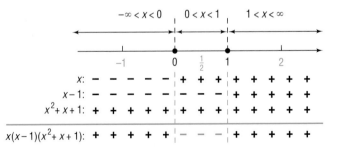

FIGURE 52

$0 \le x \le 1$ or [0, 1]

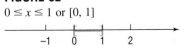

Since we want to know where $x^4 \le x$ or, equivalently, where $x(x - 1)(x^2 + x + 1) \le 0$, we conclude from Figure 51 that the solution set is $\{x \mid 0 \le x \le 1\}$, that is, all x in [0, 1]. See Figure 52. ∎

■ Now work Problem 17.

Let's solve a rational inequality.

E X A M P L E   4   *Solving a Rational Inequality*

Solve the inequality $\dfrac{4x + 5}{x + 2} \geq 3$, and graph the solution set.

Graphing Solution   We first note that the domain of the variable consists of all real numbers except -2. We rearrange the inequality so that 0 is on the right side. Thus, we subtract 3 from both sides of the inequality and obtain

$$\frac{4x + 5}{x + 2} - 3 \geq 0$$

Figure 53 shows the graph of $Y_1 = \dfrac{4x + 5}{x + 2} - 3$. Notice that we choose to graph this function in the **dot mode** to avoid having a vertical line drawn at $x = -2$, where Y_1 is not defined. The x-intercept is 1. From the graph we conjecture that $Y_1 \geq 0$ when $x < -2$ or $x \geq 1$. Therefore, the solution set is $\{x | \infty < x < -2$ or $1 \leq x < \infty\}$ or, using interval notation, $(\infty, -2)$ or $[1, \infty)$.

FIGURE 53

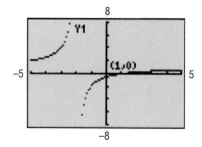

Algebraic Solution   We first note that the domain of the variable consists of all real numbers except -2. We rearrange terms so that 0 is on the right side:

$$\frac{4x + 5}{x + 2} \geq 3$$

$$\frac{4x + 5}{x + 2} - 3 \geq 0$$

$$\frac{4x + 5 - 3(x + 2)}{x + 2} \geq 0 \qquad \text{Rewrite using } x + 2 \text{ as the denominator.}$$

$$\frac{x - 1}{x + 2} \geq 0 \qquad \text{Simplify.}$$

The sign of a rational expression depends on the sign of its numerator and the sign of its denominator. Thus, for a rational expression, we separate the real number line into intervals using the numbers obtained by setting the numerator and the denominator equal to 0. For this example, they are -2 and 1. See Figure 54.

FIGURE 54

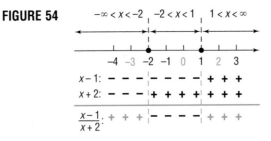

FIGURE 55

$-\infty < x < -2$ or $1 \le x < \infty$

$(-\infty, -2)$ or $[1, \infty)$

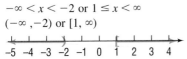

The bottom line in Figure 54 reveals the numbers x for which $(x - 1)/(x + 2)$ is positive. However, we want to know where the expression $(x - 1)/(x + 2)$ is positive or 0. Since $(x - 1)/(x + 2) = 0$ only if $x = 1$, we conclude that the solution set is $\{x | -\infty < x < -2$ or $1 \le x < \infty\}$, that is, all x in $(-\infty, -2)$ or $[1, \infty)$. See Figure 55. ∎

In the algebraic solution to Example 4, you may wonder why we did not first multiply both sides of the inequality by $x + 2$ to clear the denominator. The reason is that we do not know whether $x + 2$ is positive or negative and, as a result, we do not know whether to reverse the direction of the inequality symbol after multiplying by $x + 2$. However, there is nothing to prevent us from multiplying both sides by $(x + 2)^2$, which is always positive, since $x \ne -2$. (Do you see why?)
Then

$$\frac{4x + 5}{x + 2} \ge 3 \qquad x \ne -2$$

$$\frac{4x + 5}{x + 2}(x + 2)^2 \ge 3(x + 2)^2$$

$$(4x + 5)(x + 2) \ge 3(x^2 + 4x + 4)$$

$$4x^2 + 13x + 10 \ge 3x^2 + 12x + 12$$

$$x^2 + x - 2 \ge 0$$

$$(x + 2)(x - 1) \ge 0 \qquad x \ne -2$$

This last expression leads to the same solution set obtained in Example 4.

■ Now work Problem 29.

For our next example, we need the following equivalence property of inequalities.
If a and b are real numbers and if $a \ge 0$ and $b \ge 0$, then

$$a \le b \quad \text{is equivalent to} \quad \sqrt{a} \le \sqrt{b} \qquad (1)$$

The proof of this property is left as an exercise.

E X A M P L E   5   *Physics*

An object is dropped from the roof of a building 150 feet tall. After t seconds, the object will be $150 - 16t^2$ feet above the ground. During what interval of time will the object be between 54 and 118 feet above the ground?

Graphing Solution

Let $Y_1 =$ the height of the object and $t = x$. Figure 56 shows the graph of $Y_1 = 150 - 16x^2$. Since x represents time, we ignore negative values of x. The y-intercept of the graph represents the initial height of the object, and the x-intercept represents the time at which the object strikes the ground. Using BOX, we estimate that, when $y = 118$, $x = 1.41$ and when $y = 54$, $x = 2.44$ correct to two decimal places. Therefore, the object will be 118 feet above the ground after approximately 1.41 seconds, and the object will be 54 feet above the ground after about 2.44 seconds. So the solution set is $\{t \,|\, 1.41 \leq t \leq 2.44\}$.

FIGURE 56

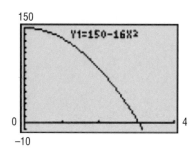

150

Y1=150-16X²

0

−10

4

Algebraic Solution

The inequality to be solved has the form

$$54 \leq 150 - 16t^2 \leq 118$$
$$-96 \leq \quad -16t^2 \quad \leq -32 \quad \text{Subtract 150 from each part.}$$
$$6 \geq \quad t^2 \quad \geq 2 \quad \text{Divide each part by } -16.$$
$$2 \leq \quad t^2 \quad \leq 6$$

For the situation described, the time t is assumed to be a nonnegative real number. With this restriction, by equation (1), the inequality

$$2 \leq t^2 \leq 6 \qquad t \geq 0$$

is equivalent to

$$\sqrt{2} \leq t \leq \sqrt{6}$$

which may be approximated by

$$1.41 \leq t \leq 2.44$$

For times ranging from approximately 1.41 to 2.44 seconds, the object will be between 54 and 118 feet above the ground. ◼

3.7

Exercise 3.7

In Problems 1–46, solve each inequality (a) graphically and (b) algebraically.

1. $(x - 5)(x + 2) < 0$

2. $(x - 5)(x + 2) > 0$

3. $x^2 - 4x > 0$

4. $x^2 + 8x > 0$

5. $x^2 - 9 < 0$

6. $x^2 - 1 < 0$

7. $x^2 + x > 12$

8. $x^2 + 7x < -12$

9. $2x^2 < 5x + 3$

10. $6x^2 < 6 + 5x$

11. $x(x - 7) > 8$

12. $x(x + 1) > 20$

13. $4x^2 + 9 < 6x$

14. $25x^2 + 16 < 40x$

15. $6(x^2 - 1) > 5x$

16. $2(2x^2 - 3x) > -9$

17. $(x - 1)(x^2 + x + 1) > 0$

18. $(x + 2)(x^2 - x + 1) > 0$

19. $(x - 1)(x - 2)(x - 3) < 0$

20. $(x + 1)(x + 2)(x + 3) < 0$

21. $x^3 - 2x^2 - 3x > 0$

22. $x^3 + 2x^2 - 3x > 0$

23. $x^4 > x^2$

24. $x^4 < 4x^2$

25. $x^3 > x^2$

26. $x^3 < 3x^2$

27. $x^4 > 1$

28. $x^3 > 1$

29. $\dfrac{x + 1}{x - 1} > 0$

30. $\dfrac{x - 3}{x + 1} > 0$

31. $\dfrac{(x - 1)(x + 1)}{x} < 0$

32. $\dfrac{(x - 3)(x + 2)}{x - 1} < 0$

33. $\dfrac{(x - 2)^2}{x^2 - 1} \geq 0$

34. $\dfrac{(x + 5)^2}{x^2 - 4} \geq 0$

35. $6x - 5 < \dfrac{6}{x}$

36. $x + \dfrac{12}{x} < 7$

37. $\dfrac{x + 4}{x - 2} \leq 1$

38. $\dfrac{x + 2}{x - 4} \geq 1$

39. $\dfrac{3x - 5}{x + 2} \leq 2$

40. $\dfrac{x - 4}{2x + 4} \geq 1$

41. $\dfrac{1}{x - 2} < \dfrac{2}{3x - 9}$

42. $\dfrac{5}{x - 3} > \dfrac{3}{x + 1}$

43. $\dfrac{2x + 5}{x + 1} > \dfrac{x + 1}{x - 1}$

44. $\dfrac{1}{x + 2} > \dfrac{3}{x + 1}$

45. $\dfrac{x^2(3 + x)(x + 4)}{(x + 5)(x - 1)} > 0$

46. $\dfrac{x(x^2 + 1)(x - 2)}{(x - 1)(x + 1)} > 0$

For Problems 47–58, first determine whether solving the inequality algebraically or graphically is easier; then proceed to solve the inequality.

47. $x^2 - 7x - 8 < 0$

48. $x^2 + 12x + 32 \geq 0$

49. $x^3 + x - 12 \geq 0$

50. $x^3 - 3x + 1 \leq 0$

51. $x^4 - 3x^2 - 4 > 0$

52. $x^4 - 5x^2 + 6 < 0$

53. $\dfrac{2x^2 - x - 1}{x - 4} \leq 0$

54. $\dfrac{3x^2 + 2x - 1}{x + 2} > 0$

55. $\dfrac{x^2 + 3x - 1}{x + 3} > 0$

56. $\dfrac{x^2 - 5x + 3}{x - 5} < 0$

57. $x^3 - 4 \geq 3x^2 + 5x - 3$

58. $x^4 - 4x \leq -x^2 + 2x + 1$

59. For what positive numbers will the cube of a number exceed 4 times its square?

60. For what positive numbers will the square of a number exceed twice the number?

61. What is the domain of the variable in the expression $\sqrt{x^2 - 16}$?

62. What is the domain of the variable in the expression $\sqrt{x^3 - 3x^2}$?

63. What is the domain of the variable in the expression $\sqrt{\dfrac{x - 2}{x + 4}}$?

64. What is the domain of the variable in the expression $\sqrt{\dfrac{x - 1}{x + 4}}$?

65. *Physics* A ball is thrown vertically upward with an initial velocity of 80 feet per second. The distance s (in feet) of the ball from the ground after t seconds is $s = 80t - 16t^2$.
  (a) For what time interval is the ball more than 96 feet above the ground? (See the figure.)
  (b) Using a graphing utility, graph the relation between s and t.
  (c) What is the maximum height of the ball?
  (d) After how many seconds does the ball reach the maximum height?

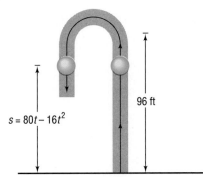

$s = 80t - 16t^2$

96 ft

66. *Physics* A ball is thrown vertically upward with an initial velocity of 96 feet per second. The distance s (in feet) of the ball from the ground after t seconds is $s = 96t - 16t^2$.
(a) For what time interval is the ball more than 112 feet above the ground?
(b) Using a graphing utility, graph the relation between s and t.
(c) What is the maximum height of the ball?
(d) After how many seconds does the ball reach the maximum height?

67. *Business* The monthly revenue achieved by selling x wristwatches is figured to be $x(40 - 0.2x)$ dollars. The wholesale cost of each watch is $32.
(a) How many watches must be sold each month to achieve a profit (revenue − cost) of at least $50?
(b) Using a graphing utility, graph the revenue function.
(c) What is the maximum revenue this firm could earn?
(d) How many wristwatches should the firm sell to maximize revenue?
(e) Using a graphing utility, graph the profit function.
(f) What is the maximum profit this firm can earn?
(g) How many watches should the firm sell to maximize profit?
(h) Provide a reasonable explanation as to why the answers found in parts (d) and (g) differ. Is the shape of the revenue function reasonable in your opinion? Why?

68. *Business* The monthly revenue achieved by selling x boxes of candy is figured to be $x(5 - 0.05x)$ dollars. The wholesale cost of each box of candy is $1.50.
(a) How many boxes must be sold each month to achieve a profit of at least $60?
(b) Using a graphing utility, graph the revenue function.
(c) What is the maximum revenue this firm could earn?
(d) How many boxes of candy should the firm sell to maximize revenue?
(e) Using a graphing utility, graph the profit function.
(f) What is the maximum profit this firm can earn?
(g) How many boxes of candy should the firm sell to maximize profit?
(h) Provide a reasonable explanation as to why the answers found in parts (d) and (g) differ. Is the shape of the revenue function reasonable in your opinion? Why?

69. Prove that if a, b are real numbers and $a \geq 0$, $b \geq 0$ then

$$a \leq b \quad \text{is equivalent to} \quad \sqrt{a} \leq \sqrt{b}$$

[*Hint:* $b - a = (\sqrt{b} - \sqrt{a})(\sqrt{b} + \sqrt{a})$]

70. Find k such that the equation $x^2 + kx + 1 = 0$ has no real solution.

71. Find k such that the equation $kx^2 + 2x + 1 = 0$ has two distinct real solutions.

72. Make up an inequality that has no solution. Make up one that has exactly one solution.

73. The inequality $x^2 + 1 < -5$ has no solution. Explain why.

3.8

Equations and Inequalities Involving Absolute Value

The absolute value of a real number a has been defined as

$$|a| = \begin{cases} a & \text{if } a \geq 0 \\ -a & \text{if } a < 0 \end{cases}$$

Also, recall that, geometrically, the absolute value of a equals the distance from the origin to the point whose coordinate is a.

In this section, we shall discuss equations and inequalities involving absolute value. In solving such equations and inequalities, the following properties will prove useful.

Theorem

1. The absolute value of any real number a is always nonnegative; that is,

$$|a| \geq 0 \qquad (1)$$

2. The absolute value of any real number a equals the principal square root of the number squared; that is,

$$|a| = \sqrt{a^2} \qquad (2)$$

3. The absolute value of the product of two real numbers a and b equals the product of their absolute values; that is,

$$|ab| = |a| \cdot |b| \qquad (3)$$

■

Proof

Property (1) follows directly from the definition of absolute value. Property (2) is given in the Appendix, Section 1. Property (3) is proved by using property (2):

$$|ab| = \sqrt{(ab)^2} = \sqrt{a^2 b^2} = \sqrt{a^2} \cdot \sqrt{b^2} = |a| \cdot |b| \qquad ■$$

We shall now take up the problem of solving equations and inequalities that contain absolute values.

Because there are two points whose distance from the origin is 5 units, -5 and 5, the equation $|x| = 5$ will have the solution set $\{-5, 5\}$. We are thus led to the following result:

Theorem

If the absolute value of an expression equals some positive number a, then the expression itself equals either a or $-a$. Thus,

$$|u| = a \quad \text{is equivalent to} \quad u = a \quad \text{or} \quad u = -a \qquad (4)$$

■

E X A M P L E   1

Solving an Equation Involving Absolute Value

Solve the equation: $|x + 4| = 13$

Graphing Solution

We first write the equation in the form $f(x) = 0$. After subtracting 13 from both sides of the equation, we have

$$|x + 4| - 13 = 0$$

Figure 57 shows the graph of $f(x) = |x + 4| - 13$. The two x-intercepts are -17 and 9. Therefore, the solution set is $\{-17, 9\}$.

FIGURE 57

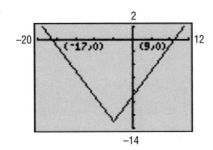

Algebraic Solution   This follows the form of equation (4), where $u = x + 4$. Thus, there are two possibilities:

$$x + 4 = 13 \quad \text{or} \quad x + 4 = -13$$
$$x = 9 \qquad\qquad x = -17$$

Thus, the solution set is $\{-17, 9\}$.   ■

■ Now work Problem 3.

Let's look at an inequality involving absolute value.

E X A M P L E   2   *Solving an Inequality Involving Absolute Value*

Solve the inequality: $|x| < 4$

Graphing Solution   Again, we write the inequality as $f(x) = |x| - 4 < 0$. Figure 58 shows the graph of $f(x) = |x| - 4$. The two x-intercepts are -4 and 4. Since the graph is below the x-axis between the two x-intercepts, the solution set is $\{x | -4 < x < 4\}$ or, using interval notation, $(-4, 4)$.

FIGURE 58

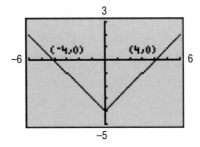

Algebraic Solution   We are looking for all points whose coordinate x is a distance less than 4 units from the origin. See Figure 59 for an illustration. Because any x between -4 and 4 satisfies the condition $|x| < 4$, the solution set consists of all numbers x for which $-4 < x < 4$; that is, all x in $(-4, 4)$.

FIGURE 59
$-4 < x < 4$ or $(-4, 4)$

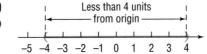

Less than 4 units from origin

                                                                          ■

We are led to the following results:

Theorem   If a is any positive number, then

$$|u| < a \quad \text{is equivalent to} \quad -a < u < a \qquad (5)$$

$$|u| \le a \quad \text{is equivalent to} \quad -a \le u \le a \qquad (6)$$

In other words, $|u| < a$ is equivalent to $-a < u$ and $u < a$.   ◼

E X A M P L E   3   *Solving an Inequality Involving Absolute Value*

Solve the inequality $|2x + 4| \le 3$, and graph the solution set.

Graphing Solution   We write the inequality as $f(x) = |2x + 4| - 3 \le 0$. Figure 60 shows the graph of $f(x) = |2x + 4| - 3$. The x-intercepts are -3.50 and -0.50 correct to two decimal places. Since the graph is below the x-axis between the two x-intercepts, the solution set is $\{x \mid -3.50 \le x \le -0.50\}$ or, using interval notation, $[-3.50, -0.50]$.

FIGURE 60

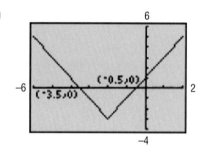

Algebraic Solution

$$|2x + 4| \le 3$$ This follows the form of statement (6); the expression $u = 2x + 4$ is inside the absolute value bars.

$$-3 \le \quad 2x + 4 \quad \le 3$$ Apply statement (6).

$$-3 - 4 \le 2x + 4 - 4 \le 3 - 4$$ Subtract 4 from each part.

$$-7 \le \quad 2x \quad \le -1$$ Simplify.

$$\frac{-7}{2} \le \quad \frac{2x}{2} \quad \le \frac{-1}{2}$$ Divide each part by 2.

$$-\frac{7}{2} \le \quad x \quad \le -\frac{1}{2}$$ Simplify.

FIGURE 61

$-\frac{7}{2} \le x \le -\frac{1}{2}$ or $[-\frac{7}{2}, -\frac{1}{2}]$

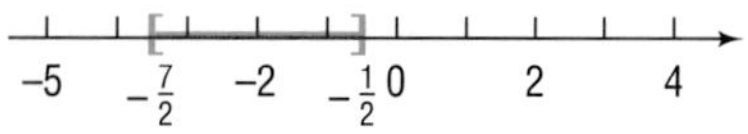

```
   |  [        ]  |  |  |  |  |
  -5  -7/2 -2  -1/2 0   2     4
```

The solution set is $\{x \mid -\frac{7}{2} \le x \le -\frac{1}{2}\}$, that is, all x in $[-\frac{7}{2}, -\frac{1}{2}]$. See Figure 61.   ◼

E X A M P L E   4   *Solving an Inequality Involving Absolute Value*

Solve the inequality $|1 - 4x| < 5$, and graph the solution set.

Graphing Solution   We write the inequality as $f(x) = |1 - 4x| - 5 < 0$. Figure 62 shows the graph of $f(x) = |1 - 4x| - 5$. The x-intercepts are -1 and 1.50. Since the graph is below the x-axis between the two x-intercepts, the solution set is $\{x \mid -1 < x < 1.50\}$ or, using interval notation, $(-1, 1.50)$.

FIGURE 62

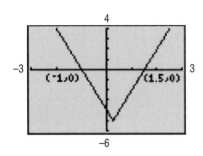

Algebraic Solution

$$|1 - 4x| < 5$$

This follows the form of statement (5); the expression $u = 1 - 4x$ is inside the absolute value bars.

$$-5 < 1 - 4x < 5$$

Apply statement (5).

$$-5 - 1 < 1 - 4x - 1 < 5 - 1$$

Subtract 1 from each part.

$$-6 < -4x < 4$$

Simplify.

$$\frac{-6}{-4} > \frac{-4x}{-4} > \frac{4}{-4}$$

Divide each part by -4, which reverses the sense of the inequality symbols.

$$\frac{3}{2} > x > -1$$

Simplify.

$$-1 < x < \frac{3}{2}$$

Rearrange the ordering.

FIGURE 63

$-1 < x < \frac{3}{2}$ or $(-1, \frac{3}{2})$

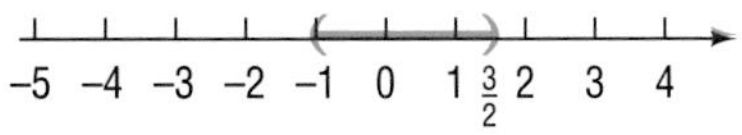

```
├──┼──┼──┼──(──┼──┼──)──┼──┼──┼──→
-5 -4 -3 -2 -1  0  1  3  2  3  4
                      2
```

The solution set is $\{x \mid -1 < x < \frac{3}{2}\}$, that is, all x in $(-1, \frac{3}{2})$. See Figure 63. ■

■   Now work Problem 27.

E X A M P L E   5   *Solving an Inequality Involving Absolute Value*

Solve the inequality $|x| > 3$, and graph the solution set.

Graphing Solution

We write the inequality as $f(x) = |x| - 3 > 0$. Figure 64 shows the graph of $f(x) = |x| - 3$. The x-intercepts are -3 and 3. Since the graph is above the x-axis left of -3 and right of 3, the solution set is $\{x \mid -\infty < x < -3 \text{ or } 3 < x < \infty\}$ or, using interval notation, $(-\infty, -3)$ or $(3, \infty)$.

FIGURE 64

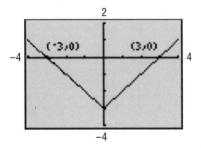

Algebraic Solution

FIGURE 65

$-\infty < x < -3$ or $3 < x < \infty$;

$(-\infty, -3)$ or $(3, \infty)$

```
←──┼──)──┼──┼──┼──┼──┼──(──┼──→
-5 -4 -3 -2 -1  0  1  2  3  4
```

We are looking for all points whose coordinate x is a distance greater than 3 units from the origin. Figure 65 illustrates the situation. We conclude that any x less than -3 or greater than 3 satisfies the condition $|x| > 3$. Consequently, the solution set consists of all numbers x for which $-\infty < x < -3$ or $3 < x < \infty$, that is, all x in $(-\infty, -3)$ or $(3, \infty)$. ■

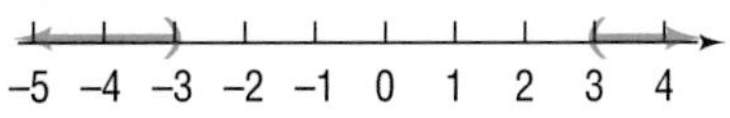

This leads us to the following results:

Theorem   If a is any positive number, then

$$|u| > a \quad \text{is equivalent to} \quad u < -a \quad \text{or} \quad u > a \qquad (7)$$

$$|u| \geq a \quad \text{is equivalent to} \quad u \leq -a \quad \text{or} \quad u \geq a \qquad (8)$$

■

EXAMPLE 6   Solving an Inequality Involving Absolute Value

Solve the inequality $|2x - 5| > 3$, and graph the solution set.

Graphing Solution   We write the inequality as $f(x) = |2x - 5| - 3 > 0$. Figure 66 shows the graph of $f(x) = |2x - 5| - 3$. The x-intercepts are 1 and 4. Since the graph is above the x-axis left of 1 and right of 4, the solution set is $\{x | -\infty < x < 1 \text{ or } 4 < x < \infty\}$ or, using interval notation, $(-\infty, 1)$ or $(4, \infty)$.

FIGURE 66

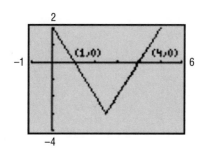

Algebraic Solution

$|2x - 5| > 3$ This follows the form of statement (7); the expression $u = 2x - 5$ is inside the absolute value bars.

$2x - 5 < -3$	or	$2x - 5 > 3$	Apply statement (7).
$2x - 5 + 5 < -3 + 5$	or	$2x - 5 + 5 > 3 + 5$	Add 5 to each part.
$2x < 2$	or	$2x > 8$	Simplify.
$\dfrac{2x}{2} < \dfrac{2}{2}$	or	$\dfrac{2x}{2} > \dfrac{8}{2}$	Divide each part by 2.
$x < 1$	or	$x > 4$	Simplify.

FIGURE 67

$-\infty < x < 1$ or $4 < x < \infty$;
$(-\infty, 1)$ or $(4, \infty)$

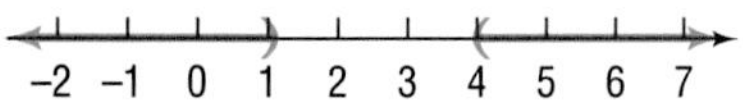

-2 -1 0 1 2 3 4 5 6 7

The solution set is $\{x | -\infty < x < 1 \text{ or } 4 < x < \infty\}$, that is, all x in $(-\infty, 1)$ or $(4, \infty)$. See Figure 67.   ■

Warning: A common error to be avoided is to attempt to write the solution $x < 1$ or $x > 4$ as $1 > x > 4$, which is incorrect, since there are no numbers x for which $1 > x$ *and* $x > 4$. Another common error is to "mix" the symbols and write $1 < x > 4$, which, of course, makes no sense.

3.8

Exercise 3.8

In Problems 1–22, solve each equation (a) graphically and (b) algebraically.

1. $|2x| = 8$
2. $|3x| = 15$
3. $|2x + 3| = 5$
4. $|3x - 1| = 2$
5. $|1 - 4t| = 5$
6. $|1 - 2z| = 3$
7. $|-2x| = 8$
8. $|-x| = 1$
9. $|-2|x = 4$
10. $|3|x = 9$
11. $\frac{2}{3}|x| = 8$
12. $\frac{3}{4}|x| = 9$
13. $\left|\frac{x}{3} + \frac{2}{5}\right| = 2$
14. $\left|\frac{x}{2} - \frac{1}{3}\right| = 1$
15. $|u - 2| = -\frac{1}{2}$
16. $|2 - v| = -1$
17. $|x^2 - 16| = 0$
18. $|x^2 - 4| = 0$
19. $|x^2 - 2x| = 3$
20. $|x^2 + x| = 12$
21. $|x^2 + x - 1| = 1$
22. $|x^2 + 3x - 2| = 2$

In Problems 23–42, solve each inequality (a) graphically and (b) algebraically. Graph the solution set.

23. $|2x| < 8$
24. $|3x| < 15$
25. $|3x| > 12$
26. $|2x| > 6$
27. $|x - 2| < 1$
28. $|x + 4| < 2$
29. $|3t - 2| \leq 4$
30. $|2u + 5| \leq 7$
31. $|x - 3| \geq 2$
32. $|x + 4| \geq 2$
33. $|1 - 4x| < 5$
34. $|1 - 2x| < 3$
35. $|1 - 2x| > 3$
36. $|2 - 3x| > 1$
37. $|x + 3| > 0$
38. $|3 - x| > 0$
39. $|x + 2| > -3$
40. $|2 - x| > -2$
41. $|2x - 1| < 0.02$
42. $|3x - 2| < 0.02$

In Problems 43–48, find a and b.

43. If $|x - 1| < 3$, then $a < x + 4 < b$.
44. If $|x + 2| < 5$, then $a < x - 2 < b$.
45. If $|x + 4| \leq 2$, then $a \leq 2x - 3 \leq b$.
46. If $|x - 3| \leq 1$, then $a \leq 3x + 1 \leq b$.
47. If $|x - 2| \leq 7$, then $a \leq \dfrac{1}{x - 10} \leq b$.
48. If $|x + 1| \leq 3$, then $a \leq \dfrac{1}{x + 5} \leq b$.

49. If $b \neq 0$, prove that $\left|\dfrac{a}{b}\right| = \dfrac{|a|}{|b|}$.
50. Show that $a \leq |a|$.

51. Prove the *triangle inequality:* $|a + b| \leq |a| + |b|$. [*Hint:* Expand $|a + b|^2 = (a + b)^2$, and use the result of Problem 50.]
52. Prove that $|a - b| \geq |a| - |b|$. [*Hint:* Apply the triangle inequality from Problem 51 to $|a| = |(a - b) + b|$.]
53. Express the fact that x differs from 3 by less than $\frac{1}{2}$ as an inequality involving an absolute value. Solve for x.
54. Express the fact that x differs from -4 by less than 1 as an inequality involving an absolute value. Solve for x.
55. Express the fact that x differs from -3 by more than 2 as an inequality involving an absolute value. Solve for x.
56. Express the fact that x differs from 2 by more than 3 as an inequality involving an absolute value. Solve for x.
57. *Body Temperature* "Normal" human body temperature is 98.6°F. If a temperature x that differs from normal by at least 1.5° is considered unhealthy, write the condition for an unhealthy temperature x as an inequality involving an absolute value, and solve for x.
58. *Household Voltage* In the United States, normal household voltage is 115 volts. However, it is not uncommon for actual voltage to differ from normal voltage by at most 5 volts. Express this situation as an inequality involving an absolute value. Use x as the actual voltage and solve for x.
59. If $a > 0$, show that the solution set of the inequality

$$x^2 < a$$

consists of all numbers x for which

$$-\sqrt{a} < x < \sqrt{a}$$

60. If $a > 0$, show that the solution set of the inequality

$$x^2 > a$$

consists of all numbers x for which

$$-\infty < x < -\sqrt{a} \quad \text{or} \quad \sqrt{a} < x < \infty$$

In Problems 61–68, use the results found in Problems 59 and 60 to solve each inequality.

61. $x^2 < 1$　　**62.** $x^2 < 4$　　**63.** $x^2 \geq 9$　　**64.** $x^2 \geq 1$

65. $x^2 \leq 16$　　**66.** $x^2 \leq 9$　　**67.** $x^2 > 4$　　**68.** $x^2 > 16$

69. Solve: $\left| 3x - |2x + 1| \right| = 4$　　**70.** Solve: $\left| x + |3x - 2| \right| = 2$

71. The equation $|x| = -2$ has no solution. Explain why.

72. The inequality $|x| > -0.5$ has all real numbers as a solution. Explain why.

Chapter Review

THINGS TO KNOW

Quadratic equation and quadratic formula

If $ax^2 + bx + c = 0$, $a \neq 0$, then $x = \dfrac{-b \pm \sqrt{b^2 - 4ac}}{2a}$.

Discriminant

If $b^2 - 4ac > 0$, there are two distinct real solutions.

If $b^2 - 4ac = 0$, there is one repeated real solution.

If $b^2 - 4ac < 0$, there are two distinct complex solutions that are not real; the solutions are conjugates of each other.

Inequality properties

Trichotomy property	$a < b$ or $a = b$ or $b < a$
Transitive property	If $a < b$ and $b < c$, then $a < c$.
	If $a > b$ and $b > c$, then $a > c$.
Addition property	If $a < b$ then $a + c < b + c$.
	If $a > b$ then $a + c > b + c$.
Multiplication properties	(a) If $a < b$ and if $c > 0$, then $ac < bc$.
	If $a < b$ and if $c < 0$, then $ac > bc$.
	(b) If $a > b$ and if $c > 0$, then $ac > bc$.
	If $a > b$ and if $c < 0$, then $ac < bc$.
Reciprocal property	If $a > 0$, then $\dfrac{1}{a} > 0$.
	If $a < 0$, then $\dfrac{1}{a} < 0$.

Absolute value

If $|u| = a$, $a > 0$, then $u = -a$ or $u = a$.

If $|u| \leq a$, $a > 0$, then $-a \leq u \leq a$.

If $|u| \geq a$, $a > 0$, then $u \leq -a$ or $u \geq a$.

HOW TO:

Solve linear equations in one variable

Solve quadratic equations

Solve radical equations

Solve linear inequalities in one variable

Solve quadratic and rational inequalities

Solve equations and inequalities involving absolute value

Solve applied problems

Solve equations and inequalities using a graphing utility

FILL-IN-THE-BLANK ITEMS

1. Two equations (or inequalities) that have precisely the same solution set are called _____.

2. An equation that is satisfied for every choice of the variable for which both sides are meaningful is called a(n) _____.

3. To complete the square of the expression $x^2 + 5x$, you would _____ the number _____.

4. The quantity $b^2 - 4ac$ is called the _____ of a quadratic equation. If it is _____, the equation has no real solution.

5. If $a < 0$, then $|a| =$ _____.

6. When a quadratic equation has a repeated solution, it is called a(n) _____ root or a root of _____.

7. When an apparent solution does not satisfy the original equation, it is called a(n) _____ solution.

8. If each side of an inequality is multiplied by a(n) _____ number, then the sense of the inequality symbol is reversed.

9. The equation $|x^2| = 4$ has two real solutions, _____ and _____, and two complex solutions, _____ and _____.

TRUE/FALSE ITEMS

T   F   1. Equations can have no solutions, one solution, or more than one solution.

T   F   2. Quadratic equations always have two real solutions.

T   F   3. If the discriminant of a quadratic equation is positive, then the equation has two complex solutions that are conjugates of each other.

T   F   4. The square of any real number is always nonnegative.

T   F   5. The expression $x^2 + x + 1$ is positive for any real number x.

6. If $a < b$ and $c < 0$, which of the following statements are true?

T   F   (a) $a \pm c < b \pm c$

T   F   (b) $a \cdot c < b \cdot c$

T   F   (c) $a/c > b/c$

7. If $a^2 < b^2$, $b \neq 0$, which of the following statements are true?

T   F   (a) $a < b$

T   F   (b) $a^2/b^2 < 1$

T   F   (c) $b^2 - a^2 > 0$

REVIEW EXERCISES

In Problems 1–34, find all real solutions, if any, of each equation (a) graphically and (b) algebraically.

1. $2 - \dfrac{x}{3} = 6$

2. $\dfrac{x}{4} - 2 = 6$

3. $-2(5 - 3x) + 8 = 4 + 5x$

4. $(6 - 3x) - 2(1 + x) = 6x$

5. $\dfrac{3x}{4} - \dfrac{x}{3} = \dfrac{1}{12}$

6. $\dfrac{4 - 2x}{3} + \dfrac{1}{6} = 2x$

7. $\dfrac{x}{x-1} = \dfrac{5}{6}, \quad x \neq 1$

8. $\dfrac{4x-5}{3-7x} = 4, \quad x \neq \frac{3}{7}$

9. $x(1-x) = 6$

10. $x(1+x) = 6$

11. $\dfrac{1}{2}\left(x - \dfrac{1}{3}\right) = \dfrac{3}{4} - \dfrac{x}{6}$

12. $\dfrac{1-3x}{4} = \dfrac{x+6}{3} + \dfrac{1}{2}$

13. $(x-1)(2x+3) = 3$

14. $x(2-x) = 3(x-4)$

15. $2x + 3 = 4x^2$

16. $1 + 6x = 4x^2$

17. $\sqrt[3]{x^2 - 1} = 2$

18. $\sqrt[3]{1 + x^3} = 3$

19. $x(x+1) + 2 = 0$

20. $3x^2 - x + 1 = 0$

21. $x^4 - 5x^2 + 4 = 0$

22. $3x^4 + 4x^2 + 1 = 0$

23. $\sqrt{2x-3} + x = 3$

24. $\sqrt{2x-1} = x - 2$

25. $x^{3/2} + 5x^{1/2} = 0$

26. $x^{2/3} + x = 0$

27. $\sqrt{x+1} + \sqrt{x-1} = \sqrt{2x+1}$

28. $\sqrt{2x-1} - \sqrt{x-5} = 3$

29. $2\sqrt[3]{x^2} + \sqrt[3]{x} = 1$

30. $4\sqrt[3]{x^2} = 1$

31. $x^{-6} - 7x^{-3} - 8 = 0$

32. $6x^{-1} - 5x^{-1/2} + 1 = 0$

33. $\sqrt{x^2 + 3x + 7} - \sqrt{x^2 - 3x + 9} + 2 = 0$

34. $\sqrt{x^2 + 3x + 7} - \sqrt{x^2 + 3x + 9} = 2$

In Problems 35–38, find all real solutions, if any, of each equation. (a, b, m, and n are positive constants.)

35. $x^2 + m^2 = 2mx + (nx)^2$

36. $b^2x^2 + 2ax = x^2 + a^2$

37. $10a^2x^2 - 2abx - 36b^2 = 0$

38. $\dfrac{1}{x-m} + \dfrac{1}{x-n} = \dfrac{2}{x}, \quad x \neq 0, m, n$

In Problems 39–58, solve each inequality (a) graphically and (b) algebraically.

39. $\dfrac{2x-3}{5} + 2 \leq \dfrac{x}{2}$

40. $\dfrac{5-x}{3} \leq 6x - 4$

41. $-9 \leq \dfrac{2x+3}{-4} \leq 7$

42. $-4 < \dfrac{2x-2}{3} < 6$

43. $6 > \dfrac{3-3x}{12} > 2$

44. $6 > \dfrac{5-3x}{2} \geq -3$

45. $2x^2 + 5x - 12 < 0$

46. $3x^2 - 2x - 1 \geq 0$

47. $\dfrac{6}{x+3} \geq 1, \quad x \neq -3$

48. $\dfrac{-2}{1-3x} < 1, \quad x \neq \frac{1}{3}$

49. $\dfrac{2x-6}{1-x} < 2, \quad x \neq 1$

50. $\dfrac{3-2x}{2x+5} \geq 2, \quad x \neq -\frac{5}{2}$

51. $\dfrac{(x-2)(x-1)}{x-3} > 0, \quad x \neq 3$

52. $\dfrac{x+1}{x(x-5)} \leq 0, \quad x \neq 0, 5$

53. $\dfrac{x^2 - 8x + 12}{x^2 - 16} > 0, \quad x \neq -4, 4$

54. $\dfrac{x(x^2 + x - 2)}{x^2 + 9x + 20} \leq 0, \quad x \neq -5, -4$

55. $|3x + 4| < \frac{1}{2}$

56. $|1 - 2x| < \frac{1}{3}$

57. $|2x - 5| \geq 9$

58. $|3x + 1| \geq 10$

In Problems 59–68, solve each equation in the complex number system.

59. $x^2 + x + 1 = 0$

60. $x^2 - x + 1 = 0$

61. $2x^2 + x - 2 = 0$

62. $3x^2 - 2x - 1 = 0$

63. $x^2 + 3 = x$

64. $2x^2 + 1 = 2x$

65. $x(1-x) = 6$

66. $x(1+x) = 2$

67. $x^4 + 2x^2 - 8 = 0$

68. $x^4 + 8x^2 - 9 = 0$

69. *Lightning and Thunder* A flash of lightning is seen, and the resulting thunderclap is heard 3 seconds later. If the speed of sound averages 1100 feet per second, how far away is the storm?

70. *Physics: Intensity of Light* The intensity I (in candlepower) of a certain light source obeys the equation $I = 900/x^2$, where x is the distance (in meters) from the light. Over what range of distances can an object be placed from this light source so that the range of intensity of light is from 1600 to 3600 candlepower, inclusive?

71. *Extent of Search and Rescue* A search plane has a cruising speed of 250 miles per hour and carries enough fuel for at most 5 hours of flying. If there is a wind that averages 30 miles per hour and the direction of search is with the wind one way and against it the other, how far can the search plane travel?

72. *Extent of Search and Rescue* Is the search plane described in Problem 71 is able to add a supplementary fuel tank that allows for an additional 2 hours of flying, how much farther can the plane extend its search?

73. *Rescue at Sea* A life raft, set adrift from a sinking ship 150 miles offshore, travels directly toward a Coast Guard station at the rate of 5 miles per hour. At the time the raft is set adrift, a rescue helicopter is dispatched from the Coast Guard station. If the helicopter's average speed is 90 miles per hour, how long will it take the helicopter to reach the life raft?

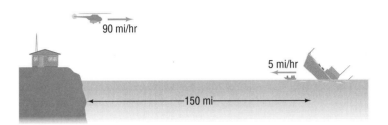

74. *Physics: Uniform Motion* Two bees leave two locations 150 meters apart and fly, without stopping, back and forth between these two locations at average speeds of 3 meters per second and 5 meters per second, respectively. How long is it until the bees meet for the first time? How long is it until they meet for the second time?

75. *Physics: Uniform Motion* A man is walking at an average speed of 4 miles per hour alongside a railroad track. A freight train, going in the same direction at an average speed of 30 miles per hour, requires 5 seconds to pass the man. How long is the freight train? Give your answer in feet.

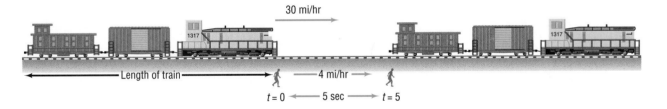

76. One formula stating the relationship between the length l and width w of a rectangle of "pleasing proportion" is $l^2 = w(l + w)$. How should a 4 foot by 8 foot sheet of plasterboard be cut so that the result is a rectangle of "pleasing proportion" with a width of 4 feet?

77. *Business: Determining the Cost of a Charter* A group of 20 senior citizens can charter a bus for a 1 day excursion trip for $15 per person. The charter company agrees to reduce the price of each ticket by 10¢ for each additional passenger in excess of 20 who goes on the trip, up to a maximum of 44 passengers (the capacity of the bus). If the final bill from the charter company was $482.40, how many seniors went on the trip, and how much did each pay?

78. A new copying machine can do a certain job in 1 hour less than an older copier. Together they can do this job in 72 minutes. How long would it take the older copier by itself to do the job?

 79. In a 100 meter race, Mike crosses the finish line 5 meters ahead of Dan. To even things up, Mike suggests to Dan that they race again, this time with Mike lining up 5 meters behind the start.
(a) Assuming Mike and Dan run at the same pace as before, does the second race end in a tie?
(b) If not, who wins?
(c) By how many meters does he win?
(d) How far back should Mike start so the race ends in a tie?

After running the race a second time, Dan, to even things up, suggests to Mike that he (Dan) line up 5 meters in front of the start.
(e) Assuming again that they run at the same pace as in the first race, does the third race result in a tie?
(f) If not, who wins?
(g) By how many meters?
(h) How far up should Dan start so the race ends in a tie?

80. Explain the difference between the following three problems. Are there any similarities in their solution?

(a) Write the expression as a single quotient: $\dfrac{x}{x-2} + \dfrac{x}{x^2-4}$

(b) Solve: $\dfrac{x}{x-2} + \dfrac{x}{x^2-4} = 0$

(c) Solve: $\dfrac{x}{x-2} + \dfrac{x}{x^2-4} < 0$

PREPARING FOR THIS CHAPTER

Before getting started on this chapter, review the following concepts:

Completing the square (Appendix, Section 5)

The discriminant of a quadratic equation (p. 187 and p. 213)

Polynomial and rational expressions (Appendix, Section 2)

Graphs of important functions (pp. 110–114)

Complex numbers (Section 3.4, pp. 206–211)

Synthetic Division (Appendix, Section 6)

POLYNOMIAL AND RATIONAL FUNCTIONS

4.1 Quadratic Functions

4.2 Polynomial Functions

4.3 Rational Functions

4.4 The Zeros of a Polynomial Function

4.5 Complex Polynomials; Fundamental Theorem of Algebra

Chapter Review

Preview   The Golden Gate Bridge

The Golden Gate Bridge, a suspension bridge, spans the entrance to San Francisco Bay. Its 746-foot-tall towers are 4200 feet apart. The bridge is suspended from two huge cables more than 3 feet in diameter; the 90-foot-wide roadway is 220 feet above the water. The cables are parabolic in shape and touch the road surface at the center of the bridge. Find the height of the cable at a distance of 1000 feet from the center.

[Example 9 in Section 4.1] ■

T n Chapters 1 and 2, we graphed linear functions $f(x) = ax + b$, $a \neq 0$; the square function $f(x) = x^2$; and the cube function $f(x) = x^3$. Each of these functions belongs to the class of functions called *polynomial functions,* which we discuss further in this chapter. We will also discuss *rational functions,* which are ratios of polynomial functions. In this chapter, we place special emphasis on the graphs of polynomial and rational functions. This emphasis will demonstrate the importance of evaluating polynomials and solving polynomial equations (Section 4.4). Section 4.5 deals with polynomials having coefficients that are complex numbers.

4.1

Quadratic Functions

A **quadratic function** is a function of the form

$$f(x) = ax^2 + bx + c \qquad (1)$$

where a, b, and c are real numbers and $a \neq 0$. The domain of a quadratic function consists of all real numbers.

Many applications require a knowledge of quadratic functions. For example, suppose that a retailer determines that the equation that relates the number x of calculators sold at the price p per calculator is given by

$$x = 15{,}000 - 750p$$

Then the revenue R derived from selling x calculators at the price p per calculator is

$$R = xp$$
$$= (15{,}000 - 750p)p$$
$$= -750p^2 + 15{,}000p$$

Figure 1 illustrates the graph of this revenue function, whose domain is $0 \leq p \leq 20$, since both x and p must be nonnegative.

FIGURE 1
Graph of a revenue function:
$R = -750p^2 + 15{,}000p$

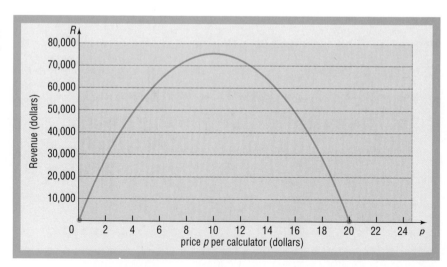

A second situation in which a quadratic function appears involves the motion of a projectile. Based on Newton's Second Law of Motion (force equals mass times acceleration, $F = ma$), it can be shown that, ignoring air resistance, the path of a projectile propelled upward at an inclination to the horizontal is the graph of a quadratic function. See Figure 2 for an illustration.

FIGURE 2
Path of a cannonball

Graphing Quadratic Functions

We know how to graph quadratic functions of the form $f(x) = ax^2$, $a \neq 0$, based on prior discussions. Figure 3 shows the graph of three functions of the form $f(x) = ax^2$, $a > 0$, for $a = 1$, $a = \frac{1}{2}$, and $a = 3$. Notice that the larger the value of a, the "narrower" the graph is and the smaller the value of a the "wider" the graph is.

FIGURE 3

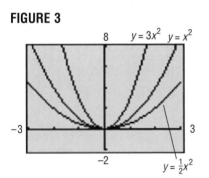

FIGURE 4

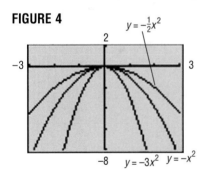

Figure 4 shows the graphs of $f(x) = ax^2$ for $a < 0$. Notice these graphs are reflections about the x-axis of the graphs in Figure 3. Based on the results of these two figures, we can draw some general conclusions about the graph of $f(x) = ax^2$. First, as $|a|$ increases, the graph becomes "narrower" and as $|a|$ gets closer to zero, the graph gets "wider." Second, if a is positive, then the graph opens "up" and if a is negative, then the graph opens "down."

The graphs in Figures 3 and 4 are typical of the graphs of all quadratic functions, which we call **parabolas.*** Refer to Figure 5, where two parabolas are pictured. The one on the left **opens up** and has a lowest point; the one on the right **opens down** and has a highest point. The lowest or highest point of a parabola is called the **vertex.** The vertical line passing through the vertex in each parabola in Figure 5 is called the **axis of symmetry** (usually abbreviated to **axis**) of the parabola. Because the parabola is symmetric about its axis, the axis of symmetry of a parabola can be used to advantage in graphing the parabola by hand.

FIGURE 5

Graphs of a quadratic function,
$f(x) = ax^2 + bx + c$, $a \neq 0$

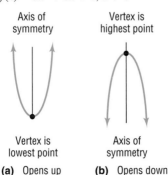

Axis of symmetry

Vertex is highest point

Vertex is lowest point

Axis of symmetry

(a) Opens up

(b) Opens down

*We shall study parabolas using a geometric definition presented later in this book.

The parabolas shown in Figure 5 are the graphs of a quadratic function $f(x) = ax^2 + bx + c$, $a \neq 0$. Notice that the coordinate axes are not included in the figure. Depending on the values of a, b, and c, the axes could be placed anywhere. The important fact is that, except possibly for compression or stretching, the shape of the graph of a quadratic function will look like one of the parabolas in Figure 5.

E X A M P L E   1

Graphing a Quadratic Function

Graph the function: $f(x) = 2x^2 + 8x + 5$. Find the vertex and line of symmetry.

Graphing Solution

Before graphing, notice the leading coefficient, 2, is positive and, therefore, the graph will open up and have a lowest point. Now graph f. See Figure 6. We observe the graph does in fact open up. To estimate the vertex of the parabola, we use the BOX function. Correct to two decimal places, the vertex is $(-2.00, -3.00)$. Therefore, the axis of symmetry is $x = -2.00$.

FIGURE 6

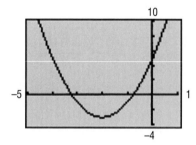

Algebraic Solution

We begin by completing the square* on the right side:

$$f(x) = 2x^2 + 8x + 5$$
$$= 2(x^2 + 4x) + 5 \qquad \text{Factor out the 2 from } 2x^2 + 8x.$$
$$= 2(x^2 + 4x + 4) + 5 - 8 \qquad \text{Complete the square of } 2(x^2 + 4x).$$
$$\qquad\qquad\qquad\qquad\qquad \text{Notice that the factor of 2 requires}$$
$$= 2(x + 2)^2 - 3 \qquad \text{that 8 be added and subtracted.} \qquad (2)$$

The graph of f can be obtained in three stages, as shown in Figure 7. Now compare this graph to the graph in Figure 5(a). The graph of $f(x) = 2x^2 + 8x + 5$ is a parabola that opens up and has its vertex (lowest point) at $(-2, -3)$. Its axis of symmetry is the line $x = -2$.

FIGURE 7

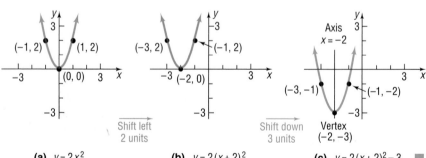

(a) $y = 2x^2$

(b) $y = 2(x + 2)^2$

(c) $y = 2(x + 2)^2 - 3$

*Refer to the Appendix, Section 5, for a review.

The method used in the Algebraic Solution to Example 1 can be used to graph any quadratic function $f(x) = ax^2 + bx + c$, $a \neq 0$, as follows:

$$f(x) = ax^2 + bx + c$$

$$= a\left(x^2 + \frac{b}{a}x\right) + c \qquad \text{Factor out } a \text{ from } ax^2 + bx.$$

$$= a\left(x^2 + \frac{b}{ax} + \frac{b^2}{4a^2}\right) + c - a\left(\frac{b^2}{4a^2}\right) \quad \text{Complete the square by adding and subtracting } a(b^2/4a^2). \text{ Look closely at this step!}$$

$$= a\left(x + \frac{b}{2a}\right)^2 + c - \frac{b^2}{4a}$$

$$= a\left(x + \frac{b}{2a}\right)^2 + \frac{4ac - b^2}{4a}$$

If we let $h = -b/2a$ and let $k = (4ac - b^2)/4a$, this last equation can be rewritten in the form

$$f(x) = a(x - h)^2 + k \qquad\qquad (3)$$

The graph of f is the parabola $y = ax^2$ shifted horizontally h units and vertically k units. As a result, the vertex is at (h, k), and the graph opens up if $a > 0$ and down if $a < 0$. The axis is the vertical line $x = h$.

For example, compare equation (3) with equation (2) of Example 1.

$$f(x) = 2(x + 2)^2 - 3$$

$$f(x) = a(x - h)^2 + k$$

We conclude that $a = 2$, so the graph opens up. Also, we find that $h = -2$ and $k = -3$, so its vertex is at $(-2, -3)$.

It is not required to complete the square to obtain the vertex. In almost every case, it is easier to obtain the vertex of a quadratic function f by remembering that its x-coordinate is $h = -b/2a$. The y-coordinate can then be found by evaluating f at $-b/2a$.

These results are summarized next:

Characteristics of the Graph of a Quadratic Function

$$f(x) = ax^2 + bx + c$$

$$\text{Vertex} = \left(\frac{-b}{2a}, f\left(\frac{-b}{2a}\right)\right) \qquad \text{Axis: The line } x = \frac{-b}{2a} \qquad (4)$$

Parabola opens up if $a > 0$. Parabola opens down if $a < 0$.

E X A M P L E   2   *Locating the Vertex without Graphing*

Without graphing, locate the vertex and axis of the parabola defined by $f(x) = -3x^2 + 6x + 1$. Does it open up or down?

Solution   For this quadratic function, $a = -3$, $b = 6$, and $c = 1$. The x-coordinate of the vertex is

$$\frac{-b}{2a} = \frac{-6}{-6} = 1$$

The y-coordinate of the vertex is therefore

$$f\left(\frac{-b}{2a}\right) = f(1) = -3 + 6 + 1 = 4$$

The vertex is located at the point (1, 4). The axis of symmetry is the line $x = 1$. Finally, because $a = -3 < 0$, the parabola opens down. ∎

The information we gathered in Example 2, together with the location of the intercepts, usually provides enough information to graph $f(x) = ax^2 + bx + c$, $a \neq 0$, by hand. The y-intercept is the value of f at $x = 0$, that is, $f(0) = c$. The x-intercepts, if there are any, are found by solving the equation

$$f(x) = ax^2 + bx + c = 0$$

This equation has two, one, or no real solutions, depending on whether the discriminant $b^2 - 4ac$ is positive, 0, or negative. Thus, it has corresponding x-intercepts, as follows:

The x-Intercepts of a Quadratic Function

1. If the discriminant $b^2 - 4ac > 0$, the graph of $f(x) = ax^2 + bx + c$ has two distinct x-intercepts and so will cross the x-axis in two places.
2. If the discriminant $b^2 - 4ac = 0$, the graph of $f(x) = ax^2 + bx + c$ has one x-intercept and touches the x-axis at its vertex.
3. If the discriminant $b^2 - 4ac < 0$, the graph of $f(x) = ax^2 + bx + c$ has no x-intercept and so will not cross or touch the x-axis.

Figure 8 illustrates these possibilities for parabolas that open up.

FIGURE 8
$f(x) = ax^2 + bx + c,\ a > 0$

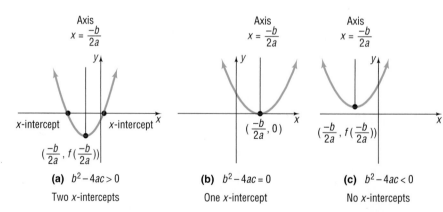

(a) $b^2 - 4ac > 0$
Two x-intercepts

(b) $b^2 - 4ac = 0$
One x-intercept

(c) $b^2 - 4ac < 0$
No x-intercepts

E X A M P L E   3   *Graphing a Quadratic Function By Hand Using Its Vertex, Axis, and Intercepts*

Use the information from Example 2 and the locations of the intercepts to graph $f(x) = -3x^2 + 6x + 1$.

FIGURE 9

$f(x) = -3x^2 + 6x + 1$

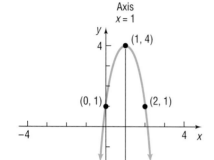

Solution In Example 2, we found the vertex to be at (1, 4) and the axis to be $x = 1$. The y-intercept is found by letting $x = 0$. Thus, the y-intercept is $f(0) = 1$. The x-intercepts are found by letting $f(x) = 0$. This results in the equation

$$-3x^2 + 6x + 1 = 0$$

The discriminant $b^2 - 4ac = (6)^2 - 4(-3)(1) = 36 + 12 = 48 > 0$, so the equation has two real solutions and the graph has two x-intercepts.
Using the quadratic formula, we find

$$x = \frac{-b + \sqrt{b^2 - 4ac}}{2a} = \frac{-6 + \sqrt{48}}{-6} = \frac{-6 + 4\sqrt{3}}{-6} \approx -0.15$$

and

$$x = \frac{-b - \sqrt{b^2 - 4ac}}{2a} = \frac{-6 - \sqrt{48}}{-6} = \frac{-6 - 4\sqrt{3}}{-6} \approx 2.15$$

The x-intercepts are approximately -0.15 and 2.15.

The graph is illustrated in Figure 9. Notice how we used the y-intercept and the axis of symmetry, $x = 1$, to obtain the additional point (2, 1) on the graph. ■

E X A M P L E 4 *Analyzing the Graph of a Quadratic Function*

Graph $f(x) = x^2 - 6x + 9$. Determine whether the graph opens up or down. Find its vertex, axis of symmetry, y-intercept, and x-intercepts, if any.

Solution Figure 10 shows the graph of $f(x) = x^2 - 6x + 9$.

FIGURE 10

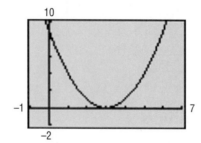

From the graph, we conjecture there is one y-intercept near 9 and one x-intercept near 3. The parabola opens up.

Now we confirm these conclusions using algebra. The parabola opens up since $a = 1 > 0$. Since $a = 1$, $b = -6$, and $c = 9$, the x-coordinate of the vertex is

$$\frac{-b}{2a} = \frac{-(-6)}{2(1)} = 3$$

The y-coordinate of the vertex is

$$f(3) = (3)^2 - 6(3) + 9 = 0$$

So the vertex is at (3, 0). The axis of symmetry is the line $x = 3$. The y-intercept is $f(0) = 9$. Since the vertex (3, 0) lies on the x-axis, the graph touches the x-axis at the x-intercept. ■

■ Now work Problem 25.

E X A M P L E   5   *Analyzing the Graph of a Quadratic Function*

Graph $f(x) = 2x^2 + x + 1$. Determine whether the graph opens up or down. Find its vertex, axis of symmetry, y-intercept, and x-intercepts, if any.

Solution   Graph the function. See Figure 11. From the graph we can see the graph opens up. We also observe the graph has no x-intercepts and one y-intercept 1.

Now we confirm these conclusions. Since $a = 2$, $b = 1$, and $c = 1$, the x-coordinate of the vertex is

$$\frac{-b}{2a} = -\frac{1}{4}$$

The y-coordinate of the vertex is

$$f\left(-\frac{1}{4}\right) = 2\left(\frac{1}{16}\right) + \left(-\frac{1}{4}\right) + 1 = \frac{7}{8}$$

So the vertex is at $(-\frac{1}{4}, \frac{7}{8})$. The axis of symmetry is the line $x = -\frac{1}{4}$. The y-intercept is $f(0) = 1$. Since the discriminant $b^2 - 4ac = (1)^2 - 4(2)(1) = -7 < 0$, the equation $2x^2 + x + 1 = 0$ has no real solutions and therefore the graph has no x-intercepts, confirming the conclusions reached earlier.

FIGURE 11

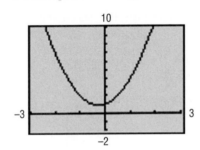

■   Now work Problem 31.

Applications

We have already seen that the graph of a quadratic function $f(x) = ax^2 + bx + c$ is a parabola with vertex at $(-b/2a, f(-b/2a))$. This vertex is the highest point on the graph if $a < 0$ and the lowest point on the graph if $a > 0$. If the vertex is the highest point ($a < 0$), then $f(-b/2a)$ is the **maximum value** of f. If the vertex is the lowest point ($a > 0$), then $f(-b/2a)$ is the **minimum value** of f. These ideas give rise to many applications.

E X A M P L E   6   *Maximizing Revenue*

A store selling calculators has found that, when the calculators are sold at a price of p dollars per unit, the revenue R (in dollars) as a function of the price p is

$$R(p) = -750p^2 + 15{,}000p$$

What unit price should be established in order to maximize revenue? If this price is charged, what is the maximum revenue?

Solution   The revenue R is

$$R(p) = -750p^2 + 15{,}000p = ap^2 + bp + c$$

The function R is a quadratic function with $a = -750$, $b = 15{,}000$, and $c = 0$. Because $a < 0$, the vertex is the highest point of the parabola. The revenue R is therefore a maximum when the price p is

$$p = \frac{-b}{2a} = \frac{-15{,}000}{2(-750)} = \frac{-15{,}000}{-1500} = \$10$$

The maximum revenue R is

$$R(10) = -750(10)^2 + 15{,}000(10) = \$75{,}000$$

See Figure 12 for an illustration.

FIGURE 12

$R(p) = -750p^2 + 15{,}000p$

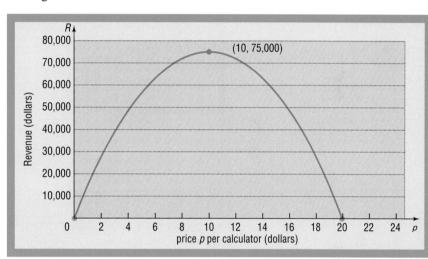

E X A M P L E   7   *Analyzing the Motion of a Projectile*

A projectile is fired from a cliff 500 feet above the water at an inclination of $45°$ to the horizontal, with a muzzle velocity of 400 feet per second. The height h of the projectile above the water is given by

$$h(x) = \frac{-32x^2}{(400)^2} + x + 500$$

where x is the horizontal distance of the projectile from the base of the cliff. See Figure 13.

FIGURE 13

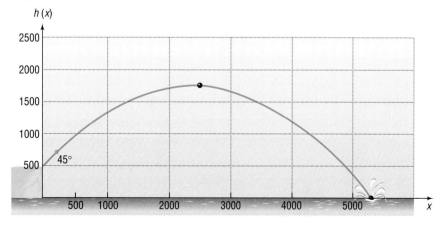

(a) Using a graphing utility, graph the function h, $0 \leq x \leq 6000$

(b) Find the maximum height of the projectile.

(c) How far from the base of the cliff will the projectile strike the water?

(d) TRACE the graph and compare the results to the solutions found in (b) and (c).

Solution (a) The graph of h is shown in Figure 14.

FIGURE 14

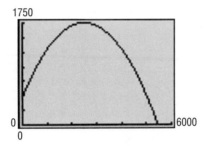

(b) The height of the projectile is given by a quadratic function:

$$h(x) = \frac{-32x^2}{(400)^2} + x + 500 = \frac{-1}{5000}x^2 + x + 500$$

We are looking for the maximum value of h. Since the maximum value is obtained at the vertex, we compute

$$x = \frac{-b}{2a} = \frac{-1}{2(-1/5000)} = \frac{5000}{2} = 2500$$

The maximum height of the projectile is

$$h(2500) = \frac{-1}{5000}(2500)^2 + 2500 + 500 = -1250 + 2500 + 500 = 1750 \text{ ft}$$

(c) The projectile will strike the water when its height is zero. To find the distance x travelled, we need to solve the equation

$$h(x) = \frac{-1}{5000}x^2 + x + 500 = 0$$

We use the quadratic formula with

$$b^2 - 4ac = 1 - 4\left(\frac{-1}{5000}\right)(500) = 1.4$$

$$x = \frac{-1 \pm \sqrt{1.4}}{2(-1/5000)} = \begin{cases} -458 \\ 5458 \end{cases}$$

We discard the negative solution and find that the projectile will strike the water a distance of 5458 feet from the base of the cliff.

(d) Figure 15 shows a TRACE of $h(x)$. The function is maximum for $x = 2489.36$ and $y = 1749.97$. These estimates could be improved using BOX or the maximum function on your graphing utility.

FIGURE 15

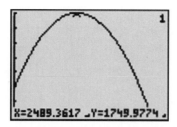

X=2489.3617 _Y=1749.9774

■   Now work Problem 61.

E X A M P L E   8

Navigation

A cruise ship leaves the Port of Miami heading due east at a constant speed of 5 knots (1 knot = 1 nautical mile per hour). At 5:00 PM, the cruise ship is 5 nautical miles due south of a cabin cruiser that is moving south at a constant speed of 10 knots. At what time are the two ships closest?

Solution   We begin with an illustration depicting the relative position of each ship at 5:00 PM. See Figure 16(a) on page 266. After a time t (in hours) has passed, the cruise ship has moved east $5t$ nautical miles, and the cabin cruiser has moved south $10t$ nautical miles. Figure 16(b) illustrates the relative position of each ship after t hours. Figure 16(c) shows a right triangle extracted from Figure 16(b). By the Pythagorean Theorem, the square of the distance d between the ships after time t is

$$d^2 = (5 - 10t)^2 + (5t)^2$$
$$= 125t^2 - 100t + 25$$

Now, the distance d is a minimum when d^2 is a minimum. Because d^2 is a quadratic function of t, it follows that d^2, and hence d, is a minimum when

$$t = \frac{-b}{2a} = \frac{100}{2(125)} = \frac{2}{5} \text{ hour}$$

Thus, the ships are closest after $\frac{2}{5}(60) = 24$ minutes, that is, at 5:24 PM.   ■

In a suspension bridge, the main cables are of parabolic shape because, if the total weight of a bridge is uniformly distributed along its length, the only cable shape that will bear the load evenly is that of a parabola. The Golden Gate Bridge in San Francisco is an example of a suspension bridge.

E X A M P L E   9

The Golden Gate Bridge

The Golden Gate Bridge, a suspension bridge, spans the entrance to San Francisco Bay. Its 746-foot-tall towers are 4200 feet apart. The bridge is suspended from two huge cables more than 3 feet in diameter; the 90-foot-wide roadway is 220 feet above the water. The cables are parabolic in shape and touch the road surface at the center of the bridge. Find the height of the cable at a distance of 1000 feet from the center.

FIGURE 16

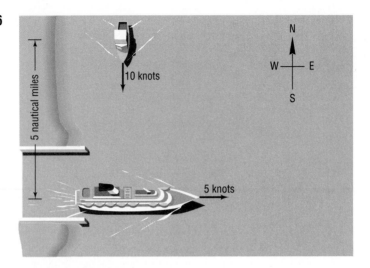

(a) Position at 5:00 PM

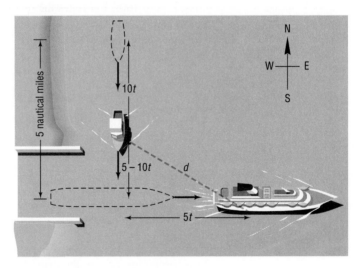

(b) Position at time t

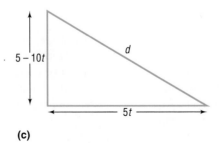

(c)

Solution   We begin by choosing the placement of the coordinate axes so that the *x*-axis co-incides with the road surface and the origin coincides with the center of the bridge. As a result, the twin towers will be vertical (height $746 - 220 = 526$ feet above the road) and located 2100 feet from the center. Also, the cable, which has the shape of a parabola, will extend from the towers, open up, and have its vertex at $(0, 0)$. As illustrated in Figure 17, the choice of placement of the axes enables us to identify the equation of the parabola as $y = ax^2$, $a > 0$. We also can see that the points $(-2100, 526)$ and $(2100, 526)$ are on the graph.

FIGURE 17

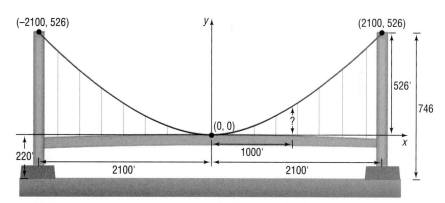

Based on these facts, we can find the value of a in $y = ax^2$:

$$y = ax^2$$
$$526 = a(2100)^2$$
$$a = \frac{526}{(2100)^2}$$

The equation of the parabola is therefore

$$y = \frac{526}{(2100)^2}x^2$$

The height of the cable when $x = 1000$ is

$$y = \frac{526}{(2100)^2}(1000)^2 \approx 119.3 \text{ feet}$$

Thus, the cable is 119.3 feet high at a distance of 1000 feet from the center of the bridge.   ■

4.1

Exercise 4.1

In Problems 1–8, match each graph on page 268 to one of the following functions:

A. $y = x^2 - 1$

B. $y = -x^2 - 1$

C. $y = x^2 - 2x + 1$

D. $y = x^2 + 2x + 1$

E. $y = x^2 - 2x + 2$

F. $y = x^2 + 2x$

G. $y = x^2 - 2x$

H. $y = x^2 + 2x + 2$

1.

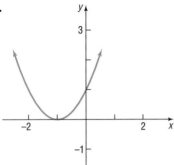

2.

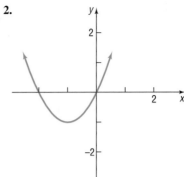

3.

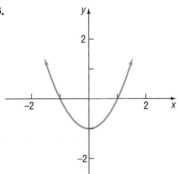

4.

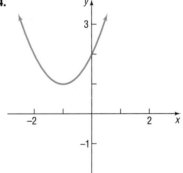

5.

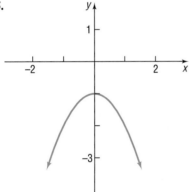

6.

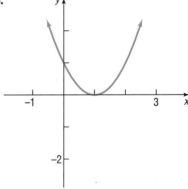

7.

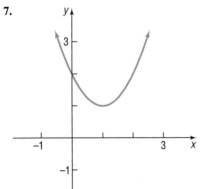

8.

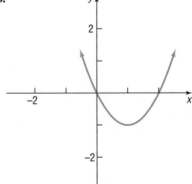

In Problems 9–12, match each graph to one of the functions listed below.

A. $y = (1/3)x^2 + 2$

B. $y = 3x^2 + 2$

C. $y = x^2 + 5x + 1$

D. $y = -x^2 + 5x + 1$

9.

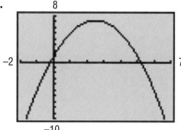

10.

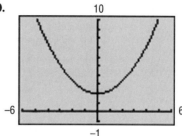

11.

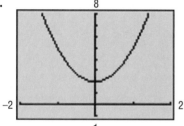

12.

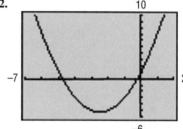

In Problems 13–42, graph each quadratic function using a graphing utility. Determine whether the graph opens up or down. Find its vertex, axis of symmetry, y-intercepts, and x-intercepts, if any.

13. $f(x) = \dfrac{1}{4}x^2$

14. $f(x) = 2x^2$

15. $f(x) = \dfrac{1}{4}x^2 - 2$

16. $f(x) = 2x^2 - 3$

17. $f(x) = \dfrac{1}{4}x^2 + 2$

18. $f(x) = 2x^2 + 4$

19. $f(x) = -\dfrac{1}{4}x^2 + 1$

20. $f(x) = -2x^2 - 2$

21. $f(x) = x^2 + 4x + 2$

22. $f(x) = x^2 - 6x - 1$

23. $f(x) = 2x^2 - 4x + 1$

24. $f(x) = 3x^2 + 6x$

25. $f(x) = -x^2 - 2x$

26. $f(x) = -2x^2 + 6x + 2$

27. $f(x) = \dfrac{1}{2}x^2 + x - 1$

28. $f(x) = \dfrac{2}{3}x^2 + \dfrac{4}{3}x - 1$

29. $f(x) = x^2 + 2x - 8$

30. $f(x) = x^2 - 2x - 3$

31. $f(x) = -x^2 - 3x + 4$

32. $f(x) = -x^2 + x + 2$

33. $f(x) = x^2 + 2x + 1$

34. $f(x) = -x^2 + 4x - 4$

35. $f(x) = 2x^2 - x + 2$

36. $f(x) = 4x^2 - 2x + 1$

37. $f(x) = -2x^2 + 2x - 3$

38. $f(x) = -3x^2 + 3x - 2$

39. $f(x) = 3x^2 + 6x + 2$

40. $f(x) = 2x^2 + 5x + 3$

41. $f(x) = -4x^2 - 6x + 2$

42. $f(x) = 3x^2 - 8x + 2$

In Problems 43–48, determine whether the given quadratic function has a maximum value or a minimum value and then find the value.

43. $f(x) = 2x^2 + 12x - 3$

44. $f(x) = 4x^2 - 8x + 3$

45. $f(x) = -x^2 + 10x - 4$

46. $f(x) = -2x^2 + 8x + 3$

47. $f(x) = -3x^2 + 12x + 1$

48. $f(x) = 4x^2 - 4x$

49. On one set of coordinate axes, graph the family of parabolas $f(x) = x^2 + 2x + c$ for $c = -3$, $c = 0$, and $c = 1$. Describe the characteristics of a member of this family.

50. On one set of coordinate axes, graph the family of parabolas $f(x) = x^2 + cx + 1$ for $c = -4$, $c = 0$, and $c = 4$. Describe the general characteristics of this family.

51. Graph $y = x^2 + 1$. Then graph $y = x^2 + x + 1$, followed by $y = x^2 + 2x + 1$, followed by $y = x^2 + 3x + 1$. What is happening? Do you see any pattern?

52. Graph $y = x^2 + x + 1$. Then graph $y = 2x^2 + x + 1$, followed by $y = 3x^2 + x + 1$, followed by $y = 4x^2 + x + 1$. What is happening? Do you see any pattern?

53. *Maximizing Revenue* Suppose that the manufacturer of a gas clothes dryer has found that, when the unit price is p dollars, the revenue R (in dollars) is

$$R = -4p^2 + 4000p$$

What unit price should be established for the dryer to maximize revenue? What is the maximum revenue?

54. *Maximizing Revenue* A tractor company has found that the revenue from sales of heavy-duty tractors is a function of the unit price p it charges. If the revenue R is

$$R = -\frac{1}{2}p^2 + 1900p$$

what unit price p should be charged to maximize revenue? What is the maximum revenue?

55. *Rectangles with Fixed Perimeter* What is the largest rectangular area that can be enclosed with 400 feet of fencing? What are the dimensions of the rectangle?

56. *Rectangles with Fixed Perimeter* What are the dimensions of a rectangle of a fixed perimeter P that result in the largest area?

57. *Enclosing the Most Area with a Fence* A farmer with 4000 meters of fencing wants to enclose a rectangular plot that borders on a river. If the farmer does not fence the side along the river, what is the largest area that can be enclosed? (See the figure.)

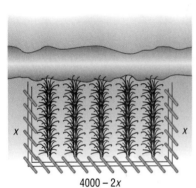

$4000 - 2x$

58. *Enclosing the Most Area with a Fence* A farmer with 2000 meters of fencing wants to enclose a rectangular plot that borders on a straight highway. If the farmer does not fence the side along the highway, what is the largest area that can be enclosed?

59. *Enclosing the Most Area with a Fence* A farmer with 10,000 meters of fencing wants to enclose a rectangular field and then divide it into two plots with a fence parallel to one of the sides (see the figure). What is the largest area that can be enclosed?

60. *Enclosing the Most Area with a Fence* A farmer with 10,000 meters of fencing wants to enclose a rectangular field and then divide it into three plots with two fences parallel to one of the sides. What is the largest area that can be enclosed?

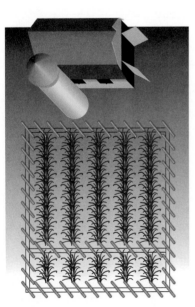

61. *Analyzing the Motion of a Projectile* A projectile is fired from a cliff 200 feet above the water at an inclination of 45° to the horizontal, with a muzzle velocity of 50 feet per second. The height h of the projectile above the water is given by

$$h(x) = \frac{-32x^2}{(50)^2} + x + 200$$

where x is the horizontal distance of the projectile from the base of the cliff.

(a) Using a graphing utility, graph the function h, $0 \le x \le 200$.

(b) Find the maximum height of the projectile.

(c) How far from the base of the cliff will the projectile strike the water?

(d) TRACE the path of the projectile, noting its maximum height and the distance from the base of the cliff at which it strikes the water. Compare your results with those obtained in parts (b) and (c). When the height of the projectile is 100 feet above the water, how far is it from the cliff?

62. *Analyzing the Motion of a Projectile* A projectile is fired at an inclination of 45° to the horizontal, with a muzzle velocity of 100 feet per second. The height h of the projectile is given by

$$h(x) = \frac{-32x^2}{(100)^2} + x$$

where x is the horizontal distance of the projectile from the firing point.

(a) Using a graphing utility, graph the function h, $0 \le x \le 700$.
(b) Find the maximum height of the projectile.
(c) How far from the firing point will the projectile strike the ground?
(d) TRACE the path of the projectile, noting its maximum height and the distance from the firing point to the point at which it strikes the ground. Compare your results with those obtained in parts (a) and (b). When the height of the projectile is 50 feet above the ground, how far has it traveled horizontally?

63. *Navigation* An aircraft carrier maintains a constant speed of 10 knots heading due north. At 4:00 PM, the ship's radar detects a destroyer 100 nautical miles due east of the carrier. If the destroyer is heading due west at 20 knots, when will the two ships be the closest? (1 knot = 1 nautical mile per hour)

64. *Air Traffic Control* An air traffic controller sees two aircraft flying at the same altitude on his screen. One, a Piper Cub, is headed due west at 150 miles per hour. The other, a Lear jet, is 15 miles due north of the Piper and is headed due south at 400 miles per hour. How close will the two aircraft come to each other?

65. *Suspension Bridge* A suspension bridge with weight uniformly distributed along its length has twin towers that extend 75 meters above the road surface and are 400 meters apart. The cables are parabolic in shape and are suspended from the tops of the towers. The cables touch the road surface at the center of the bridge. Find the height of the cables at a point 100 meters from the center. (Assume that the road is level.)

66. *Architecture* A parabolic arch has a span of 120 feet and a maximum height of 25 feet. Choose suitable rectangular coordinate axes and find the equation of the parabola. Then calculate the height of the arch at points 10 feet, 20 feet, and 40 feet from the center.

67. *Constructing Rain Gutters* A rain gutter is to be made of aluminum sheets that are 12 inches wide by turning up the edges 90°. What depth will provide maximum cross-sectional area and hence allow the most water to flow?

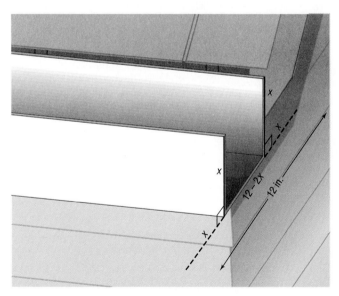

68. *Navigation* At 4 PM a cruise ship leaves the Port of Miami heading due east at a constant speed of 15 knots. At the same time, a pleasure boat located 100 nautical miles northeast of the Port of Miami is headed due south at a constant speed of 12 knots. When are the two ships closest? How close do they get to each other? (Express your answer in nautical miles; 1 knot = 1 nautical mile per hour.)

69. *Norman Windows* A Norman window has the shape of a rectangle surmounted by a semicircle of diameter equal to the width of the rectangle (see the figure). If the perimeter of the window is 20 feet, what dimensions will admit the most light (maximize the area)? [*Hint:* Circumference of circle $= 2\pi r$; Area of circle $= \pi r^2$, where r is the radius of the circle.]

70. *Constructing a Stadium* A track and field playing area is in the shape of a rectangle with semicircles at each end (see the figure). The inside perimeter of the track is to be 1500 meters. What should the dimensions of the rectangle be so that the area of the rectangle is a maximum?

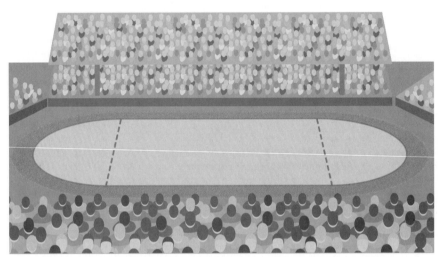

71. *Architecture* A special window has the shape of a rectangle surmounted by an equilateral triangle (see the figure). If the perimeter of the window is 16 feet, what dimensions will admit the most light? [*Hint:* Area of an equilateral triangle $= (\sqrt{3}/4)x^2$, where x is the length of a side of the triangle.]

72. *Analyzing the Motion of a Projectile* A projectile is fired at an inclination of 45° to the horizontal with an initial velocity of v_0 feet per second. If the starting point is the origin, the x-axis is horizontal, and the y-axis is vertical, then the height y (in feet) after a horizontal distance x has been traversed is approximately

$$y = \frac{-32x^2}{v_0^2} + x$$

(a) Find the maximum height in terms of the initial velocity v_0.
(b) If the initial velocity is doubled, what happens to the maximum height?
(c) Assuming the ground is flat, how far from the starting point will the projectile land if the initial velocity is 64 feet per second?

73. *Charter Club Memberships* A charter flight club charges its members $400 per year. But, for each new member in excess of 60, the charge for every member is reduced by $5. What number of members leads to a maximum revenue?

74. *Car Rentals* A car rental agency has 24 identical cars. The owner of the agency finds that all the cars can be rented at a price of $10 per day. However, for each $2 increase in rental, one of the cars is not rented. What should be charged to maximize income?

75. The graph of the function $f(x) = ax^2 + bx + c$ has vertex at $x = 0$ and passes through the points $(0, 2)$ and $(1, 8)$. Find a, b, and c.

76. The graph of the function $f(x) = ax^2 + bx + c$ has vertex at $x = 1$ and passes through the points $(0, 1)$ and $(-1, -8)$. Find a, b, and c.

77. *Chemical Reactions* A self-catalytic chemical reaction results in the formation of a compound that causes the formation ratio to increase. If the reaction rate V is given by

$$V(x) = kx(a - x) \qquad 0 \le x \le a$$

where k is a positive constant, a is the initial amount of the compound, and x is the variable amount of the compound, for what value of x is the reaction rate a maximum?

78. A rectangle has one vertex on the line $y = 10 - x$, $x > 0$, another at the origin, one on the positive x-axis, and one on the positive y-axis. Find the largest area A that can be enclosed by the rectangle.

79. *Calculus: Simpson's Rule* The figure shows the graph of $y = ax^2 + bx + c$. Suppose the points $(-h, y_0)$, $(0, y_1)$, and (h, y_2) are on the graph. It can be shown that the area enclosed by the parabola, the x-axis and the lines $x = -h$ and $x = h$ is

$$\text{Area} = \frac{h}{3}(2ah^2 + 6c)$$

Show that this area may also be given by

$$\text{Area} = \frac{h}{3}(y_0 + 4y_1 + y_2)$$

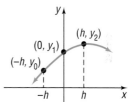

80. Let $f(x) = ax^2 + bx + c$, where a, b, and c are odd integers. If x is an integer, show that $f(x)$ must be an odd integer. [*Hint:* x is either an even integer or an odd integer.]

 81. Make up a quadratic function that opens down and has only one x-intercept. Compare yours with others in the class. What are the similarities? What are the differences?

82. The figure illustrates actual data on enrollment in all public schools (both elementary and high schools) for the academic years 1980–1981 to 1988–89. Assume that the points are the points on a parabola. Further assume that a minimum enrollment of 39 million occurred between 1984–1985 and 1985–1986. Find an equation for this parabola. Assume that the trend continues and use the equation to predict public school enrollment in 1991–1992. Determine what the actual enrollment in 1991–1992 was. How does your projection compare?

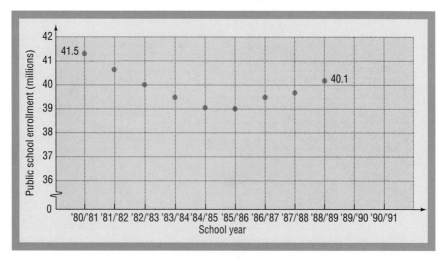

83. *CBL Experiment* As a ball bounces up and down, the maximum height the ball attains continually decreases from one bounce to the next. For a given bounce, plotting the height of the ball against time results in a parabola. The motion of the ball will be analyzed. (Activity 9, Real-World Math with the CBL System, 1994.)

84. *CBL Experiment* A cart is pushed up a ramp and allowed to return down the ramp. Plotting the distance the cart is from a motion detector against time results in the graph of a parabola. The parabola will be analyzed to determine the characteristics of the cart's motion. (Activity 8, Real-World Math with the CBL System, 1994.)

4.2

Polynomial Functions

Polynomial Function

A **polynomial function** is a function of the form

$$f(x) = a_n x^n + a_{n-1} x^{n-1} + \ldots + a_1 x + a_0 \qquad (1)$$

where $a_n, a_{n-1}, \ldots, a_1, a_0$ are real numbers and n is a non-negative integer. The domain consists of all real numbers.

Thus, a polynomial function is a function whose rule is given by a polynomial in one variable.* The **degree** of a polynomial function is the degree of the polynomial in one variable.

E X A M P L E 1   *Identifying Polynomial Functions*

Determine which of the following are polynomial functions. For those that are, state the degree; for those that are not, tell why not.

(a) $f(x) = 2 - 3x^4$ (b) $g(x) = \sqrt{x}$ (c) $h(x) = \dfrac{x^2 - 2}{x^3 - 1}$

(d) $F(x) = 0$ (e) $G(x) = 8$

Solution   (a) f is a polynomial function of degree 4.

(b) g is not a polynomial function. The variable x is raised to the $\frac{1}{2}$ power, which is not a nonnegative integer.

(c) h is not a polynomial function. It is the ratio of two polynomials, and the polynomial in the denominator is of positive degree.

(d) F is the zero polynomial function; it is not assigned a degree.

(e) G is a nonzero constant function, a polynomial function of degree 0.   ■

■ Now work Problems 1 and 5.

We have already discussed in detail polynomial functions of degrees 0, 1, and 2. See Table 1 for a summary of the characteristics of the graphs of these polynomial functions.

TABLE 1

DEGREE	FORM	NAME	GRAPH
No degree	$f(x) = 0$	Zero function	The x-axis
0	$f(x) = a_0, \quad a_0 \neq 0$	Constant function	Horizontal line with y-intercept a_0
1	$f(x) = a_1 x + a_0, \quad a_1 \neq 0$	Linear function	Nonvertical, nonhorizontal line with slope a_1 and y-intercept a_0
2	$f(x) = a_2 x^2 + a_1 x + a_0, \quad a_2 \neq 0$	Quadratic function	Parabola: graph opens up if $a_2 > 0$; graph opens down if $a_2 < 0$

*A review of polynomials may be found in the Appendix, Section 2.

Power Functions

First, we consider a special kind of polynomial function called a *power function*.

Power Function of Degree *n*

A **power function of degree *n*** is a function of the form

$$f(x) = ax^n \tag{2}$$

where a is a real number, $a \neq 0$, and $n > 0$ is an integer.

The graph of a power function of degree 1, $f(x) = ax$, is a straight line, with slope a, that passes through the origin. The graph of a power function of degree 2, $f(x) = ax^2$, is a parabola, with vertex at the origin, that opens up if $a > 0$ and down if $a < 0$.

If we know how to graph a power function of the form $f(x) = x^n$, then a compression or stretch and, perhaps, a reflection about the *x*-axis will enable us to obtain the graph of $g(x) = ax^n$. Consequently, we shall concentrate on graphing power functions of the form $f(x) = x^n$.

We begin with power functions of even degree of the form $f(x) = x^n$, $n \geq 2$ and *n even*. Using your graphing utility and a RANGE of $-2 \leq x \leq 2$; $-4 \leq y \leq 16$, graph the function $f(x) = x^4$. Can you conjecture the domain of this function from the graph? The range? On the same screen, graph $f(x) = x^8$. Now, also on the same screen, graph $f(x) = x^{12}$. What do you notice about the graphs as the magnitude of the exponent increases? Repeat the procedure given above for a RANGE of $-1 \leq x \leq 1$; $0 \leq y \leq 1$. What do you notice? See Figure 18 (a) and (b).

From the graphs, we can conjecture a few characteristics of even power functions. First, even power functions are symmetric with respect to the *y*-axis. Second, the domain is the set of all real numbers and the range is the set of nonnegative real numbers. Third, the graph always contains the points $(0, 0)$, $(1, 1)$, and $(-1, 1)$. Fourth, as the exponent increases in magnitude, the graph increases very rapidly as *x* increases, but for *x* near the origin the graph tends to flatten out and lie closer to the origin. Note that for large *n*, it appears the graph coincides with the *x*-axis near the origin, but it does not; the graph actually touches the *x*-axis only at the origin (see Table 2). Also, for large *n*, it may appear that for $x < -1$ or for $x > 1$, the graph is vertical, but it is not; it is only increasing very rapidly. If you use the TRACE function along one of the graphs these distinctions would be clear.

FIGURE 18

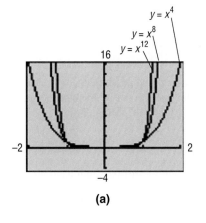

(a)

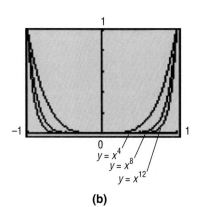

(b)

TABLE 2

	$x = 0.1$	$x = 0.3$	$x = 0.5$
$f(x) = x^8$	10^{-8}	0.0000656	0.0039063
$f(x) = x^{20}$	10^{-20}	$3.487 \cdot 10^{-11}$	0.000001
$f(x) = x^{40}$	10^{-40}	$1.216 \cdot 10^{-21}$	$9.095 \cdot 10^{-13}$

Now we consider power functions of odd degree of the form $f(x) = x^n$, n *odd*. Using your graphing utility and a RANGE of $-2 \le x \le 2$; $-16 \le y \le 16$, graph the function $f(x) = x^3$. You have seen this graph several times before. Based on the graph, can you conjecture the domain and range of this function? On the same screen, graph $f(x) = x^7$ and $f(x) = x^{11}$. What do you notice about the graphs as the magnitude of the exponent increases? Repeat the procedure given above for a RANGE of $-1 \le x \le 1$; $-1 \le y \le 1$. What do you notice? The graphs on your screen should look like Figure 19 (a) and (b).

From the graphs in Figure 19 (a) and (b), we can conjecture a few characteristics of odd power functions. First, odd power functions are symmetric with respect to the origin. Second, the domain and range of odd power functions are the set of all real numbers. Third, the graph of an odd power function always contains the points $(0, 0)$, $(1, 1)$, and $(-1, -1)$. Fourth, as the exponent increases in magnitude, the graphs become more vertical when $x > 1$ or $x < -1$ and the graphs tend to flatten out and lie closer to the x-axis when x is near the origin. Note that it appears the graph coincides with the x-axis near the origin, but it does not; the graph actually touches the x-axis only at the origin. Also it appears that as x increases, the graph is vertical, but it is not, it is increasing very rapidly. Use the TRACE function along the graphs to verify these distinctions.

FIGURE 19

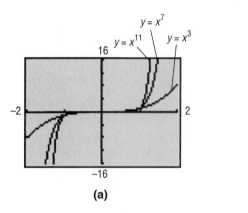

(a)

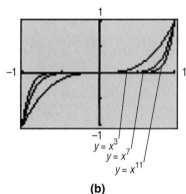

(b)

The methods of shifting, compression, stretching, and reflection studied in Section 2.3, when used with the facts just presented, will enable us to graph and analyze a variety of polynomials.

E X A M P L E   2   *Graphing In Stages*

Using a graphing utility, show each stage to obtain the graph of $f(x) = 1 - x^5$.

Solution   Stage 1: Graph: $y = x^5$. See Figure 20.

FIGURE 20

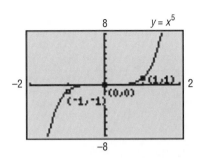

Stage 2: Graph: $y = -x^5$ (Reflection about y-axis). See Figure 21.

FIGURE 21

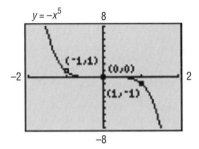

Stage 3: Graph: $y = 1 - x^5$ (Shift up 1 unit). See Figure 22.

FIGURE 22

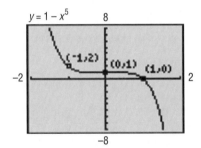

By following key points as we shift, reflect, etc., we determine the intercepts of f are $(0, 1)$ and $(1, 0)$. ∎

E X A M P L E   3   *Graphing In Stages*

Using a graphing utility, show each stage to obtain the graph of $f(x) = \frac{1}{2}(x - 1)^4$.

Solution   Stage 1: Graph: $y = x^4$. See Figure 23.

FIGURE 23

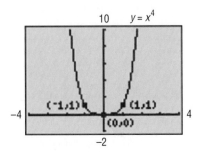

Stage 2:   Graph $y = \frac{1}{2}x^4$ (Compression). See Figure 24.

FIGURE 24

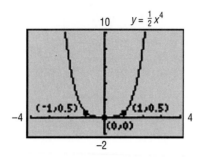

Stage 3:   Graph $y = \frac{1}{2}(x - 1)^4$ (Shift right 1 unit). See Figure 25.

FIGURE 25

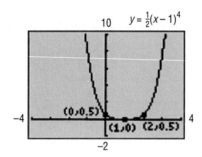

The intercepts of f are $(0, \frac{1}{2})$ and $(1, 0)$.   ■

Graphing Other Polynomials

FIGURE 26

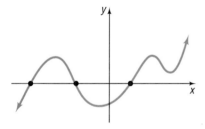

(a) Graph of a polynomial function: smooth, continuous

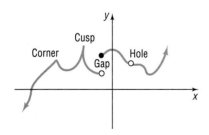

(b) Cannot be the graph of a polynomial function

To graph most polynomial functions of degree 3 or higher by hand requires techniques beyond the scope of this text. If you take a course in calculus you will learn that the graph of every polynomial function is both smooth and continuous. By **smooth,** we mean that the graph contains no sharp corners or cusps; by **continuous,** we mean that the graph has no gaps or holes and can be drawn without lifting pencil from paper. See Figures 26(a) and 26(b).

Figure 27 shows the graph of a polynomial function with four x-intercepts. Notice that at the x-intercepts the graph must either cross the x-axis or touch the x-axis. Consequently, between consecutive x-intercepts the graph is either above the x-axis or below the x-axis. We will make use of this characteristic of the graph of a polynomial shortly.

If a polynomial function f is factored completely, it is easy to solve the equation $f(x) = 0$ and locate the x-intercepts of the graph. For example, if $f(x) = (x - 1)^2(x + 3)$, then the solutions of the equation

$$f(x) = (x - 1)^2(x + 3) = 0$$

are easily identified as 1 and -3. In general, if f is a polynomial function and r is a real number for which $f(r) = 0$, then r is called a (real) **zero of** f, or **root of** f. *Thus, the real zeros of a polynomial function are the x-intercepts of its graph.* Also, if $x - r$ is a factor of a polynomial f, then $f(r) = 0$ and so r is a zero of f. If the

FIGURE 27

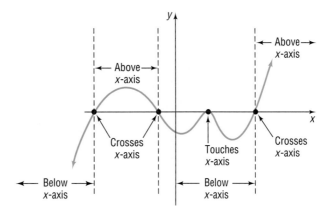

same factor $x - r$ occurs more than once, then r is called a **repeated**, or **multiple, zero of f.** More precisely, we have the following definition.

Zero of Multiplicity m	If $(x - r)^m$ is a factor of a polynomial f and $(x - r)^{m+1}$ is not a factor of f, then r is called a **zero of multiplicity m of f.**

E X A M P L E   4

Identifying Zeros and Their Multiplicities

For the polynomial

$$f(x) = 5(x - 2)(x + 3)^2\left(x - \frac{1}{2}\right)^4$$

2 is a zero of multiplicity 1

-3 is a zero of multiplicity 2

$\frac{1}{2}$ is a zero of multiplicity 4 ∎

Suppose it is possible to factor completely a polynomial function and, as a result, locate all the x-intercepts of its graph (the real zeros of the function). The following example illustrates the role the multiplicity of the x-intercept plays.

E X A M P L E   5

Investigating the Role of Multiplicity

For the polynomial: $f(x) = x^2(x - 2)$

(a) Find the x- and y-intercepts of the graph.

(b) Using a graphing utility, graph the polynomial.

(c) For each x-intercept, determine whether it is of odd or even multiplicity.

Solution

(a) The y-intercept is $f(0) = 0^2(0 - 2) = 0$. The x-intercepts satisfy the equation

$$f(x) = x^2(x - 2) = 0$$

from which we find

$$x^2 = 0 \quad \text{or} \quad x - 2 = 0$$
$$x = 0 \qquad \qquad x = 2$$

The x-intercepts are 0 and 2.

(b) See Figure 28.

FIGURE 28

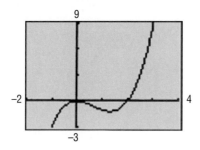

(c) We can see from the factored form of f that 0 is a zero or root of multiplicity 2 and 2 is a zero or root of multiplicity 1, so 0 is of even multiplicity and 2 is of odd multiplicity. ■

Notice from the graph in (b) that the graph of the polynomial at the root 0 just touches the x-axis so that the sign of $f(x)$ is the same on each side of the root. At the root 2, the graph of the polynomial crosses the x-axis so that the sign of $f(x)$ changes from one side of the root to the other. This observation leads us to the following result:

If r Is a Zero of Even Multiplicity	Sign of $f(x)$ does not change from one side to the other side of r.	Graph **touches** x-axis at r.

If r Is a Zero of Odd Multiplicity	Sign of $f(x)$ changes from one side to the other side of r.	Graph **crosses** x-axis at r.

■   Now work Problem 19.

Look again at Figure 28. We can use a graphing utility to determine the graph's minimum point in the interval $0 < x < 2$. After utilizing the BOX function (or function minimum if your utility has this feature), we find the graph's minimum point in the interval $0 < x < 2$ is $(1.33, -1.18)$ correct to two decimal places. In calculus, a method is presented for locating minimum and maximum points. There, such points are called *local minima* or *local maxima*. *Local minima* and *local maxima* are points where the graph changes direction (i.e., changes from an increasing function to a decreasing function or vice versa). We call these points **turning points.**

Look again at Figure 28. The graph of $f(x) = x^2(x - 2) = x^3 - 2x^2$, a polynomial of degree 3, has two turning points. Graph $y = x^3$. How many turning points do you see? How does the number of turning points relate to the degree? Graph $y = x^4$, $y = x^4 - \frac{4}{3}x^3$, and $y = x^4 - 2x^2$. How many turning points do you see? How does the number of turning points compare to the degree? The following theorem from calculus supplies the answer.

Theorem   If f is a polynomial function of degree n, then f has at most $n - 1$ turning points. ■

One last remark about Figure 28. Notice that the graph of $f(x) = x^2(x - 2)$ looks somewhat like the graph of $y = x^3$. In fact, for very large values of x, either positive or negative, there is little difference. To see for yourself, use your calculator to compare the values of $f(x) = x^2(x - 2)$ and $y = x^3$ for $x = -100,000$ and $x = 100,000$.

Theorem   For large values of x, either positive or negative, the graph of the polynomial

$$f(x) = a_n x^n + a_{n-1} x^{n-1} + \cdots + a_1 x + a_0$$

resembles the graph of the power function

$$y = a_n x^n$$ ■

The following display summarizes some features of the graph of a polynomial function.

Summary: Graph of a Polynomial Function
$f(x) = a_n x^n + a_{n-1} x^{n-1} + \cdots + a_1 x + a_0,$
$a_n \neq 0$

Degree of the polynomial f: n
Maximum number of turning points: $n - 1$
At zero of even multiplicity: graph of f touches x-axis
At zero of odd multiplicity: graph of f crosses x-axis
Between zeros, graph of f is either above the x-axis or below it.
For large x, graph of f behaves like graph of $y = a_n x^n$.

E X A M P L E   6   *Analyzing the Graph of a Polynomial Function*

For the polynomial: $f(x) = x^4 - 3x^3 - 4x^2$

(a) Using a graphing utility, graph f.

(b) Find the x- and y-intercepts.

(c) Determine whether each x-intercept is of odd or even multiplicity.

(d) Find the power function that the graph of f resembles for large values of x.

(e) Determine the number of turning points on the graph of f.

(f) Determine the *local maxima* and *local minima*, if any exist, correct to two decimal places.

Solution   (a) See Figure 29.

FIGURE 29

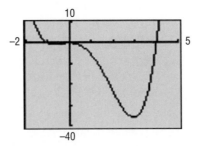

(b) The y-intercept is $f(0) = 0$. Factoring, we find $f(x) = x^4 - 3x^3 - 4x^2 = x^2(x - 4)(x + 1)$. We find the x-intercepts by solving the equation

$$f(x) = x^2(x - 4)(x + 1) = 0$$

So

$$x^2 = 0 \qquad \text{or} \qquad x - 4 = 0 \qquad \text{or} \qquad x + 1 = 0$$
$$x = 0 \qquad\qquad\qquad x = 4 \qquad\qquad\qquad x = -1$$

The x-intercepts are -1, 0, and 4.

(c) The intercept 0 is a zero of even multiplicity, 2, so the graph of f will touch the x-axis at 0; 4 and -1 are zeros of odd multiplicity, 1, so the graph of f will cross the x-axis at 4 and -1. This means the graph will cross the x-axis at -1 and 4, and touch the x-axis at 0. Look again at Figure 29. The graph is below the x-axis for $-1 < x < 0$!

(d) The graph of f behaves like $f(x) = x^4$ for large x.

(e) Since f is of degree 4, the graph can have at most three turning points. From the graph, we see the graph has three turning points: one between -1 and 0, one near $(0, 0)$, and one between 2 and 4.

(f) Correct to two decimal places, the *local maximum* is $(0.00, 0.00)$ and the *local minima* are $(-0.68, -0.69)$ and $(2.93, -36.10)$. ■

E X A M P L E   7   *Analyzing the Graph of a Polynomial Function*

Follow the instructions of Example 6 for the polynomial below:

$$f(x) = x^3 + 2.48x^2 - 4.3155x + 1.484406$$

Solution   (a) See Figure 30.

FIGURE 30

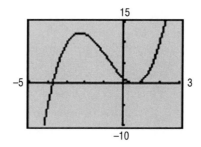

(b) The y-intercept is $f(0) = 1.484406$. In Example 6 we could easily factor $f(x)$ to find the x-intercepts. However, it is not readily apparent how $f(x)$ factors in this example. Therefore, we use a graphing utility and find the x-intercepts to be -3.74 and 0.63.

(c) The x-intercept -3.74 is of odd multiplicity since the graph of f crosses the x-axis at -3.74; the x-intercept 0.63 is of even multiplicity since the graph of f touches the x-axis at 0.63.

(d) The graph behaves like $f(x) = x^3$ for large x.

(e) Since f is of degree 3, the graph can have at most two turning points. From the graph we see the graph has two turning points: one between -3 and -2; the other at $(0.63, 0)$.

(f) Correct to two decimal places, the *local maximum* is $(-2.28, 12.36)$ and the *local minimum* is $(0.63, 0)$. ■

4.2

Exercise 4.2

In Problems 1–10, determine which functions are polynomial functions. For those that are, state the degree. For those that are not, tell why not.

1. $f(x) = 4x + x^3$

2. $f(x) = 5x^2 + 4x^4$

3. $g(x) = \dfrac{1 - x^2}{2}$

4. $h(x) = 3 - \dfrac{1}{2}x$

5. $f(x) = 1 - \dfrac{1}{x}$

6. $f(x) = x(x - 1)$

7. $g(x) = x^{3/2} - x^2 + 2$

8. $h(x) = \sqrt{x}(\sqrt{x} - 1)$

9. $F(x) = 5x^4 - \pi x^3 + \dfrac{1}{2}$

10. $F(x) = \dfrac{x^2 - 5}{x^3}$

In Problems 11–18, using a graphing utility, show the steps required to graph each function.

11. $f(x) = (x + 1)^4$

12. $f(x) = x^4 + 2$

13. $f(x) = \dfrac{1}{2}x^4$

14. $f(x) = -x^4$

15. $f(x) = 2(x + 1)^4 + 1$

16. $f(x) = 3 - (x + 2)^4$

17. $f(x) = -\dfrac{1}{2}(x - 2)^4 - 1$

18. $f(x) = 1 - 2(x + 1)^4$

In Problems 19–28, for each polynomial function, list each real zero and its multiplicity. Determine whether the graph crosses or touches the x-axis at each x-intercept.

19. $f(x) = 3(x - 7)(x + 3)^2$

20. $f(x) = 4(x + 4)(x + 3)^3$

21. $f(x) = 4(x^2 + 1)(x - 2)^3$

22. $f(x) = 2(x - 3)(x + 4)^3$

23. $f(x) = -2\left(x + \dfrac{1}{2}\right)^2(x^2 + 4)^2$

24. $f(x) = \left(x - \dfrac{1}{3}\right)^2(x - 1)^3$

25. $f(x) = (x - 5)^3(x + 4)^2$

26. $f(x) = (x + \sqrt{3})^2(x - 2)^4$

27. $f(x) = 3(x^2 + 8)(x^2 + 9)^2$

28. $f(x) = -2(x^2 + 3)^3$

In Problems 29–64, for each polynomial function f:

(a) Graph f.
(b) Find the x- and y-intercepts.
(c) Determine whether each x-intercept is of odd or even multiplicity.
(d) Find the power function that the graph of f resembles for large values of x.
(e) Determine the number of turning points on the graph of f.
(f) Determine the *local maxima* and *local minima*, if any exist, correct to two decimal places.

29. $f(x) = (x - 1)^2$

30. $f(x) = (x - 2)^3$

31. $f(x) = x^2(x - 3)$

32. $f(x) = x(x + 2)^2$

33. $f(x) = 6x^3(x + 4)$

34. $f(x) = 5x(x - 1)^3$

35. $f(x) = -4x^2(x + 2)$

36. $f(x) = -\dfrac{1}{2}x^3(x + 4)$

37. $f(x) = x(x - 2)(x + 4)$

38. $f(x) = x(x + 4)(x - 3)$

39. $f(x) = 4x - x^3$

40. $f(x) = x - x^3$

41. $f(x) = x^2(x - 2)(x + 2)$

42. $f(x) = x^2(x - 3)(x + 4)$

43. $f(x) = x^2(x - 2)^2$

44. $f(x) = x^3(x - 3)$

45. $f(x) = x^2(x - 3)(x + 1)$

46. $f(x) = x^2(x - 3)(x - 1)$

47. $f(x) = x(x + 2)(x - 4)(x - 6)$

48. $f(x) = x(x - 2)(x + 2)(x + 4)$

49. $f(x) = x^2(x - 2)(x^2 + 3)$

50. $f(x) = x^2(x^2 + 1)(x + 4)$

51. $f(x) = x^3 + 0.2x^2 - 1.5876x - 0.31752$

52. $f(x) = x^3 - 0.8x^2 - 4.6656x + 3.73248$

53. $f(x) = x^3 + 2.56x^2 - 3.31x + 0.89$

54. $f(x) = x^3 - 2.91x^2 - 7.668x - 3.8151$

55. $f(x) = x^4 - 2.5x^2 + 0.5625$

56. $f(x) = x^4 - 18.5x^2 + 50.2619$

57. $f(x) = x^4 + 0.65x^3 - 16.6319x^2 + 14.209335x - 3.1264785$

58. $f(x) = x^4 + 3.45x^3 - 11.6639x^2 - 5.864241x - 0.69257738$

59. $y = \pi x^3 + \sqrt{2}x^2 - x - 2$

60. $y = -2x^3 + \pi x^2 + \sqrt{3}x + 1$

61. $y = 2x^4 - \pi x^3 + \sqrt{5}x - 4$

62. $y = -1.2x^4 + 0.5x^2 - \sqrt{3}x + 2$

63. $y = -2x^5 - \sqrt{2}x^2 - x - \sqrt{2}$

64. $y = \pi x^5 + \pi x^4 + \sqrt{3}x + 1$

65. Consult the illustration. Which of the following polynomial functions might have this graph? (More than one answer may be possible.)

(a) $y = -4x(x - 1)(x - 2)$

(b) $y = x^2(x - 1)^2(x - 2)$

(c) $y = 3x(x - 1)(x - 2)$

(d) $y = x(x - 1)^2(x - 2)^2$

(e) $y = x^3(x - 1)(x - 2)$

(f) $y = -x(1 - x)(x - 2)$

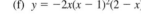

66. Consult the illustration. Which of the following polynomial functions might have this graph? (More than one answer may be possible.)

(a) $y = 2x^3(x - 1)(x - 2)^2$

(b) $y = x^2(x - 1)(x - 2)$

(c) $y = x^3(x - 1)^2(x - 2)$

(d) $y = x^2(x - 1)^2(x - 2)^2$

(e) $y = 5x(x - 1)^2(x - 2)$

(f) $y = -2x(x - 1)^2(2 - x)$

67. Consult the illustration. Which of the following polynomial functions might have this graph? (More than one answer may be possible.)

(a) $y = \dfrac{1}{2}(x^2 - 1)(x - 2)$

(b) $y = -\dfrac{1}{2}(x^2 + 1)(x - 2)$

(c) $y = (x^2 - 1)\left(1 - \dfrac{x}{2}\right)$

(d) $y = -\dfrac{1}{2}(x^2 - 1)^2(x - 2)$

(e) $y = \left(x^2 + \dfrac{1}{2}\right)(x^2 - 1)(2 - x)$

(f) $y = -(x - 1)(x - 2)(x + 1)$

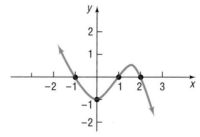

68. Consult the illustration. Which of the following polynomial functions might have this graph? (More than one answer may be possible.)

(a) $y = -\dfrac{1}{2}(x^2 - 1)(x - 2)(x + 1)$

(b) $y = -\dfrac{1}{2}(x^2 + 1)(x - 2)(x + 1)$

(c) $y = -\dfrac{1}{2}(x + 1)^2(x - 1)(x - 2)$

(d) $y = (x - 1)^2(x + 1)\left(1 - \dfrac{x}{2}\right)$

(e) $y = -(x - 1)^2(x - 2)(x + 1)$

(f) $y = -\left(x^2 + \dfrac{1}{2}\right)(x - 1)^2(x + 1)(x - 2)$

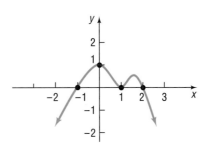

 69. Can the graph of a polynomial function have no y-intercept? Can it have no x-intercepts? Explain.

70. Write a few paragraphs that provide a general strategy for graphing a polynomial function. Be sure to mention the following: degree, intercepts, and turning points.

71. Make up a polynomial that has the following characteristics: crosses the x-axis at -1 and 4, touches the x-axis at 0 and 2, and is above the x-axis between 0 and 2. Give your polynomial to a fellow classmate and ask for a written critique of your polynomial.

72. Make up two polynomials, not of the same degree, with the following characteristics: crosses the x-axis at -2, touches the x-axis at 1, and is above the x-axis between -2 and 1. Give your polynomials to a fellow classmate and ask for a written critique of your polynomials.

73. The graph of a polynomial function is always smooth and continuous. Name a function studied earlier that is smooth and not continuous. Name one that is continuous, but not smooth.

74. Which of the following statements are true regarding the graph of the polynomial $f(x) = x^3 + bx^2 + cx + d$? (Give reasons for your conclusions.)

(a) It intersects the y-axis in one and only one point.
(b) It intersects the x-axis in at most three points.
(c) It intersects the x-axis at least once.
(d) For x very large, it behaves like the graph of $y = x^3$.
(e) It is symmetric with respect to the origin.
(f) It contains the origin.

4.3

Rational Functions

Ratios of integers are called *rational numbers*. Similarly, ratios of polynomial functions are called *rational functions*.

Rational Function

A **rational function** is a function of the form

$$R(x) = \frac{p(x)}{q(x)}$$

where p and q are polynomial functions and q is not the zero polynomial. The domain consists of all real numbers except those for which the denominator q is 0.

E X A M P L E   1   *Finding the Domain of a Rational Function*

(a) The domain of $R(x) = \dfrac{2x^2 - 4}{x + 5}$ consists of all real numbers x except -5.

(b) The domain of $R(x) = \dfrac{1}{x^2 - 4}$ consists of all real numbers x except -2 and 2.

(c) The domain of $R(x) = \dfrac{x^3}{x^2 + 1}$ consists of all real numbers.

(d) The domain of $R(x) = \dfrac{-x^2 + 2}{3}$ consists of all real numbers.

(e) The domain of $R(x) = \dfrac{x^2 - 1}{x - 1}$ consists of all real numbers x except 1.   ∎

It is important to observe that the functions

$$R(x) = \frac{x^2 - 1}{x - 1} \quad \text{and} \quad f(x) = x + 1$$

are not equal, since the domain of R is $\{x \mid x \neq 1\}$ and the domain of f is all real numbers.

∎   Now work Problem 3.

If $R(x) = p(x)/q(x)$ is a rational function and if p and q have no common factors, then the rational function R is said to be in **lowest terms.** For a rational function $R(x) = p(x)/q(x)$ in lowest terms, the zeros, if any, of the numerator are the x-intercepts of the graph of R and so will play a major role in the graph of R. The zeros of the denominator of R [that is, the numbers x, if any, for which $q(x) = 0$], although not in the domain of R, also play a major role in the graph of R. We will discuss this role shortly.

We have already discussed the characteristics of the rational function $f(x) = 1/x$. (Refer to Example 16, page 30. The next rational function we take up is $H(x) = 1/x^2$.

E X A M P L E   2   *Graphing $y = \dfrac{1}{x^2}$*

Analyze the graph: $H(x) = \dfrac{1}{x^2}$

Solution   See Figure 31.

FIGURE 31

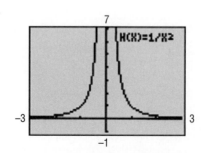

The domain of $H(x) = 1/x^2$ consists of all real numbers x except 0. Thus, the graph has no y-intercept, because x can never equal 0. The graph has no x-intercept because the equation $H(x) = 0$ has no solution. Therefore, the graph of H will not cross either of the coordinate axes.

Because

$$H(-x) = \frac{1}{(-x)^2} = \frac{1}{x^2} = H(x)$$

H is an even function, so its graph is symmetric with respect to the y-axis.

TABLE 3

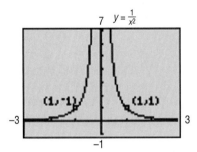

Look again at Figure 31. What happens to the function as the values of x get closer and closer to 0? We use a TABLE to answer the question. See Table 3. There, we see the values of $H(x)$ become larger and larger positive numbers. When this happens we say that H is **unbounded in the positive direction.** We symbolize this by writing $H \to \infty$ (read as "H **approaches infinity**"). In calculus, this idea is symbolized by writing $\lim_{x \to 0} H = \infty$, read as "the limit of $H(x)$ as x approaches 0 is infinity.

Look again at Table 3. As $x \to \infty$, the values of $H(x)$ approach 0 (symbolized in calculus by writing $\lim_{x \to \infty} H(x) = 0$). ∎

E X A M P L E   3   *Graphing in Stages*

Using a graphing utility, show each step to graph $R(x) = \dfrac{1}{(x - 2)^2} + 1$.

Solution   Stage 1: Graph $y = \dfrac{1}{x^2}$. See Figure 32.

FIGURE 32

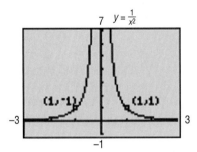

Stage 2: Graph $y = \dfrac{1}{(x - 2)^2}$. (Shift to the right 2 units.) See Figure 33.

FIGURE 33

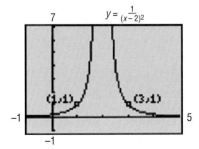

Stage 3: Graph $y = \dfrac{1}{(x-2)^2} + 1$. (Shift up 1 unit.) See Figure 34.

FIGURE 34

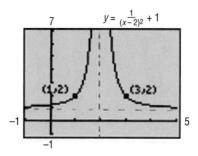

$y = \frac{1}{(x-2)^2} + 1$

Notice that the y-axis in Figure 32 is represented by the vertical line $x = 2$ in Figure 34 and the y-axis in Figure 32 is represented by the horizontal line $y = 1$ in Figure 34.

◼   Now work Problem 21.   ◼

Asymptotes

In Figure 34, notice that as the values of x become more negative, that is, as x becomes **unbounded in the negative direction** ($x \to -\infty$, read as "x **approaches negative infinity**"), the values $R(x)$ approach 1. In fact, we can conclude the following from Figure 34:

1. As $x \to -\infty$, the values $R(x)$ approach 1. $[\lim\limits_{x \to -\infty} R(x) = 1]$

2. As x approaches 2, the values $R(x) \to \infty$. $[\lim\limits_{x \to 2} R(x) = \infty]$

3. As $x \to \infty$, the values $R(x)$ approach 1. $[\lim\limits_{x \to \infty} R(x) = 1]$

Using a graphing utility and the TRACE function, verify the results discussed above for the graph shown in Figure 34. The behavior of the graph is depicted by the dashed vertical line $x = 2$ and the dashed horizontal line $y = 1$. These lines are called *asymptotes* of the graph, which we define as follows:

Let R denote a function:

Horizontal Asymptote

> If, as $x \to -\infty$ or as $x \to \infty$, the values of $R(x)$ approach some fixed number L, then the line $y = L$ is a **horizontal asymptote** of the graph of R.

Vertical Asymptote

> If, as x approaches some number c, the values $|R(x)| \to \infty$, then the line $x = c$ is a **vertical asymptote** of the graph of R.

Even though asymptotes of a function are not part of the graph of the function, they provide information about the way the graph looks. Figure 35 illustrates some of the possibilities.

Thus, an asymptote is a line that a certain part of the graph of a function gets closer and closer to, but never touches. However, other parts of the graph of

FIGURE 35

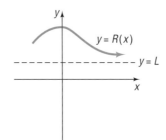

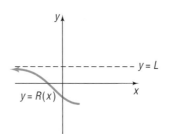

(a) As $x \to \infty$, the values of $R(x)$ approach L. That is, the points on the graph of R are getting closer to the line $y = L$; $y = L$ is a horizontal asymptote.

(b) As $x \to -\infty$, the values of $R(x)$ approach L. That is, the points on the graph of R are getting closer to the line $y = L$; $y = L$ is a horizontal asymptote.

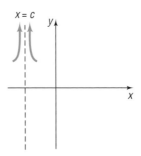

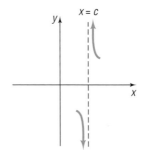

(c) As x approaches c, the values of $|R(x)| \to \infty$, That is, the points on the graph of R are getting closer to the line $x = c$; $x = c$ is a vertical asymptote.

(d) As x approaches c, the values of $|R(x)| \to \infty$, That is, the points on the graph of R are getting closer to the line $x = c$; $x = c$ is a vertical asymptote.

FIGURE 36

Oblique asymptote

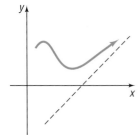

the function may intersect a nonvertical asymptote. The graph of the function will never intersect a vertical asymptote. Notice that a horizontal asymptote, when it occurs, describes a certain behavior of the graph as $x \to \infty$ or as $x \to -\infty$, while a vertical asymptote, when it occurs, describes a certain behavior of the graph when x is close to some number c.

If an asymptote is neither horizontal nor vertical, it is called **oblique.** Figure 36 shows an oblique asymptote.

Finding Asymptotes

The vertical asymptotes, if any, of a rational function $R(x) = p(x)/q(x)$ are found by factoring the denominator $q(x)$. Suppose that $x - r$ is a factor of the denominator. Now, as x approaches r, symbolized as $x \to r$, the values of $x - r$ approach 0, causing the ratio to become unbounded, that is, causing $|R(x)| \to \infty$. Based on the definition, we conclude that the line $x = r$ is a vertical asymptote.

Theorem
Locating Vertical Asymptotes

A rational function $R(x) = p(x)/q(x)$, in lowest terms, will have a vertical asymptote $x = r$, if $x - r$ is a factor of the denominator q.

Thus, if r is a zero of the denominator of a rational function $R(x) = p(x)/q(x)$, in lowest terms, then R will have the vertical asymptote $x = r$.

Warning: If a rational function is not in lowest terms, an application of this theorem may result in an incorrect listing of vertical asymptotes.

E X A M P L E   4   *Finding Vertical Asymptotes*

Find the vertical asymptotes, if any, of the graph of each rational function.

(a) $R(x) = \dfrac{x}{x^2 - 4}$ (b) $F(x) = \dfrac{x + 3}{x - 1}$ (c) $H(x) = \dfrac{x^2}{x^2 + 1}$

Solution   (a)   The zeros of the denominator $x^2 - 4$ are -2 and 2. Hence, the lines $x = -2$ and $x = 2$ are the vertical asymptotes of the graph of R.

(b)   The only zero of the denominator is 1. Hence, the line $x = 1$ is the only vertical asymptote of the graph of F.

(c)   The denominator has no zeros. Hence, the graph of H has no vertical asymptotes. ∎

As Example 4 points out, rational functions can have no vertical asymptotes, one vertical asymptote, or more than one vertical asymptote. However, the graph of a rational function will never intersect any of its vertical asymptotes. (Do you know why?)

Exploration: Graph each of the following rational functions:

$$y = \frac{1}{x - 1} \qquad y = \frac{x}{x - 1} \qquad y = \frac{x^2}{x - 1} \qquad y = \frac{x^3}{x - 1}$$

Each has the vertical asymptote $x = 1$. Use TRACE to see what happens as x approaches 1. Be sure to look at both sides of $x = 1$. ∎

The procedure for finding horizontal and oblique asymptotes is somewhat more involved. To find such asymptotes, we need to know how the values of a function behave as $x \to -\infty$ or as $x \to \infty$.

If a rational function $R(x)$ is **proper,** that is, if the degree of the numerator is less than the degree of the denominator, then as $x \to -\infty$ or as $x \to \infty$, the values of $R(x)$ approach 0. Consequently, the line $y = 0$ (the x-axis) is a horizontal asymptote of the graph.

Theorem   If a rational function is proper, the line $y = 0$ is a horizontal asymptote of its graph. ∎

E X A M P L E   5   *Finding Horizontal Asymptotes*

Find the horizontal asymptotes, if any, of the graph of

$$R(x) = \frac{x - 12}{4x^2 + x + 1}$$

Solution   The rational function R is proper, since the degree of the numerator (1) is less than the degree of the denominator (2). We conclude that the line $y = 0$ is a horizontal asymptote of the graph of R. ∎

To see why $y = 0$ is a horizontal asymptote of the function R in Example 5, we need to investigate the behavior of R for x unbounded. When x is unbounded, the numerator of R, which is $x - 12$, can be approximated by the power function $y = x$, while the denominator of R, which is $4x^2 + x + 1$, can be approximated by the power function $y = 4x^2$.

Thus,

$$R(x) = \frac{x - 12}{4x^2 + x + 1} \approx \frac{x}{4x^2} = \frac{1}{4x} \to 0$$

$$\underset{\text{For } x \text{ unbounded}}{\uparrow} \qquad \underset{\text{As } x \to -\infty \text{ or } x \to \infty}{\uparrow}$$

This shows that the line $y = 0$ is a horizontal asymptote of the graph of R.

If a rational function $R(x)$ is **improper,** that is, if the degree of the numerator is greater than or equal to the degree of the denominator, we must use long division to write the rational function as the sum of a polynomial $f(x)$ plus a proper rational function $r(x)$. That is, we write

$$R(x) = \frac{p(x)}{q(x)} = f(x) + r(x)$$

where $f(x)$ is a polynomial and $r(x)$ is a proper rational function. Since $r(x)$ is proper, then $r(x) \to 0$ as $x \to -\infty$ or as $x \to \infty$. Thus,

$$R(x) = \frac{p(x)}{q(x)} \to f(x) \qquad \text{as} \qquad x \to -\infty \text{ or as } x \to \infty$$

The possibilities are listed next:

1. If $f(x) = b$, a constant, then the line $y = b$ is a horizontal asymptote of the graph of R.
2. If $f(x) = ax + b$, $a \neq 0$, then the line $y = ax + b$ is an oblique asymptote of the graph of R.
3. In all other cases, the graph of R approaches the graph of f, and there are no horizontal or oblique asymptotes.

The following examples demonstrate these conclusions.

E X A M P L E   6

Finding Horizontal or Oblique Asymptotes

Find the horizontal or oblique asymptotes, if any, of the graph of

$$H(x) = \frac{3x^4 - x^2}{x^3 - x^2 + 1}$$

Solution

The rational function H is improper, since the degree of the numerator (4) is larger than the degree of the denominator (3). To find any horizontal or oblique asymptotes, we use long division:

$$
\begin{array}{r}
3x + 3 \\
x^3 - x^2 + 1\overline{)3x^4 \qquad\quad - x^2} \\
\underline{3x^4 - 3x^3 \qquad + 3x} \\
3x^3 - x^2 - 3x \\
\underline{3x^3 - 3x^2 \qquad + 3} \\
2x^2 - 3x - 3
\end{array}
$$

Thus,

$$H(x) = \frac{3x^4 - x^2}{x^3 - x^2 + 1} = 3x + 3 + \frac{2x^2 - 3x - 3}{x^3 - x^2 + 1}$$

Then, as $x \to -\infty$ or as $x \to \infty$, the remainder will behave as follows:

$$\frac{2x^2 - 3x - 3}{x^3 - x^2 + 1} \approx \frac{2x^2}{x^3} = \frac{2}{x} \to 0$$

Thus, as $x \to -\infty$ or as $x \to \infty$, we have $H(x) \to 3x + 3$. We conclude that the graph of the rational function H has an oblique asymptote $y = 3x + 3$. ∎

E X A M P L E   7   *Finding Horizontal or Oblique Asymptotes*

Find the horizontal or oblique asymptotes, if any, of the graph of

$$R(x) = \frac{8x^2 - x + 2}{4x^2 - 1}$$

Solution   The rational function R is improper, since the degree of the numerator (2) equals the degree of the denominator (2). To find any horizontal or oblique asymptotes, we use long division:

$$\begin{array}{r} 2 \\ 4x^2 - 1 \overline{) 8x^2 - x + 2} \\ \underline{8x^2 - 2} \\ -x + 4 \end{array}$$

Thus,

$$R(x) = \frac{8x^2 - x + 2}{4x^2 - 1} = 2 + \frac{-x + 4}{4x^2 - 1}$$

Then, as $x \to -\infty$ or as $x \to \infty$, the remainder will behave as follows:

$$\frac{-x + 4}{4x^2 - 1} \approx \frac{-x}{4x^2} = \frac{-1}{4x} \to 0$$

Thus, as $x \to -\infty$ or as $x \to \infty$, we have $R(x) \to 2$. We conclude that $y = 2$ is a horizontal asymptote of the graph. ∎

In Example 7, note that the quotient 2 obtained by long division is the quotient of the leading coefficients of the numerator polynomial and the denominator polynomial $\left(\frac{8}{4}\right)$. This means we can avoid the long division process for rational functions whose numerator and denominator *are of the same degree* and conclude that the quotient of the leading coefficients will give us the horizontal asymptote.

■ Now work Problem 35.

E X A M P L E   8   *Finding Horizontal or Oblique Asymptotes*

Find the horizontal or oblique asymptotes, if any, of the graph of

$$G(x) = \frac{2x^5 - x^3 + 2}{x^3 - 1}$$

Solution The rational function G is improper, since the degree of the numerator (5) is larger than the degree of the denominator (3). To find any horizontal or oblique asymptotes, we use long division:

$$
\begin{array}{r}
2x^2 - 1 \\
x^3 - 1\overline{)\,2x^5 - x^3 + 2} \\
\underline{2x^5 - 2x^2 } \\
-x^3 + 2x^2 + 2 \\
\underline{-x^3 + 1} \\
2x^2 + 1
\end{array}
$$

Thus,

$$
G(x) = \frac{2x^5 - x^3 + 2}{x^3 - 1} = 2x^2 - 1 + \frac{2x^2 + 1}{x^3 - 1}
$$

Then, as $x \to -\infty$ or as $x \to \infty$, the remainder will behave as follows:

$$
\frac{2x^2 + 1}{x^3 - 1} \approx \frac{2x^2}{x^3} = \frac{2}{x} \to 0
$$

Thus, as $x \to -\infty$ or as $x \to \infty$, we have $G(x) \to 2x^2 - 1$. We conclude that, for large values of x, the graph of G approaches the graph of $y = 2x^2 - 1$. ■

We summarize next the procedure for finding horizontal and oblique asymptotes.

Finding Horizontal and Oblique Asymptotes of a Rational Function R

Consider the rational function

$$
R(x) = \frac{p(x)}{q(x)} = \frac{a_n x^n + a_{n-1}x^{n-1} + \cdots + a_1 x + a_0}{b_m x^m + b_{m-1}x^{m-1} + \cdots + b_1 x + b_0}
$$

in which the degree of the numerator is n and the degree of the denominator is m.

1. If R is a proper rational function ($n < m$), the graph will have the horizontal asymptote $y = 0$ (the x-axis).
2. If R is improper ($n \geq m$), use long division.
   (a) If $n = m$, the quotient obtained will be a number L ($= a_n/b_m$) and the line $y = L$ ($= a_n/b_m$) is a horizontal asymptote.
   (b) If $n = m + 1$, the quotient obtained is of the form $ax + b$ (a polynomial of degree 1), then the line $y = ax + b$ is an oblique asymptote.
   (c) If $n > m + 1$, the quotient obtained is a polynomial of degree 2 or higher, then R has neither a horizontal nor an oblique asymptote. In this case, for x unbounded, the graph of R will behave like the graph of the quotient.

Note: The graph of a rational function either has one horizontal or one oblique asymptote or else has no horizontal and no oblique asymptote.

Now we are ready to graph rational functions.

Graphing Rational Functions

With a graphing utility, the task of graphing a rational function is as simple as entering the function into the utility and pressing graph. However, without appropriate knowledge of the behavior of rational functions, it is dangerous to draw conclusions from the graph. The analysis of a rational function $R(x) = p(x)/q(x)$ requires the following steps:

Analyzing the Graph of a Rational Function

STEP 1: Locate the intercepts, if any, of the graph. The x-intercepts, if any, of $R(x) = p(x)/q(x)$ satisfy the equation $p(x) = 0$. The y-intercept, if there is one, is $R(0)$.

STEP 2: Test for symmetry. Replace x by $-x$ in $R(x)$. If $R(-x) = R(x)$, there is symmetry with respect to the y-axis; if $R(-x) = -R(x)$, there is symmetry with respect to the origin.

STEP 3: Locate the vertical asymptotes, if any, by factoring the denominator $q(x)$ of $R(x)$ and identifying its zeros.

STEP 4: Locate the horizontal or oblique asymptotes, if any, using the procedure given earlier. Determine points, if any, at which the graph of R intersects these asymptotes.

STEP 5: Graph R.

After graphing, it is important to confirm that the graph is consistent with the conclusions found in Steps 1–4. Let's look at an example.

E X A M P L E 9   *Analyzing the Graph of a Rational Function*

Analyze the graph of the rational function: $R(x) = \dfrac{x-1}{x^2-4}$

Solution   First, we factor both the numerator and the denominator of R:

$$R(x) = \frac{x-1}{(x+2)(x-2)}$$

STEP 1: We locate the x-intercepts by finding the zeros of the numerator. By inspection, 1 is the only x-intercept. The y-intercept is $R(0) = \frac{1}{4}$.

STEP 2: Because

$$R(-x) = \frac{-x-1}{x^2-4}$$

we conclude that R is neither even nor odd. Thus, there is no symmetry with respect to the y-axis or the origin.

STEP 3: We locate the vertical asymptotes by factoring the denominator: $x^2 - 4 = (x+2)(x-2)$. The graph of R thus has two vertical asymptotes: the lines $x = -2$ and $x = 2$.

STEP 4: The degree of the numerator is less than the degree of the denominator, so R is proper and the line $y = 0$ (the x-axis) is a horizontal asymptote of the graph. To determine if the graph of R intersects the horizontal asymptote, we solve the equation $R(x) = 0$. The only solution is $x = 1$, so the graph of R intersects the horizontal asymptote at $(1, 0)$.

STEP 5: The analysis just completed helps us set the viewing rectangle to obtain a complete graph. Figure 37(a) shows the graph of $R(x) = \dfrac{x-1}{x^2-4}$ in connected mode and Figure 37(b) shows it in dot mode. Notice in Figure 37(a), the graph has vertial lines at $x = -2$ and $x = 2$. This is due to the fact that when the graphing utility is in connected mode, it will "connect the dots" between consecutive pixels. We know the graph of R does not cross the lines $x = -2$ and $x = 2$, since R is not defined at $x = -2$ or $x = 2$. Thus, when graphing rational functions, dot mode should be used if repeated experimenting with the viewing rectangle dimensions does not remove the extraneous vertical lines. You should confirm that all of the algebraic conclusions we arrived at in STEPS 1–4 are part of the graph. For example, the graph has vertical asymptotes at $x = -2$ and $x = 2$ and the graph has a horizontal asymptote at $y = 0$. The y-intercept is $\frac{1}{4}$, the x-intercept is 1.

FIGURE 37

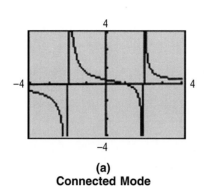

(a)
Connected Mode

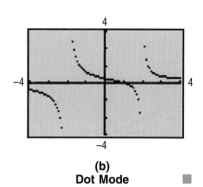

(b)
Dot Mode   ◼

◼ Now work Problem 41.

E X A M P L E   1 0   *Analyzing the Graph of a Rational Function*

Analyze the graph of the rational function: $R(x) = \dfrac{x^2 - 1}{x}$.

Solution   **STEP 1:** The graph has two x-intercepts: -1 and 1. There is no y-intercept.

STEP 2: Since $R(-x) = -R(x)$, the function is odd and the graph is symmetric with respect to the origin.

STEP 3: The graph of $R(x)$ has the line $x = 0$ (the y-axis) as a vertical asymptote.

STEP 4: The rational function R is improper since the degree of the numerator (2) is larger than the degree of the denominator (1). To find any horizontal or oblique asymptotes, we use long division:

$$
\begin{array}{r}
x \phantom{{}-1} \\
x\,)\overline{x^2 - 1} \\
\underline{x^2 \phantom{{}- 1}} \\
-\,1
\end{array}
$$

The quotient is x, so the line $y = x$ is an oblique asymptote of the graph. To determine whether the graph of R intersects the asymptote $y = x$, we solve the equation $R(x) = x$:

$$
\begin{aligned}
R(x) = \frac{x^2 - 1}{x} &= x \\
x^2 - 1 &= x^2 \\
-1 &= 0 \quad \text{Impossible}
\end{aligned}
$$

FIGURE 38

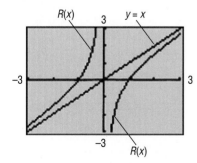

We conclude that the equation $(x^2 - 1)/x = x$ has no solution, so the graph of $R(x)$ does not intersect the line $y = x$.

STEP 5: See Figure 38. We see from the graph that there is no y-intercept and two x-intercepts, -1 and 1. The symmetry with respect to the origin is also evident. We can also see that there is a vertical asymptote at $x = 0$ and an oblique asymptote, $y = x$. Finally, it is not necessary to graph this function in dot mode since there are no extraneous vertical lines present.   ◼

◼ Now work Problem 43.

E X A M P L E   1 1   *Analyzing the Graph of a Rational Function*

Analyze the graph of the rational function: $R(x) = \dfrac{3x^2 - 3x}{x^2 + x - 12}$

Solution   We factor R to get

$$R(x) = \frac{3x(x-1)}{(x+4)(x-3)}$$

STEP 1:   The graph has two x-intercepts: 0 and 1. The y-intercept is $R(0) = 0$.

STEP 2:   There is no symmetry with respect to the y-axis or the origin.

STEP 3:   The graph of R has two vertical asymptotes: $x = -4$ and $x = 3$.

STEP 4:   Since the degree of the numerator equals the degree of the denominator, the graph has a horizontal asymptote. To find it, we form the quotient of the leading coefficient of the numerator (3) and the leading coefficient of the denominator (1). Thus, the graph of R has the horizontal asymptote $y = 3$. To find out whether the graph of R intersects the asymptote, we solve the equation $R(x) = 3$.

$$R(x) = \frac{3x^2 - 3x}{x^2 + x - 12} = 3$$
$$3x^2 - 3x = 3x^2 + 3x - 36$$
$$-6x = -36$$
$$x = 6$$

Thus, the graph intersects the line $y = 3$ only at $x = 6$, and $(6, 3)$ is a point on the graph of R.

STEP 5:   Figure 39(a) shows the graph of R in connected mode. Notice the extraneous vertical lines. As a result, we will also graph R in dot mode. See Figure 39(b). The graph indicates vertical asymptotes at $x = -4$ and $x = 3$. We use TRACE to verify $y = 3$ is a horizontal asymptote by TRACEing for large values of x and observing y gets closer to 3. We can also use TRACE to verify that the graph of R crosses the horizontal asymptote, $y = 3$, at $(6, 3)$.

FIGURE 39

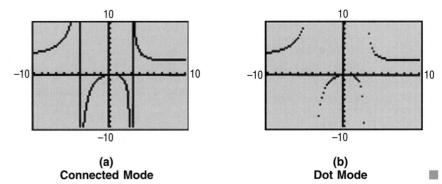

(a)
Connected Mode

(b)
Dot Mode

Figure 39 does not display the graph between the two x-intercepts, 0 and 1. Nor does it show the graph crossing the horizontal asymptote at $(6, 3)$. To see these parts better, we graph R for $0 \le x \le 1$ (Figure 40(a)), and for $5 \le x \le 100$ (Figure 40(b)).

FIGURE 40

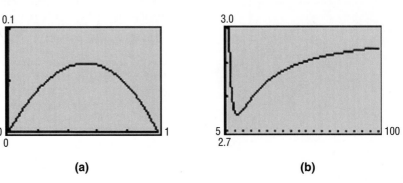

(a)

(b)

These graphs now reflect the behavior produced by the analysis. Further, we observe two turning points, one between 0 and 1; the other to the right of 3. Correct to two decimal places, these turning points are (0.52, 0.06) and (11.47, 2.74).

E X A M P L E   1 2   *Analyzing the Graph of a Rational Function*

Analyze the graph of the rational function

$$R(x) = \frac{0.5x^3 - 2x^2 + 3}{0.1x^3 + x - 3}$$

Solution   STEP 1:   The x-intercepts obey the equation

$$0.5x^3 - 2x^2 + 3 = 0$$

The x-intercepts, correct to two decimal places, are -1.08, 1.57 and 3.51. The y-intercept is $R(0) = -1$.

STEP 2:   There is no symmetry with respect to the x-axis, y-axis, or origin.

STEP 3:   The vertical asymptotes obey the equation

$$0.1x^3 + x - 3 = 0$$

The only solution, correct to two decimal places, is 2.08. The only vertical asymptote is the line $x = 2.08$.

STEP 4:   The graph of R has $y = 5$ as a horizontal asymptote. To see if the graph of R intersects $y = 5$, we solve the equation

$$\frac{0.5x^3 - 2x^2 + 3}{0.1x^3 + x - 3} = 5$$

$$0.5x^3 - 2x^2 + 3 = 0.5x^3 + 5x - 15$$

$$-2x^2 - 5x + 18 = 0$$

The solutions are $x = -4.5$ and $x = 2$.

STEP 5:   The graph of R is shown in Figure 41.

FIGURE 41

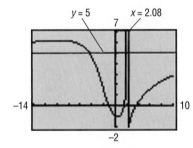

The two turning points, correct to two decimal places, are at $(-9.54, 6.17)$ and $(0.30, -1.05)$.   ■

E X A M P L E   1 3   *Analyzing the Graph of a Rational Function*

Analyze the graph of the rational function: $R(x) = \dfrac{2x^2 - 5x + 2}{x^2 - 4}$

Solution   We factor R and obtain

$$R(x) = \frac{(2x - 1)(x - 2)}{(x + 2)(x - 2)} = \frac{2x - 1}{x + 2}$$

STEP 1:   The graph has one x-intercept: 0.5. The y-intercept is $R(0) = -0.5$.

STEP 2:   There is no symmetry with respect to the y-axis or the origin.

STEP 3:   The graph has one vertical asymptote: $x = -2$, since $x + 2$ is the only factor of the denominator of $R(x)$ *in lowest terms.* However, the rational function is undefined at both $x = 2$ and $x = -2$.

STEP 4:   Since the degree of the numerator equals the degree of the denominator, the graph has a horizontal asymptote. To find it, we form the quotient of the leading coefficient of the numerator (2) and the leading coefficient of the denominator (1). Thus, the graph of R has the horizontal asymptote $y = 2$. To find out whether the graph of R intersects the asymptote, we solve the equation $R(x) = 2$.

$$R(x) = \frac{2x - 1}{x + 2} = 2$$
$$2x - 1 = 2(x + 2)$$
$$2x - 1 = 2x + 4$$
$$-1 = 4 \qquad \text{impossible}$$

Thus, the graph does not intersect the line $y = 2$.

STEP 5:   Figure 42 shows the graph of $R(x)$. Notice the graph has one vertical asymptote at $x = -2$ and the function appears to be continuous at $x = 2$. However, looking at Table 4, we see R is undefined at $x = 2$. Thus, the graph leads us to the incorrect conclusion that the graph of R is continuous at $x = 2$. In fact, there is a hole in the graph where $x = 2$. So, we must be careful when using a graphing utility to analyze rational functions.

FIGURE 42   **TABLE 4**

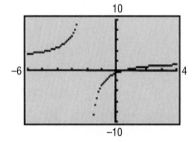

Exercise 4.3

In Problems 1–10, find the domain of each rational function.

1. $R(x) = \dfrac{4x}{x - 3}$

2. $R(x) = \dfrac{5x^2}{3 + x}$

3. $H(x) = \dfrac{-4x^2}{(x - 2)(x + 4)}$

4. $G(x) = \dfrac{6}{(x + 3)(4 - x)}$

5. $F(x) = \dfrac{3x(x - 1)}{2x^2 - 5x - 3}$

6. $Q(x) = \dfrac{-x(1 - x)}{3x^2 + 5x - 2}$

7. $R(x) = \dfrac{x}{x^3 - 8}$

8. $R(x) = \dfrac{x}{x^4 - 1}$

9. $H(x) = \dfrac{3x^2 + x}{x^2 + 4}$

10. $G(x) = \dfrac{x - 3}{x^4 + 1}$

In Problems 11–20, use the graph shown to find:

(a) The domain and range of each function

(b) The intercepts, if any

(c) Horizontal asymptotes, if any

(d) Vertical asymptotes, if any

(e) Oblique asymptotes, if any

11.

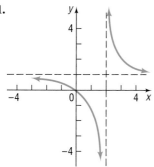

12.

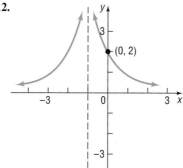

13.

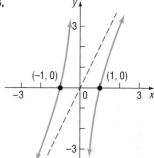

14.

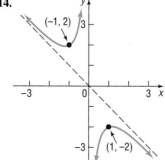

15.

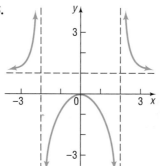

16.

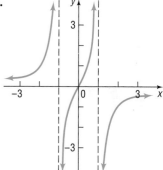

17.

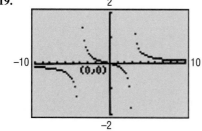

18.

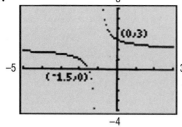

19.

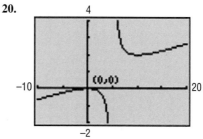

20.

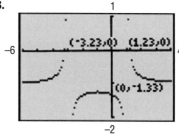

In Problems 21–30, using a graphing utility, show the steps required to graph each rational function.

21. $R(x) = \dfrac{1}{(x-1)^2}$

22. $R(x) = \dfrac{3}{x}$

23. $H(x) = \dfrac{-2}{x+1}$

24. $G(x) = \dfrac{2}{(x+2)^2}$

25. $R(x) = \dfrac{1}{x^2 + 4x + 4}$

26. $R(x) = \dfrac{1}{x-1} + 1$

27. $F(x) = 1 - \dfrac{1}{x}$

28. $Q(x) = 1 + \dfrac{1}{x}$

29. $R(x) = \dfrac{x^2 - 4}{x^2}$

30. $R(x) = \dfrac{x-4}{x}$

In Problems 31–40, find the vertical, horizontal, and oblique asymptotes, if any, of each rational function, without graphing.

31. $R(x) = \dfrac{3x}{x+4}$

32. $R(x) = \dfrac{3x+5}{x-6}$

33. $H(x) = \dfrac{x^4 + 2x^2 + 1}{x^2 - x + 1}$

34. $G(x) = \dfrac{-x^2 + 1}{x+5}$

35. $T(x) = \dfrac{x^3}{x^4 - 1}$

36. $P(x) = \dfrac{4x^5}{x^3 - 1}$

37. $Q(x) = \dfrac{5 - x^2}{3x^4}$

38. $F(x) = \dfrac{-2x^2 + 1}{2x^3 + 4x^2}$

39. $R(x) = \dfrac{3x^4 + 4}{x^3 + 3x}$

40. $R(x) = \dfrac{6x^2 + x + 12}{3x^2 - 5x - 2}$

In Problems 41–84, follow Steps 1 through 5 on page 294 to analyze the graph of each function.

41. $R(x) = \dfrac{x+1}{x(x+4)}$

42. $R(x) = \dfrac{x}{(x-1)(x+2)}$

43. $R(x) = \dfrac{3x+3}{2x+4}$

44. $R(x) = \dfrac{2x+4}{x-1}$

45. $R(x) = \dfrac{3}{x^2 - 4}$

46. $R(x) = \dfrac{6}{x^2 - x - 6}$

47. $P(x) = \dfrac{x^4 + x^2 + 1}{x^2 - 1}$

48. $Q(x) = \dfrac{x^4 - 1}{x^2 - 4}$

49. $H(x) = \dfrac{x^3 - 1}{x^2 - 9}$

50. $G(x) = \dfrac{x^3 + 1}{x^2 + 2x}$

51. $R(x) = \dfrac{x^2}{x^2 + x - 6}$

52. $R(x) = \dfrac{x^2 + x - 12}{x^2 - 4}$

53. $G(x) = \dfrac{x}{x^2 - 4}$

54. $G(x) = \dfrac{3x}{x^2 - 1}$

55. $R(x) = \dfrac{3}{(x-1)(x^2 - 4)}$

56. $R(x) = \dfrac{-4}{(x+1)(x^2 - 9)}$

57. $H(x) = 4\dfrac{x^2 - 1}{x^4 - 16}$

58. $H(x) = \dfrac{x^2 + 4}{x^4 - 1}$

59. $F(x) = \dfrac{x^2 - 3x - 4}{x+2}$

60. $F(x) = \dfrac{x^2 + 3x + 2}{x-1}$

61. $R(x) = \dfrac{x^2 + x - 12}{x-4}$

62. $R(x) = \dfrac{x^2 - x - 12}{x+5}$

63. $F(x) = \dfrac{x^2 + x - 12}{x+2}$

64. $G(x) = \dfrac{x^2 - x - 12}{x+1}$

65. $R(x) = \dfrac{x(x-1)^2}{(x+3)^3}$

66. $R(x) = \dfrac{(x-1)(x+2)(x-3)}{x(x-4)^2}$

67. $R(x) = \dfrac{4x^3 - 0.5x + 2}{2x^3 + 0.3x^2 - 1}$

68. $R(x) = \dfrac{3x^4 - 4x^2 + 8}{x^4 - x^3 + 2}$

69. $R(x) = \dfrac{3x^3 + 5x^2 - 3}{0.1x^4 - 2x^2 + 1}$

70. $R(x) = \dfrac{x^2 - 0.9x + 2}{x^3 + 0.5x^2 + 1}$

71. $R(x) = \dfrac{.5x^4 - x^3 + 1}{.1x^3 - \pi x + 1}$

72. $R(x) = \dfrac{2x^3 + \pi x + 3}{-x^2 + .9x + 1}$

73. $R(x) = \dfrac{x^2 + x - 12}{x^2 - x - 6}$

74. $R(x) = \dfrac{x^2 + 3x - 10}{x^2 + 8x + 15}$

75. $R(x) = \dfrac{6x^2 - 7x - 3}{2x^2 - 7x + 6}$

76. $R(x) = \dfrac{8x^2 + 26x + 15}{2x^2 - x - 15}$

77. $R(x) = \dfrac{x^3 + 2x^2 - 5x - 6}{x^3 + 7x^2 + 7x - 15}$

78. $R(x) = \dfrac{x^3 - 6x^2 - x + 30}{x^3 - 4x^2 + x + 6}$

79. $f(x) = x + \dfrac{1}{x}$

80. $f(x) = 2x + \dfrac{9}{x}$

81. $f(x) = x^2 + \dfrac{1}{x}$

82. $f(x) = 2x^2 + \dfrac{9}{x}$

83. $f(x) = x + \dfrac{1}{x^3}$

84. $f(x) = 2x + \dfrac{9}{x^3}$

85. If the graph of a rational function R has the vertical asymptote $x = 4$, then the factor $x - 4$ must be present in the denominator of R. Explain why.

86. If the graph of a rational function R has the horizontal asymptote $y = 2$, then the degree of the numerator of R equals the degree of the denominator of R. Explain why.

87. Consult the illustration. Which of the following rational functions might have this graph? (More than one answer might be possible.)

(a) $y = \dfrac{4x^2}{x^2 - 4}$

(b) $y = \dfrac{x}{x^2 - 4}$

(c) $y = \dfrac{x^2}{x^2 - 4}$

(d) $y = \dfrac{x^2(x^2 + 1)}{(x^2 + 4)(x^2 - 4)}$

(e) $y = \dfrac{x^3}{x^2 - 4}$

(f) $y = \dfrac{x^2 - 4}{x^2}$

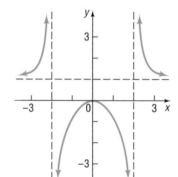

88. Consult the illustration. Which of the following rational functions might have this graph? (More than one answer may be possible.)

(a) $y = \dfrac{2x}{x^2 - 1}$

(b) $y = \dfrac{-3x}{x^2 - 1}$

(c) $y = \dfrac{x^3}{x^2 - 1}$

(d) $y = \dfrac{x^2 - 1}{-3x}$

(e) $y = \dfrac{-x^3}{(x^2 + 1)(x^2 - 1)}$

(f) $y = \dfrac{-x^2}{(x^2 + 1)(x^2 - 1)}$

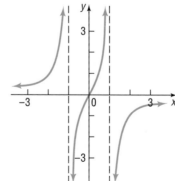

89. Consult the illustration. Make up a rational function that might have this graph. Compare yours with a friend's. What similarities do you see? What differences?

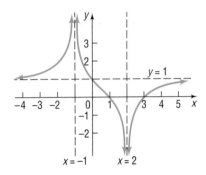

90. Can the graph of a rational function have both a horizontal and an oblique asymptote? Explain.

91. Write a few paragraphs that provide a general strategy for graphing a rational function. Be sure to mention the following: proper, improper, intercepts, and asymptotes.

92. Make up a rational function that has the following characteristics: crosses the x-axis at 3; touches the x-axis at -2; one vertical asymptote, $x = 1$; and one horizontal asymptote, $y = 2$. Give your rational function to a fellow classmate and ask for a written critique of your rational function.

93. Make up a rational function that has $y = 2x + 1$ as an oblique asymptote. Explain the methodology you used.

4.4

The Zeros of a Polynomial Function

Remainder and Factor Theorems

Recall that when we divide one polynomial (the dividend) by another (the divisor) we obtain a quotient polynomial and a remainder, the remainder being either the zero polynomial or a polynomial whose degree is less than the degree of the divisor. To check our work, we verify that

(Divisor)(Quotient) + Remainder = Dividend

This checking routine is the basis for a famous theorem called the **division algorithm* for polynomials,** which we now state without proof.

Theorem
Division Algorithm
for Polynomials

If $f(x)$ and $g(x)$ denote polynomial functions and if $g(x)$ is not the zero polynomial, then there are unique polynomial functions $q(x)$ and $r(x)$ such that

$$\frac{f(x)}{g(x)} = q(x) + \frac{r(x)}{g(x)} \quad \text{or} \quad f(x) = g(x)q(x) + r(x) \qquad (1)$$

$$\underset{\text{dividend}}{\uparrow} \quad \underset{\text{divisor}}{\uparrow} \ \underset{\text{quotient}}{\uparrow} \quad \underset{\text{remainder}}{\uparrow}$$

where $r(x)$ is either the zero polynomial or a polynomial of degree less than that of $g(x)$. ∎

In equation (1), $f(x)$ is the **dividend,** $g(x)$ is the **divisor,** $q(x)$ is the **quotient,** and $r(x)$ is the **remainder.**

---

*A systematic process in which certain steps are repeated a finite number of times is called an **algorithm.** Thus, long division is an algorithm.

ISSION POSSIBLE

Chapter 4

RESPONDING TO A CHALLENGE BY THE CARDASSIANS.

While playing with virtual reality in your school's computer lab, you are suddenly faced with some very real and very nasty-looking Cardassians who have burst into your school through your computer network. As usual, they are very scornful of the intelligence of Earthlings, and when you try to protest, they throw this challenge at you. "Find a rational function defined by this graph," they say, "and we promise to support your reputation for intellectual achievement in every inter-galactic community this side of the Macklin Nebula. Otherwise, consider yourselves the laughing stock of the universe." And they departed back through the computer screen, laughing at what they perceived to be their extreme cleverness.

As you look at the computer screen now, all you can see is this graph:

You decide to tackle this as a team project to save the reputation of all the humans on this planet! The Cardassians mentioned "rational function," so you know the solution will have the form

$$R(x) = \frac{a_n x^n + a_{n-1} x^{n-1} + \ldots + a_0}{b_m x^m + b_{m-1} x^{m-1} + \ldots + b_0}$$

Sometimes it is easier to think of it in factored form.

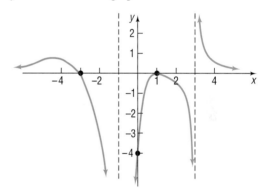

1. You notice some things right away. There are two vertical asymptotes. How do they fit into the solution? Put them in.
2. There are two x-intercepts. Where do they show up in the solution? Put them in.
3. Use a graphing utility to check what you have so far. Does it look like the original graph? Notice that 1 and -3 appear to be the only real zeros. What can you tell about the multiplicity of each one? Do you need to change the multiplicity of either zero in your solution?
4. How can you account for the fact that the graph of the function goes to $-\infty$ on both sides of one vertical asymptote and at the other vertical asymptote it goes to $-\infty$ on one side and to ∞ on the other? Can you change the denominator in some way to account for this?
5. Is the information you have so far consistent with the horizontal asymptote $y = 0$? If not, how should you adjust your function so that it is consistent?
6. Is the information you have so far consistent with the y-intercept? If not, how should you adjust your function so that it is consistent?
7. Are you finished? Have you checked your solution on a graphing utility? Are you satisfied you have it right?
8. Are there other functions whose graph might look like the one given? If so, what common characteristics do they have? Will the Cardassians be more impressed if more than one correct answer is sent?
9. Compare your group's solution to those of other groups before typing the solution into the computer. Then send all the correct solutions off to cardass@slime.uni.

If the divisor $g(x)$ is a first-degree polynomial of the form

$$g(x) = x - c \qquad c \text{ a real number}$$

then the remainder $r(x)$ is either the zero polynomial or a polynomial of degree 0. Thus, for such divisors, the remainder is some number, say, R, and we may write

$$f(x) = (x - c)q(x) + R \qquad (2)$$

This equation is an identity in x and is true for all real numbers x. In particular, it is true when $x = c$. Thus, if $x = c$, then equation (2) becomes

$$f(c) = (c - c)q(c) + R$$
$$f(c) = R$$

and equation (2) takes the form

$$f(x) = (x - c)q(x) + f(c) \qquad (3)$$

We have now proved the following result, called the **Remainder Theorem:**

Remainder Theorem   Let f be a polynomial function. If $f(x)$ is divided by $x - c$, then the remainder is $f(c)$. ■

E X A M P L E 1   *Using the Remainder Theorem*

Find the remainder if $f(x) = x^3 - 4x^2 + 2x - 5$ is divided by:

(a) $x - 3$ (b) $x + 2$

Solution   (a) We could use long division. However, it is much easier here to use the Remainder Theorem, which says the remainder is

$$f(3) = (3)^3 - 4(3)^2 + 2(3) - 5 = 27 - 36 + 6 - 5 = -8$$

(b) To find the remainder when $f(x)$ is divided by $x + 2 = x - (-2)$, we evaluate

$$f(-2) = (-2)^3 - 4(-2)^2 + 2(-2) - 5 = -8 - 16 - 4 - 5 = -33$$

Thus, the remainder is -33. ■

An important and useful consequence of the Remainder Theorem is the **Factor Theorem.**

Factor Theorem   Let f be a polynomial function. Then $x - c$ is a factor of $f(x)$ if and only if $f(c) = 0$. ■

The Factor Theorem actually consists of two separate statements:

1. If $f(c) = 0$, then $x - c$ is a factor of $f(x)$.
2. If $x - c$ is a factor of $f(x)$, then $f(c) = 0$.

Thus, the proof requires two parts.

Proof 1. Suppose that $f(c) = 0$. Then, by equation (3), we have

$$f(x) = (x - c)q(x)$$

for some polynomial $q(x)$. That is, $x - c$ is a factor of $f(x)$.

2. Suppose that $x - c$ is a factor of $f(x)$. Then there is a polynomial function q such that

$$f(x) = (x - c)q(x)$$

Replacing x by c, we find that

$$f(c) = (c - c)q(c) = 0 \cdot q(c) = 0$$

This completes the proof. ∎

One use of the Factor Theorem is to determine whether a polynomial has a particular factor.

E X A M P L E   2 *Using the Factor Theorem*

Use the Factor Theorem to determine whether the function $f(x) = 2x^3 - x^2 + 2x - 3$ has the factor:

(a) $x - 1$ (b) $x + 3$

Solution (a) Because $x - 1$ is of the form $x - c$ with $c = 1$, we find the value of $f(1)$:

$$f(1) = 2(1)^3 - (1)^2 + 2(1) - 3 = 2 - 1 + 2 - 3 = 0$$

By the Factor Theorem, $x - 1$ is a factor of $f(x)$.

(b) To test the factor $x + 3$, we first need to write it in the form $x - c$. Since $x + 3 = x - (-3)$, we find the value of $f(-3)$:

$$f(-3) = 2(-3)^3 - (-3)^2 + 2(-3) - 3 = -54 - 9 - 6 - 3 = -72$$

Because $f(-3) \neq 0$, we conclude from the Factor Theorem that $x - (-3) = x + 3$ is not a factor of $f(x)$. ∎

■ Now work Problem 1.

Descartes' Rule of Signs

The real zeros of a polynomial function f are the real solutions, if any, of the equation $f(x) = 0$. They are also the x-intercepts of the graph of f. For polynomial and rational functions, we have seen the importance of the zeros for graphing. In most cases, however, the zeros of a polynomial function are difficult to find using algebraic methods. There are no nice formulas like the quadratic formula available to help us for polynomials of degree higher than 2. Although formulas do exist for solving any third- or fourth-degree polynomial equation, they are somewhat complicated. (If you are interested in learning about them, see Problems 85–92 for solving cubic equations; for 4th degree polynomial equations, consult a book on the theory of equations.) It has been proved that no general formulas exist for polynomial equations of degree 5 or higher. In this section, we shall learn some ways of detecting information about the character of the zeros, which, in turn, will help us use our graphing utility more effectively.

Our first theorem concerns the number of zeros a polynomial function may have. In counting the zeros of a polynomial, we count each zero as many times as its multiplicity.

Theorem
Number of Zeros

A polynomial function cannot have more zeros than its degree.

Proof

The proof is based on the Factor Theorem. If r is a zero of a polynomial function f, then $f(r) = 0$ and, hence, $x - r$ is a factor of $f(x)$. Thus, each zero corresponds to a factor of degree 1. Because f cannot have more first-degree factors than its degree, the result follows.

The next theorem, called **Descartes' Rule of Signs,** provides information about the number and location of the zeros of a polynomial function, so we know where to look for zeros. Descartes' Rule of Signs assumes that the polynomial is written in descending powers of x and requires that we count the number of variations in sign of the coefficients of $f(x)$ and $f(-x)$.

For example, the following polynomial function has two variations in the signs of coefficients:

$$f(x) = -3x^7 + 4x^4 + 3x^2 - 2x - 1$$
$$= -3x^7 + 0x^6 + 0x^5 + 4x^4 + 0x^3 + 3x^2 - 2x - 1$$

$$\underbrace{\qquad\qquad\qquad}_{-\text{ to }+} \qquad \underbrace{\qquad}_{+\text{ to }-}$$

Notice that we ignored the zero coefficients in $0x^6$, $0x^5$ and $0x^3$ in counting the number of variations in sign of $f(x)$. Replacing x by $-x$, we get

$$f(-x) = 3x^7 + 4x^4 + 3x^2 + \underbrace{2x - 1}_{+\text{ to }-}$$

which has one variation in sign.

Theorem
Descartes' Rule of Signs

Let f denote a polynomial function.

The number of positive zeros of f either equals the number of variations in sign of the nonzero coefficients of $f(x)$ or else equals that number less an even integer.

The number of negative zeros of f either equals the number of variations in sign of the nonzero coefficients of $f(-x)$ or else equals that number less an even integer.

We shall not prove Descartes' Rule of Signs. Let's see how it is used.

E X A M P L E   3

Using Descartes' Rule of Signs to Locate Zeros

Discuss the zeros of: $f(x) = 3x^6 - 4x^4 + 3x^3 + 2x^2 - x - 3$

Solution

There are at most six zeros, because the polynomial is of degree 6. Since there are three variations in sign of the nonzero coefficients of $f(x)$, by Descartes' Rule of Signs we expect either three or one positive zero(s). To continue, we look at $f(-x)$:

$$f(-x) = 3x^6 - 4x^4 - 3x^3 + 2x^2 + x - 3$$

There are three variations in sign, so we expect either three or one negative zero(s). Equivalently, we now know that the graph of f has either three or one positive x-intercept(s) and three or one negative x-intercept(s).

We see in Figure 43 that f has one negative zero near -1.3 and one positive zero near 1. Thus, the conclusions of Descarte's Rule of Signs are confirmed by the graph.

FIGURE 43

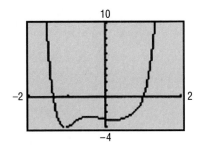

■   Now work Problem 11.

Rational Zeros Theorem

The next result, which you are asked to prove in Problem 84, is called the **Rational Zeros Theorem.** It provides information about the rational zeros of a polynomial with integer coefficients.

Theorem
Rational Zeros Theorem

Let f be a polynomial function of degree 1 or higher of the form

$$f(x) = a_n x^n + a_{n-1} x^{n-1} + \cdots + a_1 x + a_0 \qquad a_n \neq 0, \, a_0 \neq 0$$

where each coefficient is an integer. If p/q, in lowest terms, is a rational zero of f, then p must be a factor of a_0 and q must be a factor of a_n.   ■

E X A M P L E   4

Listing Potential Rational Zeros

List the potential rational zeros of

$$f(x) = 2x^3 + 11x^2 - 7x - 6$$

Solution

Because f has integer coefficients, we may use the Rational Zeros Theorem. First, we list all the integers p that are factors of $a_0 = -6$ and all the integers q that are factors of $a_3 = 2$:

$$p: \quad \pm 1, \, \pm 2, \, \pm 3, \, \pm 6$$
$$q: \quad \pm 1, \, \pm 2$$

Now we form all possible ratios p/q:

$$\frac{p}{q}: \quad \pm 1, \, \pm 2, \, \pm 3, \, \pm 6, \, \pm \frac{1}{2}, \, \pm \frac{3}{2}$$

If f has a rational zero, it will be found in this list, which contains 12 possibilities.

■

■   Now work Problem 23.

Be sure you understand what the Rational Zeros Theorem says: For a polynomial with integer coefficients, *if* there is a rational zero, it is one of those listed. There may not be any rational zeros.

The Rational Zeros Theorem provides a list of potential rational zeros of a function f. If we graph f, we can get a better sense of the location of the x-intercepts and test to see if they are rational.

The next example shows how.

E X A M P L E   5   *Finding the Rational Zeros of a Polynomial Function*

Continue working with Example 4 to find the rational zeros of

$$f(x) = 2x^3 + 11x^2 - 7x - 6$$

Solution   We gather all the information we can about the zeros:

STEP 1:   There are at most three zeros.
STEP 2:   By Descartes' Rule of Signs, there is one positive zero. Also, because

$$f(-x) = -2x^3 + 11x^2 + 7x - 6$$

there are two or no negative zeros.
STEP 3:   Now we use the list of potential rational zeros obtained in Example 4: ± 1, ± 2, ± 3, ± 6, $\pm \frac{1}{2}$, $\pm \frac{3}{2}$.

We could, of course, test each potential rational zero to see if the value of f there is zero. This is not very efficient. The graph of f will tell us approximately where the real zeros are. So we only need to test those rational zeros that are nearby. Figure 44 shows the graph of f. We see f has three zeros: one near -6, one between -1 and 0, and one near 1. From our original list of potential rational zeros, we will test:

near -6: test -6;   between -1 and 0: test $-\dfrac{1}{2}$;   near 1: test 1

$$f(-6) = 0 \qquad\qquad f\left(-\frac{1}{2}\right) = 0 \qquad\qquad f(1) = 0$$

The three zeros of f are -6, $-\frac{1}{2}$, and 1; each one is a rational zero.

FIGURE 44

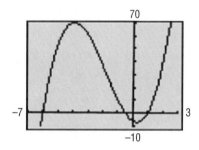

We can also use synthetic division* to determine whether a potential rational zero is, in fact, a zero. For example, to test -6, divide f by $x - (-6) = x + 6$

$$
\begin{array}{r|rrrr}
-6) & 2 & 11 & -7 & -6 \\
    &   & -12 & 6 & 6 \\
\hline
    & 2 & -1 & -1 & 0
\end{array}
$$

The fact that the remainder is zero tells us -6 is a zero and $x - (-6) = x + 6$ is a factor. Similarly, we have $x + \frac{1}{2}$ and $x - 1$ as factors of f. As a result, we can write f in factored form as

$$f(x) = 2x^3 + 11x^2 - 7x - 6 = 2\left(x^3 + \frac{11}{2}x^2 - \frac{7}{2}x - 3\right) = 2(x + 6)\left(x + \frac{1}{2}\right)(x - 1)$$

*Synthetic division is discussed in the Appendix, Section 6.

The next two examples show an advantage of using synthetic division.

E X A M P L E   6   *Finding the Real Zeros of a Polynomial Function*

Find the real zeros of: $f(x) = 3x^5 - 2x^4 - 15x^3 + 10x^2 + 12x - 8$. Then factor f over the real numbers.

Solution   We gather all the information we can about the zeros:

STEP 1: There are at most five zeros.
STEP 2: By Descartes' Rule of Signs, there are three or one positive zero(s). Also, because

$$f(-x) = -3x^5 - 2x^4 + 15x^3 + 10x^2 - 12x - 8$$

there are two or no negative zeros.
STEP 3: To obtain the list of potential rational zeros, we write the factors p of $a_0 = -8$ and the factors q of $a_5 = 3$:

$$p: \quad \pm 1, \ \pm 2, \ \pm 4, \ \pm 8$$
$$q: \quad \pm 1, \ \pm 3$$

The potential rational zeros consist of all possible quotients p/q:

$$\frac{p}{q}: \quad \pm 1, \ \pm 2, \ \pm 4, \ \pm 8, \ \pm \frac{1}{3}, \ \pm \frac{2}{3}, \ \pm \frac{4}{3}, \ \pm \frac{8}{3}$$

FIGURE 45

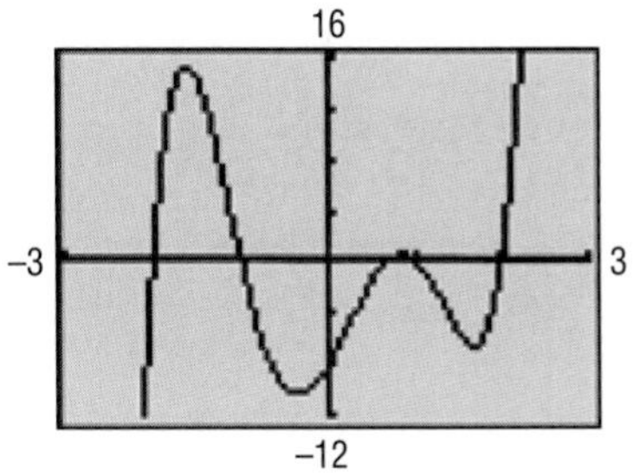

STEP 4: Figure 45 shows the graph of f. The graph has the characteristics we expect of the given polynomial of degree 5: four turning points, y-intercept -8, and behaves like $y = 3x^5$ for large x. The information found in Steps 1 and 2 is also confirmed by the graph. The polynomial has 2 or 0 negative zeros and 3 or 1 positive zeros. The graph has five x-intercepts: one near -2, one near -1, two between 0 and 1 (possibly a double root), and one near 2.

Since -2 appears to be a zero and -2 is a potential rational zero, we use synthetic division to determine if $x + 2$ is a factor of f.

$$
\begin{array}{r|rrrrrr}
-2) & 3 & -2 & -15 & 10 & 12 & -8 \\
    &   & -6 & 16 & -2 & -16 & 8 \\
\hline
    & 3 & -8 & 1 & 8 & -4 & 0
\end{array}
$$

Since the remainder is 0, we know -2 is a zero and $x + 2$ is a factor of f. We use the entries in the bottom row of the synthetic division to factor f:

$$f(x) = 3x^5 - 2x^4 - 15x^3 + 10x^2 + 12x - 8$$
$$= (x + 2)(3x^4 - 8x^3 + x^2 + 8x - 4)$$

Now any solution of the equation $3x^4 - 8x^3 + x^2 + 8x - 4 = 0$ is a zero of f. Because of this we call the equation $3x^4 - 8x^3 + x^2 + 8x - 4 = 0$ a **depressed equation** of f. Since the degree of the depressed equation of f is less than that of the original equation, we work with the depressed equation to find the zeros of f.

We check the potential rational zero -1 next using synthetic division.

$$
\begin{array}{r|rrrrr}
-1) & 3 & -8 & 1 & 8 & -4 \\
    &   & -3 & 11 & -12 & 4 \\
\hline
    & 3 & -11 & 12 & -4 & 0
\end{array}
$$

Since the remainder is 0, we know -1 is a zero and $x + 1$ is a factor of f. Thus, we can write f as

$$f(x) = (x + 2)(x + 1)(3x^3 - 11x^2 + 12x - 4)$$

We work with the second depressed equation and check the potential rational zero 1 using synthetic division,

$$
\begin{array}{r|rrrr}
1) & 3 & -11 & 12 & -4 \\
  &   & 3 & -8 & 4 \\
\hline
  & 3 & -8 & 4 & 0
\end{array}
$$

Since the remainder is 0, we conclude that 1 is a zero and $x - 1$ is a factor of f. Thus, we have

$$f(x) = (x + 2)(x + 1)(x - 1)(3x^3 - 8x + 4)$$

The new depressed equation of f, $3x^3 - 8x + 4 = 0$, is a quadratic equation with a discriminant of $b^2 - 4ac = (-8)^2 - 4(3)(4) = 16$. Therefore, this equation has two real solutions, and, in this case, we can find them by factoring:

$$3x^2 - 8x + 4 = 0$$
$$(3x - 2)(x - 2) = 0$$
$$3x - 2 = 0 \quad or \quad x - 2 = 0$$
$$x = \frac{2}{3} \qquad\qquad x = 2$$

The zeros of f are -2, -1, $\frac{2}{3}$, 1, and 2. The factored form of f is

$$f(x) = 3(x + 2)(x + 1)\left(x - \frac{2}{3}\right)(x - 1)(x - 2)$$

Now we know that f does not have a double root near 1, but rather has two nearby roots.

Based on the information gathered, we change the viewing rectangle to obtain the graph of f for $0 \le x \le 3$. See Figure 46. Now we can see the three distinct positive zeros between 0 and 3. ■

■   Now work Problem 41.

The procedure outlined in Example 6 for finding the zeros of a polynomial can also be used to solve polynomial equations.

FIGURE 46

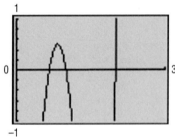

E X A M P L E   7

Solving a Polynomial Equation

Solve the equation: $x^5 - 5x^4 + 12x^3 - 24x^2 + 32x - 16 = 0$

Solution   The solutions of this equation are the zeros of the polynomial function

$$f(x) = x^5 - 5x^4 + 12x^3 - 24x^2 + 32x - 16$$

STEP 1:   There are at most five real solutions.
STEP 2:   By Descartes' Rule of Signs, there are five, three, or one positive solution(s). Because

$$f(-x) = -x^5 - 5x^4 - 12x^3 - 24x^2 - 32x - 16$$

there are no negative solutions.
STEP 3:   Because $a_5 = 1$ and there are no negative solutions, the potential rational solutions are the positive integers 1, 2, 4, 8, and 16.
STEP 4:   Figure 47 shows the graph of f. The graph has the characteristics we expect of this polynomial of degree 5: it behaves like $y = x^5$ for large x, has y-intercept -16, and has two turning points. The information obtained in Steps 1 and 2 is confirmed by the graph: no negative zeros. There is an x-intercept near 1 (where the graph crosses the x-axis) and another near 2 (which may be a double root, since the graph appears to touch the x-axis near 2).

FIGURE 47

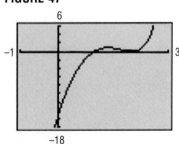

Since 1 appears to be a zero and 1 is a potential rational zero, we use synthetic division to determine if $x - 1$ is a factor of f:

$$
\begin{array}{r|rrrrr}
1) & 1 & -5 & 12 & -24 & 32 & -16 \\
  &   & 1 & -4 &   8 & -16 & 16 \\
\hline
  & 1 & -4 & 8 & -16 & 16 &   0
\end{array}
$$

Since the remainder is 0, we know 1 is a zero and $x - 1$ is a factor of f. Thus, 1 is a solution to the equation.

We now work with the first depressed equation of f:

$$x^4 - 4x^3 + 8x^2 - 16x + 16 = 0$$

We check the potential rational zero 2 next using synthetic division.

$$
\begin{array}{r|rrrrr}
2) & 1 & -4 & 8 & -16 & 16 \\
  &   & 2 & -4 &   8 & -16 \\
\hline
  & 1 & -2 & 4 & -8 & 0
\end{array}
$$

Since the remainder is 0, we conclude 2 is a zero and $x - 2$ is a factor of f. Thus, 2 is a solution to the equation.

We now work with the second depressed equation of f: $x^3 - 2x^2 + 4x - 8 = 0$. Earlier, we conjectured 2 may be a double root so we check 2 again.

$$
\begin{array}{r|rrrr}
2) & 1 & -2 & 4 & -8 \\
  &   & 2 & 0 & 8 \\
\hline
  & 1 & 0 & 4 & 0
\end{array}
$$

Since the remainder is 0, 2 is a zero and $x - 2$ is a factor of f. Thus, $x = 2$ is a zero of multiplicity 2.

The third depressed equation is $x^2 + 4 = 0$. This equation has no real solutions. Thus, the real solutions are 1 and 2 (the latter being a repeated solution).

■

Based on the solution to Example 7, the polynomial function f can be factored as follows:

$$f(x) = x^5 - 5x^4 + 12x^3 - 24x^2 + 32x - 16 = (x - 1)(x - 2)^2(x^2 + 4)$$

■ Now work Problem 61.

EXAMPLE 8

Finding the Zeros of a Polynomial

Use Descartes' Rule of Signs and the Rational Zeros Theorem to find the rational zeros of the polynomial function

$$g(x) = x^5 - 7x^4 + 8x^3 - 20x^2 + 20x - 2$$

Express any irrational zeros correct to two decimal places.

Solution **STEP 1:** There are at most five zeros.
STEP 2: There are five, three, or one positive zeros. Also, because

$$g(-x) = -x^5 - 7x^4 - 8x^3 - 20x^2 - 20x - 2$$

there are no negative zeros.
STEP 3: The potential rational zeros of g are 1 and 2.
STEP 4: Because there are no negative zeros, we graph f only for $x \geq -1$. See Figure 48.

FIGURE 48

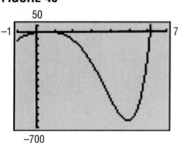

The graph has the characteristics we expect: it behaves like $y = x^5$ for large x and has y-intercept -2. The information obtained in steps 1 and 2 is confirmed. The graph has three x-intercepts: one between 0 and 1, one near 1, and one near 6. We test the potential rational zero 1 using synthetic division.

$$
\begin{array}{r|rrrrrr}
1) & 1 & -7 & 8 & -20 & 20 & -2 \\
   &   & 1 & -6 & 2 & -18 & 2 \\
\hline
   & 1 & -6 & 2 & -18 & 2 & 0
\end{array}
$$

The remainder is 0 so 1 is a zero and $x - 1$ is a factor of g.

STEP 5:   The first depressed equation of g:

$$x^4 - 6x^3 + 2x^2 - 18x + 2 = 0$$

has only 1 and 2 as potential rational zeros. Neither of these is, in fact, a zero. So the remaining zeros of g are not rational. Correct to two decimal places, these irrational zeros are: 0.10 and 6.13.   ∎

In Example 7, the quadratic factor $x^2 + 4$ that appears in the factored form of $f(x)$ is called *irreducible,* because the polynomial $x^2 + 4$ cannot be factored over the real numbers. In general, we say that a quadratic factor $ax^2 + bx + c$ is **irreducible** if it cannot be factored over the real numbers, that is, if it is prime over the real numbers.

Refer back to the polynomial function f of Example 6. We found that f has five real zeros, so, by the Factor Theorem, its factored form will contain five linear factors. The polynomial function g of Example 8 has three real zeros, so its factored form will contain three linear factors and one irreducible quadratic factor. The next result tells us what to expect when we factor a polynomial.

Theorem   Every polynomial function (with real coefficients) can be uniquely factored into a product of linear factors and/or irreducible quadratic factors.   ∎

We shall prove this result in Section 4.5, and, in fact, we shall draw several additional conclusions about the zeros of a polynomial function. One conclusion is worth noting now. If a polynomial (with real coefficients) is of odd degree, then it must contain at least one linear factor. (Do you see why?) Therefore, it will have at least one real zero.

Corollary   A polynomial function (with real coefficients) of odd degree has at least one real zero.   ∎

One of the challenges in using a graphing utility is to set the viewing window so that a complete graph is obtained. The next theorem is a tool which can be used to find bounds on the zeros. This will assure the function does not have any zeros above or below these bounds.

Upper and Lower Bounds

The search for the zeros of a polynomial function can be reduced somewhat if upper and lower bounds to the zeros can be found. A number M is an **upper bound** to the zeros of a polynomial f if no zero of f exceeds M. The number m is a **lower bound** if no zero of f is less than m.

Thus, if m is a lower bound and M is an upper bound to the zeros of a polynomial f, then

$$m \leq \text{Any zero of } f \leq M$$

One immediate advantage of knowing the values of a lower bound m and an upper bound M is that, for polynomials with integer coefficients, it may allow you to eliminate some potential rational zeros, that is, any that lie outside the interval $[m, M]$. The next result tells us how to locate lower and upper bounds.

Theorem
Bounds on Zeros

Let f denote a polynomial function whose leading coefficient is positive.

> If $M > 0$ is a real number and if the third row in the process of synthetic division of f by $x - M$ contains only numbers that are positive or 0, then M is an upper bound to the zeros of f.

> If $m < 0$ is a real number and if the third row in the process of synthetic division of f by $x - m$ contains numbers that are alternately positive (or 0) and negative (or 0), then m is a lower bound to the zeros of f.   ■

Proof (Outline)

We shall give only an outline of the proof of the first part of the theorem. Suppose that M is a positive real number, and the third row in the process of synthetic division of the polynomial f by $x - M$ contains only numbers that are positive or 0. Then there is a quotient q and a remainder R so that

$$f(x) = (x - M)q(x) + R$$

where the coefficients of $q(x)$ are positive or 0 and the remainder $R \geq 0$. Then, for any $x > M$, we must have $x - M > 0$, $q(x) > 0$, and $R \geq 0$, so that $f(x) > 0$. That is, there is no zero of f larger than M.   ■

We can use the Upper and Lower Bounds Test to determine if the RANGE we have selected for our graph is large enough to show all the real zeros of the function, ensuring no zeros lie off the viewing rectangle. Consider Example 9.

E X A M P L E   9

Using the Upper and Lower Bounds Test

Use the Upper and Lower Bounds Test on the polynomial function $f(x) = x^5 - 5x^4 + 12x^3 - 24x^2 + 32x - 16$ to verify there are no zeros larger than 2 and no zeros smaller than 1.

Solution

To get an upper bound to the zeros, the usual practice is to start with the value of the next highest integer larger than the largest zero found and continue with consecutive integers until the third row of the process of synthetic division yields only numbers that are positive or 0.

We begin with the graph of f. See Figure 49.

The largest zero we see is near 2. The next highest integer, 3, is where we begin. We use synthetic division to divide f by $x - 3$. If the third row yields only numbers that are positive or zero, then 3 is an upper bound.

FIGURE 49

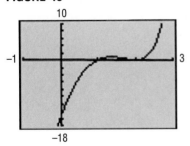

$$
\begin{array}{r|rrrrrr}
3) & 1 & -5 & 12 & -24 & 32 & -16 \\
  &   & 3 & -6 & 18 & -18 & 42 \\
\hline
  & 1 & -2 & 6 & -6 & 14 & 26 \\
\end{array}
$$

Since the third row yields some negative numbers, 3 is not an upper bound. We will try 4, the next consecutive integer.

$$
\begin{array}{r|rrrrrr}
4) & 1 & -5 & 12 & -24 & 32 & -16 \\
  &   & 4 & -4 & 32 & 32 & 256 \\
\hline
  & 1 & -1 & 8 & 8   & 64 & 240 \\
\end{array}
$$

Since the third row yields some negative numbers, 4 is not an upper bound. We will try 5.

$$
\begin{array}{r|rrrrr}
5)& 1 & -5 & 12 & -24 & 32 & -16 \\
& & 5 & 0 & 60 & 180 & 1060 \\
\hline
& 1 & 0 & 12 & 36 & 212 & 1044
\end{array}
$$

The last row yields only numbers that are positive or 0; therefore, 5 is an upper bound. We are certain there are no zeros larger than 5. This means 5 is a good choice for Xmax when we first graph f.

   To get a lower bound to the zeros, we start with the next integer smaller than the smallest zero we see.

   Look again at Figure 49. The smallest zero we see is near 1, so we begin by testing 0.

$$
\begin{array}{r|rrrrr}
0)& 1 & -5 & 12 & -24 & 32 & -16 \\
& & 0 & 0 & 0 & 0 & 0 \\
\hline
& 1 & -5 & 12 & -24 & 32 & -16
\end{array}
$$

Since the third row of the process of synthetic division yields numbers that alternate in sign, 0 is a lower bound; we are certain there are no zeros less than 0.*

   This means 0 is a good first choice for Xmin.   ∎

   ■ Now work Problem 35.

Steps for Finding the Zeros of a Polynomial	**STEP 1:** Use the degree of the polynomial to determine the maximum number of zeros.

STEP 1: Use the degree of the polynomial to determine the maximum number of zeros.
STEP 2: Use Descartes' Rule of Signs to determine the possible number of positive zeros and negative zeros.
STEP 3: If the polynomial has integer coefficients, use the Rational Zeros Theorem to identify those rational numbers that are potential zeros.
STEP 4: (a) Using a graphing utility, graph the polynomial.
   (b) Conjecture possible zeros based on the graph and the list of potential rational zeros.
   (c) Test each potential rational zero.
   (d) Approximate any irrational zeros correct to two decimal places.
STEP 5: Use the Upper and Lower Bounds Test to be sure no zeros exist off the viewing window.

Let's work one more example that shows the STEPS listed above.

E X A M P L E   1 0   *Finding the Zeros of a Polynomial*

Find all the real zeros of the polynomial function

$$
f(x) = x^5 - 1.8x^4 - 17.79x^3 + 31.672x^2 + 37.95x - 8.712
$$

Solution   **STEP 1:** There are at most 5 zeros.
STEP 2: By Descartes' Rule of Signs there are three or one positive zero(s). Also because

$$
f(-x) = -x^5 - 1.8x^4 + 17.79x^3 + 31.672x^2 - 37.95x - 8.712
$$

there are two or no negative zeros.

---

*Note: In determining lower bounds, a 0 in the bottom row following a nonzero entry may be counted as positive or negative, as needed. If the next entry is also a 0, it must be counted opposite to the way the preceding 0 was counted (i.e., if the third row yields two consecutive zeros and the first 0 was counted as positive, the next zero would count as negative).

STEP 3: Since there are noninteger coefficients, the Rational Zeros Theorem does not apply.

FIGURE 50

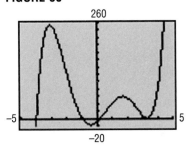

STEP 4: See Figure 50 for the graph of f. We see f appears to have four zeros, one near -4, one near -1, one between 0 and 1, and one near 3. The zero near 3 might be a zero of even multiplicity since the graph seems to touch the x-axis at the point. From Step 2, we learned the function would have two or no negative zeros and three or one positive zeros. Thus, we know there are no other negative zeros off the viewing window. However, we may only be seeing two positive zeros and Descartes' Rule of Signs says there are three.

We use the factor theorem to determine if -4, and -1 are zeros. Since $f(-4) = f(-1) = 0$, we know -4 and -1 are the two negative zeros. Correct to two decimal places, the remaining zeros are 0.20 and 3.30.

STEP 5: We will check to see if 4 is an upper bound to the zeros.

$$\begin{array}{r} 4\overline{)1 \quad -1.8 \quad -17.79 \quad 31.672 \quad 37.95 \quad -8.712} \\ \underline{4 \qquad 8.8} \\ 1 \quad 2.2 \quad -8.99 \end{array}$$

We stop the process since the Upper Bound Theorem fails. We try 5 next.

$$\begin{array}{r} 5\overline{)1 \quad -1.8 \quad -17.79 \quad 31.672 \quad 37.95 \quad -8.712} \\ \underline{5 \qquad 16} \\ 1 \quad 3.2 \quad -1.79 \end{array}$$

Again we stop the process and try 6.

$$\begin{array}{r} 6\overline{)1 \quad -1.8 \quad -17.79 \quad 31.672 \quad 37.95 \quad -8.712} \\ \underline{6 \qquad 25.2 \qquad 44.46 \qquad 456.792 \quad 2968.452} \\ 1 \quad 4.2 \quad 7.41 \quad 76.132 \quad 494.742 \quad 2959.74 \end{array}$$

So, 6 is an upper bound to the zeros. Therefore, either 3.30 is a root of multiplicity 2 or there are two zeros, both of which equal 3.30 correct to two decimal places. ∎

Historical Feature

■ Formulas for the solution of third- and fourth-degree polynomial equations exist, and, while not very practical, they do have an interesting history.

In the 1500's in Italy, mathematical contests were a popular pastime, and persons possessing methods for solving problems kept them secret. (Solutions that were published were already common knowledge.) Niccolo of Brescia (1499–1557), commonly referred to as Tartaglia ("the stammerer"), had the secret for solving cubic (third-degree) equations, which gave him a decided advantage in the contests. Girolamo Cardano (1501–1576) found out that Tartaglia had the secret, and, being interested in cubics, he requested it from Tartaglia. The reluctant Tartaglia hesitated for some time, but finally, swearing Cardano to secrecy with midnight oaths by candlelight, told him the secret. Cardano then published the solution in his book *Ars Magna* (1545), giving Tartaglia the credit but rather compromising the secrecy. Tartaglia exploded into bitter recriminations, and each wrote pamphlets that reflected on the other's mathematics, moral character, and ancestry. Tartaglia's method is discussed in Problems 85–89 in Exercise 4.4.

The quartic (fourth-degree) equation was solved by Cardano's student Lodovico Ferrari, and this solution also was included, with credit and this time with permission, in the *Ars Magna*.

Attempts were made to solve the fifth-degree equation in similar ways, all of which failed. In the early 1800's, P. Ruffini, Niels Abel, and Evariste Galois all found ways to show that it is not possible to solve fifth-degree equations by formula, but the proofs required the introduction of new methods. Galois' methods eventually developed into a large part of modern algebra. ■

4.4

Exercise 4.4

In Problems 1–10, use the Factor Theorem to determine whether $x - c$ is a factor of $f(x)$.

1. $f(x) = 4x^3 - 3x^2 - 8x + 4$; $c = 2$

2. $f(x) = -4x^3 + 5x^2 + 8$; $c = -3$

3. $f(x) = 3x^4 - 6x^3 - 5x + 10$; $c = 2$

4. $f(x) = 4x^4 - 15x^2 - 4$; $c = 2$

5. $f(x) = 3x^6 + 82x^3 + 27$; $c = -3$

6. $f(x) = 2x^6 - 18x^4 + x^2 - 9$; $c = -3$

7. $f(x) = 4x^6 - 64x^4 + x^2 - 15$; $c = -4$

8. $f(x) = x^6 - 16x^4 + x^2 - 16$; $c = -4$

9. $f(x) = 2x^4 - x^3 + 2x - 1$; $c = \dfrac{1}{2}$

10. $f(x) = 3x^4 + x^3 - 3x + 1$; $c = -\dfrac{1}{3}$

In Problems 11–22, tell the maximum number of zeros each polynomial function may have. Then, use Descartes' Rule of Signs to determine how many positive and how many negative zeros each polynomial function may have. Do not attempt to find the zeros.

11. $f(x) = -4x^7 + x^3 - x^2 + 2$

12. $f(x) = 5x^4 + 2x^2 - 6x - 5$

13. $f(x) = 2x^6 - 3x^2 - x + 1$

14. $f(x) = -3x^5 + 4x^4 + 2$

15. $f(x) = 3x^3 - 2x^2 + x + 2$

16. $f(x) = -x^3 - x^2 + x + 1$

17. $f(x) = -x^4 + x^2 - 1$

18. $f(x) = x^4 + 5x^3 - 2$

19. $f(x) = x^5 + x^4 + x^2 + x + 1$

20. $f(x) = x^5 - x^4 + x^3 - x^2 + x - 1$

21. $f(x) = x^6 - 1$

22. $f(x) = x^6 + 1$

In Problems 23–34, list the potential rational zeros of each polynomial function. Do not attempt to find the zeros.

23. $f(x) = 3x^4 - 3x^3 + x^2 - x + 1$

24. $f(x) = x^5 - x^4 + 2x^2 + 3$

25. $f(x) = x^5 - 6x^2 + 9x - 3$

26. $f(x) = 2x^5 - x^4 - x^2 + 1$

27. $f(x) = -4x^3 - x^2 + x + 2$

28. $f(x) = 6x^4 - x^2 + 2$

29. $f(x) = 3x^4 - x^2 + 2$

30. $f(x) = -4x^3 + x^2 + x + 2$

31. $f(x) = 2x^5 - x^3 + 2x^2 + 4$

32. $f(x) = 3x^5 - x^2 + 2x + 3$

33. $f(x) = 6x^4 + 2x^3 - x^2 + 2$

34. $f(x) = -6x^3 - x^2 + x + 3$

In Problems 35–40, find integer-valued upper and lower bounds to the zeros of each polynomial function.

35. $f(x) = 2x^3 + x^2 - 1$

36. $f(x) = 3x^3 - 2x^2 + x + 4$

37. $f(x) = x^3 - 5x^2 - 11x + 11$

38. $f(x) = 2x^3 - x^2 - 11x - 6$

39. $f(x) = x^4 + 3x^3 - 5x^2 + 9$

40. $f(x) = 4x^4 - 12x^3 + 27x^2 - 54x + 81$

In Problems 41–60, find the rational zeros of f. List any irrational zeros correct to two decimal places.

41. $f(x) = x^3 + 2x^2 - 5x - 6$

42. $f(x) = x^3 + 8x^2 + 11x - 20$

43. $f(x) = 2x^3 - x^2 + 2x - 1$

44. $f(x) = 2x^3 + x^2 + 2x + 1$

45. $f(x) = x^4 + x^2 - 2$

46. $f(x) = x^4 - 3x^2 - 4$

47. $f(x) = 4x^4 + 7x^2 - 2$

48. $f(x) = 4x^4 + 15x^2 - 4$

49. $f(x) = x^4 + x^3 - 3x^2 - x + 2$

50. $f(x) = x^4 - x^3 - 6x^2 + 4x + 8$

51. $f(x) = 4x^5 - 8x^4 - x + 2$

52. $f(x) = 4x^5 + 12x^4 - x - 3$

53. $f(x) = x^3 + 3.2x^2 - 16.83x - 5.31$

54. $f(x) = x^3 + 3.2x^2 - 7.25x - 6.3$

55. $f(x) = x^4 - 1.4x^3 - 33.71x^2 + 23.94x + 292.41$

56. $f(x) = x^4 + 1.2x^3 - 7.46x^2 - 4.692x + 15.2881$

57. $f(x) = \pi x^3 - (8.88\pi + 1)x^2 - (42.066\pi - 8.88)x + 42.066$

58. $f(x) = \pi x^3 - (5.63\pi + 2)x^2 - (108.392\pi - 11.26)x + 216.784$

59. $f(x) = x^3 + 19.5x^2 - 1021x + 1000.5$

60. $f(x) = x^3 + 42.2x^2 - 664.8x + 1490.4$

In Problems 61–70, solve each equation in the real number system.

61. $x^4 - x^3 + 2x^2 - 4x - 8 = 0$

62. $2x^3 + 3x^2 + 2x + 3 = 0$

63. $3x^3 + 4x^2 - 7x + 2 = 0$

64. $2x^3 - 3x^2 - 3x - 5 = 0$

65. $3x^3 - x^2 - 15x + 5 = 0$

66. $2x^3 - 11x^2 + 10x + 8 = 0$

67. $x^4 + 4x^3 + 2x^2 - x + 6 = 0$

68. $x^4 - 2x^3 + 10x^2 - 18x + 9 = 0$

69. $x^3 - \frac{2}{3}x^2 + \frac{8}{3}x + 1 = 0$

70. $x^3 + \frac{3}{2}x^2 + 3x - 2 = 0$

71. Find k such that $f(x) = x^3 - kx^2 + kx + 2$ has the factor $x - 2$.

72. Find k such that $f(x) = x^4 - kx^3 + kx^2 + 1$ has the factor $x + 2$.

73. What is the remainder when $f(x) = 2x^{20} - 8x^{10} + x - 2$ is divided by $x - 1$?

74. What is the remainder when $f(x) = -3x^{17} + x^9 - x^5 + 2x$ is divided by $x + 1$?

75. Use the Factor Theorem to prove that $x - c$ is a factor of $x^n - c^n$ for any positive integer n.

76. Use the Factor Theorem to prove that $x + c$ is a factor of $x^n + c^n$ if $n \geq 1$ is an odd integer.

77. Is $\frac{1}{3}$ a zero of $f(x) = 2x^3 + 3x^2 - 6x + 7$? Explain.

78. Is $\frac{1}{3}$ a zero of $f(x) = 4x^3 - 5x^2 - 3x + 1$? Explain.

79. Is $\frac{3}{5}$ a zero of $f(x) = 2x^6 - 5x^4 + x^3 - x + 1$? Explain.

80. Is $\frac{2}{3}$ a zero of $f(x) = x^7 + 6x^5 - x^4 + x + 2$? Explain.

81. What is the length of the edge of a cube if, after a slice 1 inch thick is cut from one side, the volume remaining is 294 cubic inches?

82. What is the length of the edge of a cube if its volume could be doubled by an increase of 6 centimeters in one edge, an increase of 12 centimeters in a second edge, and a decrease of 4 centimeters in the third edge?

83. Let $f(x)$ be a polynomial function whose coefficients are integers. Suppose that r is a real zero of f and that the leading coefficient of f is 1. Use the Rational Zeros Theorem to show that r is either an integer or an irrational number.

84. Prove the Rational Zeros Theroem. [*Hint:*   Let p/q, where p and q have no common factors except 1 and -1, be a solution of the polynomial $f(x) = a_n x^n + a_{n-1}x^{n-1} + \cdots + a_1 x + a_0$, whose coefficients are all integers. Show that $a_n p^n + a_{n-1}p^{n-1}q + \cdots + a_1 pq^{n-1} + a_0 q^n = 0$. Now, because p is a factor of the first n terms of this equation, p must also be a factor of the term $a_0 q^n$. Since p is not a factor of q (why?), p must be a factor of a_0. Similarly, q must be a factor of a_n.]

Problems 85–89 develop the Tartaglia–Cardano solution of the cubic equation and show why it is not altogether practical.

85. Show that the general cubic equation $y^3 + by^2 + cy + d = 0$ can be transformed into an equation of the form $x^3 + px + q = 0$ by using the substitution $y = x - b/3$.

86. In the equation $x^3 + px + q = 0$, replace x by $H + K$. Let $3HK = -p$, and show that $H^3 + K^3 = -q$.
[*Hint:* $3H^2K + 3HK^2 = 3HKx$.]

87. Based on Problem 86, we have the two equations

$$3HK = -p \quad \text{and} \quad H^3 + K^3 = -q$$

Solve for K in $3HK = -p$ and substitute into $H^3 + K^3 = -q$. Then show that

$$H = \sqrt[3]{\frac{-q}{2} + \sqrt{\frac{q^2}{4} + \frac{p^3}{27}}}$$

[*Hint:*   Look for an equation that is quadratic in form.]

88. Use the solution for H from Problem 87 and the equation $H^3 + K^3 = -q$ to show that

$$K = \sqrt[3]{\frac{-q}{2} - \sqrt{\frac{q^2}{4} + \frac{p^3}{27}}}$$

89. Use the results from Problems 86–88 to show that the solution of $x^3 + px + q = 0$ is

$$x = \sqrt[3]{\frac{-q}{2} + \sqrt{\frac{q^2}{4} + \frac{p^3}{27}}} + \sqrt[3]{\frac{-q}{2} - \sqrt{\frac{q^2}{4} + \frac{p^3}{27}}}$$

90. Use the result of Problem 89 to solve the equation $x^3 - 6x - 9 = 0$.

91. Use a calculator and the result of Problem 89 to solve the equation $x^3 + 3x - 14 = 0$.

92. Use the methods of this chapter to solve the equation $x^3 + 3x - 14 = 0$.

93. *Requires Complex Numbers* Show that the formula derived in Problem 89 leads to the cube root of a complex number when applied to the equation $x^3 - 6x + 4 = 0$. Use the methods of this chapter to solve the equation.

94. A function f has the property that $f(2 + x) = f(2 - x)$ for all x. If f has exactly four real zeros, find their sum.

4.5

Complex Polynomials; Fundamental Theorem of Algebra

A variable z in the complex number system is referred to as a **complex variable.** A **complex polynomial function** f of degree n is a complex function of the form

$$f(z) = a_n z^n + a_{n-1} z^{n-1} + \cdots + a_1 z + a_0 \qquad (1)$$

where $a_n, a_{n-1}, \ldots, a_1, a_0$ are complex numbers, $a_n \neq 0$, and n is a nonnegative integer. Here, a_n is called the **leading coefficient** of f. A complex number r is called a (complex) **zero** of a complex function f if $f(r) = 0$.

In Chapter 3, we discovered that some quadratic equations have no real solutions, but that in the complex number system every quadratic equation has a solution, either real or complex. The next result, proved by Karl Friedrich Gauss (1777–1855) when he was 22 years of age,* gives an extension to complex polynomials. In fact, this result is so important and useful it has become known as the **Fundamental Theorem of Algebra.**

Fundamental Theorem of Algebra   Every complex polynomial function $f(z)$ of degree $n \geq 1$ has at least one complex zero. ∎

We shall not prove this result, as the proof is beyond the scope of this book. However, using the Fundamental Theorem of Algebra and the Factor Theorem, we can prove the following result:

Theorem   Every complex polynomial function $f(z)$ of degree $n \geq 1$ can be factored into n linear factors (not necessarily distinct) of the form

$$f(z) = a_n(z - r_1)(z - r_2) \cdot \ldots \cdot (z - r_n) \qquad (2)$$

where $a_n, r_1, r_2, \ldots, r_n$ are complex numbers.

Proof   Let

$$f(z) = a_n z^n + a_{n-1} z^{n-1} + \cdots + a_1 z + a_0$$

---

*In all, Gauss gave four different proofs of this theorem, the first one in 1799 being the subject of his doctoral dissertation.

By the Fundamental Theorem of Algebra, f has at least one zero, say, r_1. Then, by the Factor Theorem, $z - r_1$ is a factor, and

$$f(z) = (z - r_1)q_1(z)$$

where $q_1(z)$ is a complex polynomial of degree $n - 1$ whose leading coefficient is a_n. Again, by the Fundamental Theorem of Algebra, the complex polynomial $q_1(z)$ has at least one zero, say, r_2. By the Factor Theorem, $q_1(z)$ has the factor $z - r_2$, so

$$q_1(z) = (z - r_2)q_2(z)$$

where $q_2(z)$ is a complex polynomial of degree $n - 2$ whose leading coefficient is a_n. Consequently,

$$f(z) = (z - r_1)(z - r_2)q_2(z)$$

Repeating this argument n times, we finally arrive at

$$f(z) = (z - r_1)(z - r_2) \cdot \ldots \cdot (z - r_n)q_n(z)$$

where $q_n(z)$ is a complex polynomial of degree $n - n = 0$ whose leading coefficient is a_n. Thus, $q_n(z) = a_n z^0 = a_n$, and so

$$f(z) = a_n(z - r_1)(z - r_2) \cdot \ldots \cdot (z - r_n)$$ ■

Complex Polynomials with Real Coefficients

We can use the Fundamental Theorem of Algebra to obtain valuable information about the zeros of complex polynomials whose coefficients are real numbers.

Theorem
Conjugate Pairs
Let $f(z)$ be a complex polynomial whose coefficients are real numbers. If $r = a + bi$ is a zero of f, then the complex conjugate $\bar{r} = a - bi$ is also a zero of f. ■

In other words, for complex polynomials whose coefficients are real numbers, the zeros occur in conjugate pairs.

Proof Let

$$f(z) = a_n z^n + a_{n-1} z^{n-1} + \cdots + a_1 z + a_0$$

where $a_n, a_{n-1}, \ldots, a_1, a_0$ are real numbers and $a_n \neq 0$. If r is a zero of f, then $f(r) = 0$, so

$$a_n r^n + a_{n-1} r^{n-1} + \cdots + a_1 r + a_0 = 0$$

We take the conjugate of both sides to get

$$\overline{a_n r^n + a_{n-1} r^{n-1} + \cdots + a_1 r + a_0} = \overline{0}$$

$$\overline{a_n r^n} + \overline{a_{n-1} r^{n-1}} + \cdots + \overline{a_1 r} + \overline{a_0} = \overline{0} \quad \text{The conjugate of a sum equals the sum of the conjugates (see Section 3.4).}$$

$$\overline{a_n}(\bar{r})^n + \overline{a_{n-1}}(\bar{r})^{n-1} + \cdots + \overline{a_1}\bar{r} + \overline{a_0} = \overline{0} \quad \text{The conjugate of a product equals the product of the conjugates.}$$

$$a_n(\bar{r})^n + a_{n-1}(\bar{r})^{n-1} + \cdots + a_1\bar{r} + a_0 = 0 \quad \text{The conjugate of a real number equals the real number.}$$

This last equation states that $f(\bar{r}) = 0$; that is, $\bar{r}$ is a zero of f. ■

The value of this result should be clear. Once we know that, say, $3 + 4i$ is a zero of a polynomial with real coefficients, then we know that $3 - 4i$ is also a zero. This result has an important corollary.

Corollary A complex polynomial f of odd degree with real coefficients has at least one real zero. ∎

Proof Because complex zeros occur as conjugate pairs in a complex polynomial with real coefficients, there will always be an even number of zeros that are not real numbers. Consequently, since f is of odd degree, one of its zeros has to be a real number. ∎

For example, the polynomial $f(z) = z^5 - 3z^4 + 4z^3 - 5$ has at least one zero that is a real number, since f is of degree 5 (odd) and has real coefficients.

Now we can prove the theorem we conjectured earlier in Section 4.4.

Theorem Every polynomial function with real coefficients can be uniquely factored over the real numbers into a product of linear factors and/or irreducible quadratic factors. ∎

Proof Every complex polynomial f of degree n has exactly n zeros and can be factored into a product of n linear factors. If its coefficients are real, then those zeros that are complex numbers will always occur as conjugate pairs. As a result, if $r = a + bi$ is a complex zero, then so is $\bar{r} = a - bi$. Consequently, when the linear factors $z - r$ and $z - \bar{r}$ of f are multiplied, we have

$$(z - r)(z - \bar{r}) = z^2 - (r + \bar{r})z + r\bar{r} = z^2 - 2az + a^2 + b^2$$

This second-degree polynomial has real coefficients and is irreducible (over the real numbers). Thus, the factors of f are either linear or irreducible quadratic factors. ∎

E X A M P L E 1 *Using the Conjugate Pairs Theorem*

A polynomial f of degree 5 whose coefficients are real numbers has the zeros 1, $5i$, and $1 + i$. Find the remaining two zeros.

Solution Since complex zeros appear as conjugate pairs, it follows that $-5i$, the conjugate of $5i$, and $1 - i$, the conjugate of $1 + i$, are the two remaining zeros. ∎

■ Now work Problem 1.

Polynomials with Complex Coefficients

The division algorithm for polynomials (see Section 4.4) is true for polynomials with complex coefficients. As a result, the Remainder Theorem and Factor Theorem are also true. In fact, the process of synthetic division also works for polynomials with complex coefficients.

E X A M P L E 2 *Verifying a Complex Zero*

Use synthetic division and the Factor Theorem to show that $1 + 2i$ is a zero of

$$f(z) = (1 + i)z^2 + (2 - i)z + (3 - 4i)$$

Solution We use synthetic division and divide $f(z)$ by $z - (1 + 2i)$:

$$1 + 2i \overline{)1 + i \quad\quad 2 - i \quad\quad 3 - 4i}$$

$$\underline{\quad\quad\quad\quad\quad -1 + 3i \quad\quad -3 + 4i}$$

$$\quad\quad 1 + i \quad\quad\quad 1 + 2i \quad\quad\quad\quad 0$$

Thus, $z - (1 + 2i)$ is a factor, and $1 + 2i$ is a zero.   ■

Of course, we could have shown that $1 + 2i$ is a zero of the polynomial $f(z)$ in Example 2 by using substitution as follows:

$$f(1 + 2i) = (1 + i)(1 + 2i)^2 + (2 - i)(1 + 2i) + (3 - 4i)$$
$$= -7 + i + 4 + 3i + 3 - 4i = 0$$

■   Now work Problem 23.

Based on equation (2), a complex polynomial f of degree n has n linear factors. These n linear factors do not have to be distinct; some may be repeated more than once. When a linear factor $z - r$ appears exactly m times in the factored form of f, then r is called a **zero of multiplicity m** of f. This leads us to the following conclusion.

Theorem   If a zero of multiplicity m of a complex polynomial f is counted m times, then a complex polynomial f of degree $n \geq 1$ will have exactly n zeros.   ■

E X A M P L E   3   *Listing all the Zeros of a Polynomial Function*

The complex polynomial function

$$f(z) = (2 + i)(z - 5)^3(z + i)^2[z - (3 + i)]^4(z - i)$$

of degree 10 has $2 + i$ as leading coefficient. Its zeros are listed next.

5:	Multiplicity	3
$-i$:	Multiplicity	2
$3 + i$:	Multiplicity	4
i:	Multiplicity	1
Degree:		10

■

E X A M P L E   4   *Using the Zeros to Form a Polynomial*

Form a polynomial $f(z)$ with complex coefficients of degree 3 and with the following zeros:

$1 + i$:	Multiplicity 1
$-i$:	Multiplicity 2

Solution   Since $1 + i$ is a zero of multiplicity 1 and $-i$ is a zero of multiplicity 2, then $z - (1 + i)$ and $(z + i)^2$ are factors of f. Thus, $f(z)$ is of the form

$$f(z) = [z - (1 + i)](z + i)^2$$
$$= [z - (1 + i)](z^2 + 2iz - 1)$$
$$= z^3 + (-1 + i)z^2 + (1 - 2i)z + 1 + i$$

Although other polynomials with complex coefficients have the three required zeros, the only ones of degree 3 will be $f(z)$ or $kf(z)$, where $k \neq 0$ is some complex number.   ■

4.5

Exercise 4.5

In Problems 1–10, information is given about a complex polynomial f(z) whose coefficients are real numbers. Find the remaining zeros of f.

1. Degree 3;   zeros: 3, $4 - i$

2. Degree 3;   zeros: 4, $3 + i$

3. Degree 4;   zeros: i, $1 + i$

4. Degree 4;   zeros: 1, 2, $2 + i$

5. Degree 5;   zeros: 1, i, $2i$

6. Degree 5;   zeros: 0, 1, 2, i

7. Degree 4;   zeros: i, 2, -2

8. Degree 4;   zeros: $2 - i$, $-i$

9. Degree 6;   zeros: 2, $2 + i$, $-3 -i$, 0

10. Degree 6;   zeros: i, $3 - 2i$, $-2 + i$

In Problems 11 and 12, tell why the facts given are contradictory.

11. $f(z)$ is a complex polynomial of degree 3 whose coefficients are real numbers; its zeros are $4 + i$, $4 - i$, and $2 + i$.

12. $f(z)$ is a complex polynomial of degree 3 whose coefficients are real numbers; its zeros are 2, i, and $3 + i$.

13. $f(z)$ is a complex polynomial of degree 4 whose coefficients are real numbers; three of its zeros are 2, $1 + 2i$, and $1 - 2i$. Explain why the remaining zero must be a real number.

14. $f(z)$ is a complex polynomial of degree 4 whose coefficients are real numbers; two of its zeros are -3 and $4 - i$. Explain why one of the remaining zeros must be a real number. Write down one of the missing zeros.

15. Find all the zeros of $f(z) = z^3 - 1$.   **16.** Find all the zeros of $f(z) = z^4 - 1$.

In Problems 17–22, evaluate each complex polynomial function f at z = 1 + i.

17. $f(z) = iz - 3$

18. $f(z) = 3z + i$

19. $f(z) = 3z^2 - z$

20. $f(z) = (4 + i)z^2 + 5 - 2i$

21. $f(z) = z^3 + iz - 1 + i$

22. $f(z) = iz^3 - 2z^2 + 1$

In Problems 23–28, use synthetic division to find the value of f(r).

23. $f(z) = 5z^5 - iz^4 + 2$; $r = 1 + i$

24. $f(z) = iz^4 + (2 + i)z^2 - z$; $r = 1 - i$

25. $f(z) = (1 + i)z^4 - z^3 + iz$; $r = 2 - i$

26. $f(z) = 2iz^3 + 8z^2 - 4iz + 1$; $r = 2 + i$

27. $f(z) = iz^5 + iz^3 + iz$; $r = 1 + 2i$

28. $f(z) = z^4 + z^2 + 1$; $r = 1 - 2i$

In Problems 29–34, form a polynomial f(z) with complex coefficients having the given degree and zeros.

29. Degree 3;   zeros: $3 + 2i$, multiplicity 1; 4, multiplicity 2

30. Degree 3;   zeros: i, multiplicity 2; $1 + 2i$, multiplicity 1

31. Degree 3;   zeros: 2, multiplicity 1; $-i$, multiplicity 1; $1 + i$, multiplicity 1

32. Degree 3;   zeros: i, multiplicity 1; $4 - i$, multiplicity 1; $2 + i$, multiplicity 1

33. Degree 4;   zeros: 3, multiplicity 2; $-i$, multiplicity 2

34. Degree 4;   zeros: 1, multiplicity 3; $1 + i$, multiplicity 1

Chapter Review

THINGS TO KNOW

Quadratic function	$f(x) = ax^2 + bx + c, a \neq 0$	Vertex: $(-b/2a, f(-b/2a))$
		Axis: The line $x = -b/2a$
		Parabola opens up if $a > 0$.
		Parabola opens down if $a < 0$.

Power function	$f(x) = x^n, n \geq 2$ even	Even function: Passes through $(-1, 1)$, $(0, 0)$, $(1, 1)$ Opens up
	$f(x) = x^n, n \geq 3$ odd	Odd function: Passes through $(-1, -1)$, $(0, 0)$, $(1, 1)$ Increasing
Polynomial function	$f(x) = a_n x^n + a_{n-1} x^{n-1} + \\ \cdots + a_1 x + a_0, a_n \neq 0$	At most $n - 1$ turning points; behaves like $y = a_n x^n$ for large x.
Rational function	$R(x) = \dfrac{p(x)}{q(x)}$, p, q are polynomial functions in lowest terms	See Steps 1 through 5 on page 294.
Zeros of a polynomial f	Numbers for which $f(x) = 0$; these are the x-intercepts of the graph of f.	
Remainder Theorem	If a polynomial $f(x)$ is divided by $x - c$, then the remainder is $f(c)$.	
Factor Theorem	$x - c$ is a factor of a polynomial $f(x)$ if and only if $f(c) = 0$.	
Descartes' Rule of Signs	Let f denote a polynomial function. The number of positive zeros of f either equals the number of variations in sign of the nonzero coefficients of $f(x)$ or else equals that number less some even integer. The number of negative zeros of f either equals the number of variations in sign of the nonzero coefficients of $f(-x)$ or else equals that number less some even integer.	
Rational Zeros Theorem	Let f be a polynomial function of degree 1 or higher of the form $$f(x) = a_n x^n + a_{n-1} x^{n-1} + \cdots + a_1 x + a_0, a_n \neq 0, a_0 \neq 0$$ where each coefficient is an integer. If p/q, in lowest terms, is a rational zero of f, then p must be a factor of a_0 and q must be a factor of a_n.	
Fundamental Theorem of Algebra	Every complex polynomial function $f(z)$ of degree $n \geq 1$ has at least one complex zero.	
Conjugate Pairs Theorem	Let $f(z)$ be a complex polynomial whose coefficients are real numbers. If $r = a + bi$ is a zero of f, then the complex conjugate $\bar{r} = a - bi$ is also a zero of f.	

How To:

Graph quadratic functions	Find the zeros of a polynomial by using Descartes' Rule of Signs, the Rational Zeros Theorem, and depressed equations
Graph polynomial functions	Solve polynomial equations using Descartes' Rule of Signs, the Rational Zeros Theorem, and depressed equations
Graph rational functions (see Steps 1 through 5, page 294)	Use the Upper and Lower Bounds Test
Use synthetic division to divide a polynomial by $x - c$	

Fill-In-The-Blank Items

1. The graph of a quadratic function is called a(n) _____. Its lowest or highest point is called the _____.

2. In the process of long division,

$$(\text{Divisor})(\text{Quotient}) + \underline{\hspace{1cm}} = \underline{\hspace{1cm}}$$

3. When a polynomial function f is divided by $x - c$, the remainder is _____.

4. A polynomial function f has the factor $x - c$ if and only if _____.

5. A number r for which $f(r) = 0$ is called a(n) _____ of the function f.

6. The polynomial function $f(x) = x^5 - 2x^3 + x^2 + x - 1$ has either _____ or _____ positive zeros; it has _____ or _____ negative zeros.

7. The possible rational zeros of $f(x) = 2x^5 - x^3 + x^2 - x + 1$ are _____.

8. The line _____ is a horizontal asymptote of: $R(x) = \dfrac{x^3 - 1}{x^3 + 1}$

9. The line _____ is a vertical asymptote of: $R(x) = \dfrac{x^3 - 1}{x^3 + 1}$

10. If $3 + 4i$ is a zero of a polynomial of degree 5 with real coefficients, then so is _____.

TRUE/FALSE ITEMS

T F **1.** Every polynomial of degree 3 with real coefficients has exactly three real zeros.

T F **2.** If $2 - 3i$ is a zero of a polynomial with real coefficients, then so is $-2 + 3i$.

T F **3.** The graph of $R(x) = \dfrac{x^2}{x - 1}$ has exactly one vertical asymptote.

T F **4.** The graph of $f(x) = x^2(x - 3)(x + 4)$ has exactly three x-intercepts.

T F **5.** If f is a polynomial function of degree 4 and if $f(2) = 5$, then

$$\frac{f(x)}{x - 2} = p(x) + \frac{5}{x - 2}$$

where $p(x)$ is a polynomial of degree 3.

REVIEW EXERCISES

In Problems 1–10, graph each quadratic function using a graphing utility. Determine whether the graph opens up or down. Find its vertex, axis of symmetry, y-intercept, and x-intercepts, if any.

1. $f(x) = (x - 2)^2 + 2$
2. $f(x) = (x + 1)^2 - 4$
3. $f(x) = \frac{1}{4}x^2 - 16$
4. $f(x) = -\frac{1}{2}x^2 + 2$
5. $f(x) = -4x^2 + 4x$
6. $f(x) = 9x^2 - 6x + 3$
7. $f(x) = \frac{9}{2}x^2 + 3x + 1$
8. $f(x) = -x^2 + x + \frac{1}{2}$
9. $f(x) = 3x^2 + 4x - 1$
10. $f(x) = -2x^2 - x + 4$

In Problems 11–16, using a graphing utility, show the stages required to graph each function.

11. $f(x) = (x + 2)^3$
12. $f(x) = -x^3 + 3$
13. $f(x) = -(x - 1)^4$
14. $f(x) = (x - 1)^4 - 2$
15. $f(x) = (x - 1)^4 + 2$
16. $f(x) = (1 - x)^3$

In Problems 17–22, determine whether the given quadratic function has a maximum value or a minimum value, and then find the value. Verify your results using a graphing utility.

17. $f(x) = 3x^2 - 6x + 4$
18. $f(x) = 2x^2 + 8x + 5$
19. $f(x) = -x^2 + 8x - 4$
20. $f(x) = -x^2 - 10x - 3$
21. $f(x) = -3x^2 + 12x + 4$
22. $f(x) = -2x^2 + 4$

In Problems 23–30:
(a) Graph f.
(b) Find the x- and y-intercepts.
(c) Determine whether each x-intercept is of odd or even multiplicity.
(d) Find the power function that the graph of f resembles for large values of x.
(e) Determine the number of turning points on the graph of f.
(f) Determine the local maxima and local minima, if any exist, correct to two decimal places.

23. $f(x) = x(x + 2)(x + 4)$
24. $f(x) = x(x - 2)(x - 4)$
25. $f(x) = (x - 2)^2(x + 4)$
26. $f(x) = (x - 2)(x + 4)^2$
27. $f(x) = x^3 - 4x^2$
28. $f(x) = x^3 + 4x$
29. $f(x) = (x - 1)^2(x + 3)(x + 1)$
30. $f(x) = (x - 4)(x + 2)^2(x - 2)$

In Problems 31–40, follow steps 1–5 on page 294 to analyze each rational function.

31. $R(x) = \dfrac{2x - 6}{x}$

32. $R(x) = \dfrac{4 - x}{x}$

33. $H(x) = \dfrac{x + 2}{x(x - 2)}$

34. $H(x) = \dfrac{x}{x^2 - 1}$

35. $R(x) = \dfrac{x^2 + x - 6}{x^2 - x - 6}$

36. $R(x) = \dfrac{x^2 - 6x + 9}{x^2}$

37. $F(x) = \dfrac{x^3}{x^2 - 4}$

38. $F(x) = \dfrac{3x^3}{(x - 1)^2}$

39. $R(x) = \dfrac{2x^4}{(x - 1)^2}$

40. $R(x) = \dfrac{x^4}{x^2 - 9}$

In Problems 41 and 42 use Descartes' Rule of Signs to determine how many positive and negative zeros each polynomial function may have. Do not attempt to find the zeros.

41. $f(x) = 12x^8 - x^7 + 8x^4 - 2x^3 + x + 3$

42. $f(x) = -6x^5 + x^4 + 5x^3 + x + 1$

43. List all the potential rational zeros of: $f(x) = 12x^8 - x^7 + 6x^4 - x^3 + x - 3$

44. List all the potential rational zeros of: $f(x) = -6x^5 + x^4 + 2x^3 - x + 1$

In Problems 45–56, use the steps listed on page 314 to find the zeros of f(x). Approximate any irrational zeros correct to two decimal places.

45. $f(x) = x^3 - 3x^2 - 6x + 8$

46. $f(x) = x^3 - x^2 - 10x - 8$

47. $f(x) = 4x^3 + 4x^2 - 7x + 2$

48. $f(x) = 4x^3 - 4x^2 - 7x - 2$

49. $f(x) = x^4 - 4x^3 + 9x^2 - 20x + 20$

50. $f(x) = x^4 + 6x^3 + 11x^2 + 12x + 18$

51. $f(x) = 2x^3 - 11.84x^2 - 9.116x + 82.46$

52. $f(x) = 12x^3 + 39.8x^2 - 4.4x - 3.4$

53. $g(x) = 15x^4 - 21.5x^3 - 1718.3x^2 + 5308x + 3796.8$

54. $g(x) = 3x^4 + 67.93x^3 + 486.265x^2 + 1121.32x + 412.195$

55. $f(x) = 3x^3 + 18.02x^2 + 11.0467x - 53.8756$

56. $f(x) = x^3 - 3.16x^2 - 39.4611x + 151.638$

In Problems 57–60, solve each equation in the real number system.

57. $2x^4 + 2x^3 - 11x^2 + x - 6 = 0$

58. $3x^4 + 3x^3 - 17x^2 + x - 6 = 0$

59. $2x^4 + 7x^3 + x^2 - 7x - 3 = 0$

60. $2x^4 + 7x^3 - 5x^2 - 28x - 12 = 0$

In Problems 61–64, find integer-valued upper and lower bounds to the zeros of each polynomial function.

61. $f(x) = 2x^3 - x^2 - 4x + 2$

62. $f(x) = 2x^3 + x^2 - 10x - 5$

63. $f(x) = 2x^3 - 7x^2 - 10x + 35$

64. $f(x) = 3x^3 - 7x^2 - 6x + 14$

In Problems 65–68, each polynomial has exactly one positive zero. Approximate the zero correct to two decimal places.

65. $f(x) = x^3 - x - 2$

66. $f(x) = 2x^3 - x^2 - 3$

67. $f(x) = 8x^4 - 4x^3 - 2x - 1$

68. $f(x) = 3x^4 + 4x^3 - 8x - 2$

In Problems 69–72, information is given about a complex polynomial f(z) whose coefficients are real numbers. Find the remaining zeros of f.

69. Degree 3; zeros: $4 + i$, 6

70. Degree 3; zeros: $3 + 4i$, 5

71. Degree 4; zeros: i, $1 + i$

72. Degree 4; zeros: 1, 2, $1 + i$

In Problems 73–78, form a polynomial f(z) with complex coefficients having the given degree and zeros.

73. Degree 4; zeros: 1, multiplicity 2; i, multiplicity 1; 3, multiplicity 1

74. Degree 4; zeros: i, multiplicity 2; 2, multiplicity 2

75. Degree 3; zeros: $1 + i$, 2, 3, each of multiplicity 1

76. Degree 3; zeros: 1, $1 + i$, $1 + 2i$, each of multiplicity 1

77. Find the quotient and remainder if $x^4 + 2x^3 - 7x^2 - 8x + 12$ is divided by $(x - 2)(x - 1)$.

78. Find the quotient and remainder if $x^4 + 2x^3 - 4x^2 - 5x - 6$ is divided by $(x - 2)(x + 3)$.

In Problems 79–82, solve each equation in the complex number system.

79. $x^3 - x^2 - 8x + 12 = 0$ **80.** $x^3 - 3x^2 - 4x + 12 = 0$

81. $3x^4 - 4x^3 + 4x^2 - 4x + 1 = 0$ **82.** $x^4 + 4x^3 + 2x^2 - 8x - 8 = 0$

83. Find integer-valued upper and lower bounds to the zeros of

$$f(x) = 4x^5 - 3x^4 + 8x^2 + x + 2$$

84. Find integer-valued upper and lower bounds to the zeros of

$$f(x) = 8x^6 - x^4 + 6x^2 + 24x + 15$$

85. Find the point on the line $y = x$ that is closest to the point $(3, 1)$. [*Hint:* Find the minimum value of the function $f(x) = d^2$, where d is the distance from $(3, 1)$ to a point on the line.]

86. Find the point on the line $y = x + 1$ that is closest to the point $(4, 1)$.

87. A horizontal bridge is in the shape of a parabolic arch. Given the information shown in the figure, what is the height h of the arch 2 feet from shore?

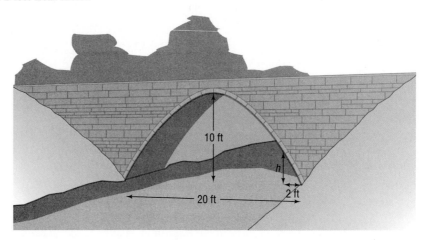

88. Find the length and width of a rectangle whose perimeter is 20 feet and whose area is 16 square feet.

89. Design a polynomial function with the following characteristics: degree 6; four real zeros, one of multiplicity 3; y-intercept 3; behaves like $y = -5x^6$ for large values of x. Is this polynomial unique? Compare your polynomial with those of other students. What terms will be the same as everyone else's? Add some more characteristics such as symmetry or naming the real zeros. How does this modify the polynomial?

90. Design a rational function with the following characteristics: three real zeros, one of multiplicity 2; y-intercept 1; vertical asymptotes $x = -2$ and $x = 3$; oblique asymptote $y = 2x + 1$. Is this rational function unique? Compare yours with those of other students. What will be the same as everyone else's? Add some more characteristics such as symmetry or naming the real zeros. How does this modify the rational function?

91. The illustration shows the graph of a polynomial function.

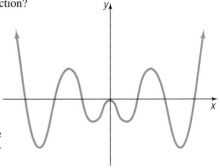

(a) Is the degree of the polynomial even or odd?

(b) Is the leading coefficient positive or negative?

(c) Is the function even, odd, or neither?

(d) Why is x^2 necessarily a factor of the polynomial?

(e) What is the minimum degree of the polynomial?

(f) Formulate five different polynomials whose graphs could look like the one shown. Compare yours to those of other students. What similarities do you see? What differences?

PREPARING FOR THIS CHAPTER

Before getting started on this chapter, review the following concepts:

Radicals; Rational Exponents (Appendix, Section 3)
Functions (Sections 2.1 and 2.2)
Inverse functions (pp. 146–152)
Simple interest (pp. 196–197)
Linear Curve Fitting (Section 1.5)

EXPONENTIAL AND LOGARITHMIC FUNCTIONS

5.1 Exponential Functions
5.2 Logarithmic Functions
5.3 Properties of Logarithms
5.4 Logarithmic and Exponential Equations
5.5 Compound Interest
5.6 Growth and Decay
5.7 Nonlinear Curve Fitting
5.8 Logarithmic Scales
     Chapter Review

Preview   Alcohol and Driving

The concentration of alcohol in a person's blood is measurable. Recent medical research suggests that the risk R (given as a percent) of having an accident while driving a car can be modeled by the equation

$$R = 6e^{kx}$$

where x is the variable concentration of alcohol in the blood and k is a constant.

(a) *Suppose that a concentration of alcohol in the blood of 0.04 results in a 10% risk (R = 10) of an accident. Find the constant k in the equation. Graph $R = 6e^{kx}$.*

(b) *Using this value of k, what is the risk if the concentration is 0.17?*

(c) *Using the same value of k, what concentration of alcohol corresponds to a risk of 100%?*

(d) *If the law asserts that anyone with a risk of having an accident of 20% or more should not have driving privileges, at what concentration of alcohol in the blood should a driver be arrested and charged with DUI (Driving Under the Influence)?*
[Example 9 in Section 5.2.]

Until now, our study of functions has concentrated primarily on polynomial and rational functions. These functions belong to the class of **algebraic functions,** that is, functions that can be expressed in terms of sums, differences, products, quotients, powers, or roots of polynomials. Functions that are not algebraic are termed **transcendental** (they transcend, or go beyond, algebraic functions).

In this chapter, we study two transcendental functions: the *exponential* and *logarithmic functions.* These functions occur frequently in a wide variety of applications.

5.1

Exponential Functions

In the Appendix, Section 3, we gave a definition for raising a real number a to a rational power. Based on that discussion, we gave meaning to expressions of the form

$$a^r$$

where the base a is a positive real number and the exponent r is a rational number.

But what is the meaning of a^x, where the base a is a positive real number and the exponent x is an irrational number? Although a rigorous definition requires methods discussed in calculus, the basis for the definition is easy to follow: Select a rational number r that is formed by truncating (removing) all but a finite number of digits from the irrational number x. Then it is reasonable to expect that

$$a^x \approx a^r$$

For example, take the irrational number $\pi = 3.14159. \ldots$ Then, an approximation to a^π is

$$a^\pi \approx a^{3.14}$$

where the digits after the hundredths position have been removed from the value for π. A better approximation would be

$$a^\pi \approx a^{3.14159}$$

where the digits after the hundred-thousandths position have been removed. Continuing in this way, we can obtain approximations to a^π to any desired degree of accuracy.

Graphing calculators can easily evaluate expressions of the form a^r as follows. Enter the base a, then press the caret key (^), enter the exponent y, and press enter.

E X A M P L E   1   *Using a Graphing Calculator to Evaluate Powers of 2*

Using a graphing calculator, evaluate:

(a) $2^{1.4}$ (b) $2^{1.41}$ (c) $2^{1.414}$ (d) $2^{1.4142}$ (e) $2^{\sqrt{2}}$

Solution   Figure 1 shows the solution to (a) using a T1-82 graphing calculator.

FIGURE 1

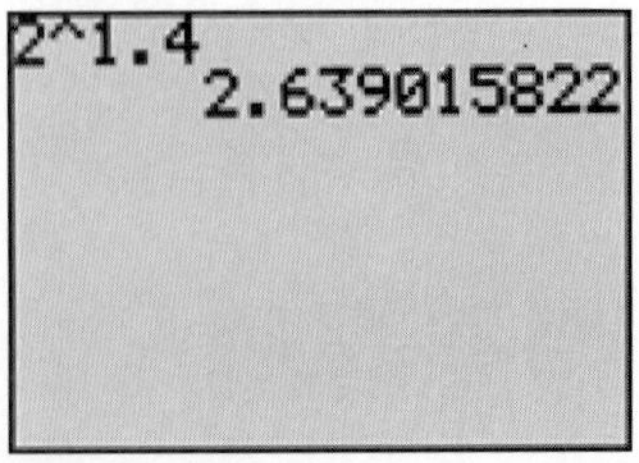

```
2^1.4
        2.639015822
```

(a) $2^{1.4} \approx 2.639015822$ (b) $2^{1.41} \approx 2.657371628$

(c) $2^{1.414} \approx 2.664749765$ (d) $2^{1.4142} \approx 2.665119089$

(e) $2^{\sqrt{2}} \approx 2.665144143$ ∎

■ Now work Problem 1.

It can be shown that the familiar laws of rational exponents hold for real exponents.

Theorem

Laws of Exponents

If *s, t, a,* and *b* are real numbers with $a > 0$ and $b > 0$, then

$$a^s \cdot a^t = a^{s+t} \qquad (a^s)^t = a^{st} \qquad (ab)^s = a^s \cdot b^s$$
$$1^s = 1 \qquad a^{-s} = \frac{1}{a^s} = \left(\frac{1}{a}\right)^s \qquad a^0 = 1 \qquad\qquad (1)$$

We are now ready for the following definition.

Exponential Function

An **exponential function** is a function of the form

$$f(x) = a^x$$

where a is a positive real number and $a \neq 1$. The domain of f is the set of all real numbers.

We exclude the base $a = 1$, because this function is simply the constant function $f(x) = 1^x = 1$. We also need to exclude bases that are negative, because, otherwise, we would have to exclude many values of x from the domain, such as $x = \frac{1}{2}$, $x = \frac{3}{4}$, and so on. [Recall that $(-2)^{1/2}$, $(-3)^{3/4}$, and so on, are not defined in the system of real numbers.]

Graphs of Exponential Functions

First, we graph the exponential function $y = 2^x$.

E X A M P L E   2   *Graphing an Exponential Function*

Graph the exponential function: $f(x) = 2^x$

Solution   The domain of $f(x) = 2^x$ consists of all real numbers. We begin by locating some points on the graph of $f(x) = 2^x$, as listed in Table 1.

Since $2^x > 0$ for all x, the range of f is $(0, \infty)$. From this, we conclude that the graph has no x-intercepts, and, in fact, the graph will lie above the x-axis. As Table 1 indicates, the y-intercept is 1. Table 1 also indicates that as $x \to -\infty$ the value of $f(x) = 2^x$ gets closer and closer to 0. Thus, the x-axis is a horizontal asymptote to the graph as $x \to -\infty$. Look again at Table 1. As $x \to \infty$, $f(x) = 2^x$ grows very quickly, causing the graph of $f(x) = 2^x$ to rise very rapidly. Thus, it is apparent that f is an increasing function and, hence, is one-to-one.

Figure 2 shows the graph of $f(x) = 2^x$. Notice that all the conclusions given earlier are confirmed by the graph.

TABLE 1

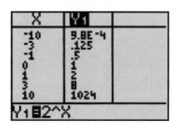

FIGURE 2

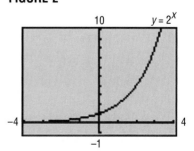

As we shall see, graphs that look like the one in Figure 2 occur very frequently in a variety of situations. For example, look at the graph in Figure 3, which illustrates the population in Ethiopia. Researchers might conclude from this graph that the population in Ethiopia is "behaving exponentially"; that is, the graph exhibits "rapid, or exponential, growth." We shall have more to say about situations that lead to exponential growth later in this chapter. For now, we continue to seek properties of the exponential functions.

FIGURE 3

Ethiopia's population is growing at one of the fastest paces in the world, exacerbating continuing shortfalls in food needed to feed its millions of people. The nation, which is only about three-quarters the size of Alaska, grows with more than 2 million births each year.

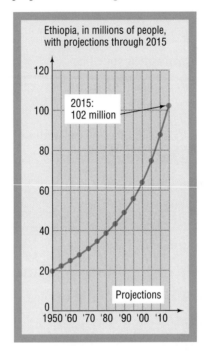

Sources: News reports, World Bank, U.S. Agency for International Development, UN Food and Agriculture Organization, Population Reference Bureau, UNICEF, U.S. Department of Agriculture, Human Nutrition Information Service.

The graph of $f(x) = 2^x$ in Figure 2 is typical of all exponential functions that have a base larger than 1. Such functions are increasing functions and hence are one-to-one. Their graphs lie above the x-axis, pass through the point $(0, 1)$, and thereafter rise rapidly as $x \to \infty$. As $x \to -\infty$, the x-axis is a horizontal asymptote. There are no vertical asymptotes. Finally, the graphs are smooth and continuous, with no corners or gaps.

Figure 4 illustrates the graphs of two more exponential functions whose bases are larger than 1. Notice that for the larger base the graph is steeper when $x > 0$. Figure 5 shows that when $x < 0$, the graph of the equation with the larger base is closer to the x-axis.

FIGURE 4

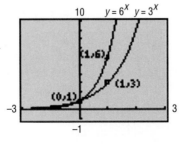

FIGURE 5

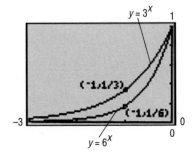

The display summarizes the information we have about $f(x) = a^x$, $a > 1$:

$$f(x) = a^x \qquad a > 1$$
Domain: $(-\infty, \infty)$ Range: $(0, \infty)$
x-intercepts: None y-intercept: 1
Horizontal asymptote: x-axis, as $x \to -\infty$
f is an increasing function
f is one-to-one passing through $(0, 1)$ and $(1, a)$

Now we consider $f(x) = a^x$ when $0 < a < 1$.

E X A M P L E   3

Graphing an Exponential Function

Graph the exponential function: $f(x) = \left(\frac{1}{2}\right)^x$

Solution   The domain of $f(x) = \left(\frac{1}{2}\right)^x$ consists of all real numbers. As before, we locate some points on the graph, as listed in Table 2. Since $\left(\frac{1}{2}\right)^x > 0$ for all x, the range of f is $(0, \infty)$. Thus, the graph lies above the x-axis and so has no x-intercepts. The y-intercept is 1. As $x \to -\infty$, $f(x) = \left(\frac{1}{2}\right)^x$ grows very quickly. As $x \to \infty$, the values of $f(x)$ approach 0. Thus, the x-axis ($y = 0$) is a horizontal asymptote as $x \to \infty$. It is apparent that f is a decreasing function and, hence, is one-to-one. Figure 6 illustrates the graph.

TABLE 2   **FIGURE 6**

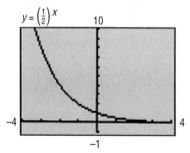

Note that we could have obtained the graph of $y = \left(\frac{1}{2}\right)^x$ from the graph of $y = 2^x$. If $f(x) = 2^x$, then $f(-x) = 2^{-x} = \frac{1}{2^x} = \left(\frac{1}{2}\right)^x$. Thus, the graph of $y = \left(\frac{1}{2}\right)^x = 2^{-x}$ is a reflection about the y-axis of the graph of $y = 2^x$. Compare Figures 2 and 6.

The graph of $f(x) = \left(\frac{1}{2}\right)^x$ in Figure 6 is typical of all exponential functions that have a base between 0 and 1. Such functions are decreasing, one-to-one functions. Their graphs lie above the x-axis and pass through the point $(0, 1)$. The graphs rise rapidly as $x \to -\infty$. As $x \to \infty$, the x-axis is a horizontal asymptote. There are no vertical asymptotes. Finally, the graphs are smooth and continuous, with no corners or gaps.

Figure 7 illustrates the graphs of two more exponential functions whose bases are between 0 and 1. Notice that the choice of a base closer to 0 results in a graph that is steeper when $x < 0$. Figure 8 shows that when $x > 0$, the graph of the equation with the smaller base is closer to the x-axis.

FIGURE 7

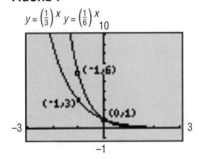

FIGURE 8

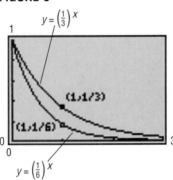

The display summarizes the information we have about $f(x) = a^x$, $0 < a < 1$:

$$f(x) = a^x \qquad 0 < a < 1$$
Domain: $(-\infty, \infty)$ \qquad Range: $(0, \infty)$
x-intercepts: None \qquad y-intercept: 1
Horizontal asymptote: x-axis, as $x \to \infty$
f is a decreasing function
f is one-to-one passing through $(0, 1)$ and $(1, a)$

The techniques of shifting, compression, stretching, and reflection may be used to graph many functions that are basically exponential functions.

E X A M P L E   4

Graphing Functions That Are Basically Exponential Using Shifts, Reflections, and the Like

Graph: $f(x) = 2^{-x} - 3$

Solution   Figure 9 shows the various steps.

FIGURE 9

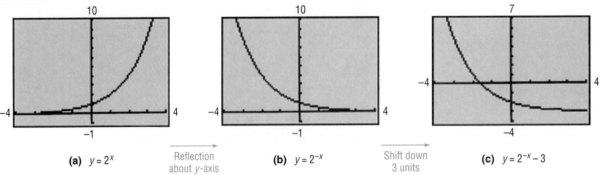

(a) $y = 2^x$ \qquad Reflection about y-axis \qquad (b) $y = 2^{-x}$ \qquad Shift down 3 units \qquad (c) $y = 2^{-x} - 3$

Note that the horizontal asymptote of $f(x) = 2^{-x} - 3$ is the line $y = -3$, as Figure 9(c) illustrates. ∎

E X A M P L E   5

Graphing Functions That Are Basically Exponential Using Shifts, Reflections, and the Like

Graph: $f(x) = -(2^{x-3})$

Solution   Figure 10 shows the various steps.

FIGURE 10

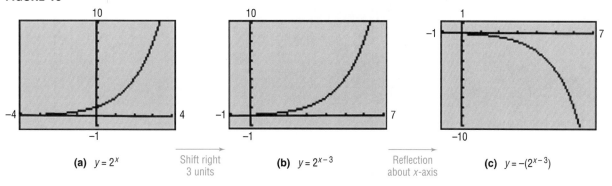

(a) $y = 2^x$ →   Shift right 3 units   (b) $y = 2^{x-3}$ →   Reflection about x-axis   (c) $y = -(2^{x-3})$

■   Now work Problems 11 and 13.

The Base e

As we shall see shortly, many problems that occur in nature require the use of an exponential function whose base is a certain irrational number, symbolized by the letter e.

Let's look now at one way of arriving at this important number e.

The Number e

> The **number e** is defined as the number that the expression
>
> $$\left(1 + \frac{1}{n}\right)^n \tag{2}$$
>
> approaches as $n \to \infty$. In calculus, this is expressed using limit notation as
>
> $$e = \lim_{n \to \infty} \left(1 + \frac{1}{n}\right)^n$$

TABLE 3

n	$\dfrac{1}{n}$	$1 + \dfrac{1}{n}$	$\left(1 + \dfrac{1}{n}\right)^n$
1	1	2	2
2	0.5	1.5	2.25
5	0.2	1.2	2.48832
10	0.1	1.1	2.59374246
100	0.01	1.01	2.704813829
1,000	0.001	1.001	2.716923932
10,000	0.0001	1.0001	2.718145926
100,000	0.00001	1.00001	2.718268237
1,000,000	0.000001	1.000001	2.718280469
1,000,000,000	10^{-9}	$1 + 10^{-9}$	2.718281828

Table 3 illustrates what happens to the defining expression (2) as n takes on increasingly large values. The last number in the last column in the table is correct to nine decimal places and is the same as the entry given for e on your calculator (if expressed correct to nine decimal places).

The exponential function $f(x) = e^x$, whose base is the number e, occurs with such frequency in applications that it is usually referred to as *the* exponential function. Indeed, graphing calculators have the key $\boxed{e^x}$ or $\boxed{exp(x),}$ which may be used to evaluate the exponential function for a given value of x. Now use your calculator to find e^x for $x = -2$, $x = -1$, $x = 0$, $x = 1$, and $x = 2$, as we have done to create Table 4. The graph of the exponential function $f(x) = e^x$ is given in Figure 11. Since $2 < e < 3$, the graph of $y = e^x$ lies between the graphs of $y = 2^x$ and $y = 3^x$. (Refer to Figures 2 and 4.)

TABLE 4

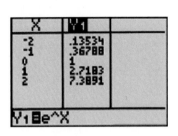

FIGURE 11

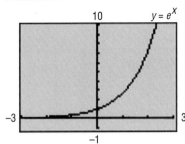

■   Now work Problem 25.

There are many applications involving the exponential function. Let's look at one.

E X A M P L E   6   *Response to Advertising*

Suppose that the percent R of people who respond to a newspaper advertisement for a new product and purchase the item advertised after t days is found using the formula

$$R = 50 - 100e^{-0.3t}$$

(a) What percent has responded and purchased after 5 days?

(b) What percent has responded and purchased after 10 days?

(c) What is the highest percent of people expected to respond and purchase?

(d) Graph $R = 50 - 100e^{-0.3t}$, $t > 0$. TRACE and compare the values of R for $t = 5$ and $t = 10$ to the ones obtained in (a) and (b). How many days are required for R to exceed 40%?

Solution   (a) After 5 days, we have $t = 5$. The corresponding percent R of people responding and purchasing is

$$R = 50 - 100e^{(-0.3)(5)} = 50 - 100e^{-1.5}$$

We evaluate this expression in Figure 12.

FIGURE 12

Thus, about 28% will have responded after 5 days.

(b) After 10 days, we have $t = 10$. The corresponding percent R of people responding and purchasing is

$$R = 50 - 100e^{(-0.3)(10)} = 50 - 100e^{-3} \approx 45.021$$

About 45% will have responded and purchased after 10 days.

(c) As time passes, more people are expected to respond and purchase. The highest percent expected is therefore found for the value of R as $t \to \infty$. Since $e^{-0.3t} = 1/e^{0.3t}$, it follows that $e^{-0.3t} \to 0$ as $t \to \infty$. Thus, the highest percent expected is 50%.

(d) See Figure 13.

FIGURE 13

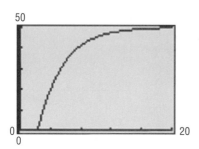

It will require 8 days to exceed 40%.

■ Now work Problem 43.

Summary

PROPERTIES OF THE EXPONENTIAL FUNCTION

$f(x) = a^x, a > 1$ Domain: $(-\infty, \infty)$; Range: $(0, \infty)$; x-intercepts: none; y-intercept: 1; horizontal asymptote: x-axis as $x \to -\infty$; increasing; one-to-one See Figure 2 for a typical graph.

$f(x) = a^x, 0 < a < 1$ Domain: $(-\infty, \infty)$; Range: $(0, \infty)$; x-intercepts: none, y-intercept: 1; horizontal asymptote: x-axis as $x \to \infty$; decreasing; one-to-one See Figure 6 for a typical graph.

5.1

Exercise 5.1

In Problems 1–10, approximate each number using a calculator. Express your answer rounded to three decimal places.

1. (a) $3^{2.2}$ (b) $3^{2.23}$ (c) $3^{2.236}$ (d) $3^{\sqrt{5}}$

2. (a) $5^{1.7}$ (b) $5^{1.73}$ (c) $5^{1.732}$ (d) $5^{\sqrt{3}}$

3. (a) $2^{3.14}$ (b) $2^{3.141}$ (c) $2^{3.1415}$ (d) 2^{π}

4. (a) $2^{2.7}$ (b) $2^{2.71}$ (c) $2^{2.718}$ (d) 2^{e}

5. (a) $3.1^{2.7}$ (b) $3.14^{2.71}$ (c) $3.141^{2.718}$ (d) π^{e}

6. (a) $2.7^{3.1}$ (b) $2.71^{3.14}$ (c) $2.718^{3.141}$ (d) e^{π}

7. $e^{1.2}$ 8. $e^{-1.3}$ 9. $e^{-0.85}$ 10. $e^{2.1}$

In Problems 11–18, the graph of an exponential function is given. Match each graph to one of the following functions:

A. $y = 3^x$ B. $y = 3^{-x}$ C. $y = -3^x$ D. $y = -3^{-x}$

E. $y = 3^x - 1$ F. $y = 3^{x-1}$ G. $y = 3^{1-x}$ H. $y = 1 - 3^x$

11.

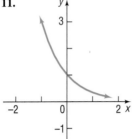

12.

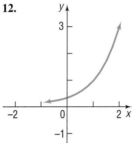

13.

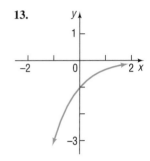

14.

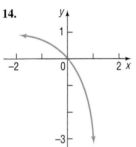

15.

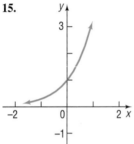

16.

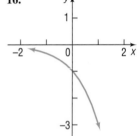

17.

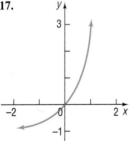

18.

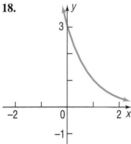

In Problems 19–24, the graph of an exponential function is given. Match each graph to one of the following functions:

A. $y = 4^x$ B. $y = 4^{-x}$ C. $y = 4^{x-1}$ D. $y = 4^x - 1$ E. $y = -4^x$ F. $y = 1 - 4^x$

19.

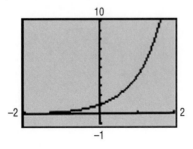

20.

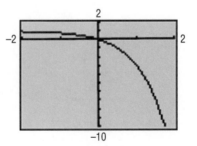

21.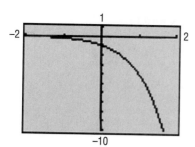

22.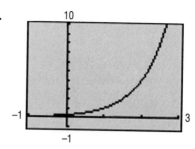

23.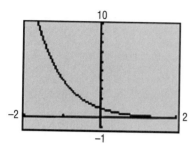

24.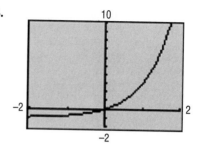

In Problems 25–32, using a graphing utility, show the stages required to graph each function.

25. $y = e^{-x}$ **26.** $y = -e^{x}$ **27.** $y = e^{x+2}$ **28.** $y = e^{x} - 1$

29. $y = 5 - e^{-x}$ **30.** $y = 9 - 3e^{-x}$ **31.** $y = 2 - e^{-x/2}$ **32.** $y = 7 - 3e^{-2x}$

33. If $4^{x} = 7$, what does 4^{-2x} equal?   **34.** If $2^{x} = 3$, what does 4^{-x} equal?

35. If $3^{-x} = 2$, what does 3^{2x} equal?   **36.** If $5^{-x} = 3$, what does 5^{3x} equal?

37. *Optics* If a single pane of glass obliterates 3% of the light passing through it, then the percent p of light that passes through n successive panes is given approximately by the equation

$$p = 100e^{-0.03n}$$

(a) What percent of light will pass through 10 panes?

(b) What percent of light will pass through 25 panes?

38. *Atmospheric Pressure* The atmospheric pressure p on a balloon or plane decreases with increasing height. This pressure, measured in millimeters of mercury, is related to the number of kilometers h above sea level by the formula

$$p = 760e^{-0.145h}$$

(a) Find the atmospheric pressure at a height of 2 kilometers (over a mile).

(b) What is it at a height of 10 kilometers (over 30,000 feet)?

39. *Space Satellites* The number of watts w provided by a space satellite's power supply over a period of d days is given by the formula

$$w = 50e^{-0.004d}$$

(a) How much power will be available after 30 days?

(b) How much power will be available after 1 year (365 days)?

40. *Healing of Wounds* The normal healing of wounds can be modeled by an exponential function. If A_0 represents the original area of the wound and if A equals the area of the wound after n days, then the formula

$$A = A_0 e^{-0.35n}$$

describes the area of a wound on the nth day following an injury when no infection is present to retard the healing. Suppose a wound initially had an area of 100 square centimeters.
    (a) If healing is taking place, how large should the area of the wound be after 3 days?
    (b) How large should it be after 10 days?

41. *Drug Medication* The formula

$$D = 5e^{-0.4h}$$

can be used to find the number of milligrams D of a certain drug that is in a patient's bloodstream h hours after the drug has been administered. How many milligrams will be present after 1 hour? After 6 hours?

42. *Spreading of Rumors* A model for the number of people N in a college community who have heard a certain rumor is

$$N = P(1 - e^{-0.15d})$$

where P is the total population of the community and d is the number of days that have elapsed since the rumor began. In a community of 1000 students, how many students will have heard the rumor after 3 days?

43. *Response to TV Advertising* The percent R of viewers who respond to a television commercial for a new product after t days is found by using the formula

$$R = 70 - 100e^{-0.2t}$$

    (a) What percent is expected to respond after 10 days?
    (b) What percent has responded after 20 days?
    (c) What is the highest percent of people expected to respond?
    (d) Graph $R = 70 - 100e^{-0.2t}$, $t > 0$. TRACE and compare the values of R for $t = 10$ and $t = 20$ to the ones obtained in parts (a) and (b). How many days are required for R to exceed 40%?

44. *Profit* The annual profit P of a company due to the sales of a particular item after it has been on the market x years is determined to be

$$P = \$100{,}000 - \$60{,}000\left(\tfrac{1}{2}\right)^x$$

    (a) What is the profit after 5 years?
    (b) What is the profit after 10 years?
    (c) What is the most profit the company can expect from this product?
    (d) Graph the profit function. TRACE and compare the values of P for $x = 5$ and $x = 10$ to the one obtained in parts (a) and (b). How many years does it take before a profit of $65,000 is obtained?

45. *Alternating Current in a RL Circuit* The equation governing the amount of current I (in amperes) after time t (in seconds) in a single RL circuit consisting of a resistance R (in ohms), an inductance L (in henrys), and an electromotive force E (in volts) is

$$I = \frac{E}{R}[1 - e^{-(R/L)t}]$$

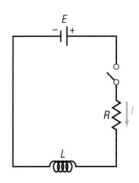

    (a) If $E = 120$ volts, $R = 10$ ohms, and $L = 5$ henrys, how much current I_1 is available after 0.3 second? After 0.5 second? After 1 second?
    (b) What is the maximum current?
    (c) Graph this function $I = I_1(t)$, measuring I along the y-axis and t along the x-axis.
    (d) If $E = 120$ volts, $R = 5$ ohms, and $L = 10$ henrys, how much current I_2 is available after 0.3 second? After 0.5 second? After 1 second?
    (e) What is the maximum current?
    (f) Graph this function $I = I_2(t)$ on the same viewing window as $I_1(t)$.

46. *Alternating Current in a RC Circuit* The equation governing the amount of current *I* (in milliamperes) after time *t* (in milliseconds) in a single *RC* circuit consisting of a resistance *R* (in ohms), a capacitance *C* (in microfarads), and an electromotive force *E* (in volts) is

$$I = \frac{E}{R} e^{-t/(RC)}$$

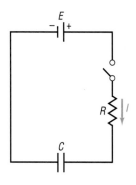

(a) If *E* = 120 volts, *R* = 2000 ohms, and *C* = 1.0 microfarad, how much current I_1 is available initially (*t* = 0)? After 1000 milliseconds? After 3000 milliseconds?

(b) What is the maximum current?

(c) Graph this function $I = I_1(t)$, measuring *I* along the *y*-axis and *t* along the *x*-axis.

(d) If *E* = 120 volts, *R* = 1000 ohms, and *C* = 2.0 microfarads, how much current I_2 is available initially? After 1000 milliseconds? After 3000 milliseconds?

(e) What is the maximum current?

(f) Graph this function $I = I_2(t)$ on the same viewing window as $I_1(t)$.

47. *The Challenger Disaster** After the *Challenger* disaster in 1986, a study of the 23 launches that preceded the fatal flight was made. A mathematical model was developed involving the relationship between the Fahrenheit temperature *x* around the O-rings and the number *y* of eroded or leaky primary O-rings. The model stated that

$$y = 6\,[1 + e^{-(5.085 - 0.1156x)}]^{-1}$$

where the number 6 indicates the 6 primary O-rings on the spacecraft.

(a) What is the predicted number of eroded or leaky primary O-rings at a temperature of 100°F?

(b) What is the predicted number of eroded or leaky primary O-rings at a temperature of 60°F?

(c) What is the predicted number of eroded or leaky primary O-rings at a temperature of 30°F?

(d) Graph the equation and TRACE. At what temperature is the predicted number of eroded or leaky O-rings 1? 3? 5?

48. *Postage Stamps†* The cumulative number *y* of different postage stamps (regular and commemorative only) issued by the U.S. Post Office can be approximated (modeled) by the exponential function

$$y = 78e^{0.025x}$$

where *x* is the number of years since 1848.

(a) What is the predicted cumulative number of stamps that will have been issued by the year 1995? Check with the Postal Service and comment on the accuracy of using the function.

(b) What is the predicted cumulative number of stamps that will have been issued by the year 1998?

(c) The cumulative number of stamps actually issued by the United States was 2 in 1848, 88 in 1868, 218 in 1888, and 341 in 1908. What conclusion can you draw about using the given function as a model over the first few decades in which stamps were issued?

49. *Another Formula for e* Use a calculator to compute the value of

$$2 + \frac{1}{2!} + \frac{1}{3!} + \ldots + \frac{1}{n!}$$

for *n* = 4, 6, 8, and 10. Compare each result with *e*.

[*Hint:* 1! = 1, 2! = 2 · 1, 3! = 3 · 2 · 1, n! = n(n − 1) · . . . · (3)(2)(1)].

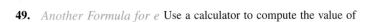

*Linda Tappin, "Analyzing Data Relating to the *Challenger* Disaster," *Mathematics Teacher,* Vol. 87, No. 6, September 1994, pp. 423–426.
†David Kullman, "Patterns of Postage-stamp Production," *Mathematics Teacher,* Vol. 85, No. 3, March 1992, pp. 188–189.

50. *Another Formula for e* Use a calculator to compute the various values of the expression

$$2 + \cfrac{1}{1 + \cfrac{1}{2 + \cfrac{2}{3 + \cfrac{3}{4 + \cfrac{4}{\text{etc.}}}}}}$$

Compare the values to e.

51. If $f(x) = a^x$, show that: $\dfrac{f(x + h) - f(x)}{h} = a^x \left(\dfrac{a^h - 1}{h} \right)$

52. If $f(x) = a^x$, show that: $f(A + B) = f(A) \cdot f(B)$

53. If $f(x) = a^x$, show that: $f(-x) = \dfrac{1}{f(x)}$

54. If $f(x) = a^x$, show that: $f(\alpha x) = [f(x)]^\alpha$

Problems 55 and 56 provide definitions for two other transcendental functions.

55. The **hyperbolic sine function,** designated by sinh x, is defined as

$$\sinh x = \frac{1}{2}(e^x - e^{-x})$$

(a) Show that $f(x) = \sinh x$ is an odd function.
(b) Graph $y = e^x$, $y = -e^{-x}$, and $f(x) = \sinh x$ on the same set of coordinate axes.
(c) How do you explain the relationship of the three graphs?

56. The **hyperbolic cosine function,** designated by cosh x, is defined as

$$\cosh x = \frac{1}{2}(e^x + e^{-x})$$

(a) Show that $f(x) = \cosh x$ is an even function.
(b) Graph $y = e^x$, $y = e^{-x}$, and $f(x) = \cosh x$ on the same set of coordinate axes.
(c) Refer to Problem 55. Show that, for every x,

$$(\cosh x)^2 - (\sinh x)^2 = 1$$

57. *Historical problem* Pierre de Fermat (1601–1665) conjectured that the function

$$f(x) = 2^{(2^x)} + 1$$

for $x = 1, 2, 3, \ldots$, would always have a value equal to a prime number. But Leonhard Euler (1707–1783) showed that this formula fails for $x = 5$. Use a calculator to determine the prime numbers produced by f for $x = 1, 2, 3, 4$. Then show that $f(5) = 641 \times 6,700,417$, which is not prime.

58. The bacteria in a 4 liter container double every minute. After 60 minutes the container is full. How long did it take to fill half the container?

59. Explain in your own words what the number e is. Provide at least two applications that require the use of this number.

60. Do you think there is a power function that increases more rapidly than an exponential function whose base is greater than 1? Explain.

5.2

Logarithmic Functions

Recall that a one-to-one function $y = f(x)$ has an inverse that is defined (implicitly) by the equation $x = f(y)$. In particular, the exponential function $y = f(x) = a^x$, $a > 0$, $a \neq 1$, is one-to-one and, hence, has an inverse that is defined implicitly by the equation

$$x = a^y \qquad a > 0, a \neq 1$$

This inverse is so important that it is given a name, the *logarithmic function*.

Logarithmic Function

The **logarithmic function to the base a,** where $a > 0$ and $a \neq 1$, is denoted by $y = \log_a x$ (read as "y is the logarithm to the base a of x") and is defined by

$$y = \log_a x \quad \text{if and only if} \quad x = a^y$$

E X A M P L E   1

Relating Logarithms to Exponents

(a) If $y = \log_3 x$, then $x = 3^y$. Thus, if $x = 9$, then $y = 2$, so $9 = 3^2$ is equivalent to $2 = \log_3 9$.

(b) If $y = \log_5 x$, then $x = 5^y$. Thus, if $x = \frac{1}{5} = 5^{-1}$, then $y = -1$, so $\frac{1}{5} = 5^{-1}$ is equivalent to $-1 = \log_5 \left(\frac{1}{5}\right)$. ∎

E X A M P L E   2

Changing Exponential Expressions to Logarithmic Expressions

Change each exponential expression to an equivalent expression involving a logarithm.

(a) $1.2^3 = m$ (b) $e^b = 9$ (c) $a^4 = 24$

Solution   We use the fact that $y = \log_a x$ and $x = a^y$, $a > 0$, $a \neq 1$, are equivalent.

(a) If $1.2^3 = m$, then $3 = \log_{1.2} m$.

(b) If $e^b = 9$, then $b = \log_e 9$.

(c) If $a^4 = 24$, then $4 = \log_a 24$. ∎

■   Now work Problem 1.

E X A M P L E   3

Changing Logarithmic Expressions to Exponential Expressions

Change each logarithmic expression to an equivalent expression involving an exponent.

(a) $\log_a 4 = 5$ (b) $\log_e b = -3$ (c) $\log_3 5 = c$

Solution   (a) If $\log_a 4 = 5$, then $a^5 = 4$.

(b) If $\log_e b = -3$, then $e^{-3} = b$.

(c) If $\log_3 5 = c$, then $3^c = 5$. ∎

■   Now work Problem 13.

To find the exact value of a logarithm, we write the logarithm in exponential notation and use the following fact:

$$\text{If } a^u = a^v, \quad \text{then} \quad u = v. \tag{1}$$

The result (1) is a consequence of the fact that exponential functions are one-to-one.

EXAMPLE 4   *Finding the Exact Value of a Logarithmic Function*

Find the exact value of:

(a) $\log_2 8$ (b) $\log_3 \frac{1}{3}$ (c) $\log_5 25$

Solution   (a) For $y = \log_2 8$, we have the equivalent exponential equation $2^y = 8 = 2^3$, so, by (1), $y = 3$. Thus, $\log_2 8 = 3$.

(b) For $y = \log_3 \frac{1}{3}$, we have $3^y = \frac{1}{3} = 3^{-1}$, so $y = -1$. Thus, $\log_3 \frac{1}{3} = -1$.

(c) For $y = \log_5 25$, we have $5^y = 25 = 5^2$, so $y = 2$. Thus, $\log_5 25 = 2$.   ■

■   Now work Problem 25.

Domain of a Logarithmic Function

The logarithmic function $y = \log_a x$ has been defined as the inverse of the exponential function $y = a^x$. That is, if $f(x) = a^x$, then $f^{-1}(x) = \log_a x$. Based on the discussion given in Section 2.5 on inverse functions, we know that for a function f and its inverse f^{-1}

$$\text{Domain } f^{-1} = \text{Range } f \quad \text{and} \quad \text{Range } f^{-1} = \text{Domain } f$$

Consequently, it follows that:

Domain of logarithmic function = Range of exponential function   = $(0, \infty)$
Range of logarithmic function   = Domain of exponential function = $(-\infty, \infty)$

In the next box, we summarize some properties of the logarithmic function:

$$y = \log_a x \quad \text{(defining equation: } x = a^y\text{)}$$
$$\text{Domain: } 0 < x < \infty \quad \text{Range: } -\infty < y < \infty$$

Notice that the domain of a logarithmic function consists of the *positive* real numbers.

EXAMPLE 5   *Finding the Domain of a Logarithmic Function*

Find the domain of each logarithmic function:

(a) $F(x) = \log_2 (1 - x)$ (b) $g(x) = \log_5 \left(\dfrac{1 + x}{1 - x} \right)$ (c) $h(x) = \log_{1/2} |x|$

Solution   (a) The domain of F consists of all x for which $(1 - x) > 0$; that is, all $x < 1$, or $(-\infty, 1)$.

(b) The domain of g is restricted to

$$\frac{1 + x}{1 - x} > 0$$

Solving this inequality, we find that the domain of g consists of all x between -1 and 1, that is, $-1 < x < 1$, or $(-1, 1)$.

(c) Since $|x| > 0$ provided $x \neq 0$, the domain of h consists of all nonzero real numbers.   ■

  ■   Now work Problem 39.

Graphs of Logarithmic Functions

Since exponential functions and logarithmic functions are inverses of each other, the graph of a logarithmic function $y = \log_a x$ is the reflection about the line $y = x$ of the graph of the exponential function $y = a^x$, as shown in Figure 14.

FIGURE 14

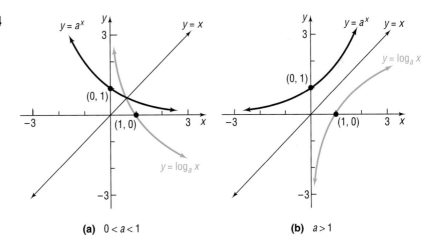

(a) $0 < a < 1$ (b) $a > 1$

Facts about the Graph
of a Logarithmic Function
$f(x) = \log_a x$

1. The x-intercept of the graph is 1. There is no y-intercept.
2. The y-axis is a vertical asymptote of the graph.
3. A logarithmic function is decreasing if $0 < a < 1$ and increasing if $a > 1$.
4. The graph is smooth and continuous, with no corners or gaps.

If the base of a logarithmic function is the number e, then we have the **natural logarithm function.** This function occurs so frequently in applications that it is given a special symbol, **ln** (from the Latin, *logarithmus naturalis*). Thus,

$$y = \ln x \quad \text{if and only if} \quad x = e^y$$

Since $y = \ln x$ and the exponential function $y = e^x$ are inverse functions, we can obtain the graph of $y = \ln x$ by reflecting the graph of $y = e^x$ about the line $y = x$. See Figure 15.

Table 5 displays other points on the graph of $f(x) = \ln x$. Notice for $x < 0$ we obtain an error message. Do you recall why?

FIGURE 15   **TABLE 5**

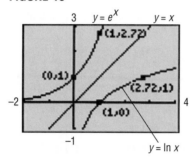

EXAMPLE   6   *Graphing Functions That Are Basically Logarithmic Using Shifting, Reflection, and the Like*

Graph $y = -\ln x$ by starting with the graph of $y = \ln x$.

Solution   The graph of $y = -\ln x$ is obtained by a reflection about the x-axis of the graph of $y = \ln x$. See Figure 16.

FIGURE 16

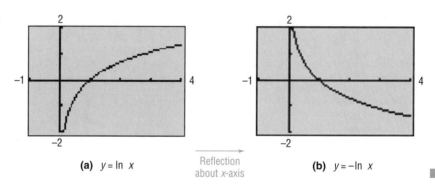

(a) $y = \ln x$ Reflection about x-axis   (b) $y = -\ln x$

EXAMPLE   7   *Graphing Functions That Are Basically Logarithmic Using Shifting, Reflection, and the Like*

Graph: $y = \ln(x + 2)$

Solution   The domain consists of all x for which

$$x + 2 > 0 \quad \text{or} \quad x > -2$$

The graph is obtained by applying a horizontal shift to the left 2 units, as shown in Figure 17. Notice that the line $x = -2$ is a vertical asymptote.

FIGURE 17

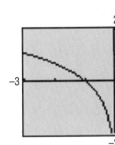
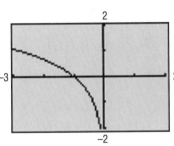
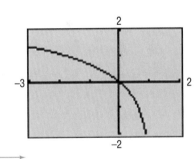

(a) $y = \ln x$ Shift left 2 units   (b) $y = \ln(x + 2)$ ■

E X A M P L E   8   *Graphing Functions That Are Basically Logarithmic Using Shifting, Reflection, and the Like*

Graph: $y = \ln(1 - x)$

Solution   The domain consists of all x for which

$$1 - x > 0 \quad \text{or} \quad x < 1$$

To obtain the graph of $y = \ln(1 - x)$, we use the steps illustrated in Figure 18. Note that the vertical asymptote of $y = \ln(1 - x)$ is $x = 1$.

FIGURE 18

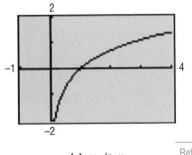
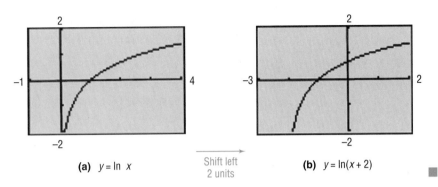

(a) $y = \ln x$ Reflection about y-axis   (b) $y = \ln(-x)$ Shift right 1 unit   (c) $y = \ln(-x + 1)$ $= \ln(1 - x)$ ■

■ Now work Problem 67.

E X A M P L E   9   *Alcohol and Driving*

The concentration of alcohol in a person's blood is measurable. Recent medical research suggests that the risk R (given as a percent) of having an accident while driving a car can be modeled by the equation

$$R = 6e^{kx}$$

where x is the variable concentration of alcohol in the blood and k is a constant.

(a) Suppose that a concentration of alcohol in the blood of 0.04 results in a 10% risk ($R = 10$) of an accident. Find the constant k in the equation. Graph $R = 6e^{kx}$.

(b) Using this value of k, what is the risk if the concentration is 0.17?

(c) Using the same value of k, what concentration of alcohol corresponds to a risk of 100%?

(d) If the law asserts that anyone with a risk of having an accident of 20% or more should not have driving privileges, at what concentration of alcohol in the blood should a driver be arrested and charged with a DUI—Driving Under the Influence?

Solution   (a) For a concentration of alcohol in the blood of 0.04 and a risk of 10%, we let $x = 0.04$ and $R = 10$ in the equation and solve for k.

$$R = 6e^{kx}$$

$$10 = 6e^{k(0.04)}$$

$$\frac{10}{6} = e^{0.04k} \qquad \text{Change to a logarithmic expression.}$$

$$0.04k = \ln\frac{10}{6} = 0.5108256$$

$$k = 12.77$$

See Figure 19 for the graph of $R = 6e^{12.77x}$.

FIGURE 19

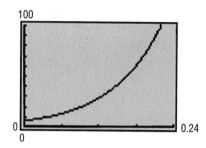

(b) Using $k = 12.77$ and $x = 0.17$ in the equation, we find the risk R to be

$$R = 6e^{kx} = 6e^{(12.77)(0.17)} = 52.6$$

For a concentration of alcohol in the blood of 0.17, the risk of an accident is about 52.6%. Verify this answer by TRACEing the graph.

(c) Using $k = 12.77$ and $R = 100$ in the equation, we find the concentration x of alcohol in the blood to be

$$R = 6e^{kx}$$

$$100 = 6e^{12.77x}$$

$$\frac{100}{6} = e^{12.77x} \qquad \text{Change to a logarithmic expression.}$$

$$12.77x = \ln\frac{100}{6} = 2.8134$$

$$x = 0.22$$

For a concentration of alcohol in the blood of 0.22, the risk of an accident is 100%. Verify this answer by TRACEing the graph.

(d) Using $k = 12.77$ and $R = 20$ in the equation, we find the concentration x of alcohol in the blood to be

$$R = 6e^{kx}$$

$$20 = 6e^{12.77x}$$

$$\frac{20}{6} = e^{12.77x}$$

$$12.77x = \ln\frac{20}{6} = 1.204$$

$$x = 0.094$$

A driver with a concentration of alcohol in the blood of 0.094 or more should be arrested and charged with DUI. Verify by TRACEing the graph. ■

Note: Most states use 0.10 as the blood alcohol content at which a DUI citation is given. A few states use 0.08.

Summary

PROPERTIES OF THE LOGARITHMIC FUNCTION

$f(x) = \log_a x, \quad a > 1$

($y = \log_a x$ means $x = a^y$)

Domain: $(0, \infty)$; Range: $(-\infty, \infty)$; x-intercept: 1; y-intercept: none; vertical asymptote: y-axis; increasing; one-to-one See Figure 14(b) for a typical graph.

$f(x) = \log_a x, \quad 0 < a < 1$

($y = \log_a x$ means $x = a^y$)

Domain: $(0, \infty)$; Range: $(-\infty, \infty)$; x-intercept: 1; y-intercept: none; vertical asymptote: y-axis; decreasing; one-to-one See Figure 14(a) for a typical graph.

5.2

Exercise 5.2

In Problems 1–12, change each exponential expression to an equivalent expression involving a logarithm.

1. $9 = 3^2$ **2.** $16 = 4^2$ **3.** $a^2 = 1.6$ **4.** $a^3 = 2.1$ **5.** $1.1^2 = M$ **6.** $2.2^3 = N$

7. $2^x = 7.2$ **8.** $3^x = 4.6$ **9.** $x^{\sqrt{2}} = \pi$ **10.** $x^\pi = e$ **11.** $e^x = 8$ **12.** $e^{2.2} = M$

In Problems 13–24, change each logarithmic expression to an equivalent expression involving an exponent.

13. $\log_2 8 = 3$ **14.** $\log_3(\frac{1}{9}) = -2$ **15.** $\log_a 3 = 6$ **16.** $\log_b 4 = 2$

17. $\log_3 2 = x$ **18.** $\log_2 6 = x$ **19.** $\log_2 M = 1.3$ **20.** $\log_3 N = 2.1$

21. $\log_{\sqrt{2}} \pi = x$ **22.** $\log_\pi x = \frac{1}{2}$ **23.** $\ln 4 = x$ **24.** $\ln x = 4$

In Problems 25–36, find the exact value of each logarithm without using a calculator.

25. $\log_2 1$ **26.** $\log_8 8$ **27.** $\log_5 25$ **28.** $\log_3(\frac{1}{9})$ **29.** $\log_{1/2} 16$ **30.** $\log_{1/3} 9$

31. $\log_{10} \sqrt{10}$ **32.** $\log_5 \sqrt[3]{25}$ **33.** $\log_{\sqrt{2}} 4$ **34.** $\log_{\sqrt{3}} 9$ **35.** $\ln \sqrt{e}$ **36.** $\ln e^3$

In Problems 37–46, find the domain of each function.

37. $f(x) = \ln(3 - x)$ **38.** $g(x) = \ln(x^2 - 1)$ **39.** $F(x) = \log_2 x^2$

40. $H(x) = \log_5 x^3$ **41.** $h(x) = \log_{1/2}(x^2 - x - 6)$ **42.** $G(x) = \log_{1/2}\left(\frac{1}{x}\right)$

43. $f(x) = \frac{1}{\ln x}$ **44.** $g(x) = \ln(x - 5)$ **45.** $g(x) = \log_5\left(\frac{x+1}{x}\right)$ **46.** $h(x) = \log_3\left(\frac{x^2}{x-1}\right)$

In Problems 47–50, use a calculator to evaluate each expression. Round your answer to three decimal places.

47. $\ln \dfrac{5}{3}$ **48.** $\dfrac{\ln 5}{3}$ **49.** $\dfrac{\ln 10/3}{0.04}$ **50.** $\dfrac{\ln 2/3}{-0.1}$

51. Find a such that the graph of $f(x) = \log_a x$ contains the point $(2, 2)$.

52. Find a such that the graph of $f(x) = \log_a x$ contains the point $(\frac{1}{2}, -4)$.

In Problems 53–60, the graph of a logarithmic function is given. Match each graph to one of the following functions:

A. $y = \log_3 x$ *B.* $y = \log_3(-x)$ *C.* $y = -\log_3 x$ *D.* $y = -\log_3(-x)$

E. $y = \log_3 x - 1$ *F.* $y = \log_3(x - 1)$ *G.* $y = \log_3(1 - x)$ *H.* $y = 1 - \log_3 x$

53.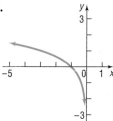

54.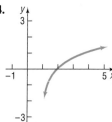

55.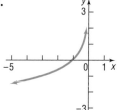

56.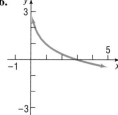

57.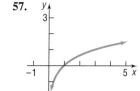

58.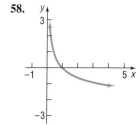

59.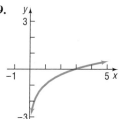

60.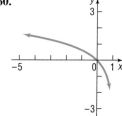

In Problems 61–67, the graph of a logarithmic function is given. Match each graph to one of the following functions:

A. $y = \log_4 x$ *B.* $y = \log_4(-x)$ *C.* $y = \log_4(x - 1)$

D. $y = -\log_4 x$ *E.* $y = 1 - \log_4 x$ *F.* $y = -\log_4(-x)$

61.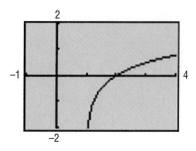

62.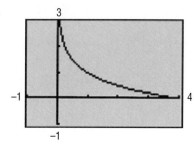

63.

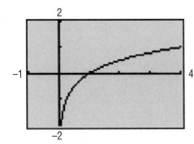

64.

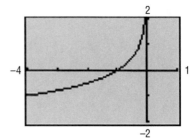

65.

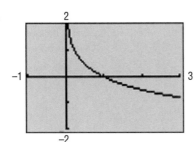

66.

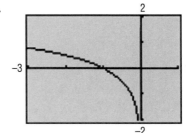

In Problems 67–76, using a graphing utility, show the stages required to graph each function.

67. $f(x) = \ln(x + 4)$ **68.** $f(x) = \ln(x - 3)$ **69.** $f(x) = \ln(-x)$ **70.** $f(x) = -\ln(-x)$

71. $g(x) = \ln 2x$ **72.** $h(x) = \ln \frac{1}{2}x$ **73.** $f(x) = 3 \ln x$ **74.** $f(x) = -2 \ln x$

75. $g(x) = \ln(3 - x)$ **76.** $h(x) = \ln(4 - x)$

77. *Optics* If a single pane of glass obliterates 10% of the light passing through it, then the percent P of light that passes through n successive panes is given approximately by the equation

$$P = 100e^{-0.1n}$$

(a) How many panes are necessary to block at least 50% of the light?

(b) How many panes are necessary to block at least 75% of the light?

78. *Chemistry* The pH of a chemical solution is given by the formula

$$pH = -\log_{10} [H^+]$$

where $[H^+]$ is the concentration of hydrogen ions in moles per liter. Values of pH range from 0 (acidic) to 14 (alkaline).

(a) Find the pH of a 1 liter container of water with 0.0000001 mole of hydrogen ion.

(b) Find the hydrogen ion concentration of a mildly acidic solution with a pH of 4.2.

79. *Space Satellites* The number of watts w provided by a space satellite's power supply after d days is given by the formula

$$w = 50e^{-0.004d}$$

(a) How long will it take for the available power to drop to 30 watts?

(b) How long will it take for the available power to drop to only 5 watts?

80. *Healing of Wounds* The normal healing of wounds can be modeled by an exponential function. If A_0 represents the original area of the wound and if A equals the area of the wound after n days, then the formula

$$A = A_0 e^{-0.35n}$$

describes the area of a wound on the nth day following an injury when no infection is present to retard the healing. Suppose a wound initially had an area of 100 square centimeters.
(a) If healing is taking place, how many days should pass before the wound is one-half its original size?
(b) How long before the wound is 10% of its original size?

81. *Drug Medication* The formula

$$D = 5e^{-0.4h}$$

can be used to find the number of milligrams D of a certain drug that is in a patient's bloodstream h hours after the drug has been administered. When the number of milligrams reaches 2, the drug is to be administered again. What is the time between injections?

82. *Spreading of Rumors* A model for the number of people N in a college community who have heard a certain rumor is

$$N = P(1 - e^{-0.15d})$$

where P is the total population of the community and d is the number of days that have elapsed since the rumor began. In a community of 1000 students, how many days will elapse before 450 students have heard the rumor?

83. *Current in a RL Circuit* The equation governing the amount of current I (in amperes) after time t (in seconds) in a simple RL circuit consisting of a resistance R (in ohms), an inductance L (in henrys), and an electromotive force E (in volts) is

$$I = \frac{E}{R}[1 - e^{-(R/L)t}]$$

If $E = 12$ volts, $R = 10$ ohms, and $L = 5$ henrys, how long does it take to obtain a current of 0.5 ampere? Of 1.0 ampere? Graph the equation.

84. *Learning Curve* Psychologists sometimes use the function

$$L(t) = A(1 - e^{-kt})$$

to measure the amount L learned at time t. The number A represents the amount to be learned, and the number k measures the rate of learning. Suppose that a student has an amount A of 200 vocabulary words to learn. A psychologist determines that the student learned 20 vocabulary words after 5 minutes.
(a) Determine the rate of learning k.
(b) Approximately how many words will the student have learned after 10 minutes?
(c) After 15 minutes?
(d) How long does it take for the student to learn 180 words?

85. *Alcohol and Driving* The concentration of alcohol in a person's blood is measurable. Suppose the risk R (given as a percent) of having an accident while driving a car can be modeled by the equation

$$R = 3e^{kx}$$

where x is the variable concentration of alcohol in the blood and k is a constant.
(a) Suppose a concentration of alcohol in the blood of 0.06 results in a 10% risk ($R = 10$) of an accident. Find the constant k in the equation.
(b) Using this value of k, what is the risk if the concentration is 0.17?
(c) Using the same value of k, what concentration of alcohol corresponds to a risk of 100%?
(d) If the law asserts that anyone with a risk of having an accident of 15% or more should not have driving privileges, at what concentration of alcohol in the blood should a driver be arrested and charged with a DUI?
(e) Compare this situation with that of Example 9. If you were a lawmaker, which situation would you support? Give your reasons.

86. Is there any function of the form $y = x^\alpha$, $0 < \alpha < 1$, that increases more slowly than a logarithmic function whose base is greater than 1? Explain.

87. *Constructing a Function* Look back at Figure 3 in Section 5.1. Assuming that the points (1950, 20) and (1990, 50) are on the graph, find an exponential equation $y = Ae^{bt}$, that fits the data. [*Hint:* Let $t = 0$ correspond to the year 1950. Then show that $A = 20$. Now find b.] Is the projection of 102 million in 2015 confirmed by your model? Try to obtain similar data about the United States birthrate and construct a function to fit those data.

88. *Critical Thinking* In buying a new car, one consideration might be how well the price of the car holds up over time. Different makes of cars have different depreciation rates. One way to compute a depreciation rate for a car is given here. Suppose the current prices of a certain Mercedes automobile are as follows:

NEW	1 YEAR OLD	2 YEARS OLD	3 YEARS OLD	4 YEARS OLD	5 YEARS OLD
$38,000	$36,600	$32,400	$28,750	$25,400	$21,200

Use the formula New = Old(e^{Rt}) to find R, the annual depreciation rate, for a specific time t. When might be the best time to trade the car in? Consult the NADA ("blue") book and compare two like models that you are interested in. Which has the better depreciation rate?

5.3

Properties of Logarithms

Logarithms have some very useful properties that can be derived directly from the definition and the laws of exponents.

EXAMPLE 1

Establishing Properties of Logarithms

(a) Show that $\log_a 1 = 0$.   (b) Show that $\log_a a = 1$.

Solution

(a) This fact was established when we graphed $y = \log_a x$ (see Figure 14). Algebraically, for $y = \log_a 1$, we have $a^y = 1 = a^0$, so $y = 0$.

(b) For $y = \log_a a$, we have $a^y = a = a^1$, so $y = 1$.

$$\log_a 1 = 0 \qquad \log_a a = 1$$

■

Theorem
Properties of Logarithms

In the properties given next, M and a are positive real numbers, with $a \neq 1$, and r is any real number.

The number $\log_a M$ is the exponent to which a must be raised to obtain M. That is,

$$a^{\log_a M} = M \tag{1}$$

The logarithm to the base a of a raised to a power equals that power. That is,

$$\log_a a^r = r \tag{2}$$

■

Proof of Property (1)

Let $x = \log_a M$. Change this logarithmic expression to the equivalent exponential expression:

$$a^x = M$$

But $x = \log_a M$, so

$$a^{\log_a M} = M$$

■

Proof of Property (2)　　Let $x = a^r$. Change this exponential expression to the equivalent logarithmic expression:

$$\log_a x = r$$

But $x = a^r$, so

$$\log_a a^r = r \qquad \blacksquare$$

E X A M P L E　2　　*Using Properties (1) and (2)*

(a) $2^{\log_2 \pi} = \pi$　　(b) $\log_{0.2} 0.2^{-\sqrt{2}} = -\sqrt{2}$　　(c) $\ln e^{kt} = kt$　　$\blacksquare$

Other useful properties of logarithms are given now.

Theorem　　In the following properties, M, N, and a are positive real numbers, with $a \neq 1$, and r is any real number.

The Log of a Product Equals
the Sum of the Logs

$$\log_a MN = \log_a M + \log_a N \qquad (3)$$

The Log of a Quotient Equals
the Difference of the Logs

$$\log_a\left(\frac{M}{N}\right) = \log_a M - \log_a N \qquad (4)$$

$$\log_a\left(\frac{1}{N}\right) = -\log_a N \qquad (5)$$

$$\log_a M^r = r \log_a M \qquad (6)$$

$\blacksquare$

We shall derive properties (3) and (6) and leave the derivations of properties (4) and (5) as exercises (see Problems 69 and 70).

Proof of Property (3)　　Let $A = \log_a M$ and let $B = \log_a N$. These expressions are equivalent to the exponential expressions

$$a^A = M \quad \text{and} \quad a^B = N$$

Now

$$\log_a MN = \log_a a^A a^B = \log_a a^{A+B} \qquad \text{Law of exponents}$$
$$= A + B \qquad \text{Property (2) of logarithms}$$
$$= \log_a M + \log_a N \qquad \blacksquare$$

Proof of Property (6)　　Let $A = \log_a M$. This expression is equivalent to

$$a^A = M$$

Now

$$\log_a M^r = \log_a (a^A)^r = \log_a a^{rA} \qquad \text{Law of exponents}$$
$$= rA \qquad \text{Property (2) of logarithms}$$
$$= r \log_a M \qquad \blacksquare$$

Logarithms can be used to transform products into sums, quotients into differences, and powers into factors. Such transformations prove useful in certain types of calculus problems.

EXAMPLE   3   *Writing a Logarithmic Expression as a Sum of Logarithms*

Write $\log_a(x\sqrt{x^2 + 1})$ as a sum of logarithms. Express all powers as factors.

Solution
$$\log_a(x\sqrt{x^2 + 1}) = \log_a x + \log_a \sqrt{x^2 + 1} \quad \text{Property (3)}$$
$$= \log_a x + \log_a(x^2 + 1)^{1/2}$$
$$= \log_a x + \tfrac{1}{2}\log_a(x^2 + 1) \quad \text{Property (6)}$$ ■

EXAMPLE   4   *Writing a Logarithmic Expression as a Difference of Logarithms*

Write
$$\log_a \frac{x^2}{(x-1)^3}$$

as a difference of logarithms. Express all powers as factors.

Solution
$$\log_a \frac{x^2}{(x-1)^3} \underset{\substack{\uparrow \\ \text{Property (4)}}}{=} \log_a x^2 - \log_a(x-1)^3 \underset{\substack{\uparrow \\ \text{Property (6)}}}{=} 2\log_a x - 3\log_a(x-1)$$ ■

■ Now work Problem 13.

EXAMPLE   5   *Writing a Logarithmic Expression as a Sum and Difference of Logarithms*

Write
$$\log_a \frac{x^3\sqrt{x^2 + 1}}{(x+1)^4}$$

as a sum and difference of logarithms. Express all powers as factors.

Solution
$$\log_a \frac{x^3\sqrt{x^2 + 1}}{(x+1)^4} = \log_a(x^3\sqrt{x^2 + 1}) - \log_a(x+1)^4$$
$$= \log_a x^3 + \log_a \sqrt{x^2 + 1} - \log_a(x+1)^4$$
$$= \log_a x^3 + \log_a(x^2 + 1)^{1/2} - \log_a(x+1)^4$$
$$= 3\log_a x + \tfrac{1}{2}\log_a(x^2 + 1) - 4\log_a(x+1)$$ ■

Another use of properties (3) through (6) is to write sums and/or differences of logarithms with the same base as a single logarithm.

EXAMPLE   6   *Writing Expressions as a Single Logarithm*

Write each of the following as a single logarithm:

(a) $\log_a 7 + 4\log_a 3$

(b) $\tfrac{2}{3}\log_a 8 - \log_a(3^4 - 8)$

(c) $\log_a x + \log_a 9 + \log_a(x^2 + 1) - \log_a 5$

Solution (a) $\log_a 7 + 4 \log_a 3 = \log_a 7 + \log_a 3^4$ Property (6)

$$= \log_a 7 + \log_a 81$$
$$= \log_a(7 \cdot 81) \qquad \text{Property (3)}$$
$$= \log_a 567$$

(b) $\frac{2}{3} \log_a 8 - \log_a(3^4 - 8) = \log_a 8^{2/3} - \log_a(81 - 8)$ Property (6)

$$= \log_a 4 - \log_a 73$$
$$= \log_a\left(\frac{4}{73}\right) \qquad \text{Property (4)}$$

(c) $\log_a x + \log_a 9 + \log_a(x^2 + 1) - \log_a 5 = \log_a 9x + \log_a(x^2 + 1) - \log_a 5$

$$= \log_a[9x(x^2 + 1)] - \log_a 5$$
$$= \log_a\left[\frac{9x(x^2 + 1)}{5}\right] \qquad \blacksquare$$

Warning: A common error made by some students is to express the logarithm of a sum as the sum of logarithms:

$$\log_a(M + N) \quad \text{is not equal to} \quad \log_a M + \log_a N$$

Correct Statement $\log_a MN = \log_a M + \log_a N$ Property (3)

Another common error is to express the difference of logarithms as the quotient of logarithms:

$$\log_a M - \log_a N \quad \text{is not equal to} \quad \frac{\log_a M}{\log_a N}$$

Correct Statement $\log_a M - \log_a N = \log_a\left(\frac{M}{N}\right)$ Property (4)

■ Now work Problem 23.

There remain two other properties of logarithms we need to know. They are a consequence of the fact that the logarithmic function $y = \log_a x$ is one-to-one.

Theorem In the following properties, M, N, and a are positive real numbers, with $a \neq 1$:

If $M = N$, then $\log_a M = \log_a N$.	(7)
If $\log_a M = \log_a N$, then $M = N$.	(8)

$\blacksquare$

Properties (7) and (8) are useful for solving *logarithmic equations*, a topic discussed in the next section.

Using a Calculator to Evaluate and Graph Logarithms with Bases Other Than e or 10

Logarithms to the base 10, called **common logarithms,** were used to facilitate arithmetic computations before the widespread use of calculators. (See the Historic Feature at the end of this section.) Natural logarithms, that is, logarithms whose base is the number e, remain very important because they arise frequently in the study of natural phenomena.

Common logarithms are usually abbreviated by writing **log,** with the base understood to be 10, just as natural logarithms are abbreviated by **ln,** with the base understood to be e.

Graphing calculators have both $\boxed{\log}$ and $\boxed{\ln}$ keys to calculate the common logarithm and natural logarithm of a number. Let's look at an example to see how to calculate logarithms having a base other than 10 or e.

E X A M P L E 7 *Evaluating Logarithms Whose Base Is Neither 10 nor e*

Evaluate: $\log_2 7$

Solution Let $y = \log_2 7$. Then $2^y = 7$, so

$$2^y = 7$$
$$\ln 2^y = \ln 7 \qquad \text{Property (7)}$$
$$y \ln 2 = \ln 7 \qquad \text{Property (6)}$$
$$y = \frac{\ln 7}{\ln 2} \qquad \text{Solve for } y.$$
$$= 2.8074 \qquad \text{Use calculator (}\boxed{\ln}\text{ key).}$$

Example 7 shows how to change the base from 2 to e. In general, to change from the base b to the base a, we use the **Change-of-Base Formula.**

Theorem If $a \neq 1$, $b \neq 1$, and M are positive real numbers, then
Change-of-Base Formula

$$\log_a M = \frac{\log_b M}{\log_b a} \qquad (9)$$

Proof We derive this formula as follows: Let $y = \log_a M$. Then $a^y = M$, so

$$\log_b a^y = \log_b M \qquad \text{Property (7)}$$
$$y \log_b a = \log_b M \qquad \text{Property (6)}$$
$$y = \frac{\log_b M}{\log_b a} \qquad \text{Solve for } y.$$
$$\log_a M = \frac{\log_b M}{\log_b a} \qquad y = \log_a M$$

Since calculators have only keys for $\boxed{\log}$ and $\boxed{\ln}$, in practice, the Change-of-Base Formula uses either $b = 10$ or $b = e$. Thus,

$$\log_a M = \frac{\log M}{\log a} \quad \text{and} \quad \log_a M = \frac{\ln M}{\ln a} \qquad (10)$$

E X A M P L E 8 *Using the Change-of-Base Formula*

Calculate:

(a) $\log_5 89$ (b) $\log_{\sqrt{2}} \sqrt{5}$

Solution   (a) $\log_5 89 = \dfrac{\log 89}{\log 5} \approx \dfrac{1.94939}{0.69897} = 2.7889$

or

$\log_5 89 = \dfrac{\ln 89}{\ln 5} \approx \dfrac{4.4886}{1.6094} = 2.7889$

(b) $\log_{\sqrt{2}} \sqrt{5} = \dfrac{\log \sqrt{5}}{\log \sqrt{2}} = \dfrac{\frac{1}{2} \log 5}{\frac{1}{2} \log 2} \approx \dfrac{0.69897}{0.30103} = 2.3219$

or

$\log_{\sqrt{2}} \sqrt{5} = \dfrac{\ln \sqrt{5}}{\ln \sqrt{2}} = \dfrac{\frac{1}{2} \ln 5}{\frac{1}{2} \ln 2} \approx \dfrac{1.6094}{0.6931} = 2.3219$ ■

■ Now work Problem 33.

We also use the Change-of-Base-Formula to graph logarithmic functions whose base is neither 10 nor e.

E X A M P L E   9   *Graphing a Logarithmic Function Whose Base Is Neither 10 Nor e*
Use a graphing utility to graph $y = \log_2 x$.

Solution   Since graphing utilities only have logarithms with the base 10 or the base e, we need to use the Change-of-Base Formula to express $y = \log_2 x$ in terms of logarithms with base 10 or base e. We can graph either $y = \ln x/\ln 2$ or $y = \log x/\log 2$ to obtain the graph of $y = \log_2 x$. See Figure 20.   ■

FIGURE 20

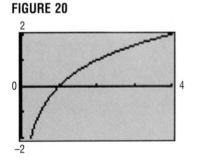

Check: Verify that $y = \ln x/\ln 2$ and $y = \log x/\log 2$ result in the same graph by graphing each one on the same screen.

■ Now work Problem 41.

Summary of Properties of Logarithms

In the list that follows, $a > 0$, $a \neq 1$, and $b > 0$, $b \neq 1$; also, $M > 0$ and $N > 0$.

Definition $y = \log_a x$ means $x = a^y$

Properties of logarithms $\log_a 1 = 0; \quad \log_a a = 1$

$a^{\log_a M} = M; \quad \log_a a^r = r$

$\log_a MN = \log_a M + \log_a N$

$\log_a\left(\dfrac{M}{N}\right) = \log_a M - \log_a N$

$\log_a\left(\dfrac{1}{N}\right) = -\log_a N$

$\log_a M^r = r \log_a M$

Change-of-Base Formula $\log_a M = \dfrac{\log_b M}{\log_b a}$

HISTORICAL FEATURE   ■ Logarithms were invented about 1590 by John Napier (1550–1617) and Jobst Bürgi (1552–1632), working independently. Napier, whose work had the greater influence, was a Scottish lord, a secretive man whose neighbors were inclined to believe him to be in league with the devil. His approach to logarithms was quite different from ours; it was based on the relationship between arithmetic and geometric sequences (see the chapter on induction and sequences), and not on the inverse function relationship of logarithms to exponential functions (described in Section 5.2). Napier's tables, published in 1614, listed what would now be called *natural logarithms* of sines and were rather difficult to use. A London professor, Henry Briggs, became interested in the tables and visited Napier. In their conversations, they developed the idea of common logarithms, and Briggs then converted Napier's tables into tables of common logarithms, which were published in 1617. Their importance for calculation was immediately recognized, and by 1650 they were being printed as far away as China. They remained an important calculation tool until the advent of the inexpensive handheld calculator about 1972, which has decreased their calculational, but not their theoretical, importance.

A side effect of the invention of logarithms was the popularization of the decimal system of notation for real numbers.   ■

5.3

Exercise 5.3

In Problems 1–12, suppose $\ln 2 = a$ *and* $\ln 3 = b$. *Use properties of logarithms to write each logarithm in terms of a and b.*

1. $\ln 6$

2. $\ln \frac{2}{3}$

3. $\ln 1.5$

4. $\ln 0.5$

5. $\ln 2e$

6. $\ln\left(\dfrac{3}{e}\right)$

7. $\ln 12$

8. $\ln 24$

9. $\ln \sqrt[5]{18}$

10. $\ln \sqrt[4]{48}$

11. $\log_2 3$

12. $\log_3 2$

In Problems 13–22, write each expression as a sum and/or difference of logarithms. Express powers as factors.

13. $\ln(x^2\sqrt{1 - x})$

14. $\ln(x\sqrt{1 + x^2})$

15. $\log_2\left(\dfrac{x^3}{x - 3}\right)$

16. $\log_5\left(\dfrac{\sqrt[3]{x^2 + 1}}{x^2 - 1}\right)$

17. $\log\left[\dfrac{x(x + 2)}{(x + 3)^2}\right]$

18. $\log\dfrac{x^3\sqrt{x + 1}}{(x - 2)^2}$

19. $\ln\left[\dfrac{x^2 - x - 2}{(x + 4)^2}\right]^{1/3}$

20. $\ln\left[\dfrac{(x - 4)^2}{x^2 - 1}\right]^{2/3}$

21. $\ln\dfrac{5x\sqrt{1 - 3x}}{(x - 4)^3}$

22. $\ln\left[\dfrac{5x^2\sqrt[3]{1 - x}}{4(x + 1)^2}\right]$

In Problems 23–32, write each expression as a single logarithm.

23. $3\log_5 u + 4\log_5 v$

24. $\log_3 u^2 - \log_3 v$

25. $\log_{1/2}\sqrt{x} - \log_{1/2} x^3$

26. $\log_2\left(\dfrac{1}{x}\right) + \log_2\left(\dfrac{1}{x^2}\right)$

27. $\ln\left(\dfrac{x}{x - 1}\right) + \ln\left(\dfrac{x + 1}{x}\right) - \ln(x^2 - 1)$

28. $\log\left(\dfrac{x^2 + 2x - 3}{x^2 - 4}\right) - \log\left(\dfrac{x^2 + 7x + 6}{x + 2}\right)$

29. $8\log_2\sqrt{3x - 2} - \log_2\left(\dfrac{4}{x}\right) + \log_2 4$

30. $21\log_3\sqrt[3]{x} + \log_3 9x^2 - \log_5 25$

31. $2\log_a 5x^3 - \frac{1}{2}\log_a(2x + 3)$

32. $\frac{1}{3}\log(x^3 + 1) + \frac{1}{2}\log(x^2 + 1)$

In Problems 33–40, use the change of base formula and a calculator to evaluate each logarithm. Round your answer to three decimal places.

33. $\log_3 21$

34. $\log_5 18$

35. $\log_{1/3} 71$

36. $\log_{1/2} 15$

37. $\log_{\sqrt{2}} 7$

38. $\log_{\sqrt{5}} 8$

39. $\log_\pi e$

40. $\log_\pi \sqrt{2}$

For Problems 41–46, graph each function using a graphing utility and the Change-of-Base Formula.

41. $y = \log_4 x$

42. $y = \log_5 x$

43. $y = \log_2(x + 2)$

44. $y = \log_4(x - 3)$

45. $y = \log_{x-1}(x + 1)$

46. $y = \log_{x+2}(x - 2)$

47. Show that: $\log_a(x + \sqrt{x^2 - 1}) + \log_a(x - \sqrt{x^2 - 1}) = 0$

48. Show that: $\log_a(\sqrt{x} + \sqrt{x - 1}) + \log_a(\sqrt{x} - \sqrt{x - 1}) = 0$

49. Show that: $\ln(1 + e^{2x}) = 2x + \ln(1 + e^{-2x})$

50. If $f(x) = \log_a x$, show that: $\dfrac{f(x + h) - f(x)}{h} = \log_a\left(1 + \dfrac{h}{x}\right)^{1/h}$, $\quad h \neq 0$

51. If $f(x) = \log_a x$, show that: $-f(x) = \log_{1/a} x$

52. If $f(x) = \log_a x$, show that: $f(1/x) = -f(x)$

53. If $f(x) = \log_a x$, show that: $f(AB) = f(A) + f(B)$

54. If $f(x) = \log_a x$, show that: $f(x^\alpha) = \alpha f(x)$

In Problems 55–64, express y as a function of x. The constant C is a positive number.

55. $\ln y = \ln x + \ln C$

56. $\ln y = \ln(x + C)$

57. $\ln y = \ln x + \ln(x + 1) + \ln C$

58. $\ln y = 2 \ln x - \ln(x + 1) + \ln C$

59. $\ln y = 3x + \ln C$

60. $\ln y = -2x + \ln C$

61. $\ln(y - 3) = -4x + \ln C$

62. $\ln(y + 4) = 5x + \ln C$

63. $3 \ln y = \frac{1}{2} \ln(2x + 1) - \frac{1}{3} \ln(x + 4) + \ln C$

64. $2 \ln y = -\frac{1}{2} \ln x + \frac{1}{3} \ln(x^2 + 1) + \ln C$

65. Find the value of $\log_2 3 \cdot \log_3 4 \cdot \log_4 5 \cdot \log_5 6 \cdot \log_6 7 \cdot \log_7 8$.

66. Find the value of $\log_2 4 \cdot \log_4 6 \cdot \log_6 8$.

67. Find the value of $\log_2 3 \cdot \log_3 4 \cdot \ldots \cdot \log_n(n + 1) \cdot \log_{n+1} 2$.

68. Find the value of $\log_2 2 \cdot \log_2 4 \cdot \ldots \cdot \log_2 2^n$.

69. Show that $\log_a(M/N) = \log_a M - \log_a N$, where a, M, and N are positive real numbers, with $a \neq 1$.

70. Show that $\log_a(1/N) = -\log_a N$, where a and N are positive real numbers, with $a \neq 1$.

 71. Find the domain of $f(x) = \log_a x^2$ and the domain of $g(x) = 2 \log_a x$. Since $\log_a x^2 = 2 \log_a x$, how do you reconcile the fact that the domains are not equal? Write a brief explanation.

5.4

Logarithmic and Exponential Equations

Logarithmic Equations

Equations that contain terms of the form $\log_a x$, where a is a positive real number, with $a \neq 1$, are often called **logarithmic equations.**

As before, our practice will be to solve equations, whenever possible, by finding exact solutions using algebraic methods. In such cases, we shall also verify the solution obtained by using a graphing utility. When algebraic methods cannot be used, approximate solutions will be obtained using a graphing utility. The reader is encouraged to pay particular attention to the form of equations for which exact solutions are possible.

$\mathcal{M}$ISSION POSSIBLE

Chapter 5

MCNEWTON'S COFFEE

Your team has been called in to solve a problem encountered by a fast food restaurant. They believe that their coffee should be brewed at 170° Fahrenheit; however, at that temperature it is too hot to drink, and a customer who accidentally spills the coffee on himself might receive third degree burns.

What they need is a special container that will heat the water to 170°, brew the coffee at that temperature, then cool it quickly to a drinkable temperature, say 140°F, and hold it there, or at least keep it at or above 120°F for a reasonable period of time without further cooking. To cool down the coffee, three companies have submitted proposals with these specifications:

(a) The CentiKeeper Company has a container that will reduce the temperature of a liquid from 200°F to 100°F in 90 minutes by maintaining a constant temperature of 70°F.

(b) The TempControl Company has a container that will reduce the temperature of a liquid from 200°F to 110°F in 60 minutes by maintaining a constant temperature of 60°F.

(c) The Hot'n'Cold, Inc., has a container that will reduce the temperature of a liquid from 210°F to 90°F in 30 minutes by maintaining a constant temperature of 50°F.

Your job is to make a recommendation as to which container to purchase. For this you will need Newton's Law of Cooling which follows:

$$u = T + (u_0 - T)e^{kt}, \quad k < 0$$

In this formula, T represents the temperature of the surrounding medium, u_0 is the initial temperature of the heated object, t is the length of time in minutes, k is a negative constant, and u represents the temperature at time t.

1. Use Newton's Law of Cooling to find the constant k of the formula for each container.
2. Use a graphing utility to graph each relation.
3. How long does it take each container to lower the coffee temperature from 170°F to 140°F?
4. How long will the coffee temperature remain between 120°F and 140°F?
5. On the basis of this information, which company should get the contract with McNewton's? What are your reasons?
6. What are "capital cost" and "operating cost"? How might they affect your choice?

E X A M P L E 1   *Solving a Logarithmic Equation*

Solve: $\log_3(4x - 7) = 2$

Solution   We can obtain an exact solution by changing the logarithm to exponential form.

$$\log_3(4x - 7) = 2$$
$$4x - 7 = 3^2$$
$$4x - 7 = 9$$
$$4x = 16$$
$$x = 4 \qquad\blacksquare$$

We verify the solution by graphing the function $f(x) = \log_3(4x - 7) - 2 = \dfrac{\ln(4x - 7)}{\ln 3} - 2$ to determine where the graph crosses the x-axis. See Figure 21. The graph of f crosses the x-axis at $x = 4$.

FIGURE 21

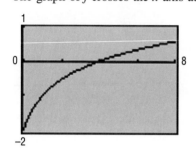

E X A M P L E 2   *Solving a Logarithmic Equation*

Solve: $2 \log_5 x = \log_5 9$

Solution   Because each logarithm is to the same base, 5, we can obtain an exact solution as follows:

$$2 \log_5 x = \log_5 9$$
$$\log_5 x^2 = \log_5 9 \qquad \text{Property (6), Section 5.3}$$
$$x^2 = 9 \qquad \text{Property (8), Section 5.3}$$
$$x = 3 \quad \text{or} \quad \cancel{x = -3} \qquad \text{Recall that logarithms of negative}$$

Recall that logarithms of negative numbers are not defined, so, in the expression $2 \log_5 x$, x must be positive. Therefore, -3 is extraneous and we discard it.

The equation has only one solution, 3. $\blacksquare$

You should verify for yourself that 3 is the only solution using a graphing utility.

  ■   Now work Problem 1.

E X A M P L E 3   *Solving a Logarithmic Equation*

Solve: $\log_4(x + 3) + \log_4(2 - x) = 1$

Solution   To obtain an exact solution, we need to express the left side as a single logarithm. Then we will change the expression to exponential form.

$$\log_4(x + 3) + \log_4(2 - x) = 1$$
$$\log_4[(x + 3)(2 - x)] = 1 \qquad \text{Property (3), Section 5.3}$$
$$(x + 3)(2 - x) = 4^1 = 4$$
$$-x^2 - x + 6 = 4$$
$$x^2 + x - 2 = 0$$
$$(x + 2)(x - 1) = 0$$
$$x = -2 \quad \text{or} \quad x = 1$$

You should verify that both of these are solutions using a graphing utility. ■

■ Now work Problem 11.

Care must be taken when solving logarithmic equations. Be sure to check each apparent solution in the original equation and discard any that are extraneous. In the expression $\log_a M$, remember that a and M are positive and $a \neq 1$.

Exponential Equations

Equations that involve terms of the form a^x, $a > 0$, $a \neq 1$, are often referred to as **exponential equations.** Such equations sometimes can be solved by appropriately applying the laws of exponents and equation (1), namely,

$$\text{If } a^u = a^v, \quad \text{then } u = v. \tag{1}$$

To use equation (1), each side of the equality must be written with the same base.

EXAMPLE   4

Solving an Exponential Equation

Solve the equation: $3^{x+1} = 81$

Solution   Since $81 = 3^4$, we can write the equation as

$$3^{x+1} = 81 = 3^4$$

Now we have the same base, 3, on each side, so we can apply (1) to obtain

$$x + 1 = 4$$
$$x = 3$$

We verify the solution by graphing $f(x) = 3^{x+1} - 81$ to determine where the graph crosses the x-axis. Figure 22 shows that the graph of f crosses the x-axis at $x = 3$. ■

FIGURE 22

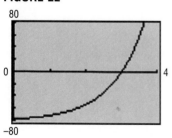

■ Now work Problem 19.

E X A M P L E   5   *Solving an Exponential Equation*

Solve the equation: $e^{-x^2} = (e^x)^2 \cdot \dfrac{1}{e^3}$

Solution   We use some laws of exponents first to get the same base e on each side:

$$e^{-x^2} = (e^x)^2 \cdot \frac{1}{e^3} = e^{2x} \cdot e^{-3} = e^{2x-3}$$

Now apply (1) to get

$$-x^2 = 2x - 3$$
$$x^2 + 2x - 3 = 0$$
$$(x + 3)(x - 1) = 0$$
$$x = -3 \quad \text{or} \quad x = 1 \qquad \blacksquare$$

You should verify these solutions using a graphing utility.

E X A M P L E   6   *Solving an Exponential Equation*

Solve the equation: $4^x - 2^x - 12 = 0$

Solution   We note that $4^x = (2^2)^x = 2^{2x} = (2^x)^2$, so the equation is actually quadratic in form, and we can rewrite it as

$$(2^x)^2 - 2^x - 12 = 0$$

Now we can factor as usual:

$$(2^x - 4)(2^x + 3) = 0$$
$$2^x - 4 = 0 \quad \text{or} \quad 2^x + 3 = 0$$
$$2^x = 4 \qquad\qquad 2^x = -3$$

The equation on the left has the solution $x = 2$, since $2^x = 4 = 2^2$; the equation on the right has no solution, since $2^x > 0$ for all x. $\blacksquare$

In each of the preceding three examples, we were able to write each exponential expression using the same base, obtaining exact solutions to the equation. When this is not possible, logarithms can sometimes be used to obtain the solution.

E X A M P L E   7   *Solving an Exponential Equation*

Solve for x: $2^x = 5$

Solution   We write the exponential equation as the equivalent logarithmic equation:

$$2^x = 5$$
$$x = \log_2 5 = \frac{\ln 5}{\underset{\uparrow}{\ln 2}}$$

Change-of-Base Formula (10)

Alternatively, we can solve the equation $2^x = 5$ by taking the natural logarithm (or common logarithm) of each side. Taking the natural logarithm,

$$2^x = 5$$
$$\ln 2^x = \ln 5$$
$$x \ln 2 = \ln 5$$
$$x = \frac{\ln 5}{\ln 2}$$

Using a calculator, the solution, rounded to three decimal places, is:

$$x = \frac{\ln 5}{\ln 2} = 2.322 \qquad \blacksquare$$

■ Now work Problem 33.

E X A M P L E   8

Solving an Exponential Equation

Solve for x: $8 \cdot 3^x = 5$

Solution

$$8 \cdot 3^x = 5 \qquad \text{Isolate } 3^x \text{ on the left side.}$$
$$3^x = \tfrac{5}{8} \qquad \text{Proceed as in Example 7.}$$
$$x = \log_3\left(\tfrac{5}{8}\right) = \frac{\ln \tfrac{5}{8}}{\ln 3}$$

Using a calculator, the solution, rounded to three decimal places, is:

$$x = \frac{\ln\left(\tfrac{5}{8}\right)}{\ln 3} = -0.428 \qquad \blacksquare$$

E X A M P L E   9

Solving an Exponential Equation

Solve for x: $5^{x-2} = 3^{3x+2}$

Solution

Because the bases are different, we take the natural logarithm of each side and apply appropriate properties of logarithms. The result is an equation in x that we can solve.

$$5^{x-2} = 3^{3x+2}$$
$$\ln 5^{x-2} = \ln 3^{3x+2} \qquad \text{Property (7)}$$
$$(x-2)\ln 5 = (3x+2)\ln 3 \qquad \text{Property (6)}$$
$$(\ln 5)x - 2\ln 5 = (3\ln 3)x + 2\ln 3$$
$$(\ln 5 - 3\ln 3)x = 2\ln 3 + 2\ln 5$$
$$x = \frac{2(\ln 3 + \ln 5)}{\ln 5 - 3\ln 3} = -3.212 \qquad \blacksquare$$

■ Now work Problem 41.

Graphing Utility Solutions

The techniques introduced in this section only apply to certain types of logarithmic and exponential equations. Solutions for other types are usually studied in calculus, using numerical methods. For example, the logarithmic equation $\log_3 x + \log_4 x = 4$ cannot be solved by previous methods. (Notice the bases are different.) We can use a graphing utility to approximate the solution.

E X A M P L E   1 0   *Solving Equations Using a Graphing Utility*

Solve: $\log_3 x + \log_4 x = 4$

Express the solution(s) correct to two decimal places.

Solution   This type of logarithmic equation cannot be solved by previous methods. However, a graphing utility can be used here. We begin by noting that the equation to be solved is equivalent to

$$\log_3 x + \log_4 x = 4$$
$$\log_3 x + \log_4 x - 4 = 0$$

The solution(s) of this equation is the same as the x-intercepts of the graph of the function $f(x) = \log_3 x + \log_4 x - 4$. (Remember to use the Change-of-Base Formula to graph f.) This function f is increasing (do you see why?) and so has at most one x-intercept. Since $f(10) \approx -0.24 < 0$ and $f(15) \approx 0.41 > 0$, it follows from the Intermediate Value Theorem (Section 3.1), that there is one x-intercept and it is between 10 and 15. So we graph the function with $10 \le x \le 15$. See Figure 23.

FIGURE 23

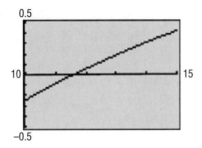

By using TRACE, ZOOM-IN, and/or BOX, we obtain the solution $x = 11.60$ correct to two decimal places. ∎

E X A M P L E   1 1   *Solving Equations Using a Graphing Utility*

Solve: $x + e^x = 2$

Express the solution(s) correct to two decimal places.

Solution   This type of exponential equation cannot be solved by previous methods. A graphing utility, though, can be used here. We begin by noting that the equation to be solved is equivalent to

$$x + e^x = 2$$
$$x + e^x - 2 = 0$$

The solution(s) of this equation is the same as the x-intercept(s) of the graph of the function $f(x) = x + e^x - 2$. This function f is increasing (do you see why?) and so has at most one x-intercept. Since $f(0) = -1 < 0$ and $f(1) = 1 + e - 2 > 0$, it follows from the Intermediate Value Theorem, that there is one x-intercept and it is between 0 and 1. So we graph the function, with $0 \le x \le 1$. See Figure 24.

FIGURE 24

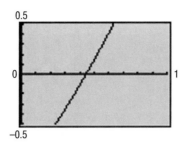

By using TRACE, ZOOM-IN, and/or BOX, we obtain the solution $x = 0.44$, correct to two decimal places. ■

5.4

Exercise 5.4

In Problems 1–58, solve each equation. Verify your solution using a graphing utility.

1. $\log_2(2x + 1) = 3$

2. $\log_3(3x - 2) = 2$

3. $\log_3(x^2 + 1) = 2$

4. $\log_5(x^2 + x + 4) = 2$

5. $\frac{1}{2}\log_3 x = 2\log_3 2$

6. $-2\log_4 x = \log_4 9$

7. $2\log_5 x = 3\log_5 4$

8. $3\log_2 x = -\log_2 27$

9. $3\log_2(x - 1) + \log_2 4 = 5$

10. $2\log_3(x + 4) - \log_3 9 = 2$

11. $\log_{10} x + \log_{10}(x + 15) = 2$

12. $\log_4 x + \log_4(x - 3) = 1$

13. $\log_x 4 = 2$

14. $\log_x\left(\frac{1}{8}\right) = 3$

15. $\log_3(x - 1)^2 = 2$

16. $\log_2(x + 4)^3 = 6$

17. $\log_{1/2}(3x + 1)^{1/3} = -2$

18. $\log_{1/3}(1 - 2x)^{1/2} = -1$

19. $2^{2x+1} = 4$

20. $5^{1-2x} = \frac{1}{5}$

21. $3^{x^3} = 9^x$

22. $4^{x^2} = 2^x$

23. $8^{x^2-2x} = \frac{1}{2}$

24. $9^{-x} = \frac{1}{3}$

25. $2^x \cdot 8^{-x} = 4^x$

26. $\left(\frac{1}{2}\right)^{1-x} = 4$

27. $2^{2x} - 2^x - 12 = 0$

28. $3^{2x} + 3^x - 2 = 0$

29. $3^{2x} + 3^{x+1} - 4 = 0$

30. $4^x - 2^x = 0$

31. $4^x = 8$

32. $9^{2x} = 27$

33. $2^x = 10$

34. $3^x = 14$

35. $8^{-x} = 1.2$

36. $2^{-x} = 1.5$

37. $3^{1-2x} = 4^x$

38. $2^{x+1} = 5^{1-2x}$

39. $\left(\frac{3}{5}\right)^x = 7^{1-x}$

40. $\left(\frac{4}{3}\right)^{1-x} = 5^x$

41. $1.2^x = (0.5)^{-x}$

42. $(0.3)^{1+x} = 1.7^{2x-1}$

43. $\pi^{1-x} = e^x$

44. $e^{x+3} = \pi^x$

45. $5(2^{3x}) = 8$

46. $0.3(4^{0.2x}) = 0.2$

47. $400e^{0.2x} = 600$

48. $500e^{0.3x} = 600$

49. $\log_a(x - 1) - \log_a(x + 6) = \log_a(x - 2) - \log_a(x + 3)$

50. $\log_a x + \log_a(x - 2) = \log_a(x + 4)$

51. $\log_{1/3}(x^2 + x) - \log_{1/3}(x^2 - x) = -1$

52. $\log_4(x^2 - 9) - \log_4(x + 3) = 3$

53. $\log_2 8^x = -3$

54. $\log_3 3^x = -1$

55. $\log_2(x^2 + 1) - \log_4 x^2 = 1$
   [*Hint:* Change $\log_4 x^2$ to base 2.]

56. $\log_2(3x + 2) - \log_4 x = 3$

57. $\log_{16} x + \log_4 x + \log_2 x = 7$

58. $\log_9 x + 3\log_3 x = 14$

In Problems 59–74, use a graphing utility to solve each equation. Express your answer correct to two decimal places.

59. $\log_5 x + \log_3 x = 1$

60. $\log_2 x + \log_6 x = 3$

61. $\log_5(x + 1) - \log_4(x - 2) = 1$

62. $\log_2(x - 1) - \log_6(x + 2) = 2$

63. $e^x = -x$

64. $e^{2x} = x + 2$

65. $e^x = x^2$

66. $e^x = x^3$

67. $\ln x = -x$

68. $\ln 2x = -x + 2$

69. $\ln x = x^3 - 1$

70. $\ln x = -x^2$

71. $e^x + \ln x = 4$

72. $e^x - \ln x = 4$

73. $e^{-x} = \ln x$

74. $e^{-x} = -\ln x$

5.5

Compound Interest

Interest is money paid for the use of money. The total amount borrowed (whether by an individual from a bank in the form of a loan or by a bank from an individual in the form of a savings account) is called the **principal.** The **rate of interest,** expressed as a percent, is the amount charged for the use of the principal for a given period of time, usually on a yearly (that is, per annum) basis.

If a principal of P dollars is borrowed for a period of t years at a per annum interest rate r, expressed as a decimal, the interest I charged is

Simple Interest Formula

$$I = Prt \qquad\qquad (1)$$

Interest charged according to formula (1) is called **simple interest.**

In working with problems involving interest we use the term **payment period** as follows:

Annually	Once per year
Semiannually	Twice per year
Quarterly	4 times per year
Monthly	12 times per year
Daily	365 times per year*

When the interest due at the end of a payment period is added to the principal so that the interest computed at the end of the next payment period is based on this new principal amount (old principal + interest), the interest is said to have been **compounded.** Thus, **compound interest** is interest paid on previously earned interest.

E X A M P L E   1

Computing Compound Interest

A credit union pays interest of 8% per annum compounded quarterly on a certain savings plan. If $1000 is deposited in such a plan and the interest is left to accumulate, how much is in the account after 1 year?

Solution

We use the simple interest formula, $I = Prt$. The principal P is $1000 and the rate of interest is $8\% = 0.08$. After the first quarter of a year, the time t is $\frac{1}{4}$ year, so the interest earned is

$$I = Prt = (\$1000)(0.08)\left(\tfrac{1}{4}\right) = \$20$$

The new principal is $P + I = \$1000 + \$20 = \$1020$. At the end of the second quarter, the interest on this principal is

$$I = (\$1020)(0.08)\left(\tfrac{1}{4}\right) = \$20.40$$

At the end of the third quarter, the interest on the new principal of $1020 + $20.40 = $1040.40 is

$$I = (\$1040.40)(0.08)\left(\tfrac{1}{4}\right) = \$20.81$$

*Some banks use a 360 day "year."

Finally, after the fourth quarter, the interest is

$$I = (\$1061.21)(0.08)\left(\tfrac{1}{4}\right) = \$21.22$$

Thus, after 1 year the account contains $1082.43.   ■

The pattern of the calculations performed in Example 1 leads to a general formula for compound interest. To fix our ideas, let P represent the principal to be invested at a per annum interest rate r, which is compounded n times per year. (For computing purposes, r is expressed as a decimal.) The interest earned after each compounding period is the principal times r/n. Thus, the amount A after one compounding period is

$$A = P + P\left(\frac{r}{n}\right) = P\left(1 + \frac{r}{n}\right)$$

After two compounding periods, the amount A, based on the new principal $P(1 + r/n)$, is

$$A = \underbrace{P\left(1 + \frac{r}{n}\right)}_{\substack{\text{New} \\ \text{principal}}} + \underbrace{P\left(1 + \frac{r}{n}\right)\left(\frac{r}{n}\right)}_{\substack{\text{Interest on} \\ \text{new principal}}} = P\left(1 + \frac{r}{n}\right)\left(1 + \frac{r}{n}\right) = P\left(1 + \frac{r}{n}\right)^2$$

After three compounding periods,

$$A = P\left(1 + \frac{r}{n}\right)^2 + P\left(1 + \frac{r}{n}\right)^2\left(\frac{r}{n}\right) = P\left(1 + \frac{r}{n}\right)^2\left(1 + \frac{r}{n}\right) = P\left(1 + \frac{r}{n}\right)^3$$

Continuing in this way, after n compounding periods (1 year),

$$A = P\left(1 + \frac{r}{n}\right)^n$$

Because t years will contain $n \cdot t$ compounding periods, after t years we have

$$A = P\left(1 + \frac{r}{n}\right)^{nt}$$

Theorem   The amount A after t years due to a principal P invested at an annual interest rate r compounded n times per year is

Compound Interest Formula

$$A = P\left(1 + \frac{r}{n}\right)^{nt} \tag{2}$$

■

E X A M P L E   2   *Comparing Investments Using Different Compounding Periods*

Investing $1000 at an annual rate of 10% compounded annually, quarterly, monthly, and daily will yield the following amounts after 1 year:

Annual compounding: $A = P(1 + r)$

$$= (\$1000)(1 + 0.10) = \$1100.00$$

Quarterly compounding: $A = P\left(1 + \dfrac{r}{4}\right)^4$

$$= (\$1000)(1 + 0.025)^4 = \$1103.81$$

Monthly compounding: $A = P\left(1 + \dfrac{r}{12}\right)^{12}$

$$= (\$1000)(1 + 0.00833)^{12} = \$1104.71$$

Daily compounding: $A = P\left(1 + \dfrac{r}{365}\right)^{365}$

$$= (\$1000)(1 + 0.000274)^{365} = \$1105.16$$ ∎

■ Now work Problem 1.

From Example 2, we can see that the effect of compounding more frequently is that the account after 1 year is higher: $1000 compounded 4 times a year at 10% results in $1103.81; $1000 compounded 12 times a year at 10% results in $1104.71; and $1000 compounded 365 times a year at 10% results in $1105.16. This leads to the following question: What would happen to the amount after 1 year if the number of times the interest is compounded were increased without bound?

Let's find the answer. Suppose P is the principal, r is the per annum interest rate, and n is the number of times the interest is compounded each year. The amount after 1 year is

$$A = P\left(1 + \frac{r}{n}\right)^n$$

Now suppose that the number n of times the interest is compounded per year gets larger and larger; that is, suppose that $n \to \infty$. Then,

$$A = P\left(1 + \frac{r}{n}\right)^n = P\left[1 + \frac{1}{n/r}\right]^n = P\left(\left[1 + \frac{1}{n/r}\right]^{n/r}\right)^r = P\left[\left(1 + \frac{1}{h}\right)^h\right]^r \quad (3)$$
$$\underset{\uparrow}{}$$
$$h = \frac{n}{r}$$

Thus, in (3), as $n \to \infty$, then $h = n/r \to \infty$, and the expression in brackets equals e, [Refer to (2) on p. 333] so $A \to Pe^r$. Table 6 compares $(1 + r/n)^n$, for large values of n, to e^r for $r = 0.05$, $r = 0.10$, $r = 0.15$, and $r = 1$. The larger n gets, the closer $(1 + r/n)^n$ gets to e^r. Thus, no matter how frequent the compounding, the amount after 1 year has the definite ceiling Pe^r.

TABLE 6

	$\left(1 + \dfrac{r}{n}\right)^n$			
	$n = 100$	$n = 1000$	$n = 10{,}000$	e^r
$r = 0.05$	1.0512579	1.05127	1.051271	1.0512711
$r = 0.10$	1.1051157	1.1051654	1.1051703	1.1051709
$r = 0.15$	1.1617037	1.1618212	1.1618329	1.1618342
$r = 1$	2.7048138	2.7169239	2.7181459	2.7182818

When interest is compounded so that the amount after 1 year is Pe^r, we say the interest is **compounded continuously.**

Theorem

The amount A after t years due to a principal P invested at an annual interest rate r compounded continuously is

Continuous Compounding

$$A = Pe^{rt} \tag{4}$$

E X A M P L E   3

Using Continuous Compounding

The amount A that results from investing a principal P of \$1000 at an annual rate r of 10% compounded continuously for a time t of 1 year is

$$A = \$1000e^{0.10} = (\$1000)(1.10517) = \$1105.17$$

■ Now work Problem 9.

The **effective rate of interest** is the equivalent annual simple rate of interest that would yield the same amount as compounding after 1 year. For example, based on Example 3, a principal of \$1000 will result in \$1105.17 at a rate of 10% compounded continuously. To get this same amount using a simple rate of interest would require that interest of \$1105.17 − \$1000.00 = \$105.17 be earned on the principal. Since \$105.17 is 10.517% of \$1000, a simple rate of interest of 10.517% is needed to equal 10% compounded continuously. Thus, the effective rate of interest of 10% compounded continuously is 10.517%.

Based on the results of Examples 2 and 3, we find the following comparisons:

	ANNUAL RATE	**EFFECTIVE RATE**
Annual compounding	10%	10%
Quarterly compounding	10%	10.381%
Monthly compounding	10%	10.471%
Daily compounding	10%	10.516%
Continuous compounding	10%	10.517%

■ Now work Problem 21.

E X A M P L E   4

Computing the Value of an IRA

On January 2, 1996, \$2000 is placed in an Individual Retirement Account (IRA) that will pay interest of 10% per annum compounded continuously. What will the IRA be worth on January 1, 2016?

Solution

The amount A after 20 years is

$$A = Pe^{rt} = \$2000e^{(0.10)(20)} = \$14,778.11$$

Check: Graph $y = 2000e^{0.1x}$ and use TRACE to verify that when $x = 20$ then $y = \$14,778.11$.

Exploration: How long will it be until $y = \$40,000$?

When people engaged in finance speak of the "time value of money," they are usually referring to the **present value** of money. The present value of A dollars to be received at a future date is the principal you would need to invest now so that

FIGURE 25

Time is Money

it would grow to A dollars in the specified time period. Thus, the present value of money to be received at a future date is always less than the amount to be received, since the amount to be received will equal the present value (money invested now) *plus* the interest accrued over the time period.

We use the compound interest formula (2) to get a formula for present value. If P is the present value of A dollars to be received after t years at a per annum interest rate r compounded n times per year, then, by formula (2),

$$A = P\left(1 + \frac{r}{n}\right)^{nt}$$

To solve for P, we divide both sides by $(1 + r/n)^{nt}$, and the result is

$$\frac{A}{(1 + r/n)^{nt}} = P \quad \text{or} \quad P = A\left(1 + \frac{r}{n}\right)^{-nt}$$

Theorem   The present value P of A dollars to be received after t years, assuming a per annum interest rate r compounded n times per year, is

Present Value Formulas

$$P = A\left(1 + \frac{r}{n}\right)^{-nt} \tag{5}$$

If the interest is compounded continuously, then

$$P = Ae^{-rt} \tag{6}$$

■

To prove (6), solve formula (4) for P.

E X A M P L E   5   *Computing the Value of a Zero-Coupon Bond*

A zero-coupon (noninterest-bearing) bond can be redeemed in 10 years for $1000. How much should you be willing to pay for it now if you want a return of:

(a) 8% compounded monthly?

(b) 7% compounded continuously?

Solution   (a) We are seeking the present value of $1000. Thus, we use formula (5) with $A = \$1000$, $n = 12$, $r = 0.08$, and $t = 10$:

$$P = A\left(1 + \frac{r}{n}\right)^{-nt}$$

$$= \$1000\left(1 + \frac{0.08}{12}\right)^{-12(10)}$$

$$= \$450.52$$

For a return of 8% compounded monthly, you should pay $450.52 for the bond.

(b)   Here, we use formula (6) with $A = \$1000$, $r = 0.07$, and $t = 10$:

$$P = Ae^{-rt}$$
$$= \$1000e^{-(0.07)(10)}$$
$$= \$496.59$$

For a return of 7% compounded continuously, you should pay $496.59 for the bond.   ■

■   Now work Problem 11.

E X A M P L E   6   *Rate of Interest Required to Double an Investment*

What annual rate of interest compounded annually should you seek if you want to double your investment in 5 years?

Algebraic Solution   If P is the principal and we want P to double, the amount A will be $2P$. We use the compound interest formula with $n = 1$ and $t = 5$ to find r:

$$2P = P(1 + r)^5$$
$$2 = (1 + r)^5$$
$$1 + r = \sqrt[5]{2}$$
$$r = \sqrt[5]{2} - 1 = 1.148698 - 1 = 0.148698$$

The annual rate of interest needed to double the principal in 5 years is 14.87%.

Graphing Solution   We solve the equation

$$2 = (1 + r)^5$$

for r by graphing the two functions $y_1 = 2$ and $y_2 = (1 + x)^5$. The x-coordinate of their point of intersection is the rate r we seek. See Figure 26. [We could also graph $y = (1 + x)^5 - 2$. Then the x-intercept is the rate r we seek.] Using the INTERSECT command*, we find the point of intersection of y_1 and y_2 is (0.14869835, 2).

FIGURE 26

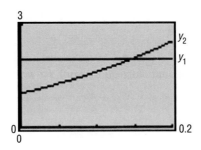

■   Now work Problem 23.

E X A M P L E   7   *Doubling and Tripling Time for an Investment*

(a)   How long will it take for an investment to double in value if it earns 5% compounded continuously?

(b)   How long will it take to triple at this rate?

*Note: If your utility does not have an INTERSECT command, you will have to use BOX or ZOOM-IN with TRACE to find the point of intersection.

Algebraic Solution

(a) If P is the initial investment and we want P to double, the amount A will be $2P$. We use formula (4) for continuously compounded interest with $r = 0.05$. Then

$$A = Pe^{rt}$$
$$2P = Pe^{0.05t}$$
$$2 = e^{0.05t}$$
$$0.05t = \ln 2$$
$$t = \frac{\ln 2}{0.05} = 13.86$$

It will take about 14 years to double the investment.

Graphing Solution

We solve the equation

$$2 = e^{0.05t}$$

for t by graphing the two functions $y_1 = 2$ and $y_2 = e^{0.05x}$. Their point of intersection is (13.86, 2). See Figure 27.

FIGURE 27

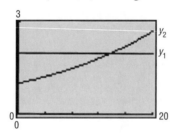

Algebraic Solution

(b) To triple the investment, we set $A = 3P$ in formula (4).

$$A = Pe^{rt}$$
$$3P = Pe^{0.05t}$$
$$3 = e^{0.05t}$$
$$0.05t = \ln 3$$
$$t = \frac{\ln 3}{0.05} = 21.97$$

It will take about 22 years to triple the investment.

Graphing Solution

We solve the equation

$$3 = e^{0.05t}$$

for t by graphing the two functions $y_1 = 3$ and $y_2 = e^{0.05x}$. Their point of intersection is (21.97, 3). See Figure 28.

FIGURE 28

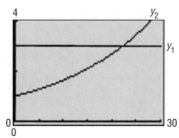

■ Now work Problem 29.

5.5

Exercise 5.5

In Problems 1–10, find the amount that results from each investment.

1. $100 invested at 4% compounded quarterly after a period of 2 years
2. $50 invested at 6% compounded monthly after a period of 3 years
3. $500 invested at 8% compounded quarterly after a period of $2\frac{1}{2}$ years
4. $300 invested at 12% compounded monthly after a period of $1\frac{1}{2}$ years
5. $600 invested at 5% compounded daily after a period of 3 years
6. $700 invested at 6% compounded daily after a period of 2 years
7. $10 invested at 11% compounded continuously after a period of 2 years
8. $40 invested at 7% compounded continuously after a period of 3 years
9. $100 invested at 10% compounded continuously after a period of $2\frac{1}{4}$ years
10. $100 invested at 12% compounded continuously after a period of $3\frac{3}{4}$ years

In Problems 11–20, find the principal needed now to get each amount; that is, find the present value.

11. To get $100 after 2 years at 6% compounded monthly
12. To get $75 after 3 years at 8% compounded quarterly
13. To get $1000 after $2\frac{1}{2}$ years at 6% compounded daily
14. To get $800 after $3\frac{1}{2}$ years at 7% compounded monthly
15. To get $600 after 2 years at 4% compounded quarterly
16. To get $300 after 4 years at 3% compounded daily
17. To get $80 after $3\frac{1}{4}$ years at 9% compounded continuously
18. To get $800 after $2\frac{1}{2}$ years at 8% compounded continuously
19. To get $400 after 1 year at 10% compounded continuously
20. To get $1000 after 1 year at 12% compounded continuously
21. Find the effective rate of interest for $5\frac{1}{4}$% compounded quarterly.
22. What interest rate compounded quarterly will give an effective interest rate of 7%?
23. What annual rate of interest is required to double an investment in 3 years? Verify your answer using a graphing utility.
24. What annual rate of interest is required to double an investment in 10 years? Verify your answer using a graphing utility.

In Problems 25–28, which of the two rates would yield the larger amount in 1 year? [Hint: Start with a principal of $10,000 in each instance.]

25. 6% compounded quarterly or $6\frac{1}{4}$% compounded annually?
26. 9% compounded quarterly or $9\frac{1}{4}$% compounded annually?
27. 9% compounded monthly or 8.8% compounded daily?
28. 8% compounded semiannually or 7.9% compounded daily?

29. How long does it take for an investment to double in value if it is invested at 8% per annum compounded monthly? Compounded continuously? Verify your answer using a graphing utility.
30. How long does it take for an investment to double in value if it is invested at 10% per annum compounded monthly? Compounded continuously? Verify your answer using a graphing utility.
31. If you have $100 to invest at 8% per annum compounded monthly, how long will it be before the amount is $150? If the compounding is continuous, how long will it be?

32. If you have $100 to invest at 10% per annum compounded monthly, how long will it be before the amount is $175? If the compounding is continuous, how long will it be?

33. How many years will it take for an initial investment of $10,000 to grow to $25,000? Assume a rate of interest of 6% compounded continuously.

34. How many years will it take for an initial investment of $25,000 to grow to $80,000? Assume a rate of interest of 7% compounded continuously.

35. What will a $90,000 house cost 5 years from now if the inflation rate over that period averages 3% compounded annually?

36. A department store charges 1.25% per month on the unpaid balance for customers with charge accounts (interest is compounded monthly). A customer charges $200 and does not pay her bill for 6 months. What is the bill at that time?

37. You will be buying a new car for $15,000 in 3 years. How much money should you ask your parents for now so that, if you invest it at 5% compounded continuously, you will have enough to buy the car?

38. You will require $3000 in 6 months to pay off a loan that has no prepayment privileges. If you have the $3000 now, how much of it should you save in an account paying 3% compounded monthly so that in 6 months you will have exactly $3000?

39. You are contemplating the purchase of 100 shares of a stock selling for $15 per share. The stock pays no dividends. The history of the stock indicates that it should grow at an annual rate of 15% per year. How much will the 100 shares of stock be worth in 5 years?

40. You are contemplating the purchase of 100 shares of a stock selling for $15 per share. The stock pays no dividends. Your broker says the stock will be worth $20 per share in 2 years. What is the annual rate of return on this investment?

41. A business purchased for $650,000 in 1994 is sold in 1997 for $850,000. What is the annual rate of return for this investment?

42. You have just inherited a diamond ring appraised at $5000. If diamonds have appreciated in value at an annual rate of 8%, what was the value of the ring 10 years ago when the ring was purchased?

43. You place $1000 in a bank account that pays 5.6% compounded continuously. After 1 year, will you have enough money to buy a computer system that costs $1060? If another bank will pay you 5.9% compounded monthly, is this a better deal?

44. On January 1, you place $1000 in a Certificate of Deposit that pays 6.8% compounded continuously and matures in 3 months. Then you place the $1000 and the interest in a passbook account that pays 5.25% compounded monthly. How much do you have in the passbook account on May 1?

45. You invest $2000 in a bond trust that pays 9% interest compounded semiannually. Your friend invests $2000 in a Certificate of Deposit (CD) that pays $8\frac{1}{2}$% compounded continuously. Who has more money after 20 years, you or your friend?

46. Suppose that you have access to an investment that will pay 10% interest compounded continuously. Which is better: To be given $1000 now so that you can take advantage of this investment opportunity or to be given $1325 after 3 years?

47. You have just purchased a house for $150,000, with the seller holding a second mortgage of $50,000. You promise to pay the seller $50,000 plus all accrued interest 5 years from now. The seller offers you three interest options on the second mortgage:
(a) Simple interest at 12% per annum   (b) $11\frac{1}{2}$% interest compounded monthly
(c) $11\frac{1}{4}$% interest compounded continuously

Which option is best; that is, which results in the least interest on the loan?

48. A bank advertises that it pays interest on saving accounts at the rate of 4.25% compounded daily. Find the effective rate if the bank uses (a) 360 days or (b) 365 days in determining the daily rate.

Problems 49–52 involve zero-coupon bonds. A zero-coupon bond is a bond that is sold now at a discount and will pay its face value at some time in the future when it matures; no interest payments are made.

49. A zero-coupon bond can be redeemed in 20 years for $10,000. How much should you be willing to pay for it now if you want a return of:
(a) 10% compounded monthly?   (b) 10% compounded continuously?

50. A child's grandparents are considering buying a $40,000 face value zero-coupon bond at birth so that she will have enough money for her college education 17 years later. If money is worth 8% compounded annually, what should they pay for the bond?

51. How much should a $10,000 face value zero-coupon bond, maturing in 10 years, be sold for now if its rate of return is to be 8% compounded annually?

52. If you pay $12,485.52 for a $25,000 face value zero-coupon bond that matures in 8 years, what is your annual rate of return?

53. Explain in your own words what the term *compound interest* means. What does *continuous compounding* mean?

54. Explain in your own words the meaning of present value.

55. Write a program that will calculate the amount after n years if a principal P is invested at r% per annum compounded quarterly. Use it to verify your answers to Problem 1 and 3.

56. Write a program that will calculate the principal needed now to get the amount A in n years at r% per annum compounded daily. Use it to verify your answer to Problem 13.

57. Write a program that will calculate the annual rate of interest required to double an investment in n years. Use it to verify your answers to Problems 23 and 24.

58. Write a program that will calculate the number of months required for an initial investment of x dollars to grow to y dollars at r% per annum compounded continuously. Use it to verify your answer to Problems 33 and 34.

59. *Time to Double or Triple an Investment* The formula

$$y = \frac{\ln m}{n \ln\left(1 + \dfrac{r}{n}\right)}$$

can be used to find the number of years y required to multiply an investment m times when r is the per annum interest rate compounded n times a year.
(a) How many years will it take to double the value of an IRA that compounds annually at the rate of 12%?
(b) How many years will it take to triple the value of a savings account that compounds quarterly at an annual rate of 6%?
(c) Give a derivation of this formula.

60. *Time to Reach an Investment Goal* The formula

$$y = \frac{\ln A - \ln P}{r}$$

can be used to find the number of years y required for an investment P to grow to a value A when compounded continuously at an annual rate r.
(a) How long will it take to increase an initial investment of $1000 to $8000 at an annual rate of 10%?
(b) What annual rate is required to increase the value of a $2000 IRA to $30,000 in 35 years?
(c) Give a derivation of this formula.

61. *Critical Thinking* You have just contracted to buy a house and will seek financing in the amount of $100,000. You go to several banks. Bank 1 will lend you $100,000 at the rate of 8.75% amortized over 30 years with a loan origination fee of 1.75%. Bank 2 will lend you $100,000 at the rate of 8.375% amortized over 15 years with a loan origination fee of 1.5%. Bank 3 will lend you $100,000 at the rate of 9.125% amortized over 30 years with no loan origination fee. Bank 4 will lend you $100,000 at the rate of 8.625% amortized over 15 years with no loan origination fee. Which loan would you take? Why? Be sure to have sound reasons for your choice. If the amount of the monthly payment does not matter to you, which loan would you take? Again, have sound reasons for your choice. Use the information in the table to assist you. Compare your final decision with others in the class. Discuss.

	MONTHLY PAYMENT	LOAN ORIGINATION FEE
Bank 1	$786.70	$1750.00
Bank 2	$977.42	$1500.00
Bank 3	$813.63	$0.00
Bank 4	$990.68	$0.00

5.6

Growth and Decay

Many natural phenomena have been found to follow the law that an amount A varies with time t according to

$$A = A_0 e^{kt} \tag{1}$$

where A_0 is the original amount ($t = 0$) and $k \neq 0$ is a constant.

If $k > 0$, then equation (1) states that the amount A is increasing over time; if $k < 0$, the amount A is decreasing over time. In either case, when an amount A varies over time according to equation (1), it is said to follow the **exponential law** or the **law of uninhibited growth** ($k > 0$) **or decay** ($k < 0$). See Figure 29.

FIGURE 29

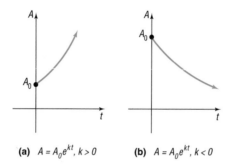

(a) $A = A_0 e^{kt}, k > 0$ (b) $A = A_0 e^{kt}, k < 0$

For example, we saw in Section 5.5 that continuously compounded interest follows the law of uninhibited growth. In this section we shall look at three additional phenomena that follow the exponential law.

Biology

Mitosis, or division of cells, is a universal process in the growth of living organisms such as amebas, plants, human skin cells, and many others. Based on an ideal situation in which no cells die and no by-products are produced, the number of cells present at a given time follows the law of uninhibited growth. Actually, however, after enough time has passed, growth at an exponential rate will cease due to the influence of factors such as lack of living space, dwindling food supply, and so on. The law of uninhibited growth accurately reflects only the early stages of the mitosis process.

The mitosis process begins with a culture containing N_0 cells. Each cell in the culture grows for a certain period of time and then divides into two identical cells. We assume that the time needed for each cell to divide in two is constant and does not change as the number of cells increases. These cells then grow, and eventually each divides in two; and so on.

A formula that gives the number N of cells in the culture after a time t has passed (in the early stages of growth) is

Uninhibited Growth of Cells

$$N(t) = N_0 e^{kt} \qquad k > 0 \tag{2}$$

where k is a positive constant.

In using equation (2) to model the growth of cells, we are using a function that yields positive real numbers, even though we are counting the number of cells, which must be an integer. This is a common practice in many applications.

E X A M P L E   1

Bacterial Growth

A colony of bacteria increases according to the law of uninhibited growth.

(a) If the number of bacteria doubles in 3 hours, how long will it take for the size of the colony to triple?

(b) Using a graphing utility, verify that the population doubles in 3 hours.

(c) Using a graphing utility, BOX and TRACE to approximate the time it takes for the population to double a second time (i.e., increase four times). Does this answer seem reasonable?

Solution   (a) Using formula (2), the number N of cells at a time t is

$$N(t) = N_0 e^{kt}$$

where N_0 is the initial number of bacteria present and k is a positive number. We first seek the number k. The number of cells doubles in 3 hours; thus, we have

$$N(3) = 2N_0$$

But $N(3) = N_0 e^{k(3)}$, so

$$N_0 e^{k(3)} = 2N_0$$
$$e^{3k} = 2$$
$$3k = \ln 2 \quad \text{Write the exponential equation as a logarithm.}$$
$$k = \tfrac{1}{3} \ln 2 \approx \tfrac{1}{3}(0.6931) = 0.2310$$

Formula (2) for this growth process is therefore

$$N(t) = N_0 e^{0.2310t}$$

The time t needed for the size of the colony to triple requires that $N = 3N_0$. Thus, we substitute $3N_0$ for N to get

$$3N_0 = N_0 e^{0.2310t}$$
$$3 = e^{0.2310t}$$
$$0.2310t = \ln 3$$
$$t = \frac{1}{0.2310} \ln 3 \approx \frac{1.0986}{0.2310} = 4.756 \text{ hours}$$

It will take about 4.756 hours for the size of the colony to triple.

(b) Figure 30 shows the graph of $y = e^{0.2310x}$ where x represents the time in hours and y represents the multiple of the original population. Using BOX and TRACE, it takes 3 hours for the population to double.

(c) Using BOX and TRACE, it takes 6 hours for the population to double again. ∎

FIGURE 30

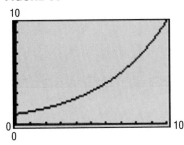

Now work Problem 1.

Radioactive Decay

Radioactive materials follow the law of uninhibited decay. Thus, the amount A of a radioactive material present at time t is given by the formula

Uninhibited Radioactive
Decay

$$A = A_0 e^{kt} \qquad k < 0 \qquad\qquad (3)$$

where A_0 is the original amount of radioactive material and k is a negative number.

All radioactive substances have a specific **half-life**, which is the time required for half of the radioactive substance to decay. In **carbon dating,** we use the fact that all living organisms contain two kinds of carbon, carbon-12 (a stable carbon) and carbon-14 (a radioactive carbon, with a half-life of 5600 years). While an organism is living, the ratio of carbon-12 to carbon-14 is constant. But when an organism dies, the original amount of carbon-12 present remains unchanged, whereas the amount of carbon-14 begins to decrease. This change in the amount of carbon-14 present relative to the amount of carbon-12 present makes it possible to calculate when an organism died.

E X A M P L E   2

Estimating the Age of Ancient Tools

Traces of burned wood found along with ancient stone tools in an archaeological dig in Chile were found to contain approximately 1.67% of the original amount of carbon-14.

(a) If the half-life of carbon-14 is 5600 years, approximately when was the tree cut and burned?

(b) Using a graphing utility, graph the relation between the percentage of carbon-14 remaining and time.

(c) Using BOX and TRACE, determine the time that elapses until half of the carbon-14 remains. This answer should equal the half-life of carbon-14.

(d) Verify the answer found in (a).

Solution   (a) Using equation (3), the amount A of carbon-14 present at time t is

$$A = A_0 e^{kt}$$

where A_0 is the original amount of carbon-14 present and k is a negative number. We first seek the number k. To find it, we use the fact that after 5600 years half of the original amount of carbon-14 remains. Thus,

$$\frac{1}{2} A_0 = A_0 e^{k(5600)}$$

$$\frac{1}{2} = e^{5600k}$$

$$5600k = \ln \frac{1}{2}$$

$$k = \frac{1}{5600} \ln \frac{1}{2} \approx -0.000124$$

Formula (3) therefore becomes

$$A = A_0 e^{-0.000124t}$$

If the amount A of carbon-14 now present is 1.67% of the original amount, it follows that

$$0.0167A_0 = A_0 e^{-0.000124t}$$
$$0.0167 = e^{-0.000124t}$$
$$-0.000124t = \ln 0.0167$$
$$t = \frac{1}{-0.000124} \ln 0.0167 \approx 33{,}000 \text{ years}$$

The tree was cut and burned about 33,000 years ago. Some archaeologists use this conclusion to argue that humans lived in the Americas 33,000 years ago, much earlier than is generally accepted.

(b) Figure 31 shows the graph of $y = e^{-0.000124x}$, where y in the fraction of carbon-14 present and x is the time.

FIGURE 31

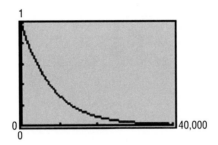

(c) Using BOX and TRACE, it takes 5600 years until half the carbon-14 remains.

(d) Using BOX and TRACE, it takes 33,000 years until 1.67% of the carbon-14 remains. ■

■ Now work Problem 3.

Newton's Law of Cooling

Newton's Law of Cooling* states that the temperature of a heated object decreases exponentially over time toward the temperature of the surrounding medium. That is, the temperature u of a heated object at a given time t satisfies the equation

Newton's Law of Cooling

$$u = T + (u_0 - T)e^{kt} \qquad k < 0 \tag{4}$$

where T is the constant temperature of the surrounding medium, u_0 is the initial temperature of the heated object, and k is a negative number.

E X A M P L E   3

Using Newton's Law of Cooling

An object is heated to 100°C and is then allowed to cool in a room whose air temperature is 30°C.

(a) If the temperature of the object is 80°C after 5 minutes, when will its temperature be 50°C?

(b) Using a graphing utility, graph the relation found between the temperature and time.

*Named after Sir Isaac Newton (1642–1727), one of the cofounders of calculus.

(c) Using BOX and TRACE, verify that when $x = 18.6$, then $y = 50$.

(d) Using BOX and TRACE, determine the elapsed time before the object is 35°C.

(e) TRACE the function until $x = 100$. What do you notice about y, the temperature?

Solution

(a) Using equation (4) with $T = 30$ and $u_0 = 100$, the temperature (in degrees Celsius) of the object at time t (in minutes) is

$$u = 30 + (100 - 30)e^{kt} = 30 + 70e^{kt} \tag{5}$$

where k is a negative number. To find k, we use the fact that $u = 80$ when $t = 5$. Then

$$80 = 30 + 70e^{k(5)}$$
$$50 = 70e^{5k}$$
$$e^{5k} = \frac{50}{70}$$
$$5k = \ln \frac{5}{7}$$
$$k = \frac{1}{5} \ln \frac{5}{7} \approx -0.0673$$

Formula (5) therefore becomes

$$u = 30 + 70e^{-0.0673t}$$

Now, we want to find t when $u = 50°C$, so

$$50 = 30 + 70e^{-0.0673t}$$
$$20 = 70e^{-0.0673t}$$
$$e^{-0.0673t} = \frac{20}{70}$$
$$-0.0673t = \ln \frac{2}{7}$$
$$t = \frac{1}{-0.0673} \ln \frac{2}{7} \approx 18.6 \text{ minutes}$$

Thus, the temperature of the object will be 50°C after about 18.6 minutes.

(b) Figure 32 shows the graph of $y = 30 + 70e^{-0.0673x}$, where y is the temperature and x is the time.

FIGURE 32

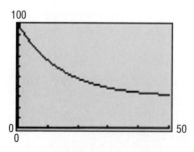

(c) Using BOX and TRACE, we find that $y = 50$ when $x = 18.6$.

(d) Using BOX and TRACE, $x = 39.21$ when $y = 35°C$.

(e) As x increases, y gets closer to 30, but never actually reaches 30. ∎

5.6

Exercise 5.6

1. *Growth of an Insect Population* The size P of a certain insect population at time t (in days) obeys the equation $P = 500e^{0.02t}$.
   (a) After how many days will the population reach 1000? 2000?
   (b) Using a graphing utility, graph the relation between P and t. Verify your answers in (a) using BOX and TRACE.

2. *Growth of Bacteria* The number N of bacteria present in a culture at time t (in hours) obeys the equation $N = 1000e^{0.01t}$.
   (a) After how many hours will the population equal 1500? 2000?
   (b) Using a graphing utility, graph the relation between N and t. Verify your answers in (a) using BOX and TRACE.

3. *Radioactive Decay* Strontium-90 is a radioactive material that decays according to the law $A = A_0e^{-0.0244t}$, where A_0 is the initial amount present and A is the amount present at time t (in years). What is the half-life of strontium-90?

4. *Radioactive Decay* Iodine-131 is a radioactive material that decays according to the law $A = A_0e^{-0.087t}$, where A_0 is the initial amount present and A is the amount present at time t (in days). What is the half-life of iodine-131?

5. (a) Use the information in Problem 3 to determine how long it takes for 100 grams of strontium-90 to decay to 10 grams.
   (b) Graph the relation $A = 100e^{-0.0244t}$ and verify your answer found in (a).

6. (a) Use the information in Problem 4 to determine how long it takes for 100 grams of iodine-131 to decay to 10 grams.
   (b) Graph the relation $A = 100e^{-0.087t}$ and verify your answer found in (a).

7. *Growth of a Colony of Mosquitoes* The population of a colony of mosquitoes obeys the law of uninhibited growth. If there are 1000 mosquitoes initially, and there are 1800 after 1 day, what is the size of the colony after 3 days? How long is it until there are 10,000 mosquitoes?

8. *Bacterial Growth* A culture of bacteria obeys the law of uninhibited growth. If 500 bacteria are present initially, and there are 800 after 1 hour, how many will be present in the culture after 5 hours? How long is it until there are 20,000 bacteria?

9. *Population Growth* The population of a southern city follows the exponential law. If the population doubled in size over an 18 month period and the current population is 10,000, what will the population be 2 years from now?

10. *Population Growth* The population of a midwestern city follows the exponential law. If the population decreased from 900,000 to 800,000 from 1993 to 1995, what will the population be in 1997?

11. *Radioactive Decay* The half-life of radium is 1690 years. If 10 grams are present now, how much will be present in 50 years?

12. *Radioactive Decay* The half-life of radioactive potassium is 1.3 billion years. If 10 grams are present now, how much will be present in 100 years? In 1000 years?

13. *Estimating the Age of A Tree* A piece of charcoal is found to contain 30% of the carbon-14 it originally had.
    (a) When did the tree from which the charcoal came die? Use 5600 years as the half-life of carbon-14.
    (b) Using a graphing utility, graph the relation between the percentage of carbon-14 remaining and time.
    (c) Using BOX and TRACE, determine the time that elapses until half of the carbon-14 remains.
    (d) Verify the answer found in (a).

14. *Estimating the Age of a Fossil* A fossilized leaf contains 70% of its normal amount of carbon-14.
    (a) How old is the fossil?
    (b) Using a graphing utility, graph the relation between the percentage of carbon-14 remaining and time.
    (c) Using BOX and TRACE, determine the time that elapses until half of the carbon-14 remains.
    (d) Verify the answer found in (a).

15. *Cooling Time of a Pizza* A pizza baked at 450°F is removed from the oven at 5:00 PM into a room that is a constant 70°F. After 5 minutes, the pizza is at 300°F.
    (a) At what time can you begin eating the pizza if you want its temperature to be 135°F?
    (b) Using a graphing utility, graph the relation between temperature and time.
    (c) Using BOX and TRACE, determine the time needed to elapse before the pizza is 160°F.
    (d) TRACE the function for large values of time. What do you notice about y, the temperature?

16. A thermometer reading 72°F is placed in a refrigerator where the temperature is a constant 38°F. If the thermometer reads 60°F after 2 minutes, what will it read after 7 minutes?
    (a) How long will it take before the thermometer reads 39°F?
    (b) Using a graphing utility, graph the relation between temperature and time.
    (c) Using BOX and TRACE, determine the time needed to elapse before the thermometer reads 45°F.
    (d) TRACE the function for large values of time. What do you notice about y, the temperature?

17. A thermometer reading 8°C is brought into a room with a constant temperature of 35°C. If the thermometer reads 15°C after 3 minutes, what will it read after being in the room for 5 minutes? For 10 minutes? [*Hint:* You need to construct a formula similar to equation (4).] Graph the relation between temperature and time. TRACE to verify your answer is correct.

18. *Thawing Time of a Steak* A frozen steak has a temperature of 28°F. It is placed in a room with a constant temperature of 70°F. After 10 minutes, the temperature of the steak has risen to 35°F. What will the temperature of the steak be after 30 minutes? How long will it take the steak to thaw to a temperature of 45°F? [See the hint given for Problem 17.] Graph the relation between temperature and time. TRACE to verify your answer is correct.

19. *Decomposition of Salt in Water* Salt (NaCl) decomposes in water into sodium (NA^+) and chloride (Cl^-) ions according to the law of uninhibited decay. If the initial amount of salt is 25 kilograms and, after 10 hours, 15 kilograms of salt are left, how much salt is left after 1 day? How long does it take until $\frac{1}{2}$ kilogram of salt is left?

20. *Voltage of a Condenser* The voltage of a certain condenser decreases over time according to the law of uninhibited decay. If the initial voltage is 40 volts, and 2 seconds later it is 10 volts, what is the voltage after 5 seconds?

21. *Radioactivity from Chernobyl* After the release of radioactive material into the atmosphere from a nuclear power plant at Chernobyl (Ukraine) in 1986, the hay in Austria was contaminated by iodine-131 (see Problem 4). If it is all right to feed the hay to cows when 10% of the iodine-131 remains, how long do the farmers need to wait to use this hay?

22. *Pig Roasts* The hotel Bora-Bora is having a pig roast. At noon, the chef put the pig in a large earthen oven. The pig's original temperature was 75°F. At 2:00 PM, the chef checked the pig's temperature and was upset because it had reached only 100°F. If the oven's temperature remains a constant 325°F, at what time may the hotel serve its guests, assuming that pork is done when it reaches 175°F?

5.7

Nonlinear Curve Fitting

In Section 1.5 we discussed how to find an equation that describes the relation that exists between two variables using data. We assumed there was a linear relationship of the form $y = ax + b$, where x is the independent variable and y is the dependent variable.

Many real-world phenomena, however, do not follow a linear relationship. For example, we saw in Section 5.6 that the uninhibited growth of bacteria follows the model $N(t) = N_0 e^{kt}$, $k > 0$. Also radioactive decay follows the model $A(t) = A_0 e^{kt}$, $k < 0$. In this section we will discuss how to use a graphing utility to find equations that describe the relation between two variables when the relation is nonlinear.

Three types of nonlinear models are popular:

1. Exponential $y = ab^x$
2. Power $y = ax^b$
3. Logarithmic $y = a + b \ln x$

Figure 33 shows scatter diagrams that typically will be observed for the three nonlinear models for different values of the parameters, a and b. Below each scatter diagram are any restrictions on the values of the independent or dependent variable.

Determining the appropriate model to use requires statistical techniques that are beyond the scope of this text. However, there are some rudimentary techniques

FIGURE 33

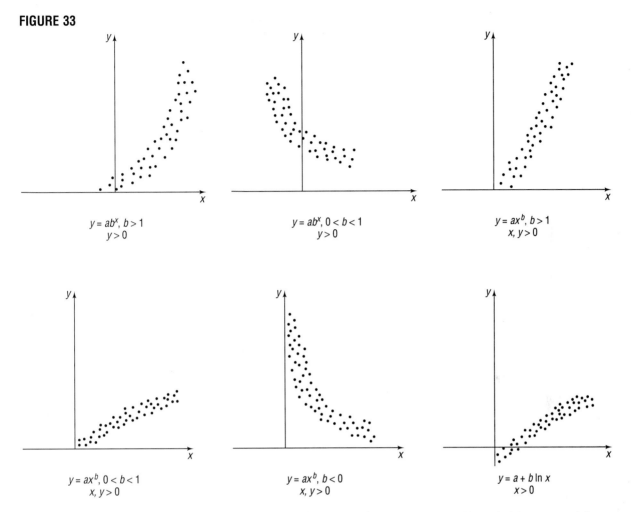

we can use that will allow us to determine which model (exponential, power, or logarithmic) is most appropriate. Before we do this, however, we need to discuss how graphing utilities are used to fit curves to data.

Most graphing utilities have REGression options that fit data to a specific type of curve. Once the data have been entered and a scatter diagram obtained, the type of curve you want to fit to the data is selected. Then that REGression option is used to obtain the curve of "best fit" of the type selected. The number r that appears measures how good the fit is. The closer $|r|$ is to 1, the better the fit.

Let's look at some examples.

Exponential Curve Fitting

As we saw in Section 5.6, many growth and decay models are exponential.

E X A M P L E   1   *Fitting a Curve to an Exponential Model*

A researcher is interested in finding a model that explains the relation between the amount of an unknown radioactive material and time. She obtains 100 grams of the radioactive material and every day for 9 days measures the amount of radioactive material in the sample and obtains the following data:

DAY	WEIGHT (IN GRAMS)
0	100
1	91.6
2	84.1
3	74.2
4	68.3
5	63.7
6	58.5
7	53.8
8	50.2
9	47.3

(a) Draw a scatter diagram.

(b) Fit an exponential curve to the data.

(c) Express the curve in the form $A = A_o e^{kt}$.

(d) Using the solution to (c), estimate the half-life of the unknown radioactive material.

(e) Predict how much of the material will remain after 30 days.

Solution   (a) After entering the data into the graphing utility, we obtain the scatter diagram shown in Figure 34.

FIGURE 34

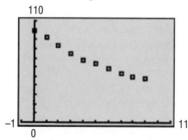

(b) A graphing utility fits the data in Figure 34 to an exponential equation of the form $y = ab^x$ by using the EXPonential REGression option.* See Figure 35.

FIGURE 35

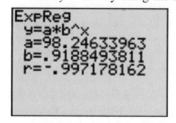

*If your utility does not have such an option but does have a LINear REGressin option, you can transform the exponential model to a linear model by the following technique:

$$y = ab^x \qquad \text{exponential form}$$
$$\ln y = \ln(ab^x) \qquad \text{take natural logs of each side}$$
$$\ln y = \ln a + \ln b^x \qquad \text{property of logarithms}$$
$$\ln y = \ln a + (\ln b)x \qquad \text{property of logarithms}$$

Now apply the LINear REGression techniques discussed in Chapter 1. Be careful though. The dependent variable is now $\ln y$, while the independent variable remains x. Once the line of best fit is obtained, you can find a and b and obtain the exponential curve $y = ab^x$.

The exponential equation of best fit to the data is

$$y = 98.24633963(0.9188493811)^x$$

Notice that $r = -0.997178162$, so $|r|$ is close to 1, indicating a good fit.

(c) To express $y = ab^x$ in the form $A = A_o e^{kt}$, we proceed as follows.

$$ab^x = A_o e^{kt}$$

$$a = A_o \qquad b^x = e^{kt} = (e^k)^t$$

$$b = e^k$$

Thus,

$$A_o = 98.2463393 \text{ and } 0.9188493811 = e^k$$

$$k = -0.0846331$$

As a result, $A = A_o e^{kt} = 98.24633963 e^{-0.0846331t}$

(d) We set $A = \frac{1}{2}A_o$ and solve the equation $\frac{1}{2}A_o = A_o e^{-0.0846331t}$ for t. The result is $t = 8.2$ days.

(e) After $t = 30$ days, the amount A of the radioactive material left is

$$A = 98.24633963 e^{-0.0846331(30)} = 7.8 \text{ grams.} \qquad \blacksquare$$

Exploration: Graph the equation found in (b) on the same screen as the scatter diagram. Does the 'fit' seem good? ◼

Logarithmic Curve Fitting

Many models do not follow uninhibited exponential growth but instead, as the independent variable increases, the growth of the dependent variable diminishes. Under these circumstances, a logarithmic model may be appropriate.

E X A M P L E   2

Fitting a Curve to a Logarithmic Model

The following data, obtained from the U.S. Census Bureau, represent the population of New York State over time. An urban economist is interested in finding a model that describes the population of New York over time.

YEAR	POPULATION
1900	7,268,894
1910	9,113,614
1920	10,385,227
1930	12,588,066
1940	13,479,142
1950	14,830,192
1960	16,782,304
1970	18,241,391
1980	17,558,165
1990	17,990,455

(a) Using a graphing utility, draw a scatter diagram of the data with time as the independent variable.

(b) Using a graphing utility, fit a logarithmic growth model to the data.

(c) Use the model found in (b) to predict the population of New York State in 1995.

Solution   (a) After entering the data into the graphing utility, we obtain the scatter diagram shown in Figure 36. Notice the data appear to follow a logarithmic growth model.

FIGURE 36

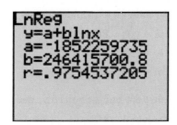

(b) A graphing utility fits the data in Figure 36 to a logarithmic equation of the form $y = a + b \ln x$ by using the Logarithm REGression option. See Figure 37.

FIGURE 37

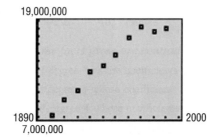

The logarithm equation of best fit to the data is

$$y = -1{,}852{,}259{,}735 + 246{,}415{,}700.8 \ln x$$

where y is the state's population and x is the year.

Notice that the value of r is close to 1, indicating a good fit.

(c) Using the equation found in (b), we predict the population of New York State in 1995 to be

$$y = -1{,}852{,}259{,}735 + 246{,}415{,}700.8 \ln(1995)$$
$$\approx 20{,}105{,}161$$ ∎

Power Curve Fitting

If the dependent variable is proportional to a power of the independent variable, then a power curve of the form

$$y = ax^b$$

where y is the dependent variable and x is the independent variable may be the appropriate model.

E X A M P L E   3

Fitting a Curve to a Power Function

The *period* of a pendulum is the time required for one oscillation; the pendulum is usually referred to as *simple* when the angle made to the vertical is less than 5°. An experiment is conducted in which simple pendulums are constructed with different lengths, *l*, and the corresponding period *T* is recorded. The following data are collected:

LENGTH *l* (IN FEET)	PERIOD *T* (IN SECONDS)
1	1.1
2	1.5
3	1.8
4	2.1
5	2.5
6	2.8
7	3.0

(a) Using a graphing utility, draw a scatter diagram of the data with length as the independent variable and the period as the dependent variable.

(b) Using a graphing utility, fit a power model to the data.

(c) If the period of a simple pendulum is known to be 2.3 seconds, what is the length of the pendulum?

Solution   (a) After entering the data into the graphing utility, we obtain the scatter diagram shown in Figure 38.

FIGURE 38

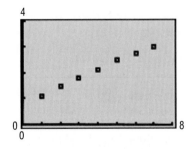

(b) A graphing utility fits the data in Figure 38 to a power curve of the form $y = ax^b$ by using the PoWeR REGression option. See Figure 39.

FIGURE 39

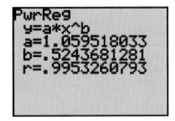

The period *T* of a simple pendulum is related to its length *l* by the power function

$$T = 1.0595 \, l^{0.524}$$

Notice that the value of r is close to l, indicating a good fit.*

(c) The length l of string corresponding to a period of 2.3 seconds obeys

$$2.3 = 1.0595 \, l^{0.524}$$
$$l = 4.39 \text{ feet} \qquad \blacksquare$$

Choosing the Best Model

We have just discussed three different nonlinear models that can be used to explain the relation between two variables. However, how can we be certain the model we used is the "best" model to explain this relation? For example, why did we use an exponential model in Example 1 and a power model in Example 3? Unfortunately, there is no such thing as *the* correct model. Modeling is not only a science but also an art form. Selecting an appropriate model requires experience and skill in the field in which you are modeling. For example, knowledge of economics is imperative when trying to determine a model to predict unemployment. The main reason for this is that there are theories in the field that can help the modeler select appropriate relations and variables. Nonetheless, there are two important tools that can be used to help determine the "best" model: scatter diagrams and the number, r. For our purposes, we will say the "best" model is the one which yields the value r for which $|r|$ is closest to 1.

When there is no theoretical basis for choosing a particular model, the best approach to follow is to first draw a scatter diagram and obtain a general idea of the shape of the data. Then fit several models to the data and determine which seems to explain the relation between the two variables best. Again, we will define "best" as the model that yields the value of $|r|$ closest to 1.

EXAMPLE 4   *Selecting the Best Model*

Suppose an astronomer claims to have discovered a tenth planet that is 4.2 billion miles from the sun. How long would it take for this planet to completely orbit the sun?

Solution   In order to solve this problem, we need to develop a model that relates the distance a planet is from the sun and the time it takes to make a complete orbit. This time is called the *sidereal year*. The data in the following table show the distances the planets are from the sun and their sidereal years.

PLANET	DISTANCE FROM SUN (MILLIONS OF MILES)	SIDEREAL YEAR
Mercury	36	0.24
Venus	67	0.62
Earth	93	1.00
Mars	142	1.88
Jupiter	483	11.9
Saturn	887	29.5
Uranus	1,785	84.0
Neptune	2,797	165.0
Pluto	3,675	248.0

*In physics, it is proven that $T = \frac{2\pi}{\sqrt{32}} \sqrt{l}$, provided friction is ignored. Since $\sqrt{l} = l^{1/2} = l^{0.5}$ and $\frac{2\pi}{\sqrt{32}} = 1.1107$, the power function obtained is reasonably close to what is expected.

The first step in determining the best model is to draw a scatter diagram. Figure 40 shows a scatter diagram for the planet data.

FIGURE 40

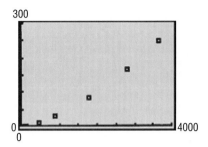

From the scatter diagram in Figure 40 it appears the data follow either a linear model, an exponential model ($k > 0$), or a power model ($b > 1$). Therefore, we shall find equations for all three models and use the value of r to determine which is best. Figure 41(a) shows the results obtained from the graphing utility for an exponential model, Figure 41(b) shows the results for a power model, and Figure 41(c) shows the results for a linear model.

FIGURE 41

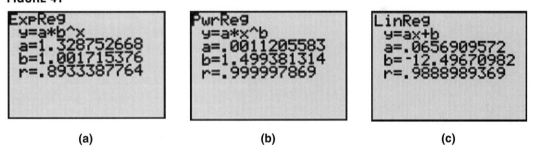

| (a) | (b) | (c) |

By looking at the value of r, we conclude that the power model is superior to both the exponential model and the linear model. Thus, we conclude the relation between the distance a planet is from the sun (x) and its sidereal year (y) is $y = 0.00112x^{1.499938}$.*

We can predict the sidereal year of the newly discovered planet by substituting $x = 4200$ into the equation just found:

$$y = 0.00112(4200)^{1.499938}$$
$$\approx 304.7 \text{ sidereal years}$$

---

*If we square both sides of this equation we obtain

$$y^2 \approx 0.0000012x^3$$

which implies the square of the sidereal year is proportional to the cube of the distance from the sun to the planet. This is Kepler's Third Law, named after Johann Kepler. Kepler discovered this relation in 1618 through experimentation without the aid of data analysis.

5.7

Exercise 5.7

1. *Biology* A certain bacteria increases according to the law of uninhibited growth. A biologist collects the following data for this bacteria:

TIME (HOURS)	POPULATION
0	1000
1	1415
2	2000
3	2828
4	4000
5	5656
6	8000

(a) Draw a scatter diagram.

(b) Fit an exponential curve to the data.

(c) Express the curve in the form $N = N_o e^{kt}$.

(d) Using the solution to (b), predict the population at $t = 7$ hours.

2. *Biology* A colony of bacteria increases according to the law of uninhibited growth. A biologist collects the following data for this bacteria:

TIME (DAYS)	POPULATION
0	50
1	153
2	234
3	357
4	547
5	839
6	1280

(a) Draw a scatter diagram.

(b) Fit an exponential curve to the data.

(c) Express the curve in the form $N = N_o e^{kt}$.

(d) Using the solution to (b), predict the population at $t = 7$ days.

3. *Economics/Marketing* A store manager collected the following data regarding price and quantity sold of a product:

PRICE ($/UNIT)	QUANTITY DEMANDED
79	10
67	20
54	30
46	40
38	50
31	60

(a) Draw a scatter diagram with quantity demanded on the x-axis and price on the y-axis.

(b) Fit an exponential curve to the data.

(c) Predict the quantity demanded if the price is $60.

4. *Economics/Marketing* A store manager collected the following data regarding price and quantity supplied of a product:

PRICE ($/UNIT)	QUANTITY SUPPLIED
25	10
32	20
40	30
46	40
60	50
74	60

(a) Draw a scatter diagram with quantity supplied on the *x*-axis and price on the *y*-axis.

(b) Fit an exponential curve to the data.

(c) Predict the quantity demanded if the price is $45.

5. *Chemistry* A chemist has a 100-gram sample of a radioactive material. He records the amount of radioactive material every week for 6 weeks and obtains the following data:

WEEK	WEIGHT (IN GRAMS)
0	100.0
1	88.3
2	75.9
3	69.4
4	59.1
5	51.8
6	45.5

(a) Draw a scatter diagram with week as the independent variable.

(b) Fit an exponential curve to the data.

(c) Express the curve in the form $A = A_o e^{kt}$.

(d) From the result found in (b), determine the half-life of the radioactive material.

(e) How much radioactive material will be left after 50 weeks?

6. *Chemistry* A chemist has a 1000-gram sample of a radioactive material. She records the amount of radioactive material remaining in the sample every day for a week and obtains the following data:

DAY	WEIGHT (IN GRAMS)
0	1000.0
1	897.1
2	802.5
3	719.8
4	651.1
5	583.4
6	521.7
7	468.3

(a) Draw a scatter diagram with day as the independent variable.

(b) Fit an exponential curve to the data.

(c) Express the curve in the form $A = A_o e^{kt}$.

(d) From the result found in (b), find the half-life of the radioactive material.

(e) How much radioactive material will be left after 20 days?

7. *Finance* The following data represent the amount of money an investor has in an investment account each year for 10 years. She wishes to determine the average annual rate of return on her investment.

YEAR	VALUE OF ACCOUNT
1985	$10,000
1986	$10,573
1987	$11,260
1988	$11,733
1989	$12,424
1990	$13,269
1991	$13,968
1992	$14,823
1993	$15,297
1994	$16,539

(a) Draw a scatter diagram with time as the independent variable and the value of the account as the dependent variable.
(b) Fit an exponential curve to the data.
(c) Based on the answer to (b), what was the average annual rate of return from this account over the past 10 years?
(d) If the investor plans on retiring in 2020, what will the predicted value of this account be?

8. *Finance* The following data show the amount of money an investor has in an investment account each year for 7 years. He wishes to determine the average annual rate of return on his investment.

YEAR	VALUE OF ACCOUNT
1988	$20,000
1989	$21,516
1990	$23,355
1991	$24,885
1992	$27,434
1993	$30,053
1994	$32,622

(a) Draw a scatter diagram with time as the independent variable and the value of the account as the dependent variable.
(b) Fit an exponential curve to the data.
(c) Based on the answer to (b), what was the average annual rate of return from this account over the past 7 years?
(d) If the investor plans on retiring in 2020, what will the predicted value of this account be?

9. *CBL Experiment* The following data were collected by placing a probe in a cup of hot water, removing the probe, and then recording temperature over time:

TIME	TEMPERATURE (°F)	TIME	TEMPERATURE (°F)
0	175.69	13	148.50
1	173.52	14	146.84
2	171.21	15	145.33
3	169.07	16	143.83
4	166.59	17	142.38
5	164.21	18	141.22
6	161.89	19	140.09
7	159.66	20	138.69
8	157.86	21	137.59
9	155.75	22	136.78
10	153.70	23	135.70
11	151.93	24	134.91
12	150.08	25	133.86

According to Newton's Law of Cooling, these data should follow an exponential model.

(a) Using a graphing utility, draw a scatter diagram for the data.
(b) Fit an exponential model to the data.
(c) Predict how long it will take for the water to reach a temperature of 110°F.

10. *CBL Experiment* The following data were collected by placing a probe in a portable heater, removing the probe, and then recording temperature over time:

TIME	TEMPERATURE (°F)	TIME	TEMPERATURE (°F)
0	165.07	8	159.35
1	164.77	9	158.61
2	163.99	10	157.89
3	163.22	11	156.83
4	162.82	12	156.11
5	161.96	13	155.08
6	161.20	14	154.40
7	160.45	15	153.72

According to Newton's Law of Cooling, these data should follow an exponential model.

(a) Using a graphing utility, draw a scatter diagram for the data.

(b) Fit an exponential model to the data.

(c) Predict how long it will take for the air blown out of the heater to reach a temperature of 110°F.

11. *Population Model* The following data obtained from the U.S. Census Bureau represent the population of Illinois. An urban economist is interested in finding a model that describes the population of Illinois.

YEAR	POPULATION
1900	4,821,550
1910	5,638,591
1920	6,485,280
1930	7,630,654
1940	7,897,241
1950	8,712,176
1960	10,081,158
1970	11,110,285
1980	11,427,409
1990	11,430,602

(a) Using a graphing utility, draw a scatter diagram of the data with time as the independent variable.

(b) Using a graphing utility, fit a logarithmic growth model to the data.

(c) Use the model found in (b) to predict the population of Illinois in 1995.

12. *Population Model* The following data obtained from the U.S. Census Bureau represent the population of Pennsylvania. An urban economist is interested in finding a model that describes the population of Pennsylvania.

YEAR	POPULATION
1900	6,302,115
1910	7,665,111
1920	8,720,017
1930	9,631,350
1940	9,900,180
1950	10,498,012
1960	11,319,366
1970	11,800,766
1980	11,864,720
1990	11,881,643

(a) Using a graphing utility, draw a scatter diagram of the data with time as the independent variable.

(b) Using a graphing utility, fit a logarithmic growth model to the data.

(c) Use the model found in (b) to predict the population of Pennsylvania in 1995.

13. *CBL Experiment* Yolanda is conducting an experiment to measure the relation between a light bulb's intensity and the distance from the light source. She measures a 100-watt light bulb's intensity 1 meter from the bulb and at 0.1-meter intervals up to 2 meters from the bulb and obtains the following data.

DISTANCE (METERS)	INTENSITY
1.0	0.29645
1.1	0.25215
1.2	0.20547
1.3	0.17462
1.4	0.15342
1.5	0.13521
1.6	0.11450
1.7	0.10243
1.8	0.09231
1.9	0.08321
2.0	0.07342

(a) Using your graphing utility, draw a scatter diagram with distance as the independent variable and intensity as the dependent variable.
(b) Using your graphing utility, fit a power model to the data.
(c) It is known that intensity I is inversely proportional to the square of the distance x, so that $I = \dfrac{a}{x^2}$. How close is the estimate of the exponent?
(d) What will the intensity of a 100-watt light bulb be if you stand 2.3 meters away?

14. *CBL Experiment* Yolanda repeats the experiment from Problem 13, but this time uses a 40-watt light bulb and obtains the following data.

DISTANCE (METERS)	INTENSITY
1.0	0.0972
1.1	0.0804
1.2	0.0674
1.3	0.0572
1.4	0.0495
1.5	0.0433
1.6	0.0384
1.7	0.0339
1.8	0.0294
1.9	0.0268
2.0	0.0224

(a) Using your graphing utility, draw a scatter diagram with distance as the independent variable and intensity as the dependent variable.
(b) Using your graphing utility, fit a power model to the data.
(c) What will the intensity of a 40-watt light bulb be if you stand 2.3 meters away?

15. *CBL Experiment* Jim is conducting an experiment in order to estimate the acceleration of an object due to gravity. Jim takes a ball and drops it from different heights and records the time it takes for the ball to hit the ground. Using an optic laser connected to a stop watch in order to determine the time, he collects the following data:

TIME (SECONDS)	DISTANCE (FEET)
1.003	16
1.365	30
1.769	50
2.093	70
2.238	80

(a) Draw a scatter diagram using time as the independent variable and distance as the dependent variable.
(b) Fit a power model to the data.
(c) Physics theory states the distance an object falls is directly proportional to the time squared. Rewrite the model found in (b) so it is of the form $s = \frac{1}{2}gt^2$, where s is distance, g is the acceleration due to gravity, and t is time. What is Jim's estimate of g? (It is known that the acceleration due to gravity is approximately 32 feet/sec².)
(d) Predict how long it will take an object to fall 100 feet.

16. *CBL Experiment* Bill, Jim's friend, doesn't believe Jim's estimate of acceleration due to gravity is correct. Bill repeats the experiment described in Problem 15 and obtains the following data:

TIME (SECONDS)	DISTANCE (FEET)
0.7907	10
1.1160	20
1.4760	35
1.6780	45
1.9380	60

(a) Draw a scatter diagram using time as the dependent variable and distance as the independent variable.
(b) Fit a power model to the data.
(c) Physics theory states the distance an object falls is directly proportional to the time squared. Rewrite the model found in (b) so it is of the form $s = \frac{1}{2}gt^2$, where s is distance, g is the acceleration due to gravity, and t is time. What is Bill's estimate of g? [It is known that the acceleration due to gravity is approximately 32 feet/sec².]
(d) Predict how long it will take an object to fall 100 feet. Why do you think this estimate is different than the one from Problem 15?

For Problems 17–22 the independent variable is x and the dependent variable is y. (a) For each set of data draw a scatter diagram of the data. (b) Use an exponential model, logarithmic model, power model, and linear model to find the "best" model that explains the relationship between the variables.

17.

x	1	2	3	4	5
y	164	269	450	740	1220

18.

x	1	2	3	4	5
y	110	245	553	1228	2728

19.

x	10	20	30	40	50
y	6.94	13.78	20.45	21.23	22.95

20.

x	10	20	30	40	50
y	4.95	6.46	7.23	7.78	8.23

21.

x	1	1.3	1.7	2.1	2.4	2.7
y	9.92	4.56	1.42	0.37	0.12	0.032

22.

x	1	1.3	1.7	2.1	2.4	2.7
y	1.72	1.94	2.48	3.28	4.01	5.07

23. *Economics* The following data are levels of the Consumer Price Index for food items (Source: Bureau of Labor Statistics).

YEAR	CPI
1984	103.2
1985	105.6
1986	109.0
1987	113.5
1988	118.2
1989	125.1
1990	132.4
1991	136.3
1992	137.9
1993	140.9

(a) Using a graphing utility, draw a scatter diagram of the data using time as the independent variable and the CPI as the dependent variable.
(b) Use an exponential model, logarithmic model, power model, and linear model to find the "best" model to describe the relation between time and the CPI.
(c) Use this model to predict the CPI for food items for 1994.

24. *Economics* The following data are levels of the Consumer Price Index for electricity (Source: Bureau of Labor Statistics).

YEAR	CPI
1984	105.3
1985	108.9
1986	110.4
1987	110.0
1988	111.5
1989	114.7
1990	117.4
1991	121.8
1992	124.2
1993	126.7

(a) Using a graphing utility, draw a scatter diagram of the data using time as the independent variable and the CPI as the dependent variable.

(b) Use an exponential model, logarithmic model, power model, and linear model to find the "best" model to describe the relation between time and the CPI.

(c) Use this model to predict the CPI for electricity for 1994.

25. *Economics* The following data represent the per capita gross domestic product (GDP) in constant 1987 dollars (this removes the effects of inflation). (Source: U.S. Bureau of Economic Analysis)

YEAR	PER CAPITA GROSS DOMESTIC PRODUCT (1987 DOLLARS)
1988	19,252
1989	19,556
1990	19,593
1991	19,238
1992	19,518
1993	19,888

(a) Using a graphing utility, draw a scatter diagram of the data using time as the independent variable and the per capita GDP as the dependent variable.

(b) Use an exponential model, logarithmic model, power model, and linear model to find the "best" model to describe the relation between time and the per capita GDP.

(c) Use this model to predict per capita GDP for 1994.

26. *Economics* The following data represent the per capita gross national product (GNP) in constant 1987 dollars (this removes the effects of inflation). (Source: U.S. Bureau of Economic Analysis)

YEAR	PER CAPITA GROSS NATIONAL PRODUCT (1987 DOLLARS)
1988	19,284
1989	19,615
1990	19,670
1991	19,290
1992	19,548
1993	19,897

(a) Using a graphing utility, draw a scatter diagram of the data using time as the independent variable and the GNP as the dependent variable.

(b) Use an exponential model, logarithmic model, power model, and linear model to find the "best" model to describe the relation between time and the per capita GNP.

(c) Use this model to predict the per capita GNP for 1994.

27. *Population Model* The following data represent the population of Florida. (Source: US Census Bureau)

YEAR	POPULATION
1900	528,542
1910	752,619
1920	968,470
1930	1,468,211
1940	1,897,414
1950	2,771,305
1960	4,951,560
1970	6,791,418
1980	9,746,961
1990	12,937,926

(a) Using a graphing utility, draw a scatter diagram using time as the independent variable and population as the dependent variable.

(b) Use an exponential model, logarithmic model, power model, and linear model to find the "best" model that describes the relation between time and population.

(c) Use the model found in (b) to predict the population of Florida in 1995.

28. *Population Model* The following data represent the population of Arizona. (Source: US Census Bureau)

YEAR	POPULATION
1900	122,931
1910	204,354
1920	334,162
1930	435,573
1940	499,261
1950	749,587
1960	1,302,161
1970	1,775,399
1980	2,716,546
1990	3,665,228

(a) Using a graphing utility, draw a scatter diagram using time as the independent variable and population as the dependent variable.

(b) Use an exponential model, logarithmic model, power model, and linear model to find the "best" model that describes the relation between time and population.

(c) Use the model found in (b) to predict the population of Arizona in 1995.

5.8

Logarithmic Scales

Common logarithms often appear in the measurement of quantities, because they provide a way to scale down positive numbers that vary from very small to very large. For example, if a certain quantity can take on values from $0.0000000001 = 10^{-10}$ to $10,000,000,000 = 10^{10}$, the common logarithms of such numbers would be between -10 and 10.

Loudness of Sound

Our first application utilizes a logarithmic scale to measure the loudness of a sound. Physicists define the **intensity of a sound wave** as the amount of energy the wave transmits through a given area. For example, the least intense sound that a human ear can detect at a frequency of 100 hertz is about 10^{-12} watt per square meter. The **loudness** $L(x)$, measured in **decibels** (named in honor of Alexander Graham Bell), of a sound of intensity x (measured in watts per square meter) is defined as

Loudness

$$L(x) = 10 \log \frac{x}{I_0} \qquad (1)$$

where $I_0 = 10^{-12}$ watt per square meter is the least intense sound that a human ear can detect. If we let $x = I_0$ in equation (1), we get

$$L(I_0) = 10 \log \frac{I_0}{I_0} = 10 \log 1 = 0$$

Thus, at the threshold of human hearing, the loudness is 0 decibels. Figure 42 gives the loudness of some common sounds.

FIGURE 42

Loudness of common sounds (in decibels)

DECIBELS

140	Shotgun blast, jet 100 feet away at takeoff	Pain
130	Motor test chamber	Human ear pain threshold
120	Firecrackers, severe thunder, pneumatic jackhammer, hockey crowd	Uncomfortably loud
110	Amplified rock music	
100	Textile loom, subway train, elevated train, farm tractor, power lawn mower, newspaper press	Loud
90	Heavy city traffic, noisy factory	
80	Diesel truck going 40 mi/hr 50 feet away, crowded restaurant, garbage disposal, average factory, vacuum cleaner	Moderately loud
70	Passenger car going 50 mi/hr 50 feet away	
60	Quiet typewriter, singing birds, window air conditioner, quiet automobile	Quiet
50	Normal conversation, average office	
40	Household refrigerator, quiet office	Very quiet
30	Average home, dripping faucet, whisper 5 feet away	
20	Light rainfall, rustle of leaves	Average person's threshold of hearing
10	Whisper across room	Just audible
0		Threshold for acute hearing

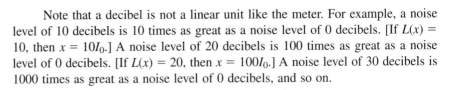

Note that a decibel is not a linear unit like the meter. For example, a noise level of 10 decibels is 10 times as great as a noise level of 0 decibels. [If $L(x) = 10$, then $x = 10I_0$.] A noise level of 20 decibels is 100 times as great as a noise level of 0 decibels. [If $L(x) = 20$, then $x = 100I_0$.] A noise level of 30 decibels is 1000 times as great as a noise level of 0 decibels, and so on.

EXAMPLE 1

Finding the Intensity of a Sound

Use Figure 42 to find the intensity of the sound of a dripping faucet.

Solution

From Figure 42, we see that the loudness of the sound of dripping water is 30 decibels. Thus, by equation (1), its intensity x may be found as follows:

$$30 = 10 \log\left(\frac{x}{I_0}\right)$$

$$3 = \log\left(\frac{x}{I_0}\right) \qquad \text{Divide by 10.}$$

$$\frac{x}{I_0} = 10^3 \qquad \text{Write in exponential form.}$$

$$x = 1000 I_0$$

where $I_0 = 10^{-12}$ watt per square meter. Thus, the intensity of the sound of a dripping faucet is 1000 times as great as a noise level of 0 decibels; that is, such a sound has an intensity of $1000 \cdot 10^{-12} = 10^{-9}$ watt per square meter. ∎

    ■   Now work Problem 5.

E X A M P L E 2   *Finding the Loudness of a Sound*

Use Figure 42 to determine the loudness of a subway train if it is known that this sound is 10 times as intense as the sound due to heavy city traffic.

Solution   The sound due to heavy city traffic has a loudness of 90 decibels. Its intensity, therefore, is the value of x in the equation

$$90 = 10 \log\left(\frac{x}{I_0}\right)$$

A sound 10 times as intense as x has loudness $L(10x)$. Thus, the loudness of the subway train is

$$L(10x) = 10 \log\left(\frac{10x}{I_0}\right) \qquad \text{Replace } x \text{ by } 10x.$$

$$= 10 \log\left(10 \cdot \frac{x}{I_0}\right)$$

$$= 10 \left[\log 10 + \log\left(\frac{x}{I_0}\right)\right] \qquad \text{Log of product = Sum of logs}$$

$$= 10 \log 10 + 10 \log\left(\frac{x}{I_0}\right) \qquad \log 10 = 1$$

$$= 10 + 90 = 100 \text{ decibels} \qquad\qquad ∎$$

Magnitude of an Earthquake

Our second application uses a logarithmic scale to measure the magnitude of an earthquake.

    The **Richter scale*** is one way of converting seismographic readings into numbers that provide an easy reference for measuring the magnitude M of an earthquake. All earthquakes are compared to a **zero-level earthquake** whose seismographic reading measures 0.001 millimeter at a distance of 100 kilometers from the epicenter. An earthquake whose seismographic reading measures x millimeters has **magnitude $M(x)$** given by

Magnitude of an Earthquake

$$M(x) = \log\left(\frac{x}{x_0}\right) \qquad\qquad (2)$$

where $x_0 = 10^{-3}$ is the reading of a zero-level earthquake the same distance from its epicenter.

E X A M P L E 3   *Finding the Magnitude of an Earthquake*

What is the magnitude of an earthquake whose seismographic reading is 0.1 millimeter at a distance of 100 kilometers from its epicenter?

---

*Named after the American scientist, C.F. Richter, who devised it in 1935.

Solution   If $x = 0.1$, the magnitude $M(x)$ of this earthquake is

$$M(0.1) = \log\left(\frac{x}{x_0}\right) = \log\left(\frac{0.1}{0.001}\right) = \log\left(\frac{10^{-1}}{10^{-3}}\right) = \log 10^2 = 2$$

This earthquake thus measures 2.0 on the Richter scale.   ■

■   Now work Problem 7.

Based on formula (2), we define the **intensity of an earthquake** as the ratio of x to x_0. For example, the intensity of the earthquake described in Example 3 is $\frac{0.1}{0.001} = 10^2 = 100$. That is, it is 100 times as intense as a zero-level earthquake.

E X A M P L E   4   *Comparing the Intensity of Two Earthquakes*

The devastating San Francisco earthquake of 1906 measured 8.9 on the Richter scale. How did the intensity of that earthquake compare to the Papua New Guinea earthquake of 1988, which measured 6.7 on the Richter scale?

FIGURE 43

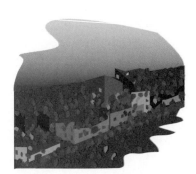

Solution   Let x_1 and x_2 denote the seismographic readings, respectively, of the 1906 San Francisco earthquake and the Papua New Guinea earthquake. Then, based on formula (2),

$$8.9 = \log\left(\frac{x_1}{x_0}\right) \qquad 6.7 = \log\left(\frac{x_2}{x_0}\right)$$

Consequently,

$$\frac{x_1}{x_0} = 10^{8.9} \qquad \frac{x_2}{x_0} = 10^{6.7}$$

The 1906 San Francisco earthquake was $10^{8.9}$ times as intense as a zero-level earthquake. The Papua New Guinea earthquake was $10^{6.7}$ times as intense as a zero-level earthquake. Thus,

$$\frac{x_1}{x_2} = \frac{10^{8.9}x_0}{10^{6.7}x_0} = 10^{2.2} \approx 158$$

$$x_1 \approx 158x_2$$

Hence, the San Francisco earthquake was 158 times as intense as the Papua New Guinea earthquake.   ■

Example 4 demonstrates that the relative intensity of two earthquakes can be found by raising 10 to a power equal to the difference of their readings on the Richter scale.

5.8

Exercise 5.8

1. *Loudness of a Dishwasher* Find the loudness of a dishwasher that operates at an intensity of 10^{-5} watt per square meter. Express your answer in decibels.

2. *Loudness of a Diesel Engine* Find the loudness of a diesel engine that operates at an intensity of 10^{-3} watt per square meter. Express your answer in decibels.

3. *Loudness of a Jet Engine* With engines at full throttle, a Boeing 727 jetliner produces noise at an intensity of 0.15 watt per square meter. Find the loudness of the engines in decibels.

4. *Loudness of a Whisper* A whisper produces noise at an intensity of $10^{-9.8}$ watt per square meter. What is the loudness of a whisper in decibels?

5. *Intensity of a Sound at Threshold of Pain* For humans, the threshold of pain due to sound averages 130 decibels. What is the intensity of such a sound in watts per square meter?

6. If one sound is 50 times as intense as another, what is the difference in the loudness of the two sounds? Express your answer in decibels.

7. *Magnitude of an Earthquake* Find the magnitude of an earthquake whose seismographic reading is 10.0 millimeters at a distance of 100 kilometers from its epicenter.

8. *Magnitude of an Earthquake* Find the magnitude of an earthquake whose seismographic reading is 1210 millimeters at a distance of 100 kilometers from its epicenter.

9. *Comparing Earthquakes* The Mexico City earthquake of 1978 registered 7.85 on the Richter scale. What would a seismograph 100 kilometers from the epicenter have measured for this earthquake? How does this earthquake compare in intensity to the 1906 San Francisco earthquake, which registered 8.9 on the Richter scale?

10. *Comparing Earthquakes* Two earthquakes differ by 1.0 when measured on the Richter scale. How would the seismographic readings differ at a distance of 100 kilometers from the epicenter? How do their intensities compare?

Chapter Review

THINGS TO KNOW

Properties of the exponential function

$f(x) = a^x, a > 1$ Domain: $(-\infty, \infty)$; Range: $(0, \infty)$; x-intercepts: none; y-intercept: 1; horizontal asymptote: x-axis as $x \to -\infty$; increasing; one-to-one
See Figure 2 for a typical graph.

$f(x) = a^x, 0 < a < 1$ Domain: $(-\infty, \infty)$; Range: $(0, \infty)$; x-intercepts: none, y-intercept: 1; horizontal asymptote: x-axis as $x \to \infty$; decreasing; one-to-one
See Figure 6 for a typical graph.

Properties of the logarithmic function

$f(x) = \log_a x, a > 1$
$(y = \log_a x$ means $x = a^y)$ Domain: $(0, \infty)$; Range: $(-\infty, \infty)$; x-intercept: 1; y-intercept: none; vertical asymptote: y-axis; increasing; one-to-one
See Figure 14(b) for a typical graph.

$f(x) = \log_a x, 0 < a < 1$
$(y = \log_a x$ means $x = a^y)$ Domain: $(0, \infty)$; Range: $(-\infty, \infty)$; x-intercept: 1; y-intercept: none; vertical asymptote: y-axis; decreasing; one-to-one
See Figure 14(a) for a typical graph.

Number e

Value approached by the expression $\left(1 + \dfrac{1}{n}\right)^n$ as $n \to \infty$; that is, $\displaystyle\lim_{n \to \infty} \left(1 + \dfrac{1}{n}\right)^n = e$

Natural logarithm

$y = \ln x$ means $x = e^y$

Properties of logarithms

$\log_a 1 = 0 \qquad \log_a a = 1 \qquad a^{\log_a M} = M \qquad \log_a a^r = r$

$\log_a MN = \log_a M + \log_a N \qquad \log_a\left(\dfrac{M}{N}\right) = \log_a M - \log_a N \qquad \log_a\left(\dfrac{1}{N}\right) = -\log_a N \qquad \log_a M^r = r \log_a M$

FORMULAS

Change-of-Base Formula	$\log_a M = \dfrac{\log_b M}{\log_b a}$
Compound interest	$A = P\left(1 + \dfrac{r}{n}\right)^{nt}$
Continuous compounding	$A = Pe^{rt}$
Present value	$P = A\left(1 + \dfrac{r}{n}\right)^{-nt}$ or $P = Ae^{-rt}$
Growth and decay	$A = A_0 e^{kt}$

HOW TO:

Graph exponential and logarithmic functions

Solve certain exponential equations

Solve certain logarithmic equations

Solve problems involving compound interest

Solve problems involving growth and decay

Solve problems involving intensity of sound and intensity of earthquakes

Fit nonlinear models to data

FILL-IN-THE-BLANK ITEMS

1. The graph of every exponential function $f(x) = a^x$, $a > 0$, $a \neq 1$, passes through the two points _____.

2. If the graph of an exponential function $f(x) = a^x$, $a > 0$, $a \neq 1$, is decreasing, then its base must be less than _____.

3. If $3^x = 3^4$, then $x = $ _____.

4. The logarithm of a product equals the _____ of the logarithms.

5. For every base, the logarithm of _____ equals 0.

6. If $\log_8 M = \log_5 7 / \log_5 8$, then $M = $ _____.

7. The domain of the logarithmic function $f(x) = \log_a x$ consists of _____.

8. The graph of every logarithmic function $f(x) = \log_a x$, $a > 0$, $a \neq 1$, passes through the two points _____.

9. If the graph of a logarithmic function $f(x) = \log_a x$, $a > 0$, $a \neq 1$, is increasing, then its base must be larger than _____.

10. If $\log_3 x = \log_3 7$, then $x = $ _____.

TRUE/FALSE ITEMS

T   F   **1.** The graph of every exponential function $f(x) = a^x$, $a > 0$, $a \neq 1$, will contain the points $(0, 1)$ and $(1, a)$.

T   F   **2.** The graphs of $y = 3^{-x}$ and $y = \left(\frac{1}{3}\right)^x$ are identical.

T   F   **3.** The present value of $1000 to be received after 2 years at 10% per annum compounded continuously is approximately $1205.

T   F   **4.** If $y = \log_a x$, then $y = a^x$.

T   F   **5.** The graph of every logarithmic function $f(x) = \log_a x$, $a > 0$, $a \neq 1$, will contain the points $(1, 0)$ and $(a, 1)$.

T   F   **6.** $a^{\log_M a} = M$, where $a > 0$, $a \neq 1$, $M > 0$

T   F   **7.** $\log_a(M + N) = \log_a M + \log_a N$, where $a > 0$, $a \neq 1$, $M > 0$, $N > 0$

T   F   **8.** $\log_a M - \log_a N = \log_a(M/N)$, where $a > 0$, $a \neq 1$, $M > 0$, $N > 0$

REVIEW EXERCISES

In Problems 1–6, evaluate each expression.

1. $\log_2\left(\frac{1}{8}\right)$ **2.** $\log_3 81$ **3.** $\ln e^{\sqrt{2}}$ **4.** $e^{\ln 0.1}$ **5.** $2^{\log_2 0.4}$ **6.** $\log_2 2^{\sqrt{3}}$

In Problems 7–12, write each expression as a single logarithm.

7. $3 \log_4 x^2 + \frac{1}{2} \log_4 \sqrt{x}$

8. $-2 \log_3\left(\frac{1}{x}\right) + \frac{1}{3} \log_3 \sqrt{x}$

9. $\ln\left(\frac{x-1}{x}\right) + \ln\left(\frac{x}{x+1}\right) - \ln(x^2 - 1)$

10. $\log(x^2 - 9) - \log(x^2 + 7x + 12)$

11. $2 \log 2 + 3 \log x - \frac{1}{2}[\log(x + 3) + \log(x - 2)]$

12. $\frac{1}{2} \ln(x^2 + 1) - 4 \ln\frac{1}{2} - \frac{1}{2}[\ln(x - 4) + \ln x]$

In Problems 13–20, find y as a function of x. The constant C is a positive number.

13. $\ln y = 2x^2 + \ln C$

14. $\ln(y - 3) = \ln 2x^2 + \ln C$

15. $\frac{1}{2} \ln y = 3x^2 + \ln C$

16. $\ln 2y = \ln(x + 1) + \ln(x + 2) + \ln C$

17. $\ln(y - 3) + \ln(y + 3) = x + C$

18. $\ln(y - 1) + \ln(y + 1) = -x + C$

19. $e^{y+C} = x^2 + 4$

20. $e^{3y-C} = (x + 4)^2$

In Problems 21–30, using a graphing utility, show the stages required to graph each function.

21. $f(x) = e^{-x}$ **22.** $f(x) = \ln(-x)$ **23.** $f(x) = 1 - e^x$ **24.** $f(x) = 3 + \ln x$

25. $f(x) = 3e^x$ **26.** $f(x) = \frac{1}{2} \ln x$ **27.** $f(x) = e^{|x|}$ **28.** $f(x) = \ln|x|$

29. $f(x) = 3 - e^{-x}$ **30.** $f(x) = 4 - \ln(-x)$

In Problems 31–50, solve each equation. Verify your result using a graphing utility.

31. $4^{1-2x} = 2$ **32.** $8^{6+3x} = 4$ **33.** $3^{x^2+x} = \sqrt{3}$

34. $4^{x-x^2} = \frac{1}{2}$ **35.** $\log_x 64 = -3$ **36.** $\log_{\sqrt{2}} x = -6$

37. $5^x = 3^{x+2}$ **38.** $5^{x+2} = 7^{x-2}$ **39.** $9^{2x} = 27^{3x-4}$

40. $25^{2x} = 5^{x^2-12}$ **41.** $\log_3 \sqrt{x - 2} = 2$ **42.** $2^{x+1} \cdot 8^{-x} = 4$

43. $8 = 4^{x^2} \cdot 2^{5x}$ **44.** $2^x \cdot 5 = 10^x$ **45.** $\log_6(x + 3) + \log_6(x + 4) = 1$

46. $\log_{10}(7x - 12) = 2 \log_{10} x$ **47.** $e^{1-x} = 5$ **48.** $e^{1-2x} = 4$

49. $2^{3x} = 3^{2x+1}$ **50.** $2^{x^3} = 3^{x^2}$

In Problems 51–54, use the following result: If x is the atmospheric pressure (measured in millimeters of mercury), then the formula for the altitude h(x) (measured in meters above sea level) is

$$h(x) = (30T + 8000)\log\left(\frac{P_0}{x}\right)$$

where T is the temperature (in degrees Celsius) and P_0 is the atmospheric pressure at sea level, which is approximately 760 millimeters of mercury.

51. *Finding the Altitude of an Airplane* At what height is an aircraft whose instruments record an outside temperature of 0°C and a barometric pressure of 300 millimeters of mercury?

52. *Finding the Height of a Mountain* How high is a mountain if instruments placed on its peak record a temperature of 5°C and a barometric pressure of 500 millimeters of mercury?

53. *Atmospheric Pressure Outside an Airplane* What is the atmospheric pressure outside an aircraft flying at an altitude of 10,000 meters if the outside air temperature is −100°C?

54. *Atmospheric Pressure at High Altitudes* What is the atmospheric pressure (in millimeters of mercury) on Mt. Everest, which has an altitude of approximately 8900 meters, if the air temperature is 5°C?

55. *Amplifying Sound* An amplifier's power output P (in watts) is related to its decibel voltage gain d by the formula $P = 25e^{0.1d}$.

(a) Find the power output for a decibel voltage gain of 4 decibels.

(b) For a power output of 50 watts, what is the decibel voltage gain?

56. *Limiting Magnitude of a Telescope* A telescope is limited in its usefulness by the brightness of the star it is aimed at and by the diameter of its lens. One measure of a star's brightness is its *magnitude:* the dimmer the star, the larger its magnitude. A formula for the limiting magnitude L of a telescope, that is, the magnitude of the dimmest star it can be used to view, is given by

$$L = 9 + 5.1 \log d$$

where d is the diameter (in inches) of the lens.

(a) What is the limiting magnitude of a 3.5 inch telescope?

(b) What diameter is required to view a star of magnitude 14?

57. *Product Demand* The demand for a new product increases rapidly at first and then levels off. The percent P of actual purchases of this product after it has been on the market t months is

$$P = 90 - 80\left(\frac{3}{4}\right)^t$$

(a) What is the percent of purchases of the product after 5 months?

(b) What is the percent of purchases of the product after 10 months?

(c) What is the maximum percent of purchases of the product?

(d) How many months does it take before 40% of purchases occurs?

(e) How many months before 70% of purchases occurs?

58. *Disseminating Information* A survey of a certain community of 10,000 residents shows that the number of residents N who have heard a piece of information after m months is given by the formula

$$m = 55.3 - 6 \ln(10,000 - N)$$

How many months will it take for half of the citizens to learn about a community program of free blood pressure readings?

59. *Salvage Value* The number of years n for a piece of machinery to depreciate to a known salvage value can be found using the formula

$$n = \frac{\log_{10} s - \log_{10} i}{\log_{10}(1 - d)}$$

where s is the salvage value of the machinery, i is its initial value, and d is the annual rate of depreciation.

(a) How many years will it take for a piece of machinery to decline in value from $90,000 to $10,000 if the annual rate of depreciation is 0.20 (20%)?

(b) How many years will it take for a piece of machinery to lose half of its value if the annual rate of depreciation is 15%?

60. *Funding a College Education* A child's grandparents purchase a $10,000 bond fund that matures in 18 years to be used for her college education. The bond fund pays 4% interest compounded semiannually. How much will the bond fund be worth at maturity?

61. *Funding a College Education* A child's grandparents wish to purchase a bond fund that matures in 18 years to be used for her college education. The bond fund pays 4% interest compounded semiannually. How much should they purchase so that the bond fund will be worth $85,000 at maturity?

62. *Funding an IRA* First Colonial Bankshares Corporation advertised the following IRA investment plans.

(a) Assuming continuous compounding, what was the annual rate of interest they offered?

(b) First Colonial Bankshares claims that $4000 invested today will have a value of over $32,000 in 20 years. Use the answer found in part (a) to find the actual value of $4000 in 20 years. Assume continuous compounding.

TARGET IRA PLANS

FOR EACH $5000 MATURITY VALUE DESIRED DEPOSIT:	AT A TERM OF:
$620.17	20 years
$1045.02	15 years
$1760.92	10 years
$2967.26	5 years

63. *Loudness of a Garbage Disposal* Find the loudness of a garbage disposal unit that operates at an intensity of 10^{-4} watt per square meter. Express your answer in decibels.

64. *Comparing Earthquakes* On September 9, 1985, the western suburbs of Chicago experienced a mild earthquake that registered 3.0 on the Richter scale. How did this earthquake compare in intensity to the great San Francisco earthquake of 1906, which registered 8.9 on the Richter scale?

65. *Estimating the Date a Prehistoric Man Died* The bones of a prehistoric man found in the desert of New Mexico contain approximately 5% of the original amount of carbon-14. If the half-life of carbon-14 is 5600 years, approximately how long ago did the man die?

66. *Temperature of a Skillet* A skillet is removed from an oven whose temperature is 450°F and placed in a room whose temperature is 70°F. After 5 minutes, the temperature of the skillet is 400°F. How long will it be until its temperature is 150°F?

In Problems 67–68, the independent variable is x and the dependent variable is y. For each set of data:
(a) draw a scatter diagram of the data; (b) use an exponential model, logarithmic model, power model, and linear model to find the "best" model that explains the relationship between the variable.

67.

x	1	4	5	7	10	12
y	1.3084	0.8932	0.8123	0.7737	0.6954	0.6659

68.

x	0.8	1.2	1.7	2.4	2.9	3.4
y	1.1843	2.8634	4.3104	5.6459	6.4683	7.2081

69. *Population Model* The following data represent the population of the United States from 1900 to 1990.

YEAR	POPULATION
1900	76,212,168
1910	92,228,496
1920	106,021,537
1930	123,202,624
1940	132,164,569
1950	151,325,798
1960	179,323,175
1970	203,302,031
1980	226,542,203
1990	248,709,873

(a) Using a graphing utility, draw a scatter diagram for the data using time as the independent variable and population as the dependent variable.

(b) Use an exponential model, logarithmic model, power model, and linear model to find the best model to describe the relationship between the variables.

(c) Using the model found in (b), predict the population of the United States for 1995.

70. *Economics* The following data represent personal consumption expenditures and per capita gross national product figures in the United States from 1984 to 1993 (Source: U.S. Bureau of Economic Analysis).

YEAR	PER CAPITA PERSONAL CONSUMPTION	PER CAPITA GROSS NATIONAL PRODUCT
1984	11,617	17,659
1985	12,015	18,007
1986	12,336	18,337
1987	12,568	18,712
1988	12,903	19,284
1989	13,029	19,615
1990	13,093	19,670
1991	12,895	19,290
1992	13,081	19,548
1993	13,372	19,897

(a) Using a graphing utility, draw a scatter diagram of the data using gross national product as the independent variable and personal consumption as the dependent variable.

(b) Use an exponential model, logarithmic model, power model, and linear model to find the best model to describe the relationship between the variables.

(c) Using the model found in (b), predict personal consumption expenditures if per capita gross national product is $20,000.

71. In a room whose constant temperature is 70°F, will a pizza heated to 450°F ever reach exactly 70°F? Newton's Law of Cooling seems to say no! What really happens? What assumptions are made in using Newton's law? Write a brief paragraph explaining your conclusions.

PREPARING FOR THIS CHAPTER

Before getting started on this chapter, review the following concepts:

For Sections 6.1–6.4: Rectangular coordinates (pp. 2–3)
Distance formula (p. 5)
Completing the square (Appendix, Section 5)
Intercepts (pp. 21–23)
Symmetry (pp. 26–31)
Circles (pp. 55–60)
For Section 6.5: Rectangular coordinates (pp. 2–3)

THE CONICS; VECTORS

6.1 Preliminaries
6.2 The Parabola
6.3 The Ellipse
6.4 The Hyperbola
6.5 Vectors
     Chapter Review

Preview   Satellite Dish

*A satellite dish is shaped like a **paraboloid of revolution,** a surface formed by rotating a parabola about its axis of symmetry. The signals that emanate from a satellite strike the surface of the dish and are reflected to a single point, where the receiver is located. If the dish is 8 feet across at its opening and is 3 feet deep at its center, at what position should the receiver be placed? [Example 10 in Section 6.2].* ■

7 his chapter contains two topics: *conics,* Sections 6.1–6.4, and *vectors,* Section 6.5. They are independent of each other and may be covered in any order. Historically, Apollonius (200 BC) was among the first to study *conics* and discover some of their interesting properties. Today, conics are still studied because of their many uses. *Paraboloids of revolution* (parabolas rotated about their axes of symmetry) are used as signal collectors (the satellite dishes used with radar and cable TV, for example), as solar energy collectors, and as reflectors (telescopes, light projection, and so on). The planets circle the Sun in approximately *elliptical* orbits. Elliptical surfaces can be used to reflect signals such as light and sound from one place to another. And *hyperbolas* can be used to determine the positions of ships at sea.

The Greeks used the methods of Euclidean geometry to study conics. We shall use the more powerful methods of analytic geometry, bringing to bear both algebra and geometry, for our study of conics. Thus, we shall give a geometric description of each conic, and then, using rectangular coordinates and the distance formula, we shall find equations that represent conics. We used this same development, you may recall, when we first defined a circle in Section 1.4.

The last section provides an introduction to *vectors* and some of its applications, an extremely important topic in engineering and physics.

6.1

Preliminaries

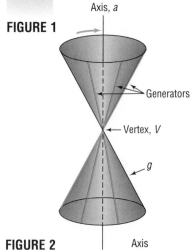

FIGURE 1

The word *conic* derives from the word *cone,* which is a geometric figure that can be constructed in the following way: Let a and g be two distinct lines that intersect at a point V. Keep the line a fixed. Now rotate the line g about a while maintaining the same angle between a and g. The collection of points swept out (generated) by the line g is called a **(right circular) cone.** See Figure 1. The fixed line a is called the **axis** of the cone; the point V is called its **vertex;** the lines that pass through V and make the same angle with a as g are called **generators** of the cone. Thus, each generator is a line that lies entirely on the cone. The cone consists of two parts, called **nappes,** that intersect at the vertex.

Conics, an abbreviation for **conic sections,** are curves that result from the intersection of a (right circular) cone and a plane. The conics we shall study arise when the plane does not contain the vertex, as shown in Figure 2. These conics are **circles** when the plane is perpendicular to the axis of the cone and intersects each generator; **ellipses** when the plane is tilted slightly so that it intersects each generator, but intersects only one nappe of the cone; **parabolas** when the plane is

FIGURE 2

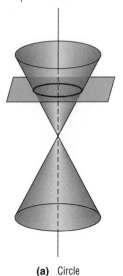

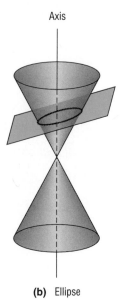

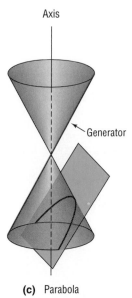

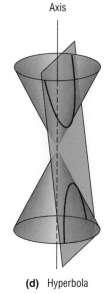

(a) Circle   (b) Ellipse   (c) Parabola   (d) Hyperbola

tilted further so that it is parallel to one (and only one) generator and intersects only one nappe of the cone; and **hyperbolas** when the plane intersects both nappes.

If the plane does contain the vertex, the intersection of the plane and the cone is a point, a line, or a pair of intersecting lines. These are usually called **degenerate conics.**

6.2

The Parabola

We stated earlier (Section 4.1) that the graph of a quadratic function is a parabola. In this section, we begin with a geometric definition of parabola and use it to obtain an equation.

Parabola

A **parabola** is defined as the collection of all points P in the plane that are the same distance from a fixed point F as they are from a fixed line D. The point F is called the **focus** of the parabola, and the line D is its **directrix.** As a result, a parabola is the set of points P for which

$$d(F, P) = d(P, D) \tag{1}$$

FIGURE 3

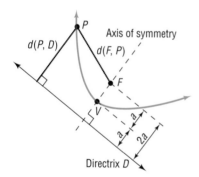

Figure 3 shows a parabola. The line through the focus F and perpendicular to the directrix D is called the **axis of symmetry** of the parabola. The point of intersection of the parabola with its axis of symmetry is called the **vertex** V.

Because the vertex V lies on the parabola, it must satisfy equation (1): $d(F, V) = d(V, D)$. Thus, the vertex is midway between the focus and the directrix. We shall let a equal the distance $d(F, V)$ from F to V. Now we are ready to derive an equation for a parabola. To do this, we use a rectangular system of coordinates, positioned so that the vertex V, focus F, and directrix D of the parabola are conveniently located. If we choose to locate the vertex V at the origin $(0, 0)$, then we can conveniently position the focus F on either the x-axis or the y-axis.

First, we consider the case where the focus F is on the positive x-axis, as shown in Figure 4. Because the distance from F to V is a, the coordinates of F will be $(a, 0)$ with $a > 0$. Similarly, because the distance from V to the directrix D is also a and because D must be perpendicular to the x-axis (since the x-axis is the axis of symmetry), the equation of the directrix D must be $x = -a$. Now, if $P = (x, y)$ is any point on the parabola, then P must obey equation (1):

$$d(F, P) = d(P, D)$$

FIGURE 4
$y^2 = 4ax$

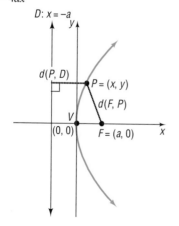

So, we have

$$
\begin{aligned}
\sqrt{(x - a)^2 + y^2} &= |x + a| && \text{Use the distance formula.} \\
(x - a)^2 + y^2 &= (x + a)^2 && \text{Square both sides.} \\
x^2 - 2ax + a^2 + y^2 &= x^2 + 2ax + a^2 \\
y^2 &= 4ax
\end{aligned}
$$

Theorem   The equation of a parabola with vertex at $(0, 0)$, focus at $(a, 0)$, and directrix $x = -a, a > 0$, is

Equation of a Parabola;
Vertex at $(0, 0)$,
Focus at $(a, 0), a > 0$

$$y^2 = 4ax \qquad (2)$$

■

E X A M P L E   1

Finding the Equation of a Parabola

Find an equation of the parabola with vertex at $(0, 0)$ and focus at $(3, 0)$. Graph the equation.

Solution   The distance from the vertex $(0, 0)$ to the focus $(3, 0)$ is $a = 3$. Based on equation (2), the equation of this parabola is

FIGURE 5
$y^2 = 12x$

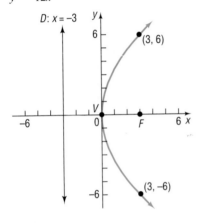

$$y^2 = 4ax$$
$$y^2 = 12x \qquad a = 3$$

To graph this parabola by hand, it is helpful to plot the two points on the graph above and below the focus. To locate them, we let $x = 3$. Then,

$$y^2 = 12x = 36$$
$$y = \pm 6$$

The points on the parabola above and below the focus are $(3, -6)$ and $(3, 6)$. See Figure 5.

■

In general, the points on a parabola $y^2 = 4ax$ that lie above and below the focus $(a, 0)$ are each at a distance $2a$ from the focus. This follows from the fact that if $x = a$ then $y^2 = 4ax = 4a^2$, or $y = \pm 2a$. The line segment joining these two points is called the **latus rectum**; its length is $4a$.

E X A M P L E   2

Graphing a Parabola Using a Graphing Utility

Graph the parabola: $y^2 = 12x$

Solution   To graph the parabola $y^2 = 12x$, we need to graph the two functions $y_1 = \sqrt{12x}$ and $y_2 = -\sqrt{12x}$. Figure 6 shows the graph of $y^2 = 12x$. Notice that the graph fails the vertical line test, so $y^2 = 12x$ is not a function.   ■

FIGURE 6
$y^2 = 12x$

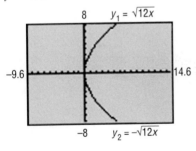

■   Now work Problem 19.

By reversing the steps we used to obtain equation (2), it follows that the graph of an equation of the form of equation (2) is a parabola; its vertex is at $(0, 0)$, its focus is at $(a, 0)$, its directrix is the line $x = -a$, and its axis of symmetry is the x-axis.

For the remainder of this section, the direction "Discuss the equation" will mean to find the vertex, focus, and directrix of the parabola and graph it.

E X A M P L E   3

Discussing the Equation of a Parabola

Discuss the equation: $y^2 = 8x$

Solution   Figure 7(a) shows the graph of $y^2 = 8x$ using a graphing utility. We now proceed to analyze the equation.

The equation $y^2 = 8x$ is of the form $y^2 = 4ax$, where $4a = 8$. Thus, $a = 2$.

Consequently, the graph of the equation is a parabola with vertex at $(0, 0)$ and focus on the positive x-axis at $(2, 0)$. The directrix is the vertical line $x = -2$. The two points defining the latus rectum are obtained by letting $x = 2$. Then $y^2 = 16$, or $y = \pm 4$. These points help in graphing the parabola by hand since they determine the "opening" of the graph. See Figure 7(b).

FIGURE 7
$y^2 = 8x$

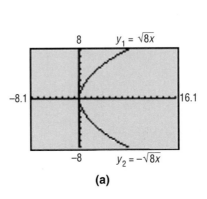

(a)

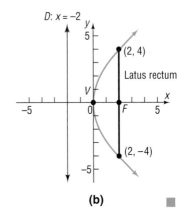

(b)

Recall that we arrived at equation (2) after placing the focus on the positive x-axis. If the focus is placed on the negative x-axis, positive y-axis, or negative y-axis, a different form of the equation for the parabola results. The four forms of the equation of a parabola with vertex at $(0, 0)$ and focus on a coordinate axis a distance a from $(0, 0)$ are given in Table 1, and their graphs are given in Figure 8. Notice that each graph is symmetric with respect to its axis of symmetry.

TABLE 1   EQUATIONS OF A PARABOLA: VERTEX AT (0, 0); FOCUS ON AXIS; $a > 0$

VERTEX	FOCUS	DIRECTRIX	EQUATION	DESCRIPTION
$(0, 0)$	$(a, 0)$	$x = -a$	$y^2 = 4ax$	Parabola, axis of symmetry is the x-axis, opens to right
$(0, 0)$	$(-a, 0)$	$x = a$	$y^2 = -4ax$	Parabola, axis of symmetry is the x-axis, opens to left
$(0, 0)$	$(0, a)$	$y = -a$	$x^2 = 4ay$	Parabola, axis of symmetry is the y-axis, opens up
$(0, 0)$	$(0, -a)$	$y = a$	$x^2 = -4ay$	Parabola, axis of symmetry is the y-axis, opens down

FIGURE 8

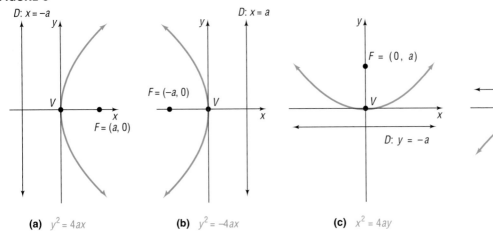

(a) $y^2 = 4ax$ (b) $y^2 = -4ax$ (c) $x^2 = 4ay$ (d) $x^2 = -4ay$

E X A M P L E   4   *Discussing the Equation of a Parabola*

Discuss the equation: $x^2 = -12y$

Solution   Figure 9(a) shows the graph of $x^2 = -12y$ using a graphing utility. We now proceed to analyze the equation.

The equation $x^2 = -12y$ is of the form $x^2 = -4ay$, with $a = 3$. Consequently,
the graph of the equation is a parabola with vertex at $(0, 0)$, focus at $(0, -3)$, and
directrix the line $y = 3$. The parabola opens down, and its axis of symmetry is the
y-axis. To obtain the points defining the latus rectum, let $y = -3$. Then $x^2 = 36$,
or $x = \pm 6$. See Figure 9(b).

FIGURE 9
$x^2 = -12y$

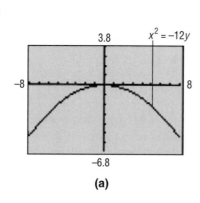

(a)

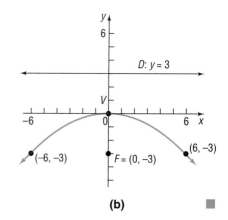

(b)

■   Now work Problem 35.

E X A M P L E   5   *Finding the Equation of a Parabola*

Find the equation of the parabola with focus at $(0, 4)$ and directrix the line
$y = -4$. Graph the equation by hand.

Solution   A parabola whose focus is at $(0, 4)$ and whose directrix is the horizontal line
$y = -4$ will have its vertex at $(0, 0)$. (Do you see why? The vertex is midway between the focus and the directrix.) Thus, the equation of this parabola is of the
form $x^2 = 4ay$, with $a = 4$; that is,

$$x^2 = 16y$$

Figure 10 shows the graph.

FIGURE 10
$x^2 = 16y$

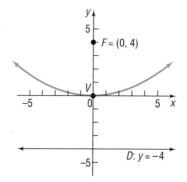

■

E X A M P L E   6   *Finding the Equation of a Parabola*

Find the equation of a parabola with vertex at $(0, 0)$ if its axis of symmetry is the x-axis and its graph contains the point $(-\frac{1}{2}, 2)$. Find its focus and directrix, and graph the equation by hand.

Solution   Because the vertex is at the origin and the axis of symmetry is the x-axis, we see from Table 1 that the form of the equation is

$$y^2 = kx$$

Because the point $(-\frac{1}{2}, 2)$ is on the parabola, the coordinates $x = -\frac{1}{2}$, $y = 2$ must satisfy the equation. Putting $x = -\frac{1}{2}$ and $y = 2$ into the equation, we find

$$4 = k(-\tfrac{1}{2})$$
$$k = -8$$

Thus, the equation of the parabola is

$$y^2 = -8x$$

Comparing this equation to $y^2 = -4ax$, we find that $a = 2$. The focus is therefore at $(-2, 0)$, and the directrix is the line $x = 2$. Letting $x = -2$, we find $y^2 = 16$ or $y = \pm 4$. The points $(-2, 4)$ and $(-2, -4)$ define the latus rectum. See Figure 11.

FIGURE 11
$y^2 = -8x$

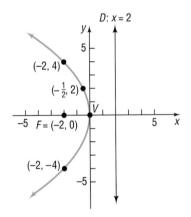

Now work Problem 27.

Vertex at (h, k)

If a parabola with vertex at the origin and axis of symmetry along a coordinate axis is shifted horizontally h units and then vertically k units, the result is a parabola with vertex at (h, k) and axis of symmetry parallel to a coordinate axis. The equations of such parabolas have the same forms as those in Table 1, but with x replaced by $x - h$ and y replaced by $y - k$. Table 2 gives the forms of the equations of such parabolas. Figure 12(a)–(d) illustrates the graphs for $h > 0$, $k > 0$.

TABLE 2   PARABOLAS WITH VERTEX AT (h, k), AXIS OF SYMMETRY PARALLEL TO A COORDINATE AXIS, $a > 0$

VERTEX	FOCUS	DIRECTRIX	EQUATION	DESCRIPTION
(h, k)	$(h + a, k)$	$x = -a + h$	$(y - k)^2 = 4a(x - h)$	Parabola, axis of symmetry parallel to x-axis, opens to right
(h, k)	$(h - a, k)$	$x = a + h$	$(y - k)^2 = -4a(x - h)$	Parabola, axis of symmetry parallel to x-axis, opens to left
(h, k)	$(h, k + a)$	$y = -a + k$	$(x - h)^2 = 4a(y - k)$	Parabola, axis of symmetry parallel to y-axis, opens up
(h, k)	$(h, k - a)$	$y = a + k$	$(x - h)^2 = -4a(y - k)$	Parabola, axis of symmetry parallel to y-axis, opens down

FIGURE 12

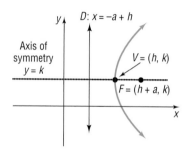

(a) $(y - k)^2 = 4a(x - h)$

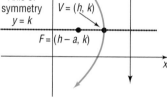

(b) $(y - k)^2 = -4a(x - h)$

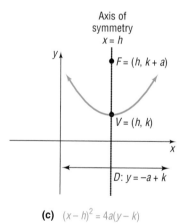

(c) $(x - h)^2 = 4a(y - k)$

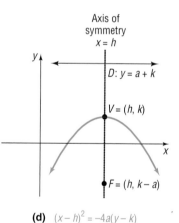

(d) $(x - h)^2 = -4a(y - k)$

E X A M P L E   7   *Finding the Equation of a Parabola, Vertex Not at Origin*

Find an equation of the parabola with vertex at $(-2, 3)$ and focus at $(0, 3)$. Graph the equation by hand.

Solution   The vertex $(-2, 3)$ and focus $(0, 3)$ both lie on the horizontal line $y = 3$ (the axis of symmetry). The distance a from $(-2, 3)$ to $(0, 3)$, is $a = 2$. Also, because the focus lies to the right of the vertex, we know the parabola opens to the right. Consequently, the form of the equation is

$$(y - k)^2 = 4a(x - h)$$

where $(h, k) = (-2, 3)$ and $a = 2$. Therefore, the equation is

$$(y - 3)^2 = 4 \cdot 2[x - (-2)]$$
$$(y - 3)^2 = 8(x + 2)$$

If $x = 0$, then $(y - 3)^2 = 16$. Thus, $y - 3 = \pm 4$ and $y = -1, y = 7$. The points $(0, -1)$ and $(0, 7)$ define the latus rectum; the line $x = -4$ is the directrix. See Figure 13.

FIGURE 13
$(y - 3)^2 = 8(x + 2)$

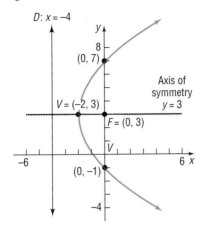

Now work Problem 25.

E X A M P L E   8

Using a Graphing Utility to Graph a Parabola, Vertex Not at Origin
Using a graphing utility, graph the equation $(y - 3)^2 = 8(x + 2)$.

Solution   First, we must solve the equation for y.

$$(y - 3)^2 = 8(x + 2)$$
$$y - 3 = \pm\sqrt{8(x + 2)} \qquad \text{Take the square root of each side.}$$
$$y = 3 \pm\sqrt{8(x + 2)} \qquad \text{Solve for } y.$$

FIGURE 14

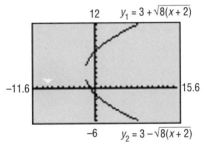

Figure 14 shows the graphs of the equations $y_1 = 3 + \sqrt{8(x + 2)}$ and $y_2 = 3 - \sqrt{8(x + 2)}$.

Now work Problem 35.

Polynomial equations define parabolas whenever they involve two variables that are quadratic in one variable and linear in the other. To discuss this type of equation, we first complete the square of the quadratic variable.

E X A M P L E   9

Discussing the Equation of a Parabola
Discuss the equation: $x^2 + 4x - 4y = 0$

Solution   Figure 15(a) shows the graph of $x^2 + 4x - 4y = 0$ using a graphing utility. We now proceed to analyze the equation.
To discuss the equation $x^2 + 4x - 4y = 0$, we complete the square involving the variable x. Thus,

$$x^2 + 4x - 4y = 0$$
$$x^2 + 4x = 4y \qquad \text{Isolate the terms involving } x \text{ on the left side.}$$
$$x^2 + 4x + 4 = 4y + 4 \qquad \text{Complete the square on the left side.}$$
$$(x + 2)^2 = 4(y + 1)$$

This equation is of the form $(x - h)^2 = 4a(y - k)$, with $h = -2$, $k = -1$, and $a = 1$. The graph is a parabola with vertex at $(h, k) = (-2, -1)$ that opens up. The focus is at $(-2, 0)$, and the directrix is the line $y = -2$. See Figure 15(b).

FIGURE 15
$x^2 + 4x - 4y = 0$

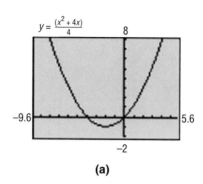

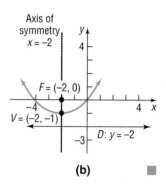

(a)   **(b)**

Parabolas find their way into many applications. For example, as we discussed in Section 4.1, suspension bridges have cables in the shape of a parabola. Another property of parabolas that is used in applications is their reflecting property.

Reflecting Property

FIGURE 16

Searchlight

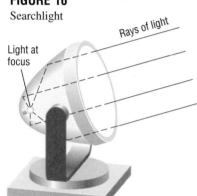

Suppose that a mirror is shaped like a **paraboloid of revolution,** a surface formed by rotating a parabola about its axis of symmetry. If a light (or any other emitting source) is placed at the focus of the parabola, all the rays emanating from the light will reflect off the mirror in lines parallel to the axis of symmetry. This principle is used in the design of searchlights, flashlights, certain automobile headlights, and other such devices. See Figure 16.

Conversely, suppose rays of light (or other signals) emanate from a distant source so that they are essentially parallel. When these rays strike the surface of a parabolic mirror whose axis of symmetry is parallel to these rays, they are reflected to a single point at the focus. This principle is used in the design of some solar energy devices, satellite dishes, and the mirrors used in some types of telescopes. See Figure 17.

FIGURE 17

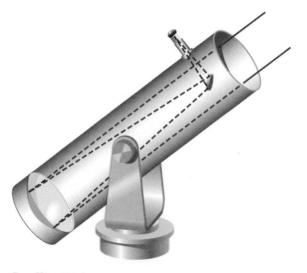

EXAMPLE 10   *Satellite Dish*

A satellite dish is shaped like a paraboloid of revolution. The signals that emanate from a satellite strike the surface of the dish and are reflected to a single point, where the receiver is located. If the dish is 8 feet across at its opening and is 3 feet deep at its center, at what position should the receiver be placed?

FIGURE 18

(a)

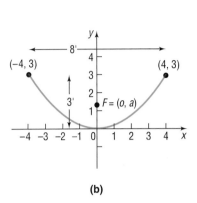

(b)

Solution   Figure 18(a) shows the satellite dish. We draw the parabola used to form the dish on a rectangular coordinate system so that the vertex of the parabola is at the origin and its focus is on the positive y-axis. See Figure 18(b). The form of the equation of the parabola is

$$x^2 = 4ay$$

and its focus is at $(0, a)$. Since $(4, 3)$ is a point on the graph, we have

$$4^2 = 4a(3)$$

$$a = \frac{4}{3}$$

The receiver should be located $1\frac{1}{3}$ feet from the base of the dish, along its axis of symmetry.   ■

6.2

Exercise 6.2

In Problems 1–8, the graph of a parabola is given. Match each graph to its equation.

A. $y^2 = 4x$
B. $x^2 = 4y$
C. $y^2 = -4x$
D. $x^2 = -4y$
E. $(y - 1)^2 = 4(x - 1)$
F. $(x + 1)^2 = 4(y + 1)$
G. $(y - 1)^2 = -4(x - 1)$
H. $(x + 1)^2 = -4(y + 1)$

1.

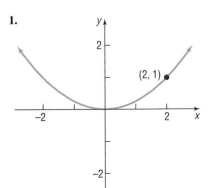

(2, 1)

2.

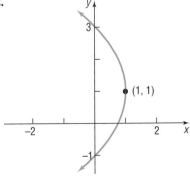

(1, 1)

3.

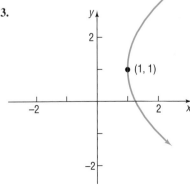

(1, 1)

4.

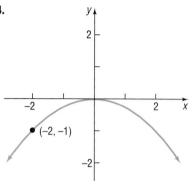

5.

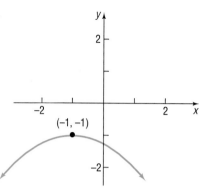

6.

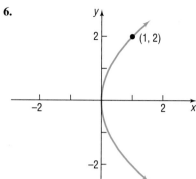

7.

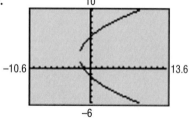

8.

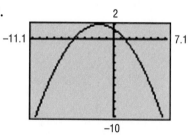

In Problems 9–16, the graph of a parabola is given. Match each graph to its equation.

A. $x^2 = 6y$

B. $x^2 = -6y$

C. $y^2 = 6x$

D. $y^2 = -6x$

E. $(y - 2)^2 = -6(x + 2)$

F. $(y - 2)^2 = 6(x + 2)$

G. $(x + 2)^2 = -6(y - 2)$

H. $(x + 2)^2 = 6(y - 2)$

9.

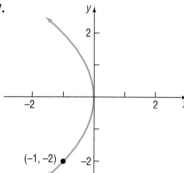

10.

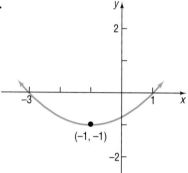

11.

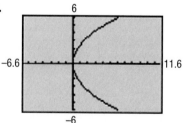

12.

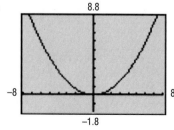

13.

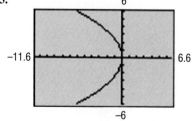

14.

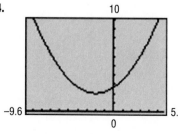

15.

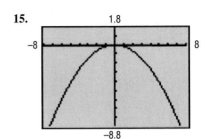

16.

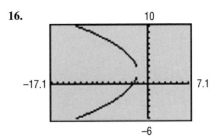

In Problems 17–32, find the equation of the parabola described. Find the two points that define the latus rectum, and graph the equation by hand.

17. Focus at (4, 0); vertex at (0, 0)

18. Focus at (0, 2); vertex at (0, 0)

19. Focus at (0, −3); vertex at (0, 0)

20. Focus at (−4, 0); vertex at (0, 0)

21. Focus at (−2, 0); directrix the line $x = 2$

22. Focus at (0, −1); directrix the line $y = 1$

23. Directrix the line $y = -\frac{1}{2}$; vertex at (0, 0)

24. Directrix the line $x = -\frac{1}{2}$; vertex at (0, 0)

25. Vertex at (2, −3); focus at (2, −5)

26. Vertex at (4, −2); focus at (6, −2)

27. Vertex at (0, 0); axis of symmetry the y-axis; containing the point (2, 3)

28. Vertex at (0, 0); axis of symmetry the x-axis; containing the point (2, 3)

29. Focus at (−3, 4); directrix the line $y = 2$

30. Focus at (2, 4); directrix the line $x = -4$

31. Focus at (−3, −2); directrix the line $x = 1$

32. Focus at (−4, 4); directrix the line $y = -2$

In Problems 33–50, find the vertex, focus, and directrix of each parabola. Graph the equation, using a graphing utility.

33. $x^2 = 4y$

34. $y^2 = 8x$

35. $y^2 = -16x$

36. $x^2 = -4y$

37. $(y - 2)^2 = 8(x + 1)$

38. $(x + 4)^2 = 16(y + 2)$

39. $(x - 3)^2 = -(y + 1)$

40. $(y + 1)^2 = -4(x - 2)$

41. $(y + 3)^2 = 8(x - 2)$

42. $(x - 2)^2 = 4(y - 3)$

43. $y^2 - 4y + 4x + 4 = 0$

44. $x^2 + 6x - 4y + 1 = 0$

45. $x^2 + 8x = 4y - 8$

46. $y^2 - 2y = 8x - 1$

47. $y^2 + 2y - x = 0$

48. $x^2 - 4x = 2y$

49. $x^2 - 4x = y + 4$

50. $y^2 + 12y = -x + 1$

In Problems 51–58, write an equation for each parabola.

51.

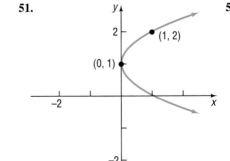

52.

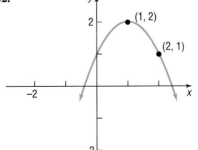

53.

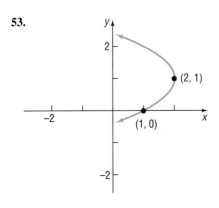

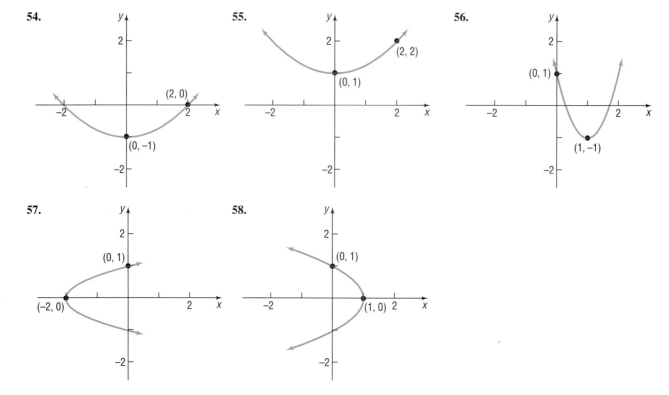

54.

55.

56.

57.

58.

59. *Satellite Dish* A satellite dish is shaped like a paraboloid of revolution. The signals that emanate from a satellite strike the surface of the dish and are reflected to a single point, where the receiver is located. If the dish is 10 feet across at its opening and is 4 feet deep at its center, at what position should the receiver be placed?

60. *Constructing a TV Dish* A cable TV receiving dish is in the shape of a paraboloid of revolution. Find the location of the receiver, which is placed at the focus, if the dish is 6 feet across at its opening and 2 feet deep.

61. *Constructing a Flashlight* The reflector of a flashlight is in the shape of a paraboloid of revolution. Its diameter is 4 inches and its depth is 1 inch. How far from the vertex should the light bulb be placed so that the rays will be reflected parallel to the axis?

62. *Constructing a Headlight* A sealed-beam headlight is in the shape of a paraboloid of revolution. The bulb, which is placed at the focus, is 1 inch from the vertex. If the depth is to be 2 inches, what is the diameter of the headlight at its opening?

63. *Suspension Bridges* The cables of a suspension bridge are in the shape of a parabola, as shown in the figure. The towers supporting the cable are 600 feet apart and 80 feet high. If the cables touch the road surface midway between the towers, what is the height of the cable at a point 150 feet from a tower?

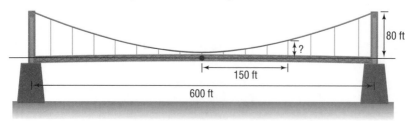

64. *Suspension Bridges* The cables of a suspension bridge are in the shape of a parabola. The towers supporting the cable are 400 feet apart and 100 feet high. If the cables are at a height of 10 feet midway between the towers, what is the height of the cable at a point 50 feet from a tower?

65. *Searchlights* A searchlight is shaped like a paraboloid of revolution. If the light source is located 2 feet from the base along the axis of symmetry and the opening is 5 feet across, how deep should the searchlight be?

66. *Searchlights* A searchlight is shaped like a paraboloid of revolution. If the light source is located 2 feet from the base along the axis of symmetry and the depth of the searchlight is 4 feet, what should the width of the opening be?

67. *Solar Heat* A mirror is shaped like a paraboloid of revolution and will be used to concentrate the rays of the sun at its focus, creating a heat source. If the mirror is 20 feet across at its opening and is 6 feet deep, where will the heat source be concentrated?

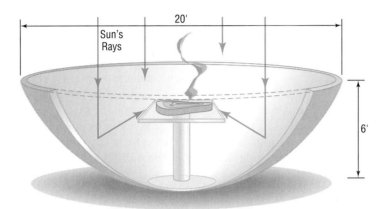

68. *Reflecting Telescopes* A reflecting telescope contains a mirror shaped like a paraboloid of revolution. If the mirror is 4 inches across at its opening and is 3 feet deep, where will the light collected be concentrated?

69. *Parabolic Arch Bridge* A bridge is built in the shape of a parabolic arch. The bridge has a span of 120 feet and a maximum height of 25 feet. See the illustration. Choose a suitable rectangular coordinate system and find the height of the arch at distances of 10, 30, and 50 feet from the center.

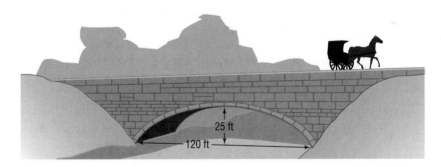

70. *Parabolic Arch Bridge* A bridge is built in the shape of a parabolic arch and is to have a span of 100 feet. The height of the arch a distance of 40 feet from the center is to be 10 feet. Find the height of the arch at its center.

71. Show that an equation of the form

$$Ax^2 + Ey = 0 \qquad A \neq 0, E \neq 0$$

is the equation of a parabola with vertex at $(0, 0)$ and axis of symmetry the y-axis. Find its focus and directrix.

72. Show that an equation of the form

$$Cy^2 + Dx = 0 \qquad C \neq 0, D \neq 0$$

is the equation of a parabola with vertex at $(0, 0)$ and axis of symmetry the x-axis. Find its focus and directrix.

73. Show that the graph of an equation of the form

$$Ax^2 + Dx + Ey + F = 0 \qquad A \neq 0$$

(a) Is a parabola if $E \neq 0$.
(b) Is a vertical line if $E = 0$ and $D^2 - 4AF = 0$.
(c) Is two vertical lines if $E = 0$ and $D^2 - 4AF > 0$.
(d) Contains no points if $E = 0$ and $D^2 - 4AF < 0$.

74. Show that the graph of an equation of the form

$$Cy^2 + Dx + Ey + F = 0 \qquad C \neq 0$$

(a) Is a parabola if $D \neq 0$.
(b) Is a horizontal line if $D = 0$ and $E^2 - 4CF = 0$.
(c) Is two horizontal lines if $D = 0$ and $E^2 - 4CF > 0$.
(d) Contains no points if $D = 0$ and $E^2 - 4CF < 0$.

6.3

The Ellipse

Ellipse

FIGURE 19

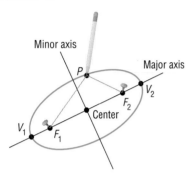

> An **ellipse** is the collection of all points in the plane the sum of whose distances from two fixed points, called the **foci,** is a constant.

The definition actually contains within it a physical means for drawing an ellipse. Find a piece of string (the length of this string is the constant referred to in the definition). Then take two thumbtacks (the foci) and stick them on a piece of cardboard so that the distance between them is less than the length of the string. Now attach the ends of the string to the thumbtacks and, using the point of a pencil, pull the string taut. Keeping the string taut, rotate the pencil around the two thumbtacks. The pencil traces out an ellipse, as shown in Figure 19.

In Figure 19, the foci are labeled F_1 and F_2. The line containing the foci is called the **major axis.** The midpoint of the line segment joining the foci is called the **center** of the ellipse. The line through the center and perpendicular to the major axis is called the **minor axis.**

The two points of intersection of the ellipse and the major axis are the **vertices,** V_1 and V_2, of the ellipse. The distance from one vertex to the other is called the **length of the major axis.** The ellipse is symmetric with respect to its major axis and with respect to its minor axis.

With these ideas in mind, we are now ready to find the equation of an ellipse in a rectangular coordinate system. First, we place the center of the ellipse at the origin. Second, we position the ellipse so that its major axis coincides with a coordinate axis. Suppose that the major axis coincides with the x-axis, as shown in Figure 20. If c is the distance from the center to a focus, then one focus will be at $F_1 = (-c, 0)$ and the other at $F_2 = (c, 0)$. As we shall see, it is convenient to let $2a$ denote the constant distance referred to in the definition. Thus, if $P = (x, y)$ is any point on the ellipse, we have

FIGURE 20
$d(F_1, P) + d(F_2, P) = 2a$

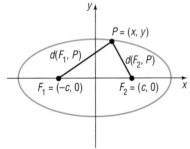

$$d(F_1, P) + d(F_2, P) = 2a \qquad \text{Sum of the distances from } P \text{ to the foci equals a constant}$$

$$\sqrt{(x + c)^2 + y^2} + \sqrt{(x - c)^2 + y^2} = 2a \qquad \text{Use the distance formula.}$$

$$\sqrt{(x + c)^2 + y^2} = 2a - \sqrt{(x - c)^2 + y^2} \qquad \text{Isolate one radical.}$$

$$(x + c)^2 + y^2 = 4a^2 - 4a\sqrt{(x - c)^2 + y^2} \qquad \text{Square both sides.}$$
$$+ (x - c)^2 + y^2$$

$$x^2 + 2cx + c^2 + y^2 = 4a^2 - 4a\sqrt{(x - c)^2 + y^2} \qquad \text{Simplify}$$
$$+ x^2 - 2cx + c^2 + y^2$$

$$4cx - 4a^2 = -4a\sqrt{(x - c)^2 + y^2} \qquad \text{Isolate the radical.}$$

$$cx - a^2 = -a\sqrt{(x - c)^2 + y^2} \qquad \text{Divide each side by 4.}$$

$$c^2x^2 - 2a^2cx + a^4 = a^2[(x - c)^2 + y^2] \qquad \text{Square both sides again.}$$

$$c^2x^2 - 2a^2cx + a^4 = a^2(x^2 - 2cx + c^2 + y^2)$$

$$(c^2 - a^2)x^2 - a^2y^2 = a^2c^2 - a^4$$

$$(a^2 - c^2)x^2 + a^2y^2 = a^2(a^2 - c^2) \qquad \text{Multiply each side by } -1; \text{ factor } a^2 \text{ on the right side.} \quad (1)$$

MISSION POSSIBLE

Chapter 6

BUILDING A BRIDGE OVER THE EAST RIVER

Your team is working for the transportation authority in New York City. You have been asked to study the construction plans for a new bridge over the East River in New York City. The space between supports needs to be 1050 feet; the height at the center of the arch needs to be 350 feet. One company has suggested the support be in the shape of a parabola; another company suggests a semi-ellipse. The engineering team will determine the relative strengths of the two plans; your job is to find out if there are any differences in the channel widths.

An empty tanker needs a 280 foot clearance to pass beneath the bridge. You need to find the width of the channel for each of the two different plans.

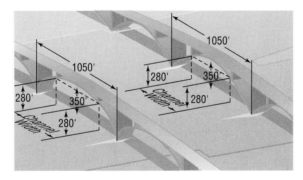

1. To determine the equation of a parabola with these characteristics, first place the parabola on coordinate axes in a convenient location and sketch it.
2. What is the equation of the parabola? (If using a decimal in the equation, you may want to carry 6 decimal places. If using a fraction, your answer will be more exact.)
3. How wide is the channel the tanker can pass through if the shape of the support is parabolic?
4. To determine the equation of a semi-ellipse with these characteristics, place the semi-ellipse on coordinate axes in a convenient location and sketch it.
5. What is the equation of the ellipse? How wide is the channel the tanker can pass through?
6. Now that you know which of the two provides the wider channel, consider some other factors. Your department is also in charge of channel depth, amount of traffic on the river, and various other factors. For example, if there were flooding in the river, and the water level rose by 10 feet, how would the clearances be affected? Make a decision about which plan you think would be the better one as far as your department is concerned, and explain why you think so.

To obtain points on the ellipse off the x-axis, it must be that $a > c$. To see why, look again at Figure 20:

$$d(F_1, P) + d(F_2, P) > d(F_1, F_2)$$ The sum of the lengths of two sides of a triangle is greater than the length of the third side.

$$2a > 2c$$ $d(F_1, P) + d(F_2, P) = 2a;$ $d(F_1, F_2) = 2c$

$$a > c$$

Since $a > c$, we also have $a^2 > c^2$, so $a^2 - c^2 > 0$. Let $b^2 = a^2 - c^2, b > 0$. Then $a > b$ and equation (1) can be written as

$$b^2x^2 + a^2y^2 = a^2b^2$$

$$\frac{x^2}{a^2} + \frac{y^2}{b^2} = 1$$ Divide each side by a^2b^2.

Theorem

An equation of the ellipse with center at $(0, 0)$ and foci at $(-c, 0)$ and $(c, 0)$ is

Equation of an Ellipse;
Center at (0, 0);
Foci at (± c, 0);
Major Axis along
the x-Axis

$$\frac{x^2}{a^2} + \frac{y^2}{b^2} = 1 \qquad \text{where } a > b > 0 \text{ and } b^2 = a^2 - c^2 \qquad (2)$$

The major axis is the x-axis.   ∎

As you can verify, the ellipse defined by equation (2) is symmetric with respect to the x-axis, y-axis, and origin.

To find the vertices of the ellipse defined by equation (2), let $y = 0$. The vertices satisfy the equation $x^2/a^2 = 1$, the solutions of which are $x = \pm a$. Consequently, the vertices of the ellipse given by equation (2) are $V_1 = (-a, 0)$ and $V_2 = (a, 0)$. The y-intercepts of the ellipse, found by letting $x = 0$, have coordinates $(0, -b)$ and $(0, b)$. These four intercepts, $(a, 0)$, $(-a, 0)$, $(0, b)$, and $(0, -b)$, are used to graph the ellipse by hand. See Figure 21.

Notice in Figure 21 the right triangle formed with the points $(0, 0)$, $(c, 0)$, and $(0, b)$. Because $b^2 = a^2 - c^2$ (or $b^2 + c^2 = a^2$), the distance from the focus at $(c, 0)$ to the point $(0, b)$ is a.

FIGURE 21

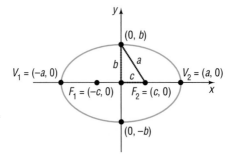

E X A M P L E   1

Finding an Equation of an Ellipse

Find an equation of the ellipse with center at the origin, one focus at $(3, 0)$, and a vertex at $(-4, 0)$. Graph the equation by hand.

Solution   The ellipse has its center at the origin, and the major axis coincides with the *x*-axis. One focus is at $(c, 0) = (3, 0)$, so $c = 3$. One vertex is at $(-a, 0) = (-4, 0)$, so $a = 4$. From equation (2), it follows that

$$b^2 = a^2 - c^2 = 16 - 9 = 7$$

so an equation of the ellipse is

$$\frac{x^2}{16} + \frac{y^2}{7} = 1$$

Figure 22 shows the graph drawn by hand.   ■

FIGURE 22

$$\frac{x^2}{16} + \frac{y^2}{7} = 1$$

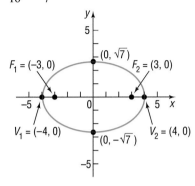

Notice in Figure 22 how we used the intercepts of the equation to graph the ellipse. Following this practice will make it easier for you to obtain an accurate graph of an ellipse when graphing by hand. It also tells you how to set the viewing rectangle when using a graphing utility.

E X A M P L E   2   *Graphing an Ellipse Using a Graphing Utility*

Use a graphing utility to graph the ellipse: $\dfrac{x^2}{16} + \dfrac{y^2}{7} = 1$.

Solution   First, we must solve $\dfrac{x^2}{16} + \dfrac{y^2}{7} = 1$ for *y*.

$$\frac{y^2}{7} = 1 - \frac{x^2}{16}$$ Subtract $\frac{x^2}{16}$ from each side.

$$y^2 = 7\left(1 - \frac{x^2}{16}\right)$$ Multiply both sides by 7.

$$y = \pm\sqrt{7\left(1 - \frac{x^2}{16}\right)}$$ Take the square root of each side.

Figure 23 shows the graphs of $y_1 = \sqrt{7\left(1 - \dfrac{x^2}{16}\right)}$ and $y_2 = -\sqrt{7\left(1 - \dfrac{x^2}{16}\right)}$.

FIGURE 23

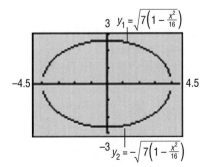

Notice in Figure 23 that we used a square screen. As with circles, this is done to avoid a distorted view of the graph.

An equation of the form of equation (2), with $a > b$, is the equation of an ellipse with center at the origin, foci on the x-axis at $(-c, 0)$ and $(c, 0)$, where $c^2 = a^2 - b^2$, and major axis along the x-axis.

For the remainder of this section, the direction "Discuss the equation" will mean to find the center, major axis, foci, and vertices of the ellipse and graph it.

E X A M P L E   3   *Discussing the Equation of an Ellipse*

Discuss the equation: $\dfrac{x^2}{25} + \dfrac{y^2}{9} = 1$

Solution   Figure 24(a) shows the graph of $\dfrac{x^2}{25} + \dfrac{y^2}{9} = 1$ using a graphing utility. We now proceed to analyze the equation. The given equation is of the form of equation (2), with $a^2 = 25$ and $b^2 = 9$. The equation is that of an ellipse with center $(0, 0)$ and major axis along the x-axis. The vertices are at $(\pm a, 0) = (\pm 5, 0)$. Because $b^2 = a^2 - c^2$, we find

$$c^2 = a^2 - b^2 = 25 - 9 = 16$$

The foci are at $(\pm c, 0) = (\pm 4, 0)$. Figure 24(b) shows the graph drawn by hand.

FIGURE 24

$\dfrac{x^2}{25} + \dfrac{y^2}{9} = 1$

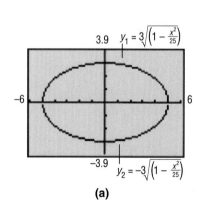

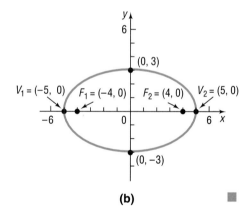

(a)   (b)

Now work Problems 9 and 19.

If the major axis of an ellipse with center at $(0, 0)$ coincides with the y-axis, then the foci are at $(0, -c)$ and $(0, c)$. Using the same steps as before, the definition of an ellipse leads to the following result:

Theorem   An equation of the ellipse with center at $(0, 0)$ and foci at $(0, -c)$ and $(0, c)$ is

**Equation of an Ellipse;
Center at (0, 0);
Foci at (0, ± c);
Major Axis
along the y-Axis**

$$\frac{x^2}{b^2} + \frac{y^2}{a^2} = 1 \qquad \text{where } a > b > 0 \text{ and } b^2 = a^2 - c^2 \qquad (3)$$

The major axis is the y-axis; the vertices are at $(0, -a)$ and $(0, a)$.

Figure 25 illustrates the graph of such an ellipse. Again, notice the right triangle with the points at $(0, 0)$, $(b, 0)$, and $(0, c)$.

FIGURE 25

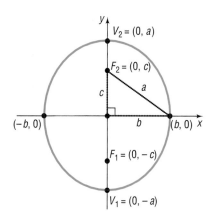

Look closely at equations (2) and (3). Although they may look alike, there is a difference! In equation (2), the larger number, a^2, is in the denominator of the x^2-term, so the major axis of the ellipse is along the x-axis. In equation (3), the larger number, a^2, is in the denominator of the y^2-term, so the major axis is along the y-axis.

E X A M P L E   4   *Discussing the Equation of an Ellipse*

Discuss the equation: $9x^2 + y^2 = 9$

Solution   Figure 26(a) shows the graph of $9x^2 + y^2 = 9$ using a graphing utility. We now proceed to analyze the equation. To put the equation in proper form, we divide each side by 9:

$$x^2 + \frac{y^2}{9} = 1$$

The larger number, 9, is in the denominator of the y^2-term so, based on equation (3), this is the equation of an ellipse with center at the origin and major axis along the y-axis. Also, we conclude that $a^2 = 9$, $b^2 = 1$, and $c^2 = a^2 - b^2 = 9 - 1 = 8$. The vertices are at $(0, \pm a) = (0, \pm 3)$, and the foci are at $(0, \pm c) = (0, \pm 2\sqrt{2})$. The graph, drawn by hand, is given in Figure 26(b).

FIGURE 26

$$x^2 + \frac{y^2}{9} = 1$$

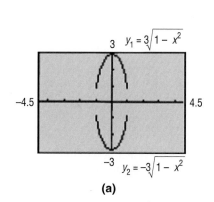

(a)

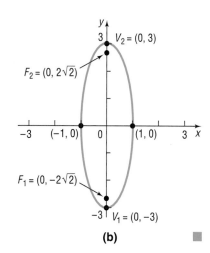

(b)

E X A M P L E   5   *Finding an Equation of an Ellipse*

Find an equation of the ellipse having one focus at $(0, 2)$ and vertices at $(0, -3)$ and $(0, 3)$. Graph the equation by hand.

Solution   Because the vertices are at $(0, -3)$ and $(0, 3)$, the center of this ellipse is at the origin. Also, its major axis coincides with the y-axis. The given information also reveals that $c = 2$ and $a = 3$, so $b^2 = a^2 - c^2 = 9 - 4 = 5$. The form of the equation of this ellipse is given by equation (3):

$$\frac{x^2}{b^2} + \frac{y^2}{a^2} = 1$$

$$\frac{x^2}{5} + \frac{y^2}{9} = 1$$

Figure 27 shows the graph.

FIGURE 27

$$\frac{x^2}{5} + \frac{y^2}{9} = 1$$

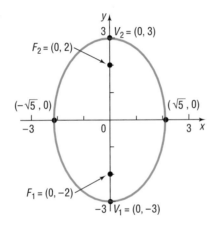

■   Now work Problems 13 and 21.

The circle may be considered a special kind of ellipse. To see why, let $a = b$ in equation (2) or in equation (3). Then

$$\frac{x^2}{a^2} + \frac{y^2}{a^2} = 1$$

$$x^2 + y^2 = a^2$$

This is the equation of a circle with center at the origin and radius a. The value of c is

$$c^2 = a^2 - b^2 = 0$$

We conclude that the closer the two foci of an ellipse are, the more the ellipse will look like a circle.

Center at *(h, k)*

If an ellipse with center at the origin and major axis coinciding with a coordinate axis is shifted horizontally h units and then vertically k units, the result is an ellipse with center at *(h, k)* and major axis parallel to a coordinate axis. Table 3 gives the forms of the equations of such ellipses, and Figure 28 shows their graphs.

TABLE 3   ELLIPSES WITH CENTER AT (h, k) AND MAJOR AXIS PARALLEL TO A COORDINATE AXIS

CENTER	MAJOR AXIS	FOCI	VERTICES	EQUATION
(h, k)	Parallel to x-axis	$(h \pm c, k)$	$(h \pm a, k)$	$\dfrac{(x - h)^2}{a^2} + \dfrac{(y - k)^2}{b^2} = 1,$ $a > b$ and $b^2 = a^2 - c^2$
(h, k)	Parallel to y-axis	$(h, k \pm c)$	$(h, k \pm a)$	$\dfrac{(x - h)^2}{b^2} + \dfrac{(y - k)^2}{a^2} = 1,$ $a > b$ and $b^2 = a^2 - c^2$

FIGURE 28

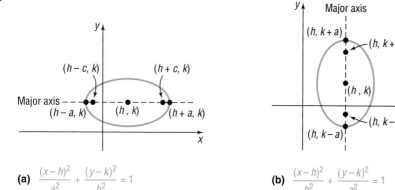

(a) $\dfrac{(x - h)^2}{a^2} + \dfrac{(y - k)^2}{b^2} = 1$

(b) $\dfrac{(x - h)^2}{b^2} + \dfrac{(y - k)^2}{a^2} = 1$

E X A M P L E   6   *Finding an Equation of an Ellipse, Center Not at the Origin*

Find an equation for the ellipse with center at $(2, -3)$, one focus at $(3, -3)$, and one vertex at $(5, -3)$. Graph the equation by hand.

Solution   The center is at $(h, k) = (2, -3)$, so $h = 2$ and $k = -3$. The major axis is parallel to the x-axis. The distance from the center $(2, -3)$ to a focus $(3, -3)$ is $c = 1$; the distance from the center $(2, -3)$ to a vertex $(5, -3)$ is $a = 3$. Thus, $b^2 = a^2 - c^2 = 9 - 1 = 8$. The form of the equation is

$$\frac{(x - h)^2}{a^2} + \frac{(y - k)^2}{b^2} = 1 \quad \text{where } h = 2, \, k = -3, \, a = 3, \, b = 2\sqrt{2}.$$

$$\frac{(x - 2)^2}{9} + \frac{(y + 3)^2}{8} = 1$$

Figure 29 shows the graph.

FIGURE 29

$\dfrac{(x - 2)^2}{9} + \dfrac{(y + 3)^2}{8} = 1$

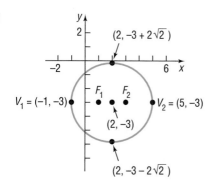

E X A M P L E   7   *Using a Graphing Utility to Graph an Ellipse, Center Not at the Origin*

Using a graphing utility, graph the ellipse: $\dfrac{(x-2)^2}{9} + \dfrac{(y+3)^2}{8} = 1$.

Solution   First, we must solve the equation $\dfrac{(x-2)^2}{9} + \dfrac{(y+3)^2}{8} = 1$ for y.

$$\frac{(y+3)^2}{8} = 1 - \frac{(x-2)^2}{9}$$ Subtract $\frac{(x-2)^2}{9}$ from each side.

$$(y+3)^2 = 8\left[1 - \frac{(x-2)^2}{9}\right]$$ Multiply each side by 8.

$$y + 3 = \pm\sqrt{8\left[1 - \frac{(x-2)^2}{9}\right]}$$ Take the square root of each side.

$$y = -3 \pm \sqrt{8\left[1 - \frac{(x-2)^2}{9}\right]}$$ Subtract 3 from each side.

FIGURE 30

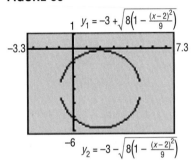

Figure 30 shows the graphs of $y_1 = -3 + \sqrt{8\left[1 - \dfrac{(x-2)^2}{9}\right]}$

and $y_2 = -3 - \sqrt{8\left[1 - \dfrac{(x-2)^2}{9}\right]}$.   ■

■   Now work Problem 33.

E X A M P L E   8   *Discussing the Equation of an Ellipse*

Discuss the equation: $4x^2 + y^2 - 8x + 4y + 4 = 0$

Solution   We proceed to complete the square in x and in y:

$$4x^2 + y^2 - 8x + 4y + 4 = 0$$
$$4x^2 - 8x + y^2 + 4y = -4$$
$$4(x^2 - 2x) + (y^2 + 4y) = -4$$
$$4(x^2 - 2x + 1) + (y^2 + 4y + 4) = -4 + 4 + 4 \quad \text{Complete each square.}$$
$$4(x-1)^2 + (y+2)^2 = 4$$
$$(x-1)^2 + \frac{(y+2)^2}{4} = 1 \qquad \text{Divide each side by 4.}$$

Figure 31(a) shows the graph of $(x-1)^2 + \dfrac{(y+2)^2}{4} = 1$ using a graphing utility. We now proceed to analyze the equation.

This is the equation of an ellipse with center at $(1, -2)$ and major axis parallel to the y-axis. Since $a^2 = 4$ and $b^2 = 1$, we have $c^2 = a^2 - b^2 = 4 - 1 = 3$. The vertices are at $(h, k \pm a) = (1, -2 \pm 2)$ or $(1, 0)$ and $(1, -4)$. The foci are at $(h, k \pm c) = (1, -2 \pm \sqrt{3})$ or $(1, -2 - \sqrt{3})$ and $(1, -2 + \sqrt{3})$. Figure 31(b) shows the graph drawn by hand.

FIGURE 31

$$(x - 1)^2 + \frac{(y + 2)^2}{4} = 1$$

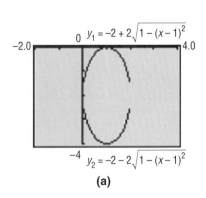

$y_1 = -2 + 2\sqrt{1 - (x-1)^2}$

$y_2 = -2 - 2\sqrt{1 - (x-1)^2}$

(a)

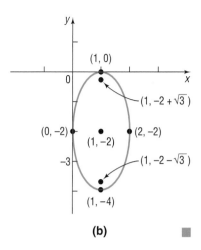

(b)

Applications

Ellipses are found in many applications in science and engineering. For example, the orbits of the planets around the Sun are elliptical, with the Sun's position at a focus. See Figure 32.

FIGURE 32

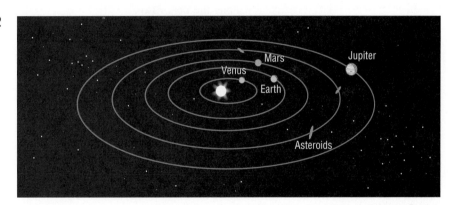

Stone and concrete bridges are often shaped as semielliptical arches. Elliptical gears are used in machinery when a variable rate of motion is required.

Ellipses also have an interesting reflection property. If a source of light (or sound) is placed at one focus, the waves transmitted by the source will reflect off the ellipse and concentrate at the other focus. This is the principal behind "whispering galleries," which are rooms designed with elliptical ceilings. A person standing at one focus of the ellipse can whisper and be heard by a person standing at the other focus, because all the sound waves that reach the ceiling are reflected to the other person.

EXAMPLE 9

Whispering Galleries

Figure 33 shows the specifications for an elliptical ceiling in a hall designed to be a whispering gallery. In a whispering gallery, a person standing at one focus of the ellipse can whisper and be heard by another person standing at the other focus, because all the sound waves that reach the ceiling from one focus are reflected to the other focus. Where in the hall are the foci located?

FIGURE 33

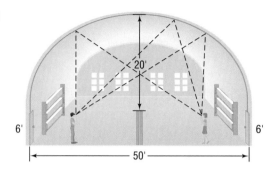

Solution   We set up a rectangular coordinate system so that the center of the ellipse is at the origin and the major axis is along the *x*-axis. See Figure 34. The equation of the ellipse is

$$\frac{x^2}{a^2} + \frac{y^2}{b^2} = 1$$

where $a = 25$ and $b = 20$. Since

$$c^2 = a^2 - b^2 = 25^2 - 20^2 = 625 - 400 = 225$$

we have $c = 15$. Thus, the foci are located 15 feet from the center of the ellipse along the major axis.

FIGURE 34

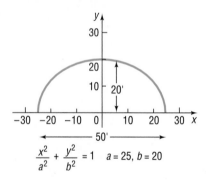

6.3

Exercise 6.3

In Problems 1–4, the graph of an ellipse is given. Match each graph to its equation.

A. $\dfrac{x^2}{4} + y^2 = 1$ B. $x^2 + \dfrac{y^2}{4} = 1$ C. $\dfrac{x^2}{16} + \dfrac{y^2}{4} = 1$ D. $\dfrac{x^2}{4} + \dfrac{y^2}{16} = 1$

1.
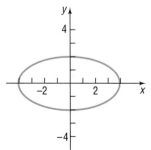

2.
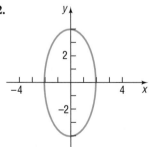

3.

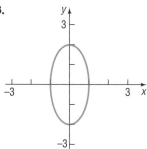

4.

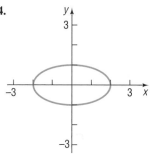

In Problems 5–8, the graph of an ellipse is given. Match each graph to its equation.

A. $\dfrac{(x+1)^2}{4} + \dfrac{(y-1)^2}{9} = 1$

B. $\dfrac{(x-1)^2}{4} + \dfrac{(y+1)^2}{9} = 1$

C. $\dfrac{(x-1)^2}{9} + \dfrac{(y+1)^2}{4} = 1$

D. $\dfrac{(x+1)^2}{9} + \dfrac{(y-1)^2}{4} = 1$

5.

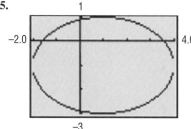

6.

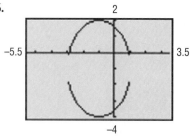

7.

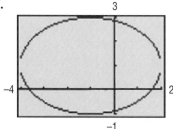

8.

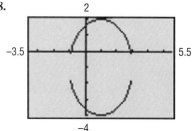

In Problems 9–18, find the vertices and foci of each ellipse. Graph each equation using a graphing utility.

9. $\dfrac{x^2}{25} + \dfrac{y^2}{4} = 1$

10. $\dfrac{x^2}{9} + \dfrac{y^2}{4} = 1$

11. $\dfrac{x^2}{9} + \dfrac{y^2}{25} = 1$

12. $\dfrac{x^2}{4} + \dfrac{y^2}{16} = 1$

13. $4x^2 + y^2 = 16$

14. $x^2 + 9y^2 = 18$

15. $4y^2 + x^2 = 8$

16. $4y^2 + 9x^2 = 36$

17. $x^2 + y^2 = 16$

18. $x^2 + y^2 = 4$

In Problems 19–28, find an equation for each ellipse. Graph the equation by hand.

19. Center at (0, 0); focus at (3, 0); vertex at (5, 0)

20. Center at (0, 0); focus at (−1, 0); vertex at (3, 0)

21. Center at (0, 0); focus at (0, −4); vertex at (0, 5)

22. Center at (0, 0); focus at (0, 1); vertex at (0, −2)

23. Foci at $(\pm 2, 0)$; length of the major axis is 6

24. Focus at $(0, -4)$; vertices at $(0, \pm 8)$

25. Foci at $(0, \pm 3)$; x-intercepts are ± 2

26. Foci at $(0, \pm 2)$; length of the major axis is 8

27. Center at $(0, 0)$; vertex at $(0, 4)$; $b = 1$

28. Vertices at $(\pm 5, 0)$; $c = 2$

In Problems 29–32, write an equation for each ellipse.

29.

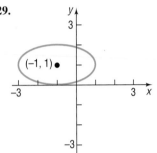

30.

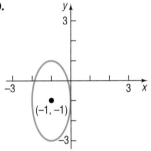

31.

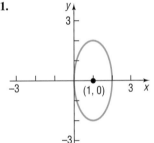

32.

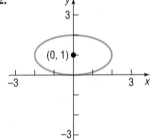

In Problems 33–44, find the center, foci, and vertices of each ellipse. Graph each equation, using a graphing utility.

33. $\dfrac{(x - 3)^2}{4} + \dfrac{(y + 1)^2}{9} = 1$

34. $\dfrac{(x + 4)^2}{9} + \dfrac{(y + 2)^2}{4} = 1$

35. $(x + 5)^2 + 4(y - 4)^2 = 16$

36. $9(x - 3)^2 + (y + 2)^2 = 18$

37. $x^2 + 4x + 4y^2 - 8y + 4 = 0$

38. $x^2 + 3y^2 - 12y + 9 = 0$

39. $2x^2 + 3y^2 - 8x + 6y + 5 = 0$

40. $4x^2 + 3y^2 + 8x - 6y = 5$

41. $9x^2 + 4y^2 - 18x + 16y - 11 = 0$

42. $x^2 + 9y^2 + 6x - 18y + 9 = 0$

43. $4x^2 + y^2 + 4y = 0$

44. $9x^2 + y^2 - 18x = 0$

In Problems 45–54, find an equation for each ellipse. Graph the equation by hand.

45. Center at $(2, -2)$; vertex at $(7, -2)$; focus at $(4, -2)$

46. Center at $(-3, 1)$; vertex at $(-3, 3)$; focus at $(-3, 0)$

47. Vertices at $(4, 3)$ and $(4, 9)$; focus at $(4, 8)$

48. Foci at $(1, 2)$ and $(-3, 2)$; vertex at $(-4, 2)$

49. Foci at $(5, 1)$ and $(-1, 1)$; length of the major axis is 8

50. Vertices at $(2, 5)$ and $(2, -1)$; $c = 2$

51. Center at $(1, 2)$; focus at $(4, 2)$; contains the point $(1, 3)$

52. Center at $(1, 2)$; focus at $(1, 4)$; contains the point $(2, 2)$

53. Center at $(1, 2)$; vertex at $(4, 2)$; contains the point $(1, 3)$

54. Center at $(1, 2)$; vertex at $(1, 4)$; contains the point $(2, 2)$

In Problems 55–58, graph each function by hand. Use a graphing utility to verify your results. [Hint: Notice that each function is half an ellipse.]

55. $f(x) = \sqrt{16 - 4x^2}$ **56.** $f(x) = \sqrt{9 - 9x^2}$

57. $f(x) = -\sqrt{64 - 16x^2}$ **58.** $f(x) = -\sqrt{4 - 4x^2}$

59. *Semielliptical Arch Bridge* An arch in the shape of the upper half of an ellipse is used to support a bridge that is to span a river 20 meters wide. The center of the arch is 6 meters above the center of the river (see the figure). Write an equation for the ellipse in which the x-axis coincides with the water level and the y-axis passes through the center of the arch.

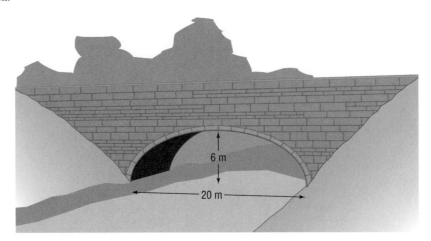

60. *Semielliptical Arch Bridge* The arch of a bridge is a semiellipse with a horizontal major axis. The span is 30 feet, and the top of the arch is 10 feet above the major axis. The roadway is horizontal and is 2 feet above the top of the arch. Find the vertical distance from the roadway to the arch at 5 foot intervals along the roadway.

61. *Whispering Galleries* A hall 100 feet in length is to be designed as a whispering gallery. If the foci are located 25 feet from the center, how high will the ceiling be at the center?

62. *Whispering Galleries* A person, standing at one focus of a whispering gallery, is 6 feet from the nearest wall. Her friend is standing at the other focus, 100 feet away. What is the length of this whispering gallery? How high is its elliptical ceiling at the center?

63. *Semielliptical Arch Bridge* A bridge is built in the shape of a semielliptical arch. The bridge has a span of 120 feet and a maximum height of 25 feet. Choose a suitable rectangular coordinate system and find the height of the arch at distances of 10, 30, and 50 feet from the center.

64. *Semielliptical Arch Bridge* A bridge is built in the shape of a semielliptical arch and is to have a span of 100 feet. The height of the arch, a distance of 40 feet from the center, is to be 10 feet. Find the height of the arch at its center.

65. *Semielliptical Arch* An arch in the form of half an ellipse is 40 feet wide and 15 feet high at the center. Find the height of the arch at intervals of 10 feet along its width.

66. *Semielliptical Arch Bridge* An arch for a bridge over a highway is in the form of half an ellipse. The top of the arch is 20 feet above the ground level (the major axis). The highway has four lanes, each 12 feet wide; a center safety strip 8 feet wide; and two side strips, each 4 feet wide. What should the span of the bridge be (the length of its major axis) if the height 28 feet from the center is to be 13 feet?

*In Problems 67–70, use the fact that the orbit of a planet about the Sun is an ellipse, with the Sun at one focus. The **aphelion** of a planet is its greatest distance from the Sun and the **perihelion** is its shortest distance. The **mean distance** of a planet from the Sun is the length of the semimajor axis of the elliptical orbit. See the illustration.*

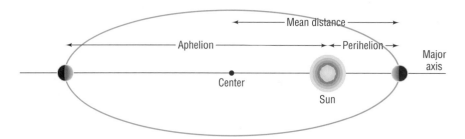

67. *Earth* The mean distance of Earth from the Sun is 93 million miles. If the aphelion of Earth is 94.5 million miles, what is the perihelion? Write an equation for the orbit of Earth around the Sun.

68. *Mars* The mean distance of Mars from the Sun is 142 million miles. If the perihelion of Mars is 128.5 million miles, what is the aphelion? Write an equation for the orbit of Mars about the Sun.

69. *Jupiter* The aphelion of Jupiter is 507 million miles. If the distance from the Sun to the center of its elliptical orbit is 23.2 million miles, what is the perihelion? What is the mean distance? Write an equation for the orbit of Jupiter around the Sun.

70. *Pluto* The perihelion of Pluto is 4551 million miles and the distance of the Sun from the center of its elliptical orbit is 897.5 million miles. Find the aphelion of Pluto. What is the mean distance of Pluto from the Sun? Write an equation for the orbit of Pluto about the Sun.

71. Consult the figure. A racetrack is in the shape of an ellipse, 100 feet long and 50 feet wide. What is the width 10 feet from the side?

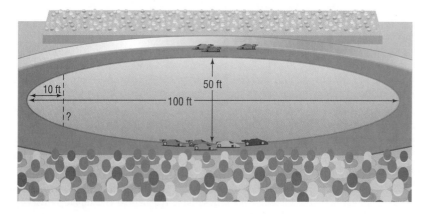

72. A racetrack is in the shape of an ellipse 80 feet long and 40 feet wide. What is the width 10 feet from the side?

73. Show that an equation of the form

$$Ax^2 + Cy^2 + F = 0 \qquad A \neq 0, C \neq 0, F \neq 0$$

where A and C are of the same sign and F is of opposite sign:

(a) Is the equation of an ellipse with center at $(0, 0)$ if $A \neq C$.

(b) Is the equation of a circle with center at $(0, 0)$ if $A = C$.

74. Show that the graph of an equation of the form

$$Ax^2 + Cy^2 + Dx + Ey + F = 0 \qquad A \neq 0, C \neq 0$$

where A and C are of the same sign:

(a) Is an ellipse if $(D^2/4A) + (E^2/4C) - F$ is the same sign as A.

(b) Is a point if $(D^2/4A) + (E^2/4C) - F = 0$.

(c) Contains no points if $(D^2/4A) + (E^2/4C) - F$ is of opposite sign to A.

75. The **eccentricity** e of an ellipse is defined as the number c/a, where a and c are the numbers given in equation (2). Because $a > c$, it follows that $e < 1$. Write a brief paragraph about the general shape of each of the following ellipses. Be sure to justify your conclusions.

(a) Eccentricity close to 0   (b) Eccentricity $= 0.5$ (c) Eccentricity close to 1

76. *CBL Experiment* The motion of a swinging pendulum is examined by plotting the velocity against the position of the pendulum. The result is the graph of an ellipse. (Activity 17, Real-World Math with the CBL System)

6.4

The Hyperbola

Hyperbola

> A **hyperbola** is the collection of all points in the plane the difference of whose distances from two fixed points, called the **foci,** is a constant.

Figure 35 illustrates a hyperbola with foci F_1 and F_2. The line containing the foci is called the **transverse axis.** The midpoint of the line segment joining the foci is called the **center** of the hyperbola. The line through the center and perpendicular to the transverse axis is called the **conjugate axis.** The hyperbola consists of two separate curves, called **branches,** that are symmetric with respect to the transverse axis, conjugate axis, and center. The two points of intersection of the hyperbola and the transverse axis are the **vertices,** V_1 and V_2, of the hyperbola.

FIGURE 35

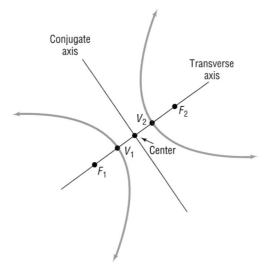

With these ideas in mind, we are now ready to find the equation of a hyperbola in a rectangular coordinate system. First, we place the center at the origin. Next, we position the hyperbola so that its transverse axis coincides with a coordinate axis. Suppose that the transverse axis coincides with the x-axis, as shown in Figure 36.

FIGURE 36
$d(F_1, P) - d(F_2, P) = \pm 2a$

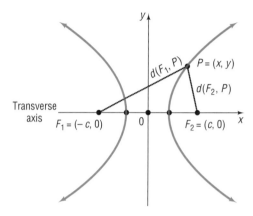

If c is the distance from the center to a focus, then one focus will be at $F_1 = (-c, 0)$ and the other at $F_2 = (c, 0)$. Now we let the constant difference of the distances from any point $P = (x, y)$ on the hyperbola to the foci F_1 and F_2 be denoted by $\pm 2a$. (If P is on the right branch, the $+$ sign is used; if P is on the left branch, the $-$ sign is used.) The coordinates of P must satisfy the equation

$$d(F_1, P) - d(F_2, P) = \pm 2a \qquad \text{Difference of the distances from } P \text{ to the foci equals } \pm 2a.$$

$$\sqrt{(x + c)^2 + y^2} - \sqrt{(x - c)^2 + y^2} = \pm 2a \qquad \text{Use the distance formula.}$$

$$\sqrt{(x + c)^2 + y^2} = \pm 2a + \sqrt{(x - c)^2 + y^2} \qquad \text{Isolate one radical.}$$

$$(x + c)^2 + y^2 = 4a^2 \pm 4a\sqrt{(x - c)^2 + y^2} \qquad \text{Square both sides.}$$
$$+ (x - c)^2 + y^2$$

Next, we remove the parentheses:

$$x^2 + 2cx + c^2 + y^2 = 4a^2 \pm 4a\sqrt{(x - c)^2 + y^2} + x^2 - 2cx + c^2 + y^2$$

$$4cx - 4a^2 = \pm 4a\sqrt{(x - c)^2 + y^2} \qquad \text{Isolate the radical.}$$

$$cx - a^2 = \pm a\sqrt{(x - c)^2 + y^2} \qquad \text{Divide each side by 4.}$$

$$(cx - a^2)^2 = a^2[(x - c)^2 + y^2] \qquad \text{Square both sides.}$$

$$c^2x^2 - 2ca^2x + a^4 = a^2(x^2 - 2cx + c^2 + y^2)$$

$$c^2x^2 + a^4 = a^2x^2 + a^2c^2 + a^2y^2$$

$$(c^2 - a^2)x^2 - a^2y^2 = a^2c^2 - a^4$$

$$(c^2 - a^2)x^2 - a^2y^2 = a^2(c^2 - a^2) \qquad (1)$$

To obtain points on the hyperbola off the x-axis, it must be that $a < c$. To see why, look again at Figure 36.

$$d(F_1, P) < d(F_2, P) + d(F_1, F_2) \qquad \text{Use triangle } F_1PF_2.$$

$$d(F_1, P) - d(F_2, P) < d(F_1, F_2) \qquad \begin{array}{l} P \text{ is on the right branch,} \\ \text{so } d(F_1, P) - d(F_2, P) = 2a. \end{array}$$

$$2a < 2c$$

$$a < c$$

Since $a < c$, we also have $a^2 < c^2$, so $c^2 - a^2 > 0$. Let $b^2 = c^2 - a^2$, $b > 0$. Then equation (1) can be written as

$$b^2x^2 - a^2y^2 = a^2b^2$$

$$\frac{x^2}{a^2} - \frac{y^2}{b^2} = 1$$

To find the vertices of the hyperbola defined by this equation, let $y = 0$. The vertices satisfy the equation $x^2/a^2 = 1$, the solutions of which are $x = \pm a$. Consequently, the vertices of the hyperbola are $V_1 = (-a, 0)$ and $V_2 = (a, 0)$.

Theorem An equation of the hyperbola with center at $(0, 0)$, foci at $(-c, 0)$ and $(c, 0)$, and vertices at $(-a, 0)$ and $(a, 0)$ is

Equation of a Hyperbola; Center at (0, 0); Foci at (±c, 0); Vertices at (±a, 0); Transverse Axis along x-Axis

$$\frac{x^2}{a^2} - \frac{y^2}{b^2} = 1 \qquad \text{where } b^2 = c^2 - a^2 \qquad (2)$$

The transverse axis is the x-axis. ∎

As you can verify, the hyperbola defined by equation (2) is symmetric with respect to the x-axis, y-axis, and origin. To find the y-intercepts, if any, let $x = 0$ in equation (2). This results in the equation $y^2/b^2 = -1$, which has no solution. We conclude that the hyperbola defined by equation (2) has no y-intercepts. In fact, since $x^2/a^2 - 1 = y^2/b^2 \geq 0$, it follows that $x^2/a^2 \geq 1$. Thus, there are no points on the graph for $-a < x < a$. See Figure 37.

FIGURE 37
$$\frac{x^2}{a^2} - \frac{y^2}{b^2} = 1,$$
$$b^2 = c^2 - a^2$$

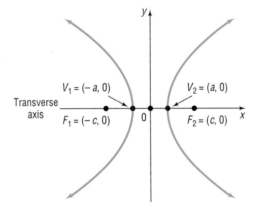

EXAMPLE 1   *Finding an Equation of a Hyperbola*

Find an equation of the hyperbola with center at the origin, one focus at $(3, 0)$, and one vertex at $(-2, 0)$. Graph the equation by hand.

Solution   The hyperbola has its center at the origin, and the transverse axis coincides with the x-axis. One focus is at $(c, 0) = (3, 0)$, so $c = 3$. One vertex is at $(-a, 0) = (-2, 0)$, so $a = 2$. From equation (2), it follows that $b^2 = c^2 - a^2 = 9 - 4 = 5$, so an equation of the hyperbola is

$$\frac{x^2}{4} - \frac{y^2}{5} = 1$$

See Figure 38.

FIGURE 38

$$\frac{x^2}{4} - \frac{y^2}{5} = 1$$

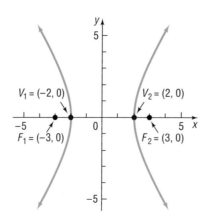

E X A M P L E **2**

Using a Graphing Utility to Graph a Hyperbola

Use a graphing utility to graph the hyperbola: $\dfrac{x^2}{4} - \dfrac{y^2}{5} = 1$

Solution   To graph the hyperbola $\dfrac{x^2}{4} - \dfrac{y^2}{5} = 1$, we need to graph the two functions

$y_1 = \sqrt{5}\sqrt{\dfrac{x^2}{4} - 1}$ and $y_2 = -\sqrt{5}\sqrt{\dfrac{x^2}{4} - 1}$. As with graphing circles and ellipses on a graphing utility, we use a square screen setting so that the graph is not distorted. Figure 39 shows the graph of the hyperbola.

FIGURE 39

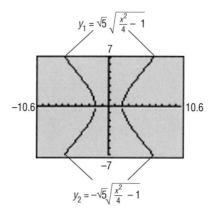

■ Now work Problem 19.

An equation of the form of equation (2) is the equation of a hyperbola with center at the origin, foci on the x-axis at $(-c, 0)$ and $(c, 0)$, where $c^2 = a^2 + b^2$, and transverse axis along the x-axis.

For the remainder of this section, the direction "Discuss the equation" will mean to find the center, transverse axis, vertices, and foci of the hyperbola and graph it.

E X A M P L E **3**

Discussing the Equation of a Hyperbola

Discuss the equation: $\dfrac{x^2}{16} - \dfrac{y^2}{4} = 1$

Solution Figure 40(a) shows the graph of $\dfrac{x^2}{16} - \dfrac{y^2}{4} = 1$ using a graphing utility. We now proceed to analyze the equation.

The given equation is of the form of equation (2), with $a^2 = 16$ and $b^2 = 4$. Thus, the graph of the equation is a hyperbola with center at (0, 0) and transverse axis along the x-axis. Also, we know that $c^2 = a^2 + b^2 = 16 + 4 = 20$. The vertices are at $(\pm a, 0) = (\pm 4, 0)$, and the foci are at $(\pm c, 0) = (\pm 2\sqrt{5}, 0)$. Figure 40(b) shows the graph drawn by hand.

FIGURE 40

$\dfrac{x^2}{16} - \dfrac{y^2}{4} = 1$

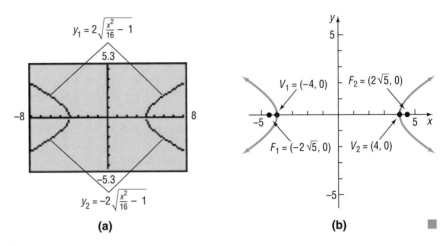

$y_1 = 2\sqrt{\dfrac{x^2}{16} - 1}$

$y_2 = -2\sqrt{\dfrac{x^2}{16} - 1}$

(a)

$V_1 = (-4, 0)$ $F_2 = (2\sqrt{5}, 0)$

$F_1 = (-2\sqrt{5}, 0)$ $V_2 = (4, 0)$

(b)

The next result gives the form of the equation of a hyperbola with center at the origin and transverse axis along the y-axis.

Theorem An equation of the hyperbola with center at (0, 0), foci at $(0, -c)$ and $(0, c)$, and vertices at $(0, -a)$ and $(0, a)$ is

$$\frac{y^2}{a^2} - \frac{x^2}{b^2} = 1 \qquad \text{where } b^2 = c^2 - a^2 \qquad (3)$$

Equation of a Hyperbola; Center at (0, 0); Foci at (0, ±c); Vertices at (0, ±a); Transverse Axis along y-Axis

The transverse axis is the y-axis.

Figure 41 shows the graph of a typical hyperbola defined by equation (3).

FIGURE 41

$\dfrac{y^2}{a^2} - \dfrac{x^2}{b^2} = 1,$

$b^2 = c^2 - a^2$

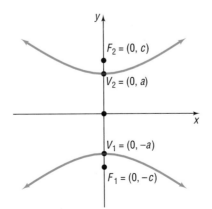

$F_2 = (0, c)$

$V_2 = (0, a)$

$V_1 = (0, -a)$

$F_1 = (0, -c)$

Notice the difference in the form of equations (2) and (3). When the y^2-term is subtracted from the x^2-term, the transverse axis is the x-axis. When the x^2-term is subtracted from the y^2-term, the transverse axis is the y-axis.

EXAMPLE   4

Discussing the Equation of a Hyperbola

Discuss the equation: $y^2 - 4x^2 = 4$

Solution   Figure 42(a) shows the graph of $y^2 - 4x^2 = 4$ using a graphing utility. We now proceed to analyze the equation.

To put the equation in proper form, we divide each side by 4:

$$\frac{y^2}{4} - x^2 = 1$$

Since the x^2-term is subtracted from the y^2-term, the equation is that of a hyperbola with center at the origin and transverse axis along the y-axis. Also, comparing the above equation to equation (3), we find $a^2 = 4$, $b^2 = 1$, and $c^2 = a^2 + b^2 = 5$. The vertices are at $(0, \pm a) = (0, \pm 2)$, and the foci are at $(0, \pm c) = (0, \pm\sqrt{5})$. The graph, drawn by hand, is given in Figure 42(b).

FIGURE 42

$$\frac{y^2}{4} - x^2 = 1$$

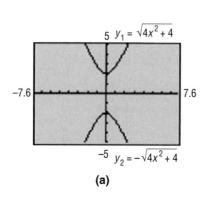

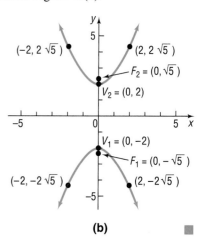

(a)   (b)

EXAMPLE   5

Finding an Equation of a Hyperbola

Find an equation of the hyperbola having one vertex at $(0, 2)$ and foci at $(0, -3)$ and $(0, 3)$. Graph the equation by hand.

Solution   Since the foci are at $(0, -3)$ and $(0, 3)$, the center of the hyperbola is at the origin. Also, the transverse axis is along the y-axis. The given information also reveals that $c = 3$, $a = 2$, and $b^2 = c^2 - a^2 = 9 - 4 = 5$. The form of the equation of the hyperbola is given by equation (3):

$$\frac{y^2}{a^2} - \frac{x^2}{b^2} = 1$$

$$\frac{y^2}{4} - \frac{x^2}{5} = 1$$

See Figure 43.

FIGURE 43

$$\frac{y^2}{4} - \frac{x^2}{5} = 1$$

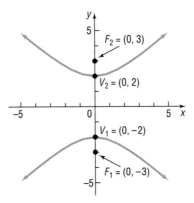

■ Now work Problem 11.

Look at the equations of the hyperbolas in Examples 4 and 5. For the hyperbola in Example 4, $a^2 = 4$ and $b^2 = 1$, so $a > b$; for the hyperbola in Example 5, $a^2 = 4$ and $b^2 = 5$, so $a < b$. We conclude that, for hyperbolas, there are no requirements involving the relative size of a and b. Contrast this situation to the case of an ellipse, in which the relative sizes of a and b dictate which axis is the major axis. Hyperbolas have another feature to distinguish them from ellipses and parabolas: Hyperbolas have asymptotes.

Asymptotes

Recall from Section 4.3 that a horizontal or oblique asymptote of a graph is a line with the property that the distance from the line to points on the graph approaches 0 as $x \to -\infty$ or as $x \to \infty$.

Theorem   The hyperbola $\dfrac{x^2}{a^2} - \dfrac{y^2}{b^2} = 1$

has the two oblique asymptotes

Asymptotes of a Hyperbola

$$y = \frac{b}{a}x \quad \text{and} \quad y = -\frac{b}{a}x$$

Proof   We begin by solving for y in the equation of the hyperbola:

$$\frac{x^2}{a^2} - \frac{y^2}{b^2} = 1$$

$$\frac{y^2}{b^2} = \frac{x^2}{a^2} - 1$$

$$y^2 = b^2\left(\frac{x^2}{a^2} - 1\right)$$

If $x \neq 0$, we can rearrange the right side in the form

$$y^2 = \frac{b^2 x^2}{a^2}\left(1 - \frac{a^2}{x^2}\right)$$

$$y = \pm\frac{bx}{a}\sqrt{1 - \frac{a^2}{x^2}}$$

Now, as $x \to -\infty$ or as $x \to \infty$, the term a^2/x^2 approaches 0, so the expression under the radical approaches 1. Thus, as $x \to -\infty$ or as $x \to \infty$, the value of y approaches $\pm bx/a$; that is, the graph of the hyperbola approaches the lines

$$y = -\frac{b}{a}x \quad \text{and} \quad y = \frac{b}{a}x$$

Thus, these lines are oblique asymptotes of the hyperbola. ■

The asymptotes of a hyperbola are not part of the hyperbola, but they do serve as a guide for graphing a hyperbola by hand. For example, suppose that we want to graph the equation

$$\frac{x^2}{a^2} - \frac{y^2}{b^2} = 1$$

We begin by plotting the vertices $(-a, 0)$ and $(a, 0)$. Then we plot the points $(0, -b)$ and $(0, b)$ and use these four points to construct a rectangle, as shown in Figure 44. The diagonals of this rectangle have slopes b/a and $-b/a$, and their extensions are the asymptotes $y = (b/a)x$ and $y = -(b/a)x$ of the hyperbola.

FIGURE 44

$$\frac{x^2}{a^2} - \frac{y^2}{b^2} = 1$$

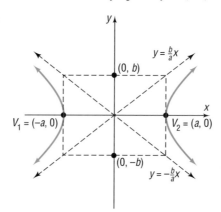

Theorem   The hyperbola $\dfrac{y^2}{a^2} - \dfrac{x^2}{b^2} = 1$

has the two oblique asymptotes

Asymptotes of a Hyperbola

$$y = \frac{a}{b}x \quad \text{and} \quad y = -\frac{a}{b}x$$

■

You are asked to prove this result in Problem 64.

E X A M P L E   6   *Discussing the Equation of a Hyperbola*
Discuss the equation: $9x^2 - 4y^2 = 36$

Solution   Figure 45(a) shows the graph of $\dfrac{x^2}{4} - \dfrac{y^2}{9} = 1$ and its asymptotes, $y = \dfrac{3}{2}x$ and

$y = -\dfrac{3}{2}x$, using a graphing utility. We now proceed to analyze the equation.

First, we divide each side by 36 to put the equation in proper form:

$$\frac{x^2}{4} - \frac{y^2}{9} = 1$$

This is the equation of a hyperbola with center at the origin and transverse axis along the x-axis. Using $a^2 = 4$ and $b^2 = 9$, we find $c^2 = a^2 + b^2 = 13$. The vertices are at $(\pm a, 0) = (\pm 2, 0)$, the foci are at $(\pm c, 0) = (\pm\sqrt{13}, 0)$, and the asymptotes have the equations

$$y = \frac{3}{2}x \quad \text{and} \quad y = -\frac{3}{2}x$$

Now form the rectangle containing the points $(\pm a, 0)$ and $(0, \pm b)$, that is, $(-2, 0)$ $(2, 0)$, $(0, -3)$, and $(0, 3)$. The extensions of the diagonals of this rectangle are the asymptotes. See Figure 45(b) for the graph drawn by hand.

FIGURE 45

$$\frac{x^2}{4} - \frac{y^2}{9} = 1$$

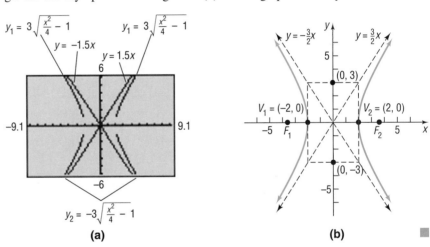

(a)   (b)

Exploration: Use TRACE to see what happens as x becomes unbounded in the positive direction for both the upper and the lower portions of the hyperbola.

Now work Problem 21.

Center at (h, k)

If a hyperbola with center at the origin and transverse axis coinciding with a coordinate axis is shifted horizontally h units and then vertically k units, the result is a hyperbola with center at (h, k) and transverse axis parallel to a coordinate axis. Table 4 gives the forms of the equations of such hyperbolas. See Figure 46 for the graphs.

TABLE 4   HYPERBOLAS WITH CENTER AT (h,k) AND TRANSVERSE AXIS PARALLEL TO A COORDINATE AXIS

CENTER	TRANSVERSE AXIS	FOCI	VERTICES	EQUATION	ASYMPTOTES
(h, k)	Parallel to x-axis	$(h \pm c, k)$	$(h \pm a, k)$	$\dfrac{(x-h)^2}{a^2} - \dfrac{(y-k)^2}{b^2} = 1,$ $b^2 = c^2 - a^2$	$y - k = \pm\dfrac{b}{a}(x - h)$
(h, k)	Parallel to y-axis	$(h, k \pm c)$	$(h, k \pm a)$	$\dfrac{(y-k)^2}{a^2} - \dfrac{(x-h)^2}{b^2} = 1,$ $b^2 = c^2 - a^2$	$y - k = \pm\dfrac{a}{b}(x - h)$

FIGURE 46

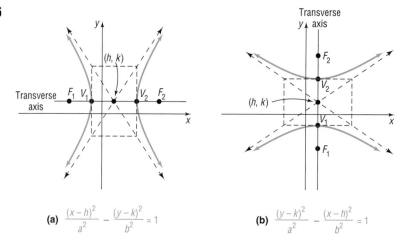

(a) $\dfrac{(x-h)^2}{a^2} - \dfrac{(y-k)^2}{b^2} = 1$ (b) $\dfrac{(y-k)^2}{a^2} - \dfrac{(x-h)^2}{b^2} = 1$

E X A M P L E   7   *Finding an Equation of a Hyperbola, Center Not at the Origin*

Find an equation for the hyperbola with center at $(1, -2)$, one focus at $(4, -2)$, and one vertex at $(3, -2)$. Graph the equation by hand.

Solution   The center is at $(h, k) = (1, -2)$, so $h = 1$ and $k = -2$. The transverse axis is parallel to the x-axis. The distance from the center $(1, -2)$ to the focus $(4, -2)$ is $c = 3$; the distance from the center $(1, -2)$ to the vertex $(3, -2)$ is $a = 2$. Thus, $b^2 = c^2 - a^2 = 9 - 4 = 5$. The equation is

$$\frac{(x-h)^2}{a^2} - \frac{(y-k)^2}{b^2} = 1$$

$$\frac{(x-1)^2}{4} - \frac{(y+2)^2}{5} = 1$$

See Figure 47.

FIGURE 47

$\dfrac{(x-1)^2}{4} - \dfrac{(y+2)^2}{5} = 1$

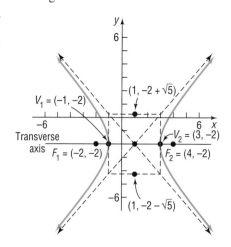

■   Now work Problem 31.

E X A M P L E   8   *Using a Graphing Utility to Graph a Hyperbola, Center Not at the Origin*

Use a graphing utility to graph the hyperbola $\dfrac{(x-1)^2}{4} - \dfrac{(y+2)^2}{5} = 1$

Solution First, we must solve the equation for y:

$$\frac{(x-1)^2}{4} - \frac{(y+2)^2}{5} = 1$$

FIGURE 48

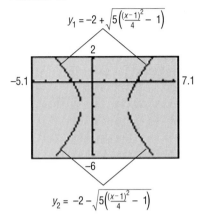

$y_1 = -2 + \sqrt{5\left(\frac{(x-1)^2}{4} - 1\right)}$

2

−5.1   7.1

−6

$y_2 = -2 - \sqrt{5\left(\frac{(x-1)^2}{4} - 1\right)}$

$$\frac{(y+2)^2}{5} = \frac{(x-1)^2}{4} - 1$$

$$(y+2)^2 = 5\left[\frac{(x-1)^2}{4} - 1\right]$$

$$y + 2 = \pm\sqrt{5\left[\frac{(x-1)^2}{4} - 1\right]}$$

$$y = -2 \pm\sqrt{5\left[\frac{(x-1)^2}{4} - 1\right]}$$

Figure 48 shows the graph of $y_1 = -2 + \sqrt{5\left[\frac{(x-1)^2}{4} - 1\right]}$ and

$y_2 = -2 - \sqrt{5\left[\frac{(x-1)^2}{4} - 1\right]}$. ∎

E X A M P L E   9   *Discussing the Equation of a Hyperbola*

Discuss the equation: $-x^2 + 4y^2 - 2x - 16y + 11 = 0$

Solution We complete the squares in x and in y:

$$-x^2 + 4y^2 - 2x - 16y + 11 = 0$$
$$-(x^2 + 2x) + 4(y^2 - 4y) = -11 \qquad \text{Group terms.}$$
$$-(x^2 + 2x + 1) + 4(y^2 - 4y + 4) = -1 + 16 - 11 \qquad \text{Complete each square.}$$
$$-(x+1)^2 + 4(y-2)^2 = 4$$
$$(y-2)^2 - \frac{(x+1)^2}{4} = 1 \qquad \text{Divide by 4.}$$

Figure 49(a) shows the graph of $(y-2)^2 - \dfrac{(x+1)^2}{4} = 1$ using a graphing utility. We now proceed to analyze the equation.

It is the equation of a hyperbola with center at $(-1, 2)$ and transverse axis parallel to the y-axis. Also, $a^2 = 1$ and $b^2 = 4$, so $c^2 = a^2 + b^2 = 5$. The vertices are at $(h, k \pm a) = (-1, 2 \pm 1)$, or $(-1, 1)$ and $(-1, 3)$. The foci are at $(h, k \pm c) = (-1, 2 \pm \sqrt{5})$. The asymptotes are $y - 2 = \frac{1}{2}(x+1)$ and $y - 2 = -\frac{1}{2}(x+1)$. Figure 49(b) shows the graph drawn by hand.

FIGURE 49

$(y-2)^2 - \dfrac{(x+1)^2}{4} = 1$

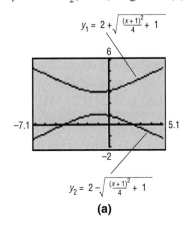

$y_1 = 2 + \sqrt{\frac{(x+1)^2}{4} + 1}$

6

−7.1   5.1

−2

$y_2 = 2 - \sqrt{\frac{(x+1)^2}{4} + 1}$

(a)

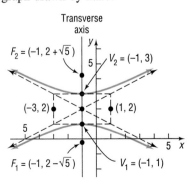

Transverse axis

$F_2 = (-1, 2 + \sqrt{5})$ $V_2 = (-1, 3)$

$(-3, 2)$ $(1, 2)$

$F_1 = (-1, 2 - \sqrt{5})$ $V_1 = (-1, 1)$

(b) ∎

FIGURE 50

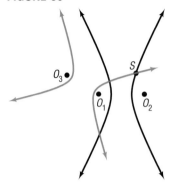

Applications

Suppose that a gun is fired from an unknown source S. An observer at O_1 hears the report (sound of gun shot) 1 second after another observer at O_2. Because sound travels at about 1100 feet per second, it follows that the point S must be 1100 feet closer to O_2 than to O_1. Thus, S lies on one branch of a hyperbola with foci at O_1 and O_2. (Do you see why? The difference of the distances from S to O_1 and from S to O_2 is the constant 1100.) If a third observer at O_3 hears the same report 2 seconds after O_1 hears it, then S will lie on a branch of a second hyperbola with foci at O_1 and O_3. The intersection of the two hyperbolas will pinpoint the location of S. See Figure 50 for an illustration.

LORAN

In the LOng RAnge Navigation system (LORAN), a master radio sending station and a secondary sending station emit signals that can be received by a ship at sea. (See Figure 51.) Because a ship monitoring the two signals will usually be nearer to one of the two stations, there will be a difference in the distance the two signals travel, which will register as a slight time difference between the signals. As long as the time difference remains constant, the difference of the two distances will also be constant. If the ship follows a path corresponding to the fixed time difference, it will follow the path of a hyperbola whose foci are located at the positions of the two sending stations. So for each time difference a different hyperbolic path results, each bringing the ship to a different shore location. Navigation charts show the various hyperbolic paths corresponding to different time differences.

FIGURE 51

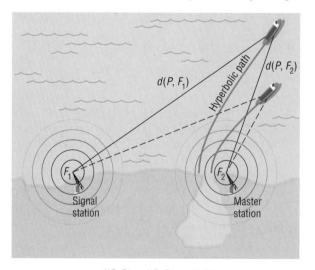

$d(P, F_1)$ $d(P, F_2)$

$d(P, F_1) - d(P, F_2) = \text{constant}$

EXAMPLE 10   *LORAN*

Two LORAN stations are positioned 250 miles apart along a straight shore.

(a) A ship records a time difference of 0.00086 second between the LORAN signals. Set up an appropriate rectangular coordinate system to determine where the ship would reach shore if it were to follow the hyperbola corresponding to this time difference.

(b) If the ship wants to enter a harbor located between the two stations 25 miles from the master station, what time difference should it be looking for?

(c) If the ship is 80 miles off shore when the desired time difference is obtained, what is the exact location of the ship? [*Note:* The speed of each radio signal is 186,000 miles per second.]

Solution   (a) We set up a rectangular coordinate system so that the two stations lie on the *x*-axis and the origin is midway between them. See Figure 52.

FIGURE 52

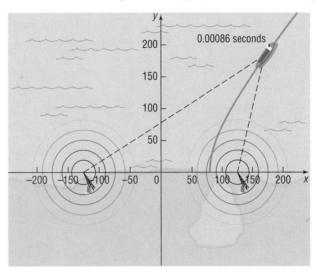

The ship lies on a hyperbola whose foci are the locations of the two stations. The reason for this is that the constant time difference of the signals from each station results in a constant difference in the distance of the ship from each station. Since the time difference is 0.00086 second and the speed of the signal is 186,000 miles per second, the difference of the distances from the ship to each station (foci) is

Distance = Speed × Time = 186,000 × 0.00086 = 160 miles

The difference of the distances from the ship to each station, 160, equals 2*a,* so *a* = 80 and the vertex of the corresponding hyperbola is at (80, 0). Since the focus is at (125, 0), following this hyperbola the ship would reach shore 45 miles from the master station.

(b) To reach shore 25 miles from the master station, the ship should follow a hyperbola with vertex at (100, 0). For this hyperbola, *a* = 100, so the constant difference of the distances from the ship to each station is 200. The time difference the ship should look for is

$$\text{Time} = \frac{\text{Distance}}{\text{Speed}} = \frac{200}{186,000} = 0.001075 \text{ second}$$

(c) To find the exact location of the ship, we need to find the equation of the hyperbola with vertex at (100, 0) and a focus at (125, 0). The form of the equation of this hyperbola is

$$\frac{x^2}{a^2} - \frac{y^2}{b^2} = 1$$

where *a* = 100. Since *c* = 125, we have

$$b^2 = c^2 - a^2 = 125^2 - 100^2 = 5625$$

The equation of the hyperbola is

$$\frac{x^2}{100^2} - \frac{y^2}{5625} = 1$$

Since the ship is 80 miles from shore, we use $y = 80$ in the equation and solve for x.

$$\frac{x^2}{100^2} - \frac{80^2}{5625} = 1$$

$$\frac{x^2}{100^2} = 1 + \frac{80^2}{5625} = 2.14$$

$$x^2 = 100^2(2.14)$$

$$x = 146$$

The ship is at the position (146, 80). See Figure 53.

FIGURE 53

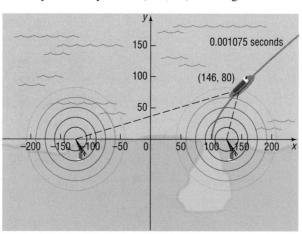

6.4

Exercise 6.4

In Problems 1–4, the graph of a hyperbola is given. Match each graph to its equation.

A. $\dfrac{x^2}{4} - y^2 = 1$ B. $x^2 - \dfrac{y^2}{4} = 1$ C. $\dfrac{y^2}{4} - x^2 = 1$ D. $y^2 - \dfrac{x^2}{4} = 1$

1.

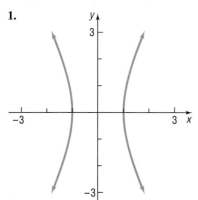

2.

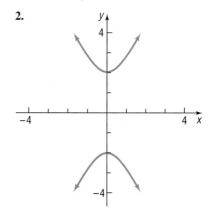

3.

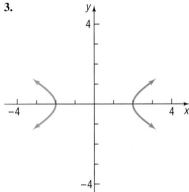

4.

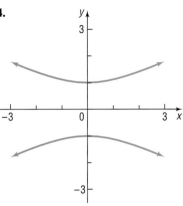

In Problems 5–8, the graph of a hyperbola is given. Match each graph to its equation.

A. $\dfrac{x^2}{16} - \dfrac{y^2}{9} = 1$ B. $\dfrac{x^2}{9} - \dfrac{y^2}{16} = 1$ C. $\dfrac{y^2}{16} - \dfrac{x^2}{9} = 1$ D. $\dfrac{y^2}{9} - \dfrac{x^2}{16} = 1$

5.

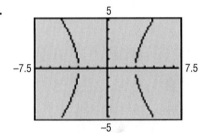

6.

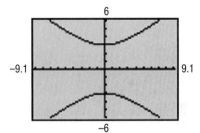

7.

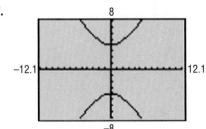

8.

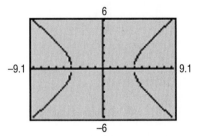

In Problems 9–18, find an equation for the hyperbola described. Graph the equation by hand.

9. Center at $(0, 0)$; focus at $(3, 0)$; vertex at $(1, 0)$
10. Center at $(0, 0)$; focus at $(0, 5)$; vertex at $(0, 3)$
11. Center at $(0, 0)$; focus at $(0, -6)$; vertex at $(0, 4)$
12. Center at $(0, 0)$; focus at $(-3, 0)$; vertex at $(2, 0)$
13. Foci at $(-5, 0)$ and $(5, 0)$; vertex at $(3, 0)$
14. Focus at $(0, 6)$; vertices at $(0, -2)$ and $(0, 2)$
15. Vertices at $(0, -6)$ and $(0, 6)$; asymptote the line $y = 2x$
16. Vertices at $(-4, 0)$ and $(4, 0)$; asymptote the line $y = 2x$
17. Foci at $(-4, 0)$ and $(4, 0)$; asymptote the line $y = -x$
18. Foci at $(0, -2)$ and $(0, 2)$; asymptote the line $y = -x$

In Problems 19–26, find the center, transverse axis, vertices, foci, and asymptotes. Graph each equation using a graphing utility.

19. $\dfrac{x^2}{16} - \dfrac{y^2}{4} = 1$ **20.** $\dfrac{y^2}{16} - \dfrac{x^2}{4} = 1$ **21.** $4x^2 - y^2 = 16$ **22.** $y^2 - 4x^2 = 16$

23. $y^2 - 9x^2 = 9$ **24.** $x^2 - y^2 = 4$ **25.** $y^2 - x^2 = 25$ **26.** $2x^2 - y^2 = 4$

In Problems 27–30, write an equation for each hyperbola.

27.

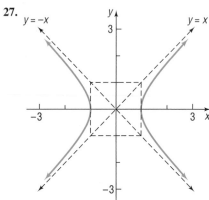

28.

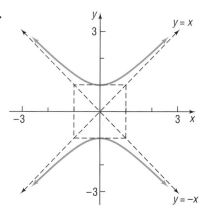

29.

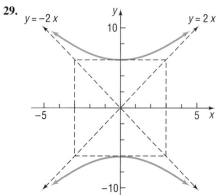

30.

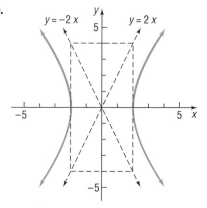

In Problems 31–38, find an equation for the hyperbola described. Graph the equation by hand.

31. Center at $(4, -1)$; focus at $(7, -1)$; vertex at $(6, -1)$

32. Center at $(-3, 1)$; focus at $(-3, 6)$; vertex at $(-3, 4)$

33. Center at $(-3, -4)$; focus at $(-3, -8)$; vertex at $(-3, -2)$

34. Center at $(1, 4)$; focus at $(-2, 4)$; vertex at $(0, 4)$

35. Foci at $(3, 7)$ and $(7, 7)$; vertex at $(6, 7)$

36. Focus at $(-4, 0)$; vertices at $(-4, 4)$ and $(-4, 2)$

37. Vertices at $(-1, -1)$ and $(3, -1)$; asymptote the line $(x - 1)/2 = (y + 1)/3$

38. Vertices at $(1, -3)$ and $(1, 1)$; asymptote the line $(x - 1)/2 = (y + 1)/3$

In Problems 39–52, find the center, transverse axis, vertices, foci, and asymptotes. Graph each equation using a graphing utility.

39. $\dfrac{(x - 2)^2}{4} - \dfrac{(y + 3)^2}{9} = 1$ **40.** $\dfrac{(y + 3)^2}{4} - \dfrac{(x - 2)^2}{9} = 1$

41. $(y - 2)^2 - 4(x + 2)^2 = 4$ **42.** $(x + 4)^2 - 9(y - 3)^2 = 9$ **43.** $(x + 1)^2 - (y + 2)^2 = 4$

44. $(y - 3)^2 - (x + 2)^2 = 4$ **45.** $x^2 - y^2 - 2x - 2y - 1 = 0$ **46.** $y^2 - x^2 - 4y + 4x - 1 = 0$

47. $y^2 - 4x^2 - 4y - 8x - 4 = 0$ **48.** $2x^2 - y^2 + 4x + 4y - 4 = 0$ **49.** $4x^2 - y^2 - 24x - 4y + 16 = 0$

50. $2y^2 - x^2 + 2x + 8y + 3 = 0$ **51.** $y^2 - 4x^2 - 16x - 2y - 19 = 0$ **52.** $x^2 - 3y^2 + 8x - 6y + 4 = 0$

In Problems 53–56, graph each function by hand. Verify your answer using a graphing utility.
[Hint: Notice that each function is "half" a hyperbola.]

53. $f(x) = \sqrt{16 + 4x^2}$ **54.** $f(x) = -\sqrt{9 + 9x^2}$

55. $f(x) = -\sqrt{-25 + x^2}$ **56** $f(x) = \sqrt{-1 + x^2}$

57. *LORAN* Two LORAN stations are positioned 200 miles apart along a straight shore.
   (a) A ship records a time difference of 0.00038 second between the LORAN signals. Set up an appropriate rectangular coordinate system to determine where the ship would reach shore if it were to follow the hyperbola corresponding to this time difference.
   (b) If the ship wants to enter a harbor located between the two stations 20 miles from the master station, what time difference should it be looking for?
   (c) If the ship is 50 miles off shore when the desired time difference is obtained, what is the exact location of the ship? [*Note:* The speed of each radio signal is 186,000 miles per second.]

58. *LORAN* Two LORAN stations are positioned 100 miles apart along a straight shore.
   (a) A ship records a time difference of 0.00032 second between the LORAN signals. Set up an appropriate rectangular coordinate system to determine where the ship would reach shore if it were to follow the hyperbola corresponding to this time difference.
   (b) If the ship wants to enter a harbor located between the two stations 10 miles from the master station, what time difference should it be looking for?
   (c) If the ship is 20 miles off shore when the desired time difference is obtained, what is the exact location of the ship? [*Note:* The speed of each radio signal is 186,000 miles per second.]

59. *Calibrating Instruments* In a test of their recording devices a team of seismologists positioned two of the devices 2000 feet apart, with the device at point *A* to the west of the device at point *B*. At a point between the devices and 200 feet from point *B*, a small amount of explosive was detonated and a note made of the time at which the sound reached each device. A second explosion is to be carried out at a point directly north of point *B*.
   (a) How far north should the site of the second explosion be chosen so that the measured time difference recorded by the devices for the second detonation is the same as that recorded for the first detonation?
   (b) Explain why this experiment can be used to calibrate the instruments.

60. Explain in your own words the LORAN system of navigation.

61. The **eccentricity** *e* of a hyperbola is defined as the number c/a, where *a* and *c* are the numbers given in equation (2). Because $c > a$, it follows that $e > 1$. Describe the general shape of a hyperbola whose eccentricity is close to 1. What is the shape if *e* is very large?

62. A hyperbola for which $a = b$ is called an **equilateral hyperbola.** Find the eccentricity *e* of an equilateral hyperbola. [*Note:* The eccentricity of a hyperbola is defined in Problem 61.]

63. Two hyperbolas that have the same set of asymptotes are called **conjugate.** Show that the hyperbolas

$$\frac{x^2}{4} - y^2 = 1 \quad \text{and} \quad y^2 - \frac{x^2}{4} = 1$$

are conjugate. Graph each hyperbola on the same set of coordinate axes.

64. Prove that the hyperbola

$$\frac{y^2}{a^2} - \frac{x^2}{b^2} = 1$$

has the two oblique asymptotes

$$y = \frac{a}{b}x \quad \text{and} \quad y = -\frac{a}{b}x$$

65. Show that the graph of an equation of the form

$$Ax^2 + Cy^2 + F = 0 \qquad A \neq 0, C \neq 0, F \neq 0$$

where A and C are of opposite sign, is a hyperbola with center at $(0, 0)$.

66. Show that the graph of an equation of the form

$$Ax^2 + Cy^2 + Dx + Ey + F = 0 \qquad A \neq 0, C \neq 0$$

where A and C are of opposite sign:
(a) Is a hyperbola if $(D^2/4A) + (E^2/4C) - F \neq 0$.
(b) Is two intersecting lines if $(D^2/4A) + (E^2/4C) - F = 0$.

6.5

Vectors

In simple terms, a **vector** (derived from the Latin *vehere*, meaning "to carry") is a quantity that has both magnitude and direction. For a vector in the plane, which is the only type we shall discuss, it is convenient to represent a vector by using an arrow. The length of the arrow represents the **magnitude** of the vector, and the arrowhead indicates the **direction** of the vector.

FIGURE 54

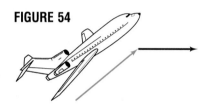

Many quantities in physics can be represented by vectors. For example, the velocity of an aircraft can be represented by an arrow that points in the direction of movement; the length of the arrow represents speed. Thus, if the aircraft speeds up, we lengthen the arrow; if the aircraft changes direction, we introduce an arrow in the new direction. See Figure 54. Based on this representation, it is not surprising that vectors and directed line segments are somehow related.

Directed Line Segments

If P and Q are two distinct points in the xy-plane, there is exactly one line containing both P and Q. The points on that part of the line that joins P to Q, including P and Q, form what is called the **line segment** $\overline{PQ}$. If we order the points so that they proceed from P to Q, we have a **directed line segment** from P to Q, which we denote by $\overrightarrow{PQ}$. In a directed line segment $\overrightarrow{PQ}$, we call P the **initial point** and Q the **terminal point,** as indicated in Figure 55.

FIGURE 55

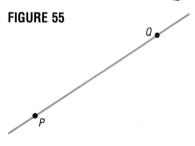

(a) Line containing P and Q

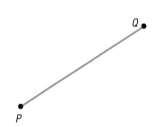

(b) Line segment $\overline{PQ}$

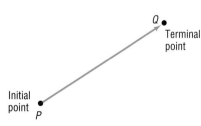

Terminal point

Initial point

(c) Directed line segment $\overrightarrow{PQ}$

The magnitude of the directed line segment $\overrightarrow{PQ}$ is the distance from the point P to the point Q, that is, it is the length of the line segment. The direction of $\overrightarrow{PQ}$ is from P to Q. If a vector $\mathbf{v}$* has the same magnitude and the same direction as the directed line segment $\overrightarrow{PQ}$, then we write

$$\mathbf{v} = \overrightarrow{PQ}$$

*Boldface letters will be used to denote vectors, in order to distinguish them from numbers. For handwritten work, an arrow is placed over the letter to signify a vector.

FIGURE 56

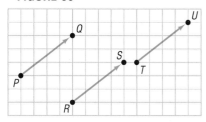

The vector **v** whose magnitude is 0 is called the **zero vector, 0.** The zero vector is assigned no direction.

Two vectors **v** and **w** are **equal,** written

$$\mathbf{v} = \mathbf{w}$$

if they have the same magnitude and the same direction.

For example, the vectors shown in Figure 56 have the same magnitude and the same direction, so they are equal, even though they have different initial points and different terminal points. As a result, we find it useful to think of a vector simply as an arrow, keeping in mind that two arrows (vectors) are equal if they have the same direction and the same magnitude (length).

Adding Vectors

FIGURE 57

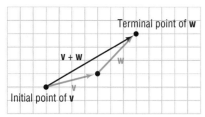

The **sum v + w** of two vectors is defined as follows: We position the vectors **v** and **w** so that the terminal point of **v** coincides with the initial point of **w**, as shown in Figure 57. The vector **v + w** is then the unique vector whose initial point coincides with the initial point of **v** and whose terminal point coincides with the terminal point of **w**.

Vector addition is **commutative.** That is, if **v** and **w** are any two vectors, then

Commutative Property
FIGURE 58

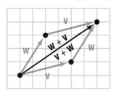

$$\mathbf{v} + \mathbf{w} = \mathbf{w} + \mathbf{v}$$

Figure 58 illustrates this fact. (Observe that the commutative property is another way of saying that opposite sides of a parallelogram are equal and parallel.)

Vector addition is also **associative.** That is, if **u**, **v**, and **w** are vectors, then

Associative Property

$$\mathbf{u} + (\mathbf{v} + \mathbf{w}) = (\mathbf{u} + \mathbf{v}) + \mathbf{w}$$

Figure 59 illustrates the associative property for vectors.

FIGURE 59
$(\mathbf{u} + \mathbf{v}) + \mathbf{w} = \mathbf{u} + (\mathbf{v} + \mathbf{w})$

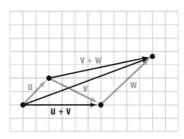

The zero vector has the property that

Identity Property

$$\mathbf{v} + \mathbf{0} = \mathbf{0} + \mathbf{v} = \mathbf{v}$$

for any vector **v**.

If **v** is a vector, then −**v** is the vector having the same magnitude as **v**, but whose direction is opposite to **v**, as shown in Figure 60.

FIGURE 60

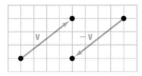

Furthermore,

Additive Inverse Property

$$\mathbf{v} + (-\mathbf{v}) = \mathbf{0}$$

If $\mathbf{v}$ and $\mathbf{w}$ are two vectors, we define the **difference $\mathbf{v} - \mathbf{w}$** as

FIGURE 61

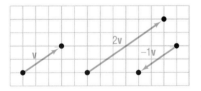

$$\mathbf{v} - \mathbf{w} = \mathbf{v} + (-\mathbf{w})$$

Figure 61 illustrates the relationships among $\mathbf{v}$, $\mathbf{w}$, $\mathbf{v} + \mathbf{w}$, and $\mathbf{v} - \mathbf{w}$.

Multiplying Vectors by Numbers

When dealing with vectors, we refer to real numbers as **scalars.** Scalars are quantities that have only magnitude. Examples from physics of scalar quantities are temperature, speed, and time. We now define how to multiply a vector by a scalar.

Scalar Product

If α is a scalar and **v** is a vector, the **scalar product** $\alpha\mathbf{v}$ is defined as:

1. If $\alpha > 0$, the product $\alpha\mathbf{v}$ is the vector whose magnitude is α times the magnitude of $\mathbf{v}$ and whose direction is the same as $\mathbf{v}$.
2. If $\alpha < 0$, the product $\alpha\mathbf{v}$ is the vector whose magnitude is $|\alpha|$ times the magnitude of $\mathbf{v}$ and whose direction is opposite that of $\mathbf{v}$.
3. If $\alpha = 0$ or if $\mathbf{v} = \mathbf{0}$, then $\alpha\mathbf{v} = \mathbf{0}$.

FIGURE 62

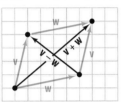

See Figure 62 for some illustrations.

For example, if **a** is the acceleration of an object of mass m due to a force **F** being exerted on it, then by Newton's Second Law of Motion, $\mathbf{F} = m\mathbf{a}$. Here, $m\mathbf{a}$ is the product of the scalar m and the vector **a**.

Scalar products have the following properties:

Properties of Scalar Products

$$0\mathbf{v} = \mathbf{0} \qquad 1\mathbf{v} = \mathbf{v} \qquad -1\mathbf{v} = -\mathbf{v}$$
$$(\alpha + \beta)\mathbf{v} = \alpha\mathbf{v} + \beta\mathbf{v} \qquad \alpha(\mathbf{v} + \mathbf{w}) = \alpha\mathbf{v} + \alpha\mathbf{w}$$
$$\alpha(\beta\mathbf{v}) = (\alpha\beta)\mathbf{v}$$

E X A M P L E   1   *Graphing Vectors*

Use the vectors illustrated in Figure 63 to graph each expression.

FIGURE 63

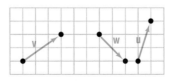

(a) $\mathbf{v} - \mathbf{w}$ (b) $2\mathbf{v} + 3\mathbf{w}$ (c) $2\mathbf{v} - \mathbf{w} + 3\mathbf{u}$

Solution   Figure 64 illustrates each graph.

FIGURE 64

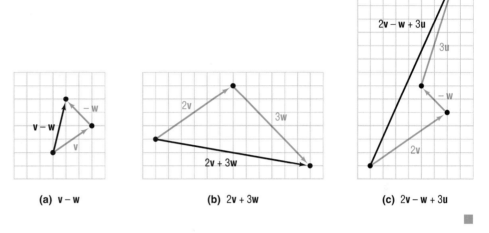

(a) v − w   (b) 2v + 3w   (c) 2v − w + 3u

■   Now work Problems 1 and 3.

Magnitudes of Vectors

If **v** is a vector, we use the symbol $\|\mathbf{v}\|$ to represent the **magnitude** of **v**. Since $\|\mathbf{v}\|$ equals the length of a directed line segment, it follows that $\|\mathbf{v}\|$ has the following properties:

Theorem   If **v** is a vector and if α is a scalar, then:
Properties of $\|v\|$

(a) $\|\mathbf{v}\| \geq 0$ (b) $\|\mathbf{v}\| = 0$ if and only if $\mathbf{v} = \mathbf{0}$

(c) $\|-\mathbf{v}\| = \|\mathbf{v}\|$ (d) $\|\alpha\mathbf{v}\| = |\alpha|\|\mathbf{v}\|$ ■

Property (a) is a consequence of the fact that distance is a nonnegative number. Property (b) follows, because the length of the directed line segment $\overrightarrow{PQ}$ is positive unless P and Q are the same point, in which case the length is 0. Property (c) follows, because the length of the line segment $\overline{PQ}$ equals the length of the line segment $\overline{QP}$. Property (d) is a direct consequence of the definition of a scalar product.

Unit Vector   A vector **u** for which $\|\mathbf{u}\| = 1$ is called a **unit vector.**

To compute the magnitude and direction of a vector, we need an algebraic way of representing vectors.

Representing Vectors in the Plane

We use a rectangular coordinate system to represent vectors in the plane. Let **i** denote a unit vector whose direction is along the positive x-axis; let **j** denote a unit vector whose direction is along the positive y-axis. If **v** is a vector with initial point at the origin O and terminal point at $P = (a, b)$, then we can represent $\mathbf{v}$ in terms of the vectors $\mathbf{i}$ and $\mathbf{j}$ as

FIGURE 65

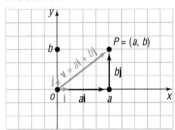

$$\mathbf{v} = a\mathbf{i} + b\mathbf{j}$$

See Figure 65. The scalars a and b are called the **components** of the vector $\mathbf{v} = a\mathbf{i} + b\mathbf{j}$, with a being the component in the direction $\mathbf{i}$ and b being the component in the direction $\mathbf{j}$.

A vector whose initial point is at the origin is called a **position vector.** The next result states that any vector whose initial point is not at the origin is equal to a unique position vector.

Theorem   Suppose that $\mathbf{v}$ is a vector with initial point $P_1 = (x_1, y_1)$, not necessarily the origin, and terminal point $P_2 = (x_2, y_2)$. If $\mathbf{v} = \overrightarrow{P_1P_2}$, then $\mathbf{v}$ is equal to the position vector

$$\mathbf{v} = (x_2 - x_1)\mathbf{i} + (y_2 - y_1)\mathbf{j} \qquad \blacksquare$$

To see why this is true, look at Figure 66. Triangle OPA and triangle P_1P_2Q are congruent. (Do you see why?) The line segments have the same magnitude, so $d(O, P) = d(P_1, P_2)$; and they have the same direction, so $\angle POA = \angle P_2P_1Q$. Since the triangles are right triangles, we have Angle–Side–Angle. Thus, it follows that corresponding sides are equal. As a result, $x_2 - x_1 = a$ and $y_2 - y_1 = b$, so $\mathbf{v}$ may be written as

$$\mathbf{v} = a\mathbf{i} + b\mathbf{j} = (x_2 - x_1)\mathbf{i} + (y_2 - y_1)\mathbf{j} \qquad (1)$$

FIGURE 66

$$\mathbf{v} = a\mathbf{i} + b\mathbf{j} = (x_2 - x_1)\mathbf{i} + (y_2 - y_1)\mathbf{j}$$

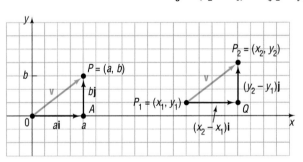

E X A M P L E   2   *Finding a Position Vector*

Find the position vector of the vector $\mathbf{v} = \overrightarrow{P_1P_2}$ if $P_1 = (-1, 2)$ and $P_2 = (4, 6)$.

Solution   By equation (1), the position vector equal to $\mathbf{v}$ is

$$\mathbf{v} = [4 - (-1)]\mathbf{i} + (6 - 2)\mathbf{j} = 5\mathbf{i} + 4\mathbf{j}$$

FIGURE 67

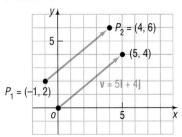

See Figure 67. $\blacksquare$

■   Now work Problem 21.

Two position vectors $\mathbf{v}$ and $\mathbf{w}$ are equal if and only if the terminal point of $\mathbf{v}$ is the same as the terminal point of $\mathbf{w}$. This leads to the following result:

Theorem

Two vectors **v** and **w** are equal if and only if their corresponding components are equal. That is:

Equality of Vectors

If $\mathbf{v} = a_1\mathbf{i} + b_1\mathbf{j}$ and $\mathbf{w} = a_2\mathbf{i} + b_2\mathbf{j}$,

then $\mathbf{v} = \mathbf{w}$ if and only if $a_1 = a_2$ and $b_1 = b_2$.

Because of the above result, we can replace any vector (directed line segment) by a unique position vector, and vice versa. This flexibility is one of the main reasons for the wide use of vectors. Unless otherwise specified, from now on the term *vector* will mean the unique position vector equal to it.

Next, we define addition, subtraction, scalar product, and magnitude in terms of the components of a vector.

Let $\mathbf{v} = a_1\mathbf{i} + b_1\mathbf{j}$ and $\mathbf{w} = a_2\mathbf{i} + b_2\mathbf{j}$ be two vectors, and let α be a scalar. Then:

$$\mathbf{v} + \mathbf{w} = (a_1 + a_2)\mathbf{i} + (b_1 + b_2)\mathbf{j} \qquad (2)$$

$$\mathbf{v} - \mathbf{w} = (a_1 - a_2)\mathbf{i} + (b_1 - b_2)\mathbf{j} \qquad (3)$$

$$\alpha\mathbf{v} = (\alpha a_1)\mathbf{i} + (\alpha b_1)\mathbf{j} \qquad (4)$$

$$\|\mathbf{v}\| = \sqrt{a_1^2 + b_1^2} \qquad (5)$$

These definitions are compatible with the geometric ones given earlier in this section. See Figure 68.

FIGURE 68

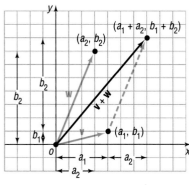

(a)   Illustration of property (2)

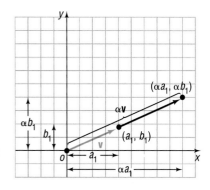

(b)   Illustration of property (4)

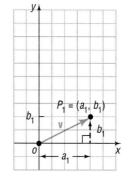

(c)   Illustration of property (5):
$\|\mathbf{v}\| = $ Distance of O to P_1
$\|\mathbf{v}\| = \sqrt{a_1^2 + b_1^2}$

Thus, to add two vectors, simply add corresponding components. To subtract two vectors, subtract corresponding components.

EXAMPLE 3   *Adding and Subtracting Vectors*

If $\mathbf{v} = 2\mathbf{i} + 3\mathbf{j}$ and $\mathbf{w} = 3\mathbf{i} - 4\mathbf{j}$; find:

(a) $\mathbf{v} + \mathbf{w}$ (b) $\mathbf{v} - \mathbf{w}$

Solution   (a) $\mathbf{v} + \mathbf{w} = (2\mathbf{i} + 3\mathbf{j}) + (3\mathbf{i} - 4\mathbf{j}) = (2 + 3)\mathbf{i} + (3 - 4)\mathbf{j}$
$$= 5\mathbf{i} - \mathbf{j}$$
(b) $\mathbf{v} - \mathbf{w} = (2\mathbf{i} + 3\mathbf{j}) - (3\mathbf{i} - 4\mathbf{j}) = (2 - 3)\mathbf{i} + [3 - (-4)]\mathbf{j}$
$$= -\mathbf{i} + 7\mathbf{j}$$ ■

EXAMPLE 4   *Forming Scalar Products*

If $\mathbf{v} = 2\mathbf{i} + 3\mathbf{j}$ and $\mathbf{w} = 3\mathbf{i} - 4\mathbf{j}$, find:

(a) $3\mathbf{v}$ (b) $2\mathbf{v} - 3\mathbf{w}$ (c) $\|\mathbf{v}\|$

Solution   (a) $3\mathbf{v} = 3(2\mathbf{i} + 3\mathbf{j}) = 6\mathbf{i} + 9\mathbf{j}$

(b) $2\mathbf{v} - 3\mathbf{w} = 2(2\mathbf{i} + 3\mathbf{j}) - 3(3\mathbf{i} - 4\mathbf{j}) = 4\mathbf{i} + 6\mathbf{j} - 9\mathbf{i} + 12\mathbf{j}$
$$= -5\mathbf{i} + 18\mathbf{j}$$
(c) $\|\mathbf{v}\| = \|2\mathbf{i} + 3\mathbf{j}\| = \sqrt{2^2 + 3^2} = \sqrt{13}$ ■

■   Now work Problems 27 and 33.

Recall that a unit vector $\mathbf{u}$ is one for which $\|\mathbf{u}\| = 1$. In many applications, it is useful to be able to find a unit vector $\mathbf{u}$ that has the same direction as a given vector $\mathbf{v}$.

Theorem   For any nonzero vector $\mathbf{v}$, the vector

Unit Vector in Direction of $\mathbf{v}$

$$\mathbf{u} = \frac{\mathbf{v}}{\|\mathbf{v}\|}$$

is a unit vector that has the same direction as $\mathbf{v}$. ■

Proof   Let $\mathbf{v} = a\mathbf{i} + b\mathbf{j}$. Then $\|\mathbf{v}\| = \sqrt{a^2 + b^2}$ and

$$\mathbf{u} = \frac{\mathbf{v}}{\|\mathbf{v}\|} = \frac{a\mathbf{i} + b\mathbf{j}}{\sqrt{a^2 + b^2}} = \frac{a}{\sqrt{a^2 + b^2}}\mathbf{i} + \frac{b}{\sqrt{a^2 + b^2}}\mathbf{j}$$

The vector $\mathbf{u}$ is in the same direction as $\mathbf{v}$, since $\|\mathbf{v}\| > 0$, and

$$\|\mathbf{u}\| = \sqrt{\frac{a^2}{a^2 + b^2} + \frac{b^2}{a^2 + b^2}} = \sqrt{\frac{a^2 + b^2}{a^2 + b^2}} = 1$$

Thus, $\mathbf{u}$ is a unit vector in the direction of $\mathbf{v}$. ■

As a consequence of this theorem, if $\mathbf{u}$ is a unit vector in the same direction as a vector $\mathbf{v}$, then $\mathbf{v}$ may be expressed as

$$\mathbf{v} = \|\mathbf{v}\|\,\mathbf{u} \tag{6}$$

This way of expressing a vector is useful in many applications.

E X A M P L E   5   *Finding a Unit Vector*

Find a unit vector in the same direction as $\mathbf{v} = 4\mathbf{i} - 3\mathbf{j}$.

Solution   We find $\|\mathbf{v}\|$ first:

$$\|\mathbf{v}\| = \|4\mathbf{i} - 3\mathbf{j}\| = \sqrt{16 + 9} = 5$$

Now we multiply $\mathbf{v}$ by the scalar $1/\|\mathbf{v}\| = \frac{1}{5}$. The result is

$$\frac{\mathbf{v}}{\|\mathbf{v}\|} = \frac{4\mathbf{i} - 3\mathbf{j}}{5} = \frac{4}{5}\mathbf{i} - \frac{3}{5}\mathbf{j}$$

Check: This vector is, in fact, a unit vector because
$\left(\frac{4}{5}\right)^2 + \left(-\frac{3}{5}\right)^2 = \frac{16}{25} + \frac{9}{25} = \frac{25}{25} = 1$.

■   Now work Problem 43.

Applications

FIGURE 69

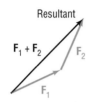

Resultant

$\mathbf{F}_1 + \mathbf{F}_2$

$\mathbf{F}_2$

$\mathbf{F}_1$

Forces provide an example of physical quantities that may be conveniently represented by vectors; two forces "combine" the way vectors "add." How do we know this? Laboratory experiments bear it out. Thus, if $\mathbf{F}_1$ and $\mathbf{F}_2$ are two forces simultaneously acting on an object, the vector sum $\mathbf{F}_1 + \mathbf{F}_2$ is the force that produces the same effect on the object as that obtained when the forces $\mathbf{F}_1$ and $\mathbf{F}_2$ act on the object. The force $\mathbf{F}_1 + \mathbf{F}_2$ is sometimes called the **resultant** of $\mathbf{F}_1$ and $\mathbf{F}_2$. See Figure 69.

Two important applications of the resultant of two vectors occur with aircraft flying in the presence of a wind and with boats cruising across a river with a current. For example, consider the velocity of wind acting on the velocity of an airplane (see Figure 70). Suppose $\mathbf{w}$ is a vector describing the velocity of the wind; that is, $\mathbf{w}$ represents the direction and speed of the wind. If $\mathbf{v}$ is the velocity of the airplane in the absence of wind (called its **velocity relative to the air**), then $\mathbf{v} + \mathbf{w}$ is the vector equal to the actual velocity of the airplane (called its **velocity relative to the ground**).

FIGURE 70

(a) Velocity **w** of wind
relative to ground

(b) Velocity **v** of airplane
relative to air

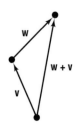

(c) Resultant **w** + **v** equals
velocity of airplane
relative to the ground

Our next example illustrates this use of vectors in navigation.

E X A M P L E   6

Finding the Actual Speed of an Aircraft

A Boeing 737 aircraft maintains a constant airspeed of 500 miles per hour in the direction due south. The velocity of the jet stream is 80 miles per hour in a northeasterly direction.

(a) Find a unit vector having northeast as direction.

(b) Find a vector 80 units in magnitude having the same direction as the unit vector found in part (a).

(c) Find the actual speed of the aircraft relative to the ground.

Solution   We set up a coordinate system in which north (N) is along the positive y-axis. See Figure 71. Let

$$\mathbf{v}_a = \text{Velocity of aircraft relative to the air} = -500\mathbf{j}$$
$$\mathbf{v}_w = \text{Velocity of jet stream}$$
$$\mathbf{v}_g = \text{Velocity of aircraft relative to ground}$$

FIGURE 71

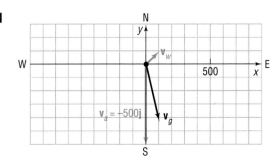

(a) A vector having northeast as direction is $\mathbf{i} + \mathbf{j}$. The unit vector in this direction is

$$\frac{\mathbf{i} + \mathbf{j}}{\|\mathbf{i} + \mathbf{j}\|} = \frac{\mathbf{i} + \mathbf{j}}{\sqrt{1 + 1}} = \frac{1}{\sqrt{2}}(\mathbf{i} + \mathbf{j})$$

(b) The velocity $\mathbf{v}_w$ of the jet stream is a vector with magnitude 80 in the direction of the unit vector $(1/\sqrt{2})(\mathbf{i} + \mathbf{j})$. Thus, from (6),

$$\mathbf{v}_w = 80\left[\frac{1}{\sqrt{2}}(\mathbf{i} + \mathbf{j})\right] = 40\sqrt{2}(\mathbf{i} + \mathbf{j})$$

(c) The velocity $\mathbf{v}_g$ of the aircraft relative to the ground is the resultant of the vectors $\mathbf{v}_a$ and $\mathbf{v}_w$. Thus,

$$\mathbf{v}_g = \mathbf{v}_a + \mathbf{v}_w = -500\mathbf{j} + 40\sqrt{2}(\mathbf{i} + \mathbf{j})$$
$$= 40\sqrt{2}\mathbf{i} + (40\sqrt{2} - 500)\mathbf{j}$$

The actual speed (speed relative to the ground) of the aircraft is

$$\|\mathbf{v}_g\| = \sqrt{(40\sqrt{2})^2 + (40\sqrt{2} - 500)^2} \approx 447 \text{ miles per hour}$$ ■

HISTORICAL FEATURE   ■ The history of vectors is surprisingly complicated for such a natural concept. In the *xy*-plane, complex numbers do a good job of imitating vectors. About 1840, mathematicians became interested in finding a system that would do for three dimensions what the complex numbers do for two dimensions. Hermann Grassmann (1809–1877), in Germany, and William Rowan Hamilton (1805–1865), in Ireland, both attempted to find solutions.

Hamilton's system was the *quaternions,* which are best thought of as a real number plus a vector, and do for four dimensions what complex numbers do for two dimensions. In this system the order of multiplication matters; that is, **ab** ≠ **ba**. Hamilton spent the rest of his life working out quaternion theory and trying to get it accepted in applied mathematics, but he encountered fierce resistance due to the complicated nature of quaternion multiplication. In the work with quaternions, two products of vectors emerged, the scalar (or dot) and the vector (or cross) products.

Grassmann fared even worse than Hamilton; if people did not like Hamilton's work, at least they understood it. Grassmann's abstract style, though easily read today, was almost impenetrable during the previous century, and only a few of his ideas were appreciated. Among those few were the same scalar and vector products that Hamilton had found.

About 1880, the American physicist Josiah Willard Gibbs (1839–1903) worked out an algebra involving only the simplest concepts—the vectors and the two products. He then added some calculus, and the resulting system was simple, flexible, and well-adapted to expressing a large number of physical laws. This system remains in use essentially unchanged. Hamilton's and Grassmann's more extensive systems each gave birth to much interesting mathematics, but little of this mathematics is seen at elementary levels.   ■

6.5

Exercise 6.5

In Problems 1–8, use the vectors in the figure to the right to graph each expression.

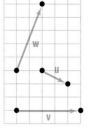

1. **v** + **w**
2. **u** + **v**
3. 3**v**
4. 4**w**
5. **v** − **w**
6. **u** − **v**
7. 3**v** + **u** − 2**w**
8. 2**u** − 3**v** + **w**

In Problems 9–16, use the figure below to find each vector.

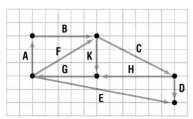

9. **x**, if **x** + **B** = **F**
10. **x**, if **x** + **D** = **E**
11. **C** in terms of **E**, **D**, and **F**
12. **G** in terms of **C**, **D**, **E**, and **K**
13. **E** in terms of **G**, **H**, and **D**
14. **E** in terms of **A**, **B**, **C**, and **D**
15. **x**, if **x** = **A** + **B** + **K** + **G**
16. **x**, if **x** = **A** + **B** + **C** + **H** + **G**
17. If ‖**v**‖ = 4, what is ‖3**v**‖
18. If ‖**v**‖ = 2, what is ‖−4**v**‖?

*In Problems 19–26, the vector **v** has initial point P and terminal point Q. Write **v** in the form a**i** + b**j**; that is, find its position vector.*

19. $P = (0, 0); Q = (3, 4)$

20. $P = (0, 0); Q = (-3, -5)$

21. $P = (3, 2); Q = (5, 6)$

22. $P = (-3, 2); Q = (6, 5)$

23. $P = (-2, -1); Q = (6, -2)$

24. $P = (-1, 4); Q = (6, 2)$

25. $P = (1, 0); Q = (0, 1)$

26. $P = (1, 1); Q = (2, 2)$

In Problems 27–32, find $\|\mathbf{v}\|$.

27. $\mathbf{v} = 3\mathbf{i} - 4\mathbf{j}$ **28.** $\mathbf{v} = -5\mathbf{i} + 12\mathbf{j}$ **29.** $\mathbf{v} = \mathbf{i} - \mathbf{j}$

30. $\mathbf{v} = -\mathbf{i} - \mathbf{j}$ **31.** $\mathbf{v} = -2\mathbf{i} + 3\mathbf{j}$ **32.** $\mathbf{v} = 6\mathbf{i} + 2\mathbf{j}$

In Problems 33–38, find each quantity if $\mathbf{v} = 3\mathbf{i} - 5\mathbf{j}$ and $\mathbf{w} = -2\mathbf{i} + 3\mathbf{j}$.

33. $2\mathbf{v} + 3\mathbf{w}$ **34.** $3\mathbf{v} - 2\mathbf{w}$ **35.** $\|\mathbf{v} - \mathbf{w}\|$

36. $\|\mathbf{v} + \mathbf{w}\|$ **37.** $\|\mathbf{v}\| - \|\mathbf{w}\|$ **38.** $\|\mathbf{v}\| + \|\mathbf{w}\|$

*In Problems 39–44, find the unit vector having the same direction as **v**.*

39. $\mathbf{v} = 5\mathbf{i}$ **40.** $\mathbf{v} = -3\mathbf{j}$ **41.** $\mathbf{v} = 3\mathbf{i} - 4\mathbf{j}$

42. $\mathbf{v} = -5\mathbf{i} + 12\mathbf{j}$ **43.** $\mathbf{v} = \mathbf{i} - \mathbf{j}$ **44.** $\mathbf{v} = 2\mathbf{i} - \mathbf{j}$

45. Find a vector **v** whose magnitude is 4 and whose component in the **i** direction is twice the component in the **j** direction.

46. Find a vector **v** whose magnitude is 3 and whose component in the **i** direction is equal to the component in the **j** direction.

47. If $\mathbf{v} = 2\mathbf{i} - \mathbf{j}$ and $\mathbf{w} = x\mathbf{i} + 3\mathbf{j}$, find all numbers x for which $\|\mathbf{v} + \mathbf{w}\| = 5$.

48. If $P = (-3, 1)$ and $Q = (x, 4)$, find all numbers x such that the vector represented by $\overrightarrow{PQ}$ has length 5.

49. *Finding Ground Speed* An airplane has an airspeed of 500 kilometers per hour in an easterly direction. If the wind velocity is 60 kilometers per hour in a northwesterly direction, find the speed of the airplane relative to the ground.

50. *Finding Air Speed* After 1 hour in the air, an airplane arrives at a point 200 miles due south of its departure point. If there was a steady wind of 30 miles per hour from the northwest during the entire flight, what was the average airspeed of the airplane?

51. *Finding Speed Without Wind* An airplane travels in a northwesterly direction at a constant ground speed of 250 miles per hour, due to an easterly wind of 50 miles per hour. How fast would the plane have gone if there had been no wind?

52. *Finding the True Speed of a Motorboat* A small motorboat in still water maintains a speed of 10 miles per hour. In heading directly across a river (that is, perpendicular to the current) whose current is 4 miles per hour, what will be the true speed of the motorboat? (See the figure).

53. Show on the graph below the force needed to prevent an object at P from moving.

54. Explain in your own words what a vector is. Give an example of a vector.

55. Write a brief paragraph comparing the algebra of complex numbers and the algebra of vectors.

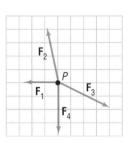

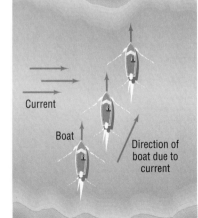

Current

Boat

Direction of boat due to current

Chapter Review

Equations of Conics

Parabola	See Tables 1 and 2.
Ellipse	See Table 3.
Hyperbola	See Table 4.
Vector	Quantity having magnitude and direction; equivalent to a directed line segment $\overrightarrow{PQ}$
Position vector	Vector whose initial point is at the origin
Unit vector	Vector whose magnitude is 1

HOW TO:

Find the vertex, focus, and directrix of a parabola given its equation

Graph a parabola given its equation by hand and using a graphing utility

Find an equation of a parabola given certain information about the parabola

Find the center, foci, and vertices of an ellipse given its equation

Graph an ellipse given its equation by hand and using a graphing utility

Find an equation of an ellipse given certain information about the ellipse

Find the center, foci, vertices, and asymptotes of a hyperbola given its equation

Graph a hyperbola given its equation by hand and using a graphing utility

Find an equation of a hyperbola given certain information about the hyperbola

Add and subtract vectors

Form scalar multiples of vectors

Find the magnitude of a vector

Solve problems involving vectors

FILL-IN-THE-BLANK ITEMS

1. A(n) _____ is the collection of all points in the plane such that the distance from each point to a fixed point equals its distance to a fixed line.

2. A(n) _____ is the collection of all points in the plane the sum of whose distances from two fixed points is a constant.

3. A(n) _____ is the collection of all points in the plane the difference of whose distances from two fixed points is a constant.

4. For an ellipse, the foci lie on the _____ axis, for a hyperbola, the foci lie on the _____ axis.

5. For the ellipse $(x^2/9) + (y^2/16) = 1$, the major axis is along the _____.

6. The equations of the asymptotes of the hyperbola $(y^2/9) - (x^2/4) = 1$ are _____ and _____.

7. A vector whose magnitude is 1 is called a(n) _____ vector.

TRUE/FALSE ITEMS

T   F   **1.** On a parabola, the distance from any point to the focus equals the distance from that point to the directrix.

T   F   **2.** The foci of an ellipse lie on its minor axis.

T   F   **3.** The foci of a hyperbola lie on its transverse axis.

T   F   **4.** Hyperbolas always have asymptotes, and ellipses never have asymptotes.

T   F   **5.** A hyperbola never intersects its conjugate axis.

T   F   **6.** A hyperbola always intersects its transverse axis.

T   F   **7.** Vectors are quantities that have magnitude and direction.

T   F   **8.** Force is a physical example of a vector.

REVIEW EXERCISES

In Problems 1–20, identify each equation. If it is a parabola, give its vertex, focus, and directrix; if it is an ellipse, give its center, vertices, and foci; if it is a hyperbola, give its center, vertices, foci, and asymptotes.

1. $y^2 = -16x$

2. $16x^2 = y$

3. $\dfrac{x^2}{25} - y^2 = 1$

4. $\dfrac{y^2}{25} - x^2 = 1$

5. $\dfrac{y^2}{25} + \dfrac{x^2}{16} = 1$

6. $\dfrac{x^2}{9} + \dfrac{y^2}{16} = 1$

7. $x^2 + 4y = 4$

8. $3y^2 - x^2 = 9$

9. $4x^2 - y^2 = 8$

10. $9x^2 + 4y^2 = 36$

11. $x^2 - 4x = 2y$

12. $2y^2 - 4y = x - 2$

13. $y^2 - 4y - 4x^2 + 8x = 4$

14. $4x^2 + y^2 + 8x - 4y + 4 = 0$

15. $4x^2 + 9y^2 - 16x - 18y = 11$

16. $4x^2 + 9y^2 - 16x + 18y = 11$

17. $4x^2 - 16x + 16y + 32 = 0$

18. $4y^2 + 3x - 16y + 19 = 0$

19. $9x^2 + 4y^2 - 18x + 8y = 23$

20. $x^2 - y^2 - 2x - 2y = 1$

In Problems 21–36, obtain an equation of the conic described. Graph the equation by hand.

21. Parabola; focus at $(-2, 0)$; directrix the line $x = 2$

22. Ellipse; center at $(0, 0)$; focus at $(0, 3)$; vertex at $(0, 5)$

23. Hyperbola; center at $(0, 0)$; focus at $(0, 4)$; vertex at $(0, -2)$

24. Parabola; vertex at $(0, 0)$; directrix the line $y = -3$

25. Ellipse; foci at $(-3, 0)$ and $(3, 0)$; vertex at $(4, 0)$

26. Hyperbola; vertices at $(-2, 0)$ and $(2, 0)$; focus at $(4, 0)$

27. Parabola; vertex at $(2, -3)$; focus at $(2, -4)$

28. Ellipse; center at $(-1, 2)$; focus at $(0, 2)$; vertex at $(2, 2)$

29. Hyperbola; center at $(-2, -3)$; focus at $(-4, -3)$; vertex at $(-3, -3)$

30. Parabola; focus at $(3, 6)$; directrix the line $y = 8$

31. Ellipse; foci at $(-4, 2)$ and $(-4, 8)$; vertex at $(-4, 10)$

32. Hyperbola; vertices at $(-3, 3)$ and $(5, 3)$; focus at $(7, 3)$

33. Center at $(-1, 2)$; $a = 3$; $c = 4$; transverse axis parallel to the x-axis

34. Center at $(4, -2)$; $a = 1$; $c = 4$; transverse axis parallel to y-axis

35. Vertices at $(0, 1)$ and $(6, 1)$; asymptote the line $3y + 2x - 9 = 0$

36. Vertices at $(4, 0)$ and $(4, 4)$; asymptote the line $y + 2x - 10 = 0$

37. Find an equation of the hyperbola whose foci are the vertices of the ellipse $4x^2 + 9y^2 = 36$ and whose vertices are the foci of this ellipse.

38. Find an equation of the ellipse whose foci are the vertices of the hyperbola $x^2 - 4y^2 = 16$ and whose vertices are the foci of this hyperbola.

39. Describe the collection of points in a plane so that the distance from each point to the point $(3, 0)$ is three-fourths of its distance from the line $x = \frac{16}{3}$.

40. Describe the collection of points in a plane so that the distance from each point to the point $(5, 0)$ is five-fourths of its distance from the line $x = \frac{16}{5}$.

In Problems 41–48, use the vectors $\mathbf{v} = -2\mathbf{i} + \mathbf{j}$ *and* $\mathbf{w} = 4\mathbf{i} - 3\mathbf{j}$.

41. Find $4\mathbf{v} - 3\mathbf{w}$.   **42.** Find $-\mathbf{v} + 2\mathbf{w}$.   **43.** Find $\|\mathbf{v}\|$.   **44.** Find $\|\mathbf{v} + \mathbf{w}\|$.

45. Find $\|\mathbf{v}\| + \|\mathbf{w}\|$.   **46.** Find $\|2\mathbf{v}\| - 3\|\mathbf{w}\|$.

47. Find a unit vector having the same direction as $\mathbf{v}$.

48. Find a unit vector having the opposite direction of $\mathbf{w}$.

49. *Actual Speed of a Swimmer* A swimmer can maintain a constant speed of 5 miles per hour. If the swimmer heads directly across a river that has a current moving at the rate of 2 miles per hour, what is the actual speed of the swimmer? (See the figure.)

50. *Speed of a Motorboat Relative to Water* A small motorboat is moving at a true speed of 11 miles per hour in a southerly direction. The current is known to be from the northeast at 3 miles per hour. What is the speed of the motorboat relative to the water?

51. *Mirrors* A mirror is shaped like a paraboloid of revolution. If a light source is located 1 foot from the base along the axis of symmetry and the opening is 2 feet across, how deep should the mirror be?

52. *Parabolic Arch Bridge* A bridge is built in the shape of a parabolic arch. The bridge has a span of 60 feet and a maximum height of 20 feet. Find the height of the arch at distances of 5, 10, and 20 feet from the center.

53. *Semielliptical Arch Bridge* A bridge is built in the shape of a semielliptical arch. The bridge has a span of 60 feet and a maximum height of 20 feet. Find the height of the arch at distances of 5, 10, and 20 feet from the center.

54. *Whispering Galleries* The figure shows the specifications for an elliptical ceiling in a hall designed to be a whispering gallery. Where in the hall are the foci located?

55. *LORAN* Two LORAN stations are positioned 150 miles apart along a straight shore.
(a) A ship records a time difference of 0.00032 second between the LORAN signals. Set up an appropriate rectangular coordinate system to determine where the ship would reach shore if it were to follow the hyperbola corresponding to this time difference.
(b) If the ship wants to enter a harbor located between the two stations 15 miles from the master station, what time difference should it be looking for?
(c) If the ship is 20 miles off shore when the desired time difference is obtained, what is the exact location of the ship? [*Note:* The speed of each radio signal is 186,000 miles per second.]

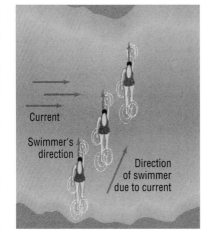

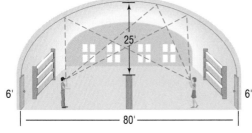
Current

Swimmer's direction

Direction of swimmer due to current

 56. Use the definition of a parabola and the idea of the eccentricity of an ellipse and hyperbola to construct a unifying definition for all three conics. [Refer to Problem 75 in Exercise 6.3 and Problem 61 in Exercise 6.4.]

57. Formulate a strategy for discussing and graphing an equation of the form $Ax^2 + Cy^2 + Dx + Ey + F = 0$.

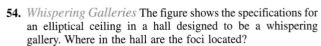

Chapter 7

PREPARING FOR THIS CHAPTER

Before getting started on this chapter, review the following concepts:

For Sections 7.1, 7.2, 7.7, 7.8: The Straight Line (Section 1.3)

For Section 7.5: Identities (p. 680)

Factoring Polynomials (Appendix, Section 2, pp. 665–667)

Proper and Improper Rational Functions (p. 290)

Fundamental Theorem of Algebra (pp. 312 and 318)

For Section 7.6:   Circles (pp. 55–60)

Parabolas (pp. 409–417)

Ellipses (pp. 422–432)

Hyperbolas (pp. 437–450)

7.1 Systems of Linear Equations: Substitution; Elimination

7.2 Systems of Linear Equations: Matrices

7.3 Systems of Linear Equations: Determinants

7.4 Matrix Algebra

7.5 Partial Fraction Decomposition

7.6 Systems of Nonlinear Equations

7.7 Systems of Inequalities

7.8 Linear Programming
     Chapter Review

Preview   Running a Marathon

In a 50-mile marathon, the winner crosses the finish line 1 mile ahead of the second-place runner and 4 miles ahead of the third-place runner. Assuming that each runner maintains a constant speed throughout the race, by how many miles does the second-place runner beat the third-place runner? [Example 8, Section 7.6] ■

In this chapter we take up the problem of solving equations and inequalities containing two or more variables. As the section titles suggest, there are various ways to solve such problems.

The *method of substitution* for solving equations in several unknowns goes back to ancient times.

The *method of elimination,* although it had existed for centuries, was put into systematic order by Karl Friedrich Gauss (1777–1855) and by Camille Jordan (1838–1922). This method is now used for solving large systems by computer.

The theory of *matrices* was developed in 1857 by Arthur Cayley (1821–1895), although only later were matrices used as we use them in this chapter. Matrices have become a very flexible instrument, useful in almost all areas of mathematics.

The method of *determinants* was invented by Seki Kōwa (1642–1708) in 1683 in Japan and by Gottfried Wilhelm von Leibniz (1646–1716) in 1693 in Germany. Both used them only in relation to linear equations. *Cramer's Rule* is named after Gabriel Cramer (1704–1752) of Switzerland, who popularized the use of determinants for solving linear systems.

Section 7.5, *partial fraction decomposition,* provides an application of systems of equations. This particular application is one that is used in integral calculus.

Section 7.6 introduces *linear programming,* a modern application of linear inequalities to certain types of problems. This topic is particularly useful for students interested in operations research.

7.1

Systems of Linear Equations: Substitution; Elimination

We begin with an example.

E X A M P L E   1   *Movie Theater Ticket Sales*

A movie theater sells tickets for $8.00 each, with Seniors receiving a discount of $2.00. One evening the theater took in $3580 in revenue. If x represents the number of tickets sold at $8.00 and y the number of tickets sold at the discounted price of $6.00, write an equation that relates these variables.

Solution   Each nondiscounted ticket brings in $8.00, so x tickets will bring in $8x$ dollars. Similarly, y discounted tickets bring in $6y$ dollars. If the total brought in is $3580, we must have

$$8x + 6y = 3580$$ ■

In Example 1, suppose we also know that 525 tickets were sold that evening. Then we have another equation relating the variables x and y:

$$x + y = 525$$

The two equations

$$8x + 6y = 3580$$
$$x + \ y = \ 525$$

form a *system* of equations.

In general, a **system of equations** is a collection of two or more equations, each containing one or more variables. Example 2 gives some samples of systems of equations.

E X A M P L E   2   *Examples of Systems of Equations*

(a) $\begin{cases} 2x + \ y = \ \ 5 & \text{(1)} \\ -4x + 6y = -2 & \text{(2)} \end{cases}$ Two equations containing two variables, x and y

(b) $\begin{cases} x + y^2 = 5 & \text{(1)} \\ 2x + y = 4 & \text{(2)} \end{cases}$ (1) Two equations containing two variables, x and y

(c) $\begin{cases} x + y + z = 6 & \text{(1)} \\ 3x - 2y + 4z = 9 & \text{(2)} \\ x - y - z = 0 & \text{(3)} \end{cases}$ (1) Three equations containing three variables, x, y, and z

(d) $\begin{cases} x + y + z = 5 & \text{(1)} \\ x - y = 2 & \text{(2)} \end{cases}$ (1) Two equations containing three variables, x, y, and z

(e) $\begin{cases} x + y + z = 6 & \text{(1)} \\ 2x + 2z = 4 & \text{(2)} \\ y + z = 2 & \text{(3)} \\ x = 4 & \text{(4)} \end{cases}$ (1) Four equations containing three variables, x, y, and z ■

We use a brace, as shown above, to remind us that we are dealing with a system of equations. We also will find it convenient to number each equation in the system.

A **solution** of a system of equations consists of values for the variables that reduce each equation of the system to a true statement. To **solve** a system of equations means to find all solutions of the system.

For example, $x = 2$, $y = 1$ is a solution of the system in Example 2(a), because

$$2(2) + 1 = 5 \quad \text{and} \quad -4(2) + 6(1) = -2$$

A solution of the system in Example 2(b) is $x = 1$, $y = 2$, because

$$1 + 2^2 = 5 \quad \text{and} \quad 2(1) + 2 = 4$$

Another solution of the system in Example 2(b) is $x = \frac{11}{4}$, $y = -\frac{3}{2}$, which you can check for yourself. A solution of the system in Example 2(c) is $x = 3$, $y = 2$, $z = 1$, because

$$\begin{cases} 3 + 2 + 1 = 6 & \text{(1)} \quad x = 3, y = 2, z = 1 \\ 3(3) - 2(2) + 4(1) = 9 & \text{(2)} \\ 3 - 2 - 1 = 0 & \text{(3)} \end{cases}$$

Note that $x = 3$, $y = 3$, $z = 0$ is not a solution of the system in Example 2(c):

$$\begin{cases} 3 + 3 + 0 = 6 & \text{(1)} \quad x = 3, y = 3, z = 0 \\ 3(3) - 2(3) + 4(0) = 3 \neq 9 & \text{(2)} \\ 3 - 3 - 0 = 0 & \text{(3)} \end{cases}$$

Although these values satisfy equations (1) and (3), they do not satisfy equation (2). Any solution of the system must satisfy *each* equation of the system.

■ Now work Problem 3.

When a system of equations has at least one solution, it is said to be **consistent;** otherwise, it is called **inconsistent.**

An equation in n variables is said to be **linear** if it is equivalent to an equation of the form

$$a_1x_1 + a_2x_2 + \cdots + a_nx_n = b$$

where $x_1, x_2, \ldots, x_n$ are n distinct variables, $a_1, a_2, \ldots, a_n, b$ are constants, and at least one of the a's is not 0.

Some examples of linear equations are

$$2x + 3y = 2 \qquad 5x - 2y + 3z = 10 \qquad 8x + 8y - 2z + 5w = 0$$

If each equation in a system of equations is linear, then we have a **system of linear equations.** Thus, the systems in Examples 2(a), (c), (d), and (e) are linear, whereas the system in Example 2(b) is nonlinear. We concentrate on solving linear systems in Sections 7.1–7.3, and we will take up nonlinear systems in Section 7.6.

Two Linear Equations Containing Two Variables

We can view the problem of solving a system of two linear equations containing two variables as a geometry problem. The graph of each equation in such a system is a straight line. Thus, a system of two equations containing two variables represents a pair of lines. The lines either (1) intersect or (2) are parallel or (3) are **coincident** (that is, identical).

1. If the lines intersect, then the system of equations has one solution, given by the point of intersection. The system is **consistent** and the equations are **independent.**
2. If the lines are parallel, then the system of equations has no solution, because the lines never intersect. The system is **inconsistent.**
3. If the lines are coincident, then the system of equations has infinitely many solutions, represented by the totality of points on the line. The system is **consistent** and the equations are **dependent.**

Figure 1 illustrates these conclusions.

FIGURE 1

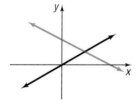

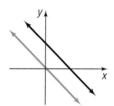

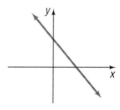

(a) Intersecting lines; system has one solution

(b) Parallel lines; system has no solution

(c) Coincident lines; system has infinitely many solutions

E X A M P L E   3

Solving a System of Equations Using a Graphing Utility

Solve: $\begin{cases} 2x + y = 5 & (1) \\ -4x + 6y = 12 & (2) \end{cases}$

Solution   First, we solve each equation for y. This is equivalent to writing each equation in slope-intercept form. Equation (1) in slope–intercept form is $y = -2x + 5$. Equation (2) in slope–intercept form is $y = \frac{2}{3}x + 2$. Figure 2 shows the graphs using a graphing utility. From the graph in Figure 2, we see that the lines intersect, so the system is consistent. Using BOX and TRACE,* we obtain the solution (1.12, 2.75) correct to two decimal places.

*Some graphing utilities have an INTERSECT command. Using this command, we obtain the solution (1.125, 2.75).

FIGURE 2

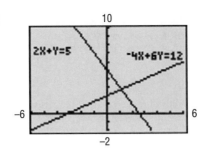

To obtain exact solutions we use algebraic methods. The first algebraic method we discuss is the *method of substitution.*

Method of Substitution

We illustrate the **method of substitution** by solving the system given in Example 3.

E X A M P L E   4

Solving a System of Equations by Substitution

Solve: $\begin{cases} 2x + y = 5 & (1) \\ -4x + 6y = 12 & (2) \end{cases}$

Solution   We solve the first equation for y, obtaining

$$y = 5 - 2x$$

We substitute this result for y in the second equation. The result is an equation containing just the variable x, which we can then solve for:

$$-4x + 6y = 12$$
$$-4x + 6(5 - 2x) = 12$$
$$-4x + 30 - 12x = 12$$
$$-16x = -18$$
$$x = \frac{-18}{-16} = \frac{9}{8}$$

Once we know that $x = \frac{9}{8}$, we can easily find the value of y by **back-substitution,** that is, by substituting $\frac{9}{8}$ for x in one of the original equations. We use the first one:

$$2x + y = 5$$
$$2\left(\frac{9}{8}\right) + y = 5$$
$$\frac{9}{4} + y = 5$$
$$y = 5 - \frac{9}{4} = \frac{20}{4} - \frac{9}{4} = \frac{11}{4}$$

The solution of the system is $x = \frac{9}{8} = 1.125$, $y = \frac{11}{4} = 2.75$.

The method used to solve the system in Example 4 is called **substitution.** The steps to be used are outlined next.

Steps for Solving by Substitution

STEP 1: Pick one of the equations and solve for one of the variables in terms of the remaining variables.

STEP 2: Substitute the result in the remaining equations.

STEP 3: If one equation in one variable results, solve this equation. Otherwise, repeat Step 1 until a single equation with one variable remains.

STEP 4: Find the values of the remaining variables by back-substitution.

STEP 5: Check the solution found.

E X A M P L E   5

Solving a System of Equations by Substitution

Solve: $\begin{cases} 3x - 2y = 5 & (1) \\ 5x - y = 6 & (2) \end{cases}$

Solution   STEP 1: After looking at the two equations, we conclude that it is easiest to solve for the variable y in equation (2):

$$5x - y = 6$$
$$y = 5x - 6$$

STEP 2: We substitute this result into equation (1) and simplify:

$$3x - 2y = 5$$
$$3x - 2(5x - 6) = 5$$
$$-7x + 12 = 5$$
$$-7x = -7$$
$$x = 1$$

STEP 3: Because we now have one solution, $x = 1$, we proceed to Step 4.

STEP 4: Knowing $x = 1$, we can find y from the equation

$$y = 5x - 6 = 5(1) - 6 = -1$$

STEP 5: *Check:* $\begin{cases} 3(1) - 2(-1) = 3 + 2 = 5 \\ 5(1) - (-1) = 5 + 1 = 6 \end{cases}$

The solution of the system is $x = 1$, $y = -1$.   ∎

E X A M P L E   6

Solving a System of Equations by Substitution

Solve: $\begin{cases} 2x - 3y = 7 & (1) \\ 4x + 5y = 3 & (2) \end{cases}$

Solution   STEP 1: In looking over the system, we conclude that there is no way to solve for one of the variables without introducing fractions. We solve for the variable x in equation (1):

$$2x - 3y = 7$$
$$2x = 3y + 7$$
$$x = \frac{3}{2}y + \frac{7}{2}$$

STEP 2: We substitute this result for x in equation (2) and simplify:

$$4x + 5y = 3$$
$$4\left(\frac{3}{2}y + \frac{7}{2}\right) + 5y = 3$$
$$6y + 14 + 5y = 3$$
$$11y + 14 = 3$$
$$11y = -11$$

STEP 3: $y = -1$

STEP 4: $x = \dfrac{3}{2}y + \dfrac{7}{2} = \dfrac{3}{2}(-1) + \dfrac{7}{2} = \dfrac{4}{2} = 2$

STEP 5: *Check:* $\begin{cases} 2(2) - 3(-1) = 4 + 3 = 7 \\ 4(2) + 5(-1) = 8 - 5 = 3 \end{cases}$

The solution is $x = 2$, $y = -1$.   ■

■   Now use substitution to work Problem 13.

Method of Elimination

A second method for solving a system of linear equations is the *method of elimination*. This method is usually preferred over substitution if substitution leads to fractions or if the system contains more than two variables. Elimination also provides the necessary motivation for solving systems using matrices (the subject of the next section).

The idea behind the method of elimination is to keep replacing the original equations in the system with equivalent equations until a system of equations with an obvious solution is reached. When we proceed in this way, we obtain **equivalent systems of equations.** The rules for obtaining equivalent equations are the same as those studied earlier. However, we may also interchange any two equations of the system and/or replace any equation in the system by the sum (or difference) of that equation and any other equation in the system.

Rules for Obtaining an Equivalent System of Equations

1. Interchange any two equations of the system.
2. Multiply (or divide) each side of an equation by the same nonzero constant.
3. Replace any equation in the system by the sum (or difference) of that equation and any other equation in the system.

An example will give you the idea. As you work through the example, pay particular attention to the pattern being followed.

E X A M P L E 7   *Solving a System of Equations by Elimination*

Solve: $\begin{cases} 2x + 3y = 1 & (1) \\ -x + y = -3 & (2) \end{cases}$

Solution   We multiply each side of equation (2) by 2 so that the coefficients of x in the two equations are negatives of one another. The result is the equivalent system

$$\begin{cases} 2x + 3y = 1 & (1) \\ -2x + 2y = -6 & (2) \end{cases}$$

If we now replace equation (2) of this system by the sum of the two equations, we obtain an equation containing just the variable y, which we can solve for:

$$\begin{cases} 2x + 3y = 1 & (1) \\ -2x + 2y = -6 & (2) \end{cases}$$

$$5y = -5$$

$$y = -1$$

We back-substitute by using this value for y in equation (1) and simplify to get

$$2x + 3(-1) = 1$$

$$2x = 4$$

$$x = 2$$

Thus, the solution of the original system is $x = 2$, $y = -1$. We leave it to you to check the solution. ∎

The procedure used in Example 7 is called the **method of elimination.** Notice the pattern of the solution. First, we eliminated the variable x from the second equation. Then we back-substituted; that is, we substituted the value found for y back into the first equation to find x.

Let's return to the movie theater example (Example 1).

E X A M P L E   8   *Movie Theater Ticket Sales*

A movie theater sells tickets for $8.00 each, with Seniors receiving a discount of $2.00. One evening the theater sold 525 tickets and took in $3580 in revenue. How many of each type of ticket was sold?

Solution   If x represents the number of tickets sold at $8.00 and y the number of tickets sold at the discounted price of $6.00, then the given information results in the system of equations

$$\begin{cases} 8x + 6y = 3580 \\ x + y = 525 \end{cases}$$

We use elimination and multiply the second equation by -6 and then add the equations.

$$\begin{cases} 8x + 6y = 3580 \\ -6x - 6y = -3150 \end{cases}$$

$$2x = 430$$

$$x = 215$$

Since $x + y = 525$, then $y = 525 - x = 525 - 215 = 310$. Thus, 215 nondiscounted tickets and 310 Senior discount tickets were sold. ∎

■ Now use elimination to work Problem 13.

The previous examples dealt with consistent systems of equations that had a unique solution. The next two examples deal with two other possibilities that may occur, the first being a system that has no solution.

E X A M P L E   9

An Inconsistent System of Equations

Solve: $\begin{cases} 2x + y = 5 & (1) \\ 4x + 2y = 8 & (2) \end{cases}$

Solution   We choose to use the method of substitution and solve equation (1) for *y:*

$$2x + y = 5$$
$$y = 5 - 2x$$

Substituting in equation (2), we get

$$4x + 2y = 8$$
$$4x + 2(5 - 2x) = 8$$
$$4x + 10 - 4x = 8$$
$$0 \cdot x = -2$$

FIGURE 3

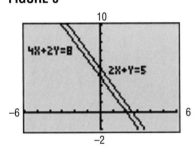

This equation has no solution. Thus, we conclude that the system itself has no solution and is therefore inconsistent.   ■

Figure 3 illustrates the pair of lines whose equations form the system in Example 9. Notice that the graphs of the two equations are lines, each with slope -2; one has a *y*-intercept of 5, the other a *y*-intercept of 4. Thus, the lines are parallel and have no point of intersection. This geometric statement is equivalent to the algebraic statement that the system has no solution.

The next example is an illustration of a system with infinitely many solutions.

E X A M P L E   1 0

Solving a System of Equations with Infinitely Many Solutions

Solve: $\begin{cases} 2x + y = 4 & (1) \\ -6x - 3y = -12 & (2) \end{cases}$

Solution   We choose to use the method of elimination:

$$\begin{cases} 2x + y = 4 & (1) \\ -6x - 3y = -12 & (2) \end{cases}$$

$$\begin{cases} 6x + 3y = 12 & (1) \text{ Multiply each side of equation (1) by 3.} \\ -6x - 3y = -12 & (2) \end{cases}$$

$$\begin{cases} 6x + 3y = 12 & (1) \text{ Replace equation (2) by the sum of} \\ 0 = 0 & (2) \text{ equations (1) and (2).} \end{cases}$$

The original system is thus equivalent to a system containing one equation, so the equations are dependent. This means that any values of *x* and *y* for which $6x + 3y = 12$ or, equivalently, $2x + y = 4$ are solutions. For example, $x = 2, y = 0$; $x = 0, y = 4$; $x = -2, y = 8$; $x = 4, y = -4$; and so on, are solutions. There are, in fact, infinitely many values of *x* and *y* for which $2x + y = 4$, so the original system has infinitely many solutions. We will write the solutions of the original systems either as

$$y = 4 - 2x$$

where x can be any real number, or as

$$x = 2 - \frac{1}{2}y$$

where y can be any real number. ■

FIGURE 4

$$\begin{cases} 2x + y = 4 \\ -6x - 3y = -12 \end{cases}$$

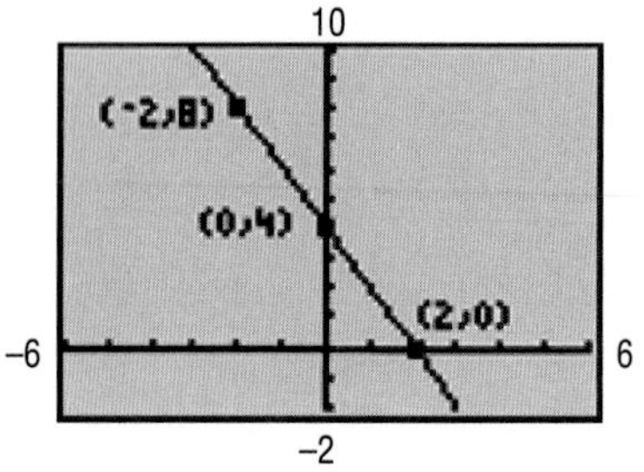

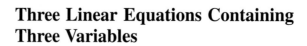

Figure 4 illustrates the situation presented in Example 10. Notice that the graphs of the two equations are lines, each with slope -2 and each with y-intercept 4. Thus, the lines are coincident. Notice also that equation (2) in the original system is just -3 times equation (1), indicating that the two equations are dependent.

For the system in Example 10, we can write down some of the infinite number of solutions by assigning values to x and then finding $y = 4 - 2x$. Thus,

If $x = -2$, then $y = 8$.

If $x = 0$, then $y = 4$.

If $x = 2$, then $y = 0$.

The pairs (x, y) are points on the line in Figure 4.

■ Now work Problems 19 and 23.

Three Linear Equations Containing Three Variables

Just as with a system of two linear equations containing two variables, a system of three linear equations containing three variables also has either exactly one solution, or no solution, or infinitely many solutions.

Let's see how elimination works on a system of three equations containing three variables.

E X A M P L E   1 1   *Solving a System of Three Linear Equations with Three Variables*

Use the method of elimination to solve the system of equations;

$$\begin{cases} x + y - z = -1 & (1) \\ 4x - 3y + 2z = 16 & (2) \\ 2x - 2y - 3z = 5 & (3) \end{cases}$$

Solution   For a system of three equations, we attempt to eliminate one variable at a time, using pairs of equations. Our plan of attack on this system will be to first eliminate the variable x from equations (1) and (2) and then from equations (1) and (3). Next, we will eliminate the variable y from the resulting equations, leaving one equation containing only the variable z. Back-substitution can then be used to obtain the values of y and then x.

We begin by multiplying each side of equation (1) by -2, in anticipation of eliminating the variable x from equation (3) by adding equations (1) and (3):

$$\begin{cases} -2x - 2y + 2z = 2 & (1) \text{ Multiply each side of equation (1)} \\ 4x - 3y + 2z = 16 & (2) \text{ by } -2 \\ 2x - 2y - 3z = 5 & (3) \end{cases}$$

$$\begin{cases} -2x - 2y + 2z = 2 & (1) \\ 4x - 3y + 2z = 16 & (2) \\ -4y - z = 7 & (3) \text{ Replace equation (3) by the sum of equations (1) and (3).} \end{cases}$$

We now eliminate the variable x from equation (2):

$$\begin{cases} -4x - 4y + 4z = 4 & (1) \text{ Multiply each side of equation (1) by 2.} \\ 4x - 3y + 2z = 16 & (2) \\ -4y - z = 7 & (3) \end{cases}$$

$$\begin{cases} -4x - 4y + 4z = 4 & (1) \\ -7y + 6z = 20 & (2) \text{ Replace equation (2) by the sum of equations (1) and (2).} \\ -4y - z = 7 & (3) \end{cases}$$

We now eliminate y from equation (3):

$$\begin{cases} -4x - 4y + 4z = 4 & (1) \\ -28y + 24z = 80 & (2) \text{ Multiply each side of equation (2) by 4.} \\ 28y + 7z = -49 & (3) \text{ Multiply each side of equation (3) by } -7. \end{cases}$$

$$\begin{cases} -4x - 4y + 4z = 4 & (1) \\ -28y + 24z = 80 & (2) \\ 31z = 31 & (3) \text{ Replace equation (3) by the sum of equations (2) and (3).} \end{cases}$$

$$\begin{cases} -4x - 4y + 4z = 4 & (1) \\ -28y + 24z = 80 & (2) \\ z = 1 & (3) \text{ Multiply each side of equation (3) by } \frac{1}{31}. \end{cases}$$

$$\begin{cases} -4x - 4y + 4 = 4 & (1) \text{ Back–substitute; replace } z \text{ by 1 in equations (1) and (2).} \\ -28y + 24 = 80 & (2) \\ z = 1 & (3) \end{cases}$$

$$\begin{cases} -4x - 4y = 0 & (1) \\ y = -2 & (2) \text{ Solve equation (2) for } y. \\ z = 1 & (3) \end{cases}$$

$$\begin{cases} -4x + 8 = 0 & (1) \text{ Back–substitute; replace } y \text{ by } -2. \\ y = -2 & (2) \\ z = 1 & (3) \end{cases}$$

$$\begin{cases} x = 2 & (1) \\ y = -2 & (2) \\ z = 1 & (3) \end{cases}$$

The solution of the original system is $x = 2$, $y = -2$, $z = 1$. (You should check this.) ■

Look back over the solution given in Example 11. Note the pattern of making equation (3) contain only the variable z, followed by making equation (2) contain only the variable y and equation (1) contain only the variable x. Although which variables to isolate is your choice, the methodology remains the same for all systems.

Note: Graphing utilities only graph in two divisions, so they can only be used with systems containing two variables.

7.1

Exercise 7.1

In Problems 1–10, verify that the values of the variables listed are solutions of the system of equations.

1. $\begin{cases} 2x - y = 5 \\ 5x + 2y = 8 \end{cases}$

$x = 2, y = -1$

2. $\begin{cases} 3x + 2y = 2 \\ x - 7y = -30 \end{cases}$

$x = -2, y = 4$

3. $\begin{cases} 3x - 4y = 4 \\ \frac{1}{2}x - 3y = -\frac{1}{2} \end{cases}$

$x = 2, y = \frac{1}{2}$

4. $\begin{cases} 2x + \frac{1}{2}y = 0 \\ 3x - 4y = -\frac{19}{2} \end{cases}$

$x = -\frac{1}{2}, y = 2$

5. $\begin{cases} x^2 - y^2 = 3 \\ xy = 2 \end{cases}$

$x = 2, y = 1$

6. $\begin{cases} x^2 - y^2 = 3 \\ xy = 2 \end{cases}$

$x = -2, y = -1$

7. $\begin{cases} \dfrac{x}{1 + x} + 3y = 6 \\ x + 9y^2 = 36 \end{cases}$

$x = 0, y = 2$

8. $\begin{cases} \dfrac{x}{x - 1} + y = 5 \\ 3x - y = 3 \end{cases}$

$x = 2, y = 3$

9. $\begin{cases} 3x + 3y + 2z = 4 \\ x - y - z = 0 \\ 2y - 3z = -8 \end{cases}$

$x = 1, y = -1, z = 2$

10. $\begin{cases} 4x - z = 7 \\ 8x + 5y - z = 0 \\ -x - y + 5z = 6 \end{cases}$

$x = 2, y = -3, z = 1$

In Problems 11–46, solve each system of equations. If the system has no solution, say it is inconsistent. Use either substitution or elimination. Verify your solution using a graphing utility (when possible).

11. $\begin{cases} x + y = 8 \\ x - y = 4 \end{cases}$

12. $\begin{cases} x + 2y = 5 \\ x + y = 3 \end{cases}$

13. $\begin{cases} 5x - y = 13 \\ 2x + 3y = 12 \end{cases}$

14. $\begin{cases} x + 3y = 5 \\ 2x - 3y = -8 \end{cases}$

15. $\begin{cases} 3x = 24 \\ x + 2y = 0 \end{cases}$

16. $\begin{cases} 4x + 5y = -3 \\ -2y = -4 \end{cases}$

17. $\begin{cases} 3x - 6y = 2 \\ 5x + 4y = 1 \end{cases}$

18. $\begin{cases} 2x + 4y = \frac{2}{3} \\ 3x - 5y = -10 \end{cases}$

19. $\begin{cases} 2x + y = 1 \\ 4x + 2y = 3 \end{cases}$

20. $\begin{cases} x - y = 5 \\ -3x + 3y = 2 \end{cases}$

21. $\begin{cases} 2x - y = 0 \\ 3x + 2y = 7 \end{cases}$

22. $\begin{cases} 3x + 3y = -1 \\ 4x + y = \frac{8}{3} \end{cases}$

23. $\begin{cases} x + 2y = 4 \\ 2x + 4y = 8 \end{cases}$

24. $\begin{cases} 3x - y = 7 \\ 9x - 3y = 21 \end{cases}$

25. $\begin{cases} 2x - 3y = -1 \\ 10x + y = 11 \end{cases}$

26. $\begin{cases} 3x - 2y = 0 \\ 5x + 10y = 4 \end{cases}$

27. $\begin{cases} 2x + 3y = 6 \\ x - y = \frac{1}{2} \end{cases}$

28. $\begin{cases} \frac{1}{2}x + y = -2 \\ x - 2y = 8 \end{cases}$

29. $\begin{cases} \frac{1}{2}x + \frac{1}{3}y = 3 \\ \frac{1}{4}x - \frac{2}{3}y = -1 \end{cases}$

30. $\begin{cases} \frac{1}{3}x - \frac{3}{2}y = -5 \\ \frac{3}{4}x + \frac{1}{3}y = 11 \end{cases}$

31. $\begin{cases} 3x - 5y = 3 \\ 15x + 5y = 21 \end{cases}$

32. $\begin{cases} 2x - y = -1 \\ x + \frac{1}{2}y = \frac{3}{2} \end{cases}$

33. $\begin{cases} x - y = 6 \\ 2x - 3z = 16 \\ 2y + z = 4 \end{cases}$

34. $\begin{cases} 2x + y = -4 \\ -2y + 4z = 0 \\ 3x - 2z = -11 \end{cases}$

35. $\begin{cases} x - 2y + 3z = 7 \\ 2x + y + z = 4 \\ -3x + 2y - 2z = -10 \end{cases}$

36. $\begin{cases} 2x + y - 3z = 0 \\ -2x + 2y + z = -7 \\ 3x - 4y - 3z = 7 \end{cases}$

37. $\begin{cases} x - y - z = 1 \\ 2x + 3y + z = 2 \\ 3x + 2y = 0 \end{cases}$

38. $\begin{cases} 2x - 3y - z = 0 \\ -x + 2y + z = 5 \\ 3x - 4y - z = 1 \end{cases}$

39. $\begin{cases} x - y - z = 1 \\ -x + 2y - 3z = -4 \\ 3x - 2y - 7z = 0 \end{cases}$

40. $\begin{cases} 2x - 3y - z = 0 \\ 3x + 2y + 2z = 2 \\ x + 5y + 3z = 2 \end{cases}$

41. $\begin{cases} 2x - 2y + 3z = 6 \\ 4x - 3y + 2z = 0 \\ -2x + 3y - 7z = 1 \end{cases}$

42. $\begin{cases} 3x - 2y + 2z = 6 \\ 7x - 3y + 2z = -1 \\ 2x - 3y + 4z = 0 \end{cases}$

43. $\begin{cases} x + y - z = 6 \\ 3x - 2y + z = -5 \\ x + 3y - 2z = 14 \end{cases}$

44. $\begin{cases} x - y + z = -4 \\ 2x - 3y + 4z = -15 \\ 5x + y - 2z = 12 \end{cases}$

45. $\begin{cases} x + 2y - z = -3 \\ 2x - 4y + z = -7 \\ -2x + 2y - 3z = 4 \end{cases}$

46. $\begin{cases} x + 4y - 3z = -8 \\ 3x - y + 3z = 12 \\ x + y + 6z = 1 \end{cases}$

47. Solve: $\begin{cases} \dfrac{1}{x} + \dfrac{1}{y} = 8 \\ \dfrac{3}{x} - \dfrac{5}{y} = 0 \end{cases}$

48. Solve: $\begin{cases} \dfrac{4}{x} - \dfrac{3}{y} = 0 \\ \dfrac{6}{x} + \dfrac{3}{2y} = 2 \end{cases}$

[Hint: Let $u = 1/x$ and $v = 1/y$, and solve for u and v. Then $x = 1/u$ and $y = 1/v$.]

In Problems 49–54, use a graphing utility to solve each system of equations. Approximate the solution correct to two decimal places.

49. $\begin{cases} y = \sqrt{2}x - 20\sqrt{7} \\ y = -0.1x + 20 \end{cases}$

50. $\begin{cases} y = -\sqrt{3}x + 100 \\ y = 0.2x + \sqrt{19} \end{cases}$

51. $\begin{cases} \sqrt{2}x + \sqrt{3}y + \sqrt{6} = 0 \\ \sqrt{3}x - \sqrt{2}y + 60 = 0 \end{cases}$

52. $\begin{cases} \sqrt{5}x - \sqrt{6}y + 60 = 0 \\ 0.2x + 0.3y + \sqrt{5} = 0 \end{cases}$

53. $\begin{cases} \sqrt{3}x + \sqrt{2}y = \sqrt{0.3} \\ 100x - 95y = 20 \end{cases}$

54. $\begin{cases} \sqrt{6}x - \sqrt{5}y + \sqrt{1.1} = 0 \\ y = -0.2x + 0.1 \end{cases}$

55. The sum of two numbers is 81. The difference of twice one and three times the other is 62. Find the two numbers.

56. The difference of two numbers is 40. Six times the smaller one less the larger one is 5. Find the two numbers.

57. The perimeter of a rectangular floor is 90 feet. Find the dimensions of the floor if the length is twice the width.

58. The length of fence required to enclose a rectangular field is 3000 meters. What are the dimensions of the field if it is known that the difference between its length and width is 50 meters?

59. *Cost of Fast Food* Four large cheeseburgers and two chocolate shakes cost a total of $7.90. Two shakes cost 15¢ more than one cheeseburger. What is the cost of a cheeseburger? A shake?

60. *Movie Theatre Tickets* A movie theater charges $9.00 for adults and $7.00 for senior citizens. On a day when 325 people paid an admission, the total receipts were $2495. How many who paid were adults? How many were seniors?

61. *Mixing Nuts* A store sells cashews for $5.00 per pound and peanuts for $1.50 per pound. The manager decides to mix 30 pounds of peanuts with some cashews and sell the mixture for $3.00 per pound. How many pounds of cashews should be mixed with the peanuts so that the mixture will produce the same revenue as would selling the nuts separately?

62. *Financial Planning* A recently retired couple need $12,000 per year to supplement their Social Security. They have $150,000 to invest to obtain this income. They have decided on two investment options: AA bonds yielding 10% per annum and a Bank Certificate yielding 5%.
(a) How much should be invested in each to realize exactly $12,000?
(b) If, after 2 years, the couple requires $14,000 per year in income, how should they reallocate their investment to achieve the new amount?

63. *Computing Wind Speed* With a tail wind, a small Piper aircraft can fly 600 miles in 3 hours. Against this same wind, the Piper can fly the same distance in 4 hours. Find the average wind speed and the average airspeed of the Piper.

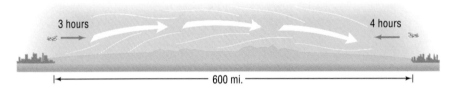

3 hours   4 hours

|← ——————————— 600 mi. ——————————— →|

64. *Computing Wind Speed* The average airspeed of a single-engine aircraft is 150 miles per hour. If the aircraft flew the same distance in 2 hours with the wind as it flew in 3 hours against the wind, what was the wind speed?

65. *Restaurant Management* A restaurant manager wants to purchase 200 sets of dishes. One design costs $25 per set, while another costs $45 per set. If she only has $7400 to spend, how many of each design should be ordered?

66. *Cost of Fast Food* One group of people purchased 10 hot dogs and 5 soft drinks at a cost of $12.50. A second bought 7 hot dogs and 4 soft drinks at a cost of $9.00. What is the cost of a single hot dog? A single soft drink?

We paid $12.50.
How much is one hot dog?
How much is one cola?

We paid $9.00.
How much is one hot dog?
How much is one cola?

67. *Computing a Refund* The grocery store we use does not mark prices on its goods. My wife went to this store, bought three 1 pound packages of bacon and two cartons of eggs, and paid a total of $7.45. Not knowing she went to the store, I also went to the same store, purchased two 1 pound packages of bacon and three cartons of eggs, and paid a total of $6.45. Now we want to return two 1 pound packages of bacon and two cartons of eggs. How much will be refunded?

68. *Finding the Current of a Stream* A swimmer requires 3 hours to swim 15 miles downstream. The return trip upstream takes 5 hours. Find the average speed of the swimmer in still water. How fast is the current of the stream? (Assume that the speed of the swimmer is the same in each direction.)

69. The sum of three numbers is 48. The sum of the two larger numbers is three times the smallest. The sum of the two smaller numbers is 6 more than the largest. Find the numbers.

70. A coin collection consists of 37 coins (nickels, dimes, and quarters). If the collection has a face value of $3.25 and there are 5 more dimes than there are nickels, how many of each coin are in the collection?

71. *Electricity: Kirchhoff's Rules* An application of *Kirchhoff's Rules* to the circuit shown results in the following system of equations:

$$\begin{cases} I_2 = I_1 + I_3 \\ 5 - 3I_1 - 5I_2 = 0 \\ 10 - 5I_2 - 7I_3 = 0 \end{cases}$$

Find the currents I_1, I_2, and I_3.[*]

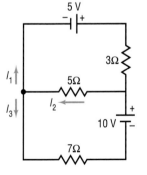

*Source: Based on Raymond Serway, Physics, 3rd ed. Philadelphia: Saunders, 1990, Prob. 26, p. 790.

72. *Electricity: Kirchhoff's Rules* An application of Kirchhoff's Rules to the circuit shown results in the following system of equations:

$$\begin{cases} I_3 = I_1 + I_2 \\ 8 = 4I_3 + 6I_2 \\ 8I_1 = 4 + 6I_2 \end{cases}$$

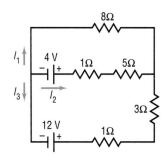

Find the currents I_1, I_2, and I_3.[†]

73. *Theater Revenues* A Broadway theater has 500 seats, divided into orchestra, main, and balcony seating. Orchestra seats sell for $50, main seats for $35, and balcony seats for $25. If all the seats are sold, the gross revenue to the theater is $17,100. If all the main and balcony seats are sold, but only half the orchestra seats are sold, the gross revenue is $14,600. How many are there of each kind of seat?

74. *Laboratory Work Stations* A chemistry laboratory can be used by 38 students at one time. The laboratory has 16 work stations, some set up for 2 students each and the others set up for 3 students each. How many are there of each kind of work station?

75. Make up three systems of two linear equations containing two variables that have:

(a) No solution   (b) Exactly one solution   (c) Infinitely many solutions

Give them to a friend to solve and critique.

76. Write a brief paragraph outlining your strategy for solving a system of two linear equations containing two variables.

77. Do you prefer the method of substitution or the method of elimination for solving a system of two linear equations containing two variables? What about for solving a system of three linear equations containing three variables? Give reasons.

78. *Curve Fitting* Find real numbers b and c such that the parabola $y = x^2 + bx + c$ passes through the points $(1, 3)$ and $(3, 5)$.

79. *Curve Fitting* Find real numbers b and c such that the parabola $y = x^2 + bx + c$ passes through the points $(1, 2)$ and $(-1, 3)$.

80. *Curve Fitting* Find real numbers b and c such that the parabola $y = x^2 + bx + c$ passes through the points (x_1, y_1) and (x_2, y_2).

81. *Curve Fitting* Find real numbers a, b, and c such that the parabola $y = ax^2 + bx + c$ passes through the points $(-1, 4)$, $(2, 3)$, and $(0, 1)$.

82. *Curve Fitting* Find real numbers a, b, and c such that the parabola $y = ax^2 + bx + c$ passes through the points $(-1, -2)$, $(1, -4)$, and $(2, 4)$.

83. Solve: $\begin{cases} y = m_1x + b_1 \\ y = m_2x + b_2 \end{cases}$

where $m_1 \neq m_2$.

84. Solve: $\begin{cases} y = m_1x + b_1 \\ y = m_2x + b_2 \end{cases}$

where $m_1 = m_2 = m$ and $b_1 \neq b_2$.

85. Solve: $\begin{cases} y = m_1x + b_1 \\ y = m_2x + b_2 \end{cases}$

where $m_1 = m_2 = m$ and $b_1 = b_2 = b$.

86. *CBL Experiment* Two students walk toward each other at a *constant* rate. Plotting distance against time on the same viewing window for each student results in a system of linear equations. By finding the intersection point of the two graphs, the time and location where the students pass each other is determined. (Activity 4, Real-World Math with the CBL System, 1994.)

[†]*Source:* Ibid., Prob. 27, p. 790.

7.2

Systems of Linear Equations: Matrices

The systematic approach of the method of elimination for solving a system of linear equations provides another method of solution that involves a simplified notation.

Consider the following system of linear equations:

$$\begin{cases} x + 4y = 14 \\ 3x - 2y = 0 \end{cases}$$

If we choose not to write the symbols used for the variables, we can represent this system as

$$\begin{bmatrix} 1 & 4 & | & 14 \\ 3 & -2 & | & 0 \end{bmatrix}$$

where it is understood that the first column represents the coefficients of the variable x, the second column the coefficients of y, and the third column the constants on the right side of the equal signs. The vertical line serves as a reminder of the equal signs. The large square brackets are the traditional symbols used to denote a *matrix* in algebra.

Matrix

A **matrix** is defined as a rectangular array of numbers,

	Column 1	Column 2		Column j		Column n
Row 1	a_{11}	a_{12}	$\cdots$	a_{1j}	$\cdots$	a_{1n}
Row 2	a_{21}	a_{22}	$\cdots$	a_{2j}	$\cdots$	a_{2n}
$\cdot$	$\cdot$	$\cdot$		$\cdot$		$\cdot$
$\cdot$	$\cdot$	$\cdot$		$\cdot$		$\cdot$
$\cdot$	$\cdot$	$\cdot$		$\cdot$	$\cdot$	$\cdot$
Row i	a_{i1}	a_{i2}	$\cdots$	a_{ij}	$\cdots$	a_{in}
$\cdot$	$\cdot$			$\cdot$		$\cdot$
$\cdot$	$\cdot$	$\cdot$		$\cdot$		$\cdot$
$\cdot$	$\cdot$	$\cdot$		$\cdot$		$\cdot$
Row m	a_{m1}	a_{m2}	$\cdots$	a_{mj}	$\cdots$	a_{mn}

(1)

Each number a_{ij} of the matrix has two indices: the **row index** i and the **column index** j. The matrix shown in display (1) has m rows and n columns. The numbers a_{ij} are usually referred to as the **entries** of the matrix.

Now we will use matrix notation to represent a system of linear equations. The matrices used to represent systems of linear equations are called **augmented matrices**.

E X A M P L E   1

Writing the Augmented Matrix of a System of Equations

Write the augmented matrix of each system of equations.

(a) $\begin{cases} 3x - 4y = -6 & (1) \\ 2x - 3y = -5 & (2) \end{cases}$

(b) $\begin{cases} 2x - y + z = 0 & (1) \\ x + z - 1 = 0 & (2) \\ x + 2y - 8 = 0 & (3) \end{cases}$

Solution

(a) The augmented matrix is

$$\begin{bmatrix} 3 & -4 & | & -6 \\ 2 & -3 & | & -5 \end{bmatrix}$$

(b) Care must be taken that the system is written with the coefficients of all variables present (if any variable is missing, its coefficient is 0) and with all constants to the right of the equal signs. Thus, we need to rearrange the given system as follows:

$$\begin{cases} 2x - y + z = 0 & (1) \\ x + z - 1 = 0 & (2) \\ x + 2y - 8 = 0 & (3) \end{cases}$$

$$\begin{cases} 2x - \quad y + \quad z = 0 & (1) \\ x + 0 \cdot y + \quad z = 1 & (2) \\ x + \quad 2y + 0 \cdot z = 8 & (3) \end{cases}$$

The augmented matrix is

$$\begin{bmatrix} 2 & -1 & 1 & \bigm| & 0 \\ 1 & 0 & 1 & \bigm| & 1 \\ 1 & 2 & 0 & \bigm| & 8 \end{bmatrix}$$

$\blacksquare$

If we do not include the constants to the right of the equal sign, that is, to the right of the vertical bar in the augmented matrix, of a system of equations, the resulting matrix is called the **coefficient matrix** of the system. For the systems discussed in Example 1, the coefficient matrices are

$$\begin{bmatrix} 3 & -4 \\ 2 & -3 \end{bmatrix} \quad \text{and} \quad \begin{bmatrix} 2 & -1 & 1 \\ 1 & 0 & 1 \\ 1 & 2 & 0 \end{bmatrix}$$

$\blacksquare$ Now work Problem 3.

E X A M P L E   2

Writing the System of Linear Equations from the Augmented Matrix

Write the system of linear equations corresponding to each augmented matrix.

(a) $\begin{bmatrix} 5 & 2 & \bigm| & 13 \\ -3 & 1 & \bigm| & -10 \end{bmatrix}$ (b) $\begin{bmatrix} 3 & -1 & -1 & \bigm| & 7 \\ 2 & 0 & 2 & \bigm| & 8 \\ 0 & 1 & 1 & \bigm| & 0 \end{bmatrix}$

Solution (a) The matrix has two rows and so represents a system of two equations. The two columns to the left of the vertical bar indicate that the system has two variables. If x and y are used to denote these variables, the system of equations is

$$\begin{cases} 5x + 2y = 13 & (1) \\ -3x + y = -10 & (2) \end{cases}$$

(b) This matrix represents a system of three equations containing three variables. If x, y, and z are the three variables, this system is

$$\begin{cases} 3x - y - z = 7 & (1) \\ 2x \quad + 2z = 8 & (2) \\ y + z = 0 & (3) \end{cases}$$

$\blacksquare$

Row Operations on a Matrix

E X A M P L E   3

Solving a System of Equations

Solve the system of equations

$$\begin{cases} 4x - 3y = 11 & (1) \\ 3x + 2y = \ \ 4 & (2) \end{cases}$$

Solution   We shall use a variation of the method of elimination to solve the system. First, multiply each side of equation (2) by -1 and add it to equation (1). Replace equation (1) with the result:

$$\begin{cases} \ \ x - 5y = 7 & (1) \\ 3x + 2y = 4 & (2) \end{cases}$$

Multiply each side of equation (1) by -3 and add it to equation (2). Replace equation (2) with the result:

$$\begin{cases} \ \ x - \ \ 5y = \ \ \ \ 7 & (1) \\ 0 \cdot x + 17y = -17 & (2) \end{cases}$$

Multiply each side of equation (2) by $\frac{1}{17}$:

$$\begin{cases} x - 5y = \ \ \ 7 & (1) \\ \ \ \ \ \ \ \ \ y = -1 & (2) \end{cases}$$

Now we back-substitute $y = -1$ into equation (1) to get

$$x - 5y = 7$$
$$x - 5(-1) = 7$$
$$x = 2$$

The solution of the system is $x = 2$, $y = -1$.   ■

The pattern of solution shown above provides a systematic way to solve any system of equations. The idea is to start with the augmented matrix of the system,

$$\begin{bmatrix} 4 & -3 & | & 11 \\ 3 & 2 & | & 4 \end{bmatrix} \qquad \begin{cases} 4x - 3y = 11 & (1) \\ 3x + 2y = \ \ 4 & (2) \end{cases}$$

and eventually arrive at the matrix,

$$\begin{bmatrix} 1 & -5 & | & 7 \\ 0 & 1 & | & -1 \end{bmatrix} \qquad \begin{cases} x - 5y = \ \ \ 7 & (1) \\ \ \ \ \ \ \ \ y = -1 & (2) \end{cases}$$

Let's go through the procedure again, this time starting with the augmented matrix and keeping the final augmented matrix given above in mind:

$$\begin{bmatrix} 4 & -3 & | & 11 \\ 3 & 2 & | & 4 \end{bmatrix} \qquad \begin{cases} 4x - 3y = 11 & (1) \\ 3x + 2y = \ \ 4 & (2) \end{cases}$$

As before, we start by multiplying each side of equation (2) by -1 and adding it to equation (1). This is equivalent to multiplying each entry in the second row of the matrix by -1, adding the result to the corresponding entries in row 1, and replacing row 1 by these entries. The result of this step is that the number 1 appears in row 1, column 1:

$$\begin{bmatrix} 1 & -5 & | & 7 \\ 3 & 2 & | & 4 \end{bmatrix} \qquad \begin{cases} x - 5y = 7 & (1) \\ 3x + 2y = 4 & (2) \end{cases}$$

Multiply each entry in the first row by -3, add the result to the entries in the second row, and replace the second row by these entries. The result of this step is that the number 0 appears in row 2, column 1:

$$\begin{bmatrix} 1 & -5 & | & 7 \\ 0 & 17 & | & -17 \end{bmatrix} \qquad \begin{cases} x - 5y = 7 & (1) \\ 0 \cdot x + 17y = -17 & (2) \end{cases}$$

Multiply each entry in the second row by $\dfrac{1}{17}$. The result of this step is that the number 1 appears in row 2, column 2:

$$\begin{bmatrix} 1 & -5 & | & 7 \\ 0 & 1 & | & -1 \end{bmatrix} \qquad \begin{cases} x - 5y = 7 & (1) \\ y = -1 & (2) \end{cases}$$

Now that we know that $y = -1$, we can back-substitute to find that $x = 2$.

The manipulations just performed on the augmented matrix are called **row operations.** There are three basic row operations:

Row Operations

1. Interchange any two rows.
2. Replace a row by a nonzero multiple of that row.
3. Replace a row by the sum of that row and a constant multiple of some other row.

These three row operations correspond to the three rules given earlier for obtaining an equivalent system of equations. Thus, when a row operation is performed on a matrix, the resulting matrix represents a system of equations equivalent to the system represented by the original matrix.

For example, consider the augmented matrix

$$\begin{bmatrix} 1 & 2 & | & 3 \\ 4 & -1 & | & 2 \end{bmatrix}$$

Suppose we want to apply a row operation to this matrix that results in a matrix whose entry in row 2, column 1 is a 0. The row operation to use is

$$\text{Multiply each entry in row 1 by } -4 \text{ and add the result} \\ \text{to the corresponding entries in row 2.} \qquad (2)$$

If we use R_2 to represent the new entries in row 2 and we use r_1 and r_2 to represent the original entries in rows 1 and 2, respectively, then we can represent the row operation in statement (2) by

$$R_2 = -4r_1 + r_2$$

Then

$$\begin{bmatrix} 1 & 2 & | & 3 \\ 4 & -1 & | & 2 \end{bmatrix} \underset{\substack{\uparrow \\ R_2 = -4r_1 + r_2}}{\rightarrow} \begin{bmatrix} 1 & 2 & | & 3 \\ -4(1)+4 & -4(2)+(-1) & | & -4(3)+2 \end{bmatrix} = \begin{bmatrix} 1 & 2 & | & 3 \\ 0 & -9 & | & -10 \end{bmatrix}$$

As desired, we now have the entry 0 in row 2, column 1.

E X A M P L E   4   *Applying a Row Operation to an Augmented Matrix*

Apply the row operation $R_2 = -3r_1 + r_2$ to the augmented matrix

$$\begin{bmatrix} 1 & -2 & | & 2 \\ 3 & -5 & | & 9 \end{bmatrix}$$

Solution   The row operation $R_2 = -3r_1 + r_2$ tells us that the entries in row 2 are to be replaced by the entries obtained after multiplying each entry in row 1 by -3 and adding the result to the corresponding entries in row 2. Thus,

$$\begin{bmatrix} 1 & -2 & | & 2 \\ 3 & -5 & | & 9 \end{bmatrix} \underset{\substack{\uparrow \\ R_2 = -3r_1 + r_2}}{\rightarrow} \begin{bmatrix} 1 & -2 & | & 2 \\ -3(1)+3 & (-3)(-2)+(-5) & | & -3(2)+9 \end{bmatrix} = \begin{bmatrix} 1 & -2 & | & 2 \\ 0 & 1 & | & 3 \end{bmatrix}$$

■

E X A M P L E   5   *Finding a Particular Row Operation*

Using the matrix

$$\begin{bmatrix} 1 & -2 & | & 2 \\ 0 & 1 & | & 3 \end{bmatrix}$$

find a row operation that will result in a matrix with a 0 in row 1, column 2.

Solution   We want a 0 in row 1, column 2. This result can be accomplished by multiplying row 2 by 2 and adding the result to row 1. That is, we apply the row operation $R_1 = 2r_2 + r_1$:

$$\begin{bmatrix} 1 & -2 & | & 2 \\ 0 & 1 & | & 3 \end{bmatrix} \underset{\substack{\uparrow \\ R_1 = 2r_2 + r_1}}{\rightarrow} \begin{bmatrix} 2(0)+1 & 2(1)+(-2) & | & 2(3)+2 \\ 0 & 1 & | & 3 \end{bmatrix} = \begin{bmatrix} 1 & 0 & | & 8 \\ 0 & 1 & | & 3 \end{bmatrix}.$$

■

A word about the notation we have introduced. A row operation such as $R_1 = 2r_2 + r_1$ changes the entries in row 1. Note also that to change the entries in a given row we multiply the entries in some other row by an appropriate number and add the results to the original entries of the row to be changed.

Now let's see how we use row operations to solve a system of linear equations.

E X A M P L E   6   *Solving a System of Equations Using Matrices*

Solve: $\begin{cases} 4x + 3y = 11 & (1) \\ x - 3y = -1 & (2) \end{cases}$

Solution First, we write the augmented matrix that represents this system:

$$\begin{bmatrix} 4 & 3 & | & 11 \\ 1 & -3 & | & -1 \end{bmatrix}$$

The first step requires getting the entry 1 in row 1, column 1. An interchange of rows 1 and 2 is the easiest way to do this:

$$\begin{bmatrix} 1 & -3 & | & -1 \\ 4 & 3 & | & 11 \end{bmatrix}$$

Next, we want a 0 under the entry 1 in column 1. We use the row operation $R_2 = -4r_1 + r_2$:

$$\begin{bmatrix} 1 & -3 & | & -1 \\ 4 & 3 & | & 11 \end{bmatrix} \rightarrow \begin{bmatrix} 1 & -3 & | & -1 \\ 0 & 15 & | & 15 \end{bmatrix}$$
$$\uparrow$$
$$R_2 = -4r_1 + r_2$$

Now we want the entry 1 in row 2, column 2. We use $R_2 = \frac{1}{15}r_2$:

$$\begin{bmatrix} 1 & -3 & | & -1 \\ 0 & 15 & | & 15 \end{bmatrix} \rightarrow \begin{bmatrix} 1 & -3 & | & -1 \\ 0 & 1 & | & 1 \end{bmatrix}$$
$$\uparrow$$
$$R_2 = \frac{1}{15}r_2$$

The second row of the matrix on the right represents the equation $y = 1$. Thus, using $y = 1$, we back-substitute into the equation $x - 3y = -1$ (from the first row) to get

$$x - 3(1) = -1 \quad y = 1$$
$$x = 2$$

The solution of the system is $x = 2$, $y = 1$. ∎

■ Now work Problem 31.

The steps we used to solve the system of linear equations in Example 6 can be summarized as follows:

Matrix Method for Solving a System of Linear Equations

STEP 1: Write the augmented matrix that represents the system.
STEP 2: Perform row operations that place the entry 1 in row 1, column 1.
STEP 3: Perform row operations that leave the entry 1 in row 1, column 1 unchanged, while causing 0's to appear below it in column 1.
STEP 4: Perform row operations that place the entry 1 in row 2, column 2 and leave the entries in columns to the left unchanged. If it is impossible to place a 1 in row 2, column 2, then proceed to place a 1 in row 2, column 3. Once a 1 is in place, perform row operations to place 0's under it.
STEP 5: Now repeat Step 4, placing a 1 in the next row, but one column to the right. Continue until the bottom row or the vertical bar is reached.
STEP 6: If any rows are obtained that contain only 0's on the left side of the vertical bar, then place such rows at the bottom of the matrix.

After Steps 1 to 6 have been completed, the matrix is said to be in **echelon form.** A little thought should convince you that a matrix is in echelon form when:

1. The entry in row 1, column 1 is a 1, and 0's appear below it
2. The first nonzero entry in each row after the first row is a 1, 0's appear below it, and it appears to the right of the first nonzero entry in any row above.
3. Any rows that contain all 0's to the left of the vertical bar appear at the bottom.

Two advantages of solving a system of equations by writing the augmented matrix in echelon form are the following:

1. The process is algorithmic; that is, it consists of repetitive steps so that it can be programmed on a computer.
2. The process works on any system of linear equations, no matter how many equations or variables are present.

The next example shows how to write a matrix in echelon form.

E X A M P L E   7   *Solving a System of Equations Using Matrices*

$$\text{Solve:} \quad \begin{cases} x - y + z = 8 & (1) \\ 2x + 3y - z = -2 & (2) \\ 3x - 2y - 9z = 9 & (3) \end{cases}$$

Solution   **STEP 1:** The augmented matrix of the system is

$$\begin{bmatrix} 1 & -1 & 1 & | & 8 \\ 2 & 3 & -1 & | & -2 \\ 3 & -2 & -9 & | & 9 \end{bmatrix}$$

STEP 2: Because the entry 1 is already present in row 1, column 1, we can go to Step 3.

STEP 3: Perform the row operations $R_2 = -2r_1 + r_2$ and $R_3 = -3r_1 + r_3$. Each of these leaves the entry 1 in row 1, column 1 unchanged, while causing 0's to appear under it:

$$\begin{bmatrix} 1 & -1 & 1 & | & 8 \\ 2 & 3 & -1 & | & -2 \\ 3 & -2 & -9 & | & 9 \end{bmatrix} \xrightarrow{\uparrow} \begin{bmatrix} 1 & -1 & 1 & | & 8 \\ 0 & 5 & -3 & | & -18 \\ 0 & 1 & -12 & | & -15 \end{bmatrix}$$
$$R_2 = -2r_1 + r_2$$
$$R_3 = -3r_1 + r_3$$

STEP 4: The easiest way to obtain the entry 1 in row 2, column 2 without altering column 1 is to interchange rows 2 and 3 (another way would be to multiply row 2 by $\frac{1}{5}$, but this introduces fractions):

$$\begin{bmatrix} 1 & -1 & 1 & | & 8 \\ 0 & 1 & -12 & | & -15 \\ 0 & 5 & -3 & | & -18 \end{bmatrix}$$

To get 0's under the 1 in row 2, column 2, perform the row operation $R_3 = -5r_2 + r_3$:

$$\begin{bmatrix} 1 & -1 & 1 & | & 8 \\ 0 & 1 & -12 & | & -15 \\ 0 & 5 & -3 & | & -18 \end{bmatrix} \xrightarrow{\uparrow} \begin{bmatrix} 1 & -1 & 1 & | & 8 \\ 0 & 1 & -12 & | & -15 \\ 0 & 0 & 57 & | & 57 \end{bmatrix}$$
$$R_3 = -5r_2 + r_3$$

STEP 5: Continuing, we place a 1 in row 3, column 3 by using $R_3 = \frac{1}{57}r_3$:

$$\begin{bmatrix} 1 & -1 & 1 & \bigm| & 8 \\ 0 & 1 & -12 & \bigm| & -15 \\ 0 & 0 & 57 & \bigm| & 57 \end{bmatrix} \rightarrow \begin{bmatrix} 1 & -1 & 1 & \bigm| & 8 \\ 0 & 1 & -12 & \bigm| & -15 \\ 0 & 0 & 1 & \bigm| & 1 \end{bmatrix}$$

$$\uparrow$$
$$R_3 = \frac{1}{57}r_3$$

Because we have reached the bottom row, the matrix is in echelon form and we can stop.

The system of equations represented by the matrix in echelon form is

$$\begin{cases} x - y + \quad z = \quad 8 \\ \qquad y - 12z = -15 \\ \qquad\qquad z = \quad 1 \end{cases}$$

Using $z = 1$, we back-substitute to get

$$\begin{cases} x - y + 1 = \quad 8 & \text{From row 1 of the matrix} \\ y - 12(1) = -15 & \text{From row 2 of the matrix} \end{cases}$$

Thus, we get $y = -3$, and back-substituting into $x - y = 7$, we find $x = 4$. The solution of the system is $x = 4$, $y = -3$, $z = 1$. ∎

Sometimes, it is advantageous to write a matrix in **reduced echelon form.** In this form, row operations are used to obtain entries that are 0 above (as well as below) the leading 1 in a row. For example, the echelon form obtained in the solution to Example 7 is

$$\begin{bmatrix} 1 & -1 & 1 & \bigm| & 8 \\ 0 & 1 & -12 & \bigm| & -15 \\ 0 & 0 & 1 & \bigm| & 1 \end{bmatrix}$$

To write this matrix in reduced echelon form, we proceed as follows:

$$\begin{bmatrix} 1 & -1 & 1 & \bigm| & 8 \\ 0 & 1 & -12 & \bigm| & -15 \\ 0 & 0 & 1 & \bigm| & 1 \end{bmatrix} \rightarrow \begin{bmatrix} 1 & 0 & -11 & \bigm| & -7 \\ 0 & 1 & -12 & \bigm| & -15 \\ 0 & 0 & 1 & \bigm| & 1 \end{bmatrix} \rightarrow \begin{bmatrix} 1 & 0 & 0 & \bigm| & 4 \\ 0 & 1 & 0 & \bigm| & -3 \\ 0 & 0 & 1 & \bigm| & 1 \end{bmatrix}$$

$$\uparrow \qquad\qquad\qquad \uparrow$$
$$R_1 = r_2 + r_1 \qquad R_1 = 11r_3 + r_1$$
$$R_2 = 12r_3 + r_2$$

The matrix is now written in reduced echelon form. The advantage of writing the matrix in this form is that the solution to the system, $x = 4$, $y = -3$, $z = 1$, is readily found, without the need to back-substitute. Another advantage will be seen in Section 7.4, where the inverse of a matrix is discussed.

■ Now work Problem 49.

The matrix method for solving a system of linear equations also identifies systems that have infinitely many solutions and systems that are inconsistent. Let's see how.

E X A M P L E   8   *Solving a System of Equations Using Matrices*

$$
\text{Solve:} \quad
\begin{cases}
6x - y - z = 4 & (1) \\
-12x + 2y + 2z = -8 & (2) \\
5x + y - z = 3 & (3)
\end{cases}
$$

Solution   We start with the augmented matrix of the system:

$$
\left[\begin{array}{ccc|c}
6 & -1 & -1 & 4 \\
-12 & 2 & 2 & -8 \\
5 & 1 & -1 & 3
\end{array}\right]
\rightarrow
\left[\begin{array}{ccc|c}
1 & -2 & 0 & 1 \\
-12 & 2 & 2 & -8 \\
5 & 1 & -1 & 3
\end{array}\right]
\rightarrow
\left[\begin{array}{ccc|c}
1 & -2 & 0 & 1 \\
0 & -22 & 2 & 4 \\
0 & 11 & -1 & -2
\end{array}\right]
$$

$$
\underset{R_1 = -1r_3 + r_1}{} \qquad\qquad \underset{\substack{R_2 = 12r_1 + r_2. \\ R_3 = -5r_1 + r_3}}{}
$$

Obtaining a 1 in row 2, column 2 without altering column 1 can be accomplished only by $R_2 = -\frac{1}{22}r_2$ or by $R_3 = \frac{1}{11}r_3$. (Do you see why?) We shall use the first of these:

$$
\left[\begin{array}{ccc|c}
1 & -2 & 0 & 1 \\
0 & -22 & 2 & 4 \\
0 & 11 & -1 & -2
\end{array}\right]
\rightarrow
\left[\begin{array}{ccc|c}
1 & -2 & 0 & 1 \\
0 & 1 & -\frac{1}{11} & -\frac{2}{11} \\
0 & 11 & -1 & -2
\end{array}\right]
\rightarrow
\left[\begin{array}{ccc|c}
1 & -2 & 0 & 1 \\
0 & 1 & -\frac{1}{11} & -\frac{2}{11} \\
0 & 0 & 0 & 0
\end{array}\right]
$$

$$
\underset{R_2 = -\frac{1}{22}r_2}{} \qquad\qquad \underset{R_3 = -11r_2 + r_3}{}
$$

This matrix is in echelon form. Because the bottom row consists entirely of 0's, the system actually consists of only two equations:

$$
\begin{cases}
x - 2y = 1 & (1) \\
y - \frac{1}{11}z = -\frac{2}{11} & (2)
\end{cases}
$$

We shall back-substitute the solution for y from the second equation, $y = \frac{1}{11}z - \frac{2}{11}$, into the first equation to get

$$
x = 2y + 1 = 2\left(\tfrac{1}{11}z - \tfrac{2}{11}\right) + 1 = \tfrac{2}{11}z + \tfrac{7}{11}
$$

Thus, the original system is equivalent to the system

$$
\begin{cases}
x = \frac{2}{11}z + \frac{7}{11} & (1) \\
y = \frac{1}{11}z - \frac{2}{11} & (2)
\end{cases}
$$

where z can be any real number.

Let's look at the situation. The original system of three equations is equivalent to a system containing two equations. This means any values of x, y, z that satisfy both

$$
x = \tfrac{2}{11}z + \tfrac{7}{11} \quad \text{and} \quad y = \tfrac{1}{11}z - \tfrac{2}{11}
$$

will be solutions. For example, $z = 0$, $x = \frac{7}{11}$, $y = -\frac{2}{11}$; $z = 1$, $x = \frac{9}{11}$, $y = -\frac{1}{11}$; and $z = -1$, $x = \frac{5}{11}$, $y = -\frac{3}{11}$ are some of the solutions of the original system. There are, in fact, infinitely many values of x, y, and z for which the two equations are satisfied. That is, the original system has infinitely many solutions. We will write the solution of the original system as

$$
\begin{cases}
x = \frac{2}{11}z + \frac{7}{11} \\
y = \frac{1}{11}z - \frac{2}{11}
\end{cases}
$$

where z can be any real number. ∎

We can also find the solution by writing the augmented matrix in reduced echelon form. Starting with the echelon form, we have

$$
\begin{bmatrix}
1 & -2 & 0 & | & 1 \\
0 & 1 & -\frac{1}{11} & | & -\frac{2}{11} \\
0 & 0 & 0 & | & 0
\end{bmatrix}
\rightarrow
\begin{bmatrix}
1 & 0 & -\frac{2}{11} & | & \frac{7}{11} \\
0 & 1 & -\frac{1}{11} & | & -\frac{2}{11} \\
0 & 0 & 0 & | & 0
\end{bmatrix}
$$

$$\uparrow$$
$$R_1 = 2r_2 + r_1$$

The matrix on the right is in reduced echelon form. The corresponding system of equations is

$$
\begin{cases}
x - \frac{2}{11}z = \frac{7}{11} & (1) \\
y - \frac{1}{11}z = -\frac{2}{11} & (2)
\end{cases}
$$

or, equivalently,

$$
\begin{cases}
x = \frac{2}{11}z + \frac{7}{11} & (1) \\
y = \frac{1}{11}z - \frac{2}{11} & (2)
\end{cases}
$$

where z can be any real number.

■ Now work Problem 53.

E X A M P L E   9   *Solving a System of Equations Using Matrices*

Solve: $\begin{cases} x + y + z = 6 \\ 2x - y - z = 3 \\ x + 2y + 2z = 0 \end{cases}$

Solution   The augmented matrix is

$$
\begin{bmatrix}
1 & 1 & 1 & | & 6 \\
2 & -1 & -1 & | & 3 \\
1 & 2 & 2 & | & 0
\end{bmatrix}
\rightarrow
\begin{bmatrix}
1 & 1 & 1 & | & 6 \\
0 & -3 & -3 & | & -9 \\
0 & 1 & 1 & | & -6
\end{bmatrix}
\rightarrow
\begin{bmatrix}
1 & 1 & 1 & | & 6 \\
0 & 1 & 1 & | & -6 \\
0 & -3 & -3 & | & -9
\end{bmatrix}
\rightarrow
\begin{bmatrix}
1 & 1 & 1 & | & 6 \\
0 & 1 & 1 & | & -6 \\
0 & 0 & 0 & | & -27
\end{bmatrix}
$$

$$\uparrow$$
$$R_2 = -2r_1 + r_2$$
$$R_3 = -1r_1 + r_3$$
Interchange rows 2 and 3.
$$R_3 = 3r_2 + r_3.$$

This matrix is in echelon form. The bottom row is equivalent to the equation

$$0x + 0y + 0z = -27$$

which has no solution. Hence, the original system is inconsistent.   ■

■ Now work Problems 23 and 29.

The matrix method is especially effective for systems of equations for which the number of equations and the number of variables are unequal. Here, too, such a system is either inconsistent or consistent. If it is consistent, it will have either exactly one solution or infinitely many solutions.

Let's look at a system of four equations containing three variables.

E X A M P L E   1 0   *Solving a System of Equations Using Matrices*

Solve:
$$\begin{cases} x - 2y + z = 0 & (1) \\ 2x + 2y - 3z = -3 & (2) \\ y - z = -1 & (3) \\ -x + 4y + 2z = 13 & (4) \end{cases}$$

Solution   The augmented matrix is

$$\begin{bmatrix} 1 & -2 & 1 & | & 0 \\ 2 & 2 & -3 & | & -3 \\ 0 & 1 & -1 & | & -1 \\ -1 & 4 & 2 & | & 13 \end{bmatrix} \rightarrow \begin{bmatrix} 1 & -2 & 1 & | & 0 \\ 0 & 6 & -5 & | & -3 \\ 0 & 1 & -1 & | & -1 \\ 0 & 2 & 3 & | & 13 \end{bmatrix} \rightarrow \begin{bmatrix} 1 & -2 & 1 & | & 0 \\ 0 & 1 & -1 & | & -1 \\ 0 & 6 & -5 & | & -3 \\ 0 & 2 & 3 & | & 13 \end{bmatrix}$$

$\uparrow$
$R_2 = -2r_1 + r_2$
$R_4 = r_1 + r_4$

$\uparrow$ Interchange rows 2 and 3.

$$\rightarrow \begin{bmatrix} 1 & -2 & 1 & | & 0 \\ 0 & 1 & -1 & | & -1 \\ 0 & 0 & 1 & | & 3 \\ 0 & 0 & 5 & | & 15 \end{bmatrix} \rightarrow \begin{bmatrix} 1 & -2 & 1 & | & 0 \\ 0 & 1 & -1 & | & -1 \\ 0 & 0 & 1 & | & 3 \\ 0 & 0 & 0 & | & 0 \end{bmatrix}$$

$\uparrow$
$R_3 = -6r_2 + r_3$
$R_4 = -2r_2 + r_4$

$\uparrow$
$R_4 = -5r_3 + r_4$

We could stop here, since the matrix is in echelon form, and back-substitute $z = 3$ to find x and y. Or we can continue to obtain the reduced echelon form:

$$\rightarrow \begin{bmatrix} 1 & 0 & -1 & | & -2 \\ 0 & 1 & -1 & | & -1 \\ 0 & 0 & 1 & | & 3 \\ 0 & 0 & 0 & | & 0 \end{bmatrix} \rightarrow \begin{bmatrix} 1 & 0 & 0 & | & 1 \\ 0 & 1 & 0 & | & 2 \\ 0 & 0 & 1 & | & 3 \\ 0 & 0 & 0 & | & 0 \end{bmatrix}$$

$\uparrow$
$R_1 = 2r_2 + r_1$

$\uparrow$
$R_1 = r_3 + r_1$
$R_2 = r_3 + r_2$

The matrix is now in reduced echelon form, and we can see that the solution is $x = 1$, $y = 2$, $z = 3$. ∎

E X A M P L E   1 1   *Mixing Acids*

A chemistry laboratory has three containers of nitric acid, HNO_3. One container holds a solution with a concentration of 10% HNO_3, the second holds 20% HNO_3, and the third holds 40% HNO_3. How many liters of each solution should be mixed to obtain 100 liters of a solution whose concentration is 25% HNO_3?

Solution   Let x, y, and z represent the number of liters of 10%, 20%, and 40% concentrations of HNO_3, respectively. We want 100 liters in all, and the concentration of HNO_3 from each solution must sum to 25% of 100 liters. Thus, we find that

$$\begin{cases} x + y + z = 100 \\ 0.10x + 0.20y + 0.40z = 0.25(100) \end{cases}$$

Now, the augmented matrix is

$$\begin{bmatrix} 1 & 1 & 1 & | & 100 \\ 0.10 & 0.20 & 0.40 & | & 25 \end{bmatrix} \rightarrow \begin{bmatrix} 1 & 1 & 1 & | & 100 \\ 0 & 0.10 & 0.30 & | & 15 \end{bmatrix}$$

$$\uparrow$$
$$R_2 = -0.10r_1 + r_2$$

$$\rightarrow \begin{bmatrix} 1 & 1 & 1 & | & 100 \\ 0 & 1 & 3 & | & 150 \end{bmatrix} \rightarrow \begin{bmatrix} 1 & 0 & -2 & | & -50 \\ 0 & 1 & 3 & | & 150 \end{bmatrix}$$

$$\uparrow \qquad\qquad\qquad \uparrow$$
$$R_2 = 10r_2 \qquad\qquad R_1 = -1r_2 + r_1$$

The matrix is now in reduced echelon form. The final matrix represents the system

$$\begin{cases} x - 2z = -50 & (1) \\ y + 3z = 150 & (2) \end{cases}$$

which has infinitely many solutions given by

$$\begin{cases} x = 2z - 50 & (1) \\ y = -3z + 150 & (2) \end{cases}$$

where z is any real number. However, the practical considerations of this problem require us to restrict the solutions to $x \geq 0$, $y \geq 0$, $z \geq 0$. Furthermore, we require $25 \leq z \leq 50$, because otherwise $x < 0$ or $y < 0$. Some of the possible solutions are given in Table 1. The final determination of what solution the laboratory will pick very likely depends on availability, cost differences, and other considerations.

TABLE 1

LITERS OF 10% SOLUTION	LITERS OF 20% SOLUTION	LITERS OF 40% SOLUTION	LITERS OF 25% SOLUTION
0	75	25	100
10	60	30	100
12	57	31	100
16	51	33	100
26	36	38	100
38	18	44	100
46	6	48	100
50	0	50	100

Row Operations on a Graphing Utility

Graphing utilities can be used to perform row operations on an augmented matrix in order to obtain a matrix in reduced echelon form.

E X A M P L E   1 2   *Using a Graphing Utility and Matrices to Solve a System of Equations*

Solve the system of Example 7 using a graphing utility:
$$\begin{cases} x - y + z = 8 & (1) \\ 2x + 3y - z = -2 & (2) \\ 3x - 2y - 9z = 9 & (3) \end{cases}$$

Solution First, we enter the augmented matrix into our graphing utility. See Figure 5.

FIGURE 5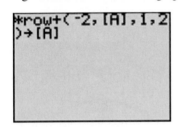

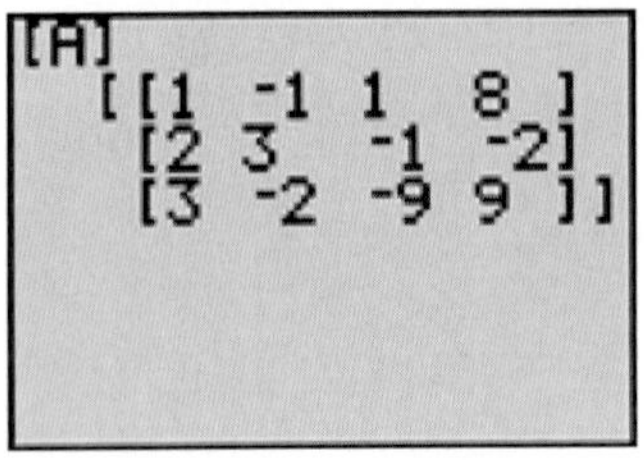

Now perform the row operations from Example 7 until the augmented matrix is in reduced echelon form. They were:

1. $R_2 = -2r_1 + r_2$
2. $R_3 = -3r_1 + r_3$
3. Interchange rows 2 and 3
4. $R_3 = -5r_2 + r_3$
5. $R_3 = \frac{1}{57}r_3$

For example, to perform the row operation $R_2 = -2r_1 + r_2$, we enter the following command into a T1-82 graphing calculator:*

FIGURE 6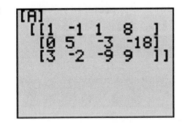

From the command listed in Figure 6, we obtain the augmented matrix given in Figure 7.

FIGURE 7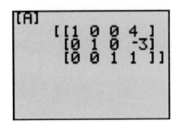

Notice row 2 is the same as row 2 in Step 3 from Example 7. We continue to enter row operations until we obtain a matrix in reduced echelon form. See Figure 8.

FIGURE 8

```
[A]
 [[1 0 0 4 ]
  [0 1 0 -3]
  [0 0 1 1 ]]
```

From the matrix shown in Figure 8, we see the solution to the system is $x = 4$, $y = -3$, $z = 1$. ∎

*Check your manual to see how it is done on your utility.

7.2

Exercise 7.2

In Problems 1–10, write the augmented matrix of the given system of equations.

1. $\begin{cases} x - 5y = 5 \\ 4x + 3y = 6 \end{cases}$

2. $\begin{cases} 3x + 4y = 7 \\ 4x - 2y = 5 \end{cases}$

3. $\begin{cases} 2x + 3y - 6 = 0 \\ 4x - 6y + 2 = 0 \end{cases}$

4. $\begin{cases} 9x - y = 0 \\ 3x - y - 4 = 0 \end{cases}$

5. $\begin{cases} 0.01x - 0.03y = 0.06 \\ 0.13x + 0.10y = 0.20 \end{cases}$

6. $\begin{cases} \frac{4}{3}x - \frac{3}{2}y = \frac{3}{4} \\ -\frac{1}{4}x + \frac{1}{3}y = \frac{2}{3} \end{cases}$

7. $\begin{cases} x - y + z = 10 \\ 3x + 2y = 5 \\ x + y + 2z = 2 \end{cases}$

8. $\begin{cases} 5x - y - z = 0 \\ x + y = 5 \\ 2x - 3z = 2 \end{cases}$

9. $\begin{cases} x + y - z = 2 \\ 3x - 2y = 2 \end{cases}$

10. $\begin{cases} 2x + 3y - 4z = 0 \\ x - 5z + 2 = 0 \end{cases}$

In Problems 11–20 use a graphing utility to perform, in order (a), followed by (b), followed by (c) on the given augmented matrix.

11. $\begin{bmatrix} 1 & -3 & -5 & | & -2 \\ 2 & -5 & -4 & | & 5 \\ -3 & 5 & 4 & | & 6 \end{bmatrix}$ (a) $R_2 = -2r_1 + r_2$ (b) $R_3 = 3r_1 + r_3$ (c) $R_3 = 4r_2 + r_3$

12. $\begin{bmatrix} 1 & -3 & -3 & | & -3 \\ 2 & -5 & 2 & | & -4 \\ -3 & 2 & 4 & | & 6 \end{bmatrix}$ (a) $R_2 = -2r_1 + r_2$ (b) $R_3 = 3r_1 + r_3$ (c) $R_3 = 7r_2 + r_3$

13. $\begin{bmatrix} 1 & -3 & 4 & | & 3 \\ 2 & -5 & 6 & | & 6 \\ -3 & 3 & 4 & | & 6 \end{bmatrix}$ (a) $R_2 = -2r_1 + r_2$ (b) $R_3 = 3r_1 + r_3$ (c) $R_3 = 6r_2 + r_3$

14. $\begin{bmatrix} 1 & -3 & 3 & | & -5 \\ 2 & -5 & -3 & | & -5 \\ -3 & -2 & 4 & | & 6 \end{bmatrix}$ (a) $R_2 = -2r_1 + r_2$ (b) $R_3 = 3r_1 + r_3$ (c) $R_3 = 11r_2 + r_3$

15. $\begin{bmatrix} 1 & -3 & 2 & | & -6 \\ 2 & -5 & 3 & | & -4 \\ -3 & -6 & 4 & | & 6 \end{bmatrix}$ (a) $R_2 = -2r_1 + r_2$ (b) $R_3 = 3r_1 + r_3$ (c) $R_3 = 15r_2 + r_3$

16. $\begin{bmatrix} 1 & -3 & -4 & | & -6 \\ 2 & -5 & 6 & | & -6 \\ -3 & 1 & 4 & | & 6 \end{bmatrix}$ (a) $R_2 = -2r_1 + r_2$ (b) $R_3 = 3r_1 + r_3$ (c) $R_3 = 8r_2 + r_3$

17. $\begin{bmatrix} 1 & -3 & 1 & | & -2 \\ 2 & -5 & 6 & | & -2 \\ -3 & 1 & 4 & | & 6 \end{bmatrix}$ (a) $R_2 = -2r_1 + r_2$ (b) $R_3 = 3r_1 + r_3$ (c) $R_3 = 8r_2 + r_3$

18. $\begin{bmatrix} 1 & -3 & -1 & | & 2 \\ 2 & -5 & 2 & | & 6 \\ -3 & -6 & 4 & | & 6 \end{bmatrix}$ (a) $R_2 = -2r_1 + r_2$ (b) $R_3 = 3r_1 + r_3$ (c) $R_3 = 15r_2 + r_3$

19. $\begin{bmatrix} 1 & -3 & -2 & | & 3 \\ 2 & -5 & 2 & | & -1 \\ -3 & -2 & 4 & | & 6 \end{bmatrix}$ (a) $R_2 = -2r_1 + r_2$ (b) $R_3 = 3r_1 + r_3$ (c) $R_3 = 11r_2 + r_3$

20. $\begin{bmatrix} 1 & -3 & 5 & | & -3 \\ 2 & -5 & 1 & | & -4 \\ -3 & 3 & 4 & | & 6 \end{bmatrix}$ (a) $R_2 = -2r_1 + r_2$ (b) $R_3 = 3r_1 + r_3$ (c) $R_3 = 6r_2 + r_3$

In Problems 21–30, the reduced echelon form of a system of linear equations is given. Write the system of equations corresponding to the given matrix. Use x, y, or x, y, z or x_1, x_2, x_3, x_4 as variables. Determine whether the system is consistent or inconsistent. If it is consistent, give the solution.

21. $\begin{bmatrix} 1 & 0 & | & 5 \\ 0 & 1 & | & -1 \end{bmatrix}$

22. $\begin{bmatrix} 1 & 0 & | & -4 \\ 0 & 1 & | & 0 \end{bmatrix}$

23. $\begin{bmatrix} 1 & 0 & 0 & | & 1 \\ 0 & 1 & 0 & | & 2 \\ 0 & 0 & 0 & | & 3 \end{bmatrix}$

24. $\begin{bmatrix} 1 & 0 & 0 & | & 0 \\ 0 & 1 & 0 & | & 0 \\ 0 & 0 & 0 & | & 2 \end{bmatrix}$

25. $\begin{bmatrix} 1 & 0 & 2 & | & -1 \\ 0 & 1 & -4 & | & -2 \\ 0 & 0 & 0 & | & 0 \end{bmatrix}$

26. $\begin{bmatrix} 1 & 0 & 4 & | & 4 \\ 0 & 1 & 3 & | & 2 \\ 0 & 0 & 0 & | & 0 \end{bmatrix}$

27. $\begin{bmatrix} 1 & 0 & 0 & 0 & | & 1 \\ 0 & 1 & 0 & 1 & | & 2 \\ 0 & 0 & 1 & 2 & | & 3 \end{bmatrix}$ **28.** $\begin{bmatrix} 1 & 0 & 0 & 0 & | & 1 \\ 0 & 1 & 0 & 2 & | & 2 \\ 0 & 0 & 1 & 3 & | & 0 \end{bmatrix}$ **29.** $\begin{bmatrix} 1 & 0 & 0 & 4 & | & 2 \\ 0 & 1 & 1 & 3 & | & 3 \\ 0 & 0 & 0 & 0 & | & 0 \end{bmatrix}$

30. $\begin{bmatrix} 1 & 0 & 0 & 0 & | & 1 \\ 0 & 1 & 0 & 0 & | & 2 \\ 0 & 0 & 1 & 2 & | & 3 \end{bmatrix}$

In Problems 31–72, solve each system of equations using matrices (row operations). If the system has no solution, say it is inconsistent.

31. $\begin{cases} x + y = 8 \\ x - y = 4 \end{cases}$

32. $\begin{cases} x + 2y = 5 \\ x + y = 3 \end{cases}$

33. $\begin{cases} x - 5y = -13 \\ 3x + 2y = 12 \end{cases}$

34. $\begin{cases} x + 3y = 5 \\ 2x - 3y = -8 \end{cases}$

35. $\begin{cases} 3x - 6y = 24 \\ 5x + 4y = 12 \end{cases}$

36. $\begin{cases} 2x + 4y = 16 \\ 3x - 5y = -9 \end{cases}$

37. $\begin{cases} 2x + y = 1 \\ 4x + 2y = 6 \end{cases}$

38. $\begin{cases} x - y = 5 \\ -3x + 3y = 2 \end{cases}$

39. $\begin{cases} 2x - 4y = -2 \\ 3x + 2y = 3 \end{cases}$

40. $\begin{cases} 3x + 3y = 3 \\ 4x + 2y = \frac{8}{3} \end{cases}$

41. $\begin{cases} x + 2y = 4 \\ 2x + 4y = 8 \end{cases}$

42. $\begin{cases} 3x - y = 7 \\ 9x - 3y = 21 \end{cases}$

43. $\begin{cases} 2x + 3y = 6 \\ x - y = \frac{1}{2} \end{cases}$

44. $\begin{cases} \frac{1}{2}x + y = -2 \\ x - 2y = 8 \end{cases}$

45. $\begin{cases} 3x - 5y = 3 \\ 15x + 5y = 21 \end{cases}$

46. $\begin{cases} 2x - y = -1 \\ x + \frac{1}{2}y = \frac{3}{2} \end{cases}$

47. $\begin{cases} x - y = 6 \\ 2x - 3z = 16 \\ 2y + z = 4 \end{cases}$

48. $\begin{cases} 2x + y = -4 \\ -2y + 4z = 0 \\ 3x - 2z = -11 \end{cases}$

49. $\begin{cases} x - 2y + 3z = 7 \\ 2x + y + z = 4 \\ -3x + 2y - 2z = -10 \end{cases}$

50. $\begin{cases} 2x + y - 3z = 0 \\ -2x + 2y + z = -7 \\ 3x - 4y - 3z = 7 \end{cases}$

51. $\begin{cases} 2x - 2y - 2z = 2 \\ 2x + 3y + z = 2 \\ 3x + 2y = 0 \end{cases}$

52. $\begin{cases} 2x - 3y - z = 0 \\ -x + 2y + z = 5 \\ 3x - 4y - z = 1 \end{cases}$

53. $\begin{cases} -x + y + z = -1 \\ -x + 2y - 3z = -4 \\ 3x - 2y - 7z = 0 \end{cases}$

54. $\begin{cases} 2x - 3y - z = 0 \\ 3x + 2y + 2z = 2 \\ x + 5y + 3z = 2 \end{cases}$

55. $\begin{cases} 2x - 2y + 3z = 6 \\ 4x - 3y + 2z = 0 \\ -2x + 3y - 7z = 1 \end{cases}$

56. $\begin{cases} 3x - 2y + 2z = 6 \\ 7x - 3y + 2z = -1 \\ 2x - 3y + 4z = 0 \end{cases}$

57. $\begin{cases} x + y - z = 6 \\ 3x - 2y + z = -5 \\ x + 3y - 2z = 14 \end{cases}$

58. $\begin{cases} x - y + z = -4 \\ 2x - 3y + 4z = -15 \\ 5x + y - 2z = 12 \end{cases}$

59. $\begin{cases} x + 2y - z = -3 \\ 2x - 4y + z = -7 \\ -2x + 2y - 3z = 4 \end{cases}$

60. $\begin{cases} x + 4y - 3z = -8 \\ 3x - y + 3z = 12 \\ x + y + 6z = 1 \end{cases}$

61. $\begin{cases} 3x + y - z = \frac{2}{3} \\ 2x - y + z = 1 \\ 4x + 2y = \frac{8}{3} \end{cases}$

62. $\begin{cases} x + y = 1 \\ 2x - y + z = 1 \\ x + 2y + z = \frac{8}{3} \end{cases}$

63. $\begin{cases} x + y + z + w = 4 \\ 2x - y + z = 0 \\ 3x + 2y + z - w = 6 \\ x - 2y - 2z + 2w = -1 \end{cases}$

64. $\begin{cases} x + y + z + w = 4 \\ -x + 2y + z = 0 \\ 2x + 3y + z - w = 6 \\ -2x + y - 2z + 2w = -1 \end{cases}$

65. $\begin{cases} x + 2y + z = 1 \\ 2x - y + 2z = 2 \\ 3x + y + 3z = 3 \end{cases}$

66. $\begin{cases} x + 2y - z = 3 \\ 2x - y + 2z = 6 \\ x - 3y + 3z = 4 \end{cases}$

67. $\begin{cases} x - y + z = 5 \\ 3x + 2y - 2z = 0 \end{cases}$

68. $\begin{cases} 2x + y - z = 4 \\ -x + y + 3z = 1 \end{cases}$

69. $\begin{cases} 2x + 3y - z = 3 \\ x - y - z = 0 \\ -x + y + z = 0 \\ x + y + 3z = 5 \end{cases}$

70. $\begin{cases} x - 3y + z = 1 \\ 2x - y - 4z = 0 \\ x - 3y + 2z = 1 \\ x - 2y = 5 \end{cases}$

71. $\begin{cases} 4x + y + z - w = 4 \\ x - y + 2z + 3w = 3 \end{cases}$

72. $\begin{cases} -4x + y = 5 \\ 2x - y + z - w = 5 \\ z + w = 4 \end{cases}$

73. *Curve Fitting* Find the parabola $y = ax^2 + bx + c$ that passes through the points $(1, 2)$, $(-2, -7)$, and $(2, -3)$.

74. *Curve Fitting* Find the parabola $y = ax^2 + bx + c$ that passes through the points $(1, -1)$, $(3, -1)$, and $(-2, 14)$.

75. *Curve Fitting* Find the function $f(x) = ax^3 + bx^2 + cx + d$ for which $f(-3) = -112, f(-1) = -2, f(1) = 4$, and $f(2) = 13$.

76. *Curve Fitting* Find the function $f(x) = ax^3 + bx^2 + cx + d$ for which $f(-2) = -10, f(-1) = 3, f(1) = 5$, and $f(3) = 15$.

77. *Mixing Acids* A chemistry laboratory has three containers of sulfuric acid, H_2SO_4. One container holds a solution with a concentration of 15% H_2SO_4, the second holds 25% H_2SO_4, and the third holds 50% H_2SO_4. How many liters of each solution should be mixed to obtain 100 liters of a solution with a concentration of 40% H_2SO_4? Construct a table similar to Table 1 illustrating some of the possible combinations.

78. *Painting a House* Three painters, Mike, Dan, and Katy, working together can paint the exterior of a home in 10 hours. Dan and Katy together have painted a similar house in 15 hours. One day, all three worked on this same kind of house for 4 hours, after which Katy left. Mike and Dan required 8 more hours to finish. Assuming no gain or loss in efficiency, how long should it take each person to complete such a job alone?

79. *Prices of Fast Food* One group of customers bought 8 deluxe hamburgers, 6 orders of large fries, and 6 large colas for $26.10. A second group ordered 10 deluxe hamburgers, 6 large fries, and 8 large colas and paid $31.60. Is there sufficient information to determine the price of each food item? If not, construct a table showing the various possibilities. Assume that the hamburgers cost between $1.75 and $2.25, the fries between $0.75 and $1.00, and the colas between $0.60 and $0.90.

80. *Prices of Fast Food* Use the information given in Problem 79, and suppose that a third group purchased 3 deluxe hamburgers, 2 large fries, and 4 large colas for $10.95. Now is there sufficient information to determine the price of each food item?

81. *Financial Planning* Three retired couples each require an additional annual income of $2000 per year. As their financial consultant, you recommend that they invest some money in Treasury bills that yield 7%, some money in corporate bonds that yield 9%, and some money in junk bonds that yield 11%. Prepare a table for each couple showing the various ways their goals can be achieved:

(a) If the first couple has $20,000 to invest.

(b) If the second couple has $25,000 to invest.

(c) If the third couple has $30,000 to invest.

(d) What advice would you give each couple regarding the amount to invest and the choices available? [Higher yields generally carry more risk].

82. *Financial Planning* A retired couple has $25,000 to invest. As their financial consultant, you recommend that they invest some money in Treasury bills that yield 7%, some money in corporate bonds that yield 9%, and some money in junk bonds that yield 11%. Prepare a table showing the various ways this couple can achieve the following goals:

    (a) The couple wants $1500 per year in income.

    (b) The couple wants $2000 per year in income.

    (c) The couple wants $2500 per year in income.

    (d) What advice would you give this couple regarding the income they require and the choices available? [Higher yields generally carry more risk].

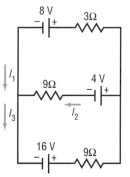

83. *Electricity: Kirchhoff's Rules* An application of Kirchhoff's Rules to the circuit shown results in the following system of equations:

$$\begin{cases} I_1 + I_2 = I_3 \\ 16 - 8 - 9I_3 - 3I_1 = 0 \\ 16 - 4 - 9I_3 - 9I_2 = 0 \\ 8 - 4 - 9I_2 + 3I_1 = 0 \end{cases}$$

    Find the currents I_1, I_2, and I_3.*

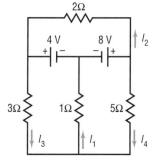

84. *Electricity: Kirchhoff's Rules* An application of Kirchhoff's Rules to the circuit shown results in the following system of equations:

$$\begin{cases} -4 + 8 - 2I_2 = 0 \\ 8 = 5I_4 + I_1 \\ 4 = 3I_3 + I_1 \\ I_3 + I_4 = I_1 \end{cases}$$

    Find the currents I_1, I_2, I_3, and I_4.†

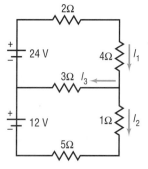

85. *Electricity: Kirchhoff's Rules* An application of Kirchhoff's Rules to the circuit shown results in the following system of equations:

$$\begin{cases} I_1 = I_3 + I_2 \\ 24 - 6I_1 - 3I_3 = 0 \\ 12 + 24 - 6I_1 - 6I_2 = 0 \end{cases}$$

    Find the currents I_1, I_2, and I_3.‡

86. Write a brief paragraph or two that outlines your strategy for solving a system of linear equations using matrices.

87. When solving a system of linear equations using matrices, do you prefer to place the augmented matrix in echelon form or in reduced echelon form? Give reasons for your choice.

88. Make up three systems of three linear equations containing three variables that have:

    (a) No solution   (b) Exactly one solution   (c) Infinitely many solutions

    Give them to a friend to solve and critique.

*Source: Based on Raymond Serway, *Physics,* 3rd ed. Philadelphia: Saunders, 1990, Prob. 31, p. 790.
†Source: Ibid., Prob. 34, p. 791.
‡Source: Ibid., Prob. 38, p. 791.

89. Consider the system of equations

$$\begin{cases} a_1x + b_1y = c_1 \\ a_2x + b_2y = c_2 \end{cases}$$

If $D = a_1b_2 - a_2b_1 \neq 0$, use matrices to show that the solution is

$$x = \frac{1}{D}(c_1b_2 - c_2b_1), \qquad y = \frac{1}{D}(a_1c_2 - a_2c_1)$$

90. For the system in Problem 89, suppose that $D = a_1b_2 - a_2b_1 = 0$. Use matrices to show that the system is inconsistent if either $a_1c_2 \neq a_2c_1$ or $b_1c_2 \neq b_2c_1$ and has infinitely many solutions if both $a_1c_2 = a_2c_1$ and $b_1c_2 = b_2c_1$.

91. The graph of a linear equation containing three variables is a plane. Give a geometrical argument for what can result when solving a system of two linear equations containing three variables. [*Hint:* Two planes in a three-dimensional space are either coincident (the same), parallel, or intersect in a line.]

92. Refer to Problem 91. Give a geometrical argument for what can result when solving a system of three linear equations containing three variables.

93. Refer to Problem 91. Give a geometrical argument for what can result when solving a system of four linear equations containing three variables.

7.3

Systems of Linear Equations: Determinants

In the preceding section, we described a method of using matrices to solve any system of linear equations. This section deals with yet another method for solving systems of linear equations; however, it can be used only when the number of equations equals the number of variables. Although the method will work for any system (provided the number of equations equals the number of variables), it is most often used for systems of two equations containing two variables or three equations containing three variables. This method, called *Cramer's Rule*, is based on the concept of a *determinant*.

2 by 2 Determinants

2 by 2 Determinant

If a, b, c, and d are four real numbers, the symbol

$$D = \begin{vmatrix} a & b \\ c & d \end{vmatrix}$$

is called a **2 by 2 determinant.** Its value is the number $ad - bc$; that is,

$$D = \begin{vmatrix} a & b \\ c & d \end{vmatrix} = ad - bc \tag{1}$$

A device that may be helpful for remembering the value of a 2 by 2 determinant is the following:

E X A M P L E   1   *Evaluating a 2 × 2 Determinant*

$$\begin{vmatrix} 3 & -2 \\ 6 & 1 \end{vmatrix} = (3)(1) - (6)(-2) = 3 - (-12) = 15$$ ∎

Graphing utilities can also be used to evaluate determinants.

E X A M P L E   2   *Using a Graphing Utility to Evaluate a 2 × 2 Determinant*

Use a graphing utility to evaluate the determinant from Example 1: $\begin{vmatrix} 3 & -2 \\ 6 & 1 \end{vmatrix}$.

Solution   First, we enter the matrix whose entries are those of the determinant into the graphing utility and name it A. Using the determinant command, we obtain the result shown in Figure 9.

FIGURE 9

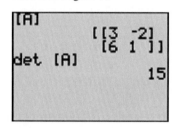

∎

■ Now work Problem 3.

Let's now see the role that a 2 by 2 determinant plays in the solution of a system of two equations containing two variables. Consider the system

$$\begin{cases} ax + by = s & (1) \\ cx + dy = t & (2) \end{cases} \qquad (2)$$

We shall use the method of elimination to solve this system.

Provided $d \neq 0$ and $b \neq 0$, this system is equivalent to the system

$$\begin{cases} adx + bdy = sd & (1) \quad \text{Multiply by } d. \\ bcx + bdy = tb & (2) \quad \text{Multiply by } b. \end{cases}$$

On subtracting the second equation from the first equation, we get

$$\begin{cases} (ad - bc)x + 0 \cdot y = sd - tb & (1) \\ \quad bcx \quad + bdy = tb & (2) \end{cases}$$

Now, the first equation can be rewritten using determinant notation:

$$\begin{vmatrix} a & b \\ c & d \end{vmatrix} x = \begin{vmatrix} s & b \\ t & d \end{vmatrix}$$

If $D = \begin{vmatrix} a & b \\ c & d \end{vmatrix} = ad - bc \neq 0$, we can solve for x to get

$$x = \frac{\begin{vmatrix} s & b \\ t & d \end{vmatrix}}{\begin{vmatrix} a & b \\ c & d \end{vmatrix}} = \frac{\begin{vmatrix} s & b \\ t & d \end{vmatrix}}{D} \qquad (3)$$

Return now to the original system (2). Provided $a \neq 0$ and $c \neq 0$, the system is equivalent to

$$\begin{cases} acx + bcy = cs & \text{(1)} \quad \text{Multiply by } c. \\ acx + ady = at & \text{(2)} \quad \text{Multiply by } a. \end{cases}$$

On subtracting the first equation from the second equation, we get

$$\begin{cases} acx + \quad bcy \quad = \quad cs & \text{(1)} \\ 0 \cdot x + (ad - bc)y = at - cs & \text{(2)} \end{cases}$$

The second equation now can be rewritten using determinant notation:

$$\begin{vmatrix} a & b \\ c & d \end{vmatrix} y = \begin{vmatrix} a & s \\ c & t \end{vmatrix}$$

If $D = \begin{vmatrix} a & b \\ c & d \end{vmatrix} = ad - bc \neq 0$, we can solve for y to get

$$y = \frac{\begin{vmatrix} a & s \\ c & t \end{vmatrix}}{\begin{vmatrix} a & b \\ c & d \end{vmatrix}} = \frac{\begin{vmatrix} a & s \\ c & t \end{vmatrix}}{D} \tag{4}$$

Equations (3) and (4) lead us to the following result, called **Cramer's Rule:**

Theorem
Cramer's Rule for Two Equations
Containing Two Variables

The solution to the system of equations

$$\begin{cases} ax + by = s & \text{(1)} \\ cx + dy = t & \text{(2)} \end{cases} \tag{5}$$

is given by

$$x = \frac{\begin{vmatrix} s & b \\ t & d \end{vmatrix}}{\begin{vmatrix} a & b \\ c & d \end{vmatrix}}, \qquad y = \frac{\begin{vmatrix} a & s \\ c & t \end{vmatrix}}{\begin{vmatrix} a & b \\ c & d \end{vmatrix}} \tag{6}$$

provided that

$$D = \begin{vmatrix} a & b \\ c & d \end{vmatrix} = ad - bc \neq 0 \qquad \blacksquare$$

In the derivation given for Cramer's Rule above, we assumed that none of the numbers a, b, c, and d were 0. In Problem 58 at the end of this section you will be asked to complete the proof under the less stringent conditions that $D = ad - bc \neq 0$.

Now look carefully at the pattern in Cramer's Rule. The denominator in the solution (6) is the determinant of the coefficients of the variables:

$$\begin{cases} ax + by = s \\ cx + dy = t \end{cases} \qquad D = \begin{vmatrix} a & b \\ c & d \end{vmatrix}$$

In the solution for x, the numerator is the determinant, denoted by D_x, formed by replacing the entries in the first column (the coefficients of x) in D by the constants on the right side of the equal sign:

$$D_x = \begin{vmatrix} s & b \\ t & d \end{vmatrix}$$

In the solution for y, the numerator is the determinant, denoted by D_y, formed by replacing the entries in the second column (the coefficients of y) in D by the constants on the right side of the equal sign:

$$D_y = \begin{vmatrix} a & s \\ c & t \end{vmatrix}$$

Cramer's Rule then states that, if $D \neq 0$,

$$x = \frac{D_x}{D}, \qquad y = \frac{D_y}{D} \tag{7}$$

E X A M P L E   3   *Solving a System of Equations Using Determinants*

Use Cramer's Rule, if applicable, to solve the system

$$\begin{cases} 3x - 2y = 4 & (1) \\ 6x + y = 13 & (2) \end{cases}$$

Solution   The determinant D of the coefficients of the variables is

$$D = \begin{vmatrix} 3 & -2 \\ 6 & 1 \end{vmatrix} = (3)(1) - (6)(-2) = 15$$

Because $D \neq 0$, Cramer's Rule (7) can be used:

$$x = \frac{D_x}{D} = \frac{\begin{vmatrix} 4 & -2 \\ 13 & 1 \end{vmatrix}}{15} = \frac{30}{15} = 2, \qquad y = \frac{D_y}{D} = \frac{\begin{vmatrix} 3 & 4 \\ 6 & 13 \end{vmatrix}}{15} = \frac{15}{15} = 1$$

The solution is $x = 2$, $y = 1$. ■

If, in attempting to use Cramer's Rule, the determinant of D of the coefficients of the variables is found to equal 0 (so that Cramer's Rule is not applicable), then the system either is inconsistent or has infinitely many solutions. (Refer to Problem 90 in Exercise 7.2.)

We can also use a graphing utility to solve a system of equations using Cramer's Rule.

E X A M P L E   4   *Solving a System of Equations Using Determinants and a Graphing Utility*

Use Cramer's Rule, if applicable, to solve the system.

$$\begin{cases} 3x - 2y = 12 & (1) \\ 2x + 4y = 24 & (2) \end{cases}$$

MISSION POSSIBLE

Chapter 7

KEEPING SECRETS

Your group has discovered that a rival consulting firm has been intercepting and reading the reports you are sending one another. You need a way to encode your final decisions so that this rival firm does not take business away from you by stealing your ideas. You have decided to find a way to encode your most important messages to each other so that prying eyes are not able to read them.

Your method of coding begins with the simple one used by children, that is, assigning to each letter of the alphabet the number which represents its position in the order. For example, A = 1, B = 2, . . . , M =13, N = 14, . . . , Z = 26. A space would be represented by 27; a period by 0. Then the message is translated into 2 × 1 matrices as in the example which follows.

Example: Suppose you wish to send the message; *Choose Dealer C.* You would write this out first as letters in groups of two, then as matrices.

$$\begin{matrix} Ch & oo & se & D & ea & le & r & C \\ \begin{bmatrix} 3 \\ 8 \end{bmatrix} & \begin{bmatrix} 15 \\ 15 \end{bmatrix} & \begin{bmatrix} 19 \\ 5 \end{bmatrix} & \begin{bmatrix} 27 \\ 4 \end{bmatrix} & \begin{bmatrix} 5 \\ 1 \end{bmatrix} & \begin{bmatrix} 12 \\ 5 \end{bmatrix} & \begin{bmatrix} 18 \\ 27 \end{bmatrix} & \begin{bmatrix} 3 \\ 0 \end{bmatrix} \end{matrix}$$

Next you would use a coding matrix, a 2 × 2 square matrix with numbers chosen so that the inverse exists. We'll use $\begin{bmatrix} 6 & 4 \\ -2 & 7 \end{bmatrix}$ as our coding matrix. To encode a message, you multiply the coding matrix times each matrix of the message, as follows:.

$$\begin{bmatrix} 6 & 4 \\ -2 & 7 \end{bmatrix} \cdot \begin{bmatrix} 3 \\ 8 \end{bmatrix} = \begin{bmatrix} 50 \\ 50 \end{bmatrix}, \begin{bmatrix} 6 & 4 \\ -2 & 7 \end{bmatrix} \cdot \begin{bmatrix} 15 \\ 15 \end{bmatrix} = \begin{bmatrix} 150 \\ 75 \end{bmatrix}, \text{etc.}$$

The resulting set of matrices would be sent along to your partner as a sequence of numbers: 50, 50, 150, 75, etc. Your partner would use the inverse of the encoding matrix to decipher your message. Obviously you would have to guard the encoding matrix and its inverse very carefully!

1. Find the inverse of the coding matrix above. Make sure it is correct by multiplying the two encoded matrices times your matrix to see if you can recover the original two matrices.
2. Next, create your own message (not too long!) and encrypt it using the coding matrix above. Copy the resulting code onto a separate piece of paper as a sequence of numbers.
3. Exchange coded messages with another group.
4. Use the inverse matrix to decode the message sent to you by the other group.
5. Cryptography is a science practiced by governments at war, or businesses which are competitive, in situations where secrecy is crucial. If your consulting firm is serious about keeping secrets, you may need to upgrade your coding procedures. For example, the message *Choose Dealer C,* given above, could be broken up into four 4 × 1 matrices. Then the coding matrix would have to be 4 × 4 with an inverse to match. A government might use a 20 × 20 coding matrix to decrease the probability that it could be deciphered. Create your own 4 × 4 coding matrix, find its inverse, and encode a message of your choosing in 4 × 1 matrices. Then pass the code and the key to a neighboring group for decoding.

Solution   We enter the coefficient matrix into our graphing utility and call it A. We also enter the matrices D_x and D_y into our graphing utility and call them B and C, respectively. Finally, we find x by calculating $\det[B]/\det[A]$ and y by calculating $\det[C]/\det[A]$. The results are shown in Figure 10.

FIGURE 10

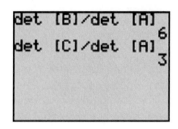

The solution is $x = 6$, $y = 3$.   ■

■   Now work Problem 11.

3 by 3 Determinants

In order to use Cramer's Rule to solve a system of three equations containing three variables, we need to define a 3 by 3 determinant.

A **3 by 3 determinant** is symbolized by

$$\begin{vmatrix} a_{11} & a_{12} & a_{13} \\ a_{21} & a_{22} & a_{23} \\ a_{31} & a_{32} & a_{33} \end{vmatrix} \qquad (8)$$

in which $a_{11}, a_{12}, \ldots$ are real numbers.

As with matrices, we use a double subscript to identify an entry by indicating its row and column numbers. For example, the entry a_{23} is in row 2, column 3.

The value of a 3 by 3 determinant may be defined in terms of 2 by 2 determinants by the following formula:

$$\begin{vmatrix} a_{11} & a_{12} & a_{13} \\ a_{21} & a_{22} & a_{23} \\ a_{31} & a_{32} & a_{33} \end{vmatrix} = a_{11}\overset{\text{Minus}}{\underset{\downarrow}{\begin{vmatrix} a_{22} & a_{23} \\ a_{32} & a_{33} \end{vmatrix}}} - a_{12}\begin{vmatrix} a_{21} & a_{23} \\ a_{31} & a_{33} \end{vmatrix} + a_{13}\begin{vmatrix} a_{21} & a_{22} \\ a_{31} & a_{32} \end{vmatrix} \qquad (9)$$

<table>
<tr><td>2 by 2 determinant left after removing row and column containing a_{11}</td><td>2 by 2 determinant left after removing row and column containing a_{12}</td><td>2 by 2 determinant left after removing row and column containing a_{13}</td></tr>
</table>

Be sure to take note of the minus sign that appears with the second term—it's easy to forget it! Formula (9) is best remembered by noting that each entry in row 1 is multiplied by the 2 by 2 determinant that remains after the row and column containing the entry have been removed, as follows:

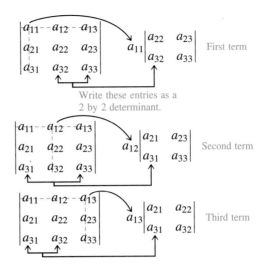

Now insert the minus sign before the middle expression and add:

$$\begin{vmatrix} a_{11} & a_{12} & a_{13} \\ a_{21} & a_{22} & a_{23} \\ a_{31} & a_{32} & a_{33} \end{vmatrix} = a_{11}\begin{vmatrix} a_{22} & a_{23} \\ a_{32} & a_{33} \end{vmatrix} \overset{\text{Minus}}{-} a_{12}\begin{vmatrix} a_{21} & a_{23} \\ a_{31} & a_{33} \end{vmatrix} + a_{13}\begin{vmatrix} a_{21} & a_{22} \\ a_{31} & a_{32} \end{vmatrix}$$

Formula (9) exhibits one way to find the value of a 3 by 3 determinant, *by expanding across row 1.* In fact, the expansion can take place across any row or down any column. The terms to be added or subtracted consist of the row (or column) entry times the value of the 2 by 2 determinant that remains after removing the row and column entry. The value of the determinant is found by adding or subtracting the terms according to the following scheme:

$$\begin{matrix} + & - & + \\ - & + & - \\ + & - & + \end{matrix}$$

For example, if we choose to expand down column 2, we obtain

$$\begin{vmatrix} a_{11} & a_{12} & a_{13} \\ a_{21} & a_{22} & a_{23} \\ a_{31} & a_{32} & a_{33} \end{vmatrix} = -a_{12}\begin{vmatrix} a_{21} & a_{23} \\ a_{31} & a_{33} \end{vmatrix} + a_{22}\begin{vmatrix} a_{11} & a_{13} \\ a_{31} & a_{33} \end{vmatrix} - a_{32}\begin{vmatrix} a_{11} & a_{13} \\ a_{21} & a_{23} \end{vmatrix}$$

Expand down column 2 (− , + , −)

If we choose to expand across row 3, we obtain

$$\begin{vmatrix} a_{11} & a_{12} & a_{13} \\ a_{21} & a_{22} & a_{23} \\ a_{31} & a_{32} & a_{33} \end{vmatrix} = a_{31}\begin{vmatrix} a_{12} & a_{13} \\ a_{22} & a_{23} \end{vmatrix} - a_{32}\begin{vmatrix} a_{11} & a_{13} \\ a_{21} & a_{23} \end{vmatrix} + a_{33}\begin{vmatrix} a_{11} & a_{12} \\ a_{21} & a_{22} \end{vmatrix}$$

Expand across row 3 (+ , − , +)

It can be shown that the value of a determinant does not depend on the choice of the row or column used in the expansion.

E X A M P L E 5 *Evaluating a 3 × 3 Determinant*

Find the value of the 3 by 3 determinant: $\begin{vmatrix} 3 & 4 & -1 \\ 4 & 6 & 2 \\ 8 & -2 & 3 \end{vmatrix}$

Solution We choose to expand across row 1.

Remember the minus sign.
↓

$$\begin{vmatrix} 3 & 4 & -1 \\ 4 & 6 & 2 \\ 8 & -2 & 3 \end{vmatrix} = 3\begin{vmatrix} 6 & 2 \\ -2 & 3 \end{vmatrix} - 4\begin{vmatrix} 4 & 2 \\ 8 & 3 \end{vmatrix} + (-1)\begin{vmatrix} 4 & 6 \\ 8 & -2 \end{vmatrix}$$

$$= 3(18 + 4) - 4(12 - 16) + (-1)(-8 - 48)$$

$$= 3(22) - 4(-4) + (-1)(-56)$$

$$= 66 + 16 + 56 = 138$$ ∎

We could also find the value of the 3 by 3 determinant in Example 5 by expanding down column 3 (the signs are +, −, +):

$$\begin{vmatrix} 3 & 4 & -1 \\ 4 & 6 & 2 \\ 8 & -2 & 3 \end{vmatrix} = (-1)\begin{vmatrix} 4 & 6 \\ 8 & -2 \end{vmatrix} - 2\begin{vmatrix} 3 & 4 \\ 8 & -2 \end{vmatrix} + 3\begin{vmatrix} 3 & 4 \\ 4 & 6 \end{vmatrix}$$

$$= -1(-8 - 48) - 2(-6 - 32) + 3(18 - 16)$$

$$= 56 + 76 + 6 = 138$$

Evaluating 3 × 3 determinants on a graphing utility follows the same procedure as evaluating 2 × 2 determinants.

■ Now work Problem 7.

Systems of Three Equations Containing Three Variables

Consider the following system of three equations containing three variables:

$$\begin{cases} a_{11}x + a_{12}y + a_{13}z = c_1 \\ a_{21}x + a_{22}y + a_{23}z = c_2 \\ a_{31}x + a_{32}y + a_{33}z = c_3 \end{cases} \qquad (10)$$

It can be shown that if the determinant D of the coefficients of the variables is not 0, that is, if

$$D = \begin{vmatrix} a_{11} & a_{12} & a_{13} \\ a_{21} & a_{22} & a_{23} \\ a_{31} & a_{32} & a_{33} \end{vmatrix} \neq 0$$

then the unique solution of system (10) is given by

Cramer's Rule for Three
Equations Containing Three
Variables

$$x = \frac{D_x}{D}, \qquad y = \frac{D_y}{D}, \qquad z = \frac{D_z}{D}$$

where

$$D_x = \begin{vmatrix} c_1 & a_{12} & a_{13} \\ c_2 & a_{22} & a_{23} \\ c_3 & a_{32} & a_{33} \end{vmatrix} \quad D_y = \begin{vmatrix} a_{11} & c_1 & a_{13} \\ a_{21} & c_2 & a_{23} \\ a_{31} & c_3 & a_{33} \end{vmatrix} \quad D_z = \begin{vmatrix} a_{11} & a_{12} & c_1 \\ a_{21} & a_{22} & c_2 \\ a_{31} & a_{32} & c_3 \end{vmatrix}$$

The similarity of this pattern and the pattern observed earlier for a system of two equations containing two variables should be apparent.

E X A M P L E   6   *Using Cramer's Rule*

Use Cramer's Rule, if applicable, to solve the following system:

$$\begin{cases} 2x + y - z = 3 & (1) \\ -x + 2y + 4z = -3 & (2) \\ x - 2y - 3z = 4 & (3) \end{cases}$$

Solution   The value of the determinant D of the coefficients of the variables is

$$D = \begin{vmatrix} 2 & 1 & -1 \\ -1 & 2 & 4 \\ 1 & -2 & -3 \end{vmatrix} = 2\begin{vmatrix} 2 & 4 \\ -2 & -3 \end{vmatrix} - 1\begin{vmatrix} -1 & 4 \\ 1 & -3 \end{vmatrix} + (-1)\begin{vmatrix} -1 & 2 \\ 1 & -2 \end{vmatrix}$$

$$= 2(2) - 1(-1) + (-1)(0)$$
$$= 4 + 1 = 5$$

Because $D \neq 0$, we proceed to find the value of D_x, D_y, and D_z:

$$D_x = \begin{vmatrix} 3 & 1 & -1 \\ -3 & 2 & 4 \\ 4 & -2 & -3 \end{vmatrix} = 3\begin{vmatrix} 2 & 4 \\ -2 & -3 \end{vmatrix} - 1\begin{vmatrix} -3 & 4 \\ 4 & -3 \end{vmatrix} + (-1)\begin{vmatrix} -3 & 2 \\ 4 & -2 \end{vmatrix}$$

$$= 3(2) - 1(-7) + (-1)(-2) = 15$$

$$D_y = \begin{vmatrix} 2 & 3 & -1 \\ -1 & -3 & 4 \\ 1 & 4 & -3 \end{vmatrix} = 2\begin{vmatrix} -3 & 4 \\ 4 & -3 \end{vmatrix} - 3\begin{vmatrix} -1 & 4 \\ 1 & -3 \end{vmatrix} + (-1)\begin{vmatrix} -1 & -3 \\ 1 & 4 \end{vmatrix}$$

$$= 2(-7) - 3(-1) + (-1)(-1)$$
$$= -14 + 3 + 1 = -10$$

$$D_z = \begin{vmatrix} 2 & 1 & 3 \\ -1 & 2 & -3 \\ 1 & -2 & 4 \end{vmatrix} = 2\begin{vmatrix} 2 & -3 \\ -2 & 4 \end{vmatrix} - 1\begin{vmatrix} -1 & -3 \\ 1 & 4 \end{vmatrix} + 3\begin{vmatrix} -1 & 2 \\ 1 & -2 \end{vmatrix}$$

$$= 2(2) - 1(-1) + 3(0) = 5$$

As a result,

$$x = \frac{D_x}{D} = \frac{15}{5} = 3, \qquad y = \frac{D_y}{D} = \frac{-10}{5} = -2, \qquad z = \frac{D_z}{D} = \frac{5}{5} = 1$$

The solution is $x = 3$, $y = -2$, $z = 1$. ∎

If the determinant of the coefficients of the variables of a system of three linear equations containing three variables is 0, then Cramer's Rule is not applicable. In such a case, the system either is inconsistent or has infinitely many solutions.

Solving systems of three equations containing three variables using Cramer's Rule on a graphing utility follows the same procedure as that for solving systems of two equations containing two variables.

■ Now work Problem 29.

More about Determinants

Determinants have several properties that are sometimes helpful for obtaining their value. We list some of them here.

Theorem   The value of a determinant changes sign if any two rows (or any two columns) are interchanged.   (11) ■

Proof for 2 by 2 Determinants

$$\begin{vmatrix} a & b \\ c & d \end{vmatrix} = ad - bc \quad \text{and} \quad \begin{vmatrix} c & d \\ a & b \end{vmatrix} = bc - ad = -(ad - bc)$$ ■

E X A M P L E   7   *Demonstrating a Theorem (11)*

$$\begin{vmatrix} 3 & 4 \\ 1 & 2 \end{vmatrix} = 6 - 4 = 2 \qquad \begin{vmatrix} 1 & 2 \\ 3 & 4 \end{vmatrix} = 4 - 6 = -2$$ ■

Theorem   If all the entries in any row (or any column) equal 0, the value of the determinant is 0.   (12) ■

Proof   Merely expand across the row (or down the column) containing the 0's. ■

Theorem   If any two rows (or any two columns) of a determinant have corresponding entries that are equal, the value of the determinant is 0.   (13) ■

You are asked to prove this result for a 3 by 3 determinant in which the entries in column 1 equal the entries in column 3 in Problem 61 at the end of this section.

E X A M P L E   8   *Demonstrating a Theorem (13)*

$$\begin{vmatrix} 1 & 2 & 3 \\ 1 & 2 & 3 \\ 4 & 5 & 6 \end{vmatrix} = 1\begin{vmatrix} 2 & 3 \\ 5 & 6 \end{vmatrix} - 2\begin{vmatrix} 1 & 3 \\ 4 & 6 \end{vmatrix} + 3\begin{vmatrix} 1 & 2 \\ 4 & 5 \end{vmatrix}$$

$$= 1(-3) - 2(-6) + 3(-3)$$

$$= -3 + 12 - 9 = 0$$ ■

Theorem   If any row (or any column) of a determinant is multiplied by a nonzero number k, the value of the determinant is also changed by a factor of k.   (14) ■

You are asked to prove this result for a 3 by 3 determinant using row 2 in Problem 60 at the end of this section.

EXAMPLE 9   *Demonstrating a Theorem (14)*

$$\begin{vmatrix} 1 & 2 \\ 4 & 6 \end{vmatrix} = 6 - 8 = -2$$

$$\begin{vmatrix} k & 2k \\ 4 & 6 \end{vmatrix} = 6k - 8k = -2k = k(-2) = k\begin{vmatrix} 1 & 2 \\ 4 & 6 \end{vmatrix}$$ ■

Theorem   If the entries of any row (or any column) of a determinant are multiplied by a nonzero number k and the result is added to the corresponding entries of another row (or column), the value of the determinant remains unchanged.   (15)   ■

In Problem 62 at the end of this section, you are asked to prove this result for a 3 by 3 determinant using rows 1 and 2.

EXAMPLE 10   *Demonstrating a Theorem (15)*

$$\begin{vmatrix} 3 & 4 \\ 5 & 2 \end{vmatrix} = \begin{vmatrix} -7 & 0 \\ 5 & 2 \end{vmatrix} = -14$$

Multiply row 2 by -2 and add to row 1.   ■

7.3

Exercise 7.3

In Problems 1–10, find the value of each determinant: (a) by hand; (b) using a graphing utility.

1. $\begin{vmatrix} 3 & 1 \\ 4 & 2 \end{vmatrix}$ **2.** $\begin{vmatrix} 6 & 1 \\ 5 & 2 \end{vmatrix}$ **3.** $\begin{vmatrix} 6 & 4 \\ -1 & 3 \end{vmatrix}$ **4.** $\begin{vmatrix} 8 & -3 \\ 4 & 2 \end{vmatrix}$ **5.** $\begin{vmatrix} -3 & -1 \\ 4 & 2 \end{vmatrix}$

6. $\begin{vmatrix} -4 & 2 \\ -5 & 3 \end{vmatrix}$ **7.** $\begin{vmatrix} 3 & 4 & 2 \\ 1 & -1 & 5 \\ 1 & 2 & -2 \end{vmatrix}$ **8.** $\begin{vmatrix} 1 & 3 & -2 \\ 6 & 1 & -5 \\ 8 & 2 & 3 \end{vmatrix}$ **9.** $\begin{vmatrix} 4 & -1 & 2 \\ 6 & -1 & 0 \\ 1 & -3 & 4 \end{vmatrix}$ **10.** $\begin{vmatrix} 3 & -9 & 4 \\ 1 & 4 & 0 \\ 8 & -3 & 1 \end{vmatrix}$

In Problems 11–38, solve each system of equations using Cramer's Rule if it is applicable: (a) by hand; (b) using a graphing utility. If Cramer's Rule is not applicable, say so.

11. $\begin{cases} x + y = 8 \\ x - y = 4 \end{cases}$ **12.** $\begin{cases} x + 2y = 5 \\ x + y = 3 \end{cases}$ **13.** $\begin{cases} 5x - y = 13 \\ 2x + 3y = 12 \end{cases}$ **14.** $\begin{cases} x + 3y = 5 \\ 2x - 3y = -8 \end{cases}$

15. $\begin{cases} 3x = 24 \\ x + 2y = 0 \end{cases}$ **16.** $\begin{cases} 4x + 5y = -3 \\ -2y = -4 \end{cases}$ **17.** $\begin{cases} 3x - 6y = 24 \\ 5x + 4y = 12 \end{cases}$ **18.** $\begin{cases} 2x + 4y = 16 \\ 3x - 5y = -9 \end{cases}$

19. $\begin{cases} 3x - 2y = 4 \\ 6x - 4y = 0 \end{cases}$ **20.** $\begin{cases} -x + 2y = 5 \\ 4x - 8y = 6 \end{cases}$ **21.** $\begin{cases} 2x - 4y = -2 \\ 3x + 2y = 3 \end{cases}$ **22.** $\begin{cases} 3x + 3y = 3 \\ 4x + 2y = \frac{8}{3} \end{cases}$

23. $\begin{cases} 2x - 3y = -1 \\ 10x + 10y = 5 \end{cases}$ **24.** $\begin{cases} 3x - 2y = 0 \\ 5x + 10y = 4 \end{cases}$ **25.** $\begin{cases} 2x + 3y = 6 \\ x - y = \frac{1}{2} \end{cases}$ **26.** $\begin{cases} \frac{1}{2}x + y = -2 \\ x - 2y = 8 \end{cases}$

27. $\begin{cases} 3x - 5y = 3 \\ 15x + 5y = 21 \end{cases}$ **28.** $\begin{cases} 2x - y = -1 \\ x + \frac{1}{2}y = \frac{3}{2} \end{cases}$ **29.** $\begin{cases} x + y - z = 6 \\ 3x - 2y + z = -5 \\ x + 3y - 2z = 14 \end{cases}$

30. $\begin{cases} x - y + z = -4 \\ 2x - 3y + 4z = -15 \\ 5x + y - 2z = 12 \end{cases}$

31. $\begin{cases} x + 2y - z = -3 \\ 2x - 4y + z = -7 \\ -2x + 2y - 3z = 4 \end{cases}$

32. $\begin{cases} x + 4y - 3z = -8 \\ 3x - y + 3z = 12 \\ x + y + 6z = 1 \end{cases}$

33. $\begin{cases} x - 2y + 3z = 1 \\ 3x + y - 2z = 0 \\ 2x - 4y + 6z = 2 \end{cases}$

34. $\begin{cases} x - y + 2z = 5 \\ 3x + 2y = 4 \\ -2x + 2y - 4z = -10 \end{cases}$

35. $\begin{cases} x + 2y - z = 0 \\ 2x - 4y + z = 0 \\ -2x + 2y - 3z = 0 \end{cases}$

36. $\begin{cases} x + 4y - 3z = 0 \\ 3x - y + 3z = 0 \\ x + y + 6z = 0 \end{cases}$

37. $\begin{cases} x - 2y + 3z = 0 \\ 3x + y - 2z = 0 \\ 2x - 4y + 6z = 0 \end{cases}$

38. $\begin{cases} x - y + 2z = 0 \\ 3x + 2y = 0 \\ -2x + 2y - 4z = 0 \end{cases}$

39. Solve: $\begin{cases} \dfrac{1}{x} + \dfrac{1}{y} = 8 \\ \dfrac{3}{x} - \dfrac{5}{y} = 0 \end{cases}$

40. Solve: $\begin{cases} \dfrac{4}{x} - \dfrac{3}{y} = 0 \\ \dfrac{6}{x} + \dfrac{3}{2y} = 2 \end{cases}$

[*Hint:*   Let $u = 1/x$ and $v = 1/y$ and solve for u and v.]

In Problems 41–46, solve for x.

41. $\begin{vmatrix} x & x \\ 4 & 3 \end{vmatrix} = 5$

42. $\begin{vmatrix} x & 1 \\ 3 & x \end{vmatrix} = -2$

43. $\begin{vmatrix} x & 1 & 1 \\ 4 & 3 & 2 \\ -1 & 2 & 5 \end{vmatrix} = 2$

44. $\begin{vmatrix} 3 & 2 & 4 \\ 1 & x & 5 \\ 0 & 1 & -2 \end{vmatrix} = 0$

45. $\begin{vmatrix} x & 2 & 3 \\ 1 & x & 0 \\ 6 & 1 & -2 \end{vmatrix} = 7$

46. $\begin{vmatrix} x & 1 & 2 \\ 1 & x & 3 \\ 0 & 1 & 2 \end{vmatrix} = -4x$

In Problems 47–54, use properties of determinants to find the value of each determinant if it is known that

$$\begin{vmatrix} x & y & z \\ u & v & w \\ 1 & 2 & 3 \end{vmatrix} = 4$$

47. $\begin{vmatrix} 1 & 2 & 3 \\ u & v & w \\ x & y & z \end{vmatrix}$

48. $\begin{vmatrix} x & y & z \\ u & v & w \\ 2 & 4 & 6 \end{vmatrix}$

49. $\begin{vmatrix} x & y & z \\ -3 & -6 & -9 \\ u & v & w \end{vmatrix}$

50. $\begin{vmatrix} 1 & 2 & 3 \\ x - u & y - v & z - w \\ u & v & w \end{vmatrix}$

51. $\begin{vmatrix} 1 & 2 & 3 \\ x - 3 & y - 6 & z - 9 \\ 2u & 2v & 2w \end{vmatrix}$

52. $\begin{vmatrix} x & y & z - x \\ u & v & w - u \\ 1 & 2 & 2 \end{vmatrix}$

53. $\begin{vmatrix} 1 & 2 & 3 \\ 2x & 2y & 2z \\ u - 1 & v - 2 & w - 3 \end{vmatrix}$

54. $\begin{vmatrix} x + 3 & y + 6 & z + 9 \\ 3u - 1 & 3v - 2 & 3w - 3 \\ 1 & 2 & 3 \end{vmatrix}$

55. *Geometry: Equation of a Line* An equation of the line containing the two points (x_1, y_1) and (x_2, y_2) may be expressed as the determinant

$$\begin{vmatrix} x & y & 1 \\ x_1 & y_1 & 1 \\ x_2 & y_2 & 1 \end{vmatrix} = 0$$

Prove this result by expanding the determinant and comparing the result to the two-point form of the equation of a line.

56. *Geometry: Collinear Points* Using the result obtained in Problem 55, show that three distinct points (x_1, y_1), (x_2, y_2), and (x_3, y_3) are collinear (lie on the same line) if and only if

$$\begin{vmatrix} x_1 & y_1 & 1 \\ x_2 & y_2 & 1 \\ x_3 & y_3 & 1 \end{vmatrix} = 0$$

57. Show that $\begin{vmatrix} x^2 & x & 1 \\ y^2 & y & 1 \\ z^2 & z & 1 \end{vmatrix} = (y - z)(x - y)(x - z)$.

58. Complete the proof of Cramer's Rule for two equations containing two variables. [*Hint:* In system (5), page 503, if $a = 0$, then $b \neq 0$ and $c \neq 0$, since $D = -bc \neq 0$. Now show that equations (6) provide a solution of the system when $a = 0$. There are then three remaining cases: $b = 0$, $c = 0$, and $d = 0$.]

59. Interchange columns 1 and 3 of a 3 by 3 determinant. Show that the value of the new determinant is -1 times the value of the original determinant.

60. Multiply each entry in row 2 of a 3 by 3 determinant by the number k, $k \neq 0$. Show that the value of the new determinant is k times the value of the original determinant.

61. Prove that a 3 by 3 determinant in which the entries in column 1 equal those in column 3 has the value 0.

62. Prove that, if row 2 of a 3 by 3 determinant is multiplied by k, $k \neq 0$, and the result is added to the entries in row 1, then there is no change in the value of the determinant.

7.4

Matrix Algebra

In Section 7.2, we defined a matrix as an array of real numbers and used an augmented matrix to represent a system of linear equations. There is, however, a branch of mathematics, called **linear algebra,** that deals with matrices in such a way that an algebra of matrices is permitted. In this section, we provide a survey of how this **matrix algebra** is developed.

Before getting started, we restate the definition of a matrix.

Matrix

A **matrix** is defined as a rectangular array of numbers:

jth column

$$\begin{bmatrix} a_{11} & a_{12} & \cdots & a_{1j} & \cdots & a_{1n} \\ a_{21} & a_{22} & \cdots & a_{2j} & \cdots & a_{2n} \\ \vdots & \vdots & & \vdots & & \vdots \\ a_{i1} & a_{i2} & \cdots & a_{ij} & \cdots & a_{in} \\ \vdots & \vdots & & \vdots & & \vdots \\ a_{m1} & a_{m2} & \cdots & a_{mj} & \cdots & a_{mn} \end{bmatrix} \quad i\text{th row}$$

Each number a_{ij} of the matrix has two indices: the **row index** i and the **column index** j. The matrix shown above has m rows and n columns. The $m \cdot n$ numbers a_{ij} are usually referred to as the **entries** of the matrix.

Let's begin with an example that illustrates how matrices can be used to conveniently represent an array of information.

E X A M P L E   1   *Arranging Data in a Matrix*

In a survey of 1000 people, the following information was obtained:

200 males	Thought federal defense spending was too high
150 males	Thought federal defense spending was too low
45 males	Had no opinion
315 females	Thought federal defense spending was too high
125 females	Thought federal defense spending was too low
165 females	Had no opinion

We can arrange the above data in a rectangular array as follows:

	TOO HIGH	TOO LOW	NO OPINION
Males	200	150	45
Females	315	125	165

or as

$$\begin{bmatrix} 200 & 150 & 45 \\ 315 & 125 & 165 \end{bmatrix}$$

This matrix has two rows (representing males and females) and three columns (representing "too high," "too low," and "no opinion"). ■

The matrix we developed in Example 1 has 2 rows and 3 columns. In general, a matrix with m rows and n columns is called an ***m* by *n* matrix.** Thus, the matrix we developed in Example 1 is a 2 by 3 matrix. Notice that an m by n matrix will contain $m \cdot n$ entries.

If an m by n matrix has the same number of rows as columns, that is, if $m = n$, then the matrix is referred to as a **square matrix.**

E X A M P L E   2   *Examples of Matrices*

(a) $\begin{bmatrix} 5 & 0 \\ -6 & 1 \end{bmatrix}$ A 2 by 2 square matrix   (b) $\begin{bmatrix} 1 & 0 & 3 \end{bmatrix}$ A 1 by 3 matrix

(c) $\begin{bmatrix} 6 & -2 & 4 \\ 4 & 3 & 5 \\ 8 & 0 & 1 \end{bmatrix}$ A 3 by 3 square matrix ■

Equality and Addition of Matrices

We begin our discussion of matrix algebra by first defining what is meant by two matrices being equal and then defining the operations of addition and subtraction. It is important to note that these definitions require each matrix to have the same number of rows *and* the same number of columns as a prerequisite for equality and for addition and subtraction.

We usually represent matrices by capital letters, such as *A, B, C,* and so on.

Equal Matrices

Two m by n matrices A and B are said to be **equal,** written as

$$A = B$$

provided each entry a_{ij} in A is equal to the corresponding entry b_{ij} in B.

For example,

$$\begin{bmatrix} 2 & 1 \\ 0.5 & -1 \end{bmatrix} = \begin{bmatrix} \sqrt{4} & 1 \\ \frac{1}{2} & -1 \end{bmatrix} \quad \text{and} \quad \begin{bmatrix} 3 & 2 & 1 \\ 0 & 1 & -2 \end{bmatrix} = \begin{bmatrix} \sqrt{9} & \sqrt{4} & 1 \\ 0 & 1 & \sqrt[3]{-8} \end{bmatrix}$$

$$\begin{bmatrix} 4 & 1 \\ 6 & 1 \end{bmatrix} \neq \begin{bmatrix} 4 & 0 \\ 6 & 1 \end{bmatrix} \qquad \text{Because the entries in row 1, column 2 are not equal}$$

$$\begin{bmatrix} 4 & 1 & 2 \\ 6 & 1 & 2 \end{bmatrix} \neq \begin{bmatrix} 4 & 1 & 2 & 3 \\ 6 & 1 & 2 & 4 \end{bmatrix} \qquad \text{Because the matrix on the left is 2 by 3 and the matrix on the right is 2 by 4}$$

If each of A and B is an m by n matrix (and each therefore contains $m \cdot n$ entries), the statement $A = B$ actually represents a system of $m \cdot n$ ordinary equations. We will make use of this fact a little later.

Suppose A and B represent two m by n matrices. We define their **sum $A + B$** to be the m by n matrix formed by adding the corresponding entries a_{ij} of A and b_{ij} of B. The **difference $A - B$** is defined as the m by n matrix formed by subtracting the entries b_{ij} in B from the corresponding entries a_{ij} in A. Addition and subtraction of matrices are allowed only for matrices having the same number m of rows and the same number n of columns. Thus, for example, a 2 by 3 matrix and a 2 by 4 matrix cannot be added or subtracted.

The advent of graphing utilities has made the sometimes tedious process of matrix algebra easy. Let's compare how a graphing utility adds and subtracts matrices with doing it by hand.

E X A M P L E   3

Adding and Subtracting Matrices

Suppose that

$$A = \begin{bmatrix} 2 & 4 & 8 & -3 \\ 0 & 1 & 2 & 3 \end{bmatrix} \quad \text{and} \quad B = \begin{bmatrix} -3 & 4 & 0 & 1 \\ 6 & 8 & 2 & 0 \end{bmatrix}$$

Find: (a) $A + B$ (b) $A - B$

Graphing Solution Enter the matrices into a graphing utility. Name them $[A]$ and $[B]$. Figure 11 shows the results of adding and subtracting $[A]$ and $[B]$.

Algebraic Solution (a) $A + B = \begin{bmatrix} 2 & 4 & 8 & -3 \\ 0 & 1 & 2 & 3 \end{bmatrix} + \begin{bmatrix} -3 & 4 & 0 & 1 \\ 6 & 8 & 2 & 0 \end{bmatrix}$

FIGURE 11

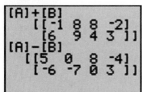

$$= \begin{bmatrix} 2 + (-3) & 4 + 4 & 8 + 0 & -3 + 1 \\ 0 + 6 & 1 + 8 & 2 + 2 & 3 + 0 \end{bmatrix} \qquad \begin{array}{l}\text{Add corresponding} \\ \text{entries.}\end{array}$$

$$= \begin{bmatrix} -1 & 8 & 8 & -2 \\ 6 & 9 & 4 & 3 \end{bmatrix}$$

(b) $A - B = \begin{bmatrix} 2 & 4 & 8 & -3 \\ 0 & 1 & 2 & 3 \end{bmatrix} - \begin{bmatrix} -3 & 4 & 0 & 1 \\ 6 & 8 & 2 & 0 \end{bmatrix}$

$= \begin{bmatrix} 2 - (-3) & 4 - 4 & 8 - 0 & -3 - 1 \\ 0 - 6 & 1 - 8 & 2 - 2 & 3 - 0 \end{bmatrix}$ Subtract corresponding entries.

$= \begin{bmatrix} 5 & 0 & 8 & -4 \\ -6 & -7 & 0 & 3 \end{bmatrix}$ ■

■ Now work Problem 1.

Many of the algebraic properties of sums of real numbers are also true for sums of matrices. Suppose that A, B, and C are m by n matrices. Then matrix addition is **commutative.** That is,

Commutative Property

$$A + B = B + A$$

Matrix addition is also **associative.** That is,

Associative Property

$$(A + B) + C = A + (B + C)$$

Although we shall not prove these results, the proofs, as the following example illustrates, are based on the commutative and associative properties for real numbers.

E X A M P L E   4   *Demonstrating the Commutative Property*

$\begin{bmatrix} 2 & 3 & -1 \\ 4 & 0 & 7 \end{bmatrix} + \begin{bmatrix} -1 & 2 & 1 \\ 5 & -3 & 4 \end{bmatrix} = \begin{bmatrix} 2 + (-1) & 3 + 2 & -1 + 1 \\ 4 + 5 & 0 + (-3) & 7 + 4 \end{bmatrix}$

$= \begin{bmatrix} -1 + 2 & 2 + 3 & 1 + (-1) \\ 5 + 4 & -3 + 0 & 4 + 7 \end{bmatrix}$

$= \begin{bmatrix} -1 & 2 & 1 \\ 5 & -3 & 4 \end{bmatrix} + \begin{bmatrix} 2 & 3 & -1 \\ 4 & 0 & 7 \end{bmatrix}$ ■

A matrix whose entries are all equal to 0 is called a **zero matrix.** Each of the following matrices is a zero matrix:

$\begin{bmatrix} 0 & 0 \\ 0 & 0 \end{bmatrix}$ 2 by 2 square zero matrix $\begin{bmatrix} 0 & 0 & 0 \\ 0 & 0 & 0 \end{bmatrix}$ 2 by 3 zero matrix $\begin{bmatrix} 0 & 0 & 0 \end{bmatrix}$ 1 by 3 zero matrix

Zero matrices have properties similar to the real number 0. Thus, if A is an m by n matrix and 0 is an m by n zero matrix, then

$$A + 0 = A$$

In other words, the zero matrix is the additive identity in matrix algebra.

We also can multiply a matrix by a real number. If k is a real number and A is an m by n matrix, the matrix kA is the m by n matrix formed by multiplying each entry in A by k. The number k is sometimes referred to as a **scalar,** and the matrix kA is called a **scalar multiple** of A.

E X A M P L E   5   *Operations Using Matrices*

Suppose that

$$A = \begin{bmatrix} 3 & 1 & 5 \\ -2 & 0 & 6 \end{bmatrix} \qquad B = \begin{bmatrix} 4 & 1 & 0 \\ 8 & 1 & -3 \end{bmatrix} \qquad C = \begin{bmatrix} 9 & 0 \\ -3 & 6 \end{bmatrix}$$

Find: (a) $4A$ (b) $\frac{1}{3}C$ (c) $3A - 2B$

Graphing Solution   Enter the matrices $[A]$, $[B]$, and $[C]$ into a graphing utility. Figure 12 shows the required computations.

FIGURE 12

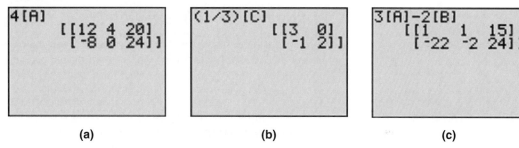

(a)   (b)   (c)

Algebraic Solution   (a) $4A = 4\begin{bmatrix} 3 & 1 & 5 \\ -2 & 0 & 6 \end{bmatrix} = \begin{bmatrix} 4 \cdot 3 & 4 \cdot 1 & 4 \cdot 5 \\ 4(-2) & 4 \cdot 0 & 4 \cdot 6 \end{bmatrix} = \begin{bmatrix} 12 & 4 & 20 \\ -8 & 0 & 24 \end{bmatrix}$

(b) $\frac{1}{3}C = \frac{1}{3}\begin{bmatrix} 9 & 0 \\ -3 & 6 \end{bmatrix} = \begin{bmatrix} \frac{1}{3} \cdot 9 & \frac{1}{3} \cdot 0 \\ \frac{1}{3}(-3) & \frac{1}{3} \cdot 6 \end{bmatrix} = \begin{bmatrix} 3 & 0 \\ -1 & 2 \end{bmatrix}$

(c) $3A - 2B = 3\begin{bmatrix} 3 & 1 & 5 \\ -2 & 0 & 6 \end{bmatrix} - 2\begin{bmatrix} 4 & 1 & 0 \\ 8 & 1 & -3 \end{bmatrix}$

$= \begin{bmatrix} 3 \cdot 3 & 3 \cdot 1 & 3 \cdot 5 \\ 3(-2) & 3 \cdot 0 & 3 \cdot 6 \end{bmatrix} - \begin{bmatrix} 2 \cdot 4 & 2 \cdot 1 & 2 \cdot 0 \\ 2 \cdot 8 & 2 \cdot 1 & 2(-3) \end{bmatrix}$

$= \begin{bmatrix} 9 & 3 & 15 \\ -6 & 0 & 18 \end{bmatrix} - \begin{bmatrix} 8 & 2 & 0 \\ 16 & 2 & -6 \end{bmatrix}$

$= \begin{bmatrix} 9 - 8 & 3 - 2 & 15 - 0 \\ -6 - 16 & 0 - 2 & 18 - (-6) \end{bmatrix}$

$= \begin{bmatrix} 1 & 1 & 15 \\ -22 & -2 & 24 \end{bmatrix}$

■ Now work Problem 5.

We list next some of the algebraic properties of scalar multiplication. Let h and k be real numbers, and let A and B be m by n matrices. Then

Properties of Scalar
Multiplication

$$k(hA) = (kh)A$$
$$(k + h)A = kA + hA$$
$$k(A + B) = kA + kB$$

The proofs of these properties are based on properties of real numbers. For example, if A and B are 2 by 2 matrices, then

$$k(A + B) = k\left(\begin{bmatrix} a_{11} & a_{12} \\ a_{21} & a_{22} \end{bmatrix} + \begin{bmatrix} b_{11} & b_{12} \\ b_{21} & b_{22} \end{bmatrix}\right) = k\begin{bmatrix} a_{11} + b_{11} & a_{12} + b_{12} \\ a_{21} + b_{21} & a_{22} + b_{22} \end{bmatrix}$$

$$= \begin{bmatrix} k(a_{11} + b_{11}) & k(a_{12} + b_{12}) \\ k(a_{21} + b_{21}) & k(a_{22} + b_{22}) \end{bmatrix} = \begin{bmatrix} ka_{11} + kb_{11} & ka_{12} + kb_{12} \\ ka_{21} + kb_{21} & ka_{22} + kb_{22} \end{bmatrix}$$

$$= \begin{bmatrix} ka_{11} & ka_{12} \\ ka_{21} & ka_{22} \end{bmatrix} + \begin{bmatrix} kb_{11} & kb_{12} \\ kb_{21} & kb_{22} \end{bmatrix} = k\begin{bmatrix} a_{11} & a_{12} \\ a_{21} & a_{22} \end{bmatrix} + k\begin{bmatrix} b_{11} & b_{12} \\ b_{21} & b_{22} \end{bmatrix} = kA + kB$$

Multiplication of Matrices

Unlike the straightforward definition for adding two matrices, the definition for multiplying two matrices is not what we might expect. In preparation for this definition, we need the following definitions:

A **row vector** R is a 1 by n matrix

$$R = [r_1 \quad r_2 \quad \cdots \quad r_n]$$

A **column vector** C is an n by 1 matrix

$$C = \begin{bmatrix} c_1 \\ c_2 \\ \cdot \\ \cdot \\ \cdot \\ c_n \end{bmatrix}$$

The **product** RC of R times C is defined as the number

$$RC = [r_1 \quad r_2 \quad \cdots \quad r_n]\begin{bmatrix} c_1 \\ c_2 \\ \cdot \\ \cdot \\ \cdot \\ c_n \end{bmatrix} = r_1c_1 + r_2c_2 + \cdots + r_nc_n$$

Notice that a row vector and a column vector can be multiplied only if they contain the same number of entries.

EXAMPLE 6   *The Product of a Row Vector by a Column Vector*

If $R = [3 \quad -5 \quad 2]$ and $C = \begin{bmatrix} 3 \\ 4 \\ -5 \end{bmatrix}$, then

$$RC = [3 \quad -5 \quad 2]\begin{bmatrix} 3 \\ 4 \\ -5 \end{bmatrix} = 3 \cdot 3 + (-5)4 + 2(-5)$$

$$= 9 - 20 - 10 = -21$$

Let's look at an application of the product of a row vector by a column vector.

E X A M P L E   7   *Using Matrices to Compute Revenue*

A clothing store sells men's shirts for $25, silk ties for $8, and wool suits for $300. Last month, the store had sales consisting of 100 shirts, 200 ties, and 50 suits. What was the total revenue due to these sales?

Solution   We set up a row vector R to represent the prices of each item and a column vector C to represent the corresponding number of items sold.

Then

$$
R = \begin{matrix} \text{Shirts} \ \ \text{Ties} \ \ \text{Suits} \\ [\ 25 \quad 8 \quad 300\ ] \end{matrix} \qquad
C = \begin{matrix} \text{Number} \\ \text{sold} \\ \begin{bmatrix} 100 \\ 200 \\ 50 \end{bmatrix} \end{matrix} \begin{matrix} \\ \\ \text{Shirts} \\ \text{Ties} \\ \text{Suits} \end{matrix}
$$

The total revenue obtained is the product RC. That is,

$$
RC = \begin{bmatrix} 25 & 8 & 300 \end{bmatrix} \begin{bmatrix} 100 \\ 200 \\ 50 \end{bmatrix}
$$

$$
= \underbrace{25 \cdot 100}_{\text{Shirt revenue}} + \underbrace{8 \cdot 200}_{\text{Tie revenue}} + \underbrace{300 \cdot 50}_{\text{Suit revenue}} = \underbrace{\$19{,}100}_{\text{Total revenue}}
$$

The definition for multiplying two matrices is based on the definition of a row vector times a column vector.

Product

Let A denote an m by r matrix and let B denote an r by n matrix. The **product** AB is defined as the m by n matrix whose entry in row i, column j is the product of the ith row of A and the jth column of B.

The definition of the product AB of two matrices A and B, in this order, requires that the number of columns of A equal the number of rows of B; otherwise, no product is defined:

$$
\begin{matrix} A & & B \\ m \text{ by } r & & r \text{ by } n \end{matrix}
$$

Must be same
for AB to be defined

AB is m by n

An example will help clarify the definition.

E X A M P L E   8   *Multiplying Two Matrices*

Find the product AB if

$$
A = \begin{bmatrix} 2 & 4 & -1 \\ 5 & 8 & 0 \end{bmatrix} \quad \text{and} \quad B = \begin{bmatrix} 2 & 5 & 1 & 4 \\ 4 & 8 & 0 & 6 \\ -3 & 1 & -2 & -1 \end{bmatrix}
$$

Graphing Solution   Enter the matrices A and B into a graphing utility. Figure 13 shows the product AB.

FIGURE 13

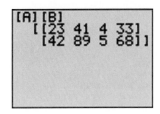

Algebraic Solution   First, we note that A is 2 by 3 and B is 3 by 4, so the product AB is defined and will be a 2 by 4 matrix. Suppose we want the entry in row 2, column 3 of AB. To find it, we find the product of the row vector from row 2 of A and the column vector from column 3 of B:

$$\underset{\text{Row 2 of } A}{[5 \quad 8 \quad 0]} \overset{\text{Column 3 of } B}{\begin{bmatrix} 1 \\ 0 \\ -2 \end{bmatrix}} = 5 \cdot 1 + 8 \cdot 0 + 0(-2) = 5$$

So far, we have

$$AB = \begin{bmatrix} \underline{\quad} & \underline{\quad} & \overset{\text{Column 3}}{\underset{\downarrow}{\underline{\quad}}} & \underline{\quad} \\ \underline{\quad} & \underline{\quad} & 5 & \underline{\quad} \end{bmatrix} \quad \leftarrow\text{Row 2}$$

Now, to find the entry in row 1, column 4 of AB, we find the product of row 1 of A and column 4 of B:

$$\underset{\text{Row 1 of } A}{[2 \quad 4 \quad -1]} \overset{\text{Column 4 of } B}{\begin{bmatrix} 4 \\ 6 \\ -1 \end{bmatrix}} = 2 \cdot 4 + 4 \cdot 6 + (-1)(-1) = 33$$

Continuing in this fashion, we find AB:

$$AB = \begin{bmatrix} 2 & 4 & -1 \\ 5 & 8 & 0 \end{bmatrix} \begin{bmatrix} 2 & 5 & 1 & 4 \\ 4 & 8 & 0 & 6 \\ -3 & 1 & -2 & -1 \end{bmatrix}$$

$$= \begin{bmatrix} \begin{matrix}\text{Row 1 of } A \\ \text{times} \\ \text{column 1 of } B\end{matrix} & \begin{matrix}\text{Row 1 of } A \\ \text{times} \\ \text{column 2 of } B\end{matrix} & \begin{matrix}\text{Row 1 of } A \\ \text{times} \\ \text{column 3 of } B\end{matrix} & \begin{matrix}\text{Row 1 of } A \\ \text{times} \\ \text{column 4 of } B\end{matrix} \\ \\ \begin{matrix}\text{Row 2 of } A \\ \text{times} \\ \text{column 1 of } B\end{matrix} & \begin{matrix}\text{Row 2 of } A \\ \text{times} \\ \text{column 2 of } B\end{matrix} & \begin{matrix}\text{Row 2 of } A \\ \text{times} \\ \text{column 3 of } B\end{matrix} & \begin{matrix}\text{Row 2 of } A \\ \text{times} \\ \text{column 4 of } B\end{matrix} \end{bmatrix}$$

$$= \begin{bmatrix} 2 \cdot 2 + 4 \cdot 4 + (-1)(-3) & 2 \cdot 5 + 4 \cdot 8 + (-1)1 & 2 \cdot 1 + 4 \cdot 0 + (-1)(-2) & 33 \text{ (from earlier)} \\ 5 \cdot 2 + 8 \cdot 4 + 0(-3) & 5 \cdot 5 + 8 \cdot 8 + 0 \cdot 1 & 5 \text{ (from earlier)} & 5 \cdot 4 + 8 \cdot 6 + 0(-1) \end{bmatrix}$$

$$= \begin{bmatrix} 23 & 41 & 4 & 33 \\ 42 & 89 & 5 & 68 \end{bmatrix}$$

■

■ Now work Problem 17.

Notice that for the matrices given in Example 8 the product BA is not defined, because B is 3 by 4 and A is 2 by 3. Try calculating BA on a graphing utility. What do you notice?

Another result that can occur when multiplying two matrices is illustrated in the next example.*

E X A M P L E   9   *Multiplying Two Matrices*

If

$$A = \begin{bmatrix} 2 & 1 & 3 \\ 1 & -1 & 0 \end{bmatrix} \quad \text{and} \quad B = \begin{bmatrix} 1 & 0 \\ 2 & 1 \\ 3 & 2 \end{bmatrix}$$

find: (a) AB (b) BA

Solution   (a) $AB = \begin{bmatrix} 2 & 1 & 3 \\ 1 & -1 & 0 \end{bmatrix} \begin{bmatrix} 1 & 0 \\ 2 & 1 \\ 3 & 2 \end{bmatrix} = \begin{bmatrix} 13 & 7 \\ -1 & -1 \end{bmatrix}$

　　　　　　　　2 by 3　　　　3 by 2　　　　2 by 2

(b) $BA = \begin{bmatrix} 1 & 0 \\ 2 & 1 \\ 3 & 2 \end{bmatrix} \begin{bmatrix} 2 & 1 & 3 \\ 1 & -1 & 0 \end{bmatrix} = \begin{bmatrix} 2 & 1 & 3 \\ 5 & 1 & 6 \\ 8 & 1 & 9 \end{bmatrix}$

　　　　　　　3 by 2　　　2 by 3　　　　3 by 3   ■

Notice in Example 9 that AB is 2 by 2 and BA is 3 by 3. Thus, it is possible for both AB and BA to be defined, yet be unequal. In fact, even if A and B are both n by n matrices, so that AB and BA are each defined and n by n, AB and BA will nearly always be unequal.

E X A M P L E   1 0   *Multiplying Two Square Matrices*

If

$$A = \begin{bmatrix} 2 & 1 \\ 0 & 4 \end{bmatrix} \quad \text{and} \quad B = \begin{bmatrix} -3 & 1 \\ 1 & 2 \end{bmatrix}$$

find: (a) AB (b) BA

Solution   (a) $AB = \begin{bmatrix} 2 & 1 \\ 0 & 4 \end{bmatrix} \begin{bmatrix} -3 & 1 \\ 1 & 2 \end{bmatrix}$

$= \begin{bmatrix} 2(-3) + 1 \cdot 1 & 2 \cdot 1 + 1 \cdot 2 \\ 0(-3) + 4 \cdot 1 & 0 \cdot 1 + 4 \cdot 2 \end{bmatrix} = \begin{bmatrix} -5 & 4 \\ 4 & 8 \end{bmatrix}$

(b) $BA = \begin{bmatrix} -3 & 1 \\ 1 & 2 \end{bmatrix} \begin{bmatrix} 2 & 1 \\ 0 & 4 \end{bmatrix}$

$= \begin{bmatrix} (-3)2 + 1 \cdot 0 & (-3)1 + 1 \cdot 4 \\ 1 \cdot 2 + 2 \cdot 0 & 1 \cdot 1 + 2 \cdot 4 \end{bmatrix} = \begin{bmatrix} -6 & 1 \\ 2 & 9 \end{bmatrix}$ ■

*For most of the examples that follow, we will multiply matrices by hand. You should verify each result using a graphing utility.

The preceding examples demonstrate that an important property of real numbers, the commutative property of multiplication, is not shared by matrices. Thus, in general:

Theorem   Matrix multiplication is not commutative.   ■

■ Now work Problems 7 and 9.

Next we give two of the properties of real numbers that are shared by matrices. Assuming each product and sum is defined, we have:

Associative Property

$$A(BC) = (AB)C$$

Distributive Property

$$A(B + C) = AB + AC$$

The Identity Matrix

For an n by n square matrix, the entries located in row i, column i, $1 \leq i \leq n$, are called the **diagonal entries.** An n by n square matrix whose diagonal entries are 1's, while all other entries are 0's, is called the **identity matrix I_n.** For example,

$$I_2 = \begin{bmatrix} 1 & 0 \\ 0 & 1 \end{bmatrix} \qquad I_3 = \begin{bmatrix} 1 & 0 & 0 \\ 0 & 1 & 0 \\ 0 & 0 & 1 \end{bmatrix}$$

and so on.

E X A M P L E   1 1   *Multiplication with an Identity Matrix*

Let

$$A = \begin{bmatrix} -1 & 2 & 0 \\ 0 & 1 & 3 \end{bmatrix} \quad \text{and} \quad B = \begin{bmatrix} 3 & 2 \\ 4 & 6 \\ 5 & 2 \end{bmatrix}$$

Find:

(a) AI_3 (b) I_2A (c) BI_2

Solution   (a) $AI_3 = \begin{bmatrix} -1 & 2 & 0 \\ 0 & 1 & 3 \end{bmatrix} \begin{bmatrix} 1 & 0 & 0 \\ 0 & 1 & 0 \\ 0 & 0 & 1 \end{bmatrix} = \begin{bmatrix} -1 & 2 & 0 \\ 0 & 1 & 3 \end{bmatrix} = A$

(b) $I_2A = \begin{bmatrix} 1 & 0 \\ 0 & 1 \end{bmatrix} \begin{bmatrix} -1 & 2 & 0 \\ 0 & 1 & 3 \end{bmatrix} = \begin{bmatrix} -1 & 2 & 0 \\ 0 & 1 & 3 \end{bmatrix} = A$

(c) $BI_2 = \begin{bmatrix} 3 & 2 \\ 4 & 6 \\ 5 & 2 \end{bmatrix} \begin{bmatrix} 1 & 0 \\ 0 & 1 \end{bmatrix} = \begin{bmatrix} 3 & 2 \\ 4 & 6 \\ 5 & 2 \end{bmatrix} = B$ ■

Example 11 demonstrates the following property:

If A is an m by n matrix, then

Identity Property

$$I_m A = A \quad \text{and} \quad AI_n = A$$

If A is an n by n square matrix, then $AI_n = I_n A = A$.

Thus, an identity matrix has properties analogous to those of the real number 1. In other words, the identity matrix is a multiplicative identity in matrix algebra.

The Inverse of a Matrix

Inverse Matrix

Let A be a square n by n matrix. If there exists an n by n matrix A^{-1}, read "A inverse," for which

$$AA^{-1} = A^{-1}A = I_n$$

then A^{-1} is called the **inverse** of the matrix A.

As we shall soon see, not every square matrix has an inverse. When a matrix A does have an inverse A^{-1}, then A is said to be **nonsingular.** If a matrix A has no inverse, it is called **singular.**

E X A M P L E   1 2   *Multiplying a Matrix by Its Inverse*

Show that the inverse of

$$A = \begin{bmatrix} 3 & 1 \\ 2 & 1 \end{bmatrix} \quad \text{is} \quad A^{-1} = \begin{bmatrix} 1 & -1 \\ -2 & 3 \end{bmatrix}$$

Solution   We need to show that $AA^{-1} = A^{-1}A = I_2$.

$$AA^{-1} = \begin{bmatrix} 3 & 1 \\ 2 & 1 \end{bmatrix}\begin{bmatrix} 1 & -1 \\ -2 & 3 \end{bmatrix} = \begin{bmatrix} 3 \cdot 1 + 1(-2) & 3(-1) + 1 \cdot 3 \\ 2 \cdot 1 + 1(-2) & 2(-1) + 1 \cdot 3 \end{bmatrix}$$

$$= \begin{bmatrix} 1 & 0 \\ 0 & 1 \end{bmatrix} = I_2$$

$$A^{-1}A = \begin{bmatrix} 1 & -1 \\ -2 & 3 \end{bmatrix}\begin{bmatrix} 3 & 1 \\ 2 & 1 \end{bmatrix} = \begin{bmatrix} 3 - 2 & 1 - 1 \\ -6 + 6 & -2 + 3 \end{bmatrix} = \begin{bmatrix} 1 & 0 \\ 0 & 1 \end{bmatrix} = I_2 \quad \blacksquare$$

We now show one way to find the inverse of

$$A = \begin{bmatrix} 3 & 1 \\ 2 & 1 \end{bmatrix}$$

Suppose that A^{-1} is given by

$$A^{-1} = \begin{bmatrix} x & y \\ z & w \end{bmatrix} \tag{1}$$

where x, y, z, and w are four variables. Based on the definition of an inverse, if, indeed, A has an inverse, then we have

$$AA^{-1} = I_2$$

$$\begin{bmatrix} 3 & 1 \\ 2 & 1 \end{bmatrix} \begin{bmatrix} x & y \\ z & w \end{bmatrix} = \begin{bmatrix} 1 & 0 \\ 0 & 1 \end{bmatrix}$$

$$\begin{bmatrix} 3x + z & 3y + w \\ 2x + z & 2y + w \end{bmatrix} = \begin{bmatrix} 1 & 0 \\ 0 & 1 \end{bmatrix}$$

Because corresponding entries must be equal, it follows that this matrix equation is equivalent to four ordinary equations:

$$\begin{cases} 3x + z = 1 \\ 2x + z = 0 \end{cases} \qquad \begin{cases} 3y + w = 0 \\ 2y + w = 1 \end{cases}$$

The augmented matrix of each system is

$$\begin{bmatrix} 3 & 1 & | & 1 \\ 2 & 1 & | & 0 \end{bmatrix} \qquad \begin{bmatrix} 3 & 1 & | & 0 \\ 2 & 1 & | & 1 \end{bmatrix} \tag{2}$$

The usual procedure would be to transform each augmented matrix into reduced echelon form. Notice, though, that the left sides of the augmented matrices are equal, so the same row operations (see Section 7.2) can be used to reduce each one. Thus, we find it more efficient to combine the two augmented matrices (2) into a single matrix, as shown next, and then transform it into reduced echelon form:

$$\begin{bmatrix} 3 & 1 & | & 1 & 0 \\ 2 & 1 & | & 0 & 1 \end{bmatrix}$$

Now we attempt to transform the left side into an identity matrix:

$$\begin{bmatrix} 3 & 1 & | & 1 & 0 \\ 2 & 1 & | & 0 & 1 \end{bmatrix} \rightarrow \begin{bmatrix} 1 & 0 & | & 1 & -1 \\ 2 & 1 & | & 0 & 1 \end{bmatrix}$$

$$\uparrow \\ R_1 = -1r_2 + r_1$$

$$\rightarrow \begin{bmatrix} 1 & 0 & | & 1 & -1 \\ 0 & 1 & | & -2 & 3 \end{bmatrix} \tag{3}$$

$$\uparrow \\ R_2 = -2r_1 + r_2$$

Matrix (3) is in reduced echelon form. Now we reverse the earlier step of combining the two augmented matrices in (2) and write the single matrix (3) as two augmented matrices:

$$\begin{bmatrix} 1 & 0 & | & 1 \\ 0 & 1 & | & -2 \end{bmatrix} \text{ and } \begin{bmatrix} 1 & 0 & | & -1 \\ 0 & 1 & | & 3 \end{bmatrix}$$

We conclude from these matrices that $x = 1$, $z = -2$, and $y = -1$, $w = 3$. Substituting these values into matrix (1), we find that

$$A^{-1} = \begin{bmatrix} 1 & -1 \\ -2 & 3 \end{bmatrix}$$

Notice in matrix (3) that the 2 by 2 matrix to the right of the vertical bar is, in fact, the inverse of A. Also notice that the identity matrix I_2 is the matrix that appears to the left of the vertical bar. These observations and the procedures followed above will work in general.

Procedure for Finding the Inverse of a Nonsingular Matrix

To find the inverse of an n by n nonsingular matrix A, proceed as follows:

STEP 1: Form the matrix $[A|I_n]$.

STEP 2: Transform the matrix $[A|I_n]$ into reduced echelon form.

STEP 3: The reduced echelon form of $[A|I_n]$ will contain the identity matrix I_n on the left of the vertical bar; the n by n matrix on the right of the vertical bar is the inverse of A.

In other words, if A is nonsingular, we begin with the matrix $[A|I_n]$ and, after transforming it into reduced echelon form, we end up with the matrix $[I_n|A^{-1}]$.

Let's look at another example.

E X A M P L E   1 3   *Finding the Inverse of a Matrix*

The matrix

$$A = \begin{bmatrix} 1 & 1 & 0 \\ -1 & 3 & 4 \\ 0 & 4 & 3 \end{bmatrix}$$

is nonsingular. Find its inverse.

Graphing Solution   Enter the matrix A into a graphing utility. Figure 14 shows A^{-1}.

FIGURE 14

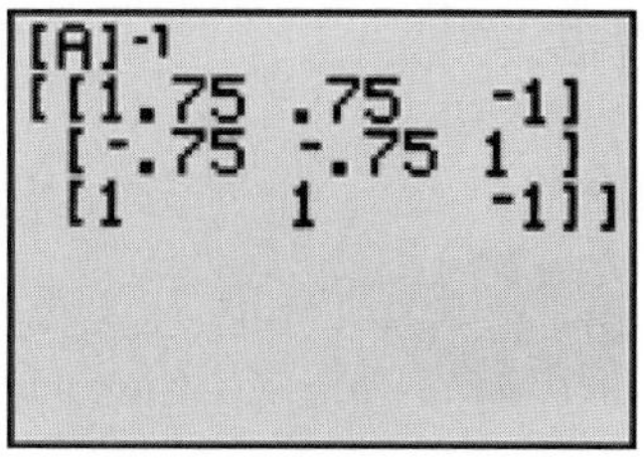

```
[A]-1
[[1.75 .75 -1]
 [-.75 -.75 1 ]
 [1    1    -1]]
```

Algebraic Solution   First, we form the matrix

$$[A|I_3] = \begin{bmatrix} 1 & 1 & 0 & | & 1 & 0 & 0 \\ -1 & 3 & 4 & | & 0 & 1 & 0 \\ 0 & 4 & 3 & | & 0 & 0 & 1 \end{bmatrix}$$

Next, we use row operations to transform $[A|I_3]$ into reduced echelon form:

$$
\begin{bmatrix}
1 & 1 & 0 & | & 1 & 0 & 0 \\
-1 & 3 & 4 & | & 0 & 1 & 0 \\
0 & 4 & 3 & | & 0 & 0 & 1
\end{bmatrix}
\rightarrow
\underset{\substack{\uparrow \\ R_2 = r_2 + r_1}}{
\begin{bmatrix}
1 & 1 & 0 & | & 1 & 0 & 0 \\
0 & 4 & 4 & | & 1 & 1 & 0 \\
0 & 4 & 3 & | & 0 & 0 & 1
\end{bmatrix}}
\rightarrow
\underset{\substack{\uparrow \\ R_2 = \frac{1}{4} r_2}}{
\begin{bmatrix}
1 & 1 & 0 & | & 1 & 0 & 0 \\
0 & 1 & 1 & | & \frac{1}{4} & \frac{1}{4} & 0 \\
0 & 4 & 3 & | & 0 & 0 & 1
\end{bmatrix}}
$$

$$
\rightarrow
\underset{\substack{\uparrow \\ R_1 = -1r_2 + r_1 \\ R_3 = -4r_2 + r_3}}{
\begin{bmatrix}
1 & 0 & -1 & | & \frac{3}{4} & -\frac{1}{4} & 0 \\
0 & 1 & 1 & | & \frac{1}{4} & \frac{1}{4} & 0 \\
0 & 0 & -1 & | & -1 & -1 & 1
\end{bmatrix}}
\rightarrow
\underset{\substack{\uparrow \\ R_3 = -1r_3}}{
\begin{bmatrix}
1 & 0 & -1 & | & \frac{3}{4} & -\frac{1}{4} & 0 \\
0 & 1 & 1 & | & \frac{1}{4} & \frac{1}{4} & 0 \\
0 & 0 & 1 & | & 1 & 1 & -1
\end{bmatrix}}
$$

$$
\rightarrow
\underset{\substack{\uparrow \\ R_1 = r_1 + r_3 \\ R_2 = -1r_3 + r_2}}{
\begin{bmatrix}
1 & 0 & 0 & | & \frac{7}{4} & \frac{3}{4} & -1 \\
0 & 1 & 0 & | & -\frac{3}{4} & -\frac{3}{4} & 1 \\
0 & 0 & 1 & | & 1 & 1 & -1
\end{bmatrix}}
$$

The matrix $[A|I_3]$ is now in reduced echelon form, and the identity matrix I_3 is on the left of the vertical bar. Hence, the inverse of A is

$$
A^{-1} = \begin{bmatrix}
\frac{7}{4} & \frac{3}{4} & -1 \\
-\frac{3}{4} & -\frac{3}{4} & 1 \\
1 & 1 & -1
\end{bmatrix}
$$
■

You can (and should) verify that this is the correct inverse by showing that $AA^{-1} = A^{-1}A = I_3$.

■   Now work Problem 23.

If transforming the matrix $[A|I_n]$ into reduced echelon form does not result in the identity matrix I_n to the left of the vertical bar, then A has no inverse. The next example demonstrates such a matrix.

E X A M P L E   1 4   *Showing a Matrix Has No Inverse*

Show that the following matrix has no inverse.

$$
A = \begin{bmatrix} 4 & 6 \\ 2 & 3 \end{bmatrix}
$$

Graphing Solution   Enter the matrix A. Figure 15 shows the result when we try to find its inverse. The ERRor comes about because A is singular.

FIGURE 15

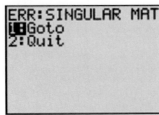

Algebraic Solution   Proceeding as in Example 13, we form the matrix

$$[A|I_2] = \begin{bmatrix} 4 & 6 & | & 1 & 0 \\ 2 & 3 & | & 0 & 1 \end{bmatrix}$$

Then we use row operations to transform $[A \quad | \quad I_2]$ into reduced echelon form:

$$[A|I_2] = \begin{bmatrix} 4 & 6 & | & 1 & 0 \\ 2 & 3 & | & 0 & 1 \end{bmatrix} \underset{\underset{R_1 = \frac{1}{4}r_1}{\uparrow}}{\rightarrow} \begin{bmatrix} 1 & \frac{3}{2} & | & \frac{1}{4} & 0 \\ 2 & 3 & | & 0 & 1 \end{bmatrix} \underset{\underset{R_2 = -2r_1 + r_2}{\uparrow}}{\rightarrow} \begin{bmatrix} 1 & \frac{3}{2} & | & \frac{1}{4} & 0 \\ 0 & 0 & | & -\frac{1}{2} & 1 \end{bmatrix}$$

The matrix $[A|I_2]$ is sufficiently reduced for us to see the identity matrix cannot appear to the left of the vertical bar. We conclude that A has no inverse. ■

■ Now work Problem 51.

Solving Systems of Equations

Inverse matrices can be used to solve systems of equations in which the number of equations is the same as the number of variables.

E X A M P L E   1 5   *Using the Inverse Matrix to Solve a System of Equations*

Solve the system of equations: $\begin{cases} x + y = 3 \\ -x + 3y + 4z = -3 \\ 4y + 3z = 2 \end{cases}$

Solution   If we let

$$A = \begin{bmatrix} 1 & 1 & 0 \\ -1 & 3 & 4 \\ 0 & 4 & 3 \end{bmatrix}, \quad X = \begin{bmatrix} x \\ y \\ z \end{bmatrix}, \quad \text{and} \quad B = \begin{bmatrix} 3 \\ -3 \\ 2 \end{bmatrix}$$

then the original system of equations can be written compactly as the matrix equation

$$AX = B \tag{4}$$

We know from Example 13 that the matrix A has the inverse A^{-1}, so we multiply each side of equation (4) by A^{-1}:

$$AX = B$$
$$A^{-1}(AX) = A^{-1}B$$
$$(A^{-1}A)X = A^{-1}B \quad \text{Associative property for multiplication}$$
$$I_3X = A^{-1}B \quad \text{Definition of inverse matrix}$$
$$X = A^{-1}B \quad \text{Property of identity matrix} \tag{5}$$

Now we use (5) to find $X = \begin{bmatrix} x \\ y \\ z \end{bmatrix}$:

Graphing Solution   Enter the matrices A and B into a graphing utility. Figure 16 shows the solution to the system of equations.

FIGURE 16

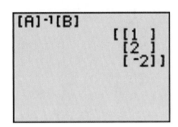

Algebraic Solution

$$X = \begin{bmatrix} x \\ y \\ z \end{bmatrix} = A^{-1}B = \begin{bmatrix} \frac{7}{4} & \frac{3}{4} & -1 \\ -\frac{3}{4} & -\frac{3}{4} & 1 \\ 1 & 1 & -1 \end{bmatrix} \begin{bmatrix} 3 \\ -3 \\ 2 \end{bmatrix} = \begin{bmatrix} 1 \\ 2 \\ -2 \end{bmatrix}$$

Example 13

Thus, $x = 1$, $y = 2$, $z = -2$. ∎

The method used in Example 15 to solve a system of equations is particularly useful when it is necessary to solve several systems of equations in which the constants appearing to the right of the equal signs change, while the coefficients of the variables on the left side remain the same. See Problems 31–50 for some illustrations.

HISTORICAL FEATURE   ∎ Matrices were invented in 1857 by Arthur Cayley (1821–1895) as a way of efficiently computing the result of substituting one linear system into another (see Historical Problem 2). The resulting system had incredible richness, in the sense that a very wide variety of mathematical systems could be mimicked by the matrices. Cayley and his friend J.J. Sylvester (1814–1897) spent much of the rest of their lives elaborating the theory. The torch was then passed to G. Frobenius (1848–1917), whose deep investigations established a central place for matrices in modern mathematics. In 1924, rather to the surprise of physicists, it was found that matrices (with complex numbers in them) were exactly the right tool for describing the behavior of atomic systems. Today, matrices are used in a wide variety of applications. ∎

HISTORICAL PROBLEMS   ∎ 1. *Matrices and Complex Numbers* Frobenius emphasized in his research how matrices could be used to mimic other mathematical systems. Here, we mimic the behavior of complex numbers using matrices. Mathematicians call such a relationship an *isomorphism*.

Complex number⟷ Matrix

$$a + bi \quad \longleftrightarrow \quad \begin{bmatrix} a & b \\ -b & a \end{bmatrix}$$

Note that the complex number can be read off the top line of the matrix. Thus,

$$2 + 3i \longleftrightarrow \begin{bmatrix} 2 & 3 \\ -3 & 2 \end{bmatrix} \quad \text{and} \quad \begin{bmatrix} 4 & -2 \\ 2 & 4 \end{bmatrix} \longleftrightarrow 4 - 2i$$

(a) Find the matrices corresponding to $2 - 5i$ and $1 + 3i$.

(b) Multiply the two matrices.

(c) Find the corresponding complex number for the matrix found in part (b).

(d) Multiply $2 - 5i$ by $1 + 3i$. The result should be the same as that found in part (c).

The process also works for addition and subtraction. Try it for yourself.

2. *Cayley's Definition of Matrix Multiplication*   Cayley invented matrix multiplication to simplify the following problem:

$$\begin{cases} u = ar + bs \\ v = cr + ds \end{cases} \qquad \begin{cases} x = ku + lv \\ y = mu + nv \end{cases}$$

(a) Find x and y in terms of r and s by substituting u and v from the first system of equations into the second system of equations.

(b) Use the result of part (a) to find the 2 by 2 matrix A in

$$\begin{bmatrix} x \\ y \end{bmatrix} = A \begin{bmatrix} r \\ s \end{bmatrix}$$

(c) Now look at the following way to do it: Write the equations in matrix form

$$\begin{bmatrix} u \\ v \end{bmatrix} = \begin{bmatrix} a & b \\ c & d \end{bmatrix} \begin{bmatrix} r \\ s \end{bmatrix} \qquad \begin{bmatrix} x \\ y \end{bmatrix} = \begin{bmatrix} k & l \\ m & n \end{bmatrix} \begin{bmatrix} u \\ v \end{bmatrix}$$

So

$$\begin{bmatrix} x \\ y \end{bmatrix} = \begin{bmatrix} k & l \\ m & n \end{bmatrix} \begin{bmatrix} a & b \\ c & d \end{bmatrix} \begin{bmatrix} r \\ s \end{bmatrix}$$

Do you see how Cayley defined matrix multiplication?   ∎

7.4

Exercise 7.4

In Problems 1–16, use the following matrices to compute the given expression: (a) by hand; (b) using a graphing utility.

$$A = \begin{bmatrix} 0 & 3 & -5 \\ 1 & 2 & 6 \end{bmatrix} \qquad B = \begin{bmatrix} 4 & 1 & 0 \\ -2 & 3 & -2 \end{bmatrix} \qquad C = \begin{bmatrix} 4 & 1 \\ 6 & 2 \\ -2 & 3 \end{bmatrix}$$

1. $A + B$	**2.** $A - B$	**3.** $4A$	**4.** $-3B$
5 $3A - 2B$	**6.** $2A + 4B$	**7.** AC	**8.** BC
9. CA	**10.** CB	**11.** $C(A + B)$	**12.** $(A + B)C$
13. $AC - 3I_2$	**14.** $CA + 5I_3$	**15.** $CA - CB$	**16.** $AC + BC$

In Problems 17–20, compute each product: (a) by hand; (b) using a graphing utility.

17. $\begin{bmatrix} 2 & -2 \\ 1 & 0 \end{bmatrix} \begin{bmatrix} 2 & 1 & 4 & 6 \\ 3 & -1 & 3 & 2 \end{bmatrix}$

18. $\begin{bmatrix} 4 & 1 \\ 2 & 1 \end{bmatrix} \begin{bmatrix} -6 & 6 & 1 & 0 \\ 2 & 5 & 4 & -1 \end{bmatrix}$

19. $\begin{bmatrix} 1 & 0 & 1 \\ 2 & 4 & 1 \\ 3 & 6 & 1 \end{bmatrix} \begin{bmatrix} 1 & 3 \\ 6 & 2 \\ 8 & -1 \end{bmatrix}$

20. $\begin{bmatrix} 4 & -2 & 3 \\ 0 & 1 & 2 \\ -1 & 0 & 1 \end{bmatrix} \begin{bmatrix} 2 & 6 \\ 1 & -1 \\ 0 & 2 \end{bmatrix}$

In Problems 21–30, each matrix is nonsingular. Find the inverse of each matrix. Be sure to check your answer using a graphing utility (when possible).

21. $\begin{bmatrix} 2 & 1 \\ 1 & 1 \end{bmatrix}$

22. $\begin{bmatrix} 3 & -1 \\ -2 & 1 \end{bmatrix}$

23. $\begin{bmatrix} 6 & 5 \\ 2 & 2 \end{bmatrix}$

24. $\begin{bmatrix} -4 & 1 \\ 6 & -2 \end{bmatrix}$

25. $\begin{bmatrix} 2 & 1 \\ a & a \end{bmatrix}, \quad a \neq 0$

26. $\begin{bmatrix} b & 3 \\ b & 2 \end{bmatrix}, \quad b \neq 0$

27. $\begin{bmatrix} 1 & -1 & 1 \\ 0 & -2 & 1 \\ -2 & -3 & 0 \end{bmatrix}$

28. $\begin{bmatrix} 1 & 0 & 2 \\ -1 & 2 & 3 \\ 1 & -1 & 0 \end{bmatrix}$

29. $\begin{bmatrix} 1 & 1 & 1 \\ 3 & 2 & -1 \\ 3 & 1 & 2 \end{bmatrix}$

30. $\begin{bmatrix} 3 & 3 & 1 \\ 1 & 2 & 1 \\ 2 & -1 & 1 \end{bmatrix}$

In Problems 31–50, use the inverses found in Problems 21–30 to solve each system of equations by hand.

31. $\begin{cases} 2x + y = 8 \\ x + y = 5 \end{cases}$

32. $\begin{cases} 3x - y = 8 \\ -2x + y = 4 \end{cases}$

33. $\begin{cases} 2x + y = 0 \\ x + y = 5 \end{cases}$

34. $\begin{cases} 3x - y = 4 \\ -2x + y = 5 \end{cases}$

35. $\begin{cases} 6x + 5y = 7 \\ 2x + 2y = 2 \end{cases}$

36. $\begin{cases} -4x + y = 0 \\ 6x - 2y = 14 \end{cases}$

37. $\begin{cases} 6x + 5y = 13 \\ 2x + 2y = 5 \end{cases}$

38. $\begin{cases} -4x + y = 5 \\ 6x - 2y = -9 \end{cases}$

39. $\begin{cases} 2x + y = -3 \\ ax + ay = -a \end{cases} \quad a \neq 0$

40. $\begin{cases} bx + 3y = 2b + 3 \\ bx + 2y = 2b + 2 \end{cases} \quad b \neq 0$

41. $\begin{cases} 2x + y = 7/a \\ ax + ay = 5 \end{cases} \quad a \neq 0$

42. $\begin{cases} bx + 3y = 14 \\ bx + 2y = 10 \end{cases} \quad b \neq 0$

43. $\begin{cases} x - y + z = 0 \\ -2y + z = -1 \\ -2x - 3y = -5 \end{cases}$

44. $\begin{cases} x + 2z = 6 \\ -x + 2y + 3z = -5 \\ x - y = 6 \end{cases}$

45. $\begin{cases} x - y + z = 2 \\ -2y + z = 2 \\ -2x - 3y = \frac{1}{2} \end{cases}$

46. $\begin{cases} x + 2z = 2 \\ -x + 2y + 3z = -\frac{3}{2} \\ x - y = 2 \end{cases}$

47. $\begin{cases} x + y + z = 9 \\ 3x + 2y - z = 8 \\ 3x + y + 2z = 1 \end{cases}$

48. $\begin{cases} 3x + 3y + z = 8 \\ x + 2y + z = 5 \\ 2x - y + z = 4 \end{cases}$

49. $\begin{cases} x + y + z = 2 \\ 3x + 2y - z = \frac{7}{3} \\ 3x + y + 2z = \frac{10}{3} \end{cases}$

50. $\begin{cases} 3x + 3y + z = 1 \\ x + 2y + z = 0 \\ 2x - y + z = 4 \end{cases}$

In Problems 51–56, by hand, show that each matrix has no inverse.

51. $\begin{bmatrix} 4 & 2 \\ 2 & 1 \end{bmatrix}$

52. $\begin{bmatrix} -3 & \frac{1}{2} \\ 6 & 1 \end{bmatrix}$

53. $\begin{bmatrix} 15 & 3 \\ 10 & 2 \end{bmatrix}$

54. $\begin{bmatrix} -3 & 0 \\ 4 & 0 \end{bmatrix}$

55. $\begin{bmatrix} -3 & 1 & -1 \\ 1 & -4 & -7 \\ 1 & 2 & 5 \end{bmatrix}$

56. $\begin{bmatrix} 1 & 1 & -3 \\ 2 & -4 & 1 \\ -5 & 7 & 1 \end{bmatrix}$

In Problems 57–60, use a graphing utility to find the inverse, if it exists, of each matrix.

57. $\begin{bmatrix} 25 & 61 & -12 \\ 18 & -2 & 4 \\ 8 & 35 & 21 \end{bmatrix}$

58. $\begin{bmatrix} 18 & -3 & 4 \\ 6 & -20 & 14 \\ 10 & 25 & -15 \end{bmatrix}$

59. $\begin{bmatrix} 44 & 21 & 18 & 6 \\ -2 & 10 & 15 & 5 \\ 21 & 12 & -12 & 4 \\ -8 & -16 & 4 & 9 \end{bmatrix}$

60. $\begin{bmatrix} 16 & 22 & -3 & 5 \\ 21 & -17 & 4 & 8 \\ 2 & 8 & 27 & 20 \\ 5 & 15 & -3 & -10 \end{bmatrix}$

In Problems 61–64, use the idea behind Example 15 with a graphing utility to solve the following systems of equations.

61. $\begin{cases} 25x + 61y - 12z = 10 \\ 18x - 12y + 7z = -9 \\ 3x + 4y - z = 12 \end{cases}$

62. $\begin{cases} 25x + 61y - 12z = 15 \\ 18x - 12y + 7z = -3 \\ 3x + 4y - z = 12 \end{cases}$

63. $\begin{cases} 25x + 61y - 12z = 21 \\ 18x - 12y + 7z = 7 \\ 3x + 4y - z = -2 \end{cases}$

64. $\begin{cases} 25x + 61y - 12z = 25 \\ 18x - 12y + 7z = 10 \\ 3x + 4y - z = -4 \end{cases}$

65. *Computing the Cost of Production* The Acme Steel Company is a producer of stainless steel and aluminum containers. On a certain day, the following stainless steel containers were manufactured: 500 with 10 gallon capacity, 350 with 5 gallon capacity, and 400 with 1 gallon capacity. On the same day, the following aluminum containers were manufactured: 700 with 10 gallon capacity, 500 with 5 gallon capacity, and 850 with 1 gallon capacity.
(a) Find a 2 by 3 matrix representing the above data. Find a 3 by 2 matrix to represent the same data.
(b) If the amount of material used in the 10 gallon containers is 15 pounds, the amount used in the 5 gallon containers is 8 pounds, and the amount used in the 1 gallon containers is 3 pounds, find a 3 by 1 matrix representing the amount of material.
(c) Multiply the 2 by 3 matrix found in part (a) and the 3 by 1 matrix found in part (b) to get a 2 by 1 matrix showing the day's usage of material.
(d) If stainless steel costs Acme $0.10 per pound and aluminum costs $0.05 per pound, find a 1 by 2 matrix representing cost.
(e) Multiply the matrices found in parts (c) and (d) to determine what the total cost of the day's production was.

66. *Computing Profit* A car dealership has two locations, one in a city and the other in the suburbs. In January, the city location sold 400 subcompacts, 250 intermediate-size cars, and 50 station wagons; in February, it sold 350 subcompacts, 100 intermediates, and 30 station wagons. At the suburban location in January, 450 subcompacts, 200 intermediates, and 140 station wagons were sold. In February, the suburban location sold 350 subcompacts, 300 intermediates, and 100 station wagons.
(a) Find 2 by 3 matrices that summarize the sales data for each location for January and February (one matrix for each month).
(b) Use matrix addition to obtain total sales for the 2 month period.
(c) The profit on each kind of car is $100 per subcompact, $150 per intermediate, and $200 per station wagon. Find a 3 by 1 matrix representing this profit.
(d) Multiply the matrices found in parts (b) and (c) to get a 2 by 1 matrix showing the profit at each location.

67. Consider the 2 by 2 square matrix

$$A = \begin{bmatrix} a & b \\ c & d \end{bmatrix}$$

If $D = ad - bc \neq 0$, show that A is nonsingular and that

$$A^{-1} = \frac{1}{D} \begin{bmatrix} d & -b \\ -c & a \end{bmatrix}$$

68. Make up a situation different from any found in the text that can be represented by a matrix.

7.5

Partial Fraction Decomposition

Consider the problem of adding the two fractions

$$\frac{3}{x+4} \quad \text{and} \quad \frac{2}{x-3}$$

The result is

$$\frac{3}{x+4} + \frac{2}{x-3} = \frac{3(x-3) + 2(x+4)}{(x+4)(x-3)} = \frac{5x-1}{x^2+x-12}$$

The reverse procedure, of starting with the rational expression $(5x-1)/(x^2+x-12)$ and writing it as the sum (or difference) of the two simpler fractions $3/(x+4)$ and $2/(x-3)$ is referred to as **partial fraction decomposition,** and the two simpler fractions are called **partial fractions.** Decomposing a rational expression into a sum of partial fractions is important in solving certain types of calculus problems. This section presents a systematic way to decompose rational expressions.

We begin by recalling that a rational expression is the ratio of two polynomials, say, P and $Q \neq 0$, that have no common factors. Recall also that a rational expression P/Q is called **proper** if the degree of the polynomial in the numerator is less than the degree of the polynomial in the denominator. Otherwise, the rational expression is termed **improper.**

Because any improper rational expression can be reduced by long division to a mixed form consisting of the sum of a polynomial and a proper rational expression, we shall restrict the discussion that follows to proper rational expressions.

The partial fraction decomposition of the rational expression P/Q depends on the factors of the denominator Q. Recall (from Section 4.4) that any polynomial whose coefficients are real numbers can be factored (over the real numbers) into products of linear and/or irreducible quadratic factors. Thus, the denominator Q of the rational expression P/Q will contain only factors of one or both of the following types:

1. *Linear factors* of the form $x - a$, where a is a real number.
2. *Irreducible quadratic factors* of the form $ax^2 + bx + c$, where a, b, and c are real numbers, $a \neq 0$, and $b^2 - 4ac < 0$ (which guarantees that $ax^2 + bx + c$ cannot be written as the product of two linear factors with real coefficients).

As it turns out, there are four cases to be examined. We begin with the case for which Q has only nonrepeated linear factors.

CASE 1: Q has only nonrepeated linear factors.

Under the assumption that Q has only nonrepeated linear factors, the polynomial Q has the form

$$Q(x) = (x - a_1)(x - a_2) \cdot \ldots \cdot (x - a_n)$$

where none of the numbers $a_1, a_2, \ldots, a_n$ are equal. In this case, the partial fraction decomposition of P/Q is of the form

$$\frac{P(x)}{Q(x)} = \frac{A_1}{x - a_1} + \frac{A_2}{x - a_2} + \cdots + \frac{A_n}{x - a_n} \tag{1}$$

where the numbers $A_1, A_2, \ldots, A_n$ are to be determined.
We show how to find these numbers in the example that follows.

E X A M P L E   1   *Nonrepeated Linear Factors*

Write the partial fraction decomposition of: $\dfrac{x}{x^2 - 5x + 6}$

Solution   First, we factor the denominator,

$$x^2 - 5x + 6 = (x - 2)(x - 3)$$

and conclude that the denominator contains only nonrepeated linear factors. Then we decompose the rational expression according to equation (1):

$$\frac{x}{x^2 - 5x + 6} = \frac{A}{x - 2} + \frac{B}{x - 3} \tag{2}$$

where A and B are to be determined. To find A and B, we clear the fractions by multiplying each side by $(x - 2)(x - 3) = x^2 - 5x + 6$. The result is

$$x = A(x - 3) + B(x - 2) \tag{3}$$

or

$$x = (A + B)x + (-3A - 2B)$$

This equation is an identity in x. Thus, we may equate the coefficients of like powers of x to get

$$\begin{cases} 1 = A + B & \text{Equate coefficients of } x\text{: } 1x = (A + B)x \\ 0 = -3A - 2B & \text{Equate coefficients of } x^0, \text{ the constants:} \\ & \quad 0x^0 = (-3A - 2B)x^0 \end{cases}$$

This system of two equations containing two variables, A and B, can be solved using whatever method you wish. Solving it, we get

$$A = -2 \qquad B = 3$$

Thus, from equation (2), the partial fraction decomposition is

$$\frac{x}{x^2 - 5x + 6} = \frac{-2}{x - 2} + \frac{3}{x - 3}$$

Check: The decomposition can be checked by adding the fractions:

$$\frac{-2}{x - 2} + \frac{3}{x - 3} = \frac{-2(x - 3) + 3(x - 2)}{(x - 2)(x - 3)} = \frac{x}{(x - 2)(x - 3)}$$

$$= \frac{x}{x^2 - 5x + 6}$$

■   Now work Problem 9.

The numbers to be found in the partial fraction decomposition can sometimes be found more readily by using suitable choices for x (which may include complex numbers) in the identity obtained after fractions have been cleared. In Example 1, the identity after clearing fractions, equation (3), is

$$x = A(x - 3) + B(x - 2)$$

If we let $x = 2$ in this expression, the term containing B drops out, leaving $2 = A(-1)$, or $A = -2$. Similarly, if we let $x = 3$, the term containing A drops out, leaving $3 = B$. Thus, as before $A = -2$ and $B = 3$.

We use this method in the next example.

CASE 2: Q has repeated linear factors.

If the polynomial Q has a repeated factor, say, $(x - a)^n$, $n \geq 2$ an integer, then, in the partial fraction decomposition of P/Q, we allow for the terms

$$\frac{A_1}{x - a} + \frac{A_2}{(x - a)^2} + \cdots + \frac{A_n}{(x - a)^n}$$

where the numbers $A_1, A_2, \ldots, A_n$ are to be determined.

E X A M P L E   2   *Repeated Linear Factors*

Write the partial fraction decomposition of: $\dfrac{x + 2}{x^3 - 2x^2 + x}$

Solution   First, we factor the denominator,

$$x^3 - 2x^2 + x = x(x^2 - 2x + 1) = x(x - 1)^2$$

and find that the denominator has the nonrepeated linear factor x and the repeated linear factor $(x - 1)^2$. By Case 1, we must allow for the term A/x in the decomposition; and, by Case 2, we must allow for the terms $B/(x - 1) + C/(x - 1)^2$ in the decomposition.

Thus, we write

$$\frac{x + 2}{x^3 - 2x^2 + x} = \frac{A}{x} + \frac{B}{x - 1} + \frac{C}{(x - 1)^2} \tag{4}$$

Again, we clear fractions by multiplying each side by $x^3 - 2x^2 + x = x(x - 1)^2$. The result is the identity

$$x + 2 = A(x - 1)^2 + Bx(x - 1) + Cx \tag{5}$$

If we let $x = 0$ in this expression, the terms containing B and C drop out leaving $2 = A(-1)^2$, or $A = 2$. Similarly, if we let $x = 1$, the terms containing A and B drop out, leaving $3 = C$. Thus, equation (5) becomes

$$x + 2 = 2(x - 1)^2 + Bx(x - 1) + 3x$$

Now, let $x = 2$ (any choice other than 0 or 1 will work as well). The result is

$$4 = 2(1)^2 + B(2)(1) + 3(2)$$
$$2B = 4 - 2 - 6 = -4$$
$$B = -2$$

Thus, we have $A = 2$, $B = -2$, and $C = 3$.

From equation (4), the partial fraction decomposition is

$$\frac{x + 2}{x^3 - 2x^2 + x} = \frac{2}{x} + \frac{-2}{x - 1} + \frac{3}{(x - 1)^2}$$ ∎

E X A M P L E 3   *Repeated Linear Factors*

Write the partial fraction decomposition of: $\dfrac{x^3 - 8}{x^2(x - 1)^3}$

Solution   The denominator contains the repeated linear factor x^2 and the repeated linear factor $(x - 1)^3$. Thus, the partial fraction decomposition takes the form

$$\frac{x^3 - 8}{x^2(x - 1)^3} = \frac{A}{x} + \frac{B}{x^2} + \frac{C}{x - 1} + \frac{D}{(x - 1)^2} + \frac{E}{(x - 1)^3} \qquad (6)$$

As before, we clear fractions and obtain the identity

$$x^3 - 8 = Ax(x - 1)^3 + B(x - 1)^3 + Cx^2(x - 1)^2 + Dx^2(x - 1) + Ex^2 \qquad (7)$$

Let $x = 0$. (Do you see why this choice was made?) Then,

$$-8 = B(-1)$$
$$B = 8$$

Now let $x = 1$ in equation (7). Then

$$-7 = E$$

Use $B = 8$ and $E = -7$ in equation (7) and collect like terms:

$$x^3 - 8 = Ax(x - 1)^3 + 8(x - 1)^3$$
$$+ Cx^2(x - 1)^2 + Dx^2(x - 1) - 7x^2$$
$$x^3 - 8 - 8(x^3 - 3x^2 + 3x - 1) + 7x^2 = Ax(x - 1)^3 + Cx^2(x - 1)^2 + Dx^2(x - 1)$$
$$-7x^3 + 31x^2 - 24x = x(x - 1)[A(x - 1)^2 + Cx(x - 1) + Dx]$$
$$x(x - 1)(-7x + 24) = x(x - 1)[A(x - 1)^2 + Cx(x - 1) + Dx]$$
$$-7x + 24 = A(x - 1)^2 + Cx(x - 1) + Dx \qquad (8)$$

We now work with equation (8). Let $x = 0$. Then

$$24 = A$$

Now, let $x = 1$ in equation (8). Then,

$$17 = D$$

Use $A = 24$ and $D = 17$ in equation (8) and collect like terms:

$$-7x + 24 = 24(x - 1)^2 + Cx(x - 1) + 17x$$
$$-24x^2 + 48x - 24 - 17x - 7x + 24 = Cx(x - 1)$$
$$-24x^2 + 24x = Cx(x - 1)$$
$$-24x(x - 1) = Cx(x - 1)$$
$$-24 = C$$

We now know all the numbers A, B, C, D, and E, so, from equation (6), we have the decomposition

$$\frac{x^3 - 8}{x^2(x - 1)^3} = \frac{24}{x} + \frac{8}{x^2} + \frac{-24}{x - 1} + \frac{17}{(x - 1)^2} + \frac{-7}{(x - 1)^3} \quad \blacksquare$$

The method employed in Example 3, although somewhat tedious, is still preferable to solving the system of five equations containing five variables that the expansion of equation (6) leads to.

■ Now work Problem 15.

The final two cases involve irreducible quadratic factors. As mentioned in Section 4.4, a quadratic factor is irreducible if it cannot be factored into linear factors with real coefficients. A quadratic expression $ax^2 + bx + c$ is irreducible whenever $b^2 - 4ac < 0$. For example, $x^2 + x + 1$ and $x^2 + 4$ are irreducible.

CASE 3: Q contains a nonrepeated irreducible quadratic factor.

If Q contains a nonrepeated irreducible quadratic factor of the form $ax^2 + bx + c$, then, in partial fraction decomposition of P/Q, allow for the term

$$\frac{Ax + B}{ax^2 + bx + c}$$

where the numbers A and B are to be determined.

EXAMPLE 4 *Nonrepeated Irreducible Quadratic Factor*

Write the partial fraction decomposition of: $\dfrac{3x - 5}{x^3 - 1}$

Solution We factor the denominator,

$$x^3 - 1 = (x - 1)(x^2 + x + 1)$$

and find that it has a nonrepeated linear factor $x - 1$ and a nonrepeated irreducible quadratic factor $x^2 + x + 1$. Thus, we allow for the term $A/(x - 1)$ by Case 1, and we allow for the term $(Bx + C)/(x^2 + x + 1)$ by Case 3. Hence, we write

$$\frac{3x - 5}{x^3 - 1} = \frac{A}{x - 1} + \frac{Bx + C}{x^2 + x + 1} \tag{9}$$

We clear fractions by multiplying each side of equation (9) by $x^3 - 1 = (x - 1)(x^2 + x + 1)$ to obtain the identity

$$3x - 5 = A(x^2 + x + 1) + (Bx + C)(x - 1) \tag{10}$$

Now let $x = 1$. Then equation (10) gives $-2 = A(3)$, or $A = -\frac{2}{3}$. We use this value of A in equation (10) and simplify:

$$3x - 5 = -\frac{2}{3}(x^2 + x + 1) + (Bx + C)(x - 1)$$

$$3(3x - 5) = -2(x^2 + x + 1) + 3(Bx + C)(x - 1) \quad \text{Multiply each side by 3.}$$

$$9x - 15 = -2x^2 - 2x - 2 + 3(Bx + C)(x - 1)$$

$$2x^2 + 11x - 13 = 3(Bx + C)(x - 1) \qquad \text{Collect terms.}$$

$$(2x + 13)(x - 1) = 3(Bx + C)(x - 1) \qquad \text{Factor the left side.}$$

$$2x + 13 = 3Bx + 3C$$

$$2 = 3B \quad \text{and} \quad 13 = 3C \qquad \text{Equate coefficients.}$$

$$B = \frac{2}{3} \qquad\qquad C = \frac{13}{3}$$

Thus, from equation (9), we see that

$$\frac{3x - 5}{x^3 - 1} = \frac{-\frac{2}{3}}{x - 1} + \frac{\frac{2}{3}x + \frac{13}{3}}{x^2 + x + 1}$$

■ Now work Problem 17.

CASE 4: *Q* contains repeated irreducible quadratic factors.

If the polynomial Q contains a repeated irreducible quadratic factor $(ax^2 + bx + c)^n$, $n \geq 2$, n an integer, then, in the partial fraction decomposition of P/Q, allow for the terms

$$\frac{A_1x + B_1}{ax^2 + bx + c} + \frac{A_2x + B_2}{(ax^2 + bx + c)^2} + \cdots + \frac{A_nx + B_n}{(ax^2 + bx + c)^n}$$

where the numbers $A_1, B_1, A_2, B_2, \ldots, A_n, B_n$ are to be determined.

EXAMPLE 5   *Repeated Irreducible Quadratic Factor*

Write the partial fraction decomposition of: $\dfrac{x^3 + x^2}{(x^2 + 4)^2}$

Solution   The denominator contains the repeated irreducible quadratic factor $(x^2 + 4)^2$, so we write

$$\frac{x^3 + x^2}{(x^2 + 4)^2} = \frac{Ax + B}{x^2 + 4} + \frac{Cx + D}{(x^2 + 4)^2} \tag{11}$$

We clear fractions to obtain

$$x^3 + x^2 = (Ax + B)(x^2 + 4) + Cx + D$$

Collecting like terms yields

$$x^3 + x^2 = Ax^3 + Bx^2 + (4A + C)x + D + 4B$$

Equating coefficients, we arrive at the system

$$\begin{cases} A = 1 \\ B = 1 \\ 4A + C = 0 \\ D + 4B = 0 \end{cases}$$

The solution is $A = 1$, $B = 1$, $C = -4$, $D = -4$. Hence, from equation (11),

$$\frac{x^3 + x^2}{(x^2 + 4)^2} = \frac{x + 1}{x^2 + 4} + \frac{-4x - 4}{(x^2 + 4)^2} \qquad\blacksquare$$

7.5

Exercise 7.5

In Problems 1–8, tell whether the given rational expression is proper or improper. If improper, rewrite it as the sum of a polynomial and a proper rational expression.

1. $\dfrac{x}{x^2 - 1}$

2. $\dfrac{5x + 2}{x^3 - 1}$

3. $\dfrac{x^2 + 5}{x^2 - 4}$

4. $\dfrac{3x^2 - 2}{x^2 - 1}$

5. $\dfrac{5x^3 + 2x - 1}{x^2 - 4}$

6. $\dfrac{3x^4 + x^2 - 2}{x^3 + 8}$

7. $\dfrac{x(x - 1)}{(x + 4)(x - 3)}$

8. $\dfrac{2x(x^2 + 4)}{x^2 + 1}$

In Problems 9–42, write the partial fraction decomposition of each rational expression.

9. $\dfrac{4}{x(x - 1)}$

10. $\dfrac{3x}{(x + 2)(x - 1)}$

11. $\dfrac{1}{x(x^2 + 1)}$

12. $\dfrac{1}{(x + 1)(x^2 + 4)}$

13. $\dfrac{x}{(x - 1)(x - 2)}$

14. $\dfrac{3x}{(x + 2)(x - 4)}$

15. $\dfrac{x^2}{(x - 1)^2(x + 1)}$

16. $\dfrac{x + 1}{x^2(x - 2)}$

17. $\dfrac{1}{x^3 - 8}$

18. $\dfrac{2x + 4}{x^3 - 1}$

19. $\dfrac{x^2}{(x - 1)^2(x + 1)^2}$

20. $\dfrac{x + 1}{x^2(x - 2)^2}$

21. $\dfrac{x - 3}{(x + 2)(x + 1)^2}$

22. $\dfrac{x^2 + x}{(x + 2)(x - 1)^2}$

23. $\dfrac{x + 4}{x^2(x^2 + 4)}$

24. $\dfrac{10x^2 + 2x}{(x - 1)^2(x^2 + 2)}$

25. $\dfrac{x^2 + 2x + 3}{(x + 1)(x^2 + 2x + 4)}$

26. $\dfrac{x^2 - 11x - 18}{x(x^2 + 3x + 3)}$

27. $\dfrac{x}{(3x - 2)(2x + 1)}$

28. $\dfrac{1}{(2x + 3)(4x - 1)}$

29. $\dfrac{x}{x^2 + 2x - 3}$

30. $\dfrac{x^2 - x - 8}{(x + 1)(x^2 + 5x + 6)}$

31. $\dfrac{x^2 + 2x + 3}{(x^2 + 4)^2}$

32. $\dfrac{x^3 + 1}{(x^2 + 16)^2}$

33. $\dfrac{7x + 3}{x^3 - 2x^2 - 3x}$

34. $\dfrac{x^5 + 1}{x^6 - x^4}$

35. $\dfrac{x^2}{x^3 - 4x^2 + 5x - 2}$

36. $\dfrac{x^2 + 1}{x^3 + x^2 - 5x + 3}$

37. $\dfrac{x^3}{(x^2 + 16)^3}$

38. $\dfrac{x^2}{(x^2 + 4)^3}$

39. $\dfrac{4}{2x^2 - 5x - 3}$

40. $\dfrac{4x}{2x^2 + 3x - 2}$

41. $\dfrac{2x + 3}{x^4 - 9x^2}$

42. $\dfrac{x^2 + 9}{x^4 - 2x^2 - 8}$

7.6

Systems of Nonlinear Equations

There is no general methodology for solving a system of nonlinear equations by hand. There are times when substitution is best; other times, elimination is best; and there are times when neither of these methods works. Experience and a certain degree of imagination are your allies here.

Before we begin, two comments are in order:

1. If the system contains two variables, then graph them. By graphing each equation in the system, we can get an idea of how many solutions a system has and approximately where they are located.
2. Extraneous solutions can creep in when solving nonlinear systems, so it is imperative that all apparent solutions be checked.

EXAMPLE 1

Solving a System of Nonlinear Equations Using Substitution

Solve the following system of equations:

$$\begin{cases} 3x - y = -2 & (1) \quad \text{A line} \\ 2x^2 - y = 0 & (2) \quad \text{A parabola} \end{cases}$$

Graphing Solution   We use a graphing utility to graph $y = 3x + 2$ and $y = 2x^2$. From Figure 17, we see that the system apparently has two solutions. Using BOX and TRACE (or INTERSECT), the solutions to the system of equations are $(-0.50, 0.50)$ and $(2.00, 8.00)$.

FIGURE 17

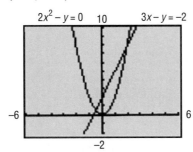

Algebraic Solution   First, we notice that the system contains two variables and that we know how to graph each equation by hand. In Figure 18, we see that the system apparently has two solutions.

FIGURE 18

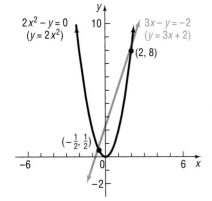

We shall use substitution to solve the system. Equation (1) is easily solved for y:

$$3x - y = -2$$
$$y = 3x + 2$$

We substitute this expression for y in equation (2). The result is an equation containing just the variable x, which we can then solve:

$$2x^2 - y = 0$$
$$2x^2 - (3x + 2) = 0$$
$$2x^2 - 3x - 2 = 0$$
$$(2x + 1)(x - 2) = 0$$
$$2x + 1 = 0 \quad \text{or} \quad x - 2 = 0$$
$$x = -\frac{1}{2} \qquad\qquad x = 2$$

Using these values for x in $y = 3x + 2$, we find

$$y = 3\left(-\frac{1}{2}\right) + 2 = \frac{1}{2} \quad \text{or} \quad y = 3(2) + 2 = 8$$

The apparent solutions are $x = -\frac{1}{2}$, $y = \frac{1}{2}$ and $x = 2$, $y = 8$.

Check: For $x = -\frac{1}{2}$, $y = \frac{1}{2}$:

$$\begin{cases} 3\left(-\frac{1}{2}\right) - \frac{1}{2} = -\frac{3}{2} - \frac{1}{2} = -2 & (1) \\ 2\left(-\frac{1}{2}\right)^2 - \frac{1}{2} = 2\left(\frac{1}{4}\right) - \frac{1}{2} = 0 & (2) \end{cases}$$

For $x = 2$, $y = 8$:

$$\begin{cases} 3(2) - 8 = 6 - 8 = -2 & (1) \\ 2(2)^2 - 8 = 2(4) - 8 = 0 & (2) \end{cases}$$

Each solution checks. Now we know the graphs in Figure 18 intersect at $\left(-\frac{1}{2}, \frac{1}{2}\right)$ and at $(2, 8)$. ■

■   Now work Problem 11.

Our next example illustrates how the method of elimination works for non-linear systems.

E X A M P L E   2

Solving a System of Nonlinear Equations Using Elimination

Solve: $\begin{cases} x^2 + y^2 = 13 & \text{(1)} \quad \text{A circle} \\ x^2 - y = 7 & \text{(2)} \quad \text{A parabola} \end{cases}$

Graphing Solution

We use a graphing utility to graph $x^2 + y^2 = 13$ and $x^2 - y = 7$. [Remember that to graph $x^2 + y^2 = 13$ requires two functions, $y = \sqrt{13 - x^2}$ and $y = -\sqrt{13 - x^2}$, and a square screen.] From Figure 19 we see that the system apparently has four solutions. Using BOX and TRACE (or INTERSECT), the solutions to the system of equations are $(-3.00, 2.00)$, $(3.00, 2.00)$, $(-2.00, -3.00)$, and $(2.00, -3.00)$.

FIGURE 19

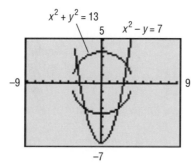

Algebraic Solution

First, we graph each equation, as shown in Figure 20. Based on the graph, we expect four solutions. By subtracting equation (2) from equation (1), the variable x is eliminated, leaving

$$y^2 + y = 6$$

This quadratic equation in y is easily solved by factoring:

FIGURE 20

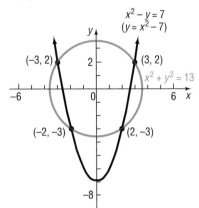

$$y^2 + y - 6 = 0$$
$$(y + 3)(y - 2) = 0$$
$$y = -3 \quad \text{or} \quad y = 2$$

We use these values for y in equation (2) to find x. If $y = 2$, then $x^2 = y + 7 = 9$ and $x = 3$ or -3. If $y = -3$, then $x^2 = y + 7 = 4$ and $x = 2$ or -2. Thus, we have four solutions: $x = 3, y = 2; x = -3, y = 2; x = 2, y = -3;$ and $x = -2, y = -3$. You should verify that, in fact, these four solutions also satisfy equation (1), so that all four are solutions of the system. The four points, $(3, 2)$, $(-3, 2)$, $(2, -3)$, and $(-2, -3)$, are the points of intersection of the graphs. Look again at Figure 20. ∎

■   Now work Problem 9.

E X A M P L E   3   *Solving a System of Nonlinear Equations Using Elimination*

Solve: $\begin{cases} x^2 - y^2 = 1 & (1) \\ x^3 - y^2 = x & (2) \end{cases}$

Graphing Solution   We use a graphing utility to graph $x^2 - y^2 = 1$ and $x^3 - y^2 = x$. [You will need to graph four functions: $y = \sqrt{x^2 - 1}$, $y = -\sqrt{x^2 - 1}$, $y = \sqrt{x^3 - x}$, and $y = -\sqrt{x^3 - x}$]. From Figure 21 we see that the system apparently has two solutions. Using BOX and TRACE (or INTERSECT), the solutions to the system of equations are $(-1.00, 0.00)$ and $(1.00, 0.00)$.

FIGURE 21

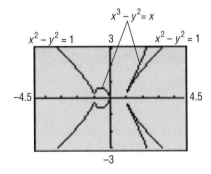

Algebraic Solution   Because the second equation is not so easy to graph by hand, we omit the graphing step. We use elimination, subtracting equation (2) from equation (1), to obtain

$$x^2 - x^3 = 1 - x$$
$$x^2(1 - x) = 1 - x$$
$$x^2(1 - x) - (1 - x) = 0$$
$$(x^2 - 1)(1 - x) = 0$$
$$x^2 - 1 = 0 \quad \text{or} \quad 1 - x = 0$$
$$x = \pm 1 \qquad\qquad x = 1$$

We now use equation (1) to get y. If $x = 1$, then $1 - y^2 = 1$ and $y = 0$. If $x = -1$, then $1 - y^2 = 1$ and $y = 0$. There are two apparent solutions: $x = 1$, $y = 0$ and $x = -1$, $y = 0$. Because each of these solutions also satisfies equation (2), the system has two solutions: $x = 1$, $y = 0$ and $x = -1$, $y = 0$. ∎

E X A M P L E   4   *Solving a System of Nonlinear Equations Using Elimination*

Solve: $\begin{cases} x^2 + x + y^2 - 3y + 2 = 0 & (1) \\ x + 1 + \dfrac{y^2 - y}{x} = 0 & (2) \end{cases}$

Graphing Solution   First, we multiply equation (2) by x to eliminate the fraction. The result is an equivalent system because x cannot be 0 [look at equation (2) to see why]:

$$\begin{cases} x^2 + x + y^2 - 3y + 2 = 0 & (1) \\ x^2 + x + y^2 - y = 0 & (2) \end{cases}$$

We need to solve each equation for y. First, we solve equation (1) for y:

$$x^2 + x + y^2 - 3y + 2 = 0$$

$$y^2 - 3y = -x^2 - x - 2 \qquad \text{Rearrange so terms involving } y \text{ are on left side.}$$

$$y^2 - 3y + \frac{9}{4} = -x^2 - x - 2 + \frac{9}{4} \qquad \text{Complete the square involving } y.$$

$$\left(y - \frac{3}{2}\right)^2 = -x^2 - x + \frac{1}{4}$$

$$y - \frac{3}{2} = \pm\sqrt{-x^2 - x + \frac{1}{4}} \qquad \text{Solve for the squared term.}$$

$$y = \frac{3}{2} \pm \sqrt{-x^2 - x + \frac{1}{4}} \qquad \text{Solve for } y.$$

Now we solve equation (2) for y:

$$x^2 + x + y^2 - y = 0$$

$$y^2 - y = -x^2 - x \qquad \text{Rearrange so terms involving } y \text{ are on left side.}$$

$$y^2 - y + \frac{1}{4} = -x^2 - x + \frac{1}{4} \qquad \text{Complete the square involving } y.$$

$$\left(y - \frac{1}{2}\right)^2 = -x^2 - x + \frac{1}{4}$$

$$y - \frac{1}{2} = \pm\sqrt{-x^2 - x + \frac{1}{4}} \qquad \text{Solve for the squared term.}$$

$$y = \frac{1}{2} \pm \sqrt{-x^2 - x + \frac{1}{4}} \qquad \text{Solve for } y.$$

FIGURE 22

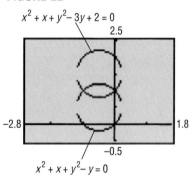

$x^2 + x + y^2 - 3y + 2 = 0$

$x^2 + x + y^2 - y = 0$

Now graph each equation using a graphing utility. See Figure 22.

Using BOX and TRACE (or INTERSECT), the points of intersection are $(-1.00, 1.00)$ and $(0.00, 1.00)$. Because x cannot be 0, the value $x = 0$ is extraneous, and we discard it. Thus, the only solution is $x = -1.00$ and $y = 1.00$.

Algebraic Solution   First, we multiply equation (2) by x to eliminate the fraction. The result is an equivalent system because x cannot be 0 [look at equation (2) to see why]:

$$\begin{cases} x^2 + x + y^2 - 3y + 2 = 0 & (1) \\ x^2 + x + y^2 - y = 0 & (2) \end{cases}$$

Now subtract equation (2) from equation (1) to eliminate x. The result is

$$-2y + 2 = 0$$

$$y = 1$$

To find x, we back-substitute $y = 1$ in equation (1):

$$x^2 + x + 1 - 3 + 2 = 0$$
$$x^2 + x = 0$$
$$x(x + 1) = 0$$
$$x = 0 \quad \text{or} \quad x = -1$$

Because x cannot be 0, the value $x = 0$ is extraneous, and we discard it. Thus, the solution is $x = -1$, $y = 1$.

Check: We now check $x = -1$, $y = 1$:

$$\begin{cases} (-1)^2 + (-1) + 1^2 - 3(1) + 2 = 1 - 1 + 1 - 3 + 2 = 0 & (1) \\ \\ -1 + 1 + \dfrac{1^2 - 1}{-1} = 0 + \dfrac{0}{-1} = 0 & (2) \end{cases}$$

Thus, the only solution to the system is $x = -1$, $y = 1$. ■

■   Now work Problems 25 and 49.

E X A M P L E   5   *Solving a System of Nonlinear Equations*

Solve: $\begin{cases} x^2 - y^2 = 4 & (1) \quad \text{A hyperbola} \\ \quad\quad\; y = x^2 & (2) \quad \text{A parabola} \end{cases}$

Graphing Solution   We graph $y = x^2$ and $x^2 - y^2 = 4$ in Figure 23(a). [You will need to graph $x^2 - y^2 = 4$ as two functions:

$$y = \sqrt{x^2 - 4} \quad \text{and} \quad y = -\sqrt{x^2 - 4}$$

From Figure 23(a) we see that the graphs of these two equations do not intersect. Thus, the system is inconsistent.

FIGURE 23(a)

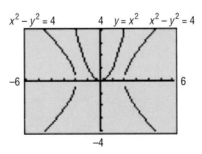

Algebraic Solution   Either substitution or elimination can be used here. We use substitution and replace x^2 by y in equation (1). The result is

$$y - y^2 = 4$$
$$y^2 - y + 4 = 0$$

This is a quadratic equation whose discriminant is $1 - 4 \cdot 4 = -15 < 0$. Thus, the equation has no real solutions, and hence, the system is inconsistent. The graphs of these two equations will not intersect. See Figure 23(b).

FIGURE 23(b)

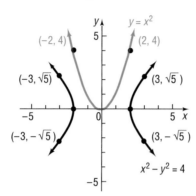

The following examples illustrate two of the more imaginative ways to solve systems of nonlinear equations algebraically.

E X A M P L E   6   *Solving a System of Nonlinear Equations*

Solve: $\begin{cases} 4x^2 - 9xy - 28y^2 = 0 & (1) \\ 16x^2 - 4xy = 16 & (2) \end{cases}$

Graphing Solution   Instead of solving for y by completing the square, we notice that equation (1) can be factored:

$$4x^2 - 9xy - 28y^2 = 0$$
$$(4x + 7y)(x - 4y) = 0$$

This results in the two equations:

$$4x + 7y = 0 \qquad \text{or} \quad x - 4y = 0$$
$$y = -\frac{4}{7}x \qquad\qquad y = \frac{1}{4}x$$

We graph these two equations and equation (2): $y = \dfrac{4(x^2 - 1)}{x}$. See Figure 24.

FIGURE 24

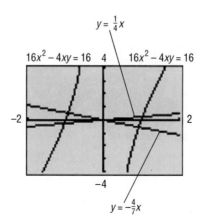

Using BOX and TRACE (or INTERSECT), we find the intersection points to be $(-0.93, 0.53)$, $(-1.03, -0.25)$, $(0.93, -0.53)$, and $(1.03, 0.25)$, each correct to two decimal places.

Algebraic Solution

We take note of the fact that equation (1) can be factored:

$$4x^2 - 9xy - 28y^2 = 0$$

$$(4x + 7y)(x - 4y) = 0$$

This results in the two equations

$$4x + 7y = 0 \quad \text{or} \quad x - 4y = 0$$
$$x = -\tfrac{7}{4}y \qquad\qquad x = 4y$$

We substitute each of these values for x in equation (2):

$$16x^2 - 4xy = 16 \qquad\qquad 16x^2 - 4xy = 16$$
$$16(-\tfrac{7}{4}y)^2 - 4(-\tfrac{7}{4}y)y = 16 \qquad 16(4y)^2 - 4(4y)y = 16$$
$$49y^2 + 7y^2 = 16 \qquad\qquad 16(16y^2) - 16y^2 = 16$$
$$56y^2 = 16 \qquad\qquad 15y^2 = 1$$
$$7y^2 = 2 \qquad\qquad y^2 = \tfrac{1}{15}$$
$$y^2 = \tfrac{2}{7}$$

Thus, we have

$$y = \pm\sqrt{\frac{2}{7}} = \pm\frac{\sqrt{14}}{7} \qquad y = \pm\frac{\sqrt{15}}{15}$$

$$x = -\frac{7}{4}y = \mp\frac{\sqrt{14}}{4} \qquad x = 4y = \pm\frac{4\sqrt{15}}{15}$$

You should verify for yourself that, in fact, the four solutions $x = -\sqrt{14}/4$, $y = \sqrt{14}/7$; $x = \sqrt{14}/4$, $y = -\sqrt{14}/7$; $x = 4\sqrt{15}/15$, $y = \sqrt{15}/15$; and $x = -4\sqrt{15}/15$, $y = -\sqrt{15}/15$ are actually solutions of the system. ∎

E X A M P L E   7

Solving a System of Nonlinear Equations

Solve: $\begin{cases} 3xy - 2y^2 = -2 & (1) \\ 9x^2 + 4y^2 = 10 & (2) \end{cases}$

Graphing Solution

To graph $3xy - 2y^2 = -2$, we need to solve for y. In this instance, it is easier to view the equation as a quadratic equation in the variable y.

$$3xy - 2y^2 = -2$$
$$2y^2 - 3xy - 2 = 0 \qquad \text{Place in standard form.}$$
$$y = \frac{-(-3x) \pm \sqrt{(-3x)^2 - 4(2)(-2)}}{2(2)} \qquad \begin{array}{l}\text{Use the quadratic formula} \\ a = 2,\ b = -3x,\ c = -2.\end{array}$$
$$y = \frac{3x \pm \sqrt{9x^2 + 16}}{4} \qquad \text{Simplify.}$$

Using a graphing utility, we graph $y = \dfrac{3x + \sqrt{9x^2 + 16}}{4}$, and $y = \dfrac{3x - \sqrt{9x^2 + 16}}{4}$.

From (2), We graph $y = \dfrac{\sqrt{10 - 9x^2}}{2}$ and $y = \dfrac{-\sqrt{10 - 9x^2}}{2}$. See Figure 25.

FIGURE 25

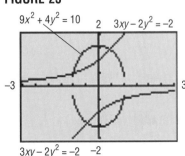

$9x^2 + 4y^2 = 10$ $3xy - 2y^2 = -2$

-3 ⋯⋯ 3

$3xy - 2y^2 = -2$

Using BOX and TRACE (or INTERSECT), the solutions to the system of equations are $(-1.00, 0.50)$, $(0.47, 1.41)$, $(1.00, 0.50)$, and $(-0.47, -1.41)$, each correct to two decimal places.

Algebraic Solution We multiply equation (1) by 2 and add the result to equation (2) to eliminate the y^2-terms:

$$\begin{cases} 6xy - 4y^2 = -4 & \text{(1)} \\ 9x^2 + 4y^2 = 10 & \text{(2)} \end{cases}$$

$$9x^2 + 6xy = 6$$

$$3x^2 + 2xy = 2 \qquad \text{Divide each side by 3.}$$

Since $x \neq 0$ (do you see why?), we can solve for y in this equation to get

$$y = \frac{2 - 3x^2}{2x}, \qquad x \neq 0 \tag{1}$$

Now substitute for y in equation (2) of the system:

$$9x^2 + 4y^2 = 10$$

$$9x^2 + 4\left(\frac{2 - 3x^2}{2x}\right)^2 = 10$$

$$9x^2 + \frac{4 - 12x^2 + 9x^4}{x^2} = 10$$

$$9x^4 + 4 - 12x^2 + 9x^4 = 10x^2$$

$$18x^4 - 22x^2 + 4 = 0$$

$$9x^4 - 11x^2 + 2 = 0$$

This quadratic equation (in x^2) can be factored:

$$(9x^2 - 2)(x^2 - 1) = 0$$

$$\begin{array}{ccc} 9x^2 - 2 = 0 & \text{or} & x^2 - 1 = 0 \\ x^2 = \frac{2}{9} & & x^2 = 1 \\ x = \pm \dfrac{\sqrt{2}}{3} & & x = \pm 1 \end{array}$$

To find y, we use equation (1):

$$\text{If } x = \frac{\sqrt{2}}{3}: \qquad y = \frac{2 - 3x^2}{2x} = \frac{2 - \frac{2}{3}}{2(\sqrt{2}/3)} = \frac{4}{2\sqrt{2}} = \sqrt{2}$$

$$\text{If } x = -\frac{\sqrt{2}}{3}: \qquad y = \frac{2 - 3x^2}{2x} = \frac{2 - \frac{2}{3}}{-2(\sqrt{2}/3)} = \frac{4}{-2\sqrt{2}} = -\sqrt{2}$$

$$\text{If } x = 1: \qquad y = \frac{2 - 3x^2}{2x} = \frac{2 - 3}{2} = -\frac{1}{2}$$

$$\text{If } x = -1: \qquad y = \frac{2 - 3x^2}{2x} = \frac{2 - 3}{-2} = \frac{1}{2}$$

The system has four solutions. Check them for yourself. ∎

■ Now work Problem 45.

E X A M P L E   8

Running a Marathon

In a 50 mile marathon race, the winner crosses the finish line 1 mile ahead of the second place runner and 4 miles ahead of the third place runner. Assuming that each runner maintains a constant speed throughout the race, by how many miles does the second place runner beat the third place runner?

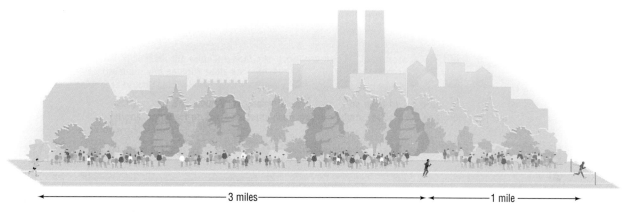

$\xleftarrow{\hspace{2cm}}$ 3 miles $\xrightarrow{\hspace{2cm}}$ $\xleftarrow{\hspace{1cm}}$ 1 mile $\xrightarrow{\hspace{1cm}}$

Solution

Let v_1, v_2, v_3 denote the speeds of the first, second, and third place runners, respectively. Let t_1 and t_2 denote the times (in hours) required for the first place runner and second place runner to finish the race. Then we have the system of equations

$$\begin{cases} 50 = v_1t_1 & \text{(1)} \quad \text{First place runner goes 50 miles in } t_1 \text{ hours.} \\ 49 = v_2t_1 & \text{(2)} \quad \text{Second place runner goes 49 miles in } t_1 \text{ hours.} \\ 46 = v_3t_1 & \text{(3)} \quad \text{Third place runner goes 46 miles in } t_1 \text{ hours.} \\ 50 = v_2t_2 & \text{(4)} \quad \text{Second place runner goes 50 miles in } t_2 \text{ hours.} \end{cases}$$

We seek the distance of the third place runner from the finish at time t_2. That is, we seek

$$50 - v_3t_2 = 50 - v_3\left(t_1 \cdot \frac{t_2}{t_1}\right)$$

$$= 50 - (v_3t_1) \cdot \frac{t_2}{t_1}$$

$$= 50 - 46 \cdot \frac{50/v_2}{50/v_1} \quad \begin{cases} \text{From (3), } v_3t_1 = 46; \\ \text{from (4), } t_2 = 50/v_2; \\ \text{from (1), } t_1 = 50/v_1. \end{cases}$$

$$= 50 - 46 \cdot \frac{v_1}{v_2}$$

$$= 50 - 46 \cdot \frac{50}{49} \qquad \text{Form the quotient of (1) and (2).}$$

$$\approx 3.06 \text{ miles} \qquad\qquad\qquad ■$$

HISTORICAL FEATURE

■ Recall that, in the beginning of this section, we said imagination and experience are important in solving simultaneous nonlinear equations. Indeed, these kinds of problems lead into some of the deepest and most difficult parts of modern mathematics. Look again at the graphs in Examples 1 and 2 of this section (Figures 17 or 18 and 19 or 20). We see that Example 1 has two solutions, and Example 2 has four solutions. We might conjecture that the number of solutions is equal to the product of the degrees of the equations involved. This conjecture was indeed made by Etienne Bezout (1739–1783), but working out the details took about 150 years. It turns out that, to arrive at the correct number of intersections, we must count not only the complex number intersections, but also those intersections that, in a certain sense, lie at infinity. For example, a parabola and a line lying on the axis of the parabola intersect at the vertex and at infinity. This topic is part of the study of algebraic geometry. ∎

HISTORICAL PROBLEM   ∎ 1. A papyrus dating back to 1950 BC contains the following problem: A given surface area of 100 units of area shall be represented as the sum of two squares whose sides are to each other as $1:\frac{3}{4}$. Solve for the sides by solving the system of equations

$$\begin{cases} x^2 + y^2 = 100 \\ \quad\quad x = \frac{3}{4}y \end{cases}$$

∎

7.6

Exercise 7.6

In Problems 1–20, use a graphing utility to graph each equation of the system. Then solve the system by finding the intersection points. Express your answer correct to two decimal places. Also solve each system algebraically.

1. $\begin{cases} y = x^2 + 1 \\ y = x + 1 \end{cases}$

2. $\begin{cases} y = x^2 + 1 \\ y = 4x + 1 \end{cases}$

3. $\begin{cases} y = \sqrt{36 - x^2} \\ y = 8 - x \end{cases}$

4. $\begin{cases} y = \sqrt{4 - x^2} \\ y = 2x + 4 \end{cases}$

5. $\begin{cases} y = \sqrt{x} \\ y = 2 - x \end{cases}$

6. $\begin{cases} y = \sqrt{x} \\ y = 6 - x \end{cases}$

7. $\begin{cases} x = 2y \\ x = y^2 - 2y \end{cases}$

8. $\begin{cases} y = x - 1 \\ y = x^2 - 6x + 9 \end{cases}$

9. $\begin{cases} x^2 + y^2 = 4 \\ x^2 + 2x + y^2 = 0 \end{cases}$

10. $\begin{cases} x^2 + y^2 = 8 \\ x^2 + y^2 + 4y = 0 \end{cases}$

11. $\begin{cases} y = 3x - 5 \\ x^2 + y^2 = 5 \end{cases}$

12. $\begin{cases} x^2 + y^2 = 10 \\ y = x + 2 \end{cases}$

13. $\begin{cases} x^2 + y^2 = 4 \\ y^2 - x = 4 \end{cases}$

14. $\begin{cases} x^2 + y^2 = 16 \\ x^2 - 2y = 8 \end{cases}$

15. $\begin{cases} xy = 4 \\ x^2 + y^2 = 8 \end{cases}$

16. $\begin{cases} x^2 = y \\ xy = 1 \end{cases}$

17. $\begin{cases} x^2 + y^2 = 4 \\ y = x^2 - 9 \end{cases}$

18. $\begin{cases} xy = 1 \\ y = 2x + 1 \end{cases}$

19. $\begin{cases} y = x^2 - 4 \\ y = 6x - 13 \end{cases}$

20. $\begin{cases} x^2 + y^2 = 10 \\ xy = 3 \end{cases}$

In Problems 21–52, solve each system. Use any method you wish.

21. $\begin{cases} 2x^2 + y^2 = 18 \\ xy = 4 \end{cases}$

22. $\begin{cases} x^2 - y^2 = 21 \\ x + y = 7 \end{cases}$

23. $\begin{cases} y = 2x + 1 \\ 2x^2 + y^2 = 1 \end{cases}$

24. $\begin{cases} x^2 - 4y^2 = 16 \\ 2y - x = 2 \end{cases}$

25. $\begin{cases} x + y + 1 = 0 \\ x^2 + y^2 + 6y - x = -5 \end{cases}$

26. $\begin{cases} 2x^2 - xy + y^2 = 8 \\ xy = 4 \end{cases}$

27. $\begin{cases} 4x^2 - 3xy + 9y^2 = 15 \\ 2x + 3y = 5 \end{cases}$

28. $\begin{cases} 2y^2 - 3xy + 6y + 2x + 4 = 0 \\ 2x - 3y + 4 = 0 \end{cases}$

29. $\begin{cases} x^2 - 4y^2 + 7 = 0 \\ 3x^2 + y^2 = 31 \end{cases}$

30. $\begin{cases} 3x^2 - 2y^2 + 5 = 0 \\ 2x^2 - y^2 + 2 = 0 \end{cases}$

31. $\begin{cases} 7x^2 - 3y^2 + 5 = 0 \\ 3x^2 + 5y^2 = 12 \end{cases}$

32. $\begin{cases} x^2 - 3y^2 + 1 = 0 \\ 2x^2 - 7y^2 + 5 = 0 \end{cases}$

33. $\begin{cases} x^2 + 2xy = 10 \\ 3x^2 - xy = 2 \end{cases}$

34. $\begin{cases} 5xy + 13y^2 + 36 = 0 \\ xy + 7y^2 = 6 \end{cases}$

35. $\begin{cases} 2x^2 + y^2 = 2 \\ x^2 - 2y^2 + 8 = 0 \end{cases}$

36. $\begin{cases} y^2 - x^2 + 4 = 0 \\ 2x^2 + 3y^2 = 6 \\ \dfrac{5}{x^2} - \dfrac{2}{y^2} + 3 = 0 \end{cases}$

37. $\begin{cases} x^2 + 2y^2 = 16 \\ 4x^2 - y^2 = 24 \\ \dfrac{2}{x^2} - \dfrac{3}{y^2} + 1 = 0 \end{cases}$

38. $\begin{cases} 4x^2 + 3y^2 = 4 \\ 2x^2 - 6y^2 = -3 \end{cases}$

39. $\begin{cases} \dfrac{3}{x^2} + \dfrac{1}{y^2} = 7 \end{cases}$

40. $\begin{cases} \dfrac{6}{x^2} - \dfrac{7}{y^2} + 2 = 0 \end{cases}$

41. $\begin{cases} \dfrac{1}{x^4} + \dfrac{6}{y^4} = 6 \\ \dfrac{2}{x^4} - \dfrac{2}{y^4} = 19 \end{cases}$

42. $\begin{cases} \dfrac{1}{x^4} - \dfrac{1}{y^4} = 1 \\ \dfrac{1}{x^4} + \dfrac{1}{y^4} = 4 \end{cases}$

43. $\begin{cases} x^2 - 3xy + 2y^2 = 0 \\ x^2 + xy = 6 \end{cases}$

44. $\begin{cases} x^2 - xy - 2y^2 = 0 \\ xy + x + 6 = 0 \end{cases}$

45. $\begin{cases} xy - x^2 + 3 = 0 \\ 3xy - 4y^2 = 2 \end{cases}$

46. $\begin{cases} 5x^2 + 4xy + 3y^2 = 36 \\ x^2 + xy + y^2 = 9 \end{cases}$

47. $\begin{cases} x^3 - y^3 = 26 \\ x - y = 2 \end{cases}$

48. $\begin{cases} x^3 + y^3 = 26 \\ x + y = 2 \end{cases}$

49. $\begin{cases} y^2 + y + x^2 - x - 2 = 0 \\ y + 1 + \dfrac{x - 2}{y} = 0 \end{cases}$

50. $\begin{cases} x^3 - 2x^2 + y^2 + 3y - 4 = 0 \\ x - 2 + \dfrac{y^2 - y}{x^2} = 0 \end{cases}$

51. $\begin{cases} \log_x y = 3 \\ \log_x (4y) = 5 \end{cases}$

52. $\begin{cases} \log_x (2y) = 3 \\ \log_x (4y) = 2 \end{cases}$

In Problems 53–60, use a graphing utility to solve each system of equations. Express the solution(s) correct to two decimal places.

53. $\begin{cases} y = x^{2/3} \\ y = e^{-x} \end{cases}$

54. $\begin{cases} y = x^{3/2} \\ y = e^{-x} \end{cases}$

55. $\begin{cases} x^2 + y^3 = 2 \\ x^3 y = 4 \end{cases}$

56. $\begin{cases} x^3 + y^2 = 2 \\ x^2 y = 4 \end{cases}$

57. $\begin{cases} x^4 + y^4 = 12 \\ xy^2 = 2 \end{cases}$

58. $\begin{cases} x^4 + y^4 = 6 \\ xy = 1 \end{cases}$

59. $\begin{cases} xy = 2 \\ y = \ln x \end{cases}$

60. $\begin{cases} x^2 + y^2 = 4 \\ y = \ln x \end{cases}$

61. The difference of two numbers is 2 and the sum of their squares is 10. Find the numbers.

62. The sum of two numbers is 7 and the difference of their squares is 21. Find the numbers.

63. The product of two numbers is 4 and the sum of their squares is 8. Find the numbers.

64. The product of two numbers is 10 and the difference of their squares is 21. Find the numbers.

65. The difference of two numbers is the same as their product, and the sum of their reciprocals is 5. Find the numbers.

66. The sum of two numbers is the same as their product, and the difference of their reciprocals is 3. Find the numbers.

67. The ratio of a to b is $\frac{2}{3}$. The sum of a and b is 10. What is the ratio of $a + b$ to $b - a$?

68. The ratio of a to b is $\frac{4}{3}$. The sum of a and b is 14. What is the ratio of $a - b$ to $a + b$?

In Problems 69–74, graph each equation by hand and find the point(s) of intersection, if any.

69. The line $x + 2y = 0$ and the circle $(x - 1)^2 + (y - 1)^2 = 5$

70. The line $x + 2y + 6 = 0$ and the circle $(x + 1)^2 + (y + 1)^2 = 5$

71. The circle $(x - 1)^2 + (y + 2)^2 = 4$ and the parabola $y^2 + 4y - x + 1 = 0$

72. The circle $(x + 2)^2 + (y - 1)^2 = 4$ and the parabola $y^2 - 2y - x - 5 = 0$

73. The graph of $y = \dfrac{4}{x - 3}$ and the circle $x^2 - 6x + y^2 + 1 = 0$

74. The graph of $y = \dfrac{4}{x + 2}$ and the circle $x^2 + 4x + y^2 - 4 = 0$

75. *Geometry* The perimeter of a rectangle is 16 inches and its area is 15 square inches. What are its dimensions?

76. *Geometry* An area of 52 square feet is to be enclosed by two squares whose sides are in the ratio of 2:3. Find the sides of the squares.

77. *Geometry* Two circles have perimeters that add up to 12π centimeters and areas that add up to 20π square centimeters. Find the radius of each circle.

78. *Geometry* The altitude of an isosceles triangle drawn to its base is 3 centimeters, and its perimeter is 18 centimeters. Find the length of its base.

79. *The Tortoise and the Hare* In a 21 meter race between a tortoise and a hare, the tortoise leaves 9 minutes before the hare. The hare, by running at an average speed of 0.5 meter per hour faster than the tortoise, crosses the finish line 3 minutes before the tortoise. What are the average speeds of the tortoise and the hare?

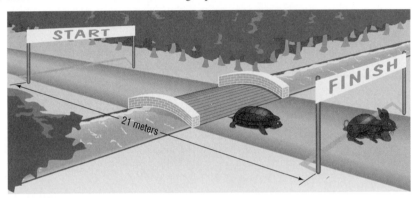

80. *Running a Race* In a 1 mile race, the winner crosses the finish line 10 feet ahead of the second place runner and 20 feet ahead of the third place runner. Assuming that each runner maintains a constant speed throughout the race, by how many feet does the second place runner beat the third place runner?

81. *Constructing a Box* A rectangular piece of cardboard, whose area is 216 square centimeters, is made into an open box by cutting a 2 centimeter square from each corner and turning up the sides. See the figure. If the box is to have a volume of 224 cubic centimeters, what size cardboard should you start with?

82. *Constructing a Cylindrical Tube* A rectangular piece of cardboard, whose area is 216 square centimeters, is made into a cylindrical tube by joining together two sides of the rectangle. (See the figure.) If the tube is to have a volume of 224 cubic centimeters, what size cardboard should you start with?

83. *Fencing* A farmer has 300 feet of fence available to enclose 4500 square feet in the shape of adjoining squares, with sides of length x and y. See the figure. Find x and y.

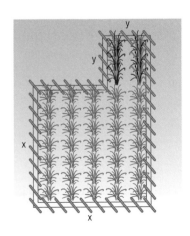

84. *Bending Wire* A wire 60 feet long is cut into two pieces. Is it possible to bend one piece into the shape of a square and the other into the shape of a circle so that the total area enclosed by the two pieces is 100 square feet? If this is possible, find the length of the side of the square and the radius of the circle.

85. *Geometry* Find formulas for the length l and width w of a rectangle in terms of its area A and perimeter P.

86. *Geometry* Find formulas for the base b and one of the equal sides l of an isosceles triangle in terms of its altitude h and perimeter P.

87. *Descartes' Method of Equal Roots* Descartes' method for finding tangents depends on the idea that, for many graphs, the tangent line at a given point is the *unique* line that intersects the graph at that point only. We will apply his method to find an equation of the tangent line to the parabola $y = x^2$ at the point $(2, 4)$; see the figure. First, we know the equation of the tangent line must be in the form $y = mx + b$. Using the fact that the point $(2, 4)$ is on the line, we can solve for b in terms of m and get the equation $y = mx + (4 - 2m)$. Now we want $(2, 4)$ to be the *unique* solution to the system

$$\begin{cases} y = x^2 \\ y = mx + 4 - 2m \end{cases}$$

From this system, we get $x^2 = mx + 4 - 2m$ or $x^2 - mx + (2m - 4) = 0$. By using the quadratic formula, we get

$$x = \frac{m \pm \sqrt{m^2 - 4(2m - 4)}}{2}$$

To obtain a unique solution for x, the two roots must be equal; in other words, the expression $m^2 - 4(2m - 4)$ must be 0. Complete the work to get m, and write an equation of the tangent line.

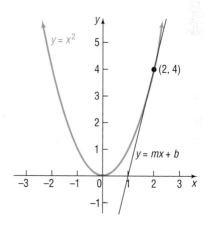

In Problems 88–94, use Descartes' method from Problem 87 to find the equation of the line tangent to each graph at the given point.

88. $x^2 + y^2 = 10$; at $(1, 3)$

89. $y = x^2 + 2$; at $(1, 3)$

90. $x^2 + y = 5$; at $(-2, 1)$

91. $2x^2 + 3y^2 = 14$; at $(1, 2)$

92. $3x^2 + y^2 = 7$; at $(-1, 2)$

93. $x^2 - y^2 = 3$; at $(2, 1)$

94. $2y^2 - x^2 = 14$; at $(2, 3)$

95. If r_1 and r_2 are two solutions of a quadratic equation $ax^2 + bx + c = 0$, then it can be shown that

$$r_1 + r_2 = -\frac{b}{a} \quad \text{and} \quad r_1 r_2 = \frac{c}{a}$$

Solve this system of equations for r_1 and r_2.

96. A circle and a line intersect at most twice. A circle and a parabola intersect at most four times. Deduce that a circle and the graph of a polynomial of degree 3 intersect at most six times. What do you conjecture about a polynomial of degree 4? What about a polynomial of degree n? Can you explain your conclusions using an algebraic argument?

97. Suppose that you are the manager of a sheet metal shop. A customer asks you to manufacture 10,000 boxes, each box being open on top. The boxes are required to have a square base and a 9 cubic foot capacity. You construct the boxes by cutting a square out from each corner of a square piece of sheet metal and folding along the edges.
(a) What are the dimensions of the square to be cut if the area of the square piece of sheet metal is 100 square feet?
(b) Could you make the box using a smaller piece of sheet metal? Make a list of the dimensions of the box for various pieces of sheet metal.

7.7

Systems of Inequalities

In Chapter 3, we discussed inequalities in one variable. In this section, we discuss inequalities in two variables. Samples are given in Example 1.

E X A M P L E   1   *Samples of Inequalities in Two Variables*

(a) $3x + y - 6 < 0$ (b) $x^2 + y^2 < 4$ (c) $y^2 \leq x$ ■

An inequality in two variables x and y is **satisfied** by an ordered pair (a, b) if, when x is replaced by a and y by b, a true statement results. A **graph of an inequality in two variables** x and y consists of all points (x, y) whose coordinates satisfy the inequality.

Let's look at an example.

E X A M P L E   2   *Graphing a Linear Inequality*

Graph the linear inequality: $3x + y - 6 \leq 0$

Solution   We begin with the associated problem of the graph of the linear equality

$$3x + y - 6 = 0$$

formed by replacing (for now) the $\leq$ symbol with an $=$ sign. The graph of the linear equation is a line. See Figure 26(a). This line is part of the graph of the inequality we seek because the inequality is nonstrict. (Do you see why? We are seeking points for which $3x + y - 6$ is less than *or equal to* 0.)

FIGURE 26

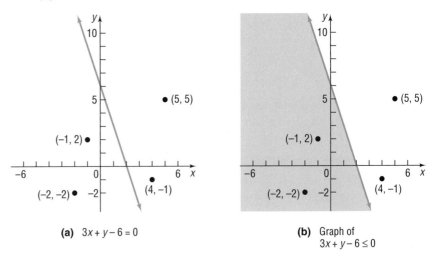

(a) $3x + y - 6 = 0$

(b) Graph of $3x + y - 6 \leq 0$

Now, let's test a few randomly selected points to see whether they belong to the graph of the inequality

	$3x + y - 6$	**CONCLUSION**
$(4, -1)$	$3(4) + (-1) - 6 = 5 > 0$	Does not belong to graph
$(5, 5)$	$3(5) + 5 - 6 = 14 > 0$	Does not belong to graph
$(-1, 2)$	$3(-1) + 2 - 6 = -7 < 0$	Belongs to graph
$(-2, -2)$	$3(-2) + (-2) - 6 = -14 < 0$	Belongs to graph

Look again at Figure 26(a). Notice that the two points that belong to the graph both lie on the same side of the line, and the two points that do not belong to the graph lie on the opposite side. As it turns out, this is always the case. Thus, the graph we seek consists of all points that lie on the same side of the line as do $(-1, 2)$ and $(-2, -2)$. The graph we seek is the shaded region in Figure 26(b). ∎

The graph of any inequality in two variables may be obtained in a like way. First, the equation corresponding to the inequality is graphed, using dashes if the inequality is strict and solid marks if it is nonstrict. This graph, in almost every case, will separate the xy-plane into two or more regions. In each region either all points satisfy the inequality or no points satisfy the inequality. Thus, the use of a single test point in each region is all that is required to determine whether the points of that region are part of the graph or not. The steps to follow are given next.

Steps for Graphing an Inequality By Hand	**STEP 1:** Replace the inequality symbol by an equal sign and graph the resulting equation. If the inequality is strict, use dashes; if it is nonstrict, use a solid mark. This graph separates the xy-plane into two or more regions.
	STEP 2: In each of the regions, select a test point P.
	(a) If the coordinates of P satisfy the inequality, then so do all the points in that region. Indicate this by shading the region.
	(b) If the coordinates of P do not satisfy the inequality, then none of the points in that region do.

Graphing utilities can also be used to graph inequalities.

E X A M P L E   3   *Graphing an Inequality Using a Graphing Utility*

Use a graphing utility to graph $3x + y - 6 \leq 0$.

Solution   We begin by graphing the equation $3x + y - 6 = 0$. See Figure 27.

FIGURE 27

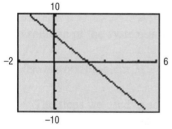

As with graphing by hand, we need to test points selected from each region and determine whether or not they satisfy the inequality. To test the point $(-1, 2)$, for example, enter $3*-1+2-6\leq0$. See Figure 28(a). The 1 that appears indicates the statement entered (the inequality) is true. When the point $(5, 5)$ is tested, a 0 appears, indicating the statement entered is false. Thus, $(-1, 2)$ is a part of the graph of the inequality and $(5, 5)$ is not. Figure 28(b) shows the graph of the inequality.

FIGURE 28

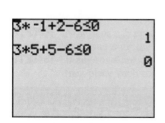

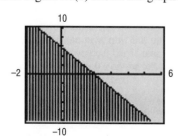

(a)   (b)   ∎

The steps to follow to graph an inequality using a graphing utility are given next.

Steps for Graphing an Inequality Using a Graphing Utility	**STEP 1:** Replace the inequality symbol by an equal sign and graph the resulting equation.
	STEP 2: In each of the regions, select a test point P.
	    (a) Use a graphing utility to determine if the test point P satisfies the inequality. If the test point satisfies the inequality, then so do all the points in the region. Indicate this by using the graphing utility to shade the region.
	    (b) If the coordinates of P do not satisfy the inequality, then none of the points in that region do.

E X A M P L E   4

Graphing an Inequality

Graph: $x^2 + y^2 \le 4$

FIGURE 29

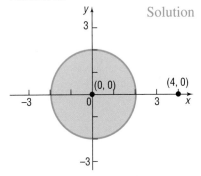

Solution   First we graph the equation $x^2 + y^2 = 4$, a circle of radius 2, center at the origin. A solid color will be used because the inequality is not strict. We use two test points, one inside the circle, the other outside;

Inside $(0, 0)$: $x^2 + y^2 = 0^2 + 0^2 = 0 \le 4$ Belongs to the graph

Outside $(4, 0)$: $x^2 + y^2 = 4^2 + 0^2 = 16 > 4$ Does not belong to graph

All the points inside and on the circle satisfy the inequality. See Figure 29.   ■

■   Now work Problem 7.

Linear inequalities are inequalities in one of the forms

$$Ax + By < C, \qquad Ax + By > C, \qquad Ax + By \le C, \qquad Ax + By \ge C$$

The graph of the corresponding equation of a linear inequality is a line, which separates the xy-plane into two regions, called **half-planes.** See Figure 30.

As shown there, if $Ax + By = C$ is the equation of the boundary line, then it divides the plane into two half-planes; one for which $Ax + By < C$ and the other for which $Ax + By > C$. Because of this, for linear inequalities, only one test point is required.

FIGURE 30

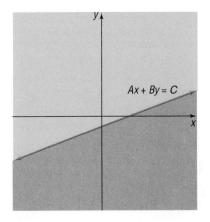

E X A M P L E   5

Graphing Linear Inequalities

Graph: (a) $y < 2$ (b) $y \ge 2x$

Solution (a) The graph of the equation $y = 2$ is a horizontal line and is not part of the graph of the inequality. Since $(0, 0)$ satisfies the inequality, the graph consists of the half-plane below the line $y = 2$. See Figure 31.

FIGURE 31 **FIGURE 32**

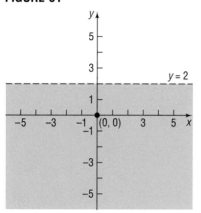 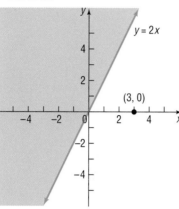

(b) The graph of the equation $y = 2x$ is a line and is part of the graph of the inequality. Using $(3, 0)$ as a test point, we find it does not satisfy the inequality $[0 < 2 \cdot 3]$. Thus, points in the half-plane on the opposite side of $y = 2x$ satisfy the inequality. See Figure 32. ■

■ Now work Problem 3.

Systems of Inequalities in Two Variables

The **graph of a system of inequalities** in two variables x and y is the set of all points (x, y) that simultaneously satisfy *each* of the inequalities in the system. Thus, the graph of a system of inequalities can be obtained by graphing each inequality individually and then determining where, if at all, they intersect.

E X A M P L E 6 *Graphing a System of Linear Inequalities Using a Graphing Utility*

Graph the system: $\begin{cases} x + y \geq 2 \\ 2x - y \leq 4 \end{cases}$

Solution First, we graph the lines $x + y = 2$ and $2x - y = 4$. See Figure 33.

FIGURE 33

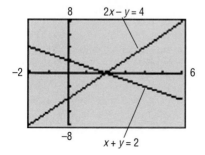

Notice the graphs divide the viewing window into four regions. We select a test point for each region and determine whether the point makes *both* inequalities true.

We choose to test $(0, 0)$, $(2, 3)$, $(4, 0)$, and $(2, -2)$. Figure 34(a) shows that $(2, 3)$ is the only point for which both inequalities are true. Thus, we obtain the graph shown in Figure 34(b).

FIGURE 34

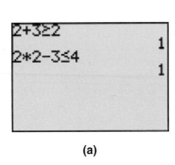

(a)

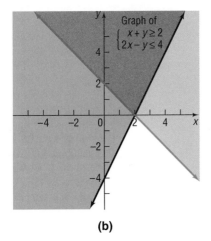

(b)

Notice in Figure 34(b) that the darker region, the graph of the system of inequalities, is the intersection of the graphs of the single inequalities $x + y \geq 2$ and $2x - y \leq 4$. Obtaining Figure 34(b) by this method is sometimes faster than using test points.

E X A M P L E   7   *Graphing a System of Linear Inequalities*

Graph the system: $\begin{cases} x + y \leq 2 \\ x + y \geq 0 \end{cases}$

Solution   See Figure 35. The overlapping, darker shaded region between the two boundary lines is the graph of the system.

FIGURE 35

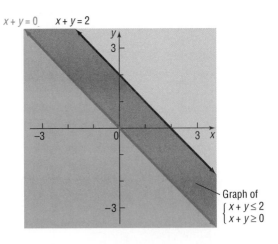

Now work Problem 25.

E X A M P L E   8   *Graphing a System of Linear Inequalities*

Graph the system: $\begin{cases} 2x - y \geq 2 \\ 2x - y \geq 0 \end{cases}$

Solution   See Figure 36. The overlapping, darker shaded region is the graph of the system. Note that the graph of the system is identical to the graph of the single inequality $2x - y \geq 2$.

FIGURE 36

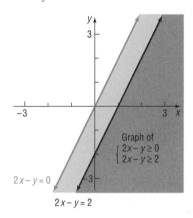

Graph of
$\begin{cases} 2x - y \geq 0 \\ 2x - y \geq 2 \end{cases}$

$2x - y = 0$

$2x - y = 2$

E X A M P L E   9   *Graphing a System of Linear Inequalities*

Graph the system: $\begin{cases} x + 2y \leq 2 \\ x + 2y \geq 6 \end{cases}$

Solution   See Figure 37. Because no overlapping region results, there are no points in the *xy*-plane that simultaneously satisfy each inequality. Hence, the system has no solution.

FIGURE 37

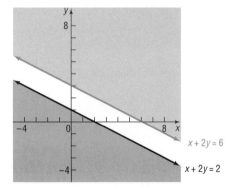

$x + 2y = 6$

$x + 2y = 2$

E X A M P L E   1 0   *Graphing a System of Inequalities*

Graph the system: $\begin{cases} y \geq x^2 - 4 \\ x + y \leq 2 \end{cases}$

Solution   Figure 38 shows the graph of the equations $y = x^2 - 4$ and $x + y = 2$. The graphs divide the viewing window into five regions.

FIGURE 38

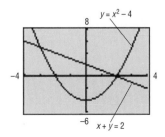

$y = x^2 - 4$

$x + y = 2$

Instead of selecting test points in each region, we determine the graph of the system consists of the intersection of the graphs of the single inequalities $y \geq x^2 - 4$ and $x + y \leq 2$. See Figure 39. The points of intersection of the two equations are found by solving the system of equations

$$\begin{cases} y = x^2 - 4 \\ x + y = 2 \end{cases}$$

Using substitution, we find

$$x + (x^2 - 4) = 2$$
$$x^2 + x - 6 = 0$$
$$(x + 3)(x - 2) = 0$$
$$x = -3, \qquad x = 2$$

The two points of intersection are $(-3, 5)$ and $(2, 0)$.

FIGURE 39

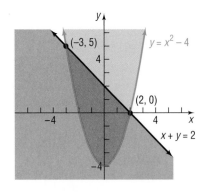

■ Now work Problem 19.

E X A M P L E   1 1   *Graphing a System of Four Linear Inequalities*

Graph the system: $\begin{cases} x + y \geq 3 \\ 2x + y \geq 4 \\ x \geq 0 \\ y \geq 0 \end{cases}$

Solution   The two inequalities $x \geq 0$ and $y \geq 0$ require that the graph be in quadrant I. Thus, we set our viewing window accordingly. Figure 40 shows the graph of $x + y = 3$ and $2x + y = 4$. The graphs divide the viewing window into four regions.

FIGURE 40

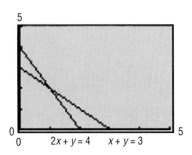

Instead of selecting test points in each region, we determine that the graph of the system is the intersection of the graphs of the single inequalities $x + y \geq 3$ and $2x + y \geq 4$ in quadrant I. See Figure 41.

FIGURE 41

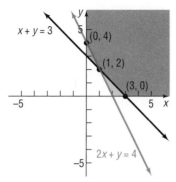

E X A M P L E   1 2   *Financial Planning*

A retired couple has up to $25,000 to invest. As their financial adviser, you recommend that they place at least $15,000 in Treasury bills yielding 6% and at most $5000 in corporate bonds yielding 9%.

(a) Using x to denote the amount of money invested in Treasury bills and y the amount invested in corporate bonds, write a system of linear inequalities that describes the possible amounts of each investment.

(b) Graph the system.

Solution (a) The system of linear inequalities is

$$\begin{cases} x \geq 0 & \text{\small x and y are nonnegative variables since they represent money invested.} \\ y \geq 0 & \\ x + y \leq 25,000 & \text{\small The total of the two investments, $x + y$, cannot exceed \$25,000.} \\ x \geq 15,000 & \text{\small At least \$15,000 in Treasury bills} \\ y \leq 5000 & \text{\small At most \$5000 in corporate bonds} \end{cases}$$

(b) See the shaded region in Figure 42. Note that the inequalities $x \geq 0$ and $y \geq 0$ again require that the graph of the system be in quadrant I.

FIGURE 42

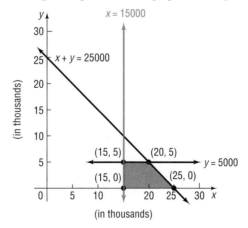

The graph of the system of linear inequalities in Figure 42 is said to be **bounded,** because it can be contained within some circle of sufficiently large radius. A graph that cannot be contained in any circle is said to be **unbounded**. For example, the graph of the system of linear inequalities in Figure 41 is unbounded, since it extends indefinitely in a particular direction.

Notice in Figures 41 and 42 that those points belonging to the graph that are also points of intersection of boundary lines have been plotted. Such points are referred to as **vertices** or **corner points** of the graph. Thus, the system graphed in Figure 41 has three corner points: (0, 4), (1, 2), and (3, 0). The system graphed in Figure 42 has four corner points: (15, 0), (25, 0), (20, 5), and (15, 5).

These ideas will be used in the next section in developing a method for solving linear programming problems, an important application of linear inequalities.

■ Now work Problem 33.

7.7

Exercise 7.7

In Problems 1–12, graph each inequality.

1. $x \geq 0$ **2.** $y \geq 0$ **3.** $x \geq 4$ **4.** $y \leq 2$

5. $2x + y \geq 6$ **6.** $3x + 2y \leq 6$ **7.** $x^2 + y^2 > 1$ **8.** $x^2 + y^2 \leq 9$

9. $y \leq x^2 - 1$ **10.** $y > x^2 + 2$ **11.** $xy \geq 4$ **12.** $xy \leq 1$

In Problems 13–30, graph each system of inequalities.

13. $\begin{cases} x + y \leq 2 \\ 2x + y \geq 4 \end{cases}$ **14.** $\begin{cases} 3x - y \geq 6 \\ x + 2y \leq 2 \end{cases}$ **15.** $\begin{cases} 2x - y \leq 4 \\ 3x + 2y \geq -6 \end{cases}$ **16.** $\begin{cases} 4x - 5y \leq 0 \\ 2x - y \geq 2 \end{cases}$

17. $\begin{cases} 2x - 3y \leq 0 \\ 3x + 2y \leq 6 \end{cases}$ **18.** $\begin{cases} 4x - y \geq 2 \\ x + 2y \geq 2 \end{cases}$ **19.** $\begin{cases} x^2 + y^2 \leq 9 \\ x + y \geq 3 \end{cases}$ **20.** $\begin{cases} x^2 + y^2 \geq 9 \\ x + y \leq 3 \end{cases}$

21. $\begin{cases} y \geq x^2 - 4 \\ y \leq x - 2 \end{cases}$ **22.** $\begin{cases} y^2 \leq x \\ y \geq x \end{cases}$ **23.** $\begin{cases} xy \geq 4 \\ y \geq x^2 + 1 \end{cases}$ **24.** $\begin{cases} y + x^2 \leq 1 \\ y \geq x^2 - 1 \end{cases}$

25. $\begin{cases} x - 2y \leq 6 \\ 2x - 4y \geq 0 \end{cases}$ **26.** $\begin{cases} x + 4y \leq 8 \\ x + 4y \geq 4 \end{cases}$ **27.** $\begin{cases} 2x + y \geq -2 \\ 2x + y \geq 2 \end{cases}$ **28.** $\begin{cases} x - 4y \leq 4 \\ x - 4y \geq 0 \end{cases}$

29. $\begin{cases} 2x + 3y \geq 6 \\ 2x + 3y \leq 0 \end{cases}$ **30.** $\begin{cases} 2x + y \geq 0 \\ 2x + y \geq 2 \end{cases}$

In Problems 31–40, graph each system of linear inequalities. Tell whether the graph is bounded or unbounded, and label the corner points.

31. $\begin{cases} x \geq 0 \\ y \geq 0 \\ 2x + y \leq 6 \\ x + 2y \leq 6 \end{cases}$ **32.** $\begin{cases} x \geq 0 \\ y \geq 0 \\ x + y \geq 4 \\ 2x + 3y \geq 6 \end{cases}$ **33.** $\begin{cases} x \geq 0 \\ y \geq 0 \\ x + y \geq 2 \\ 2x + y \geq 4 \end{cases}$ **34.** $\begin{cases} x \geq 0 \\ y \geq 0 \\ 3x + y \leq 6 \\ 2x + y \leq 2 \end{cases}$ **35.** $\begin{cases} x \geq 0 \\ y \geq 0 \\ x + y \geq 2 \\ 2x + 3y \leq 12 \\ 3x + y \leq 12 \end{cases}$

36. $\begin{cases} x \geq 0 \\ y \geq 0 \\ x + y \geq 2 \\ x + y \leq 10 \\ 2x + y \leq 3 \end{cases}$ **37.** $\begin{cases} x \geq 0 \\ y \geq 0 \\ x + y \geq 2 \\ x + y \leq 8 \\ 2x + y \leq 10 \end{cases}$ **38.** $\begin{cases} x \geq 0 \\ y \geq 0 \\ x + y \geq 2 \\ x + y \leq 8 \\ x + 2y \geq 1 \end{cases}$ **39.** $\begin{cases} x \geq 0 \\ y \geq 0 \\ x + 2y \geq 1 \\ x + 2y \leq 10 \end{cases}$ **40.** $\begin{cases} x \geq 0 \\ y \geq 0 \\ x + 2y \geq 1 \\ x + 2y \leq 10 \\ x + y \geq 2 \\ x + y \leq 8 \end{cases}$

In Problems 41–44, write a system of linear inequalities that has the given graph.

41.

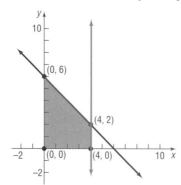

42.

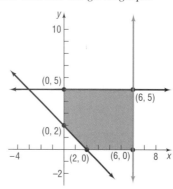

43.

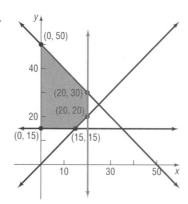

44.

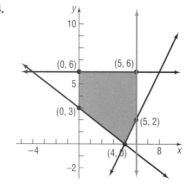

45. *Financial Planning* A retired couple has up to $50,000 to invest. As their financial adviser, you recommend they place at least $35,000 in Treasury bills yielding 7% and at most $10,000 in corporate bonds yielding 10%.
   (a) Using x to denote the amount of money invested in Treasury bills and y the amount invested in corporate bonds, write a system of linear inequalities that describes the possible amounts of each investment.
   (b) Graph the system and label the corner points.

46. *Manufacturing Trucks* Mike's Toy Truck Company manufactures two models of toy trucks, a standard model and a deluxe model. Each standard model requires 2 hours for painting and 3 hours for detail work; each deluxe model requires 3 hours for painting and 4 hours for detail work. Two painters and three detail workers are employed by the company, and each works 40 hours per week.
   (a) Using x to denote the number of standard model trucks and y to denote the number of deluxe model trucks, write a system of linear inequalities that describes the possible number of each model of truck that can be manufactured in a week.
   (b) Graph the system and label the corner points.

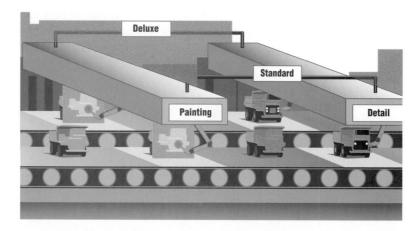

47. *Blending Coffee* A store that specializes in coffee has available 75 pounds of *A* grade coffee and 120 pounds of *B* grade coffee. These will be blended into 1 pound packages as follows: An economy blend that contains 4 ounces of *A* grade coffee and 12 ounces of *B* grade coffee and a superior blend that contains 8 ounces of *A* grade coffee and 8 ounces of *B* grade coffee.

(a) Using *x* to denote the number of packages of the economy blend and *y* to denote the number of packages of the superior blend, write a system of linear inequalities that describes the possible number of packages of each kind of blend.

(b) Graph the system and label the corner points.

48. *Mixed Nuts* A store that specializes in selling nuts has 90 pounds of cashews and 120 pounds of peanuts available. These are to be mixed in 12 ounce packages as follows: a lower priced package containing 8 ounces of peanuts and 4 ounces of cashews and a quality package containing 6 ounces of peanuts and 6 ounces of cashews.

(a) Use *x* to denote the number of lower priced packages and use *y* to denote the number of quality packages. Write a system of linear inequalities that describes the possible number of each kind of package.

(b) Graph the system and label the corner points.

49. *Transporting Goods* A small truck can carry no more than 1600 pounds of cargo or 150 cubic feet of cargo. A printer weighs 20 pounds and occupies 3 cubic feet of space. A microwave oven weighs 30 pounds and occupies 2 cubic feet of space.

(a) Using *x* to represent the number of microwave ovens and *y* to represent the number of printers, write a system of linear inequalities that describes the number of ovens and printers that can be hauled by the truck.

(b) Graph the system and label the corner points.

7.8

Linear Programming

Historically, linear programming evolved as a technique for solving problems involving resource allocation of goods and materials for the U.S. Air Force during World War II. Today, linear programming techniques are used to solve a wide variety of problems, such as optimizing airline scheduling and establishing telephone lines. Although most practical linear programming problems involve systems of several hundred linear inequalities containing several hundred variables, we shall limit our discussion to problems containing only two variables, because we can solve such problems using graphing techniques.*

We begin by returning to Example 12 of the previous section.

EXAMPLE 1   *Financial Planning*

A retired couple has up to $25,000 to invest. As their financial adviser, you recommend that they place at least $15,000 in Treasury bills yielding 6% and at most $5000 in corporate bonds yielding 9%. How much money should be placed in each investment so that income is maximized?   ∎

The problem given here is typical of a *linear programming problem.* The problem requires that a certain linear expression, the income, be maximized. If *I* represents income, *x* the amount invested in Treasury bills at 6%, and *y* the amount invested in corporate bonds at 9%, then

$$I = 0.06x + 0.09y$$

This linear expression is called the **objective function.** Furthermore, the problem requires that the maximum income be achieved under certain conditions or

---

*The **simplex method** is a way to solve linear programming problems involving many inequalities and variables. This method was developed by George Dantzig in 1946 and is particularly well suited for computerization. In 1984, Narendra Karmarkar of Bell Laboratories discovered a way of solving large linear programming problems that improves on the simplex method.

constraints, each of which is a linear inequality involving the variables. (See Example 12 in Section 7.7.) The linear programming problem given in Example 1 may be restated as

Maximize $I = 0.06x + 0.09y$

subject to the conditions that

$$x \geq 0, \qquad y \geq 0$$
$$x + y \leq 25{,}000$$
$$x \geq 15{,}000$$
$$y \leq 5000$$

In general, every linear programming problem has two components:

1. A linear objective function that is to be maximized or minimized.
2. A collection of linear inequalities that must be satisfied simultaneously.

Linear Programming Problem

> A **linear programming problem** in two variables x and y consists of maximizing (or minimizing) a linear objective function
>
> $$z = Ax + By, \qquad A \text{ and } B \text{ are real numbers, not both } 0$$
>
> subject to certain conditions, or constraints, expressible as linear inequalities in x and y.

To maximize (or minimize) the quantity $z = Ax + By$, we need to identify points (x, y) that make the expression for z the largest (or smallest) possible. But not all points (x, y) are eligible; only those that also satisfy each linear inequality (constraint) can be used. We refer to each point (x, y) that satisfies the system of linear inequalities (the constraints) as a **feasible point.** Thus, in a linear programming problem, we seek the feasible point(s) that maximizes (or minimizes) the objective function.
Let's look again at the linear programming problem in Example 1.

E X A M P L E   2   *Analyzing a Linear Programming Problem*

Consider the linear programming problem:

Maximize $I = 0.06x + 0.09y$

subject to the conditions that

$$x \geq 0, \qquad y \geq 0$$
$$x + y \leq 25{,}000$$
$$x \geq 15{,}000$$
$$y \leq 5000$$

Graph the constraints. Then graph the objective function for $I = 0, 0.9, 1.35, 1.65,$ and 1.8.

Solution   Figure 43 shows the graph of the constraints. We superimpose on this graph the graph of the objective function for the given values of I.

For $I = 0$, the objective function is the line $0 = 0.06x + 0.09y$.

For $I = 0.9$, the objective function is the line $0.9 = 0.06x + 0.09y$.

For $I = 1.35$, the objective function is the line $1.35 = 0.06x + 0.09y$.

For $I = 1.65$, the objective function is the line $1.65 = 0.06x + 0.09y$.

For $I = 1.8$, the objective function is the line $1.8 = 0.06x + 0.09y$.

FIGURE 43

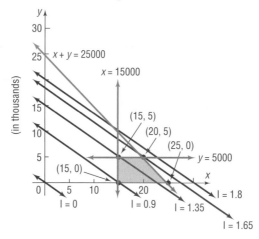

A **solution** to a linear programming problem consists of the feasible point(s) that maximizes (or minimizes) the objective function, together with the corresponding value of the objective function. One condition for a linear programming problem in two variables to have a solution is that the graph of the feasible points be bounded. (Refer to page 559.)

If none of the feasible points maximizes (or minimizes) the objective function or if there are no feasible points, then the linear programming problem has no solution.

Consider the linear programming problem stated in Example 2, and look again at Figure 43. The feasible points are the points that lie in the shaded region. For example, $(20, 3)$ is a feasible point, as is $(15, 5)$, $(20, 5)$, $(18, 4)$, etc. To find the solution of the problem requires that we find a feasible point (x, y) that makes $I = 0.06x + 0.09y$ as large as possible. Notice that as I increases in value from $I = 0$ to $I = 0.9$ to $I = 1.35$ to $I = 1.65$ to $I = 1.8$, we obtain a collection of parallel lines. Furthermore, notice that the largest value of I that can be obtained while feasible points are present is $I = 1.65$, which corresponds to the line $1.65 = 0.06x + 0.09y$. Any larger value of I results in a line that does not pass through any feasible points. Finally, notice that the feasible point that yields $I = 1.65$ is the point $(20, 5)$, a corner point. These observations form the basis of the following result, which we state without proof.

Theorem

Location of Solution
of a Linear Programming
Problem

If a linear programming problem has a solution, it is located at a corner point of the graph of the feasible points.

If a linear programming problem has multiple solutions, at least one of them is located at a corner point of the graph of the feasible points.

In either case, the corresponding value of the objective function is unique. ∎

We shall not consider here linear programming problems that have no solution. As a result, we can outline the procedure for solving a linear programming problem as follows:

Procedure for Solving a Linear Programming Problem	
STEP 1:	Write an expression for the quantity to be maximized (or minimized). This expression is the objective function.
STEP 2:	Write all the constraints as a system of linear inequalities and graph the system.
STEP 3:	List the corner points of the graph of the feasible points.
STEP 4:	List the corresponding values of the objective function at each corner point. The largest (or smallest) of these is the solution.

E X A M P L E 3

Solving a Minimum Linear Programming Problem

Minimize the expression

$$z = 2x + 3y$$

subject to the constraints

$$y \le 5, \quad x \le 6, \quad x + y \ge 2, \quad x \ge 0, \quad y \ge 0$$

FIGURE 44

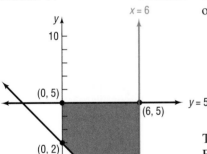

Solution The objective function is $z = 2x + 3y$. We seek the smallest value of z that can occur if x and y are solutions of the system of linear inequalities

$$\begin{cases} y \le 5 \\ x \le 6 \\ x + y \ge 2 \\ x \ge 0 \\ y \ge 0 \end{cases}$$

The graph of this system (the feasible points) is shown as the shaded region in Figure 44. We have also plotted the corner points. Table 2 lists the corner points and the corresponding values of the objective function. From the table, we can see that the minimum value of z is 4, and it occurs at the point $(2, 0)$.

TABLE 2

CORNER POINT	VALUE OF THE OBJECTIVE FUNCTION
(x, y)	$z = 2x + 3y$
$(0, 2)$	$z = 2(0) + 3(2) = 6$
$(0, 5)$	$z = 2(0) + 3(5) = 15$
$(6, 5)$	$z = 2(6) + 3(5) = 27$
$(6, 0)$	$z = 2(6) + 3(0) = 12$
$(2, 0)$	$z = 2(2) + 3(0) = 4$

■ Now work Problems 3 and 9.

E X A M P L E 4

Maximizing Profit

At the end of every month, after filling orders for its regular customers, a coffee company has some pure Colombian coffee and some special-blend coffee remaining. The practice of the company has been to package a mixture of the two coffees into 1 pound packages as follows: a low-grade mixture containing 4 ounces of Colombian coffee and 12 ounces of special-blend coffee and a high-grade mixture containing 8 ounces of Colombian and 8 ounces of special-blend coffee. A profit of $0.30 per package is made on the low-grade mixture, whereas a profit of $0.40

per package is made on the high-grade mixture. This month, 120 pounds of special-blend coffee and 100 pounds of pure Colombian coffee remain. How many packages of each mixture should be prepared to achieve a maximum profit? Assume that all packages prepared can be sold.

Solution   We begin by assigning symbols for the two variables:

$$x = \text{Number of packages of the low-grade mixture}$$
$$y = \text{Number of packages of the high-grade mixture}$$

If P denotes the profit, then

$$P = \$0.30x + \$0.40y$$

This expression is the objective function. We seek to maximize P subject to certain constraints on x and y. Because x and y represent numbers of packages, the only meaningful values for x and y are nonnegative integers. Thus, we have the two constraints

$$x \geq 0, \qquad y \geq 0 \quad \text{Nonnegative constraints}$$

We also have only so much of each type of coffee available. For example, the total amount of Colombian coffee used in the two mixtures cannot exceed 100 pounds, or 1600 ounces. Because we use 4 ounces in each low-grade package and 8 ounces in each high-grade package, we are led to the constraint

$$4x + 8y \leq 1600 \quad \text{Colombian coffee constraint}$$

Similarly, the supply of 120 pounds, or 1920 ounces, of special-blend coffee leads to the constraint

$$12x + 8y \leq 1920 \quad \text{Special-blend coffee constraint}$$

The linear programming problem may be stated as

$$\text{Maximize} \quad P = 0.3x + 0.4y$$

subject to the constraints

$$x \geq 0, \quad y \geq 0, \quad 4x + 8y \leq 1600, \quad 12x + 8y \leq 1920$$

The graph of the constraints (the feasible points) is illustrated in Figure 45. We list the corner points and evaluate the objective function at each one. In Table 3,

FIGURE 45

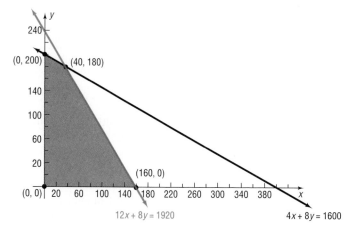

TABLE 3

CORNER POINT	VALUE OF PROFIT
(x, y)	$P = 0.3x + 0.4y$
$(0, 0)$	$P = 0$
$(0, 200)$	$P = 0.3(0) + 0.4(200) = \$80$
$(40, 180)$	$P = 0.3(40) + 0.4(180) = \$84$
$(160, 0)$	$P = 0.3(160) + 0.4(0) = \$48$

we can see that the maximum profit, $84, is achieved with 40 packages of the low-grade mixture and 180 packages of the high-grade mixture. ∎

■ Now work Problem 17.

7.8

Exercise 7.8

In Problems 1–6, find the maximum and minimum value of the given objective function.

The figure illustrates the graph of the feasible points of a linear programming problem.

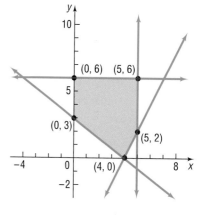

1. $z = x + y$ 2. $z = 2x + 3y$ 3. $z = x + 10y$

4. $z = 10x + y$ 5. $z = 5x + 7y$ 6. $z = 7x + 5y$

In Problems 7–16, solve each linear programming problem.

7. Maximize $z = 2x + y$ subject to $x \geq 0$, $y \geq 0$, $x + y \leq 6$, $x + y \geq 1$

8. Maximize $z = x + 3y$ subject to $x \geq 0$, $y \geq 0$, $x + y \geq 3$, $x \leq 5$, $y \leq 7$

9. Minimize $z = 2x + 5y$ subject to $x \geq 0$, $y \geq 0$, $x + y \geq 2$, $x \leq 5$, $y \leq 3$

10. Minimize $z = 3x + 4y$ subject to $x \geq 0$, $y \geq 0$, $2x + 3y \geq 6$, $x + y \leq 8$

11. Maximize $z = 3x + 5y$ subject to $x \geq 0$, $y \geq 0$, $x + y \geq 2$, $2x + 3y \leq 12$, $3x + 2y \leq 12$

12. Maximize $z = 5x + 3y$ subject to $x \geq 0$, $y \geq 0$, $x + y \geq 2$, $x + y \leq 8$, $2x + y \leq 10$

13. Minimize $z = 5x + 4y$ subject to $x \geq 0$, $y \geq 0$, $x + y \geq 2$, $2x + 3y \leq 12$, $3x + y \leq 12$

14. Minimize $z = 2x + 3y$ subject to $x \geq 0$, $y \geq 0$, $x + y \geq 3$, $x + y \leq 9$, $x + 3y \geq 6$

15. Maximize $z = 5x + 2y$ subject to $x \geq 0$, $y \geq 0$, $x + y \leq 10$, $2x + y \geq 10$, $x + 2y \geq 10$

16. Maximize $z = 2x + 4y$ subject to $x \geq 0$, $y \geq 0$, $2x + y \geq 4$, $x + y \leq 9$

17. *Maximizing Profit* A manufacturer of skis produces two types: downhill and cross-country. Use the following table to determine how many of each kind of ski should be produced to achieve a maximum profit. What is the maximum profit? What would the maximum profit be if the maximum time available for manufacturing is increased to 48 hours?

	DOWNHILL	CROSS-COUNTRY	MAXIMUM TIME AVAILABLE
Manufacturing time per ski	2 hours	1 hour	40 hours
Finishing time per ski	1 hour	1 hour	32 hours
Profit per ski	$70	$50	

18. *Farm Management* A farmer has 70 acres of land available for planting either soybeans or wheat. The cost of preparing the soil, the workdays required, and the expected profit per acre planted for each type of crop are given in the following table:

	SOYBEANS	WHEAT
Preparation cost per acre	$60	$30
Workdays required per acre	3	4
Profit per acre	$180	$100

The farmer cannot spend more than $1800 in preparation costs nor more than a total of 120 workdays. How many acres of each crop should be planted in order to maximize the profit? What is the maximum profit? What is the maximum profit if the farmer is willing to spend no more than $2400 on preparation?

19. *Farm Management* A small farm in Illinois has 100 acres of land available on which to grow corn and soybeans. The following table shows the cultivation cost per acre, the labor cost per acre, and the expected profit per acre. The column on the right shows the amount of money available for each of these expenses. Find the number of acres of each crop that should be planted in order to maximize profit.

	SOYBEANS	CORN	MONEY AVAILABLE
Cultivation cost per acre	$40	$60	$1800
Labor cost per acre	$60	$60	$2400
Profit per acre	$200	$250	

20. *Dietary Requirements* A certain diet requires at least 60 units of carbohydrates, 45 units of protein, and 30 units of fat each day. Each ounce of Supplement A provides 5 units of carbohydrates, 3 units of protein, and 4 units of fat. Each ounce of Supplement B provides 2 units of carbohydrates, 2 units of protein, and 1 unit of fat. If Supplement A costs $1.50 per ounce and Supplement B costs $1.00 per ounce, how many ounces of each supplement should be taken daily to minimize the cost of the diet?

21. *Production Scheduling* In a factory, machine 1 produces 8″ plyers at the rate of 60 units per hour, and 6″ plyers at the rate of 70 units per hour. Machine 2 produces 8″ plyers at the rate of 40 units per hour and 6″ plyers at the rate of 20 units per hour. It costs $50 per hour to operate machine 1, while machine 2 costs $30 per hour to operate. The production schedule requires that at least 240 units of 8″ plyers and at least 140 units of 6″ plyers must be produced during each 10 hour day. Which combination of machines will cost the least money to operate?

22. *Farm Management* An owner of a fruit orchard hires a crew of workers to prune at least 25 of his 50 fruit trees. Each newer tree requires one hour to prune, while each older tree needs one-and-a-half hours. The crew contracts to work for at least 30 hours and charges $15 for each newer tree and $20 for each older tree. To minimize his cost, how many of each kind of tree will the orchard owner have pruned? What will be the cost?

23. *Managing a Meat Market* A meat market combines ground beef and ground pork in a single package for meat loaf. The ground beef is 75% lean (75% beef, 25% fat) and costs the market $0.75 per pound. The ground pork is 60% lean and costs the market $0.45 per pound. The meat loaf must be at least 70% lean. If the market wants to use at least 50 lbs of its available pork, but no more than 200 lbs of its available groundbeef, how much ground beef should be mixed with ground pork so that the cost is minimized?

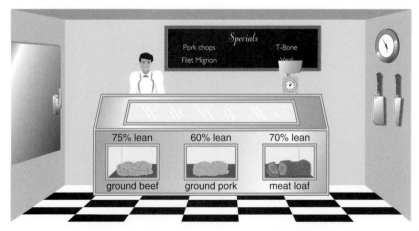

24. *Return on Investment* An investment broker is instructed by her client to invest up to $20,000, some in a junk bond yielding 9% per annum and some in Treasury bills yielding 7% per annum. The client wants to invest at least $8000 in T bills and no more than $12,000 in the junk bond.
   (a) How much should the broker recommend that the client place in each investment to maximize income if the client insists that the amount invested in T-bills must equal or exceed the amount placed in junk bonds?
   (b) How much should the broker recommend the client place in each investment to maximize income if the client insists that the amount invested in T-bills must not exceed the amount placed in junk bonds?

25. *Maximizing Profit on Ice Skates* A factory manufactures two kinds of ice skates: racing skates and figure skates. The racing skates require 6 work-hours in the fabrication department, whereas the figure skates require 4 work-hours there. The racing skates require 1 work-hour in the finishing department, whereas the figure skates require 2 work-hours there. The fabricating department has available at most 120 work-hours per day, and the finishing department has no more than 40 work-hours per day available. If the profit on each racing skate is $10 and the profit on each figure skate is $12, how many of each should be manufactured each day to maximize profit? (Assume that all skates made are sold.)

26 *Financial Planning* A retired couple has up to $50,000 to place in fixed-income securities. Their financial adviser suggests two securities to them: one is an AAA bond that yields 8% per annum; the other is a Certificate of Deposit (CD) that yields 4%. After careful consideration of the alternatives, the couple decides to place at most $20,000 in the AAA bond and at least $15,000 in the CD. They also instruct the financial adviser to place at least as much in the CD as in the AAA bond. How should the financial adviser proceed to maximize the return on their investment?

27. *Product Design* An entrepreneur is having a design group produce at least six samples of a new kind of fastener that he wants to market. It costs $9.00 to produce each metal fastener and $4.00 to produce each plastic fastener. He wants to have at least two of each version of the fastener, and needs to have all the samples 24 hours from now. It takes 4 hours to produce each metal sample and 2 hours to produce each plastic sample. To minimize the cost of the samples, how many of each kind should the entrepreneur order? What will be the cost of the samples?

28. *Animal Nutrition* Ted's dog Amadeus likes two kinds of canned dog food. "Gourmet Dog" costs 40 cents a can and has 20 units of a vitamin complex; the calorie content is 75 calories. "Chow Hound" costs 32 cents a can and has 35 units of vitamins and 50 calories. Ted likes Amadeus to have at least 1175 units of vitamins a month and at least 2375 calories during the same time period. Ted has space to store only 60 cans of dog food at a time. How much of each kind of dog food should Ted buy each month in order to minimize his cost?

29. *Airline Revenue* An airline has two classes of service: first class and coach. Management's experience has been that each aircraft should have at least 8 but not more than 16 first-class seats and at least 80 but not more than 120 coach seats.
   (a) If management decides that the ratio of first class to coach seats should never exceed 1:12, with how many of each type seat should an aircraft be configured to maximize revenue?
   (b) If management decides that the ratio of first class to coach seats should never exceed 1:8, with how many of each type seat should an aircraft be configured to maximize revenue?
   (c) If you were management, what would you do?
   {*Hint:* Assume that the airline charges $C for a coach seat and $F for a first class seat; $C > 0, F > 0$.}

30. *Minimizing Cost* A farm that specializes in raising frying chickens supplements the regular chicken feed with four vitamins. The owner wants the supplemental food to contain at least 50 units of vitamin I, 90 units of vitamin II, 60 units of vitamin III, and 100 units of vitamin IV per 100 ounces of feed. Two supplements are available: supplement A, which contains 5 units of vitamin I, 25 units of vitamin II, 10 units of vitamin III, and 35 units of vitamin IV per ounce; and supplement B, which contains 25 units of vitamin I, 10 units of vitamin II, 10 units of vitamin III, and 20 units of vitamin IV per ounce. If supplement A costs $0.06 per ounce and supplement B costs $0.08 per ounce, how much of each supplement should the manager of the farm buy to add to each 100 ounces of feed in order to keep the total cost at a minimum, while still meeting the owner's vitamin specifications?

31. Explain in your own words what a linear programming problem is and how it can be solved.

Chapter Review

THINGS TO KNOW

Systems of equations

Systems with no solutions are inconsistent.   Systems with a solution are consistent.

Consistent systems have either a unique solution or an infinite number of solutions.

Matrix	Rectangular array of numbers, called entries
m by n matrix	Matrix with m rows and n columns
Identity matrix I	Square matrix whose diagonal entries are 1's, while all other entries are 0's
Inverse of a matrix	A^{-1} is the inverse of A if $AA^{-1} = A^{-1}A = I$
Nonsingular matrix	A matrix that has an inverse

Linear programming problem

Maximize (or minimize) a linear objective function, $z = Ax + By$, subject to certain conditions, or constraints, expressible as linear inequalities in x and y.

Feasible point

A point (x, y) that satisfies the constraints of a linear programming problem

Location of solution

If a linear programming problem has a solution, it is located at a corner point of the graph of the feasible points.

If a linear programming problem has multiple solutions, at least one of them is located at a corner point of the graph of the feasible points.

In either case, the corresponding value of the objective function is unique.

How To:

Solve a system of linear equations using the method of substitution

Solve a system of linear equations using the method of elimination

Solve a system of linear equations using matrices

Solve a system of linear equations using determinants

Recognize equal matrices

Add and subtract matrices

Multiply matrices

Find the inverse of a nonsingular matrix

Solve a system of equations using the inverse of a matrix

Write the partial fraction decomposition of a rational expression

Solve a system of nonlinear equations

Graph a system of inequalities

Find the corner points of the graph of a system of linear inequalities

Solve linear programming problems

Fill-in-the-Blank Items

1. If a system of equations has no solution, it is said to be _____.

2. An m by n rectangular array of numbers is called a(n) _____.

3. Cramer's Rules uses _____ to solve a system of linear equations.

4. The matrix used to represent a system of linear equations is called a(n) _____ matrix.

5. A matrix B, for which $AB = I_n$, the identity matrix, is called the _____ of A.

6. A matrix that has the same number of rows as columns is called a(n) _____ matrix.

7. In the algebra of matrices, the matrix that has the properties similar to the number 1 is called the _____ matrix.

8. A rational function is called _____ if the degree of its numerator is less than the degree of its denominator.

9. The graph of a linear inequality is called a(n) _____.

10. A linear programming problem requires that a linear expression, called the _____ _____, be maximized or minimized.

11. Each point that satisfies the constraints of a linear programming problem is called a(n) _____ _____.

TRUE/FALSE ITEMS

T F **1.** A system of two linear equations containing two unknowns always has at least one solution.

T F **2.** The augmented matrix of a system of two equations containing three variables has two rows and four columns.

T F **3.** A 3 by 3 determinant can never equal 0.

T F **4.** A consistent system of equations will have exactly one solution.

T F **5.** Every square matrix has an inverse.

T F **6.** Matrix multiplication is commutative.

T F **7.** Any pair of matrices can be multiplied.

T F **8.** The factors of the denominator of a rational expression are used to arrive at the partial fraction decomposition.

T F **9.** The graph of a linear inequality is a half-plane.

T F **10.** The graph of a system of linear inequalities is sometimes unbounded.

T F **11.** If a linear programming problem has a solution, it is located at a corner point of the graph of the feasible points.

REVIEW EXERCISES

In Problems 1–20, solve each system of equations using the method of substitution or the method of elimination. If the system has no solution, say it is inconsistent.

1. $\begin{cases} 2x - y = 5 \\ 5x + 2y = 8 \end{cases}$

2. $\begin{cases} 2x + 3y = 2 \\ 7x - y = 3 \end{cases}$

3. $\begin{cases} 3x - 4y = 4 \\ x - 3y = \frac{1}{2} \end{cases}$

4. $\begin{cases} 2x + y = 0 \\ 5x - 4y = -\frac{13}{2} \end{cases}$

5. $\begin{cases} x - 2y - 4 = 0 \\ 3x + 2y - 4 = 0 \end{cases}$

6. $\begin{cases} x - 3y + 5 = 0 \\ 2x + 3y - 5 = 0 \end{cases}$

7. $\begin{cases} y = 2x - 5 \\ x = 3y + 4 \end{cases}$

8. $\begin{cases} x = 5y + 2 \\ y = 5x + 2 \end{cases}$

9. $\begin{cases} x - y + 4 = 0 \\ \frac{1}{2}x + \frac{1}{6}y + \frac{2}{5} = 0 \end{cases}$

10. $\begin{cases} x + \frac{1}{4}y = 2 \\ y + 4x + 2 = 0 \end{cases}$

11. $\begin{cases} x - 2y - 8 = 0 \\ 2x + 2y - 10 = 0 \end{cases}$

12. $\begin{cases} x - 3y + \frac{7}{2} = 0 \\ \frac{1}{2}x + 3y - 5 = 0 \end{cases}$

13. $\begin{cases} y - 2x = 11 \\ 2y - 3x = 18 \end{cases}$

14. $\begin{cases} 3x - 4y - 12 = 0 \\ 5x + 2y + 6 = 0 \end{cases}$

15. $\begin{cases} 2x + 3y - 13 = 0 \\ 3x - 2y = 0 \end{cases}$

16. $\begin{cases} 4x + 5y = 21 \\ 5x + 6y = 42 \end{cases}$

17. $\begin{cases} 3x - 2y = 8 \\ x - \frac{2}{3}y = 12 \end{cases}$

18. $\begin{cases} 2x + 5y = 10 \\ 4x + 10y = 15 \end{cases}$

19. $\begin{cases} x + 2y - z = 6 \\ 2x - y + 3z = -13 \\ 3x - 2y + 3z = -16 \end{cases}$

20. $\begin{cases} x + 5y - z = 2 \\ 2x + y + z = 7 \\ x - y + 2z = 11 \end{cases}$

In Problems 21–28, use the following matrices to compute each expression.

$$A = \begin{bmatrix} 1 & 0 \\ 2 & 4 \\ -1 & 2 \end{bmatrix} \quad B = \begin{bmatrix} 4 & -3 & 0 \\ 1 & 1 & -2 \end{bmatrix} \quad C = \begin{bmatrix} 3 & -4 \\ 1 & 5 \\ 5 & -2 \end{bmatrix}$$

21. $A + C$

22. $A - C$

23. $6A$

24. $-4B$

25. AB

26. BA

27. CB

28. BC

In Problems 29–34, find the inverse of each matrix, if there is one. If there is not an inverse, say that the matrix is singular.

29. $\begin{bmatrix} 4 & 6 \\ 1 & 3 \end{bmatrix}$

30. $\begin{bmatrix} -3 & 2 \\ 1 & -2 \end{bmatrix}$

31. $\begin{bmatrix} 1 & 3 & 3 \\ 1 & 2 & 1 \\ 1 & -1 & 2 \end{bmatrix}$

32. $\begin{bmatrix} 3 & 1 & 2 \\ 3 & 2 & -1 \\ 1 & 1 & 1 \end{bmatrix}$

33. $\begin{bmatrix} 4 & -8 \\ -1 & 2 \end{bmatrix}$

34. $\begin{bmatrix} -3 & 1 \\ -6 & 2 \end{bmatrix}$

In Problems 35–44, solve each system of equations using matrices. If the system has no solution, say it is inconsistent.

35. $\begin{cases} 3x - 2y = 1 \\ 10x + 10y = 5 \end{cases}$

36. $\begin{cases} 3x + 2y = 6 \\ x - y = -\frac{1}{2} \end{cases}$

37. $\begin{cases} 5x + 6y - 3z = 6 \\ 4x - 7y - 2z = -3 \\ 3x + y - 7z = 1 \end{cases}$

38. $\begin{cases} 2x + y + z = 5 \\ 4x - y - 3z = 1 \\ 8x + y - z = 5 \end{cases}$

39. $\begin{cases} x - 2z = 1 \\ 2x + 3y = -3 \\ 4x - 3y - 4z = 3 \end{cases}$

40. $\begin{cases} x + 2y - z = 2 \\ 2x - 2y + z = -1 \\ 6x + 4y + 3z = 5 \end{cases}$

41. $\begin{cases} x - y + z = 0 \\ x - y - 5z - 6 = 0 \\ 2x - 2y + z - 1 = 0 \end{cases}$

42. $\begin{cases} 4x - 3y + 5z = 0 \\ 2x + 4y - 3z = 0 \\ 6x + 2y + z = 0 \end{cases}$

43. $\begin{cases} x - y - z - t = 1 \\ 2x + y + z + 2t = 3 \\ x - 2y - 2z - 3t = 0 \\ 3x - 4y + z + 5t = -3 \end{cases}$

44. $\begin{cases} x - 3y + 3z - t = 4 \\ x + 2y - z = -3 \\ x + 3z + 2t = 3 \\ x + y + 5z = 6 \end{cases}$

In Problems 45–50, find the value of each determinant.

45. $\begin{vmatrix} 3 & 4 \\ 1 & 3 \end{vmatrix}$

46. $\begin{vmatrix} -4 & 0 \\ 1 & 3 \end{vmatrix}$

47. $\begin{vmatrix} 1 & 4 & 0 \\ -1 & 2 & 6 \\ 4 & 1 & 3 \end{vmatrix}$

48. $\begin{vmatrix} 2 & 3 & 10 \\ 0 & 1 & 5 \\ -1 & 2 & 3 \end{vmatrix}$

49. $\begin{vmatrix} 2 & 1 & -3 \\ 5 & 0 & 1 \\ 2 & 6 & 0 \end{vmatrix}$

50. $\begin{vmatrix} -2 & 1 & 0 \\ 1 & 2 & 3 \\ -1 & 4 & 2 \end{vmatrix}$

In Problems 51–56, use Cramer's Rule, if applicable, to solve each system.

51. $\begin{cases} x - 2y = 4 \\ 3x + 2y = 4 \end{cases}$

52. $\begin{cases} x - 3y = -5 \\ 2x + 3y = 5 \end{cases}$

53. $\begin{cases} 2x + 3y - 13 = 0 \\ 3x - 2y = 0 \end{cases}$

54. $\begin{cases} 3x - 4y - 12 = 0 \\ 5x + 2y + 6 = 0 \end{cases}$

55. $\begin{cases} x + 2y - z = 6 \\ 2x - y + 3z = -13 \\ 3x - 2y + 3z = -16 \end{cases}$

56. $\begin{cases} x - y + z = 8 \\ 2x + 3y - z = -2 \\ 3x - y - 9z = 9 \end{cases}$

In Problems 57–66, write the partial fraction decomposition of each rational expression.

57. $\dfrac{6}{x(x - 4)}$

58. $\dfrac{x}{(x + 2)(x - 3)}$

59. $\dfrac{x - 4}{x^2(x - 1)}$

60. $\dfrac{2x - 6}{(x - 2)^2(x - 1)}$

61. $\dfrac{x}{(x^2 + 9)(x + 1)}$

62. $\dfrac{3x}{(x - 2)(x^2 + 1)}$

63. $\dfrac{x^3}{(x^2 + 4)^2}$

64. $\dfrac{x^3 + 1}{(x^2 + 16)^2}$

65. $\dfrac{x^2}{(x^2 + 1)(x^2 - 1)}$

66. $\dfrac{4}{(x^2 + 4)(x^2 - 1)}$

In Problems 67–76, solve each system of equations.

67. $\begin{cases} 2x + y + 3 = 0 \\ \quad x^2 + y^2 = 5 \end{cases}$ **68.** $\begin{cases} x^2 + y^2 = 16 \\ 2x - y^2 = -8 \end{cases}$ **69.** $\begin{cases} 2xy + y^2 = 10 \\ 3y^2 - xy = \;\; 2 \end{cases}$ **70.** $\begin{cases} 3x^2 - y^2 = 1 \\ 7x^2 - 2y^2 - 5 = 0 \end{cases}$

71. $\begin{cases} x^2 + y^2 = 64 \\ \quad\quad x^2 = 3y \end{cases}$ **72.** $\begin{cases} 2x^2 + y^2 = 9 \\ \;\; x^2 + y^2 = 9 \end{cases}$ **73.** $\begin{cases} 3x^2 + 4xy + 5y^2 = 8 \\ \;\; x^2 + 3xy + 2y^2 = 0 \end{cases}$ **74.** $\begin{cases} 3x^2 + 2xy - 2y^2 = 6 \\ \quad xy - 2y^2 + 4 = 0 \end{cases}$

75. $\begin{cases} x^2 - 3x + y^2 + y = -2 \\ \dfrac{x^2 - x}{y} + y + 1 = \;\; 0 \end{cases}$ **76.** $\begin{cases} x^2 + x + y^2 = y + 2 \\ \quad x + 1 = \dfrac{2 - y}{x} \end{cases}$

In Problems 77–82, graph each system of inequalities. Tell whether the graph is bounded or unbounded, and label the corner points.

77. $\begin{cases} -2x + y \le 2 \\ \;\; x + y \ge 2 \end{cases}$ **78.** $\begin{cases} x - 2y \le 6 \\ 2x + \;\; y \ge 2 \end{cases}$ **79.** $\begin{cases} x \ge 0 \\ y \ge 0 \\ x + \;\; y \le 4 \\ 2x + 3y \le 6 \end{cases}$ **80.** $\begin{cases} x \ge 0 \\ y \ge 0 \\ 3x + y \ge 6 \\ 2x + y \ge 2 \end{cases}$

81. $\begin{cases} x \ge 0 \\ y \ge 0 \\ 2x + \;\; y \le 8 \\ \;\; x + 2y \ge 2 \end{cases}$ **82.** $\begin{cases} x \ge 0 \\ y \ge 0 \\ 3x + \;\; y \le 9 \\ 2x + 3y \ge 6 \end{cases}$

In Problems 83–86, graph each system of inequalities.

83. $\begin{cases} x^2 + y^2 \le 16 \\ \quad x + y \ge 2 \end{cases}$ **84.** $\begin{cases} y^2 \le x - 1 \\ x - y \le 3 \end{cases}$ **85.** $\begin{cases} y \le x^2 \\ xy \le 4 \end{cases}$ **86.** $\begin{cases} x^2 + y^2 \ge 1 \\ x^2 + y^2 \le 4 \end{cases}$

In Problems 87–92, solve each linear programming problem.

87. Maximize $z = 3x + 4y$ subject to $x \ge 0$, $y \ge 0$, $3x + 2y \ge 6$, $x + y \le 8$

88. Maximize $z = 2x + 4y$ subject to $x \ge 0$, $y \ge 0$, $x + y \le 6$, $x \ge 2$

89. Minimize $z = 3x + 5y$ subject to $x \ge 0$, $y \ge 0$, $x + y \ge 1$, $3x + 2y \le 12$, $x + 3y \le 12$

90. Minimize $z = 3x + y$ subject to $x \ge 0$, $y \ge 0$, $x \le 8$, $y \le 6$, $2x + y \ge 4$

91. Maximize $z = 5x + 4y$ subject to $x \ge 0$, $y \ge 0$, $x + 2y \ge 2$, $3x + 4y \le 12$, $y \ge x$

92. Maximize $z = 4x + 5y$ subject to $x \ge 0$, $y \ge 0$, $2x + 3y \ge 6$, $x \ge y$, $2x + y \le 12$

93. Find A such that the system of equations has infinitely many solutions.

$$\begin{cases} 2x + \;\; 5y = 5 \\ 4x + 10y = A \end{cases}$$

94. Find A such that the system in Problem 93 is inconsistent.

95. *Curve Fitting* Find the quadratic function $y = ax^2 + bx + c$ that passes through the three points $(0, 1)$, $(1, 0)$, and $(-2, 1)$.

96. *Curve Fitting* Find the general equation of the circle that passes through the three points $(0, 1)$, $(1, 0)$, and $(-2, 1)$.
[*Hint:* The general equation of a circle is $x^2 + y^2 + Dx + Ey + F = 0$.]

97. *Blending Coffee* A coffee distributor is blending a new coffee that will cost $3.90 per pound. It will consist of a blend of $3.00 per pound coffee and $6.00 per pound coffee. What amounts of each type of coffee should be mixed to achieve the desired blend? [*Hint:* Assume that the weight of the blended coffee is 100 pounds.]

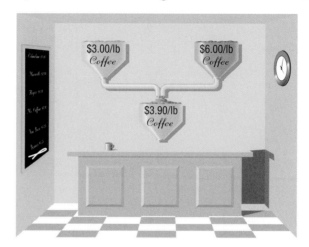

98. *Farming* A 1000 acre farm in Illinois is used to raise corn and soy beans. The cost per acre for raising corn is $65 and the cost per acre for soy beans is $45. If $54,325 has been budgeted for costs and all the acreage is to be used, how many acres should be allocated for each crop?

99. *Cookie Orders* A cookie company makes three kinds of cookies, oatmeal raisin, chocolate chip, and shortbread, packaged in small, medium, and large boxes. The small box contains 1 dozen oatmeal raisin and 1 dozen chocolate chip; the medium box has 2 dozen oatmeal raisin, 1 dozen chocolate chip, and 1 dozen shortbread; the large box contains 2 dozen oatmeal raisin, 2 dozen chocolate chip, and 3 dozen shortbread. If you require exactly 15 dozen oatmeal raisin, 10 dozen chocolate chip, and 11 dozen shortbread, how many of each size box should you buy?

100. *Mixed Nuts* A store that specializes in selling nuts has 72 pounds of cashews and 120 pounds of peanuts available. These are to be mixed in 12 ounce packages as follows: a lower-priced package containing 8 ounces of peanuts and 4 ounces of cashews and a quality package containing 6 ounces of peanuts and 6 ounces of cashews.
(a) Use x to denote the number of lower-priced packages and use y to denote the number of quality packages. Write a system of linear inequalities that describes the possible number of each kind of package.
(b) Graph the system and label the corner points.

101. A small rectangular lot has a perimeter of 68 feet. If its diagonal is 26 feet, what are the dimensions of the lot?

102. The area of a rectangular window is 4 square feet. If the diagonal measures $2\sqrt{2}$ feet, what are the dimensions of the window?

103. *Geometry* A certain right triangle has a perimeter of 14 inches. If the hypotenuse is 6 inches long, what are the lengths of the legs?

104. *Geometry* A certain isosceles triangle has a perimeter of 18 inches. If the altitude is 6 inches, what is the length of the base?

105. *Building a Fence* How much fence is required to enclose 5000 square feet by two squares whose sides are in the ratio of 1:2?

106. *Mixing Acids* A chemistry laboratory has three containers of hydrochloric acid, HCl. One container holds a solution with a concentration of 10% HCl, the second holds 25% HCl, and the third holds 40% HCl. How many liters of each should be mixed to obtain 100 liters of a solution with a concentration of 30% HCl? Construct a table showing some of the possible combinations.

107. *Calculating Allowances* Katy, Mike, Danny, and Colleen agreed to do yard work at home for $45 to be split among them. After they finished, their father determined that Mike deserves twice what Katy gets, Katy and Colleen deserve the same amount, and Danny deserves half of what Katy gets. How much does each one receive?

108. *Finding the Speed of the Jet Stream* On a flight between Midway Airport in Chicago and Ft. Lauderdale, Florida, a Boeing 737 jet maintains an airspeed of 475 miles per hour. If the trip from Chicago to Ft. Lauderdale takes 2 hours, 30 minutes and the return flight takes 2 hours, 50 minutes, what is the speed of the jet stream? (Assume the speed of the jet stream remains constant at the various altitudes of the plane.)

109. *Constant Rate Jobs* If Katy and Mike work together for 1 hour and 20 minutes, they will finish a certain job. If Mike and Danny work together for 1 hour and 36 minutes, the same job can be finished. If Danny and Katy work together, they can complete this job in 2 hours and 40 minutes. How long will it take each of them working alone to finish the job?

110. *Maximizing Profit on Figurines* A factory manufactures two kinds of ceramic figurines: a dancing girl and a mermaid, each requiring three processes—molding, painting, and glazing. The daily labor available for molding is no more than 90 work-hours, labor available for painting does not exceed 120 work-hours, and labor available for glazing is no more than 60 work-hours. The dancing girl requires 3 work-hours for molding, 6 work-hours for painting, and 2 work-hours for glazing. The mermaid requires 3 work-hours for molding, 4 work-hours for painting, and 3 work-hours for glazing. If the profit on each figurine is $25 for dancing girls and $30 for mermaids, how many of each should be produced each day to maximize profit? If management decides to produce the number of each figurine that maximizes profit, determine which of these processes has excess work-hours assigned to it.

111. *Minimizing Production Cost* A factory produces gasoline engines and diesel engines. Each week the factory is obligated to deliver at least 20 gasoline engines and at least 15 diesel engines. Due to physical limitations, however, the factory cannot make more than 60 gasoline engines nor more than 40 diesel engines. Finally, to prevent layoffs, a total of at least 50 engines must be produced. If gasoline engines cost $450 each to produce and diesel engines cost $550 each to produce, how many of each should be produced per week to minimize the cost? What is the excess capacity of the factory; that is, how many of each kind of engine are being produced in excess of the number the factory is obligated to deliver?

112. Describe four ways of solving a system of three linear equations containing three variables. Which method do you prefer? Why?

Chapter 8

SEQUENCES;
INDUCTION;
COUNTING;
PROBABILITY

8.1 Sequences
8.2 Arithmetic Sequences
8.3 Geometric Sequences;
     Geometric Series
8.4 Mathematical
     Induction
8.5 The Binomial Theorem
8.6 Sets and Counting
8.7 Permutations and
     Combinations
8.8 Probability
Chapter Review

Preview   Creating a Floor Design

A ceramic tile floor is designed in the shape of a trapezoid 20 feet wide at the base and 10 feet wide at the top. The tiles, 12 inches by 12 inches, are to be placed so that each successive row contains one less tile than the row below. How many tiles will be required? [Example 8 of Section 8.2] ■

This chapter introduces topics that are covered in more detail in courses titled *Finite Mathematics* or *Discrete Mathematics*. Applications of these topics can be found in the fields of computer science, engineering, business and economics, the social sciences, and the physical and biological sciences.

The chapter may be divided into four independent parts:

Sections 8.1–8.3, Sequences, which are functions whose domain is the set of natural numbers. Sequences form the basis for the *recursively defined* functions and *recursive procedures* used in computer programming.

Section 8.4, Mathematical Induction, a technique for proving theorems involving the natural numbers.

Section 8.5, the Binomial Theorem, a formula for the expansion of $(x + a)^n$, where n is any natural number.

Sections 8.6–8.8. The first two sections deal with techniques and formulas for counting the number of objects in a set, a part of the branch of mathematics called *combinatorics*. These formulas are used in computer science to analyze algorithms and recursive functions and to study stacks and queues. They are also used to determine *probabilities,* the likelihood that a certain outcome of a random experiment will occur.

8.1

Sequences

Sequence

A **sequence** is a function whose domain is the set of positive integers.

FIGURE 1

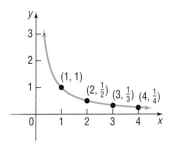

(a) $f(x) = \frac{1}{x}, x > 0$

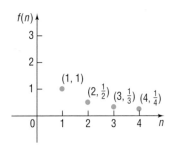

(b) $f(n) = \frac{1}{n}$

Because a sequence is a function, it will have a graph. In Figure 1(a), you will recognize the graph of the function $f(x) = 1/x$, $x > 0$. If all the points on this graph were removed except those whose x-coordinates are positive integers—that is, if all points were removed except $(1, 1)$, $(2, \frac{1}{2})$, $(3, \frac{1}{3})$ and so on—the remaining points would be the graph of the sequence $f(n) = 1/n$, as shown in Figure 1(b).

A sequence is usually represented by listing its values in order. For example, the sequence whose graph is given in Figure 1(b) might be represented as

$$f(1), f(2), f(3), f(4), \ldots \quad \text{or} \quad 1, \frac{1}{2}, \frac{1}{3}, \frac{1}{4}, \ldots$$

The list never ends, as the ellipsis dots indicate. The numbers in this ordered list are called the **terms** of the sequence.

In dealing with sequences, we usually use subscripted letters, for example, a_1 to represent the first term, a_2 for the second term, a_3 for the third term, and so on. Thus, for the sequence $f(n) = 1/n$, we write

$$a_1 = f(1) = 1$$

$$a_2 = f(2) = \frac{1}{2}$$

$$a_3 = f(3) = \frac{1}{3}$$

$$a_4 = f(4) = \frac{1}{4}$$

$$\vdots$$

$$a_n = f(n) = \frac{1}{n}$$

$$\vdots$$

In other words, we usually do not use the traditional function notation $f(n)$ for sequences. For this particular sequence, we have a rule for the nth term, namely, $a_n = 1/n$, so it is easy to find any term of the sequence.

When a formula for the nth term of a sequence is known, rather than write out the terms of the sequence, we usually represent the entire sequence by placing braces around the formula for the nth term. For example, the sequence whose nth term is $b_n = (\frac{1}{2})^n$ may be represented as

$$\{b_n\} = \left\{ \left(\frac{1}{2}\right)^n \right\}$$

or by

$$b_1 = \frac{1}{2}$$

$$b_2 = \frac{1}{4}$$

$$b_3 = \frac{1}{8}$$

$$\vdots$$

$$b_n = \left(\frac{1}{2}\right)^n$$

$$\vdots$$

E X A M P L E   1   *Writing the First Several Terms of a Sequence*

Write down the first six terms of the following sequence and graph it.

$$\{a_n\} = \left\{ \frac{n-1}{n} \right\}$$

Solution

$$a_1 = 0$$

$$a_2 = \frac{1}{2}$$

$$a_3 = \frac{2}{3}$$

$$a_4 = \frac{3}{4}$$

$$a_5 = \frac{4}{5}$$

$$a_6 = \frac{5}{6}$$

FIGURE 2

See Figure 2.   ■

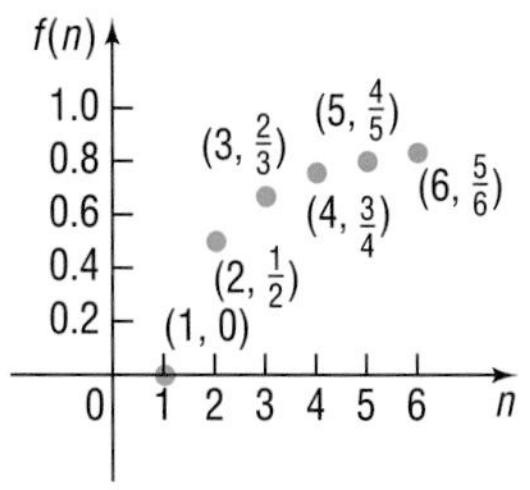

Graphing utilities can be used to write the terms of a sequence and graph them as the following example illustrates.

E X A M P L E   2   *Using a Graphing Utility to Write the First Several Terms of a Sequence*

Use a graphing utility to write the first six terms of the following sequence and graph it.

$$\{a_n\} = \left\{\frac{n-1}{n}\right\}$$

Solution   Figure 3 shows the sequence generated on a TI-82 graphing calculator. We can see the first few terms of the sequence on the viewing window. You need to press the right arrow key to scroll right in order to see the remaining terms of the sequence.

FIGURE 3

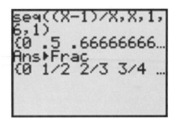

Figure 4 shows a graph of the sequence. Notice the first term of the sequence is not visible since it lies on the *x*-axis.

FIGURE 4

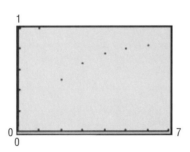

We will usually provide solutions done by hand. The reader is encouraged to verify solutions using a graphing utility. ■

E X A M P L E   3   *Writing the First Several Terms of a Sequence*

Write down the first six terms of the following sequence and graph it.

$$\{b_n\} = \left\{(-1)^{n-1}\left(\frac{2}{n}\right)\right\}$$

FIGURE 5

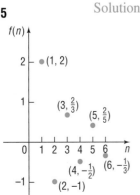

Solution

$$b_1 = 2$$

$$b_2 = -1$$

$$b_3 = \frac{2}{3}$$

$$b_4 = -\frac{1}{2}$$

$$b_5 = \frac{2}{5}$$

$$b_6 = -\frac{1}{3}$$

See Figure 5.   ■

E X A M P L E   4   *Writing the First Several Terms of a Sequence*

Write down the first six terms of the following sequence and graph it.

$$\{c_n\} = \begin{cases} n & \text{if } n \text{ is even} \\ 1/n & \text{if } n \text{ is odd} \end{cases}$$

Solution

FIGURE 6

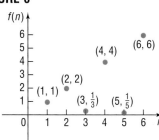

See Figure 6.

$c_1 = 1$

$c_2 = 2$

$c_3 = \dfrac{1}{3}$

$c_4 = 4$

$c_5 = \dfrac{1}{5}$

$c_6 = 6$

■ Now work Problems 3 and 5.

Sometimes a sequence is indicated by an observed pattern in the first few terms that makes it possible to infer the makeup of the nth term. In the example that follows, a sufficient number of terms of the sequence is given so that a natural choice for the nth term is suggested.

E X A M P L E   5   *Determining a Sequence from a Pattern*

(a) $e, \dfrac{e^2}{2}, \dfrac{e^3}{3}, \dfrac{e^4}{4}, \ldots$ $a_n = \dfrac{e^n}{n}$

(b) $1, \dfrac{1}{3}, \dfrac{1}{9}, \dfrac{1}{27}, \ldots$ $b_n = \dfrac{1}{3^{n-1}}$

(c) $1, 3, 5, 7, \ldots$ $c_n = 2n - 1$

(d) $1, 4, 9, 16, 25, \ldots$ $d_n = n^2$

(e) $1, -\dfrac{1}{2}, \dfrac{1}{3}, -\dfrac{1}{4}, \dfrac{1}{5}, \ldots$ $e_n = (-1)^{n+1}\left(\dfrac{1}{n}\right)$

■

Notice in the sequence $\{e_n\}$ in Example 5(e) that the signs of the terms **alternate**. When this occurs, we use factors such as $(-1)^{n+1}$, which equals 1 if n is odd and -1 if n is even, or $(-1)^n$, which equals -1 if n is odd and 1 if n is even.

■ Now work Problem 13.

The Factorial Symbol

Factorial Symbol, $n!$

If $n \geq 0$ is an integer, the **factorial symbol $n!$** is defined as follows:

$$0! = 1 \qquad 1! = 1$$

$$n! = n(n - 1) \cdot \ldots \cdot 3 \cdot 2 \cdot 1 \qquad \text{if } n \geq 2$$

For example, $2! = 2 \cdot 1 = 2$, $3! = 3 \cdot 2 \cdot 1 = 6$, $4! = 4 \cdot 3 \cdot 2 \cdot 1 = 24$, and so on. Table 1 lists the values of $n!$ for $0 \leq n \leq 6$.

TABLE 1

n	0	1	2	3	4	5	6
$n!$	1	1	2	6	24	120	720

Because

$$n! = \underbrace{n(n-1)(n-2) \cdot \ldots \cdot 3 \cdot 2 \cdot 1}_{(n-1)!}$$

we can use the formula

$$n! = n(n-1)!$$

to find successive factorials. For example, because $6! = 720$, we have

$$7! = 7 \cdot 6! = 7(720) = 5040$$

and

$$8! = 8 \cdot 7! = 8(5040) = 40{,}320$$

Comment: Your calculator may have a factorial key. Use it to see how fast factorials increase in value. Find the value of $69!$. What happens when you try to find $70!$? In fact, $70!$ is larger than 10^{100} (a *googol*), the largest number most calculators can display.

Recursion Formulas

A second way of defining a sequence is to assign a value to the first (or the first few) terms and specify the nth term by a formula or equation that involves one or more of the terms preceding it. Sequences defined this way are said to be defined **recursively,** and the rule or formula is called a **recursive formula.**

E X A M P L E   6   *Writing the Terms of a Recursively Defined Sequence*

Write down the first five terms of the following recursively defined sequence.

$$s_1 = 1, \qquad s_n = 4s_{n-1}$$

Solution   The first term is given as $s_1 = 1$. To get the second term, we use $n = 2$ in the formula to get $s_2 = 4s_1 = 4 \cdot 1 = 4$. To get the third term, we use $n = 3$ in the formula to get $s_3 = 4s_2 = 4 \cdot 4 = 16$. To get a new term requires that we know the value of the preceding term. The first five terms are

$$s_1 = 1$$
$$s_2 = 4 \cdot 1 = 4$$
$$s_3 = 4 \cdot 4 = 16$$
$$s_4 = 4 \cdot 16 = 64$$
$$s_5 = 4 \cdot 64 = 256 \qquad ■$$

Graphing utilities can be used to generate recursively defined sequences when the nth term depends upon the term preceding it.

E X A M P L E   7   *Using a Graphing Utility to Write the Terms of a Recursively Defined Sequence*

Use a graphing utility to write down the first five terms of the following recursively defined sequence

$$s_1 = 1, \qquad s_n = 4s_{n-1}$$

Solution   First, put the graphing utility into sequence mode and dot mode. Using $Y =$, enter the recursive formula into the graphing utility. Third, set up the viewing window to generate the desired sequence. Finally, graph the recursion relation and use TRACE to determine the terms in the sequence. For example, in Figure 7 we see the fourth term of the sequence is 64.

FIGURE 7

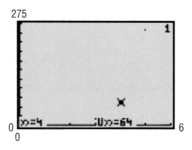

Note: If your graphing utility is capable of generating tables, the TABLE function displays the terms in the sequence. See Table 2.

TABLE 2

n	u_n	
1	1	
2	4	
3	16	
4	64	
5	256	
6	1024	
7	4096	

$U_n\boxminus 4U_{n-1}$

E X A M P L E   8   *Writing the Terms of a Recursively Defined Sequence*

Write down the first five terms of the following recursively defined sequence.

$$u_1 = 1, \qquad u_2 = 1, \qquad u_{n+2} = u_n + u_{n+1}$$

Solution   We are given the first two terms. To get the third term requires that we know each of the previous two terms. Thus,

$$u_1 = 1$$
$$u_2 = 1$$
$$u_3 = u_1 + u_2 = 2$$
$$u_4 = u_2 + u_3 = 1 + 2 = 3$$
$$u_5 = u_3 + u_4 = 2 + 3 = 5$$

The sequence defined in Example 8 is called a **Fibonacci sequence,** and the terms of this sequence are called **Fibonacci numbers.** These numbers appear in a wide variety of applications (see Problems 55 and 56).

Note: Since the terms in the sequence defined in Example 8 depend upon the previous two terms, a graphing calculator cannot be used to generate the terms of the sequence. However, a computer algebra system such as *Mathematica* could be used to generate the terms of Example 8.

E X A M P L E   9   *Writing the Terms of a Recursively Defined Sequence*

Write down the first five terms of the following recursively defined sequence.

$$f_1 = 1, \qquad f_{n+1} = (n + 1)f_n$$

Solution   Here

$$f_1 = 1$$
$$f_2 = 2f_1 = 2 \cdot 1 = 2$$
$$f_3 = 3f_2 = 3 \cdot 2 = 6$$
$$f_4 = 4f_3 = 4 \cdot 6 = 24$$
$$f_5 = 5f_4 = 5 \cdot 24 = 120$$ ■

You should recognize the nth term of the sequence in Example 9 as $n!$.

■   Now work Problems 21 and 29.

Adding the First n Terms of a Sequence; Summation Notation

It is often important to be able to find the sum of the first n terms of a sequence $\{a_n\}$, namely,

$$a_1 + a_2 + a_3 + \cdot \cdot \cdot + a_n \tag{1}$$

Rather than write down all these terms, we introduce a more concise way to express the sum, called **summation notation.** Using summation notation, we would write the sum (1) as

$$a_1 + a_2 + a_3 + \cdot \cdot \cdot + a_n = \sum_{k=1}^{n} a_k$$

The symbol $\sum$ (a stylized version of the Greek letter sigma, which is an S in our alphabet) is simply an instruction to sum, or add up, the terms. The integer k is called the **index** of the sum; it tells you where to start the sum and where to end it. Therefore, the expression

$$\sum_{k=1}^{n} a_k \tag{2}$$

is an instruction to add the terms a_k of the sequence $\{a_n\}$ from $k = 1$ through $k = n$. We read expression (2) as "the sum of a_k from $k = 1$ to $k = n$."

E X A M P L E   1 0   *Expanding Summation Notation*

Write out each sum.

(a) $\displaystyle\sum_{k=1}^{n} \frac{1}{k}$ (b) $\displaystyle\sum_{k=1}^{n} k!$

Solution (a) $\displaystyle\sum_{k=1}^{n} \frac{1}{k} = \frac{1}{1} + \frac{1}{2} + \frac{1}{3} + \cdots + \frac{1}{n}$ (b) $\displaystyle\sum_{k=1}^{n} k! = 1! + 2! + \cdots + n!$ ∎

E X A M P L E   1 1 *Writing a Sum in Summation Notation*
Express each sum using summation notation.

(a) $1^2 + 2^2 + 3^2 + \cdots + n^2$ (b) $1 + \dfrac{1}{2} + \dfrac{1}{4} + \dfrac{1}{8} + \cdots + \dfrac{1}{2^{n-1}}$

Solution (a) The sum $1^2 + 2^2 + 3^2 + \cdots + n^2$ has n terms, each of the form k^2, and starts at $k = 1$ and ends at $k = n$. Thus,

$$1^2 + 2^2 + 3^2 + \cdots + n^2 = \sum_{k=1}^{n} k^2$$

(b) The sum

$$1 + \frac{1}{2} + \frac{1}{4} + \frac{1}{8} + \cdots + \frac{1}{2^{n-1}}$$

has n terms, each of the form $1/2^{k-1}$, and starts at $k = 1$ and ends at $k = n$. Thus,

$$1 + \frac{1}{2} + \frac{1}{4} + \frac{1}{8} + \cdots + \frac{1}{2^{n-1}} = \sum_{k=1}^{n} \frac{1}{2^{k-1}}$$ ∎

The index of summation need not always begin at 1 nor end at n; for example,

$$\sum_{k=0}^{n-1} \frac{1}{2^k} = 1 + \frac{1}{2} + \frac{1}{4} + \cdots + \frac{1}{2^{n-1}}$$

Letters other than k may be used as the index. For example,

$$\sum_{j=1}^{n} j! \quad \text{and} \quad \sum_{i=1}^{n} i!$$

each represent the same sum as the one given in Example 10(b).

■ Now work Problems 37 and 47.

Next, we list some properties of sequences using summation notation.

Theorem If $\{a_n\}$ and $\{b_n\}$ are two sequences and c is a real number, then
Properties of Sequences

1. $\displaystyle\sum_{k=1}^{n} ca_k = c \sum_{k=1}^{n} a_k$

2. $\displaystyle\sum_{k=1}^{n} (a_k + b_k) = \sum_{k=1}^{n} a_k + \sum_{k=1}^{n} b_k$

3. $\displaystyle\sum_{k=1}^{n} (a_k - b_k) = \sum_{k=1}^{n} a_k - \sum_{k=1}^{n} b_k$

4. $\displaystyle\sum_{k=1}^{n} a_k = \sum_{k=1}^{j} a_k + \sum_{k=j+1}^{n} a_k, \quad \text{when } 1 < j < n$ ∎

Although we shall not prove these properties, the proofs are based on properties of real numbers.

8.1

Exercise 8.1

In Problems 1–12, write down the first five terms of each sequence by hand. Verify your results using a graphing utility.

1. $\{n\}$

2. $\{n^2 + 1\}$

3. $\left\{\dfrac{n}{n+2}\right\}$

4. $\left\{\dfrac{2n+1}{2n}\right\}$

5. $\{(-1)^{n+1}n^2\}$

6. $\left\{(-1)^{n-1}\left(\dfrac{n}{2n-1}\right)\right\}$

7. $\left\{\dfrac{2^n}{3^n+1}\right\}$

8. $\left\{\left(\dfrac{4}{3}\right)^n\right\}$

9. $\left\{\dfrac{(-1)^n}{(n+1)(n+2)}\right\}$

10. $\left\{\dfrac{3^n}{n}\right\}$

11. $\left\{\dfrac{n}{e^n}\right\}$

12. $\left\{\dfrac{n^2}{2^n}\right\}$

In Problems 13–20, the given pattern continues. Write down the nth term of each sequence suggested by the pattern.

13. $\dfrac{1}{2}, \dfrac{2}{3}, \dfrac{3}{4}, \dfrac{4}{5}, \ldots$

14. $\dfrac{1}{1 \cdot 2}, \dfrac{1}{2 \cdot 3}, \dfrac{1}{3 \cdot 4}, \dfrac{1}{4 \cdot 5}, \ldots$

15. $1, \dfrac{1}{2}, \dfrac{1}{4}, \dfrac{1}{8}, \ldots$

16. $\dfrac{2}{3}, \dfrac{4}{9}, \dfrac{8}{27}, \dfrac{16}{81}, \ldots$

17. $1, -1, 1, -1, 1, -1, \ldots$

18. $1, \dfrac{1}{2}, 3, \dfrac{1}{4}, 5, \dfrac{1}{6}, 7, \dfrac{1}{8}, \ldots$

19. $1, -2, 3, -4, 5, -6, \ldots$

20. $2, -4, 6, -8, 10, \ldots$

In Problems 21–34, a sequence is defined recursively. Write the first five terms by hand. When possible, use a graphing utility to verify your results.

21. $a_1 = 2; \ a_{n+1} = 3 + a_n$

22. $a_1 = 3; \ a_{n+1} = 4 - a_n$

23. $a_1 = -2; \ a_{n+1} = n + a_n$

24. $a_1 = 1; \ a_{n+1} = n - a_n$

25. $a_1 = 5; \ a_{n+1} = 2a_n$

26. $a_1 = 2; \ a_{n+1} = -a_n$

27. $a_1 = 3; \ a_{n+1} = \dfrac{a_n}{n}$

28. $a_1 = -2; \ a_{n+1} = n + 3a_n$

29. $a_1 = 1; \ a_2 = 2; \ a_{n+2} = a_n a_{n+1}$

30. $a_1 = -1; \ a_2 = 1; \ a_{n+2} = a_{n+1} + na_n$

31. $a_1 = A; \ a_{n+1} = a_n + d$

32. $a_1 = A; \ a_{n+1} = ra_n, \ r \neq 0$

33. $a_1 = \sqrt{2}; \ a_{n+1} = \sqrt{2 + a_n}$

34. $a_1 = \sqrt{2}; \ a_{n+1} = \sqrt{a_n/2}$

In Problems 35–44, write out each sum.

35. $\displaystyle\sum_{k=1}^{n} (k + 2)$

36. $\displaystyle\sum_{k=1}^{n} (2k + 1)$

37. $\displaystyle\sum_{k=1}^{n} \dfrac{k^2}{2}$

38. $\displaystyle\sum_{k=1}^{n} (k + 1)^2$

39. $\displaystyle\sum_{k=0}^{n} \dfrac{1}{3^k}$

40. $\displaystyle\sum_{k=0}^{n} \left(\dfrac{3}{2}\right)^k$

41. $\displaystyle\sum_{k=0}^{n-1} \dfrac{1}{3^{k+1}}$

42. $\displaystyle\sum_{k=0}^{n-1} (2k + 1)$

43. $\displaystyle\sum_{k=2}^{n} (-1)^k \ln k$

44. $\displaystyle\sum_{k=3}^{n} (-1)^{k+1} 2^k$

In Problems 45–54, express each sum using summation notation.

45. $1 + 2 + 3 + \cdots + n$

46. $1^3 + 2^3 + 3^3 + \cdots + n^3$

47. $\dfrac{1}{2} + \dfrac{2}{3} + \dfrac{3}{4} + \cdots + \dfrac{n}{n+1}$

48. $1 + 3 + 5 + 7 + \cdots + (2n - 1)$

49. $1 - \dfrac{1}{3} + \dfrac{1}{9} - \dfrac{1}{27} + \cdots + (-1)^n \left(\dfrac{1}{3^n}\right)$

50. $\dfrac{2}{3} - \dfrac{4}{9} + \dfrac{8}{27} - \cdots + (-1)^{n+1}\left(\dfrac{2}{3}\right)^n$

51. $3 + \dfrac{3^2}{2} + \dfrac{3^3}{3} + \cdots + \dfrac{3^n}{n}$

52. $\dfrac{1}{e} + \dfrac{2}{e^2} + \dfrac{3}{e^3} + \cdots + \dfrac{n}{e^n}$

53. $a + (a + d) + (a + 2d) + \cdots + (a + nd)$

54. $a + ar + ar^2 + \cdots + ar^{n-1}$

55. *Growth of a Rabbit Colony* A colony of rabbits begins with one pair of mature rabbits, which will produce a pair of offspring (one male, one female) each month. Assume that all rabbits mature in 1 month and produce a pair of offspring (one male, one female) after 2 months. If no rabbits ever die, how many pairs of mature rabbits are there after 7 months? [*Hint:* A Fibonacci sequence models this colony. Do you see why?]

1 mature pair

1 mature pair

2 mature pairs

3 mature pairs

56. *Fibonacci Sequence* Let

$$u_n = \frac{(1 + \sqrt{5})^n - (1 - \sqrt{5})^n}{2^n \sqrt{5}}$$

define the *n*th term of a sequence.

(a) Show that $u_1 = 1$ and $u_2 = 1$.   (b) Show that $u_{n+2} = u_{n+1} + u_n$.

(c) Draw the conclusion that $\{u_n\}$ is a Fibonacci sequence.

In Problems 57 and 58, we use the fact that in some programming languages it is possible to have a function subroutine include a call to itself.

57. *Programming Exercise* Write a program that accepts integer as input and prints the number and its factorial. Use a recursively defined function; that is, use a function subroutine that calls itself.

58. *Programming Exercise* Write a program that accepts a positive integer N as input and outputs the *N*th Fibonacci number. Use a recursively defined subroutine.

59. Divide the triangular array below (called Pascal's triangle) using diagonal lines as shown. Find the sum of the numbers in each of these diagonal rows. Do you recognize this sequence?

60. Investigate various applications that lead to a Fibonacci sequence. Write an essay on these applications.

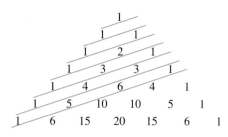

8.2

Arithmetic Sequences

When the difference between successive terms of a sequence is always the same number, the sequence is called **arithmetic.** Thus, an arithmetic sequence* may be defined recursively as $a_1 = a$, $a_{n+1} - a_n = d$, or as

Arithmetic Sequence	$$a_1 = a, \qquad a_{n+1} = a_n + d$$	(1)

where $a = a_1$ and d are real numbers. The number a is the first term, and the number d is called the **common difference.**

Thus, the terms of an arithmetic sequence with first term a and common difference d follow the pattern

$$a, \quad a + d, \quad a + 2d, \quad a + 3d, \quad \cdots$$

E X A M P L E 1   *Determining If a Sequence is Arithmetic*

The sequence

$$4, 7, 10, 13, \ldots$$

is arithmetic since the difference of successive terms is 3. The first term is 4, and the common difference is 3. ∎

E X A M P L E 2   *Determining If a Sequence is Arithmetic*

Show that the following sequence is arithmetic. Find the first term and the common difference.

$$\{s_n\} = \{3n + 5\}$$

Solution   The first term is $s_1 = 3 \cdot 1 + 5 = 8$. The $(n + 1)$st and nth terms of the sequence $\{s_n\}$ are

$$s_{n+1} = 3(n + 1) + 5 = 3n + 8 \quad \text{and} \quad s_n = 3n + 5$$

Their difference is

$$s_{n+1} - s_n = (3n + 8) - (3n + 5) = 8 - 5 = 3$$

Thus, the difference of two successive terms does not depend on n; the common difference is 3, and the sequence is arithmetic. ∎

E X A M P L E 3   *Determining If a Sequence is Arithmetic*

Show that the sequence is arithmetic. Find the first term and the common difference.

$$\{t_n\} = \{4 - n\}$$

Solution   The first term is $t_1 = 4 - 1 = 3$. The $(n + 1)$st and nth terms are

$$t_{n+1} = 4 - (n + 1) = 3 - n \quad \text{and} \quad t_n = 4 - n$$

*Sometimes called an **arithmetic progression.**

Their difference is

$$t_{n+1} - t_n = (3 - n) - (4 - n) = 3 - 4 = -1$$

The difference of two successive terms does not depend on n; it always equals the same number, -1. Hence, $\{t_n\}$ is an arithmetic sequence whose common difference is -1. ∎

■ Now work Problem 3.

Suppose that a is the first term of an arithmetic sequence whose common difference is d. We seek a formula for the nth term, a_n. To see the pattern, we write down the first few terms:

$$a_1 = a + 0 \cdot d$$
$$a_2 = a + d = a + 1 \cdot d$$
$$a_3 = a_2 + d = (a + d) + d = a + 2 \cdot d$$
$$a_4 = a_3 + d = (a + 2 \cdot d) + d = a + 3 \cdot d$$
$$a_5 = a_4 + d = (a + 3 \cdot d) + d = a + 4 \cdot d$$
$$\vdots$$
$$a_n = a_{n-1} + d = [a + (n-2)d] + d = a + (n-1)d$$

We are led to the following result:

Theorem For an arithmetic sequence $\{a_n\}$ whose first term is a and whose common difference is d, the nth term is determined by the formula

*n*th Term of an Arithmetic Sequence

$$a_n = a + (n-1)d \qquad (2)$$

E X A M P L E   4

Finding a Particular Term of an Arithmetic Sequence

Find the 13th term of the arithmetic sequence $2, 6, 10, 14, 18, \ldots$.

Solution The first term of this arithmetic sequence is $a = 2$, and the common difference is 4. By formula (2), the nth term is

$$a_n = 2 + (n-1)4$$

Hence, the 13th term is

$$a_{13} = 2 + 12 \cdot 4 = 50$$ ∎

Exploration: Use a graphing utility to find the 13th term of the sequence given in Example 4. Use it to find the 20th term and the 50th term. ∎

E X A M P L E   5

Finding a Recursive Formula for an Arithmetic Sequence

The 8th term of an arithmetic sequence is 75, and the 20th term is 39. Find the first term and the common difference. Give a recursive formula for the sequence.

Solution By equation (2), we know that

$$\begin{cases} a_8 = a + 7d = 75 \\ a_{20} = a + 19d = 39 \end{cases}$$

This is a system of two linear equations containing two variables, which we can solve by elimination. Thus, subtracting the second equation from the first equation, we get

$$-12d = 36$$
$$d = -3$$

With $d = -3$, we find $a = 75 - 7d = 75 - 7(-3) = 96$. A recursive formula for this sequence is

$$a_1 = 96, \qquad a_{n+1} = a_n - 3 \qquad \blacksquare$$

Based on formula (2), a formula for the nth term of the sequence $\{a_n\}$ in Example 5 is

$$a_n = a + (n - 1)d = 96 + (n - 1)(-3) = 99 - 3n$$

◼ Now work Problems 19 and 25.

Adding the First n Terms of an Arithmetic Sequence

The next result gives a formula for finding the sum of the first n terms of an arithmetic sequence.

Theorem   Let $\{a_n\}$ be an arithmetic sequence with first term a and common difference d. The sum S_n of the first n terms of $\{a_n\}$ is

---

Sum of n Terms of an Arithmetic Sequence

$$S_n = \frac{n}{2}[2a + (n - 1)d] = \frac{n}{2}(a + a_n) \qquad (3)$$

Proof

$$
\begin{aligned}
S_n &= a_1 + a_2 + a_3 + \cdots + a_n \\
&= a + (a + d) + (a + 2d) + \cdots + [a + (n - 1)d] \\
&= \underbrace{(a + a + \cdots + a)}_{n \text{ terms}} + [d + 2d + \cdots + (n - 1)d] \\
&= na + d[1 + 2 + \cdots + (n - 1)] \\
&= na + d\left[\frac{(n - 1)n}{2}\right] \\
&= na + \frac{n}{2}(n - 1)d \\
&= \frac{n}{2}[2a + (n - 1)d] \qquad \text{Factor out } n/2. \\
&= \frac{n}{2}[a + a + (n - 1)d] \\
&= \frac{n}{2}(a + a_n) \qquad \text{Formula (2)} \qquad \blacksquare
\end{aligned}
$$

Formula (3) provides two ways to find the sum of the first n terms of an arithmetic sequence. Notice that one involves the first term and common difference, while the other involves the first term and the nth term. Use whichever form is easier.

E X A M P L E   6   *Finding the Sum of n Terms of an Arithmetic Sequence*

Find the sum S_n of the first n terms of the sequence $\{3n + 5\}$; that is, find

$$8 + 11 + 14 + \cdots + (3n + 5)$$

Solution   The sequence $\{3n + 5\}$ is an arithmetic sequence with first term $a = 8$ and the nth term $(3n + 5)$. To find the sum S_n we use formula (3):

$$S_n = \frac{n}{2}(a + a_n) = \frac{n}{2}[8 + (3n + 5)] = \frac{n}{2}(3n + 13) \qquad \blacksquare$$

■   Now work Problem 33.

E X A M P L E   7   *Using a Graphing Utility to Find the Sum of 20 Terms of an Arithmetic Sequence*

Use a graphing utility to find the sum S_n of the first 20 terms of the sequence $\{9.5n + 2.6\}$.

Solution   Figure 8 shows the results obtained using a TI-82 graphing calculator.

FIGURE 8

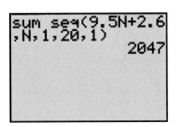

```
sum seq(9.5N+2.6
,N,1,20,1)
            2047
```

Thus, the sum of the first 20 terms of the sequence $\{9.5n + 2.6\}$ is 2047.   ■

■   Now work Problem 41.

E X A M P L E   8   *Creating a Floor Design*

A ceramic tile floor is designed in the shape of a trapezoid 20 feet wide at the base and 10 feet wide at the top. See Figure 9. The tiles, 12 inches by 12 inches, are to be placed so that each successive row contains one less tile than the row below. How many tiles will be required?

FIGURE 9

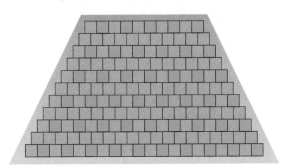

Solution   The bottom row requires 20 tiles and the top row 10 tiles. Since each successive row requires one less tile, the total number of tiles required is

$$S = 20 + 19 + 18 + \cdots + 11 + 10$$

This is the sum of an arithmetic sequence; the common difference is -1. The number of terms to be added is $n = 11$, with the first term $a = 20$ and the last term $a_{11} = 10$. The sum S is

$$S = \frac{n}{2}(a + a_{11}) = \frac{11}{2}(20 + 10) = 165$$

Thus, 165 tiles will be required.   ■

8.2

Exercise 8.2

In Problems 1–10, an arithmetic sequence is given. Find the common difference and write out the first four terms.

1. $\{n + 4\}$ **2.** $\{n - 5\}$ **3.** $\{2n - 5\}$ **4.** $\{3n + 1\}$ **5.** $\{6 - 2n\}$

6. $\{4 - 2n\}$ **7.** $\left\{\frac{1}{2} - \frac{1}{3}n\right\}$ **8.** $\left\{\frac{2}{3} + \frac{n}{4}\right\}$ **9.** $\{\ln 3^n\}$ **10.** $\{e^{\ln n}\}$

In Problems 11–18, find the nth term of the arithmetic sequence whose initial term a *and common difference* d *are given. What is the fifth term?*

11. $a = 2; d = 3$ **12.** $a = -2; d = 4$ **13.** $a = 5; d = -3$

14. $a = 6; d = -2$ **15.** $a = 0; d = \frac{1}{2}$ **16.** $a = 1; d = -\frac{1}{3}$

17. $a = \sqrt{2}; d = \sqrt{2}$ **18.** $a = 0; d = \pi$

In Problems 19–24, find the indicated term in each arithmetic sequence.

19. 12th term of 2, 4, 6, . . .   **20.** 8th term of -1, 1, 3, . . .

21. 10th term of 1, -2, -5, . . .   **22.** 9th term of 5, 0, -5, . . .

23. 8th term of a, $a + b$, $a + 2b$, . . .   **24.** 7th term of $2\sqrt{5}$, $4\sqrt{5}$, $6\sqrt{5}$, . . .

In Problems 25–32, find the first term and the common difference of the arithmetic sequence described. Give a recursive formula for the sequence.

25. 8th term is 8; 20th term is 44   **26.** 4th term is 3; 20th term is 35

27. 9th term is -5; 15th term is 31   **28.** 8th term is 4; 18th term is -96

29. 15th term is 0; 40th term is -50 **30.** 5th term is -2; 13th term is 30

31. 14th term is -1; 18th term is -9 **32.** 12th term is 4; 18th term is 28

In Problems 33–40, find the sum.

33. $1 + 3 + 5 + \cdots + (2n - 1)$ **34.** $2 + 4 + 6 + \cdots + 2n$

35. $7 + 12 + 17 + \cdots + (2 + 5n)$ **36.** $-1 + 3 + 7 + \cdots + (4n - 5)$

37. $2 + 4 + 6 + \cdots + 70$ **38.** $1 + 3 + 5 + \cdots + 59$

39. $5 + 9 + 13 + \cdots + 49$ **40.** $2 + 5 + 8 + \cdots + 41$

For Problems 41–46, use a graphing utility to find the sum of each sequence.

41. $\{3.45n + 4.12\}$ $n = 20$

42. $\{2.67n - 1.23\}$ $n = 25$

43. $2.8 + 5.2 + 7.6 + \cdots + 36.4$

44. $5.4 + 7.3 + 9.2 + \cdots + 32$

45. $4.9 + 7.48 + 10.06 + \cdots + 66.82$

46. $3.71 + 6.9 + 10.09 + \cdots + 80.27$

47. Find x so that $x + 3$, $2x + 1$, and $5x + 2$ are terms of an arithmetic sequence.

48. Find x so that $2x$, $3x + 2$, and $5x + 3$ are terms of an arithmetic sequence.

49. *Drury Lane Theater* The Drury Lane Theater has 25 seats in the first row and 30 rows in all. Each successive row contains one additional seat. How many seats are in the theater?

50. *Football Stadium* The corner section of a football stadium has 15 seats in the first row and 40 rows in all. Each successive row contains two additional seats. How many seats are in this section?

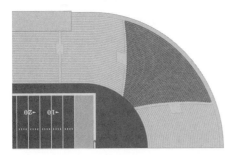

51. *Creating a Mosaic* A mosaic is designed in the shape of an equilateral triangle, 20 feet on each side. Each tile in the mosaic is in the shape of an equilateral triangle, 12 inches to a side. The tiles are to alternate in color as shown in the illustration. How many tiles of each color will be required?

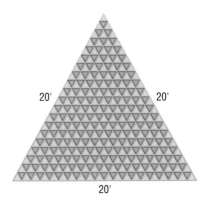

52. *Constructing a Brick Staircase* A brick staircase has a total of 30 steps. The bottom step requires 100 bricks. Each successive step requires two less bricks than the prior step.

(a) How many bricks are required for the top step?

(b) How many bricks are required to build the staircase?

 53. Make up an arithmetic sequence. Give it to a friend and ask for its 20th term.

8.3

Geometric Sequences; Geometric Series

When the ratio of successive terms of a sequence is always the same nonzero number, the sequence is called **geometric.** Thus, a geometric sequence* may be defined recursively as $a_1 = a$, $a_{n+1} / a_n = r$, or as

Geometric Sequence	$a_1 = a, \qquad a_{n+1} = ra_n$	(1)

where $a_1 = a$ and $r \neq 0$ are real numbers. The number a is the first term, and the nonzero number r is called the **common ratio.**

*Sometimes called a **geometric progression.**

Thus, the terms of a geometric sequence with first term a and common ratio r follow the pattern

$$a, \ ar, \ ar^2, \ ar^3, \ \ldots$$

E X A M P L E 1

Determining If a Sequence is Geometric

The sequence

$$2, 6, 18, 54, 162, \ldots$$

is geometric since the ratio of successive terms is 3. The first term is 2, and the common ratio is 3. ∎

E X A M P L E 2

Determining If a Sequence is Geometric

Show that the following sequence is geometric. Find the first term and the common ratio.

$$\{s_n\} = 2^{-n}$$

Solution   The first term is $s_1 = 2^{-1} = \frac{1}{2}$. The $(n + 1)$st and nth terms of the sequence $\{s_n\}$ are

$$s_{n+1} = 2^{-(n+1)} \quad \text{and} \quad s_n = 2^{-n}$$

Their ratio is

$$\frac{s_{n+1}}{s_n} = \frac{2^{-(n+1)}}{2^{-n}} = 2^{-n-1+n} = 2^{-1} = \frac{1}{2}$$

Because the ratio of successive terms is a nonzero number independent of n, the sequence $\{s_n\}$ is geometric with common ratio $\frac{1}{2}$. ∎

E X A M P L E 3

Determining If a Sequence is Geometric

Show that the following sequence is geometric. Find the first term and the common ratio.

$$\{t_n\} = \{4^n\}$$

Solution   The first term is $t_1 = 4^1 = 4$. The $(n + 1)$st and nth terms are

$$t_{n+1} = 4^{n+1} \quad \text{and} \quad t_n = 4^n$$

Their ratio is

$$\frac{t_{n+1}}{t_n} = \frac{4^{n+1}}{4^n} = 4$$

Thus, $\{t_n\}$ is a geometric sequence with common ratio 4. ∎

∎ Now work Problem 3.

Suppose a is the first term of a geometric sequence with common ratio $r \neq 0$. We seek a formula for the nth term a_n. To see the pattern, we write down the first few terms:

$$a_1 = 1 \cdot a = ar^0$$
$$a_2 = ra = ar^1$$
$$a_3 = ra_2 = r(ar) = ar^2$$
$$a_4 = ra_3 = r(ar^2) = ar^3$$
$$a_5 = ra_4 = r(ar^3) = ar^4$$
$$\vdots$$
$$a_n = ra_{n-1} = r(ar^{n-2}) = ar^{n-1}$$

We are led to the following result:

Theorem   For a geometric sequence $\{a_n\}$ whose first term is a and whose common ratio is r, the nth term is determined by the formula

---

nth Term of a Geometric Sequence

$$a_n = ar^{n-1}, \qquad r \neq 0 \tag{2}$$

■

E X A M P L E   4   *Finding a Particular Term of a Geometric Sequence*
Find the 9th term of the geometric sequence $2, \frac{2}{3}, \frac{2}{9}, \frac{2}{27}, \ldots$.

Solution   The first term of this geometric sequence is $a = 2$, and the common ratio is $\frac{1}{3}$. (Use $\frac{2}{3}/2 = \frac{1}{3}$, or $\frac{2}{9}/\frac{2}{3} = \frac{1}{3}$, or any two successive terms.) By formula (2), the nth term is

$$a_n = 2\left(\frac{1}{3}\right)^{n-1}$$

Hence, the 9th term is

$$a_9 = 2\left(\frac{1}{3}\right)^8 = \frac{2}{3^8} = \frac{2}{6561} \approx 0.0003 \qquad ■$$

Exploration: Use a graphing utility to find the 9th term of the sequence given in Example 4. Use it to find the 20th term and the 50th term.   ■

■   Now work Problems 25 and 33.

Adding the First n Terms of a Geometric Sequence

The next result gives us a formula for finding the sum of the first n terms of a geometric sequence.

Theorem Let $\{a_n\}$ be a geometric sequence with first term a and common ratio r. The sum S_n of the first n terms of $\{a_n\}$ is

Sum of n Terms
of a Geometric Sequence

$$S_n = a\frac{1 - r^n}{1 - r}, \qquad r \neq 0, 1 \tag{3}$$

Proof
$$S_n = a + ar + \cdots + ar^{n-1} \tag{4}$$

Multiply each side by r to obtain

$$rS_n = ar + ar^2 + \cdots + ar^n \tag{5}$$

Now, subtract (5) from (4). The result is

$$S_n - rS_n = a - ar^n$$
$$(1 - r)S_n = a(1 - r^n)$$

Since $r \neq 1$, we can solve for S_n:

$$S_n = a\frac{1 - r^n}{1 - r}$$ ∎

E X A M P L E   5

Finding the Sum of n Terms of a Geometric Sequence

Find the sum S_n of the first n terms of the sequence $\left\{\left(\frac{1}{2}\right)^n\right\}$; that is, find

$$\frac{1}{2} + \frac{1}{4} + \frac{1}{8} + \cdots + \left(\frac{1}{2}\right)^n$$

Solution The sequence $\left\{\left(\frac{1}{2}\right)^n\right\}$ is a geometric sequence with $a = \frac{1}{2}$ and $r = \frac{1}{2}$. The sum S_n that we seek is the sum of the first n terms of the sequence, so we use formula (3) to get

$$S_n = \sum_{k=1}^{n}\left(\frac{1}{2}\right)^k = \frac{1}{2} + \frac{1}{4} + \frac{1}{8} + \cdots + \left(\frac{1}{2}\right)^n$$

$$= \frac{1}{2}\left[\frac{1 - \left(\frac{1}{2}\right)^n}{1 - \frac{1}{2}}\right]$$

$$= \frac{1}{2}\left[\frac{1 - \left(\frac{1}{2}\right)^n}{\frac{1}{2}}\right]$$

$$= 1 - \left(\frac{1}{2}\right)^n$$ ■

■ Now work Problem 39.

E X A M P L E   6

Using a Graphing Utility to Find the Sum of a Geometric Sequence

Use a graphing utility to find the sum S_n of the first 15 terms of the sequence $\left\{\left(\frac{1}{3}\right)^n\right\}$; that is, find

$$\frac{1}{3} + \frac{1}{9} + \frac{1}{27} + \cdots + \left(\frac{1}{3}\right)^{15}$$

Solution   Figure 10 shows the result obtained using a TI-82 graphing calculator.

FIGURE 10

```
sum seq((1/3)^N,
N,1,15,1)
        .4999999652
```

Thus, the sum of the first 15 terms of the sequence $\left\{ \left(\dfrac{1}{3}\right)^n \right\}$ is 0.4999999652. ∎

■ Now work Problem 45.

Geometric Series

Infinite Geometric Series

An infinite sum of the form

$$a + ar + ar^2 + \cdots + ar^{n-1} + \cdots$$

with first term a and common ratio r, is called an **infinite geometric series** and is denoted by

$$\sum_{k=1}^{\infty} ar^{k-1}$$

Based on formula (3), the sum S_n of the first n terms of a geometric series is

$$S_n = a\frac{1 - r^n}{1 - r} = \frac{a}{1 - r} - \frac{ar^n}{1 - r} \qquad (6)$$

If this finite sum S_n approaches a number L as $n \to \infty$, then we call L the **sum of the infinite geometric series,** and we write

$$L = \sum_{k=1}^{\infty} ar^{k-1}$$

Theorem   If $|r| < 1$, the sum of the infinite geometric series $\displaystyle\sum_{k=1}^{\infty} ar^{k-1}$ is

Sum of an Infinite Geometric
Series

$$\sum_{k=1}^{\infty} ar^{k-1} = \frac{a}{1 - r} \qquad (7)$$

■

Intuitive Proof   Since $|r| < 1$, it follows that $|r^n|$ approaches 0 as $n \to \infty$. Then, based on formula (6), the sum S_n approaches $a/(1 - r)$ as $n \to \infty$. ∎

E X A M P L E   7   *Finding the Sum of a Geometric Series*

Find the sum of the geometric series $2 + \frac{4}{3} + \frac{8}{9} + \cdots$.

Solution   The first term is $a = 2$ and the common ratio is

$$r = \frac{\frac{4}{3}}{2} = \frac{4}{6} = \frac{2}{3}$$

Since $|r| < 1$, we use formula (7) to find that

$$2 + \frac{4}{3} + \frac{8}{9} + \cdots = \frac{2}{1 - \frac{2}{3}} = 6$$ ■

■   Now work Problem 51.

Exploration: Use a graphing utility to graph $U_n = 2\left(\frac{2}{3}\right)^{n-1} + U_{n-1}$ in sequence mode. TRACE the graph for large values of n. What happens to the value of U_n as n increases without bound? What can you conclude about $\sum\limits_{n=1}^{\infty} 2\left(\frac{2}{3}\right)^{n-1}$?   ■

E X A M P L E   8   *Repeating Decimals*

Show that the repeating decimal 0.999 . . . equals 1.

Solution
$$0.999 \ldots = \frac{9}{10} + \frac{9}{100} + \frac{9}{1000} + \cdots$$

Thus, 0.999 . . . is a geometric series with first term $\frac{9}{10}$ and common ratio $\frac{1}{10}$. Hence,

$$0.999 \ldots = \frac{\frac{9}{10}}{1 - \frac{1}{10}} = \frac{\frac{9}{10}}{\frac{9}{10}} = 1$$ ■

E X A M P L E   9   *Pendulum Swings*

Initially, a pendulum swings through an arc of 18 inches. See Figure 11. On each successive swing, the length of the arc is 0.98 of the previous length.

(a) What is the length of arc after 10 swings?
(b) On which swing is the length of arc first less than 12 inches?
(c) After 15 swings, what total length will the pendulum have swung?
(d) When it stops, what total length will the pendulum have swung?

Solution   (a) The length of the first swing is 18 inches. The length of the second swing is 0.98(18) inches; the length of the third swing is $0.98(0.98)(18) = 0.98^2(18)$ inches. The length of arc of the 10th swing is

$$(0.98)^9(18) = 15.007 \text{ inches}$$

FIGURE 11

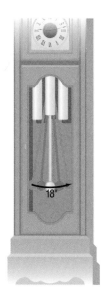

(b) The length of arc of the *n*th swing is $(0.98)^{n-1}(18)$. For this to be exactly 12 inches requires

$$(0.98)^{n-1}(18) = 12$$

$$(0.98)^{n-1} = \frac{12}{18} = \frac{2}{3}$$

$$n - 1 = \log_{0.98}\left(\frac{2}{3}\right)$$

$$n = 1 + \frac{\ln\left(\frac{2}{3}\right)}{\ln 0.98} = 1 + 20.07 = 21.07$$

The length of arc of the pendulum exceeds 12 inches on the 21st swing and is first less than 12 inches on the 22nd swing.

(c) After 15 swings, the pendulum will have swung the following total length *L*:

$$L = \underset{\text{1st}}{18} + \underset{\text{2nd}}{0.98(18)} + \underset{\text{3rd}}{(0.98)^2(18)} + \underset{\text{4th}}{(0.98)^3(18)} + \cdots + \underset{\text{15th}}{(0.98)^{14}(18)}$$

This is the sum of a geometric sequence. The common ratio is 0.98; the first term is 18. The sum has 15 terms, so

$$L = 18\,\frac{1 - 0.98^{15}}{1 - 0.98} = 18(13.07) = 235.29 \text{ inches}$$

The pendulum will have swung through 235.29 inches after 15 swings.

(d) When the pendulum stops, it will have swung the following total length *T*:

$$T = 18 + 0.98(18) + (0.98)^2(18) + (0.98)^3(18) + \cdots$$

This is the sum of a geometric series. The common ratio is $r = 0.98$; the first term is $a = 18$. The sum is

$$T = \frac{a}{1 - r} = \frac{18}{1 - 0.98} = 900$$

The pendulum will have swung a total of 900 inches when it finally stops. ■

HISTORICAL FEATURE   ■ Sequences are among the oldest objects of mathematical investigation, having been studied for over 3500 years. After the initial steps, however, little progress was made until about 1600.

Arithmetic and geometric sequences appear in the Rhind papyrus, a mathematical text containing 85 problems copied around 1650 BC by the Egyptian scribe Ahmes from an earlier work (see Historical Problems 1). Fibonacci (AD 1220) wrote about problems similar to those found in the Rhind papyrus, leading one to suspect that Fibonacci may have had material available that is now lost. This material would have been in the non-Euclidean Greek tradition of Heron (about AD 75) and Diophantus (about AD 250). One problem, again modified slightly, is still with us in the familiar puzzle rhyme "As I was going to St. Ives . . . " (see Historical Problem 2).

The Rhind papyrus indicates that the Egyptians knew how to add up the terms of an arithmetic or geometric sequence, as did the Babylonians. The rule for summing up a geometric sequence is found in Euclid's *Elements* (book IX, 35, 36), where, like all of Euclid's algebra, it is presented in a geometric form.

Investigations of other kinds of sequences began in the 1500's, when algebra became sufficiently developed to handle the more complicated problems. The development of calculus in the 1600's added a powerful new tool, especially for finding the sum of infinite series, and the subject continues to flourish today. ■

HISTORICAL PROBLEMS

■ 1. *Arithmetic sequence problem from the Rhind papyrus (statement modified slightly for clarity)* One hundred loaves of bread are to be divided among five people so that the amounts they receive form an arithmetic sequence. The first two together receive one-seventh of what the last three receive. How many does each receive? [*Partial answer:* First person receives $1\frac{2}{3}$ loaves.]

2. The following old English children's rhyme resembles one of the Rhind papyrus problems:

> As I was going to St. Ives
> I met a man with seven wives
> Each wife had seven sacks
> Each sack had seven cats
> Each cat had seven kits [kittens]
> Kits, cats, sacks, wives
> How many were going to St. Ives?

(a) Assuming that the speaker and the cat fanciers met by traveling in opposite directions, what is the answer?

(b) How many kittens are being transported?

(c) Kits, cats, sacks, wives; how many? [*Hint:* It is easier to include the man, find the sum with the formula, and then subtract 1 for the man.] ■

8.3

Exercise 8.3

In Problems 1–10, a geometric sequence is given. Find the common ratio and write out the first four terms.

1. $\{3^n\}$

2. $\{(-5)^n\}$

3. $\left\{-3\left(\frac{1}{2}\right)^n\right\}$

4. $\left\{\left(\frac{5}{2}\right)^n\right\}$

5. $\left\{\frac{2^{n-1}}{4}\right\}$

6. $\left\{\frac{3^n}{9}\right\}$

7. $\{2^{n/3}\}$

8. $\{3^{2n}\}$

9. $\left\{\frac{3^{n-1}}{2^n}\right\}$

10. $\left\{\frac{2^n}{3^{n-1}}\right\}$

In Problems 11–24, determine whether the given sequence is arithmetic, geometric, or neither. If the sequence is arithmetic, find the common difference; if it is geometric, find the common ratio.

11. $\{n + 2\}$

12. $\{2n - 5\}$

13. $\{4n^2\}$

14. $\{5n^2 + 1\}$

15. $\left\{3 - \frac{2}{3}n\right\}$

16. $\left\{8 - \frac{3}{4}n\right\}$

17. $1, 3, 6, 10, \ldots$

18. $2, 4, 6, 8, \ldots$

19. $\left\{\left(\frac{2}{3}\right)^n\right\}$

20. $\left\{\left(\frac{5}{4}\right)^n\right\}$

21. $-1, -2, -4, -8, \ldots$

22. $1, 1, 2, 3, 5, 8, \ldots$

23. $\{3^{n/2}\}$

24. $\{(-1)^n\}$

In Problems 25–32, find the fifth term and the nth term of the geometric sequence whose initial term a and common ratio r are given.

25. $a = 2; r = 3$

26. $a = -2; r = 4$

27. $a = 5; r = -1$

28. $a = 6; r = -2$

29. $a = 0; r = \frac{1}{2}$

30. $a = 1; r = -\frac{1}{3}$

31. $a = \sqrt{2}; r = \sqrt{2}$

32. $a = 0; r = 1/\pi$

In Problems 33–38, find the indicated term of each geometric sequence.

33. 7th term of $1, \frac{1}{2}, \frac{1}{4}, \ldots$

34. 8th term of $1, 3, 9, \ldots$

35. 9th term of $1, -1, 1, \ldots$

36. 10th term of $-1, 2, -4, \ldots$

37. 8th term of $0.4, 0.04, 0.004, \ldots$

38. 7th term of $0.1, 1.0, 10.0, \ldots$

In Problems 39–44, find the sum.

39. $\dfrac{1}{4} + \dfrac{2}{4} + \dfrac{2^2}{4} + \dfrac{2^3}{4} + \cdots + \dfrac{2^{n-1}}{4}$

40. $\dfrac{3}{9} + \dfrac{3^2}{9} + \dfrac{3^3}{9} + \cdots + \dfrac{3^n}{9}$

41. $\displaystyle\sum_{k=1}^{n} \left(\tfrac{2}{3}\right)^k$

42. $\displaystyle\sum_{k=1}^{n} 4 \cdot 3^{k-1}$

43. $-1 - 2 - 4 - 8 - \cdots - (2^{n-1})$

44. $2 + \dfrac{6}{5} + \dfrac{18}{25} + \cdots + 2\left(\dfrac{3}{5}\right)^n$

For Problems 45–50, use a graphing utility to find the sum of each geometric sequence.

45. $\dfrac{1}{4} + \dfrac{2}{4} + \dfrac{2^2}{4} + \dfrac{2^3}{4} + \cdots + \dfrac{2^{14}}{4}$

46. $\dfrac{3}{9} + \dfrac{3^2}{9} + \dfrac{3^3}{9} + \cdots + \dfrac{3^{15}}{9}$

47. $\displaystyle\sum_{n=1}^{15} \left(\tfrac{2}{3}\right)^n$

48. $\displaystyle\sum_{n=1}^{15} 4 \cdot 3^{n-1}$

49. $-1 - 2 - 4 - 8 - \cdots - 2^{14}$

50. $2 + \dfrac{6}{5} + \dfrac{18}{25} + \cdots + 2\left(\dfrac{3}{5}\right)^{15}$

In Problems 51–60, find the sum of each infinite geometric series.

51. $1 + \tfrac{1}{3} + \tfrac{1}{9} + \cdots$

52. $2 + \tfrac{4}{3} + \tfrac{8}{9} + \cdots$

53. $8 + 4 + 2 + \cdots$

54. $6 + 2 + \tfrac{2}{3} + \cdots$

55. $2 - \tfrac{1}{2} + \tfrac{1}{8} - \tfrac{1}{32} + \cdots$

56. $1 - \tfrac{3}{4} + \tfrac{9}{16} - \tfrac{27}{64} + \cdots$

57. $\displaystyle\sum_{k=1}^{\infty} 5\left(\tfrac{1}{4}\right)^{k-1}$

58. $\displaystyle\sum_{k=1}^{\infty} 8\left(\tfrac{1}{3}\right)^{k-1}$

59. $\displaystyle\sum_{k=1}^{\infty} 6\left(-\tfrac{2}{3}\right)^{k-1}$

60. $\displaystyle\sum_{k=1}^{\infty} 4\left(-\tfrac{1}{2}\right)^{k-1}$

61. Find x so that x, $x + 2$, and $x + 3$ are terms of a geometric sequence.

62. Find x so that $x - 1$, x, and $x + 2$ are terms of a geometric sequence.

63. *Pendulum Swings* Initially, a pendulum swings through an arc of 2 feet. On each successive swing, the length of arc is 0.9 of the previous length.

(a) What is the length of arc after 10 swings?

(b) On which swing is the length of arc first less than 1 foot?

(c) After 15 swings, what total length will the pendulum have swung?

(d) When it stops, what total length will the pendulum have swung?

64. *Bouncing Balls* A ball is dropped from a height of 30 feet. Each time it strikes the ground, it bounces up to 0.8 of the previous height.

(a) What height will the ball bounce up to after it strikes the ground for the third time?

(b) What is its height after it strikes the ground for the nth time?

(c) How many times does the ball need to strike the ground before its height is less than 6 inches?

(d) What total distance does the ball travel before it stops bouncing?

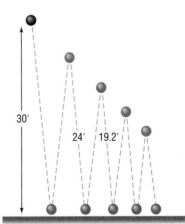

65. *Salary Increases* Suppose you have just been hired at an annual salary of $18,000 and expect to receive annual increases of 5%. What will your salary be when you begin your fifth year?

66. *Equipment Depreciation* A new piece of equipment cost a company $15,000. Each year, for tax purposes, the company depreciates the value by 15%. What value should the company give the equipment after 5 years?

67. *Critical Thinking* You have just signed a 7 year professional football league contract with a beginning salary of $2,000,000 per year. Management gives you the following options with regard to your salary over the seven years.

    (1) A bonus of $100,000 each year

    (2) An annual increase of 4.5% per year beginning after 1 year

    (3) An annual increase of $95,000 per year beginning after 1 year

Which option provides the most money over the 7 year period? Which the least? Which would you choose? Why?

68. *A Rich Man's Promise* A rich man promises to give you $1000 on September 1, 1996. Each day thereafter he will give you $\frac{9}{10}$ of what he gave you the previous day. What is the first date on which the amount you receive is less than 1¢? How much have you received when this happens?

69. *Grains of Wheat on a Chess Board* In an old fable, a commoner who had just saved the king's life was told he could ask the king for any just reward. Being a shrewd man, the commoner said, "A simple wish, sire. Place one grain of wheat on the first square of a chessboard, two grains on the second square, four grains on the third square, continuing until you have filled the board. This is all I seek." Compute the total number of grains needed to do this to see why the request, seemingly simple, could not be granted. (A chessboard consists of $8 \times 8 = 64$ squares.)

70. Can a sequence be both arithmetic and geometric? Give reasons for your answer.

71. Make up a geometric sequence. Give it to a friend and ask for its 20th term.

72. Make up two infinite geometric series, one that has a sum and one that does not. Give them to a friend and ask for the sum of each series.

73. If $x < 1$, then $1 + x + x^2 + x^3 + \cdots + x^n + \cdots = 1/(1 - x)$. Make up a table of values using $x = 0.1$, $x = 0.25$, $x = 0.5$, $x = 0.75$, and $x = 0.9$ to compute $1/(1 - x)$. Now determine how many terms are needed in the expansion $1 + x + x^2 + x^3 + \cdots + x^n + \cdots$ before it approximates $1/(1 - x)$ correct to two decimal places. For example, if $x = 0.1$, then $1/(1 - x) = 10/9 = 1.111 \ldots$. The expansion requires three terms.

74. Which of the following choices, *A* or *B*, results in more money?

    A: To receive $1000 on day 1, $999 on day 2, $998 on day 3, with the process to end after 1000 days

    B: To receive $1 on day 1, $2 on day 2, $4 on day 3, for 19 days

75. You are interviewing for a job and receive two offers:

    A: $20,000 to start with guaranteed annual increases of 6% for the first 5 years

    B: $22,000 to start with guaranteed annual increases of 3% for the first 5 years

Which offer is best if your goal is to be making as much as possible after 5 years? Which is best if your goal is to make as much money as possible over the contract (5 years)?

76. Look at the figure on the right. What fraction of the square is eventually shaded if the indicated shading process continues indefinitely?

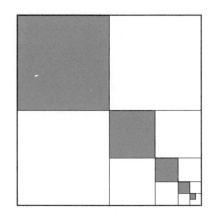

8.4

Mathematical Induction

Mathematical induction is a method for proving that statements involving natural numbers are true for all natural numbers.* For example, the statement, "$2n$ is always an even integer" can be proved true for all natural numbers by using mathematical induction. Also, the statement "the sum of the first n positive odd integers equals n^2," that is,

$$1 + 3 + 5 + \cdots + (2n - 1) = n^2 \qquad (1)$$

can be proved for all natural numbers n by using mathematical induction.

Before stating the method of mathematical induction, let's try to gain a sense of the power of the method. We shall use the statement in equation (1) for this purpose by restating it for various values of $n = 1, 2, 3, \ldots$:

$n = 1$ The sum of the first positive odd integer is 1^2; $1 = 1^2$.

$n = 2$ The sum of the first 2 positive odd integers is 2^2; $1 + 3 = 4 = 2^2$.

$n = 3$ The sum of the first 3 positive odd integers is 3^2; $1 + 3 + 5 = 9 = 3^2$.

$n = 4$ The sum of the first 4 positive odd integers is 4^2; $1 + 3 + 5 + 7 = 16 = 4^2$.

Although from this pattern we might conjecture that statement (1) is true for any choice of n, can we really be sure that it does not fail for some choice of n? The method of proof by mathematical induction will, in fact, prove that the statement is true for all n.

Theorem
The Principle of Mathematical Induction

Suppose the following two conditions are satisfied with regard to a statement about natural numbers:

CONDITION I: The statement is true for the natural number 1.

CONDITION II: If the statement is true for some natural number k, it is also true for the next natural number $k + 1$.

Then the statement is true for all natural numbers. ∎

We shall not prove this principle. However, we can provide a physical interpretation that will help us see why the principle works. Think of a collection of natural numbers obeying a statement as a collection of infinitely many dominoes (see Figure 12).

Now, suppose we are told two facts:

FIGURE 12

1. The first domino is pushed over.
2. If one of the dominoes falls over, say the kth domino, then so will the next one, the $(k + 1)$st domino.

Is it safe to conclude that *all* the dominoes fall over? The answer is yes, because, if the first one falls (Condition I), then the second one does also (by Condition II); and if the second one falls, then so does the third (by Condition II); and so on.

Now let's prove some statements about natural numbers using mathematical induction.

*Recall from the Appendix, Section 1, that the natural numbers are the numbers 1, 2, 3, 4, ... In other words, the terms *natural numbers* and *positive integers* are synonymous.

E X A M P L E   1

Using Mathematical Induction

Show that the following statement is true for all natural numbers n:

$$1 + 3 + 5 + \cdots + (2n - 1) = n^2 \tag{2}$$

Solution   We need to show first that statement (2) holds for $n = 1$. Because $1 = 1^2$, statement (2) is true for $n = 1$. Thus, Condition I holds.

Next, we need to show that Condition II holds. Suppose we know for some k that

$$1 + 3 + \cdots + (2k - 1) = k^2 \tag{3}$$

We wish to show that, based on equation (3), statement (2) holds for $k + 1$. Thus, we look at the sum of the first $k + 1$ positive odd integers to determine whether this sum equals $(k + 1)^2$:

$$1 + 3 + \cdots + (2k - 1) + (2k + 1) = \underbrace{[1 + 3 + \cdots + (2k - 1)]}_{= k^2 \text{ by equation (3)}} + (2k + 1)$$

$$= k^2 + (2k + 1)$$

$$= k^2 + 2k + 1 = (k + 1)^2$$

Conditions I and II are satisfied; thus, by the principle of mathematical induction, statement (2) is true for all natural numbers.   ■

E X A M P L E   2

Using Mathematical Induction

Show that the following statement is true for all natural numbers n:

$$2^n > n$$

Solution   First, we show that the statement $2^n > n$ holds when $n = 1$. Because $2^1 = 2 > 1$, the inequality is true for $n = 1$. Thus, Condition I holds.

Next, we assume, for some natural number k, that $2^k > k$. We wish to show that the formula holds for $k + 1$; that is, we wish to show that $2^{k+1} > k + 1$. Now,

$$2^{k+1} = 2 \cdot 2^k > 2 \cdot k = k + k \geq k + 1$$

$$\underset{\substack{\text{We know that} \\ 2^k > k.}}{\uparrow} \qquad\qquad \underset{k \geq 1}{\uparrow}$$

Thus, if $2^k > k$, then $2^{k+1} > k + 1$, so Condition II of the principle of mathematical induction is satisfied. Hence, the statement $2^n > n$ is true for all natural numbers n.   ■

E X A M P L E   3

Using Mathematical Induction

Show that the following formula is true for all natural numbers n:

$$1 + 2 + 3 + \cdots + n = \frac{n(n + 1)}{2} \tag{4}$$

Solution   First, we show that formula (4) is true when $n = 1$. Because

$$\frac{1(1 + 1)}{2} = \frac{1(2)}{2} = 1$$

Condition I of the principle of mathematical induction holds.

Next, we assume that formula (4) holds for some k, and we determine whether the formula then holds for $k + 1$. Thus, we assume that

$$1 + 2 + 3 + \cdots + k = \frac{k(k + 1)}{2} \quad \text{for some } k \tag{5}$$

Now, we need to show that

$$1 + 2 + 3 + \cdots + k + (k + 1) = \frac{(k + 1)(k + 1 + 1)}{2} = \frac{(k + 1)(k + 2)}{2}$$

We do this as follows:

$$1 + 2 + 3 + \cdots + k + (k + 1) = \underbrace{[1 + 2 + 3 + \cdots + k]}_{= \frac{k(k + 1)}{2} \text{ by equation (5)}} + (k + 1)$$

$$= \frac{k(k + 1)}{2} + (k + 1)$$

$$= \frac{k^2 + k + 2k + 2}{2}$$

$$= \frac{k^2 + 3k + 2}{2} = \frac{(k + 1)(k + 2)}{2}$$

Thus, Condition II also holds. As a result, formula (4) is true for all natural numbers. ■

■ Now work Problem 1.

E X A M P L E 4   *Using Mathematical Induction*

Show that $3^n - 1$ is divisible by 2 for all natural numbers n.

Solution   First, we show that the statement is true when $n = 1$. Because $3^1 - 1 = 3 - 1 = 2$ is divisible by 2, the statement is true when $n = 1$. Thus, Condition I is satisfied.

Next, we assume that the statement holds for some k, and we determine whether the statement then holds for $k + 1$. Thus, we assume that $3^k - 1$ is divisible by 2 for some k. We need to show that $3^{k+1} - 1$ is divisible by 2. Now,

$$3^{k+1} - 1 = 3^{k+1} - 3^k + 3^k - 1$$
$$= 3^k(3 - 1) + (3^k - 1) = 3^k \cdot 2 + (3^k - 1)$$

Because $3^k \cdot 2$ is divisible by 2 and $3^k - 1$ is divisible by 2, it follows that $3^k \cdot 2 + (3^k - 1) = 3^{k+1} - 1$ is divisible by 2. Thus, Condition II is also satisfied. As a result, the statement, "$3^n - 1$ is divisible by 2" is true for all natural numbers n. ■

Warning: The conclusion that a statement involving natural numbers is true for all natural numbers is made only after *both* Conditions I and II of the principle of mathematical induction have been satisfied. Problem 27 (below) demonstrates a statement for which only Condition I holds, but the statement is not true for all natural numbers. Problem 28 demonstrates a statement for which only Condition II holds, but the statement is *not* true for any natural number.

8.4

Exercise 8.4

In Problems 1–26, use the principle of mathematical induction to show that the given statement is true for all natural numbers.

1. $2 + 4 + 6 + \cdots + 2n = n(n + 1)$

2. $1 + 5 + 9 + \cdots + (4n - 3) = n(2n - 1)$

3. $3 + 4 + 5 + \cdots + (n + 2) = \frac{1}{2}n(n + 5)$

4. $3 + 5 + 7 + \cdots + (2n + 1) = n(n + 2)$

5. $2 + 5 + 8 + \cdots + (3n - 1) = \frac{1}{2}n(3n + 1)$

6. $1 + 4 + 7 + \cdots + (3n - 2) = \frac{1}{2}n(3n - 1)$

7. $1 + 2 + 2^2 + \cdots + 2^{n-1} = 2^n - 1$

8. $1 + 3 + 3^2 + \cdots + 3^{n-1} = \frac{1}{2}(3^n - 1)$

9. $1 + 4 + 4^2 + \cdots + 4^{n-1} = \frac{1}{3}(4^n - 1)$

10. $1 + 5 + 5^2 + \cdots + 5^{n-1} = \frac{1}{4}(5^n - 1)$

11. $\dfrac{1}{1 \cdot 2} + \dfrac{1}{2 \cdot 3} + \dfrac{1}{3 \cdot 4} + \cdots + \dfrac{1}{n(n + 1)} = \dfrac{n}{n + 1}$

12. $\dfrac{1}{1 \cdot 3} + \dfrac{1}{3 \cdot 5} + \dfrac{1}{5 \cdot 7} + \cdots + \dfrac{1}{(2n - 1)(2n + 1)} = \dfrac{n}{2n + 1}$

13. $1^2 + 2^2 + 3^2 + \cdots + n^2 = \frac{1}{6}n(n + 1)(2n + 1)$

14. $1^3 + 2^3 + 3^3 + \cdots + n^3 = \frac{1}{4}n^2(n + 1)^2$

15. $4 + 3 + 2 + \cdots + (5 - n) = \frac{1}{2}n(9 - n)$

16. $-2 - 3 - 4 - \cdots - (n + 1) = -\frac{1}{2}n(n + 3)$

17. $1 \cdot 2 + 2 \cdot 3 + 3 \cdot 4 + \cdots + n(n + 1) = \frac{1}{3}n(n + 1)(n + 2)$

18. $1 \cdot 2 + 3 \cdot 4 + 5 \cdot 6 + \cdots + (2n - 1)(2n) = \frac{1}{3}n(n + 1)(4n - 1)$

19. $n^2 + n$ is divisible by 2.

20. $n^3 + 2n$ is divisible by 3.

21. $n^2 - n + 2$ is divisible by 2.

22. $n(n + 1)(n + 2)$ is divisible by 6.

23. If $x > 1$, then $x^n > 1$.

24. If $0 < x < 1$, then $0 < x^n < 1$.

25. $a - b$ is a factor of $a^n - b^n$. [*Hint:* $a^{k+1} - b^{k+1} = a(a^k - b^k) + b^k(a - b)$]

26. $a + b$ is a factor of $a^{2n+1} + b^{2n+1}$.

27. Show that the statement "$n^2 - n + 41$ is a prime number," is true for $n = 1$, but is not true for $n = 41$.

28. Show that the formula

$$2 + 4 + 6 + \cdots + 2n = n^2 + n + 2$$

obeys Condition II of the principle of mathematical induction. That is, show that if the formula is true for some k it is also true for $k + 1$. Then show that the formula is false for $n = 1$ (or for any other choice of n).

29. Use mathematical induction to prove that if $r \neq 1$ then

$$a + ar + ar^2 + \cdots + ar^{n-1} = a\frac{1 - r^n}{1 - r}$$

30. Use mathematical induction to prove that

$$a + (a + d) + (a + 2d) + \cdots + [a + (n - 1)d] = na + d\frac{n(n - 1)}{2}$$

31. *Geometry* Use mathematical induction to show that the sum of the interior angles of a convex polygon of n sides equals $(n - 2) \cdot 180°$.

32. *The Extended Principle of Mathematical Induction* The extended principle of mathematical induction states that if conditions I and II hold, that is,

    (I)   A statement is true for a natural number j.

    (II) If the statement is true for some natural number $k > j$, then it is also true for the next natural number $k + 1$.

    Then the statement is true for *all* natural numbers $\geq j$.

    Use the extended principle of mathematical induction to show that the number of diagonals in a convex polygon of n sides is $\frac{1}{2}n(n - 3)$. [*Hint:* Begin by showing that the result is true when $n = 4$ (Condition I).]

33. How would you explain to a friend the principle of mathematical induction?

8.5

The Binomial Theorem

In the Appendix, Section 2, we listed some special products. Among these were formulas for expanding $(x + a)^n$ for $n = 2$ and $n = 3$. The *Binomial Theorem** is a formula for the expansion of $(x + a)^n$ for n any positive integer. If $n = 1, 2, 3,$ and 4, the expansion of $(x + a)^n$ is straightforward:

$$(x + a)^1 = x + a \qquad\qquad \text{2 terms, beginning with } x^1 \text{ and ending with } a^1$$

$$(x + a)^2 = x^2 + 2ax + a^2 \qquad\qquad \text{3 terms, beginning with } x^2 \text{ and ending with } a^2$$

$$(x + a)^3 = x^3 + 3ax^2 + 3a^2x + a^3 \qquad\qquad \text{4 terms, beginning with } x^3 \text{ and ending with } a^3$$

$$(x + a)^4 = x^4 + 4ax^3 + 6a^2x^2 + 4a^3x + a^4 \qquad \text{5 terms, beginning with } x^4 \text{ and ending with } a^4$$

Notice that each expansion of $(x + a)^n$ begins with x^n and ends with a^n. As you read from left to right, the powers of x are decreasing, while the powers of a are increasing. Also, the number of terms that appear equals $n + 1$. Notice, too, that the degree of each monomial in the expansion equals n. For example, in the expansion of $(x + a)^3$, each monomial $(x^3, 3ax^2, 3a^2x, a^3)$ is of degree 3. As a result, we might conjecture that the expansion of $(x + a)^n$ would look like this:

$$(x + a)^n = x^n + \_ax^{n-1} + \_a^2x^{n-2} + \cdots + \_a^{n-1}x + a^n$$

where the blanks are numbers to be found. This is, in fact, the case, as we shall see shortly.

First, we need to introduce a symbol.

The Symbol $\begin{pmatrix} n \\ j \end{pmatrix}$

We define the symbol $\begin{pmatrix} n \\ j \end{pmatrix}$, read "$n$ taken j at a time," as follows:

Symbol $\begin{pmatrix} n \\ j \end{pmatrix}$

If j and n are integers with $0 \le j \le n$, the symbol $\begin{pmatrix} n \\ j \end{pmatrix}$ is defined as

$$\begin{pmatrix} n \\ j \end{pmatrix} = \frac{n!}{j!(n - j)!} \tag{1}$$

Comment: On a graphing calculator, the symbol $\begin{pmatrix} n \\ j \end{pmatrix}$ may be denoted by the key $\boxed{nCr}$.

E X A M P L E   1   *Evaluating* $\begin{pmatrix} n \\ j \end{pmatrix}$

Find:

(a) $\begin{pmatrix} 3 \\ 1 \end{pmatrix}$ (b) $\begin{pmatrix} 4 \\ 2 \end{pmatrix}$ (c) $\begin{pmatrix} 8 \\ 7 \end{pmatrix}$ (d) $\begin{pmatrix} 65 \\ 15 \end{pmatrix}$

*The name *binomial* derives from the fact that $x + a$ is a binomial, that is, contains two terms.

Solution   (a) $\dbinom{3}{1} = \dfrac{3\ !}{1!(3-1)!} = \dfrac{3!}{1!2!} = \dfrac{3 \cdot 2 \cdot 1}{1(2 \cdot 1)} = \dfrac{6}{2} = 3$

(b) $\dbinom{4}{2} = \dfrac{4!}{2!(4-2)!} = \dfrac{4!}{2!2!} = \dfrac{4 \cdot 3 \cdot 2 \cdot 1}{(2 \cdot 1)(2 \cdot 1)} = \dfrac{24}{4} = 6$

(c) $\dbinom{8}{7} = \dfrac{8!}{7!(8-7)!} = \dfrac{8!}{7!1!} \underset{\uparrow}{=} \dfrac{8 \cdot 7!}{7! \cdot 1!} = \dfrac{8}{1} = 8$

$8! = 8 \cdot 7!$

FIGURE 13

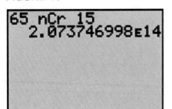

(d) Figure 13 shows the solution using a TI-82 graphing calculator.

Thus, $\dbinom{65}{15} = 2.073746998 \times 10^{14}$.   ■

■   Now work Problem 1.

Two useful formulas involving the symbol $\dbinom{n}{j}$ are

$$\dbinom{n}{0} = 1 \quad \text{and} \quad \dbinom{n}{n} = 1$$

Proof $$\dbinom{n}{0} = \dfrac{n!}{0!(n-0)!} = \dfrac{n!}{0!n!} = \dfrac{1}{1} = 1$$

You are asked to show that $\dbinom{n}{n} = 1$ in Problem 41 at the end of this section.   ■

Suppose we arrange the various values of the symbol $\dbinom{n}{j}$ in a triangular display, as shown next and in Figure 14, on page 610.

$$\dbinom{0}{0}$$

$$\dbinom{1}{0} \quad \dbinom{1}{1}$$

$$\dbinom{2}{0} \quad \dbinom{2}{1} \quad \dbinom{2}{2}$$

$$\dbinom{3}{0} \quad \dbinom{3}{1} \quad \dbinom{3}{2} \quad \dbinom{3}{3}$$

$$\dbinom{4}{0} \quad \dbinom{4}{1} \quad \dbinom{4}{2} \quad \dbinom{4}{3} \quad \dbinom{4}{4}$$

$$\dbinom{5}{0} \quad \dbinom{5}{1} \quad \dbinom{5}{2} \quad \dbinom{5}{3} \quad \dbinom{5}{4} \quad \dbinom{5}{5}$$

This display is called the **Pascal triangle,** named after Blaise Pascal (1623–1662), a French mathematician.

$\mathcal{M}$ ISSION POSSIBLE

Chapter 8

G ETTING THE MOST OUT OF YOUR CONTRACT

You and your band have just signed a recording contract with the nationally acclaimed recording studio, NASHBURG TENS. They've promised $200,000 a year for six years, plus a choice of one of the following four options:

    (a) a bonus of $10,000 per year
    (b) an annual increase of 4.5% per year (starting after the first year)
    (c) an annual increase of 6% per year (starting after the second year)
    (d) an annual increase of $9,500 per year (starting after the first year)

You have to tell them today which option you want to take. Your agent is out of town and out of cellular phone range, so you'll have to do the math yourselves.

1. For each option, find out what your payment would be every year for the six years. Round to the nearest dollar.
2. Identify which options are examples of arithmetic or geometric sequences.
3. For each option, find out how much the recording studio will have paid in total over the whole six years. Do you have formulas that will shorten your work?
4. Which option pays your band the most over all? Are there any considerations or circumstances that would make other options better choices even though they pay less money? Are there advantages to being paid more at the beginning of the contract? What are they?

FIGURE 14
Pascal triangle

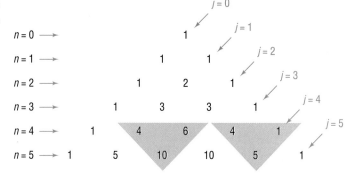

The Pascal triangle has 1's down the sides. To get any other entry, merely add the two nearest entries in the row above it. The shaded triangles in Figure 14 serve to illustrate this feature of the Pascal triangle. Based on this feature, the row corresponding to $n = 6$ is found as follows:

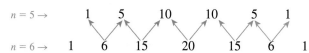

Later, we shall prove that this addition always works (see the Theorem on page 612).

Although the Pascal triangle provides an interesting and organized display of the symbol $\binom{n}{j}$, in practice it is not all that helpful. For example, if you wanted to know the value of $\binom{12}{5}$, you would need to produce twelve rows of the triangle before seeing the answer. It is much faster instead to use the definition (1).

The Binomial Theorem

Now we are ready to state the **Binomial Theorem.** A proof is given at the end of this section.

Theorem   Let x and a be real numbers. For any positive integer n, we have

$$(x + a)^n = \binom{n}{0}x^n + \binom{n}{1}ax^{n-1} + \cdots + \binom{n}{j}a^j x^{n-j} + \cdots + \binom{n}{n}a^n$$

Binomial Theorem

$$= \sum_{j=0}^{n}\binom{n}{j}x^{n-j}a^j \qquad (2)$$

Now you know why we needed to introduce the symbol $\binom{n}{j}$; these symbols are the numerical coefficients that appear in the expansion of $(x + a)^n$. Because of this, the symbol $\binom{n}{j}$ is called the **binomial coefficient.**

E X A M P L E   2   *Expanding a Binomial*

Use the Binomial Theorem to expand $(x + 2)^5$.

Solution   In the Binomial Theorem, let $a = 2$ and $n = 5$. Then

$$(x + 2)^5 = \binom{5}{0}x^5 + \binom{5}{1}2x^4 + \binom{5}{2}2^2x^3 + \binom{5}{3}2^3x^2 + \binom{5}{4}2^4x + \binom{5}{5}2^5$$

↑
Use equation (2).

$$= 1 \cdot x^5 + 5 \cdot 2x^4 + 10 \cdot 4x^3 + 10 \cdot 8x^2 + 5 \cdot 16x + 1 \cdot 32$$

↑
Use row $n = 5$ of the Pascal triangle or formula (1) for $\binom{n}{j}$.

$$= x^5 + 10x^4 + 40x^3 + 80x^2 + 80x + 32 \qquad \blacksquare$$

E X A M P L E   3   *Expanding a Binomial*

Expand $(2y - 3)^4$ using the Binomial Theorem.

Solution   First, we rewrite the expression $(2y - 3)^4$ as $[2y + (-3)]^4$. Now we use the Binomial Theorem with $n = 4$, $x = 2y$, and $a = -3$:

$$[2y + (-3)]^4 = \binom{4}{0}(2y)^4 + \binom{4}{1}(-3)(2y)^3 + \binom{4}{2}(-3)^2(2y)^2$$

$$+ \binom{4}{3}(-3)^3(2y) + \binom{4}{4}(-3)^4$$

$$= 1 \cdot 16y^4 + 4(-3)8y^3 + 6 \cdot 9 \cdot 4y^2 + 4(-27)2y + 1 \cdot 81$$

↑
Use row $n = 4$ of the Pascal triangle or formula (1) for $\binom{n}{j}$

$$= 16y^4 - 96y^3 + 216y^2 - 216y + 81$$

In this expansion, note that the signs alternate due to the fact that $a = -3 < 0$.

$\blacksquare$

$\blacksquare$ Now work Problem 17.

E X A M P L E   4   *Finding a Particular Coefficient in a Binomial Expansion*

Find the coefficient of y^8 in the expansion of $(2y + 3)^{10}$.

Solution   We write out the expansion using the Binomial Theorem:

$$(2y + 3)^{10} = \binom{10}{0}(2y)^{10} + \binom{10}{1}(2y)^9(3)^1 + \binom{10}{2}(2y)^8(3)^2 + \binom{10}{3}(2y)^7(3)^3$$

$$+ \binom{10}{4}(2y)^6(3)^4 + \cdots + \binom{10}{9}(2y)(3)^9 + \binom{10}{10}(3)^{10}$$

From the third term in the expression, the coefficient of y^8 is

$$\binom{10}{2}(2)^8(3)^2 = \frac{10!}{2!8!} \cdot 2^8 \cdot 9 = \frac{10 \cdot 9 \cdot 8!}{2 \cdot 8!} \cdot 2^8 \cdot 9 = 103,680 \qquad \blacksquare$$

As this solution demonstrates, we can use the Binomial Theorem to write a particular term in an expansion without writing the entire expansion. Based on the expansion of $(x + a)^n$, the term containing x^j is

$$\binom{n}{n-j}a^{n-j}x^j \qquad\qquad (3)$$

For example, we can solve Example 4 by using formula (3) with $n = 10$, $a = 3$, $x = 2y$, and $j = 8$. Then the term containing y^8 is

$$\binom{10}{10-8} 3^{10-8}(2y)^8 = \binom{10}{2} \cdot 3^2 \cdot 2^8 \cdot y^8 = \frac{10!}{2!8!} \cdot 9 \cdot 2^8 y^8$$

$$= \frac{10 \cdot 9 \cdot 8!}{2!8!} \cdot 9 \cdot 2^8 y^8 = 103,680 y^8$$

E X A M P L E   5   *Finding a Particular Term in a Binomial Expansion*

Find the sixth term in the expansion of $(x + 2)^9$.

Solution A   We expand using the Binomial Theorem until the sixth term is reached:

$$(x + 2)^9 = \binom{9}{0}x^9 + \binom{9}{1}x^8 \cdot 2 + \binom{9}{2}x^7 \cdot 2^2 + \binom{9}{3}x^6 \cdot 2^3 + \binom{9}{4}x^5 \cdot 2^4$$

$$+ \binom{9}{5}x^4 \cdot 2^5 + \cdots$$

The sixth term is

$$\binom{9}{5}x^4 \cdot 2^5 = \frac{9!}{5!4!} \cdot x^4 \cdot 32 = 4032x^4$$

Solution B   The sixth term in the expansion of $(x + 2)^9$, which has ten terms total, contains x^4. (Do you see why?) Thus, by formula (3), the sixth term is

$$\binom{9}{9-4}2^{9-4}x^4 = \binom{9}{5}2^5 x^4 = \frac{9!}{5!4!} \cdot 32x^4 = 4032x^4$$ ■

■ Now work Problems 25 and 31.

Next we show that the "triangular addition" feature of the Pascal triangle illustrated in Figure 14 always works.

Theorem   If n and j are integers with $1 \leq j \leq n$, then

$$\binom{n}{j-1} + \binom{n}{j} = \binom{n+1}{j} \qquad (4)$$

Proof $\binom{n}{j-1} + \binom{n}{j} = \dfrac{n!}{(j-1)![n-(j-1)]!} + \dfrac{n!}{j!(n-j)!}$ Multiply the first term by j/j and the second term by $(n-j+1)/(n-j+1)$.

$$= \frac{n!}{(j-1)!(n-j+1)!} + \frac{n!}{j!(n-j)!}$$

$$= \frac{jn!}{j(j-1)!(n-j+1)!} + \frac{(n-j+1)n!}{j!(n-j+1)(n-j)!}$$

$$= \frac{jn!}{j!(n-j+1)!} + \frac{(n-j+1)n!}{j!(n-j+1)!}$$ Now the denominators are equal.

$$= \frac{jn! + (n - j + 1)n!}{j!(n - j + 1)!}$$

$$= \frac{n!(j + n - j + 1)}{j!(n - j + 1)!}$$

$$= \frac{n!(n + 1)}{j!(n - j + 1)!} = \frac{(n + 1)!}{j![(n + 1) - j]!} = \binom{n + 1}{j}$$ ∎

Proof of the Binomial Theorem   We use mathematical induction to prove the Binomial Theorem. First, we show that formula (2) is true for $n = 1$:

$$(x + a)^1 = x + a = \binom{1}{0}x^1 + \binom{1}{1}a^1$$

Next we suppose that formula (2) is true for some k. That is, we assume that

$$(x + a)^k = \binom{k}{0}x^k + \binom{k}{1}ax^{k-1} + \cdots + \binom{k}{j-1}a^{j-1}x^{k-j+1} + \binom{k}{j}a^jx^{k-j} + \cdots + \binom{k}{k}a^k \tag{5}$$

Now we calculate $(x + a)^{k+1}$:

$$(x + a)^{k+1} = (x + a)(x + a)^k = x(x + a)^k + a(x + a)^k$$

Use ↑ equation (5).

$$\stackrel{=}{} x\left[\binom{k}{0}x^k + \binom{k}{1}ax^{k-1} + \cdots + \binom{k}{j-1}a^{j-1}x^{k-j+1} + \binom{k}{j}a^jx^{k-j} + \cdots + \binom{k}{k}a^k\right]$$

$$+ a\left[\binom{k}{0}x^k + \binom{k}{1}ax^{k-1} + \cdots + \binom{k}{j-1}a^{j-1}x^{k-j+1} + \binom{k}{j}a^jx^{k-j} + \cdots + \binom{k}{k-1}a^{k-1}x + \binom{k}{k}a^k\right]$$

$$= \binom{k}{0}x^{k+1} + \binom{k}{1}ax^k + \cdots + \binom{k}{j-1}a^{j-1}x^{k-j+2} + \binom{k}{j}a^jx^{k-j+1} + \cdots + \binom{k}{k}a^kx$$

$$+ \binom{k}{0}ax^k + \binom{k}{1}a^2x^{k-1} + \cdots + \binom{k}{j-1}a^jx^{k-j+1} + \binom{k}{j}a^{j+1}x^{k-j} + \cdots + \binom{k}{k-1}a^kx + \binom{k}{k}a^{k+1}$$

$$= \binom{k}{0}x^{k+1} + \left[\binom{k}{1} + \binom{k}{0}\right]ax^k + \cdots + \left[\binom{k}{j} + \binom{k}{j-1}\right]a^jx^{k-j+1} + \cdots + \left[\binom{k}{k} + \binom{k}{k-1}\right]a^kx + \binom{k}{k}a^{k+1}$$

Because

$$\binom{k}{0} = 1 = \binom{k+1}{0}, \quad \binom{k}{1} + \binom{k}{0} \underset{\underset{(4)}{\uparrow}}{=} \binom{k+1}{1}, \quad \cdots,$$

$$\binom{k}{j} + \binom{k}{j-1} \underset{\underset{(4)}{\uparrow}}{=} \binom{k+1}{j}, \quad \cdots, \quad \binom{k}{k} = 1 = \binom{k+1}{k+1}$$

we have

$$(x + a)^{k+1} = \binom{k+1}{0}x^{k+1} + \binom{k+1}{1}ax^k + \cdots + \binom{k+1}{j}a^jx^{k-j+1} + \cdots + \binom{k+1}{k+1}a^{k+1}$$

Thus, Conditions I and II of the principle of mathematical induction are satisfied, and formula (2) is therefore true for all n. ∎

■ The case $n = 2$ of the Binomial Theorem, $(a + b)^2$, was known to Euclid in 300 BC, but the general law seems to have been discovered by the Persian mathematician and astronomer Omar Khayyám (1044?–1123?), who is also well known as the author of the *Rubaiyat,* a collection of four-line poems making observations on the human condition. Omar Khayyám did not state the Binomial Theorem explicitly, but he claimed to have a method for extracting third, fourth, fifth roots, and so on. A little study shows that one must know the Binomial Theorem to create such a method.

The heart of the Binomial Theorem is the formula for the numerical coefficients, and, as we saw, they can be written out in a symmetric triangular form. The Pascal triangle appears first in the books of Yang Hui (about 1270) and Chu Shih-chie (1303). Pascal's name is attached to the triangle because of the many applications he made of it, especially to counting and probability. In establishing these results, he was one of the earliest users of mathematical induction.

Many people worked on the proof of the Binomial Theorem, which was finally completed for all n (including complex numbers) by Niels Abel (1802–1829). ■

8.5

Exercise 8.5

In Problems 1–12, evaluate each expression by hand. Use a graphing utility to verify your answer.

1. $\begin{pmatrix} 5 \\ 3 \end{pmatrix}$ **2.** $\begin{pmatrix} 7 \\ 3 \end{pmatrix}$ **3.** $\begin{pmatrix} 7 \\ 5 \end{pmatrix}$ **4.** $\begin{pmatrix} 9 \\ 7 \end{pmatrix}$ **5.** $\begin{pmatrix} 50 \\ 49 \end{pmatrix}$ **6.** $\begin{pmatrix} 100 \\ 98 \end{pmatrix}$

7. $\begin{pmatrix} 1000 \\ 1000 \end{pmatrix}$ **8.** $\begin{pmatrix} 1000 \\ 0 \end{pmatrix}$ **9.** $\begin{pmatrix} 55 \\ 23 \end{pmatrix}$ **10.** $\begin{pmatrix} 60 \\ 20 \end{pmatrix}$ **11.** $\begin{pmatrix} 47 \\ 25 \end{pmatrix}$ **12.** $\begin{pmatrix} 37 \\ 19 \end{pmatrix}$

In Problems 13–24, expand each expression using the Binomial Theorem.

13. $(x + 1)^5$ **14.** $(x - 1)^5$ **15.** $(x - 2)^6$ **16.** $(x + 3)^4$

17. $(3x + 1)^4$ **18.** $(2x + 3)^5$ **19.** $(x^2 + y^2)^5$ **20.** $(x^2 - y^2)^6$

21. $(\sqrt{x} + \sqrt{2})^6$ **22.** $(\sqrt{x} - \sqrt{3})^4$ **23.** $(ax + by)^5$ **24.** $(ax - by)^4$

In Problems 25–38, use the Binomial Theorem to find the indicated coefficient or term.

25. The coefficient of x^6 in the expansion of $(x + 3)^{10}$

26. The coefficient of x^3 in the expansion of $(x - 3)^{10}$

27. The coefficient of x^7 in the expansion of $(2x - 1)^{12}$

28. The coefficient of x^3 in the expansion of $(2x + 1)^{12}$

29. The coefficient of x^7 in the expansion of $(2x + 3)^9$

30. The coefficient of x^2 in the expansion of $(2x - 3)^9$

31. The fifth term in the expansion of $(x + 3)^7$

32. The third term in the expansion of $(x - 3)^7$

33. The third term in the expansion of $(3x - 2)^9$

34. The sixth term in the expansion of $(3x + 2)^8$

35. The coefficient of x^0 in the expansion of $\left(x^2 + \dfrac{1}{x}\right)^{12}$

36. The coefficient of x^0 in the expansion of $\left(x - \dfrac{1}{x^2}\right)^9$

37. The coefficient of x^4 in the expansion of $\left(x - \dfrac{2}{x}\right)^{10}$

38. The coefficient of x^2 in the expansion of $\left(\sqrt{x} + \dfrac{3}{x}\right)^8$

39. Use the Binomial Theorem to find the numerical value of $(1.001)^5$ correct to five decimal places.
[*Hint:* $(1.001)^5 = (1 + 10^{-3})^5$]

40. Use the Binomial Theorem to find the numerical value of $(0.998)^6$ correct to five decimal places.

41. Show that $\dbinom{n}{n} = 1$.

42. Show that, if n and j are integers with $0 \le j \le n$, then
$$\binom{n}{j} = \binom{n}{n-j}$$
Thus, conclude that the Pascal triangle is symmetric with respect to a vertical line drawn from the topmost entry.

43. If n is a positive integer, show that
$$\binom{n}{0} + \binom{n}{1} + \cdots + \binom{n}{n} = 2^n$$
[*Hint:* $2^n = (1 + 1)^n$; now use the Binomial Theorem.]

44. If n is a positive integer, show that
$$\binom{n}{0} - \binom{n}{1} + \binom{n}{2} - \cdots + (-1)^n\binom{n}{n} = 0$$

45. $\dbinom{5}{0}\left(\dfrac{1}{4}\right)^5 + \dbinom{5}{1}\left(\dfrac{1}{4}\right)^4\left(\dfrac{3}{4}\right) + \dbinom{5}{2}\left(\dfrac{1}{4}\right)^3\left(\dfrac{3}{4}\right)^2 + \dbinom{5}{3}\left(\dfrac{1}{4}\right)^2\left(\dfrac{3}{4}\right)^3 + \dbinom{5}{4}\left(\dfrac{1}{4}\right)\left(\dfrac{3}{4}\right)^4 + \dbinom{5}{5}\left(\dfrac{3}{4}\right)^5 = ?$

46. *Stirling's Formula* for approximating $n!$ when n is large is given by
$$n! \approx \sqrt{2n\pi}\left(\dfrac{n}{e}\right)^n\left(1 + \dfrac{1}{12n - 1}\right)$$
Calculate 12!, 20!, and 25!. Then use Stirling's formula to approximate 12!, 20!, and 25!.

8.6

Sets and Counting

Sets

A **set** is a well-defined collection of distinct objects. The objects of a set are called its **elements.** By **well-defined,** we mean that there is a rule that enables us to determine whether a given object is an element of the set. If a set has no elements, it is called the **empty set,** or **null set,** and is denoted by the symbol ∅.

Because the elements of a set are distinct, we never repeat elements. Thus, we would never write $\{1, 2, 3, 2\}$; the correct listing is $\{1, 2, 3\}$. Furthermore, because a set is a collection, the order in which the elements are listed is immaterial. Thus, $\{1, 2, 3\}$, $\{1, 3, 2\}$, $\{2, 1, 3\}$, and so on, all represent the same set.

E X A M P L E   1   *Writing the Elements of a Set*

Write the set consisting of the possible outcomes from tossing a coin twice. Use H for "heads" and T for "tails."

Solution   In tossing a coin twice, we can get heads each time, HH; or heads the first time and tails the second, HT; or tails the first time and heads the second, TH; or tails each time, TT. Because no other possibilities exist, the set of outcomes is

$$\{HH, HT, TH, TT\}$$

■

If two sets A and B have precisely the same elements, then we say that A and B are **equal** and write $A = B$.

If each element of a set A is also an element of a set B, then we say that A is a **subset** of B and write $A \subseteq B$.

If $A \subseteq B$ and $A \neq B$, then we say that A is a **proper subset** of B and write $A \subset B$.

Thus, if $A \subseteq B$, every element in set A is also in set B, but B may or may not have additional elements. If $A \subset B$, every element in A is also in B, and B has at least one element not found in A.

Finally, we agree that the empty set is a subset of every set; that is,

$$\varnothing \subseteq A \qquad \text{for any set } A$$

E X A M P L E   2   *Finding All the Subsets of a Set*

Write down all the subsets of the set $\{a, b, c\}$.

Solution   To organize our work, we write down all the subsets with no elements, then those with one element, then those with two elements, and finally those with three elements. These will give us all the subsets. Do you see why?

0 ELEMENTS	1 ELEMENT	2 ELEMENTS	3 ELEMENTS
$\varnothing$	$\{a\}, \{b\}, \{c\}$	$\{a, b\}, \{b, c\}, \{a, c\}$	$\{a, b, c\}$

■

■   Now work Problem 21.

Intersection; Union

If A and B are sets, the **intersection** of A with B, denoted $A \cap B$, is the set consisting of elements that belong to *both A and B*. The **union** of A with B, denoted $A \cup B$, is the set consisting of elements that belong to *either A or B, or both*.

E X A M P L E   3   *Finding the Intersection and Union of Sets*

Let $A = \{1, 3, 5, 8\}$, $B = \{3, 5, 7\}$, and $C = \{2, 4, 6, 8\}$. Find:

(a) $A \cap B$ (b) $A \cup B$ (c) $B \cap (A \cup C)$

Solution   (a) $A \cap B = \{1, 3, 5, 8\} \cap \{3, 5, 7\} = \{3, 5\}$
(b) $A \cup B = \{1, 3, 5, 8\} \cup \{3, 5, 7\} = \{1, 3, 5, 7, 8\}$
(c) $B \cap (A \cup C) = \{3, 5, 7\} \cap [\{1, 3, 5, 8\} \cup \{2, 4, 6, 8\}]$
$= \{3, 5, 7\} \cap \{1, 2, 3, 4, 5, 6, 8\} = \{3, 5\}$

■

■ Now work Problem 5.

Usually, in working with sets, we designate a **universal set,** the set consisting of all the elements we wish to consider. Once a universal set has been designated, we can consider elements of the universal set not found in a given set.

Complement

> If *A* is a set, the **complement** of *A*, denoted *A'*, is the set consisting of all the elements not in *A*.

E X A M P L E   4

Finding the Complement of a Set

If the universal set is $U = \{1, 2, 3, 4, 5, 6, 7, 8, 9\}$, and if $A = \{1, 3, 5, 7, 9\}$, then $A' = \{2, 4, 6, 8\}$. ■

Notice that: $A \cup A' = U$ and $A \cap A' = \varnothing.$

■ Now work Problem 13.

FIGURE 15

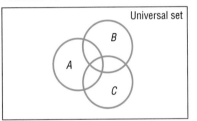

It is often helpful to draw pictures of sets. Such pictures, called **Venn diagrams,** represent sets as circles enclosed in a rectangle, which represents the universal set. Such diagrams often help us to visualize various relationships among sets. See Figure 15.

If we know that $A \subseteq B$, we might use the Venn diagram in Figure 16(a). If we know that *A* and *B* have no elements in common, that is, if $A \cap B = \varnothing$, we might use the Venn diagram in Figure 16(b).

FIGURE 16

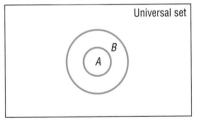

(a) $A \subseteq B$

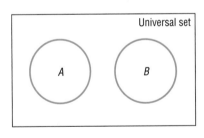

(b) $A \cap B = \varnothing$

Figures 17(a), 17(b), and 17(c) use Venn diagrams to illustrate the definitions of intersection, union, and complement, respectively.

FIGURE 17

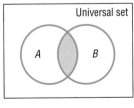

(a) $A \cap B$

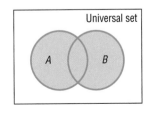

(b) $A \cup B$

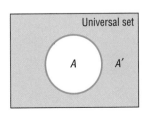

(c) A'

Counting

As you count the number of students in a classroom or the number of pennies in your pocket, what you are really doing is matching, on a one-to-one basis, each object to be counted with the counting numbers 1, 2, 3, . . . , n, for some number n. If a set A matched up in this fashion with the set $\{1, 2, \ldots, 25\}$, you would conclude that there are 25 elements in the set A. We use the notation $n(A) = 25$ to indicate that there are 25 elements in the set A.

Because the empty set has no elements, we write

$$n(\varnothing) = 0$$

If the number of elements in a set is a nonnegative integer, we say the set is **finite.** Otherwise, it is **infinite.** We shall concern ourselves only with finite sets.

From Example 2, we can see that a set with 3 elements has $2^3 = 8$ subsets. In fact, it can be shown that a set with n elements has exactly 2^n subsets. This fact has an important application to computers, which we take up at the end of this section.

E X A M P L E   5   *Analyzing Survey Data*

In a survey of 100 college students, 35 were registered in College Algebra, 52 were registered in Introduction to Computer Science, and 18 were in both courses. How many were registered in neither course?

Solution   First, we let A = Set of students in College Algebra
B = Set of students in Introduction to Computer Science

FIGURE 18

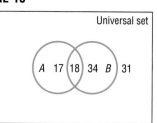

Then the information tells us that

$$n(A) = 35 \qquad n(B) = 52 \qquad n(A \cap B) = 18$$

Refer to Figure 18. Do you see how the numerical entries were determined? Based on the diagram, we conclude that $17 + 18 + 34 = 69$ students were registered in at least one of the two courses. Since 100 students were surveyed, it follows that $100 - 69 = 31$ were registered in neither course. ■

■ Now work Problem 35.

The conclusions drawn in Example 5 lead us to formulate a general counting formula. If we count the elements in each of two sets A and B, we necessarily count twice any elements that are in both A and B, that is, those elements in $A \cap B$. Thus, to count correctly the elements that are in A or B, that is, to find $n(A \cup B)$, we need to subtract those in $A \cap B$ from $n(A) + n(B)$.

Theorem   If A and B are finite sets, then

Counting Formula

$$n(A \cup B) = n(A) + n(B) - n(A \cap B) \tag{1}$$

A special case of the counting formula (1) occurs if A and B have no elements in common. In this case, $A \cap B = \varnothing$ so that $n(A \cap B) = 0$.

Theorem   If two sets A and B have no elements in common, then

Addition Principle
of Counting

$$n(A \cup B) = n(A) + n(B) \qquad (2)$$

EXAMPLE 6   *Counting the Number of Possible Codes*

A certain code is to consist of either a letter of the alphabet or a digit, but not both. How many codes are possible?

Solution   Let the sets A and B be defined as

$$A = \text{Set of letters in the alphabet}$$
$$B = \text{Set of digits } \{0, 1, 2, \ldots, 9\}$$

Then

$$n(A) = 26 \qquad n(B) = 10$$

Because letters and digits are different, $A \cap B = \varnothing$. The number of ways either a letter or a digit can be chosen is, therefore,

$$n(A \cup B) = n(A) + n(B) = 26 + 10 = 36$$

Application to Computers

Information stored in a computer may be thought of as a series of switches, which are either on or off and are denoted by either the number 0 (off) or the number 1 (on). These numbers are the binary digits, or **bits.** A **register** holds a certain fixed number of bits. For example, the Z-80 microprocessor has 8 bit registers; the PDP-11 minicomputer has 16 bit registers, and the IBM-370 computer has 32 bit registers. Thus, a Z-80 register may hold an entry that looks like this: 01111001 (8 bits). We wish to find out how many different representations are possible in a given register.

We proceed in steps, first looking at a hypothetical 3 bit register. Look again at the solution to Example 2 and arrange all the subsets of $\{a, b, c\}$ as shown in Table 3. As the table illustrates, the number of subsets of a set with 3 elements equals the number of different representations in a 3 bit register. A set with n elements has 2^n subsets; thus, an n bit register has 2^n representations. So an 8 bit register can hold $2^8 = 256$ different symbols, a 16 bit register can hold $2^{16} = 65,536$ different symbols, and a 32 bit register can hold $2^{32} \approx 4.3 \times 10^9$ different symbols.

TABLE 3

a	b	c	SUBSET
0	0	0	$\varnothing$
1	0	0	$\{a\}$
0	1	0	$\{b\}$
0	0	1	$\{c\}$
1	1	0	$\{a, b\}$
0	1	1	$\{b, c\}$
1	0	1	$\{a, c\}$
1	1	1	$\{a, b, c\}$

8.6

Exercise 8.6

In Problems 1–10, use $A = \{1, 3, 5, 7, 9\}$, $B = \{1, 5, 6, 7\}$, and $C = \{1, 2, 4, 6, 8, 9\}$ to find each set.

1. $A \cup B$
2. $A \cup C$
3. $A \cap B$
4. $A \cap C$
5. $(A \cup B) \cap C$
6. $(A \cap C) \cup (B \cap C)$
7. $(A \cap B) \cup C$
8. $(A \cup B) \cup C$
9. $(A \cup C) \cap (B \cup C)$
10. $(A \cap B) \cap C$

In Problems 11–20, use U = Universal set = {0, 1, 2, 3, 4, 5, 6, 7, 8, 9}, A = {1, 3, 4, 5, 9}, B = {2, 4, 6, 7, 8}, and C = {1, 3, 4, 6} to find each set.

11. A' **12.** C' **13.** $(A \cap B)'$ **14.** $(B \cup C)'$

15. $A' \cup B'$ **16.** $B' \cap C'$ **17.** $(A \cap C')'$ **18.** $(B' \cup C)'$

19. $(A \cup B \cup C)'$ **20.** $(A \cap B \cap C)'$

21. Write down all the subsets of $\{a, b, c, d\}$.

22. Write down all the subsets of $\{a, b, c, d, e\}$.

23. If $n(A) = 15$, $n(B) = 20$, and $n(A \cap B) = 10$, find $n(A \cup B)$.

24. If $n(A) = 20$, $n(B) = 40$, and $n(A \cup B) = 35$, find $n(A \cap B)$.

25. If $n(A \cup B) = 50$, $n(A \cap B) = 10$, and $n(B) = 20$, find $n(A)$.

26. If $n(A \cup B) = 60$, $n(A \cap B) = 40$, and $n(A) = n(B)$, find $n(A)$.

In Problems 27–34, use the information given in the figure.

27. How many are in set A?

28. How many are in set B?

29. How many are in A or B?

30. How many are in A and B?

31. How many are in A but not C?

32. How many are not in A?

33. How many are in A and B and C?

34. How many are in A or B or C?

35. *Analyzing Survey Data* In a consumer survey of 500 people, 200 indicated that they would be buying a major appliance within the next month; 150 indicated they would buy a car, and 25 said they would purchase both a major appliance and a car. How many will purchase neither? How many will purchase only a car?

36. *Analyzing Survey Data* In a student survey, 200 indicated that they would attend Summer Session I and 150 indicated Summer Session II. If 75 students plan to attend both summer sessions and 275 indicated that they would attend neither session, how many students participated in the survey?

37. *Analyzing Survey Data* In a survey of 100 investors in the stock market,

50 owned shares in IBM	20 owned shares in both IBM and GE
40 owned shares in AT&T	15 owned shares in both AT&T and GE
45 owned shares in GE	5 owned shares in all three
20 owned shares in both IBM and AT&T	

(a) How many of the investors surveyed did not have shares in any of the three companies?

(b) How many owned just IBM shares?

(c) How many owned just GE shares?

(d) How many owned neither IBM nor GE?

(e) How many owned either IBM or AT&T but no GE?

38. *Classifying Blood Types* Human blood is classified as either Rh+ or Rh−. Blood is also classified by type: A, if it contains an A antigen; B, if it contains a B antigen; AB, if it contains both A and B antigens; and O, if it contains neither antigen. Draw a Venn diagram illustrating the various blood types. Based on this classification, how many different kinds of blood are there?

39. Make up a problem different from any found in the text that requires the addition principle of counting to solve. Give it to a friend to solve and critique.

40. Investigate the notion of counting as it relates to infinite sets. Write an essay on your findings.

8.7

Permutations and Combinations

Counting plays a major role in many diverse areas, such as probability, statistics, and computer science. In this section we shall look at special types of counting problems and develop general formulas for solving them.

We begin with an example that will demonstrate a general counting principle.

E X A M P L E   1   *Counting the Number of Possible Meals*

The fixed-price dinner at a restaurant provides the following choices:

Appetizer: soup or salad

Entree: baked chicken, broiled beef patty, baby beef liver, or roast beef au jus

Dessert: ice cream or cheese cake

How many different meals can be ordered?

Solution   Ordering such a meal requires three separate decisions:

CHOOSE AN APPETIZER	**CHOOSE AN ENTREE**	**CHOOSE A DESSERT**
2 choices	4 choices	2 choices

Look at the **tree diagram** in Figure 19. We see that, for each choice of appetizer, there are 4 choices of entrees. And for each of these $2 \cdot 4 = 8$ choices, there are 2 choices for dessert. Thus, there are a total of

$$2 \cdot 4 \cdot 2 = 16$$

different meals that can be ordered.

FIGURE 19

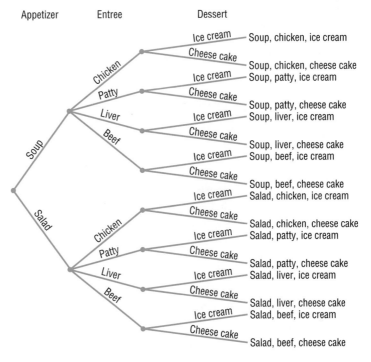

Example 1 illustrates a general counting principle.

Theorem
Multiplication Principle
of Counting

If a task consists of a sequence of choices in which there are p selections for the first choice, q selections for the second choice, r selections for the third choice, and so on, then the task of making these selections can be done in

$$p \cdot q \cdot r \cdot \ldots$$

different ways.   ■

EXAMPLE 2   *Counting Airport Codes*

The International Airline Transportation Association (IATA) assigns three-letter codes to represent airport locations. For example, JFK represents Kennedy International in New York. How many different airport codes are possible?

Solution   The task consists of making three selections. Each selection requires choosing a letter of the alphabet (26 choices). Thus, by the multiplication principle, there are

$$26 \cdot 26 \cdot 26 = 17{,}576$$

different airport codes.   ■

In Example 2, we were allowed to repeat a letter. For example, a valid airport code is FLL (Ft. Lauderdale International Airport), in which the letter L appears twice. In the next example, such repetition is not allowed.

EXAMPLE 3   *Counting without Repetition*

Suppose that we wish to establish a three-letter code using any of the 26 letters of the alphabet, but we require that no letter be used more than once. How many different three-letter codes are there?

Solution   The task consists of making three selections. The first selection requires choosing from 26 letters. Because no letter can be used more than once, the second selection requires choosing from 25 letters. The third selection requires choosing from 24 letters. (Do you see why?) By the multiplication principle, there are

$$26 \cdot 25 \cdot 24 = 15{,}600$$

different three-letter codes with no letter repeated.   ■

■ Now work Problems 25 and 29.

Example 3 illustrates a type of counting problem referred to as a *permutation*.

Permutation   A **permutation** is an ordered arrangement of n distinct objects without repetitions. The symbol $P(n, r)$ represents the number of permutations of n distinct objects, taken r at a time, where $r \leq n$.

For example, the question posed in Example 3 asks for the number of ways the 26 letters of the alphabet can be arranged using three nonrepeated letters. The answer is

$$P(26, 3) = 26 \cdot 25 \cdot 24 = 15{,}600$$

To arrive at a formula for $P(n, r)$, we note that the task of obtaining an ordered arrangement of n objects in which only $r \leq n$ of them are used, without repeating any of them, requires making r selections. For the first selection, there are n choices; for the second selection, there are $n - 1$ choices; for the third selection, there are $n - 2$ choices; . . . ; for the rth selection, there are $n - (r - 1)$ choices. By the multiplication principle, we have

$$
\begin{array}{cccc}
\text{1st} & \text{2nd} & \text{3rd} & \text{rth} \\
\end{array}
$$
$$
\begin{aligned}
P(n, r) &= n \cdot (n - 1) \cdot (n - 2) \cdot \ldots \cdot [n - (r - 1)] \\
&= n \cdot (n - 1) \cdot (n - 2) \cdot \ldots \cdot (n - r + 1)
\end{aligned}
$$

This formula for $P(n, r)$ can be compactly written using factorial notation*:

$$
\begin{aligned}
P(n, r) &= n \cdot (n - 1) \cdot (n - 2) \cdot \ldots \cdot (n - r + 1) \\
&= n \cdot (n - 1) \cdot (n - 2) \cdot \ldots \cdot (n - r + 1) \cdot \frac{(n - r) \cdot \ldots \cdot 3 \cdot 2 \cdot 1}{(n - r) \cdot \ldots \cdot 3 \cdot 2 \cdot 1} = \frac{n!}{(n - r)!}
\end{aligned}
$$

Theorem
Number of Permutations
of n Distinct Objects
Taken r at a Time

The number of different arrangements of n objects using $r \leq n$ of them, in which

1. the n objects are distinct,
2. once an object is used it cannot be repeated, and
3. order is important,

is given by the formula

$$
P(n, r) = \frac{n!}{(n - r)!} \tag{1}
$$

■

E X A M P L E   4

Evaluate: (a) $P(7, 3)$ (b) $P(6, 1)$ (c) $P(52, 5)$

Solution   We shall work parts (a) and (b) in two ways.

(a) $P(7, 3) = \underbrace{7 \cdot 6 \cdot 5}_{\text{3 factors}} = 210$

or

$$
P(7, 3) = \frac{7!}{(7 - 3)!} = \frac{7!}{4!} = \frac{7 \cdot 6 \cdot 5 \cdot 4!}{4!} = 210
$$

(b) $P(6, 1) = \underbrace{6}_{\text{1 factor}} = 6$

or

$$
P(6, 1) = \frac{6!}{(6 - 1)!} = \frac{6!}{5!} = \frac{6 \cdot 5!}{5!} = 6
$$

FIGURE 20

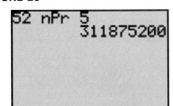

```
52 nPr 5
         311875200
```

(c) Figure 20 shows the solution using a TI-82 graphing calculator.

Thus, $P(52, 5) = 311,875,200$.

■

■   Now work Problem 1.

*Recall that $0! = 1$, $1! = 1$, $2! = 2 \cdot 1$, $\cdots$, $n! = n(n - 1) \cdot \ldots \cdot 3 \cdot 2 \cdot 1$.

E X A M P L E   5   In how many ways can 5 people be lined up?

Solution   The 5 people are obviously distinct. Once a person is in line, that person will not be repeated elsewhere in the line; and, in lining up people, order is important. Thus, we have a permutation of 5 objects taken 5 at a time. We can line up the 5 people in

$$P(5, 5) = \underbrace{5 \cdot 4 \cdot 3 \cdot 2 \cdot 1}_{5 \text{ factors}} = 5! = 120 \text{ ways}$$ ∎

■ Now work Problem 31.

Combinations

In a permutation, order is important; for example, the arrangements *ABC, CAB, BAC, . . .* are considered different arrangements of the letters *A, B,* and *C*. In many situations, though, order is unimportant. For example, in the card game of poker, the order in which the cards are received does not matter; it is the *combination* of the cards that matters.

Combination   A **combination** is an arrangement, without regard to order, of n distinct objects without repetitions. The symbol C(n, r) represents the number of combinations of n distinct objects taken r at a time, where r ≤ n.

E X A M P L E   6   *Listing Combinations*

List all the combinations of the 4 objects *a, b, c, d* taken 2 at a time. What is *C*(4, 2)?

Solution   One combination of *a, b, c, d* taken 2 at a time is

$$ab$$

The object *ba* is excluded, because order is not important in a combination. The list of all such combinations (convince yourself of this) is

$$ab, \quad ac, \quad ad, \quad bc, \quad bd, \quad cd$$

Thus,

$$C(4, 2) = 6$$ ∎

We can find a formula for *C(n, r)* by noting that the only difference between a permutation and a combination is that we disregard order in combinations. Thus, to determine *C(n, r)*, we need only eliminate from the formula for *P(n, r)* the number of permutations that were simply rearrangements of a given set of *r* objects. But that is easily determined from the formula for *P(n, r)* by calculating *P(r, r) = r!*. So, if we divide *P(n, r)* by *r!*, we will have the desired formula for *C(n, r)*:

$$C(n, r) = \frac{P(n, r)}{r!} \underset{\underset{\text{Use formula (1)}}{\uparrow}}{=} \frac{n!/(n-r)!}{r!} = \frac{n!}{(n-r)!r!}$$

We have proved the following result.

Theorem
Number of Combinations
of *n* Distinct Objects
Taken *r* at a Time

The number of different arrangements of *n* objects using $r \le n$ of them, in which

1. the *n* objects are distinct,
2. once an object is used, it cannot be repeated, and
3. order is not important

is given by the formula

$$C(n, r) = \frac{n!}{(n - r)!r!} \qquad (2)$$

Based on formula (2), we discover that the symbol $C(n, r)$ and the symbol $\binom{n}{r}$ for the binomial coefficients are, in fact, the same. Thus, the Pascal triangle (see Section 8.5) can be used to find the value of $C(n, r)$. However, because it is more practical and convenient, we shall use formula (2) instead.

E X A M P L E   7

Using Formula (2)

Use formula (2) to find the value of each expression.

(a) $C(3, 1)$ (b) $C(6, 3)$ (c) $C(n, n)$ (d) $C(n, 0)$ (e) $C(52, 5)$

Solution

(a) $C(3, 1) = \dfrac{3!}{(3 - 1)!1!} = \dfrac{3!}{2!1!} = \dfrac{3 \cdot 2 \cdot 1}{2 \cdot 1 \cdot 1} = 3$

(b) $C(6, 3) = \dfrac{6!}{(6 - 3)!3!} = \dfrac{6 \cdot 5 \cdot 4 \cdot 3!}{3! \cdot 3!} = \dfrac{6 \cdot 5 \cdot 4}{6} = 20$

(c) $C(n, n) = \dfrac{n!}{(n - n)!n!} = \dfrac{n!}{0!n!} = \dfrac{1}{1} = 1$

(d) $C(n, 0) = \dfrac{n!}{(n - 0)!0!} = \dfrac{n!}{n!0!} = \dfrac{1}{1} = 1$

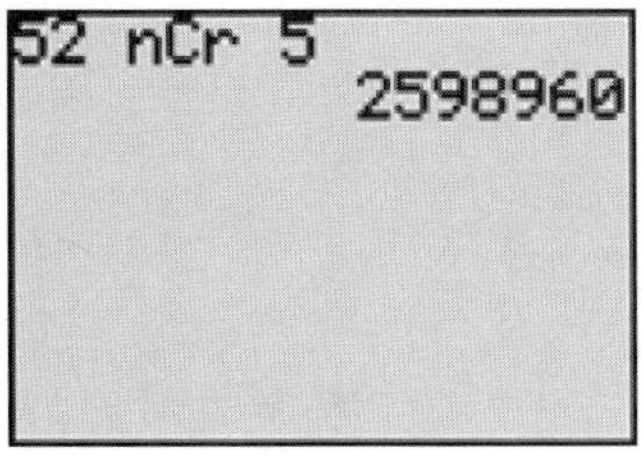

(e) Figure 21 shows the solution using a TI-82 graphing calculator.

Thus, $C(52, 5) = 2,598,960$.

FIGURE 21

52 nCr 5
             2598960

Now work Problem 9.

E X A M P L E   8

Forming Committees

How many different committees of 3 people can be formed from a pool of 7 people?

Solution

The 7 people are, of course, distinct. More important, though, is the observation that the order of being selected on a committee is not significant. Thus, the problem asks for the number of combinations of 7 objects taken 3 at a time:

$$C(7, 3) = \frac{7!}{4!3!} = \frac{7 \cdot 6 \cdot 5 \cdot 4!}{4!3!} = \frac{7 \cdot 6 \cdot 5}{6} = 35$$

E X A M P L E   9   *Forming Committees*

In how many ways can a committee consisting of 2 faculty members and 3 students be formed if there are 6 faculty members and 10 students eligible to serve on the committee?

Solution   The problem can be separated into two parts: the number of ways the faculty members can be chosen, $C(6, 2)$, and the number of ways the student members can be chosen, $C(10, 3)$. By the multiplication principle, the committee can be formed in

$$C(6, 2) \cdot C(10, 3) = \frac{6!}{4!2!} \cdot \frac{10!}{7!3!} = \frac{6 \cdot 5 \cdot 4!}{4!2!} \cdot \frac{10 \cdot 9 \cdot 8 \cdot 7!}{7!3!}$$

$$= \frac{30}{2} \cdot \frac{720}{6} = 1800 \text{ ways} \qquad \blacksquare$$

■ Now work Problem 47.

Permutations with Repetition

Recall that a permutation involves counting *distinct* objects. A permutation in which some of the objects are repeated is called a **permutation with repetition.** Some books refer to this as a **nondistinguishable permutation.**

Let's look at an example.

E X A M P L E   1 0   *Forming Different Words*

How many different words can be formed using all the letters in the word REARRANGE?

Solution   Each word formed will have 9 letters: 3 R's, 2 A's, 2 E's, 1 N, and 1 G. To construct each word, we need to fill in 9 positions with the 9 letters:

$$\overline{1} \ \overline{2} \ \overline{3} \ \overline{4} \ \overline{5} \ \overline{6} \ \overline{7} \ \overline{8} \ \overline{9}$$

The process of forming a word consists of five tasks:

Task 1: Choose the positions for the 3 R's.

Task 2: Choose the positions for the 2 A's.

Task 3: Choose the positions for the 2 E's.

Task 4: Choose the position for the 1 N.

Task 5: Choose the position for the 1 G.

Task 1 can be done in $C(9, 3)$ ways. There then remain 6 positions to be filled, so Task 2 can be done in $C(6, 2)$ ways. There remain 4 positions to be filled, so Task 3 can be done in $C(4, 2)$ ways. There remain 2 positions to be filled, so Task 4 can be done in $C(2, 1)$ ways. The last position can be filled in $C(1, 1)$ way. Using the Multiplication Principle, the number of possible words that can be found is

$$C(9, 3) \cdot C(6, 2) \cdot C(4, 2) \cdot C(2, 1) \cdot C(1, 1) = \frac{9!}{3! \cdot 6!} \frac{6!}{2! \cdot 4!} \frac{4!}{2! \cdot 2!} \frac{2!}{1! \cdot 1!} \frac{1!}{0! \cdot 1!}$$

$$= \frac{9!}{3! \cdot 2! \cdot 2! \cdot 1! \cdot 1!} \qquad \blacksquare$$

The form of the answer to Example 10 is suggestive of a general result. Had the letters in REARRANGE each been different, there would have been $P(9, 9) = 9!$ possible words formed. This is the numerator of the answer. The presence of 3 R's, 2 A's, and 2 E's reduces the number of different words, as the entries in the denominator illustrate. We are led to the following result:

Theorem

Permutations with Repetition

The number of permutations of n objects of which n_1 are of one kind, n_2 are of a second kind, ..., and n_k are of a kth kind is given by

$$\frac{n!}{n_1! \cdot n_2! \cdot \ldots \cdot n_k!} \tag{3}$$

where $n = n_1 + n_2 + \cdots + n_k$.   ■

E X A M P L E   1 1   *Arranging Flags*

How many different vertical arrangements are there of 8 flags if 4 are white, 3 are blue, and 1 is red?

Solution   We seek the number of permutations of 8 objects, of which 4 are of one kind, 3 of a second kind, and 1 of a third kind. Using formula (3), we find that there are

$$\frac{8!}{4! \cdot 3! \cdot 1!} = \frac{8 \cdot 7 \cdot 6 \cdot 5 \cdot 4!}{4! \cdot 3! \cdot 1!} = 280 \text{ different arrangements}$$ ■

■   Now work Problem 53.

8.7

Exercise 8.7

In Problems 1–8, find the value of each permutation. Verify your results using a graphing calculator.

1. $P(6, 2)$ **2.** $P(7, 2)$ **3.** $P(5, 5)$ **4.** $P(4, 4)$

5. $P(8, 0)$ **6.** $P(9, 0)$ **7.** $P(8, 3)$ **8.** $P(8, 5)$

In Problems 9–16, use formula (2) to find the value of each combination. Verify your results using a graphing calculator.

9. $C(8, 2)$ **10.** $C(8, 6)$ **11.** $C(6, 4)$ **12.** $C(6, 2)$

13. $C(15, 15)$ **14.** $C(18, 1)$ **15.** $C(26, 13)$ **16.** $C(18, 9)$

17. List all the permutations of 5 objects $a, b, c, d,$ and e taken 3 at a time. What is $P(5, 3)$?

18. List all the permutations of 5 objects $a, b, c, d,$ and e taken 2 at a time. What is $P(5, 2)$?

19. List all the permutations of 4 objects 1, 2, 3, and 4 taken 3 at a time. What is $P(4, 3)$?

20. List all the permutations of 6 objects, 1, 2, 3, 4, 5, and 6 taken 3 at a time. What is $P(6, 3)$?

21. List all the combinations of the 5 objects $a, b, c, d,$ and e taken 3 at a time. What is $C(5, 3)$?

22. List all the combinations of the 5 objects $a, b, c, d,$ and e taken 2 at a time. What is $C(5, 2)$?

23. List all the combinations of the 4 objects 1, 2, 3, and 4 taken 3 at a time. What is $C(4, 3)$?

24. List all the combinations of the 6 objects 1, 2, 3, 4, 5, and 6 taken 3 at a time. What is $C(6, 3)$?

25. A man has 5 shirts and 3 ties. How many different shirt and tie combinations can he wear?

26. A woman has 3 blouses and 5 skirts. How many different outfits can she wear?

27. *Forming Codes* How many two-letter codes can be formed using the letters *A, B, C,* and *D*? Repeated letters are allowed.

28. *Forming Codes* How many two-letter codes can be formed using the letters *A, B, C, D,* and *E*? Repeated letters are allowed.

29. *Forming Numbers* How many three-digit numbers can be formed using the digits 0 and 1? Repeated digits are allowed.

30. *Forming Numbers* How many three-digit numbers can be formed using the digits 0, 1, 2, 3, 4, 5, 6, 7, 8, and 9? Repeated digits are allowed.

31. In how many ways can 4 people be lined up?

32. In how many ways can 5 different boxes be stacked?

33. *Forming Codes* How many different three-letter codes are there if only the letters *A, B, C, D,* and *E* can be used and no letter can be used more than once?

34. *Forming Codes* How many different four-letter codes are there if only the letters *A, B, C, D, E,* and *F* can be used and no letter can be used more than once?

35. *Arranging Letters* How many arrangements are there of the letters in the word MONEY?

36. *Arranging Digits* How many arrangements are there of the digits in the number 51,342?

37. *Establishing Committees* In how many ways can a committee of 4 students be formed from a pool of 7 students?

38. *Establishing Committees* In how many ways can a committee of 3 professors be formed from a department having 8 professors?

39. *Possible Answers on a True/False Test* How many arrangements of answers are possible for a true/false test with 10 questions?

40. *Possible Answers on a Multiple-choice Test* How many arrangements of answers are possible in a multiple-choice test with 5 questions, each of which has 4 possible answers?

41. How many four-digit numbers can be formed using the digits 0, 1, 2, 3, 4, 5, 6, 7, 8, and 9 if the first digit cannot be 0? Repeated digits are allowed.

42. How many five-digit numbers can be formed using the digits 0, 1, 2, 3, 4, 5, 6, 7, 8, and 9 if the first digit cannot be 0 or 1? Repeated digits are allowed.

43. *Arranging Books* Five different mathematics books are to be arranged on a student's desk. How many arrangements are possible?

44. *Forming License Plate Numbers* How many different license plate numbers can be made using 2 letters followed by 4 digits selected from the digits 0 through 9, if

(a) Letters and digits may be repeated?

(b) Letters may be repeated, but digits are not repeated?

(c) Neither letters nor digits may be repeated?

45. *Stock Portfolios* As a financial planner, you are asked to select one stock each from the following groups: 8 DOW stocks, 15 NASDAQ stocks, and 4 global stocks. How many different portfolios are possible?

46. *Combination Locks* A combination lock has 50 numbers on it. To open it, you turn to a number, then rotate clockwise to a second number, and then counterclockwise to the third number. How many different lock combinations are there?

47. A student dance committee is to be formed consisting of 2 boys and 3 girls. If the membership is to be chosen from 4 boys and 8 girls, how many different committees are possible?

48. *Baseball Teams* A baseball team has 15 members. Four of the players are pitchers, and the remaining 11 members can play any position. How many different teams of 9 players can be formed?

49. The student relations committee of a college consists of 2 administrators, 3 faculty members, and 5 students. There are 4 administrators, 8 faculty members, and 20 students eligible to serve. How many different committees are possible?

50. *Football Teams* A defensive football squad consists of 25 players. Of these, 10 are linemen, 10 are linebackers, and 5 are safeties. How many different teams of 5 linemen, 3 linebackers, and 3 safeties can be formed?

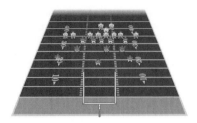

51. *Baseball* In the American Baseball League, a designated hitter may be used. How many batting orders is it possible for a manager to use? (There are 9 regular players on a team.)

52. *Baseball* In the National Baseball League, the pitcher usually bats ninth. If this is the case, how many batting orders are possible for a manager to use?

53. *Forming Words* How many different 9 letter words (real or imaginary) can be formed from the letters in the word ECONOMICS?

54. *Forming Words* How many different 11 letter words (real or imaginary) can be formed from the letters in the word MATHEMATICS?

55. *Senate Committees* The U.S. Senate has 100 members. Suppose it is desired to place each senator on exactly 1 of 7 possible committees. The first committee has 22 members, the second has 13, the third has 10, the fourth has 5, the fifth has 16, and the sixth and seventh have 17 apiece. In how many ways can these committees be formed?

56. *World Series* In the World Series the American League team (A) and the National League team (N) play until one team wins four games. If the sequence of winners is designated by letters (for example, *NAAAA* means the National League team won the first game and the American League won the next four), how many different sequences are possible?

57. *Basketball Teams* A basketball team has 6 players who play guard (2 of 5 starting positions). How many different teams are possible, assuming that the remaining 3 positions are filled and it is not possible to distinguish a left guard from a right guard?

58. *Basketball Teams* On a basketball team of 12 players, 2 only play center, 3 only play guard, and the rest play forward (5 players on a team: 2 forwards, 2 guards, and 1 center). How many different teams are possible, assuming that it is not possible to distinguish left and right guards and left and right forwards?

59. *Selecting Objects* An urn contains 7 white balls and 3 red balls. Three balls are selected. In how many ways can the 3 balls be drawn from the total of 10 balls:
(a) If 2 balls are white and 1 is red?
(b) If all 3 balls are white?
(c) If all 3 balls are red?

60. *Selecting Objects* An urn contains 15 red balls and 10 white balls. Five balls are selected. In how many ways can the 5 balls be drawn from the total of 25 balls:
(a) If all balls are red?
(b) If 3 balls are red and 2 are white?
(c) If at least 4 are red balls?

61. *Programming Exercise* When both n and r are large, finding $C(n, r)$ on a computer may lead to integers too large to compute. To avoid this, we can approximate the values of $C(n, r)$. One way to do this is the following:

$$C(40, 20) = \frac{40!}{20!20!} = \frac{40 \cdot 39 \cdot 38 \cdot \ldots \cdot 21}{20 \cdot 19 \cdot 18 \cdot \ldots \cdot 1} = \frac{40}{20} \cdot \frac{39}{19} \cdot \frac{38}{18} \cdot \ldots \cdot \frac{21}{1}$$

$$\approx 2.000 \cdot 2.053 \cdot 2.111 \cdot \ldots \cdot 21.000 = 1.3784652 \times 10^{11}$$

(a) Write a program that inputs two integers N and R and computes $C(N, R)$ using formula (1).
(b) Use the program to determine where overflow occurs on your computer.
(c) Write a program that inputs two integers N and R and computes $C(N, R)$ by the approximation technique shown above.
(d) Compare the answers found in parts (a) and (c).

62. Make up a problem different from any found in the text that requires the multiplication principle of counting to solve. Give it to a friend to solve and critique.

63. Make up a problem different from any found in the text that requires a permutation to solve. Give it to a friend to solve and critique.

64. Make up a problem different from any found in the text that requires a combination to solve. Give it to a friend to solve and critique.

65. Explain the difference between a permutation and a combination. Give an example to illustrate your explanation.

8.8

Probability

Probability is an area of mathematics that deals with experiments that yield random results yet admit a certain regularity. Such experiments do not always produce the same result or outcome, so the result of any one observation is not predictable. However, the results of the experiment over a long period do produce regular patterns that enable us to predict with remarkable accuracy.

E X A M P L E   1

Tossing a Fair Coin

In tossing a fair coin, we know that the outcome is either a head or a tail. On any particular throw, we cannot predict what will happen, but, if we toss the coin many times, we observe that the number of times a head comes up is approximately equal to the number of times we get a tail. It seems reasonable, therefore, to assign a probability of $\frac{1}{2}$ that a head comes up and a probability of $\frac{1}{2}$ that a tail comes up. ∎

Probability Models

The discussion in Example 1 constitutes the construction of a **probability model** for the experiment of tossing a fair coin once. A probability model has two components: a sample space and an assignment of probabilities. A **sample space** S is a set whose elements represent all the possibilities that can occur as a result of the experiment. Each element of S is called an **outcome.** To each outcome, we assign a number, called the **probability** of that outcome, which has two properties:

1. Each probability is nonnegative.
2. The sum of all the probabilities equals 1.

Thus, if a probability model has the sample space

$$S = \{e_1, e_2, \ldots, e_n\}$$

where $e_1, e_2, \ldots, e_n$ are the possible outcomes, and if $P(e_1), P(e_2), \ldots, P(e_n)$ denote the respective probabilities of these outcomes, then

$$P(e_1) \geq 0, \quad P(e_2) \geq 0, \quad \ldots, \quad P(e_n) \geq 0 \qquad (1)$$
$$P(e_1) + P(e_2) + \cdots + P(e_n) = 1 \qquad (2)$$

Let's look at an example.

E X A M P L E   2   *Constructing a Probability Model*

An experiment consists of rolling a fair die once.* Construct a probability model for this experiment.

Solution   A sample space S consists of all the possibilities that can occur. Because rolling the die will result in one of six faces showing, the sample space S consists of

$$S = \{1, 2, 3, 4, 5, 6\}$$

FIGURE 22

Because the die is fair, one face is no more likely to occur than another. As a result, our assignment of probabilities is

$$P(1) = \tfrac{1}{6} \qquad P(2) = \tfrac{1}{6}$$

$$P(3) = \tfrac{1}{6} \qquad P(4) = \tfrac{1}{6}$$

$$P(5) = \tfrac{1}{6} \qquad P(6) = \tfrac{1}{6}$$ ∎

Suppose that a die is loaded so that the probability assignments are

$$P(1) = 0, \quad P(2) = 0, \quad P(3) = \frac{1}{3}, \quad P(4) = \frac{2}{3}, \quad P(5) = 0, \quad P(6) = 0$$

This assignment would be made if the die was loaded so that only a 3 or a 4 could occur and the 4 is twice as likely as the 3 to occur. This assignment is consistent with the definition since each assignment is between 0 and 1, and the sum of all the probability assignments equals 1.   ∎

■   Now work Problem 13.

E X A M P L E   3   *Constructing a Probability Model*

An experiment consists of tossing a coin. The coin is weighted so that heads (H) is three times as likely to occur as tails (T). Construct a probability model for this experiment.

Solution   The sample space S is $S = \{H, T\}$. If x denotes the probability that a tail occurs, then

$$P(T) = x \quad \text{and} \quad P(H) = 3x$$

Since the sum of the probabilities of the possible outcomes must equal 1, we have

$$P(T) + P(H) = x + 3x = 1$$
$$4x = 1$$
$$x = \frac{1}{4}$$

Thus, we assign the probabilities

$$P(T) = \frac{1}{4} \qquad P(H) = \frac{3}{4}$$ ∎

■   Now work Problem 17.

*A die is a cube with each face having either 1, 2, 3, 4, 5, or 6 dots on it. See Figure 22.

E X A M P L E   4

FIGURE 23

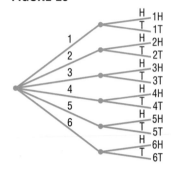

An experiment consists of tossing a fair die and then a fair coin. Construct a probability model for this experiment.

Solution   A tree diagram is helpful in listing all the possible outcomes. See Figure 23. The sample space consists of the outcomes

$$S = \{1H, 1T, 2H, 2T, 3H, 3T, 4H, 4T, 5H, 5T, 6H, 6T\}$$

The die and the coin are fair; thus, no one outcome is more likely to occur than another. As a result, we assign the probability $\frac{1}{12}$ to each of the 12 outcomes.   ■

In working with probability models, the term **event** is used to describe a set of possible outcomes of the experiment. Thus, an event E is some subset of the sample space S. The **probability of an event** E, $E \neq 0$, denoted by $P(E)$, is defined as the sum of the probabilities of the outcomes in E. If $E = \varnothing$, then $P(E) = 0$; if $E = S$, then $P(E) = P(S) = 1$.

E X A M P L E   5   *Finding the Probability of an Event*

For the experiment described in Example 4, what is the probability that an even number followed by a head occurs?

Solution   The event E, an even number followed by a head, consists of

$$E = \{2H, 4H, 6H\}$$

The probability of E is

$$P(E) = P(2H) + P(4H) + P(6H) = \tfrac{1}{12} + \tfrac{1}{12} + \tfrac{1}{12} = \tfrac{1}{4}$$ ■

The next result, called the **Additive Rule,** may be used to find the probability of the union of two events.

Theorem   For any two events E and F,
Additive rule

$$P(E \cup F) = P(E) + P(F) - P(E \cap F) \qquad (3)$$

■

If E and F are disjoint, so that $E \cap F = \varnothing$, then formula (3) takes the form

Mutually Exclusive Events

$$P(E \cup F) = P(E) + P(F) \qquad (4)$$

When formula (4) applies, we say that E and F are **mutually exclusive events.**

E X A M P L E   6   *Using Formulas (3) and (4)*

(a) If $P(E) = 0.2$, $P(F) = 0.3$, and $P(E \cap F) = 0.1$, find $P(E \cup F)$.

(b) If $P(E) = 0.2$, $P(F) = 0.3$, and E, F are mutually exclusive, find $P(E \cup F)$.

Solution   (a) We use the Additive Rule, formula (3).

$$P(E \cup F) = P(E) + P(F) - P(E \cap F) = 0.2 + 0.3 - 0.1 = 0.4$$

(b) Since E, F are mutually exclusive, we use formula (4).

$$P(E \cup F) = P(E) + P(F) = 0.2 + 0.3 = 0.5$$ ■

A Venn diagram can sometimes be used to obtain probabilities. To construct a Venn diagram representing the information in Example 6(a), we draw two sets E and F. We begin with the fact that $P(E \cap F) = 0.1$. See Figure 24(a). Then, since $P(E) = 0.2$ and $P(F) = 0.3$, we fill in E with $0.2 - 0.1 = 0.1$ and F with $0.3 - 0.1 = 0.2$. See Figure 24(b). Since $P(S) = 1$, we complete the diagram by inserting $1 - [0.1 + 0.1 + 0.2] = 0.6$. See Figure 24(c). Now it is easy to see, for example, that the probability of F, but not E, is 0.2. Also, the probability of neither E nor F is 0.6.

FIGURE 24

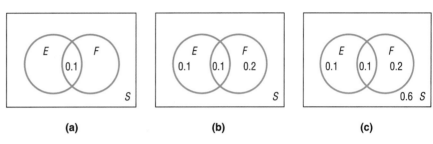

(a)	(b)	(c)

■ Now work Problem 23.

Equally Likely Outcomes

When the same probability is assigned to each outcome of the sample space, the experiment is said to have **equally likely outcomes.**

Theorem
Probability for Equally
Likely Outcomes

If an experiment has n equally likely outcomes, and if the number of ways an event E can occur is m, then the probability of E is

$$P(E) = \frac{\text{Number of ways that } E \text{ can occur}}{\text{Number of all logical possibilities}} = \frac{m}{n} \qquad (5)$$

Thus, if S is the sample space of this experiment, then

$$P(E) = \frac{n(E)}{n(S)} \qquad (6)$$

■

Based on (6), an alternative method of solution of Example 5 is

$$P(E) = \frac{n(E)}{n(S)} = \frac{3}{12} = \frac{1}{4}$$

E X A M P L E 7

Computing Probabilities for Equally Likely Outcomes

A jar contains 10 marbles; 5 are solid color, 4 are speckled, and 1 is clear.

(a) If one marble is picked at random, what is the probability it is speckled?

(b) If one marble is picked at random, what is the probability it is clear or a solid color?

Solution

The experiment is an example of one in which the outcomes are equally likely; that is, no one marble is more likely to be picked than another. If S is the sample space, then there are 10 possible outcomes in S, so $n(S) = 10$.

(a) Define the event E: speckled marble is picked. There are 4 ways E can occur. Thus,

$$P(E) = \frac{n(E)}{n(S)} = \frac{4}{10} = 0.4$$

(b) Define the events F: clear marble is picked and G: solid color marble is picked. Then there is 1 way for F to occur and 5 ways for G to occur. Thus,

$$P(F) = \frac{n(F)}{n(S)} = \frac{1}{10}, \qquad P(G) = \frac{n(G)}{n(S)} = \frac{5}{10}$$

We seek the probability of the event F or G, that is, $P(F \cup G)$. Since F, G are mutually exclusive, we use (4).

$$P(F \cup G) = P(F) + P(G) = \frac{1}{10} + \frac{5}{10} = \frac{6}{10} = 0.6 \qquad \blacksquare$$

■   Now work Problem 27.

EXAMPLE 8

The Game of Craps

In the game of "craps," two fair dice are rolled. If the total of the faces equals 7 or 11, you win. If the totals are 2, 3, or 12, you have thrown craps and you lose. In all other cases, you throw again.

(a) What is the probability that you will win?
(b) What is the probability that you will lose?
(c) What is the probability that you will need to throw again?

Solution

We begin by constructing a probability model for the experiment. The tree diagram in Figure 25 will help us to see all the possibilities.

Because the dice are fair, no one of the 36 possible outcomes in the sample space S is more likely to occur than any other. Thus, we have equally likely outcomes with $n(S) = 36$.

(a) The event E, "the dice total 7 or 11," consists of the outcomes

$$E = \{(1, 6), (2, 5), (3, 4), (4, 3), (5, 2), (6, 1), (5, 6), (6, 5)\}$$

Because $n(E) = 8$, we have

$$P(E) = \frac{n(E)}{n(S)} = \frac{8}{36} = \frac{2}{9} \approx 0.222$$

(b) The event F, "the dice total 2, 3, or 12," consists of the outcomes

$$F = \{(1, 1), (1, 2), (2, 1), (6, 6)\}$$

Because $n(F) = 4$,

$$P(F) = \frac{n(F)}{n(S)} = \frac{4}{36} = \frac{1}{9} \approx 0.111$$

FIGURE 25

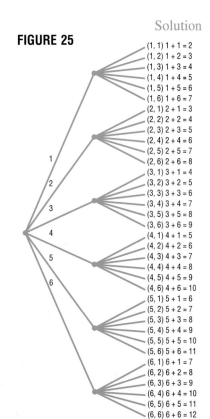

(c) The number of possibilities that require that you throw again is $36 - n(E) - n(F) = 36 - 8 - 4 = 24$. Thus, the probability that another throw is required is

$$\frac{24}{36} = \frac{2}{3} \approx 0.667$$

■

■ Now work Problem 29.

Applications Involving Permutations and Combinations

E X A M P L E 9   *Computing Probabilities*

Because of a mistake in packaging, 5 defective phones were packaged with 15 good ones. All phones look alike and have equal probability of being chosen. Three phones are selected.

(a) What is the probability that all 3 are defective?
(b) What is the probability that exactly 2 are defective?
(c) What is the probability that at least 2 are defective?

Solution   The sample space S consists of the number of ways 3 objects can be selected from 20 objects, that is, the number of combinations of 20 things taken 3 at a time.

$$n(S) = C(20, 3) = \frac{20!}{17! \cdot 3!} = \frac{20 \cdot 19 \cdot 18}{6} = 1140$$

Each of these outcomes is equally likely to occur.

(a) If E is the event "3 are defective," then the number of elements in E is the number of ways the 3 defective phones can be chosen from the 5 defective phones: $C(5, 3) = 10$. Thus, the probability of E is

$$P(E) = \frac{n(E)}{n(S)} = \frac{10}{1140} \approx 0.0088$$

(b) If F is the event "exactly 2 are defective" and 3 phones are selected, then the number of elements in F is the number of ways to select 2 defective phones from the 5 defective phones and 1 good phone from the 15 good ones. The first of these can be done in $C(5, 2)$ ways and the second in $C(15, 1)$ ways. By the Multiplication Principle, the event F can occur in

$$C(5, 2)C(15, 1) = \frac{5!}{3! \cdot 2!} \frac{15!}{14! \cdot 1!} = 10 \cdot 15 = 150 \text{ ways}$$

The probability of F is therefore

$$P(F) = \frac{n(F)}{n(S)} = \frac{150}{1140} \approx 0.1316$$

(c) The event G, "at least two are defective," when 3 are chosen, is equivalent to requiring that either exactly 2 defective are chosen or exactly 3 defective are chosen. That is, $G = E \cup F$. Since E and F are mutually exclusive (it is not possible to select 2 defective phones and, at the same time, select 3 defective phones), we find

$$P(G) = P(E) + P(F) \approx 0.0088 + 0.1316 = 0.1404$$ ■

■ Now work Problem 43.

E X A M P L E 1 0   *Tossing a Coin*

A fair coin is tossed 6 times.

(a) What is the probability of obtaining exactly 5 heads and one tail?

(b) What is the probability of obtaining between 4 and 6 heads, inclusive?

Solution   The number of elements in the sample space S is found using the Multiplication Principle. Each toss results in a head (H) or a tail (T). Since the coin is tossed 6 times, we have

$$n(S) = \underbrace{2 \cdot 2 \cdot \ldots \cdot 2}_{6 \text{ tosses}} = 2^6 = 64$$

The outcomes are equally likely since the coin is fair.

(a) Any sequence that contains 5 heads and 1 tail is determined once the position of the 5 heads (or 1 tail) is known. The number of ways we can position 5 heads in a sequence of 6 slots is $C(6, 5) = 6$. The probability of the event E: exactly 5 heads and one tail is

$$P(E) = \frac{n(E)}{n(S)} = \frac{C(6, 5)}{2^6} = \frac{6}{64} \approx 0.0938$$

(b) Let F be the event: between 4 and 6 heads, inclusive. To obtain between 4 and 6 heads is equivalent to the event: either 4 heads or 5 heads or 6 heads. Since each of these is mutually exclusive (it is impossible to obtain both 4 heads and 5 heads when tossing a coin 6 times), we have

$$P(F) = P(4 \text{ heads or 5 heads or 6 heads})$$
$$= P(4 \text{ heads}) + P(5 \text{ heads}) + P(6 \text{ heads})$$

The probabilities on the right are obtained as in part (a). Thus

$$P(F) = \frac{C(6, 4)}{2^6} + \frac{C(6, 5)}{2^6} + \frac{C(6, 6)}{2^6} = \frac{15}{64} + \frac{6}{64} + \frac{1}{64} = \frac{22}{64} \approx 0.3438$$ ■

HISTORICAL FEATURE   ■ Set theory, counting, and probability first took form as a systematic theory in the exchange of letters (1654) between Pierre de Fermat (1601–1665) and Blaise Pascal (1623–1662). They discussed the problem of how to divide the stakes in a game that is interrupted before completion, knowing how many points each player needs to win. Fermat solved the problem by listing all possibilities and counting the favorable ones, whereas Pascal made use of the triangle that now bears his name. As mentioned in the text, the entries in Pascal's triangle are equivalent to $C(n, r)$. This recognition of the role of $C(n, r)$ in counting is the foundation of all further developments.

The first book on probability, the work of Christian Huygens (1629–1695), appeared in 1657. In it, the notion of mathematical expectations is explored. This allows the calculation of the profit or loss a gambler may expect, knowing the probabilities involved in the game (see the Historical Problems that follow).

It is interesting to note that Girolamo Cardano (1501–1576) wrote a treatise on probability, but it was not published until 1663 in Cardano's collected works, and this was too late to have any effect on the development of the theory.

In 1713, the posthumously published *Ars Conjectandi* of Jacob Bernoulli gave the theory the form it would have until 1900. In the current century, both combinatorics (counting) and probability have undergone rapid development due to the use of computers.

A final comment about notation. The notations $C(n, r)$ and $P(n, r)$ are variants of a form of notation developed in England after 1830. The notation $\binom{n}{r}$ for $C(n, r)$ goes back to Leonhard Euler (1707–1783), but is now losing ground because it has no clearly related symbolism of the same type for permutations. The set symbols $\cup$ and $\cap$ were introduced by Giuseppe Peano (1858–1932) in 1888 in a slightly different context. The inclusion symbol $\subset$ was introduced by E. Schroeder (1841–1902) about 1890. The treatment of set theory in the text is due to George Boole (1815–1864), who wrote $A + B$ for $A \cup B$ and AB for $A \cap B$ (statisticians still use AB for $A \cap B$). ■

HISTORICAL PROBLEMS

■ 1. *The Problem Discussed by Fermat and Pascal* A game between two equally skilled players, A and B, is interrupted when A needs 2 points to win and B needs 3 points. In what proportion should the stakes be divided? [*Note:* If each play results in 1 point for either player, at most four more plays will decide the game.]

(a) *Fermat's solution* List all possible outcomes that will end the game to form the sample space (for example, *ABAA*, *ABBB*, etc.). The probabilities for A to win and B to win then determine how the stakes should be divided.

(b) *Pascal's solution* Use combinations to determine the number of ways the 2 points needed for A to win could occur in four plays. Then use combinations to determine the number of ways the 3 points needed for B to win could occur. This is trickier than it looks, since A can win with 2 points in either two plays, three plays, or four plays. Compute the probabilities and compare with the results in part (a).

2. *Huygens's Mathematical Expectation* In a game with n possible outcomes with probabilities $p_1, p_2, \ldots, p_n$, suppose that the *net* winnings are $w_1, w_2, \ldots, w_n$, respectively. Then the mathematical expectation is

$$E = p_1 w_1 + p_2 w_2 + \cdots + p_n w_n$$

The number E represents the profit or loss per game in the long run. The following problems are a modification of those of Huygens:

(a) A fair die is tossed. A gambler wins $3 if he throws a 6 and $6 if he throws a 5. What is his expectation? [*Note:* $w_1 = w_2 = w_3 = w_4 = 0$]

(b) A gambler plays the same game as in part (a), but now the gambler must pay $1 to play. This means $w_5 = \$5$, $w_6 = \$2$, and $w_1 = w_2 = w_3 = w_4 = -\$1$. What is the expectation? ■

8.8

Exercise 8.8

In Problems 1–6, construct a probability model for each experiment.

1. Tossing a fair coin twice

2. Tossing two fair coins once

3. Tossing two fair coins, then a fair die   **5.** Tossing three fair coins once

4. Tossing a fair coin, a fair die, and then a fair coin   **6.** Tossing one fair coin three times

In Problems 7–12, use the spinners shown below, and construct a probability model for each experiment.

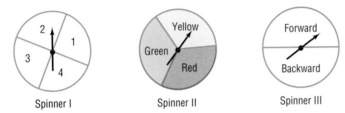

     Spinner I   Spinner II   Spinner III

7. Spin spinner I, then spinner II. What is the probability of getting a 2 or a 4, followed by Red?

8. Spin spinner III, then spinner II. What is the probability of getting Forward, followed by Yellow or Green?

9. Spin spinner I, then II, then III. What is the probability of getting a 1, followed by Red or Green, followed by Backward?

10. Spin spinner II, then I, then III. What is the probability of getting Yellow, followed by a 2 or a 4, followed by Forward?

11. Spin spinner I twice, then spinner II. What is the probability of getting a 2, followed by a 2 or a 4, followed by Red or Green?

12. Spin spinner III, then spinner I twice. What is the probability of getting Forward, followed by a 1 or a 3, followed by a 2 or a 4?

In Problems 13–16, consider the experiment of tossing a coin twice. The table lists six possible assignments of probabilities for this experiment. Using this table, answer the following questions.

13. Which of the assignments of probabilities are consistent with the definition of the probability of an outcome?

14. Which of the assignments of probabilities should be used if the coin is known to be fair?

15. Which of the assignments of probabilities should be used if the coin is known to always come up tails?

16. Which of the assignments of probabilities should be used if tails is twice as likely as heads to occur?

ASSIGNMENTS	SAMPLE SPACE			
	HH	**HT**	**TH**	**TT**
A	$\frac{1}{4}$	$\frac{1}{4}$	$\frac{1}{4}$	$\frac{1}{4}$
B	0	0	0	1
C	$\frac{3}{16}$	$\frac{5}{16}$	$\frac{5}{16}$	$\frac{3}{16}$
D	$\frac{1}{2}$	$\frac{1}{2}$	$-\frac{1}{2}$	$\frac{1}{2}$
E	$\frac{1}{8}$	$\frac{1}{4}$	$\frac{1}{4}$	$\frac{1}{8}$
F	$\frac{1}{9}$	$\frac{2}{9}$	$\frac{2}{9}$	$\frac{4}{9}$

17. *Assigning Probabilities* A coin is weighted so that heads is four times as likely as tails to occur. What probability should we assign to heads? to tails?

18. *Assigning Probabilities* A coin is weighted so that tails is twice as likely as heads to occur. What probability should we assign to heads? to tails?

19. *Assigning Probabilities* A die is weighted so that an odd-numbered face is twice as likely as an even-numbered face. What probability should we assign to each face?

20. *Assigning Probabilities* A die is weighted so that a six cannot appear. The other faces occur with the same probability. What probability should we assign to each face?

In Problems 21–24, find the probability of the indicated event if $P(A) = 0.30$ and $P(B) = 0.40$.

21. $P(A \cup B)$ if A, B are mutually exclusive

22. $P(A \cap B)$ if A, B are mutually exclusive

23. $P(A \cup B)$ if $P(A \cap B) = 0.15$

24. $P(A \cap B)$ if $P(A \cup B) = 0.6$

In Problems 25–28, a golf ball is selected at random from a container. If the container has 9 white balls, 8 green balls, and 3 orange ones, find the probability of each event.

25. The golf ball is white.

26. The golf ball is green.

27. The golf ball is white or green.

28. The golf ball is not white.

29. What is the probability of throwing a 6 or an 8 in a game of craps? (Consult Example 8.)

30. What is the probability of throwing a 5 or a 9 in a game of craps? (Consult Example 8.)

Problems 31–34 are based on a consumer survey of annual incomes in 100 households. The following table gives the data:

Income	$0–9999	$10,000–19,999	$20,000–29,999	$30,000–39,999	$40,000 or more
Number of households	5	35	30	20	10

31. What is the probability that a household has an annual income of $30,000 or more?

32. What is the probability that a household has an annual income between $10,000 and $29,999, inclusive?

33. What is the probability that a household has an annual income less than $20,000?

34. What is the probability that a household has an annual income of $20,000 or more?

35. *Surveys* In a survey about the number of TV sets in a house, the following probability table was constructed:

Number of TV sets	0	1	2	3	4 or more
Probability	0.05	0.24	0.33	0.21	0.17

Find the probability of a house having:

(a) 1 or 2 TV sets (b) 1 or more TV sets (c) 3 or fewer TV sets (d) 3 or more TV sets

(e) Less than 2 TV sets (f) Less than 1 TV set (g) 1, 2, or 3 TV sets (h) 2 or more TV sets

36. *Checkout Lines* Through observation it has been determined that the probability for a given number of people waiting in line at the "5 items or less" checkout register of a supermarket is:

Number waiting in line	0	1	2	3	4 or more
Probability	0.10	0.15	0.20	0.24	0.31

Find the probability of:
(a) At most 2 people in line (b) At least 2 people in line
(c) At least 1 person in line

37. *Winning a Lottery* In a certain lottery, there are ten balls, numbered 1, 2, 3, 4, 5, 6, 7, 8, 9, 10. Of these, five are drawn in the correct order. If you pick five numbers that match those drawn in the correct order, you win $1,000,000. What is the probability of winning such a lottery?

38. A committee of 6 people is to be chosen at random from a group of 14 people consisting of 2 supervisors, 5 skilled laborers, and 7 unskilled laborers. What is the probability that the committee chosen consists of 2 skilled and 4 unskilled laborers?

39. A fair coin is tossed 5 times.
(a) Find the probability that exactly 3 heads appear.
(b) Find the probability that no heads appear.

40. A fair coin is tossed 4 times.
(a) Find the probability that exactly 1 tail appears.
(b) Find the probability that no more than 1 tail appears.

41. A pair of fair dice are tossed 3 times.
(a) Find the probability that the sum of seven appears 3 times.
(b) Find the probability that a sum of 7 or 11 appears at least twice.

42. A pair of fair dice are tossed 5 times.
(a) Find the probability that the sum is never 2.
(b) Find the probability that the sum is never 7.

43. Through a mix-up on the production line, 5 defective TV's were shipped out with 25 good ones. If 5 are selected at random, what is the probability that all 5 are defective? What is the probability that at least 2 of them are defective?

44. In a shipment of 50 transformers, 10 are known to be defective. If 30 transformers are picked at random, what is the probability that all 30 are nondefective? Assume that all transformers look alike and have an equal probability of being chosen.

45. In a promotion, 50 silver dollars are placed in a bag, one of which is valued at more than $10,000. The winner of the promotion is given the opportunity to reach into the bag, while blindfolded, and pull out 5 coins. What is the probability that one of the 5 coins is the one valued at more than $10,000?

 46. Go to the library and look up the "birthday problem" in a book on probability. Write a brief essay about this problem and its solution.

Chapter Review

THINGS TO KNOW

Sequence	A function whose domain is the set of positive integers.
Factorials	$0! = 1$, $1! = 1$, $n! = n(n - 1) \cdot \ldots \cdot 3 \cdot 2 \cdot 1$ if $n \geq 2$
Arithmetic sequence	$a_1 = a$, $a_{n+1} = a_n + d$, where a = first term, d = common difference, $a_n = a + (n - 1)d$
Sum of the first n terms of an arithmetic sequence	$S_n = \dfrac{n}{2}[2a + (n - 1)d] = \dfrac{n}{2}(a + a_n)$
Geometric sequence	$a_1 = a$, $a_{n+1} = ra_n$; where a = first term, r = common ratio, $a_n = ar^{n-1}$, $r \neq 0$
Sum of the first n terms of a geometric sequence	$S_n = a\dfrac{1 - r^n}{1 - r}$, $r \neq 0, 1$
Infinite geometric series	$a + ar + \cdots + ar^{n-1} + \cdots = \displaystyle\sum_{k=1}^{\infty} ar^{k-1}$
Sum of an infinite geometric series	$\displaystyle\sum_{k=1}^{\infty} ar^{k-1} = \dfrac{a}{1 - r}$, $\lvert r \rvert < 1$
Principle of mathematical induction	Condition I: The statement is true for the natural number 1. Condition II: If the statement is true for some natural number k, it is also true for $k + 1$. Then the statement is true for all natural numbers.

Binomial coefficient	$\binom{n}{j} = \dfrac{n!}{j!(n-j)!}$	
Pascal triangle	See Figure 14.	
Binomial Theorem	$(x + a)^n = \binom{n}{0}x^n + \binom{n}{1}ax^{n-1} + \cdots + \binom{n}{j}a^j x^{n-j} + \cdots + \binom{n}{n}a^n$	
Set		Well-defined collection of distinct objects, called elements
Null set	$\varnothing$	Set that has no elements
Equality	$A = B$	A and B have the same elements
Subset	$A \subseteq B$	Each element of A is also an element of B.
Intersection	$A \cap B$	Set consisting of elements that belong to both A and B
Union	$A \cup B$	Set consisting of elements that belong to either A or B, or both
Universal set	U	Set consisting of all the elements we wish to consider
Complement	A'	Set consisting of elements of the universal set that are not in A
Finite set		The number of elements in the set is a nonnegative integer
Infinite set		A set that is not finite
Counting formula	$n(A \cup B) = n(A) + n(B) - n(A \cap B)$	
Addition principle		If $A \cap B = \varnothing$, then $n(A \cup B) = n(A) + n(B)$.
Multiplication principle		If a task consists of a sequence of choices in which there are p selections for the first choice, q selections for the second choice, and so on, then the task of making these selections can be done in $p \cdot q \cdots$ different ways.
Permutation	$P(n, r) = n(n-1) \cdot \ldots \cdot [n - (r-1)]$ $\qquad = \dfrac{n!}{(n-r)!}$	An ordered arrangement of n distinct objects without repetition
Combination	$C(n, r) = \dfrac{P(n, r)}{r!}$ $\qquad = \dfrac{n!}{(n-r)!r!}$	An arrangement, without regard to order, of n distinct objects without repetition
Permutations with repetition	$\dfrac{n!}{n_1! n_2! \cdots n_k!}$	The number of permutations of n objects of which n_1 are of one kind, n_2 are of a second kind, $\ldots$, and n_k are of a kth kind, where $n = n_1 + n_2 + \cdots + n_k$
Sample space		Set whose elements represent all the logical possibilities that can occur as a result of an experiment
Probability		A number assigned to each outcome of a sample space; the sum of all the probabilities of the outcomes equals 1
Additive rule	$P(E \cup F) = P(E) + P(F) - P(E \cap F)$	
Equally likely outcomes	$P(E) = \dfrac{n(E)}{n(S)}$	The same probability is assigned to each outcome.

How To:

Write down the terms of a sequence

Use summation notation

Identify an arithmetic sequence

Find the sum of the first n terms of an arithmetic sequence

Identify a geometric sequence

Find the sum of arithmetic and geometric sequences using a graphing utility

Find the sum of the first n terms of a geometric sequence

Find the sum of an infinite geometric series

Prove statements about natural numbers using mathematical induction

Apply the Binomial Theorem

Find unions, intersections, and complements of sets

Use Venn diagrams to illustrate sets

Recognize a permutation problem

Recognize a combination problem

Solve certain probability problems

Count the elements in a sample space

Draw a tree diagram

FILL-IN-THE-BLANK ITEMS

1. A(n) _____ is a function whose domain is the set of positive integers.

2. In a(n) _____ sequence, the difference between successive terms is always the same number.

3. In a(n) _____ sequence, the ratio of successive terms is always the same number.

4. The _____ _____ is a triangular display of the binomial coefficients.

5. $\binom{6}{2} =$ _____

6. The _____ of A with B consists of all elements in either A or B; the _____ of A with B consists of all elements in both A and B.

7. $P(5, 2) =$ _____; $C(5, 2) =$ _____

8. A(n) _____ is an ordered arrangement of n distinct objects.

9. A(n) _____ is an arrangement of n distinct objects without regard to order.

10. When the same probability is assigned to each outcome of a sample space, the experiment is said to have _____ _____ outcomes.

TRUE/FALSE ITEMS

T   F   **1.** A sequence is a function.

T   F   **2.** For arithmetic sequences, the difference of successive terms is always the same number.

T   F   **3.** For geometric sequences, the ratio of successive terms is always the same number.

T   F   **4.** Mathematical induction can sometimes be used to prove theorems that involve natural numbers.

T   F   **5.** $\binom{n}{j} = \dfrac{j!}{n!(n-j)!}$

T   F   **6.** The expansion of $(x + a)^n$ contains n terms.

T   F   **7.** $\displaystyle\sum_{i=1}^{n+1} i = 1 + 2 + 3 + \cdots + n$

T   F   **8.** The intersection of two sets is always a subset of their union.

T   F   **9.** $P(n, r) = \dfrac{n!}{r!}$

T   F   **10.** In a combination problem, order is not important.

T   F   **11.** In a permutation problem, once an object is used, it cannot be repeated.

T   F   **12.** The probability of an event can never equal 0.

REVIEW EXERCISES

In Problems 1–8, evaluate each expression by hand. Verify your results using a graphing utility.

1. 5!   2. 6!   3. $\binom{5}{2}$ 4. $\binom{8}{6}$ 5. $P(8, 3)$ 6. $P(7, 3)$ 7. $C(8, 3)$ 8. $C(7, 3)$

In Problems 9–16, write down the first five terms of each sequence.

9. $\left\{(-1)^n\left(\dfrac{n+3}{n+2}\right)\right\}$

10. $\{(-1)^{n+1}(2n+3)\}$

11. $\left\{\dfrac{2^n}{n^2}\right\}$

12. $\left\{\dfrac{e^n}{n}\right\}$

13. $a_1 = 3;\quad a_{n+1} = \frac{2}{3}a_n$

14. $a_1 = 4;\quad a_{n+1} = -\frac{1}{4}a_n$

15. $a_1 = 2;\quad a_{n+1} = 2 - a_n$

16. $a_1 = -3;\quad a_{n+1} = 4 + a_n$

In Problems 17–28, determine whether the given sequence is arithmetic, geometric, or neither. If the sequence is arithmetic, find the common difference and the sum of the first n terms. If the sequence is geometric, find the common ratio and the sum of the first n terms.

17. $\{n+5\}$

18. $\{4n+3\}$

19. $\{2n^3\}$

20. $\{2n^2 - 1\}$

21. $\{2^{3n}\}$

22. $\{3^{2n}\}$

23. $0, 4, 8, 12, \ldots$

24. $1, -3, -7, -11, \ldots$

25. $3, \frac{3}{2}, \frac{3}{4}, \frac{3}{8}, \frac{3}{16}, \ldots$

26. $5, -\frac{5}{3}, \frac{5}{9}, -\frac{5}{27}, \frac{5}{81}, \ldots$

27. $\frac{2}{3}, \frac{3}{4}, \frac{4}{5}, \frac{5}{6}, \ldots$

28. $\frac{3}{2}, \frac{5}{4}, \frac{7}{6}, \frac{9}{8}, \frac{11}{10}, \ldots$

In Problems 29–34, find the indicated term in each sequence: (a) by hand; (b) using a graphing utility.

29. 9th term of $3, 7, 11, 15, \ldots$

30. 8th term of $1, -1, -3, -5, \ldots$

31. 11th term of $1, \frac{1}{10}, \frac{1}{100}, \ldots$

32. 11th term of $1, 2, 4, 8, \ldots$

33. 9th term of $\sqrt{2}, 2\sqrt{2}, 3\sqrt{2}, \ldots$

34. 9th term of $\sqrt{2}, 2, 2^{3/2}, \ldots$

In Problems 35–38, find a general formula for each arithmetic sequence.

35. 7th term is 31;   20th term is 96

36. 8th term is -20;   17th term is -47

37. 10th term is 0;   18th term is 8

38. 12th term is 30;   22nd term is 50

In Problems 39–44, find the sum of each infinite geometric series.

39. $3 + 1 + \frac{1}{3} + \frac{1}{9} + \cdots$

40. $2 + 1 + \frac{1}{2} + \frac{1}{4} + \cdots$

41. $2 - 1 + \frac{1}{2} - \frac{1}{4} + \cdots$

42. $6 - 4 + \frac{8}{3} - \frac{16}{9} + \cdots$

43. $\displaystyle\sum_{k=1}^{\infty} 4\left(\frac{1}{2}\right)^{k-1}$

44. $\displaystyle\sum_{k=1}^{\infty} 3\left(-\frac{3}{4}\right)^{k-1}$

In Problems 45–50, use the principle of mathematical induction to show that the given statement is true for all natural numbers.

45. $3 + 6 + 9 + \cdots + 3n = \dfrac{3n}{2}(n+1)$

46. $2 + 6 + 10 + \cdots + (4n - 2) = 2n^2$

47. $2 + 6 + 18 + \cdots + 2 \cdot 3^{n-1} = 3^n - 1$

48. $3 + 6 + 12 + \cdots + 3 \cdot 2^{n-1} = 3(2^n - 1)$

49. $1^2 + 4^2 + 7^2 + \cdots + (3n - 2)^2 = \frac{1}{2}n(6n^2 - 3n - 1)$

50. $1 \cdot 3 + 2 \cdot 4 + 3 \cdot 5 + \cdots + n(n+2) = \dfrac{n}{6}(n+1)(2n+7)$

In Problems 51–54, expand each expression using the Binomial Theorem.

51. $(x+2)^5$ **52.** $(x-3)^4$ **53.** $(2x+3)^5$ **54.** $(3x-4)^4$

55. Find the coefficient of x^7 in the expansion of $(x+2)^9$. **57.** Find the coefficient of x^2 in the expansion of $(2x+1)^7$.

56. Find the coefficient of x^3 in the expansion of $(x-3)^8$. **58.** Find the coefficient of x^6 in the expansion of $(2x+1)^8$.

In Problems 59–66, use U = Universal set = {1, 2, 3, 4, 5, 6, 7, 8, 9}, A = {1, 3, 5, 7}, B = {3, 5, 6, 7, 8}, and C = {2, 3, 7, 8, 9} to find each set.

59. $A \cup B$ **60.** $B \cup C$ **61.** $A \cap C$ **62.** $A \cap B$

63. $A' \cup B'$ **64.** $B' \cap C'$ **65.** $(B \cap C)'$ **66.** $(A \cup B)'$

67. If $n(A) = 8$, $n(B) = 12$ and $n(A \cap B) = 3$, find $n(A \cup B)$.

68. If $n(A) = 12$, $n(A \cup B) = 30$, and $n(A \cap B) = 6$, find $n(B)$.

In Problems 69–74, use the information supplied in the figure:

69. How many are in A?

70. How many are in A or B?

71. How many are in A and C?

72. How many are not in B?

73. How many are in neither A nor C?

74. How many are in B but not in C?

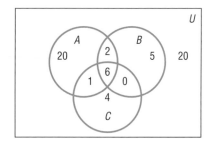

75. A clothing store sells pure wool and polyester/wool suits. Each suit comes in 3 colors and 10 sizes. How many suits are required for a complete assortment?

76. In connecting a certain electrical device, 5 wires are to be connected to 5 different terminals. How many different wirings are possible if 1 wire is connected to each terminal?

77. *Baseball* On a given day, the American Baseball League schedules 7 games. How many different outcomes are possible, assuming that each game is played to completion?

78. *Baseball* On a given day, the National Baseball League schedules 6 games. How many different outcomes are possible, assuming that each game is played to completion?

79. If 4 people enter a bus having 9 vacant seats, in how many ways can they be seated?

80. How many different arrangements are there of the letters in the word ROSE?

81. In how many ways can a squad of 4 relay runners be chosen from a track team of 8 runners?

82. A professor has 10 similar problems to put on a test with 3 problems. How many different tests can she design?

83. *Baseball* In how many different ways can the 14 baseball teams in the American League be paired without regard to which team is at home?

84. *Arranging Books on a Shelf* There are 5 different French books and 5 different Spanish books. How many ways are there to arrange them on a shelf if:
(a) Books of the same language must be grouped together, French on the left, Spanish on the right?
(b) French and Spanish books must alternate in the grouping, beginning with a French book?

85. *Telephone Numbers* Using the digits 0, 1, 2, . . . , 9, how many 7-digit numbers can be formed if the first digit cannot be 0 or 9 and if the last digit is greater than or equal to 2 and less than or equal to 3? Repeated digits are allowed.

86. *Home Choices* A contractor constructs homes with 5 different choices of exterior finish, 3 different roof arrangements, and 4 different window designs. How many different types of homes can be built?

87. *License Plate Possibilities* A license plate consists of 1 letter, excluding O and I, followed by a 4-digit number that cannot have a 0 in the lead position. How many different plates are possible?

88. Using the digits 0 and 1, how many different numbers consisting of 8 digits can be formed?

89. *Forming Different Words* How many different words can be formed using all the letters in the word MISSING?

90. *Arranging Flags* How many different vertical arrangements are there of 10 flags, if 4 are white, 3 are blue, 2 are green, and 1 is red?

91. *Forming Committees* A group of 9 people is going to be formed into committees of 4, 3, and 2 people. How many committees can be formed if:
(a) A person can serve on any number of committees?
(b) No person can serve on more than one committee?

92. *Forming Committees* A group consists of 5 men and 8 women. A committee of 4 is to be formed from this group, and policy dictates that at least 1 woman be on this committee.
(a) How many committees can be formed that contain exactly 1 man?
(b) How many committees can be formed that contain exactly 2 women?
(c) How many committees can be formed that contain at least 1 man?

93. From a box containing three 40 watt bulbs, six 60 watt bulbs, and eleven 75 watt bulbs, a bulb is drawn at random. What is the probability that the bulb is 40 watts? What is the probability that it is not a 75 watt bulb?

94. You have four $1 bills, three $5 bills, and two $10 bills in your wallet. If you pick a bill at random, what is the probability it will be a $1 bill?

95. Each of the letters in the word ROSE is written on an index card and the cards are then shuffled. What is the probability that, when the cards are dealt out, they spell the word ROSE?

96. Each of the numbers, 1, 2, . . . , 100 is written on an index card and the cards are then shuffled. If a card is selected at random, what is the probability that the number on the card is divisible by 5? What is the probability that the card selected either is a 1 or names a prime number?

97. *Computing Probabilities* Because of a mistake in packaging, a case of 12 bottles of red wine contained 5 Merlot and 7 Cabernet, each without labels. All the bottles look alike and have equal probability of being chosen. Three bottles are selected.
(a) What is the probability all 3 are Merlot?
(b) What is the probability exactly 2 are Merlot?
(c) What is the probability none is a Merlot?

98. *Tossing a Coin* A fair coin is tossed 10 times.
(a) What is the probability of obtaining exactly 5 heads?
(b) What is the probability of obtaining all heads?

99. *Constructing a Brick Staircase* A brick staircase has a total of 25 steps. The bottom step requires 80 bricks. Each successive step requires three less bricks than the prior step.
(a) How many bricks are required for the top step?
(b) How many bricks are required to build the staircase?

100. *Creating a Floor Design* A mosaic tile floor is designed in the shape of a trapezoid 30 feet wide at the base and 15 feet wide at the top. See Figure 5. The tiles, 12 inches by 12 inches, are to be placed so that each successive row contains one less tile than the row below. How many tiles will be required?

101. *Bouncing Balls* A ball is dropped from a height of 20 feet. Each time it strikes the ground, it bounces up to $\frac{3}{4}$ of the previous height.
(a) What height will the ball bounce up to after it strikes the ground for the third time?
(b) What is its height after it strikes the ground for the nth time?
(c) How many times does the ball need to strike the ground before its height is less than 6 inches?
(d) What total distance does the ball travel before it stops bouncing?

102. *Salary Increases* Your friend has just been hired at an annual salary of $20,000. If she expects to receive annual increases of 4%, what will her salary be as she begins her fifth year?

103. At the Milex tune-up and brake repair shop, the manager has found that a car will require a tune-up with a probability of 0.6, a brake job with a probability of 0.1, and both with a probability of 0.02.
(a) What is the probability that a car requires either a tune-up or a brake job?
(b) What is the probability that a car requires a tune-up but not a brake job?
(c) What is the probability that a car requires neither type of repair?

Appendix REVIEW

1. Topics From Algebra and Geometry
2. Polynomials and Rational Expressions
3. Radicals; Rational Exponents
4. Solving Equations
5. Completing the Square
6. Synthetic Division

1

Topics from Algebra and Geometry

Sets

When we want to treat a collection of similar but distinct objects as a whole, we use the idea of a **set.** For example, the set of *digits* consists of the collection of numbers 0, 1, 2, 3, 4, 5, 6, 7, 8, and 9. If we use the symbol D to denote the set of digits, then we can write

$$D = \{0, 1, 2, 3, 4, 5, 6, 7, 8, 9\}$$

In this notation, the braces { } are used to enclose the objects, or **elements,** in the set. This method of denoting a set is called the **roster method.** A second way to denote a set is to use **set-builder notation,** where the set D of digits is written as

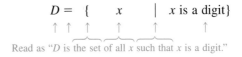

$$D = \{ \quad x \quad | \quad x \text{ is a digit}\}$$

Read as "D is the set of all x such that x is a digit."

E X A M P L E   1

Using Set-builder Notation and the Roster Method

(a) $E = \{x | x \text{ is an even digit}\} = \{0, 2, 4, 6, 8\}$

(b) $O = \{x | x \text{ is an odd digit}\} = \{1, 3, 5, 7, 9\}$ ∎

In listing the elements of a set, we do not list an element more than once because the elements of a set are distinct. Also, the order in which the elements are listed is not relevant. Thus, for example, {2, 3} and {3, 2} both represent the same set.

If every element of a set A is also an element of a set B, then we say that A is a **subset** of B. If two sets A and B have the same elements, then we say that A **equals** B. For example, {1, 2, 3} is a subset of {1, 2, 3, 4, 5}; and {1, 2, 3} equals {2, 3, 1}.

Real Numbers

Real numbers are represented by symbols such as

$$25, \quad 0, \quad -3, \quad \tfrac{1}{2}, \quad -\tfrac{5}{4}, \quad 0.125, \quad \sqrt{2}, \quad \pi, \quad \sqrt[3]{-2}, \quad 0.666\ldots$$

FIGURE 1

$\pi = \dfrac{C}{d}$

The set of **counting numbers,** or **natural numbers,** is the set $\{1, 2, 3, 4, \ldots \}$. (The three dots, called an **ellipsis,** indicate that the pattern continues indefinitely.) The set of **integers** is the set $\{ \ldots , -3, -2, -1, 0, 1, 2, 3, \ldots \}$. A **rational number** is a number that can be expressed as a quotient a/b of two integers, where the integer b cannot be 0. Examples of rational numbers are $\frac{3}{4}, \frac{5}{2}, \frac{0}{4}$, and $-\frac{2}{3}$. Since $a/1 = a$ for any integer a, every integer is also a rational number. Real numbers that are not rational are called **irrational.** Examples of irrational numbers are $\sqrt{2}$ and π (the Greek letter pi), which equals the constant ratio of the circumference to the diameter of a circle. See Figure 1.

Real numbers can be represented as **decimals.** Rational real numbers have decimal representations that either **terminate** or are nonterminating with **repeating** blocks of digits. For example, $\frac{3}{4} = 0.75$, which terminates; and $\frac{2}{3} = 0.666\ldots$, in which the digit 6 repeats indefinitely. Irrational real numbers have decimal representations that neither repeat nor terminate. For example, $\sqrt{2} = 1.414213\ldots$ and $\pi = 3.14159\ldots$. In practice, irrational numbers are generally represented by approximations. We use the symbol $\approx$ (read as "approximately equal to") to write $\sqrt{2} \approx 1.4142$ and $\pi \approx 3.1416$.

Often, letters are used to represent numbers. If the letter used is to represent *any* number from a given set of numbers, it is referred to as a **variable.** A **constant** is either a fixed number, such as 5, $\sqrt{2}$, and so on, or a letter that represents a fixed (possibly unspecified) number. In general, we will follow the practice of using letters near the beginning of the alphabet, such as a, b, and c, for constants and using those near the end, such as x, y, and z, as variables.

In working with expressions or formulas involving variables, the variables may only be allowed to take on values from a certain set of numbers, called the **domain of the variable.** For example, in the expression $1/x$, the variable x cannot take on the value 0, since division by 0 is not allowed.

It can be shown that there is a one-to-one correspondence between real numbers and points on a line. That is, every real number corresponds to a point on the line and, conversely, each point on the line has a unique real number associated with it. We establish this correspondence of real numbers with points on a line in the following manner.

We start with a line that is, for convenience, drawn horizontally. Pick a point on the line and label it O, for **origin.** Then pick another point some fixed distance to the right of O and label it U, for **unit.** The fixed distance, which may be 1 inch, 1 centimeter, 1 light-year, or any unit distance, determines the **scale.** We associate the real number 0 with the origin O and the number 1 with the point U. Refer to Figure 2. The point to the right of U that is twice as far from O as U is associated with the number 2. The point to the right of U that is three times as far from O as U is associated with the number 3. The point midway between O and U is assigned the number 0.5, or $\frac{1}{2}$. Corresponding points to the left of the origin O are assigned the numbers $-\frac{1}{2}, -1, -2, -3$, and so on. The real number x associated with a point P is called the **coordinate** of P, and the line whose points have been assigned coordinates is called the **real number line.** Notice in Figure 2 that we placed an arrowhead on the right end of the line to indicate the direction in which the assigned numbers increase. Figure 2 also shows the points associated with the irrational numbers $\sqrt{2}$ and π.

The real number line divides the real numbers into three classes: the **negative real numbers** are the coordinates of points to the left of the origin O; the real number **zero** is the coordinate of the origin O; the **positive real numbers** are the coordinates of points to the right of the origin O.

FIGURE 2

Real number line

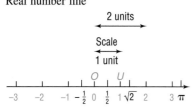

Let a and b be two real numbers. If the difference $a - b$ is positive, then we say that a is **greater than** b and write $a > b$. Alternatively, if $a - b$ is positive, we can also say that b is **less than** a and write $b < a$. Thus, $a > b$ and $b < a$ are equivalent statements.

On the real number line, if $a > b$, the point with coordinate a is to the right of the point with coordinate b. For example, $0 > -1$, $\pi > 3$, and $\sqrt{2} < 2$. Furthermore,

> $a > 0$ is equivalent to a is positive
>
> $a < 0$ is equivalent to a is negative

If the difference $a - b$ of two real numbers is positive or 0, that is, if $a > b$ or $a = b$, then we say that a is **greater than or equal to** b and write $a \geq b$. Alternatively, if $a \geq b$, we can also say that b is **less than or equal to** a and write $b \leq a$.

Statements of the form $a < b$ or $b > a$ are called **strict inequalities;** statements of the form $a \leq b$ or $b \geq a$ are called **nonstrict inequalities.** The symbols $>$, $<$, $\geq$, and $\leq$ are called **inequality signs.**

If x is a real number and $x \geq 0$, then x is either positive or 0. As a result, we describe the inequality $x \geq 0$ by saying that x is nonnegative.

Inequalities are useful in representing certain subsets of real numbers. In so doing, though, other variations of the inequality notation may be used.

E X A M P L E   2

Graphing Inequalities

(a) In the inequality $x > 4$, x is any number greater than 4. In Figure 3, we use a left parenthesis to indicate that the number 4 is not part of the graph.

FIGURE 3
$x > 4$

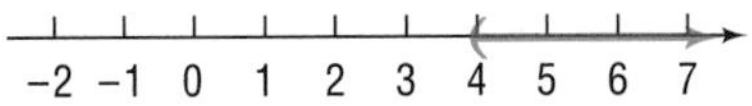

(b) In the inequality $4 < x \leq 6$, x is any number between 4 and 6, including 6 but excluding 4. In Figure 4, we use a right bracket to indicate that 6 is part of the graph.

FIGURE 4
$x > 4$ and $x \leq 6$

■ Now work Problem 15 (in the exercise set at the end of this section).

Let a and b represent two real numbers with $a < b$: A **closed interval,** denoted by **[a, b]**, consists of all real numbers x for which $a \leq x \leq b$. An **open interval,** denoted by **(a, b)**, consists of all real numbers x for which $a < x < b$. The **half-open,** or **half-closed, intervals** are **(a, b]**, consisting of all real numbers x for which $a < x \leq b$, and **[a, b)**, consisting of all real numbers x for which $a \leq x < b$. In each of these definitions, a is called the **left end point** and b the **right end point** of the interval. Figure 5 illustrates each type of interval.

FIGURE 5

(a) Closed interval   (b) Open interval   (c) Half-open (half-closed) intervals

The symbol ∞ (read as "infinity") is not a real number, but a notational device used to indicate unboundedness in the positive direction. The symbol −∞ (read as "minus infinity") also is not a real number, but a notational device used to indicate unboundedness in the negative direction. Using the symbols ∞ and −∞, we can define five other kinds of intervals:

$[a, \infty)$ consists of all real numbers x for which $a \leq x < \infty$ $(x \geq a)$

(a, ∞) consists of all real numbers x for which $a < x < \infty$ $(x > a)$

$(-\infty, a]$ consists of all real numbers x for which $-\infty < x \leq a$ $(x \leq a)$

$(-\infty, a)$ consists of all real numbers x for which $-\infty < x < a$ $(x < a)$

$(-\infty, \infty)$ consists of all real numbers x for which $-\infty < x < \infty$ (all real numbers)

Figure 6 illustrates these types of intervals.

FIGURE 6

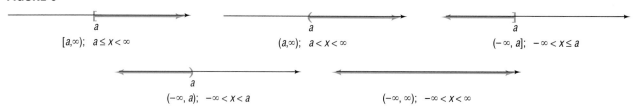

Now work Problems 21 and 25.

The *absolute value* of a number a is the distance from the point whose coordinate is a to the origin. For example, the point whose coordinate is −4 is 4 units from the origin. The point whose coordinate is 3 is 3 units from the origin. See Figure 7. Thus, the absolute value of −4 is 4, and the absolute value of 3 is 3.

FIGURE 7

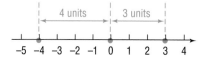

A more formal definition of absolute value is given next.

Absolute Value

The **absolute value** of a real number a, denoted by the symbol $|a|$, is defined by the rules

$$|a| = a \quad \text{if } a \geq 0 \quad \text{and} \quad |a| = -a \quad \text{if } a < 0$$

For example, since $-4 < 0$, then the second rule must be used to get $|-4| = -(-4) = 4$.

E X A M P L E   3

Computing Absolute Value

(a) $|8| = 8$ (b) $|0| = 0$ (c) $|-15| = 15$ ■

■   Now work Problem 29.

Look again at Figure 7. The distance from the point whose coordinate is -4 to the point whose coordinate is 3 is 7 units. This distance is the difference $3 - (-4)$, obtained by subtracting the smaller coordinate from the larger. However, since $|3 - (-4)| = |7| = 7$ and $|-4 - 3| = |-7| = 7$, we can use the absolute value to calculate the distance between two points without being concerned about which coordinate is smaller.

Distance between *P* and *Q*

If *P* and *Q* are two points on a real number line with coordinates *a* and *b*, respectively, the **distance between P and Q,** denoted by $d(P, Q)$, is

$$d(P, Q) = |b - a|$$

Since $|b - a| = |a - b|$, it follows that $d(P, Q) = d(Q, P)$.

E X A M P L E   4

Finding Distance on a Number Line

Let *P*, *Q*, and *R* be points on the real number line with coordinates -5, 7, and -3, respectively. Find the distance:

(a) Between *P* and *Q*   (b) Between *Q* and *R*

Solution   (a) $d(P, Q) = |7 - (-5)| = |12| = 12$ (See Figure 8.)
(b) $d(Q, R) = |-3 - 7| = |-10| = 10$

FIGURE 8

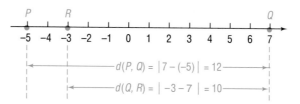

■

Exponents

Integer exponents provide a shorthand device for representing repeated multiplications of a real number.

If *a* is a real number and *n* is a positive integer, then the symbol a^n represents the product of *n* factors of *a*. That is,

$$a^n = \underbrace{a \cdot a \cdot \,\ldots\, \cdot a}_{n \text{ factors}}$$

where it is understood that $a^1 = a$. Thus, $a^2 = a \cdot a$, $a^3 = a \cdot a \cdot a$, and so on. In the expression a^n, *a* is called the **base** and *n* is called the **exponent,** or **power.** We read a^n as "a raised to the power n" or as "a to the nth power." We usually read a^2 as "a squared" and a^3 as "a cubed."

Care must be taken when parentheses are used in conjunction with exponents. For example, $-2^4 = -(2 \cdot 2 \cdot 2 \cdot 2) = -16$, whereas $(-2)^4 = (-2) \cdot (-2) \cdot (-2) \cdot (-2) = 16$. Notice the difference: The exponent applies only to the number or parenthetical expression immediately preceding it.

If $a \neq 0$, we define

$$a^0 = 1 \qquad \text{if } a \neq 0$$

If $a \neq 0$ and if n is a positive integer, then we define

$$a^{-n} = \frac{1}{a^n} \qquad \text{if } a \neq 0$$

With these definitions, the symbol a^n is defined for any integer n.

The following properties, called the **laws of exponents,** can be proved using the preceding definitions. In the list, a and b are real numbers, and m and n are integers.

Laws of Exponents

$$a^m a^n = a^{m+n} \qquad (a^m)^n = a^{mn} \qquad (ab)^n = a^n b^n$$

$$\frac{a^m}{a^n} = a^{m-n} = \frac{1}{a^{n-m}}, \qquad \text{if } a \neq 0 \qquad \left(\frac{a}{b}\right)^n = \frac{a^n}{b^n}, \qquad \text{if } b \neq 0$$

E X A M P L E   5

Using the Laws of Exponents

Write each expression so that all exponents are positive.

(a) $\dfrac{x^5 y^{-2}}{x^3 y}, \quad x \neq 0, y \neq 0$ (b) $\dfrac{xy}{x^{-1} - y^{-1}}, \quad x \neq 0, y \neq 0$

Solution   (a) $\dfrac{x^5 y^{-2}}{x^3 y} = \dfrac{x^5}{x^3} \cdot \dfrac{y^{-2}}{y} = x^{5-3} \cdot y^{-2-1} = x^2 y^{-3} = x^2 \cdot \dfrac{1}{y^3} = \dfrac{x^2}{y^3}$

(b) $\dfrac{xy}{x^{-1} - y^{-1}} = \dfrac{xy}{\dfrac{1}{x} - \dfrac{1}{y}} = \dfrac{xy}{\dfrac{y-x}{xy}} = \dfrac{(xy)(xy)}{y-x} = \dfrac{x^2 y^2}{y-x}$ ∎

▪ Now work Problem 57.

The **principal nth root of a number** a, symbolized by $\sqrt[n]{a}$, is defined as follows:

$$\sqrt[n]{a} = b \quad \text{means} \quad a = b^n \qquad \begin{array}{l} \text{where } a \geq 0 \text{ and } b \geq 0 \text{ if } n \text{ is even} \\ \text{and } a, b \text{ are any real numbers if } n \text{ is odd} \end{array}$$

Notice that if a is negative and n is even, then $\sqrt[n]{a}$ is not defined. When it is defined, the principal nth root of a number is unique.

The symbol $\sqrt[n]{a}$ for the principal nth root of a is sometimes called a **radical;** the integer n is called the **index,** and a is called the **radicand.** If the index of a radical is 2, we call $\sqrt[2]{a}$ the **square root** of a and omit the index 2 by simply writing $\sqrt{a}$. If the index is 3, we call $\sqrt[3]{a}$ the **cube root** of a.

E X A M P L E   6

Simplifying Principal nth Roots

(a) $\sqrt[3]{8} = 2$ because $8 = 2^3$ (b) $\sqrt{64} = 8$ because $64 = 8^2$

(c) $\sqrt[3]{-64} = -4$ because $-64 = (-4)^3$ (d) $\sqrt[4]{\frac{1}{16}} = \frac{1}{2}$ because $\frac{1}{16} = (\frac{1}{2})^4$

(e) $\sqrt{0} = 0$ because $0 = 0^2$ ■

These are examples of **perfect roots.** Thus, 8 and -64 are perfect cubes, since $8 = 2^3$ and $-64 = (-4)^3$; 64 and 0 are perfect squares, since $64 = 8^2$ and $0 = 0^2$; and $\frac{1}{2}$ is a perfect 4th root of $\frac{1}{16}$, since $\frac{1}{16} = (\frac{1}{2})^4$.

In general, if $n \geq 2$ is a positive integer and a is a real number,

$$\sqrt[n]{a^n} = a \qquad \text{if } n \text{ is odd} \tag{1a}$$
$$\sqrt[n]{a^n} = |a| \qquad \text{if } n \text{ is even} \tag{1b}$$

Notice the need for the absolute value in equation (1b). If n is even, then a^n is positive whether $a > 0$ or $a < 0$. But if n is even, the principal nth root must be nonnegative. Hence, the reason for using the absolute value—it gives a nonnegative result.

E X A M P L E   7

Simplifying Radicals

(a) $\sqrt{8} = \sqrt{4 \cdot 2} = \sqrt{4} \cdot \sqrt{2} = 2\sqrt{2}$
(b) $\sqrt[3]{-16} = \sqrt[3]{-8 \cdot 2} = \sqrt[3]{-8} \cdot \sqrt[3]{2} = -2\sqrt[3]{2}$
(c) $\sqrt{x^2} = |x|$ ■

■ Now work Problem 49.

Radicals are used to define **rational exponents.** If a is a real number and $n \geq 2$ is an integer, then

$$a^{1/n} = \sqrt[n]{a}$$

provided $\sqrt[n]{a}$ exists.

If a is a real number and m and n are integers containing no common factors with $n \geq 2$, then

$$a^{m/n} = \sqrt[n]{a^m} = (\sqrt[n]{a})^m \tag{2}$$

provided $\sqrt[n]{a}$ exists.

In simplifying $a^{m/n}$, either $\sqrt[n]{a^m}$ or $(\sqrt[n]{a})^m$ may be used. Generally, taking the root first, as in $(\sqrt[n]{a})^m$, is preferred.

E X A M P L E   8   *Using Equation (2)*

(a) $8^{2/3} = (\sqrt[3]{8})^2 = 2^2 = 4$ (b) $16^{3/2} = (\sqrt{16})^3 = 4^3 = 64$

(c) $(-8x^5)^{1/3} = \sqrt[3]{-8x^3 \cdot x^2} = \sqrt[3]{(-2x)^3 \cdot x^2} = \sqrt[3]{(-2x)^3}\sqrt[3]{x^2}$

$$= -2x\sqrt[3]{x^2} \qquad\blacksquare$$

A more detailed discussion of radicals and rational exponents is given in Section 3 of this Appendix.

■   Now work Problem 47.

Polynomials

Monomial

A **monomial** in one variable is the product of a constant times a variable raised to a nonnegative integer power. Thus, a monomial is of the form

$$ax^k$$

where a is a constant, x is a variable, and $k \geq 0$ is an integer.

Two monomials ax^k and bx^k, when added or subtracted, can be combined into a single monomial by using the distributive property. For example,

$$2x^2 + 5x^2 = (2 + 5)x^2 = 7x^2 \quad\text{and}\quad 8x^3 - 5x^3 = (8 - 5)x^3 = 3x^3$$

Polynomial

A **polynomial** in one variable is an algebraic expression of the form

$$a_nx^n + a_{n-1}x^{n-1} + \cdots + a_1x + a_0$$

where $a_n, a_{n-1}, \ldots, a_1, a_0$ are constants,* called the **coefficients** of the polynomial, $n \geq 0$ is an integer, and x is a variable. If $a_n \neq 0$, it is called the **leading coefficient,** and n is called the **degree** of the polynomial.

The monomials that make up a polynomial are called its **terms.** If all the coefficients are 0, the polynomial is called the **zero polynomial,** which has no degree.

---

*The notation a_n is read as "a sub n." The number n is called a **subscript** and should not be confused with an exponent. We use subscripts to distinguish one constant from another when a large or undetermined number of constants is required.

Polynomials are usually written in **standard form,** beginning with the nonzero term of highest degree and continuing with terms in descending order according to degree. Examples of polynomials are

POLYNOMIAL	COEFFICIENTS	DEGREE
$3x^2 - 5 = 3x^2 + 0 \cdot x + (-5)$	$3, 0, -5$	2
$8 - 2x + x^2 = 1 \cdot x^2 - 2x + 8$	$1, -2, 8$	2
$5x + \sqrt{2} = 5x^1 + \sqrt{2}$	$5, \sqrt{2}$	1
$3 = 3 \cdot 1 = 3 \cdot x^0$	3	0
0	0	No degree

Although we have been using x to represent the variable, letters such as y or z are also commonly used. Thus,

$3x^4 - x^2 + 2$ is a polynomial (in x) of degree 4.

$9y^3 - 2y^2 + y - 3$ is a polynomial (in y) of degree 3.

$z^5 + \pi$ is a polynomial (in z) of degree 5.

A more detailed discussion of polynomials is given in Section 2 of this Appendix.

Pythagorean Theorem

The *Pythagorean Theorem* is a statement about *right triangles*. A **right triangle** is one that contains a **right angle,** that is, an angle of 90°. The side of the triangle opposite the 90° angle is called the **hypotenuse;** the remaining two sides are called **legs.** In Figure 9 we have used c to represent the length of the hypotenuse and a and b to represent the lengths of the legs. Notice the use of the symbol $\ulcorner$ to show the 90° angle. We now state the Pythagorean Theorem.

FIGURE 9

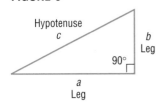

Hypotenuse
c

b
Leg

90°

a
Leg

Pythagorean Theorem

In a right triangle, the square of the length of the hypotenuse is equal to the sum of the squares of the lengths of the legs. That is, in the right triangle shown in Figure 9,

$$c^2 = a^2 + b^2 \qquad (3)$$

■

E X A M P L E   9

Finding the Hypotenuse of a Right Triangle

In a right triangle, one leg is of length 4 and the other is of length 3. What is the length of the hypotenuse?

Solution

Since the triangle is a right triangle, we use the Pythagorean Theorem with $a = 4$ and $b = 3$ to find the length c of the hypotenuse. Thus, from equation (3), we have

$$c^2 = a^2 + b^2$$
$$c^2 = 4^2 + 3^2 = 16 + 9 = 25$$
$$c = 5$$

■

■ Now work Problem 69.

The converse of the Pythagorean Theorem is also true.

Converse of the Pythagorean Theorem   In a triangle, if the square of the length of one side equals the sum of the squares of the lengths of the other two sides, then the triangle is a right triangle. The 90° angle is opposite the longest side.   ■

E X A M P L E   1 0   *Verifying That a Triangle Is a Right Triangle*

Show that a triangle whose sides are of lengths 5, 12, and 13 is a right triangle. Identify the hypotenuse.

FIGURE 10   Solution   We square the lengths of the sides:

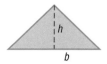

$$25, \quad 144, \quad 169$$

Notice that the sum of the first two squares (25 and 144) equals the third square (169). Hence, the triangle is a right triangle. The longest side, 13, is the hypotenuse. See Figure 10.   ■

■ Now work Problem 79.

Geometry Formulas

Certain formulas from geometry are useful in solving algebra problems. We list some of these formulas next.

For a rectangle of length l and width w,

$$\text{Area} = lw \qquad \text{Perimeter} = 2l + 2w$$

For a triangle with base b and altitude h,

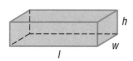

$$\text{Area} = \frac{1}{2}bh$$

For a circle of radius r (diameter $d = 2r$),

$$\text{Area} = \pi r^2 \qquad \text{Circumference} = 2\pi r = \pi d$$

For a rectangular box of length l, width w, and height h,

$$\text{Volume} = lwh$$

1

Exercise 1

In Problems 1–10, replace the question mark by $<$, $>$, or $=$, whichever is correct.

1. $\frac{1}{2}$? 0
2. 5 ? 6
3. -1 ? -2
4. -3 ? $-\frac{5}{2}$
5. π ? 3.14
6. $\sqrt{2}$? 1.41
7. $\frac{1}{2}$? 0.5
8. $\frac{1}{3}$? 0.33
9. $\frac{2}{3}$? 0.67
10. $\frac{1}{4}$? 0.25

11. On the real number line, label the points with coordinates 0, 1, -1, $\frac{5}{2}$, -2.5, $\frac{3}{4}$, and 0.25.
12. Repeat Problem 11 for the coordinates 0, -2, 2, -1.5, $\frac{3}{2}$, $\frac{1}{3}$, and $\frac{2}{3}$.

In Problems 13–20, write each statement as an inequality.

13. x is positive
14. z is negative
15. x is less than 2
16. y is greater than -5
17. x is less than or equal to 1
18. x is greater than or equal to 2
19. x is less than 5 and x is greater than 2
20. y is less than or equal to 2 and y is greater than 0

In Problems 21–24, write each inequality using interval notation, and illustrate each inequality using the real number line.

21. $0 \le x \le 4$
22. $-1 < x < 5$
23. $4 \le x < 6$
24. $-2 < x \le 0$

In Problems 25–28, write each interval as an inequality involving x, and illustrate each inequality using the real number line.

25. $[2, 5]$
26. $(1, 2)$
27. $[4, \infty)$
28. $(-\infty, 2]$

In Problems 29–32, find the value of each expression if x = 2 and y = −3.

29. $|x + y|$
30. $|x - y|$
31. $|x| + |y|$
32. $|x| - |y|$

In Problems 33–52, simplify each expression.

33. 3^0
34. 3^2
35. 4^{-2}
36. $(-3)^2$
37. $\left(\frac{2}{3}\right)^2$
38. $\left(\frac{-4}{5}\right)^3$
39. $3^{-6} \cdot 3^4$
40. $4^{-2} \cdot 4^3$
41. $\left(\frac{2}{3}\right)^{-2}$
42. $\left(\frac{3}{2}\right)^{-3}$
43. $\dfrac{2^3 \cdot 3^2}{2 \cdot 3^{-2}}$
44. $\dfrac{3^{-2} \cdot 5^3}{3 \cdot 5}$
45. $9^{3/2}$
46. $16^{3/4}$
47. $(-8)^{4/3}$
48. $(-27)^{2/3}$
49. $\sqrt{32}$
50. $\sqrt[3]{24}$
51. $\sqrt[3]{-\frac{8}{27}}$
52. $\sqrt{\frac{4}{9}}$

In Problems 53–68, simplify each expression so that all exponents are positive. Whenever an exponent is negative or 0, we assume that the base does not equal 0.

53. $x^0 y^2$
54. $x^{-1} y$
55. $x^{-2} y$
56. $x^4 y^0$
57. $\dfrac{x^{-2} y^3}{xy^4}$
58. $\dfrac{x^{-2} y}{xy^2}$
59. $\left(\dfrac{4x}{5y}\right)^{-2}$
60. $(xy)^{-2}$
61. $\dfrac{x^{-1} y^{-2} z}{x^2 y z^3}$
62. $\dfrac{3x^{-2} y z^2}{x^4 y^{-3} z}$
63. $\dfrac{(-2)^3 x^4 (yz)^2}{3^2 x y^3 z^4}$
64. $\dfrac{4x^{-2}(yz)^{-1}}{(-5)^2 x^4 y^2 z^{-2}}$
65. $\dfrac{x^{-2}}{}$
66. $\dfrac{x^{-1} + y^{-1}}{x^{-1} - y^{-1}}$
67. $\left(\dfrac{3x^{-1}}{4y^{-1}}\right)^{-2}$
68. $\left(\dfrac{5x^{-2}}{6y^{-2}}\right)^{-3}$

In Problems 69–78, a and b are the lengths of the legs of a right triangle and c is the length of the hypotenuse. Find the missing length.

69. $a = 5, b = 12, c = ?$ **70.** $a = 6, b = 8, c = ?$ **71.** $a = 10, b = 24, c = ?$

72. $a = 4, b = 3, c = ?$ **73.** $a = 7, b = 24, c = ?$ **74.** $a = 14, b = 48, c = ?$

75. $a = 3, c = 5, b = ?$ **76.** $b = 6, c = 10, a = ?$ **77.** $b = 7, c = 25, a = ?$

78. $a = 10, c = 13, b = ?$

In Problems 79–84, the lengths of the sides of a triangle are given. Determine which are right triangles. For those that are right triangles, identify the hypotenuse.

79. 3, 4, 5   **80.** 6, 8, 10   **81.** 4, 5, 6

82. 2, 2, 3   **83.** 7, 24, 25   **84.** 10, 24, 26

85. *Geometry* Find the diagonal of a rectangle whose length is 8 inches and whose width is 5 inches.

86. *Geometry* Find the length of a rectangle of width 3 inches if its diagonal is 20 inches long.

87. *Finding the Length of a Guy Wire* A radio transmission tower is 100 feet high. How long does a guy wire need to be if it is to connect a point halfway up the tower to a point 30 feet from the base?

88. Answer Problem 99 if the guy wire is attached to the top of the tower.

89. *How Far Can You See?* The tallest inhabited building in North America is the Sears Tower in Chicago.* If the observation tower is 1454 feet above ground level, use the figure to determine how far a person standing in the observation tower can see (with the aid of a telescope). Use 3960 miles for the radius of Earth. [*Note:* 1 mile = 5280 feet]

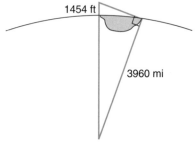

In Problems 90–92, use the fact that the radius of Earth is 3960 miles.

90. *How Far Can You See?* The conning tower of the USS *Silversides,* a World War II submarine now permanently stationed in Muskegon, Michigan, is approximately 20 feet above sea level. How far can one see from the conning tower?

91. *How Far Can You See?* A person who is 6 feet tall is standing on the beach in Fort Lauderdale, Florida and looks out onto the Atlantic Ocean. Suddenly, a ship appears on the horizon. How far is the ship from shore?

92. *How Far Can You See?* The deck of a destroyer is 100 feet above sea level. How far can a person see from the deck? How far can a person see from the bridge, which is 150 feet above sea level?

93. If $a \leq b$ and $c > 0$, show that $ac \leq bc$. [*Hint:* Since $a \leq b$, it follows that $a - b \leq 0$. Now multiply each side by c.]

94. If $a \leq b$ and $c < 0$, show that $ac \geq bc$.

95. If $a < b$, show that $a < (a + b)/2 < b$. The number $(a + b)/2$ is called the **arithmetic mean** of a and b.

96. Refer to Problem 95. Show that the arithmetic mean of a and b is equidistant from a and b.

97. Are there any real numbers that are both rational and irrational? Are there any real numbers that are neither? Explain your reasoning.

*Source: Guinness Book of World Records.

98. Explain why the sum of a rational number and an irrational number must be irrational.

99. What rational number does the repeating decimal 0.9999 . . . equal?

100. Is there a positive real number "closest" to 0?

101. I'm thinking of a number! It lies between 1 and 10; its square is rational and lies between 1 and 10. The number is larger than π. Correct to two decimal places, name the number. Now think of your own number, describe it, and challenge a fellow student to name it.

102. Write a brief paragraph that illustrates the similarities and differences between "less than" ($<$) and "less than or equal" ($\leq$).

103. *The Gibb's Hill Lighthouse, Southampton, Bermuda,* in operation since 1846, stands 117 feet high on a hill 245 feet high, so its beam of light is 362 feet above sea level. A brochure states that the light itself can be seen on the horizon about 26 miles distant. Verify the correctness of this information. The brochure further states that ships 40 miles away can see the light and planes flying at 10,000 feet can see it 120 miles away. Verify the accuracy of these statements. What assumption did the brochure make about the height of the ship?

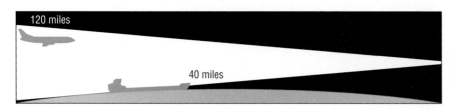

104. You have 1000 feet of flexible pool siding and wish to construct a swimming pool. Experiment with rectangular-shaped pools with perimeters of 1000 feet. How do their areas vary? What is the shape of the rectangle with the largest area? Now compute the area enclosed by a circular pool with a perimeter (circumference) of 1000 feet. What would be your choice of shape for the pool? If rectangular, what is your preference for dimensions? Justify your choice. If your only consideration is to have a pool that encloses the most area, what shape should you use?

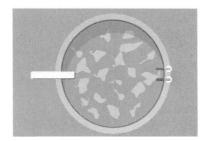

2

Polynomials and Rational Expressions

Algebra may be described as a generalization of arithmetic in which letters are used to represent real numbers. We shall use the letters at the end of the alphabet, such as x, y, and z, to represent variables and the letters at the beginning of the alphabet, such as a, b, and c, to represent constants. Thus, in the expressions $3x + 5$ and $ax + b$, it is understood that x is a variable and that a and b are constants, even though the constants a and b are unspecified. As you will find out, the context usually makes the intended meaning clear.

Now we introduce some basic vocabulary.

Monomial

A **monomial** in one variable is the product of a constant times a variable raised to a nonnegative integer power. Thus, a monomial is of the form

$$ax^k$$

where a is a constant, x is a variable, and $k \geq 0$ is an integer. The constant a is called the **coefficient** of the monomial. If $a \neq 0$, then k is called the **degree** of the monomial.

Examples of monomials are as follows:

MONOMIAL	COEFFICIENT	DEGREE	
$6x^2$	6	2	
$-\sqrt{2}x^3$	$-\sqrt{2}$	3	
3	3	0	Since $3 = 3 \cdot 1 = 3x^0$
$-5x$	-5	1	Since $-5x = -5x^1$
x^4	1	4	Since $x^4 = 1 \cdot x^4$

Two monomials ax^k and bx^k with the same degree and the same variable are called **like terms.** Such monomials when added or subtracted can be combined into a single monomial by using the distributive property. For example,

$$2x^2 + 5x^2 = (2 + 5)x^2 = 7x^2 \quad \text{and} \quad 8x^3 - 5x^3 = (8 - 5)x^3 = 3x^3$$

The sum or difference of two monomials having different degrees is called a **binomial.** The sum or difference of three monomials with three different degrees is called a **trinomial.** For example,

$x^2 - 2$ is a binomial

$x^3 - 3x + 5$ is a trinomial

$2x^2 + 5x^2 + 2 = 7x^2 + 2$ is a binomial

Polynomial

A **polynomial** in one variable is an algebraic expression of the form

$$a_n x^n + a_{n-1} x^{n-1} + \cdots + a_1 x + a_0 \tag{1}$$

where $a_n, a_{n-1}, \ldots, a_1, a_0$ are constants* called the **coefficients** of the polynomial, $n \geq 0$ is an integer, and x is a variable. If $a_n \neq 0$, it is called the **leading coefficient** and n is called the **degree** of the polynomial.

*The notation a_n is read as "a sub n." The number n is called a **subscript** and should not be confused with an exponent. We use subscripts in order to distinguish one constant from another when a large or undetermined number of constants is required.

The monomials that make up a polynomial are called its **terms.** If all the coefficients are 0, the polynomial is called the **zero polynomial,** which has no degree.

Polynomials are usually written in **standard form,** beginning with the nonzero term of highest degree and continuing with terms in descending order according to degree. Examples of polynomials are the following:

POLYNOMIAL	COEFFICIENTS	DEGREE
$3x^2 - 5 = 3x^2 + 0 \cdot x + (-5)$	$3, 0, -5$	2
$8 - 2x + x^2 = 1 \cdot x^2 - 2x + 8$	$1, -2, 8$	2
$5x + \sqrt{2} = 5x^1 + \sqrt{2}$	$5, \sqrt{2}$	1
$3 = 3 \cdot 1 = 3 \cdot x^0$	3	0
0	0	No degree

Although we have been using x to represent the variable, letters such as y or z are also commonly used. Thus,

$3x^4 - x^2 + 2$ is a polynomial (in x) of degree 4.

$9y^3 - 2y^2 + y - 3$ is a polynomial (in y) of degree 3.

$z^5 + \pi$ is a polynomial (in z) of degree 5.

Algebraic expressions such as

$$\frac{1}{x} \quad \text{and} \quad \frac{x^2 + 1}{x + 5}$$

are not polynomials. The first is not a polynomial because $1/x = x^{-1}$ has an exponent that is not a nonnegative integer. Although the second expression is the quotient of two polynomials, the polynomial in the denominator has degree greater than 0, so the expression cannot be a polynomial.

Adding and Subtracting Polynomials

Polynomials are added and subtracted by combining like terms.

E X A M P L E   1   *Finding Sums and Differences of Polynomials*

Find

(a) $(8x^3 - 2x^2 + 6x - 2) + (3x^4 - 2x^3 + x^2 + x)$

(b) $(3x^4 - 4x^3 + 6x^2 - 1) - (2x^4 - 8x^2 - 6x + 5)$

Solution   (a) The idea here is to group the like terms and then combine them.

$(8x^3 - 2x^2 + 6x - 2) + (3x^4 - 2x^3 + x^2 + x)$

$$= 3x^4 + (8x^3 - 2x^3) + (-2x^2 + x^2) + (6x + x) - 2$$

$$= 3x^4 + 6x^3 - x^2 + 7x - 2$$

(b) $(3x^4 - 4x^3 + 6x^2 - 1) - (2x^4 - 8x^2 - 6x + 5)$

$$= 3x^4 - 4x^3 + 6x^2 - 1 \underbrace{- 2x^4 + 8x^2 + 6x - 5}$$

Be sure to change the sign of each
term in the second polynomial.

$$= (3x^4 - 2x^4) + (-4x^3) + (6x^2 + 8x^2) + 6x + (-1 - 5)$$
↑
Group like terms.

$$= x^4 - 4x^3 + 14x^2 + 6x - 6 \qquad \blacksquare$$

■ Now work Problem 1.

Multiplying Polynomials

Products of polynomials are found by repeated use of the distributive property and the laws of exponents.

E X A M P L E 2

Finding the Product of Two Polynomials

Find the product: $(2x + 5)(x^2 - x + 2)$

Solution

Horizontal Multiplication

$$(2x + 5)(x^2 - x + 2) = 2x(x^2 - x + 2) + 5(x^2 - x + 2)$$
↑
Distributive property

$$= 2x \cdot x^2 - 2x \cdot x + 2x \cdot 2 + 5 \cdot x^2 - 5 \cdot x + 5 \cdot 2$$
↑
Distributive property

$$= 2x^3 - 2x^2 + 4x + 5x^2 - 5x + 10$$
↑
Law of exponents

$$= 2x^3 + 3x^2 - x + 10$$
↑
Combine like terms

Vertical Multiplication: The idea here is very much like multiplying a two-digit number by a three-digit number.

$$
\begin{array}{r}
x^2 - x + 2 \\
2x + 5 \\
\hline
2x^3 - 2x^2 + 4x \qquad \text{This line is } 2x(x^2 - x + 2). \\
(+) \qquad 5x^2 - 5x + 10 \qquad \text{This line is } 5(x^2 - x + 2). \\
\hline
2x^3 + 3x^2 - x + 10 \qquad \text{The sum of the preceding two lines.}
\end{array}
$$

$\blacksquare$

■ Now work Problem 5.

Certain products, which we call **special products,** occur frequently in algebra. In the list that follows, x, a, b, c, and d are real numbers:

Difference of Two Squares

$$(x - a)(x + a) = x^2 - a^2 \qquad (2)$$

Squares of Binomials, or Perfect Squares	
$(x + a)^2 = x^2 + 2ax + a^2$	(3a)
$(x - a)^2 = x^2 - 2ax + a^2$	(3b)

Miscellaneous Trinomials	
$(x + a)(x + b) = x^2 + (a + b)x + ab$	(4a)
$(ax + b)(cx + d) = acx^2 + (ad + bc)x + bd$	(4b)

Cubes of Binomials, or Perfect Cubes	
$(x + a)^3 = x^3 + 3ax^2 + 3a^2x + a^3$	(5a)
$(x - a)^3 = x^3 - 3ax^2 + 3a^2x - a^3$	(5b)

Difference of Two Cubes	
$(x - a)(x^2 + ax + a^2) = x^3 - a^3$	(6)

Sum of Two Cubes	
$(x + a)(x^2 - ax + a^2) = x^3 + a^3$	(7)

The formulas in equations (2) through (7) are used often and their patterns should be committed to memory. But if you forget one or are unsure of its form, you should be able to derive it as needed.

A **polynomial in two variables** x and y is the sum of one or more monomials of the form ax^ny^m, where a is a constant called the **coefficient,** x and y are variables, and n and m are nonnegative integers. The **degree** of the monomial ax^ny^m is $n + m$. The **degree** of a polynomial in two variables x and y is the highest degree of all the monomials with nonzero coefficients that appear.

Polynomials in three variables x, y, and z and polynomials in more than three variables are defined in a similar way. Here are some examples:

$3x^2 + 2x^3y + 5$ $\qquad$ $\pi x^3 - y^2$ $\qquad$ $x^4 + 4x^3y - xy^3 + y^4$

Two variables, $\qquad\qquad$ Two variables, $\qquad\qquad$ Two variables,
degree is 4 $\qquad\qquad\quad$ degree is 3 $\qquad\qquad\quad$ degree is 4

$x^2 + y^2 - z^2 + 4$ $\qquad$ x^3y^2z $\qquad$ $5x^2 - 4y^2 + z^3y + 2w^2x$

Three variables, $\qquad\qquad$ Three variables, $\qquad\qquad$ Four variables,
degree is 2 $\qquad\qquad\quad$ degree is 6 $\qquad\qquad\quad$ degree is 4

Adding and multiplying polynomials in two or more variables is handled in the same way as for polynomials in one variable.

■   Now work Problem 3.

Dividing Polynomials

The procedure for dividing two polynomials is similar to the procedure for dividing two integers. This process should be familiar to you, but we review it briefly next.

E X A M P L E   3   *Dividing Two Integers*

Divide 842 by 15.

Solution

$$
\begin{array}{r}
56 \quad \leftarrow \text{Quotient} \\
\text{Divisor} \rightarrow \quad 15\overline{)842} \quad \leftarrow \text{Dividend} \\
75 \quad \leftarrow 5 \cdot 15 \quad \text{(Subtract)} \\
\hline
92 \\
90 \quad \leftarrow 6 \cdot 15 \quad \text{(Subtract)} \\
\hline
2 \quad \leftarrow \text{Remainder}
\end{array}
$$

Thus, $\frac{842}{15} = 56 + \frac{2}{15}$. ■

In the long division process detailed in Example 3, the number 15 is called the **divisor,** the number 842 is called the **dividend,** the number 56 is called the **quotient,** and the number 2 is called the **remainder.**

To check the answer obtained in a division problem, multiply the quotient by the divisor and add the remainder. The answer should be the dividend.

$$(\text{Quotient})(\text{Divisor}) + \text{Remainder} = \text{Dividend}$$

For example, we can check the results obtained in Example 3 as follows:

$$(56)(15) + 2 = 840 + 2 = 842$$

To divide two polynomials, we first must write each polynomial in standard form. The process then follows a pattern similar to that of Example 3. The next example illustrates the procedure.

E X A M P L E   4   *Dividing Two Polynomials*

Find the quotient and the remainder when

$$3x^3 + 4x^2 + x + 7 \quad \text{is divided by} \quad x^2 + 1$$

Solution   Each polynomial is in standard form. The dividend is $3x^3 + 4x^2 + x + 7$, and the divisor is $x^2 + 1$.

STEP 1:   Divide the leading term of the dividend, $3x^3$, by the leading term of the divisor, x^2. Enter the result, $3x$, over the term $3x^3$, as follows:

$$
\begin{array}{r}
3x \\
x^2 + 1\overline{)3x^3 + 4x^2 + \ x + 7}
\end{array}
$$

STEP 2:   Multiply $3x$ by $x^2 + 1$ and enter the result below the dividend.

$$
\begin{array}{r}
3x \\
x^2 + 1\overline{)3x^3 + 4x^2 + \ x + 7} \\
\underline{3x^3 + 3x} \quad \leftarrow 3x \cdot (x^2 + 1) = 3x^3 + 3x
\end{array}
$$

↑
Notice that we align the $3x$ term under the x to make the next step easier.

STEP 3:   Subtract and bring down the remaining terms.

$$
\begin{array}{r}
3x \\
x^2 + 1\overline{)3x^3 + 4x^2 +\ \ x + 7} \\
\underline{3x^3 + 3x} \quad \leftarrow \text{Subtract.} \\
4x^2 - 2x + 7 \quad \leftarrow \text{Bring down the } 4x^2 \text{ and the } 7.
\end{array}
$$

STEP 4:   Repeat Steps 1 through 3 using $4x^2 - 2x + 7$ as the dividend.

$$
\begin{array}{r}
3x + 4 \\
x^2 + 1\overline{)3x^3 + 4x^2 +\ \ x + 7} \\
\underline{3x^3 + 3x} \\
4x^2 - 2x + 7 \quad \leftarrow \text{Divide } 4x^2 \text{ by } x^2 \text{ to get } 4. \\
\underline{4x^2 + 4} \quad \leftarrow \text{Multiply } x^2 + 1 \text{ by } 4; \text{ subtract.} \\
-2x + 3
\end{array}
$$

Since x^2 does not divide $-2x$ evenly (that is, the result is not a monomial), the process ends. The quotient is $3x + 4$, and the remainder is $-2x + 3$.

Check: (Quotient)(Divisor) + Remainder

$$
\begin{aligned}
&= (3x + 4)(x^2 + 1) + (-2x + 3) \\
&= 3x^3 + 4x^2 + 3x + 4 + (-2x + 3) \\
&= 3x^3 + 4x^2 + x + 7 = \text{Dividend}
\end{aligned}
$$

Thus,

$$
\frac{3x^3 + 4x^2 + x + 7}{x^2 + 1} = 3x + 4 + \frac{-2x + 3}{x^2 + 1} \qquad \blacksquare
$$

The process for dividing two polynomials leads to the following result:

Theorem   The remainder after dividing two polynomials is either the zero polynomial or a polynomial of degree less than the degree of the divisor. $\blacksquare$

■   Now work Problem 13.

Factoring

Consider the following product:

$$
(2x + 3)(x - 4) = 2x^2 - 5x - 12
$$

The two polynomials on the left are called **factors** of the polynomial on the right. Expressing a given polynomial as a product of other polynomials, that is, finding the factors of a polynomial, is called **factoring.**

We shall restrict our discussion here to factoring polynomials in one variable into products of polynomials in one variable, where all coefficients are integers. We call this **factoring over the integers.** There will be times, though, when we will want to **factor over the rational numbers** and even **factor over the real numbers.** Factoring over the rational numbers means to write a given polynomial whose coefficients are rational numbers as a product of polynomials whose coefficients are also rational numbers. Factoring over the real numbers means to write a given polynomial whose coefficients are real numbers as a product of polynomials whose coefficients are also real numbers. Unless specified otherwise, we will be factoring over the integers.

Any polynomial can be written as the product of 1 times itself or as -1 times its additive inverse. If a polynomial cannot be written as the product of two other polynomials (excluding 1 and -1), then the polynomial is said to be **prime.** When a polynomial has been written as a product consisting only of prime factors, then it is said to be **factored completely.** Examples of prime polynomials are

$$2, \quad 3, \quad 5, \quad x, \quad x + 1, \quad x - 1, \quad 3x + 4$$

The first factor to look for in a factoring problem is a common monomial factor present in each term of the polynomial. If one is present, use the distributive property to factor it out. For example,

POLYNOMIAL	COMMON MONOMIAL FACTOR	REMAINING FACTOR	FACTORED FORM
$2x + 4$	2	$x + 2$	$2x + 4 = 2(x + 2)$
$3x - 6$	3	$x - 2$	$3x - 6 = 3(x - 2)$
$2x^2 - 4x + 8$	2	$x^2 - 2x + 4$	$2x^2 - 4x + 8 = 2(x^2 - 2x + 4)$
$8x - 12$	4	$2x - 3$	$8x - 12 = 4(2x - 3)$
$x^2 + x$	x	$x + 1$	$x^2 + x = x(x + 1)$
$x^3 - 3x^2$	x^2	$x - 3$	$x^3 - 3x^2 = x^2(x - 3)$
$6x^2 + 9x$	$3x$	$2x + 3$	$6x^2 + 9x = 3x(2x + 3)$

The list of special products (2) through (7) given on pages 662–663 provides a list of factoring formulas when the equations are read from right to left. For example, equation (2) states that if the polynomial is the difference of two squares, $x^2 - a^2$, it can be factored into $(x - a)(x + a)$. The following example illustrates several factoring techniques.

E X A M P L E 5   *Factoring Polynomials*

Factor completely each polynomial.

(a) $x^4 - 16$ (b) $x^3 - 1$ (c) $9x^2 - 6x + 1$ (d) $x^2 + 4x - 12$

(e) $3x^2 + 10x - 8$ (f) $x^3 - 4x^2 + 2x - 8$

Solution   (a) $x^4 - 16 = (x^2 - 4)(x^2 + 4) = (x - 2)(x + 2)(x^2 + 4)$

     Difference of squares   Difference of squares

(b) $x^3 - 1 = (x - 1)(x^2 + x + 1)$

     Difference of cubes

(c) $9x^2 - 6x + 1 = (3x - 1)^2$

     Perfect square

(d) $x^2 + 4x - 12 = (x + 6)(x - 2)$

     6 and -2 are factors of -12, and the sum of 6 and -2 is 4

$$12x - 2x = 10x$$

(e) $3x^2 + 10x - 8 = (3x - 2)(x + 4)$

     $3x^2$ -8

(f) $x^3 - 4x^2 + 2x - 8 = (x^3 - 4x^2) + (2x - 8)$

$\underset{\text{Regroup}}{\uparrow}$

$= x^2(x - 4) + 2(x - 4) = (x^2 + 2)(x - 4)$

$\underset{\text{Distributive property}}{\uparrow}$ ■

The technique used in Example 5(f) is called **factoring by grouping.**

■ Now work Problem 21.

Rational Expressions

If we form the quotient of two polynomials, the result is called a **rational expression.** Some examples of rational expressions are

(a) $\dfrac{x^3 + 1}{x}$ (b) $\dfrac{3x^2 + x - 2}{x^2 + 5}$ (c) $\dfrac{x}{x^2 - 1}$ (d) $\dfrac{xy^2}{(x - y)^2}$

Expressions (a), (b), and (c) are rational expressions in one variable, x, whereas (d) is a rational expression in two variables, x and y.

Rational expressions are described in the same manner as rational numbers. Thus, in expression (a), the polynomial $x^3 + 1$ is called the **numerator,** and x is called the **denominator.** When the numerator and denominator of a rational expression contain no common factors (except 1 and −1), we say that the rational expression is **reduced to lowest terms,** or **simplified.**

A rational expression is reduced to lowest terms by completely factoring the numerator and the denominator and canceling any common factors by using the cancellation property,

$$\frac{ac}{bc} = \frac{a}{b}, \qquad b \neq 0, c \neq 0$$

We shall follow the common practice of using a slash mark to indicate cancellation. For example,

$$\frac{x^2 - 1}{x^2 - 2x - 3} = \frac{(x - 1)(x + 1)}{(x - 3)(x + 1)} = \frac{x - 1}{x - 3}$$

E X A M P L E   6   *Simplifying Rational Expressions*

Reduce each rational expression to lowest terms.

(a) $\dfrac{x^2 + 4x + 4}{x^2 + 3x + 2}$ (b) $\dfrac{x^3 - 8}{x^3 - 2x^2}$ (c) $\dfrac{8 - 2x}{x^2 - x - 12}$

Solution   (a) $\dfrac{x^2 + 4x + 4}{x^2 + 3x + 2} = \dfrac{(x + 2)(x + 2)}{(x + 2)(x + 1)} = \dfrac{x + 2}{x + 1}, \quad x \neq -2, -1$

(b) $\dfrac{x^3 - 8}{x^3 - 2x^2} = \dfrac{(x - 2)(x^2 + 2x + 4)}{x^2(x - 2)} = \dfrac{x^2 + 2x + 4}{x^2}, \quad x \neq 0, 2$

(c) $\dfrac{8 - 2x}{x^2 - x - 12} = \dfrac{2(4 - x)}{(x - 4)(x + 3)} = \dfrac{2(-1)(x - 4)}{(x - 4)(x + 3)} = \dfrac{-2}{x + 3}, \quad x \neq -3, 4$ ■

The rules for multiplying and dividing rational expressions are the same as the rules for multiplying and dividing rational numbers:

$$\frac{a}{b} \cdot \frac{c}{d} = \frac{ac}{bd}, \qquad \text{if } b \neq 0, d \neq 0 \tag{8}$$

$$\frac{\dfrac{a}{b}}{\dfrac{c}{d}} = \frac{a}{b} \cdot \frac{d}{c} = \frac{ad}{bc}, \qquad \text{if } b \neq 0, c \neq 0, d \neq 0 \tag{9}$$

In using equations (8) and (9) with rational expressions, be sure first to factor each polynomial completely so that common factors can be canceled. We shall follow the practice of leaving our answers in factored form.

E X A M P L E   7

Finding Products and Quotients of Rational Expressions

Perform the indicated operation and simplify the result. Leave your answer in factored form.

(a) $\dfrac{x^2 - 2x + 1}{x^3 + x} \cdot \dfrac{4x^2 + 4}{x^2 + x - 2}$ (b) $\dfrac{\dfrac{x + 3}{x^2 - 4}}{\dfrac{x^2 - x - 12}{x^3 - 8}}$

Solution   (a) $\dfrac{x^2 - 2x + 1}{x^3 + x} \cdot \dfrac{4x^2 + 4}{x^2 + x - 2} = \dfrac{(x - 1)^2}{x(x^2 + 1)} \cdot \dfrac{4(x^2 + 1)}{(x + 2)(x - 1)}$

$$= \frac{(x - 1)^2(4)(x^2 + 1)}{x(x^2 + 1)(x + 2)(x - 1)} = \frac{4(x - 1)}{x(x + 2)},$$

$$x \neq -2, 0, 1$$

(b) $\dfrac{\dfrac{x + 3}{x^2 - 4}}{\dfrac{x^2 - x - 12}{x^3 - 8}} = \dfrac{x + 3}{x^2 - 4} \cdot \dfrac{x^3 - 8}{x^2 - x - 12}$

$$= \frac{x + 3}{(x - 2)(x + 2)} \cdot \frac{(x - 2)(x^2 + 2x + 4)}{(x - 4)(x + 3)}$$

$$= \frac{(x + 3)(x - 2)(x^2 + 2x + 4)}{(x - 2)(x + 2)(x - 4)(x + 3)} = \frac{x^2 + 2x + 4}{(x + 2)(x - 4)},$$

$$x \neq -3, -2, 2, 4 \qquad \blacksquare$$

Note: Slanting the cancellation marks in different directions for different factors, as in Example 7, is a good practice to follow, since it will help in checking for errors.

■   Now work Problem 31.

If the denominators of two rational expressions to be added (or subtracted) are equal, we add (or subtract) the numerators and keep the common denominator. That is, if *a/b* and *c/b* are two rational expressions, then

$$\frac{a}{b} + \frac{c}{b} = \frac{a+c}{b} \qquad \frac{a}{b} - \frac{c}{b} = \frac{a-c}{b}, \qquad \text{if } b \neq 0 \qquad (10)$$

E X A M P L E   8   *Finding the Sum of Two Rational Expressions*

Perform the indicated operation and simplify the result. Leave your answer in factored form.

$$\frac{2x^2 - 4}{2x + 5} + \frac{x + 3}{2x + 5}, \qquad x \neq -\tfrac{5}{2}$$

Solution

$$\frac{2x^2 - 4}{2x + 5} + \frac{x + 3}{2x + 5} = \frac{(2x^2 - 4) + (x + 3)}{2x + 5}$$

$$= \frac{2x^2 + x - 1}{2x + 5} = \frac{(2x - 1)(x + 1)}{2x + 5} \qquad \blacksquare$$

If the denominators of two rational expressions to be added or subtracted are not equal, we can use the general formulas for adding and subtracting quotients:

$$\frac{a}{b} + \frac{c}{d} = \frac{a \cdot d}{b \cdot d} + \frac{b \cdot c}{b \cdot d} = \frac{ad + bc}{bd}, \qquad \text{if } b \neq 0, d \neq 0$$

$$\frac{a}{b} - \frac{c}{d} = \frac{a \cdot d}{b \cdot d} - \frac{b \cdot c}{b \cdot d} = \frac{ad - bc}{bd}, \qquad \text{if } b \neq 0, d \neq 0 \qquad (11)$$

E X A M P L E   9   *Finding the Difference of Two Rational Expressions*

Perform the indicated operation and simplify the result. Leave your answer in factored form.

$$\frac{x^2}{x^2 - 4} - \frac{1}{x}, \qquad x \neq -2, 0, 2$$

Solution

$$\frac{x^2}{x^2 - 4} - \frac{1}{x} = \frac{x^2(x) - (x^2 - 4)(1)}{(x^2 - 4)(x)} = \frac{x^3 - x^2 + 4}{(x - 2)(x + 2)(x)} \qquad \blacksquare$$

Least Common Multiple (LCM)

If the denominators of two rational expressions to be added (or subtracted) have common factors, we usually do not use the general rules given by equation (11), since, in doing so, we make the problem more complicated than it needs to be. Instead, just as with fractions, we apply the **least common multiple (LCM) method** by using the polynomial of least degree that contains each denominator polynomial as a factor. Then we rewrite each rational expression using the LCM as the common denominator and use equation (10) to do the addition (or subtraction).

To find the least common multiple of two or more polynomials, first factor completely each polynomial. The LCM is the product of the different prime factors of each polynomial, each factor appearing the greatest number of times it occurs in each polynomial. The next example will give you the idea.

E X A M P L E   1 0   *Finding the Least Common Multiple*

Find the least common multiple of the following pair of polynomials:

$$x(x - 1)^2(x + 1) \quad \text{and} \quad 4(x - 1)(x + 1)^3$$

Solution   The polynomials are already factored completely as

$$x(x - 1)^2(x + 1) \quad \text{and} \quad 4(x - 1)(x + 1)^3$$

Start by writing the factors of the left-hand polynomial. (Alternatively, you could start with the one on the right.)

$$x(x - 1)^2(x + 1)$$

Now look at the right-hand polynomial. Its first factor, 4, does not appear in our list, so we insert it:

$$4x(x - 1)^2(x + 1)$$

The next factor, $x - 1$, is already in our list, so no change is necessary. The final factor is $(x + 1)^3$. Since our list has $x + 1$ to the first power only, we replace $x + 1$ in the list by $(x + 1)^3$. The LCM is

$$4x(x - 1)^2(x + 1)^3$$

Notice that the LCM is, in fact, the polynomial of least degree that contains $x(x - 1)^2(x + 1)$ and $4(x - 1)(x + 1)^3$ as factors. ■

The next example illustrates how the LCM is used for adding and subtracting rational expressions.

E X A M P L E   1 1   *Using the LCM to Add Rational Expressions*

Perform the indicated operation and simplify the result. Leave your answer in factored form.

$$\frac{x}{x^2 + 3x + 2} + \frac{2x - 3}{x^2 - 1}, \quad x \neq -2, -1, 1$$

Solution   First, we find the LCM of the denominators:

$$x^2 + 3x + 2 = (x + 2)(x + 1)$$
$$x^2 - 1 = (x - 1)(x + 1)$$

The LCM is $(x + 2)(x + 1)(x - 1)$. Next, we rewrite each rational expression using the LCM as the denominator:

$$\frac{x}{x^2 + 3x + 2} = \frac{x}{(x + 2)(x + 1)} = \frac{x(x - 1)}{(x + 2)(x + 1)(x - 1)}$$

$\uparrow$ Multiply numerator and denominator by $x - 1$ to get the LCM in the denominator.

$$\frac{2x - 3}{x^2 - 1} = \frac{2x - 3}{(x - 1)(x + 1)} = \frac{(2x - 3)(x + 2)}{(x - 1)(x + 1)(x + 2)}$$

$\uparrow$ Multiply numerator and denominator by $x + 2$ to get the LCM in the denominator.

Now we can add by using equation (10).

$$\frac{x}{x^2 + 3x + 2} + \frac{2x - 3}{x^2 - 1} = \frac{x(x - 1)}{(x + 2)(x + 1)(x - 1)} + \frac{(2x - 3)(x + 2)}{(x + 2)(x + 1)(x - 1)}$$

$$= \frac{(x^2 - x) + (2x^2 + x - 6)}{(x + 2)(x + 1)(x - 1)}$$

$$= \frac{3x^2 - 6}{(x + 2)(x + 1)(x - 1)} = \frac{3(x^2 - 2)}{(x + 2)(x + 1)(x - 1)}$$

■

If we had not used the LCM technique to add the quotients in Example 11, but decided instead to use the general rule of equation (11), we would have obtained a more complicated expression, as follows:

$$\frac{x}{x^2 + 3x + 2} + \frac{2x - 3}{x^2 - 1} = \frac{x(x^2 - 1) + (x^2 + 3x + 2)(2x - 3)}{(x^2 + 3x + 2)(x^2 - 1)}$$

$$= \frac{3x^3 + 3x^2 - 6x - 6}{(x^2 + 3x + 2)(x^2 - 1)} = \frac{3(x^3 + x^2 - 2x - 2)}{(x^2 + 3x + 2)(x^2 - 1)}$$

Now we are faced with a more complicated problem of expressing this quotient in lowest terms. It is always best to first look for common factors in the denominators of expressions to be added or subtracted and to use the LCM if any common factors are found.

■   Now work Problem 35.

Mixed Quotients

When sums and/or differences of rational expressions appear as the numerator and/or denominator of a quotient, the quotient is called a **mixed quotient.** For example,

$$\frac{1 + \dfrac{1}{x}}{1 - \dfrac{1}{x}} \quad \text{and} \quad \frac{\dfrac{x^2}{x^2 - 4} - 3}{\dfrac{x - 3}{x + 2} - 1}$$

are mixed quotients. To **simplify** a mixed quotient means to write it as a rational expression reduced to lowest terms. This can be accomplished by treating the numerator and denominator of the mixed quotient separately, performing whatever operations are indicated and simplifying the results. Follow this by simplifying the resulting rational expression.

E X A M P L E   1 2   *Simplifying Mixed Quotients*

Simplify the mixed quotient: $\dfrac{1 + \dfrac{1}{x}}{1 - \dfrac{1}{x}}$

Solution

$$\frac{1 + \dfrac{1}{x}}{1 - \dfrac{1}{x}} = \frac{\dfrac{x}{x} + \dfrac{1}{x}}{\dfrac{x}{x} - \dfrac{1}{x}} = \frac{\dfrac{x + 1}{x}}{\dfrac{x - 1}{x}} = \frac{x + 1}{x} \cdot \frac{x}{x - 1}$$

$$= \frac{(x + 1)x}{x(x - 1)} = \frac{x + 1}{x - 1}$$ ■

■ Now work Problem 39.

2

Exercise 2

In Problems 1–10, perform the indicated operations. Express each answer as a polynomial.

1. $(10x^5 - 8x^2) + (3x^3 - 2x^2 + 6)$
2. $3(x^2 - 3x + 1) + 2(3x^2 + x - 4)$
3. $(x + a)^2 - x^2$
4. $(x - a)^2 - x^2$
5. $(x + 8)(2x + 1)$
6. $(2x - 1)(x + 2)$
7. $(x^2 + x - 1)(x^2 - x + 1)$
8. $(x^2 + 2x + 1)(x^2 - 3x + 4)$
9. $(x + 1)^3 - (x - 1)^3$
10. $(x + 1)^3 - (x + 2)^3$

In Problems 11–20, find the quotient and the remainder. Check your work by verifying that

$$(\text{Quotient})(\text{Divisor}) + \text{Remainder} = \text{Dividend}$$

11. $4x^3 - 3x^2 + x + 1$ divided by x
12. $3x^3 - x^2 + x - 2$ divided by x
13. $4x^3 - 3x^2 + x + 1$ divided by $x + 2$
14. $3x^3 - x^2 + x - 2$ divided by $x + 2$
15. $4x^3 - 3x^2 + x + 1$ divided by $x - 4$
16. $3x^3 - x^2 + x - 2$ divided by $x - 4$
17. $4x^3 - 3x^2 + x + 1$ divided by x^2
18. $3x^3 - x^2 + x - 2$ divided by x^2
19. $4x^3 - 3x^2 + x + 1$ divided by $x^2 + 2$
20. $3x^3 - x^2 + x - 2$ divided by $x^2 + 2$

In Problems 21–30, factor completely each polynomial. If the polynomial cannot be factored, say it is prime.

21. $x^2 - 2x - 15$
22. $x^2 - 6x - 14$
23. $ax^2 - 4a^2x - 45a^3$
24. $bx^2 + 14b^2x + 45b^3$
25. $x^3 - 27$
26. $x^3 + 27$
27. $3x^2 + 4x + 1$
28. $4x^2 + 3x - 1$
29. $x^7 - x^5$
30. $x^8 - x^5$

In Problems 31–34, perform the indicated operation and simplify the result. Leave your answer in factored form.

31. $\dfrac{3x - 6}{5x} \cdot \dfrac{x^2 - x - 6}{x^2 - 4}$
32. $\dfrac{9x - 25}{2x - 2} \cdot \dfrac{1 - x^2}{6x - 10}$
33. $\dfrac{4x^2 - 1}{x^2 - 16} \cdot \dfrac{x^2 - 4x}{2x + 1}$
34. $\dfrac{12}{x^2 - x} \cdot \dfrac{x^2 - 1}{4x - 2}$

In Problems 35–42, perform the indicated operations and simplify the result. Leave your answer in factored form.

35. $\dfrac{x}{x^2 - 7x + 6} - \dfrac{x}{x^2 - 2x - 24}$
36. $\dfrac{x}{x - 3} - \dfrac{x + 1}{x^2 + 5x - 24}$

37. $\dfrac{4}{x^2 - 4} - \dfrac{2}{x^2 + x - 6}$

38. $\dfrac{3}{x - 1} - \dfrac{x - 4}{x^2 - 2x + 1}$

39. $\dfrac{x - \dfrac{1}{x}}{x + \dfrac{1}{x}}$

40. $\dfrac{1 - \dfrac{x}{x + 1}}{2 - \dfrac{x - 1}{x}}$

41. $\dfrac{3 - \dfrac{x^2}{x + 1}}{1 + \dfrac{x}{x^2 - 1}}$

42. $\dfrac{3x - \dfrac{3}{x^2}}{\dfrac{1}{(x - 1)^2} - 1}$

3

Radicals; Rational Exponents

Square Roots

A real number is squared when it is raised to the power 2. The inverse of squaring is finding a **square root.** For example, since $6^2 = 36$ and $(-6)^2 = 36$, the numbers 6 and -6 are square roots of 36.

The symbol $\sqrt{}$, called a **radical sign,** is used to denote the **principal,** or nonnegative, square root. Thus, $\sqrt{36} = 6$.

Principal Square Root

> In general, if a is a nonnegative real number, the nonnegative number b such that $b^2 = a$ is the **principal square root** of a and is denoted by $b = \sqrt{a}$.

The following comments are noteworthy:

1. Negative numbers do not have square roots (in the real number system), because the square of any real number is *nonnegative.* For example, $\sqrt{-4}$ is not a real number, because there is no real number whose square is -4.
2. The principal square root of 0 is 0, since $0^2 = 0$. That is, $\sqrt{0} = 0$.
3. The principal square root of a positive number is positive.
4. If $c \ge 0$, then $(\sqrt{c})^2 = c$. For example, $(\sqrt{2})^2 = 2$ and $(\sqrt{3})^2 = 3$.

E X A M P L E   1   *Evaluating Square Roots*

(a) $\sqrt{64} = 8$ (b) $\sqrt{\frac{1}{16}} = \frac{1}{4}$ (c) $(\sqrt{1.4})^2 = 1.4$ ■

Examples 1(a) and (b) are examples of **perfect square roots.** Thus, 64 is a **perfect square,** since $64 = 8^2$; and $\frac{1}{16}$ is a perfect square, since $\frac{1}{16} = (\frac{1}{4})^2$.

In general, we have

$$\sqrt{a^2} = |a| \qquad (1)$$

Notice the need for the absolute value in equation (1). Since $a^2 \ge 0$, the principal square root of a^2 is defined whether $a > 0$ or $a < 0$. However, since the principal square root is nonnegative, we need the absolute value to ensure the nonnegative result.

E X A M P L E   2   *Using Equation (1)*

(a) $\sqrt{(2.3)^2} = |2.3| = 2.3$ (b) $\sqrt{(-2.3)^2} = |-2.3| = 2.3$ (c) $\sqrt{x^2} = |x|$

■

*n*th Roots

The **principal *n*th root of a real number *a*,** symbolized by $\sqrt[n]{a}$, is defined as follows:

Principal *n*th Root	$\sqrt[n]{a} = b$ means $a = b^n$, where $a \geq 0$ and $b \geq 0$ if n is even and a, b are any real numbers if n is odd

Notice that if a is negative and n is even then $\sqrt[n]{a}$ is not defined. When it is defined, the principal *n*th root of a number is unique.

The symbol $\sqrt[n]{a}$ for the principal *n*th root of a is sometimes called a **radical;** the integer n is called the **index,** and a is called the **radicand.** If the index of a radical is 2, we call $\sqrt[2]{a}$ the **square root** of a and omit the index 2 by simply writing $\sqrt{a}$. If the index is 3, we call $\sqrt[3]{a}$ the **cube root** of a.

EXAMPLE 3   *Evaluating Principal nth Roots*

$$\sqrt[3]{8} = 2 \qquad \sqrt[6]{64} = 2 \qquad \sqrt[3]{-64} = -4 \qquad \sqrt[4]{\tfrac{1}{16}} = \tfrac{1}{2}$$

because

$$8 = 2^3 \qquad 64 = 2^6 \qquad -64 = (-4)^3 \qquad \tfrac{1}{16} = (\tfrac{1}{2})^4 \qquad \blacksquare$$

These are examples of **perfect roots.** Thus, 8 and -64 are perfect cubes, since $8 = 2^3$ and $-64 = (-4)^3$; 2 is a perfect sixth root of 64, since $64 = 2^6$; and $\tfrac{1}{2}$ is a perfect fourth root of $\tfrac{1}{16}$, since $\tfrac{1}{16} = (\tfrac{1}{2})^4$.

In general, if $n \geq 2$ is a positive integer and a is a real number, we have

$\sqrt[n]{a^n} = a$, if n is odd	(1a)
$\sqrt[n]{a^n} = |a|$, if n is even	(1b)

Notice the need for the absolute value in equation (1b). If n is even, then a^n is positive whether $a > 0$ or $a < 0$. But if n is even, the principal *n*th root must be nonnegative. Hence, the reason for using the absolute value—it gives a nonnegative result.

EXAMPLE 4   *Using Equations (1a) and (1b)*

(a) $\sqrt[3]{4^3} = 4$ (b) $\sqrt[5]{(-3)^5} = -3$ (c) $\sqrt[4]{2^4} = 2$
(d) $\sqrt[4]{(-3)^4} = |-3| = 3$ (e) $\sqrt{x^2} = |x|$ $\blacksquare$

Properties of Radicals

Let $n \geq 2$ and $m \geq 2$ denote positive integers, and let a and b represent real numbers. Assuming that all radicals are defined, we have the following properties:

$$\sqrt[n]{ab} = \sqrt[n]{a}\sqrt[n]{b} \tag{2a}$$

$$\sqrt[n]{\frac{a}{b}} = \frac{\sqrt[n]{a}}{\sqrt[n]{b}} \tag{2b}$$

$$\sqrt[n]{a^m} = (\sqrt[n]{a})^m \tag{2c}$$

$$\sqrt[m]{\sqrt[n]{a}} = \sqrt[mn]{a} \tag{2d}$$

When used in reference to radicals, the direction to "simplify" will mean to remove from the radicals any perfect roots that occur as factors. Let's look at some examples of how the rules listed in the box are applied to simplify radicals.

E X A M P L E   5   *Simplifying Radicals*

Simplify each expression. Assume that all variables are positive when they appear.

(a) $\sqrt{32}$ (b) $\sqrt[3]{8x^4}$ (c) $\sqrt{\sqrt[3]{x^7}}$ (d) $\sqrt[3]{\dfrac{8x^5}{27y^2}}$ (e) $\dfrac{\sqrt{x^5y}}{\sqrt{x^3y^3}}$

Solution   (a) $\sqrt{32} \underset{\substack{\uparrow \\ (2a) \\ 16 \text{ is a perfect square.}}}{=} \sqrt{16 \cdot 2} = \sqrt{16}\sqrt{2} = 4\sqrt{2}$

(b) $\sqrt[3]{8x^4} \underset{\substack{\uparrow \\ \text{Factor out} \\ \text{perfect cube.}}}{=} \sqrt[3]{8x^3 \cdot x} = \sqrt[3]{(2x)^3 \cdot x} \underset{\substack{\uparrow \\ (2a)}}{=} \sqrt[3]{(2x)^3}\,\sqrt[3]{x} \underset{\substack{\uparrow \\ (1a)}}{=} 2x\sqrt[3]{x}$

(c) $\sqrt{\sqrt[3]{x^7}} \underset{\substack{\uparrow \\ (2d)}}{=} \sqrt[6]{x^7} = \sqrt[6]{x^6 \cdot x} = \sqrt[6]{x^6} \cdot \sqrt[6]{x} \underset{\substack{\uparrow \\ (1b)}}{=} |x|\,\sqrt[6]{x}$

(d) $\sqrt[3]{\dfrac{8x^5}{27y^2}} = \sqrt[3]{\dfrac{2^3 x^3 x^2}{3^3 y^2}} = \sqrt[3]{\left(\dfrac{2x}{3}\right)^3 \cdot \dfrac{x^2}{y^2}} = \sqrt[3]{\left(\dfrac{2x}{3}\right)^3} \cdot \sqrt[3]{\dfrac{x^2}{y^2}} = \dfrac{2x}{3}\sqrt[3]{\dfrac{x^2}{y^2}}$

(e) $\dfrac{\sqrt{x^5y}}{\sqrt{x^3y^3}} = \sqrt{\dfrac{x^5y}{x^3y^3}} = \sqrt{\dfrac{x^2}{y^2}} = \sqrt{\left(\dfrac{x}{y}\right)^2} = \left|\dfrac{x}{y}\right|$

∎

■   Now work Problems 1 and 15.

Rationalizing

When radicals occur in quotients, it has become common practice to rewrite the quotient so that the denominator contains no radicals. This process is referred to as **rationalizing the denominator.**

The idea is to find an appropriate expression so that, when it is multiplied by the radical in the denominator, the new denominator that results contains no radicals. For example,

IF RADICAL IS	MULTIPLY BY	TO GET PRODUCT FREE OF RADICALS
$\sqrt{3}$	$\sqrt{3}$	$\sqrt{9} = 3$
$\sqrt[3]{4}$	$\sqrt[3]{2}$	$\sqrt[3]{8} = 2$
$\sqrt{3} + 1$	$\sqrt{3} - 1$	$(\sqrt{3})^2 - 1^2 = 3 - 1 = 2$
$\sqrt{2} - 3$	$\sqrt{2} + 3$	$(\sqrt{2})^2 - 3^2 = 2 - 9 = -7$
$\sqrt{5} - \sqrt{3}$	$\sqrt{5} + \sqrt{3}$	$(\sqrt{5})^2 - (\sqrt{3})^2 = 5 - 3 = 2$

You are correct if you observed in this list that, after the second type of radical, the special product for differences of squares is the basis for determining by what to multiply.

E X A M P L E   6   *Rationalizing Denominators*

Rationalize the denominator of each expression.

(a) $\dfrac{4}{\sqrt{2}}$ (b) $\dfrac{\sqrt{3}}{\sqrt[3]{2}}$ (c) $\dfrac{\sqrt{x}-2}{\sqrt{x}+2}$, $x \geq 0$

Solution   (a) $\dfrac{4}{\sqrt{2}} = \dfrac{4}{\sqrt{2}} \cdot \dfrac{\sqrt{2}}{\sqrt{2}} = \dfrac{4\sqrt{2}}{(\sqrt{2})^2} = \dfrac{4\sqrt{2}}{2} = 2\sqrt{2}$

Multiply by $\dfrac{\sqrt{2}}{\sqrt{2}}$.

(b) $\dfrac{\sqrt{3}}{\sqrt[3]{2}} = \dfrac{\sqrt{3}}{\sqrt[3]{2}} \cdot \dfrac{\sqrt[3]{4}}{\sqrt[3]{4}} = \dfrac{\sqrt{3}\sqrt[3]{4}}{\sqrt[3]{8}} = \dfrac{\sqrt{3}\sqrt[3]{4}}{2}$

Multiply by $\dfrac{\sqrt[3]{4}}{\sqrt[3]{4}}$.

(c) $\dfrac{\sqrt{x}-2}{\sqrt{x}+2} = \dfrac{\sqrt{x}-2}{\sqrt{x}+2} \cdot \dfrac{\sqrt{x}-2}{\sqrt{x}-2} = \dfrac{(\sqrt{x}-2)^2}{(\sqrt{x})^2 - 2^2}$

$= \dfrac{(\sqrt{x})^2 - 4\sqrt{x} + 4}{x - 4} = \dfrac{x - 4\sqrt{x} + 4}{x - 4}$ ∎

In calculus, sometimes the numerator must be rationalized.

E X A M P L E   7   *Rationalizing Numerators*

Rationalize the numerator: $\dfrac{\sqrt{x}-2}{\sqrt{x}+1}$, $x \geq 0$

Solution   We multiply by $\dfrac{\sqrt{x}+2}{\sqrt{x}+2}$:

$\dfrac{\sqrt{x}-2}{\sqrt{x}+1} = \dfrac{\sqrt{x}-2}{\sqrt{x}+1} \cdot \dfrac{\sqrt{x}+2}{\sqrt{x}+2} = \dfrac{(\sqrt{x})^2 - 2^2}{(\sqrt{x}+1)(\sqrt{x}+2)} = \dfrac{x - 4}{x + 3\sqrt{x} + 2}$ ∎

■ Now work Problem 31.

Rational Exponents

Radicals are used to define rational exponents.

$a^{1/n}$ If a is a real number and $n \geq 2$ is an integer, then

$$a^{1/n} = \sqrt[n]{a} \tag{3}$$

provided $\sqrt[n]{a}$ exists.

E X A M P L E   8

Using Equation (3)

(a) $4^{1/2} = \sqrt{4} = 2$ (b) $(-27)^{1/3} = \sqrt[3]{-27} = -3$

(c) $8^{1/2} = \sqrt{8} = 2\sqrt{2}$ (d) $16^{1/3} = \sqrt[3]{16} = 2\sqrt[3]{2}$ ∎

$a^{m/n}$

If a is a real number and m and n are integers containing no common factors with $n \geq 2$, then

$$a^{m/n} = \sqrt[n]{a^m} = (\sqrt[n]{a})^m \tag{4}$$

provided $\sqrt[n]{a}$ exists.

We have two comments about equation (4):

1. The exponent m/n must be in lowest terms and n must be positive.
2. In simplifying $a^{m/n}$, either $\sqrt[n]{a^m}$ or $(\sqrt[n]{a})^m$ may be used. Generally, taking the root first is preferred.

It can be shown that the laws of exponents hold for rational exponents.

E X A M P L E   9

Simplifying Expressions with Rational Exponents

Simplify each expression. Express your answer so that only positive exponents occur. Assume that the variables are positive.

(a) $\left(\dfrac{2x^{1/3}}{y^{2/3}}\right)^{-3}$ (b) $(x^{2/3}y^{-3/4})(x^{-2}y)^{1/2}$

(c) $\left(\dfrac{9x^2y^{1/3}}{x^{1/3}y}\right)^{1/2}$ (d) $\dfrac{(2x+5)^{1/3}(2x+5)^{-1/2}}{(2x+5)^{-3/4}}$

Solution

(a) $\left(\dfrac{2x^{1/3}}{y^{2/3}}\right)^{-3} = \left(\dfrac{y^{2/3}}{2x^{1/3}}\right)^{3} = \dfrac{(y^{2/3})^3}{(2x^{1/3})^3} = \dfrac{y^2}{2^3(x^{1/3})^3} = \dfrac{y^2}{8x}$

(b) $(x^{2/3}y^{-3/4})(x^{-2}y)^{1/2} = (x^{2/3}y^{-3/4})[(x^{-2})^{1/2}y^{1/2}]$

$\qquad = x^{2/3}y^{-3/4}x^{-1}y^{1/2} = (x^{2/3}x^{-1})(y^{-3/4}y^{1/2})$

$\qquad = x^{-1/3}y^{-1/4} = \dfrac{1}{x^{1/3}y^{1/4}}$

(c) $\left(\dfrac{9x^2y^{1/3}}{x^{1/3}y}\right)^{1/2} = \left(\dfrac{9x^{2-(1/3)}}{y^{1-(1/3)}}\right)^{1/2} = \left(\dfrac{9x^{5/3}}{y^{2/3}}\right)^{1/2} = \dfrac{9^{1/2}(x^{5/3})^{1/2}}{(y^{2/3})^{1/2}} = \dfrac{3x^{5/6}}{y^{1/3}}$

(d) $\dfrac{(2x+5)^{1/3}(2x+5)^{-1/2}}{(2x+5)^{-3/4}} = (2x+5)^{(1/3)-(1/2)-(-3/4)}$

$\qquad\qquad\qquad = (2x+5)^{(4-6+9)/12} = (2x+5)^{7/12}$ ∎

▨   Now work Problem 53.

The next two examples illustrate some algebra that you will need to know for certain calculus problems.

E X A M P L E 1 0   *Writing an Expression as a Single Quotient*

Write the expression as a single quotient in which only positive exponents appear.

$$(x^2 + 1)^{1/2} + x \cdot \frac{1}{2}(x^2 + 1)^{-1/2} \cdot 2x$$

Solution

$$(x^2 + 1)^{1/2} + x \cdot \frac{1}{2}(x^2 + 1)^{-1/2} \cdot 2x = (x^2 + 1)^{1/2} + \frac{x^2}{(x^2 + 1)^{1/2}}$$

$$= \frac{(x^2 + 1)^{1/2}(x^2 + 1)^{1/2} + x^2}{(x^2 + 1)^{1/2}}$$

$$= \frac{(x^2 + 1) + x^2}{(x^2 + 1)^{1/2}}$$

$$= \frac{2x^2 + 1}{(x^2 + 1)^{1/2}} \qquad\blacksquare$$

E X A M P L E 1 1   *Factoring an Expression Containing Rational Exponents*

Factor: $4x^{1/3}(2x + 1) + 2x^{4/3}$

Solution

We begin by looking for factors that are common to the two terms. Notice that 2 and $x^{1/3}$ are common factors. Thus,

$$4x^{1/3}(2x + 1) + 2x^{4/3} = 2x^{1/3}[2(2x + 1) + x]$$
$$= 2x^{1/3}(5x + 2) \qquad\blacksquare$$

3

Exercise 3

In Problems 1–20, simplify each expression. Assume that all variables are positive when they appear.

1. $\sqrt{8}$ **2.** $\sqrt[4]{32}$ **3.** $\sqrt[3]{16x^4}$ **4.** $\sqrt{27x^3}$

5. $\sqrt[3]{\sqrt{x^6}}$ **6.** $\sqrt{\sqrt{x^6}}$ **7.** $\sqrt{\dfrac{32x^3}{9x}}$ **8.** $\sqrt[3]{\dfrac{x}{8x^4}}$

9. $\sqrt[4]{x^{12}y^8}$ **10.** $\sqrt[5]{x^{10}y^5}$ **11.** $\sqrt[4]{\dfrac{x^9y^7}{xy^3}}$ **12.** $\sqrt[3]{\dfrac{3xy^2}{81x^4y^2}}$

13. $\sqrt{36x}$ **14.** $\sqrt{9x^5}$ **15.** $\sqrt{3x^2}\sqrt{12x}$ **16.** $\sqrt{5x}\sqrt{20x^3}$

17. $(\sqrt{5}\sqrt[3]{9})^2$ **18.** $(\sqrt[3]{3}\sqrt{10})^4$

19. $\sqrt{\dfrac{2x-3}{2x^4+3x^3}}\sqrt{\dfrac{x}{4x^2-9}}$ **20.** $\sqrt[3]{\dfrac{x-1}{x^2+2x+1}}\sqrt[3]{\dfrac{(x-1)^2}{x+1}}$

In Problems 21–26, perform the indicated operation and simplify the result. Assume that all variables are positive when they appear.

21. $(3\sqrt{6})(2\sqrt{2})$ **22.** $(5\sqrt{8})(-3\sqrt{3})$ **23.** $(\sqrt{3} + 3)(\sqrt{3} - 1)$

24. $(\sqrt{5} - 2)(\sqrt{5} + 3)$ **25.** $(\sqrt{x} - 1)^2$ **26.** $(\sqrt{x} + \sqrt{5})^2$

In Problems 27–36, rationalize the denominator of each expression. Assume that all variables are positive when they appear.

27. $\dfrac{1}{\sqrt{2}}$ **28.** $\dfrac{6}{\sqrt[3]{4}}$ **29.** $\dfrac{-\sqrt{3}}{\sqrt{5}}$ **30.** $\dfrac{-\sqrt[3]{3}}{\sqrt{8}}$

31. $\dfrac{\sqrt{3}}{5-\sqrt{2}}$ **32.** $\dfrac{\sqrt{2}}{\sqrt{7}+2}$ **33.** $\dfrac{2-\sqrt{5}}{2+3\sqrt{5}}$ **34.** $\dfrac{\sqrt{3}-1}{2\sqrt{3}+3}$

35. $\dfrac{\sqrt{x+h}-\sqrt{x}}{\sqrt{x+h}+\sqrt{x}}$ **36.** $\dfrac{\sqrt{x+h}+\sqrt{x-h}}{\sqrt{x+h}-\sqrt{x-h}}$

In Problems 37–48, simplify each expression.

37. $8^{2/3}$ **38.** $4^{3/2}$ **39.** $(-27)^{1/3}$ **40.** $16^{3/4}$ **41.** $16^{3/2}$

42. $64^{3/2}$ **43.** $9^{-3/2}$ **44.** $25^{-5/2}$ **45.** $\left(\dfrac{9}{8}\right)^{3/2}$ **46.** $\left(\dfrac{27}{8}\right)^{2/3}$

47. $\left(\dfrac{8}{9}\right)^{-3/2}$ **48.** $\left(\dfrac{8}{27}\right)^{-2/3}$

In Problems 49–56, simplify each expression. Express your answer so that only positive exponents occur. Assume that the variables are positive.

49. $x^{5/4}x^{2/3}x^{-1/2}$ **50.** $x^{4/3}x^{1/2}x^{-1/4}$ **51.** $(x^3y^6)^{2/3}$ **52.** $(x^4y^8)^{5/4}$

53. $(x^2y)^{1/3}(xy^2)^{2/3}$ **54.** $(xy)^{1/4}(x^2y^2)^{1/2}$ **55.** $(16x^2y^{-1/3})^{3/4}$ **56.** $(4x^{-1}y^{1/3})^{3/2}$

In Problems 57–62, write each expression as a single quotient in which only positive exponents and/or radicals appear.

57. $\dfrac{x}{(1+x)^{1/2}}+2(1+x)^{1/2}$ **58.** $\dfrac{1+x}{2x^{1/2}}+x^{1/2}$

59. $\dfrac{\sqrt{1+x}-x\cdot\dfrac{1}{2\sqrt{1+x}}}{1+x}$ **60.** $\dfrac{\sqrt{x^2+1}-x\cdot\dfrac{2x}{2\sqrt{x^2+1}}}{x^2+1}$

61. $\dfrac{(x+4)^{1/2}-2x(x+4)^{-1/2}}{x+4}$ **62.** $\dfrac{(9-x^2)^{1/2}+x^2(9-x^2)^{-1/2}}{9-x^2}$

In Problems 63–66, factor each expression.

63. $(x+1)^{3/2}+x\cdot\dfrac{3}{2}(x+1)^{1/2}$ **64.** $(x^2+4)^{4/3}+x\cdot\dfrac{4}{3}(x^2+4)^{1/3}\cdot 2x$

65. $6x^{1/2}(x^2+x)-8x^{3/2}-8x^{1/2}$ **66.** $6x^{1/2}(2x+3)+x^{3/2}\cdot 8$

4

Solving Equations

An **equation in one variable** is a statement in which two expressions, at least one containing the variable, are equal. The expressions are called the **sides** of the equation. Since an equation is a statement, it may or may not be true, depending on the value of the variable. Unless otherwise restricted, the admissible values of the variable are those in the domain of the variable. Those admissible values of the variable, if any, that result in a true statement are called **solutions,** or **roots,** of the equation. To **solve an equation** means to find all the solutions of the equation.

For example, the following are all equations in one variable, x:

$$x + 5 = 9 \qquad x^2 + 5x = 2x - 2 \qquad \frac{x^2 - 4}{x + 1} = 0 \qquad x^2 + 9 = 5$$

The first of these statements, $x + 5 = 9$, is true when $x = 4$ and false for any other choice of x. Thus, 4 is a solution of the equation $x + 5 = 9$. We also say that 4 **satisfies** the equation $x + 5 = 9$, because, when x is replaced by 4, a true statement results.

Sometimes an equation will have more than one solution. For example, the equation

$$\frac{x^2 - 4}{x + 1} = 0$$

has either $x = -2$ or $x = 2$ as a solution.

Sometimes we will write the solutions of an equation in set notation. This set is called the **solution set** of the equation. For example, the solution set of the equation $x^2 - 9 = 0$ is $\{-3, 3\}$.

Unless indicated otherwise, we will limit ourselves to real solutions. Some equations have no real solution. For example, $x^2 + 9 = 5$ has no real solution, because there is no real number whose square when added to 9 equals 5.

An equation that is satisfied for every choice of the variable for which both sides are defined is called an **identity.** For example, the equation

$$3x + 5 = x + 3 + 2x + 2$$

is an identity, because this statement is true for any real number x.

Two or more equations that have precisely the same solutions are called **equivalent equations.** For example, all the following equations are equivalent, because each has only the solution $x = 5$:

$$2x + 3 = 13$$
$$2x = 10$$
$$x = 5$$

These three equations illustrate one method for solving many types of equations: Replace the original equation by an equivalent equation, and continue until an equation with an obvious solution, such as $x = 5$, is reached. The question, though, is: "How do I obtain an equivalent equation?" In general, there are five ways to do so.

Procedures That Result in Equivalent Equations

1. Interchange the two sides of the equation:

   Replace $3 = x$ by $x = 3$

2. Simplify the sides of the equation by combining like terms, eliminating parentheses, and so on:

   Replace $(x + 2) + 6 = 2x + (x + 1)$

   by $x + 8 = 3x + 1$

3. Add or subtract the same expression on both sides of the equation:

$$\text{Replace} \qquad 3x - 5 = 4$$
$$\text{by} \qquad (3x - 5) + 5 = 4 + 5$$

4. Multiply or divide both sides of the equation by the same nonzero expression:

$$\text{Replace} \qquad \frac{3x}{x - 1} = \frac{6}{x - 1} \qquad x \neq 1$$

$$\text{by} \qquad \frac{3x}{x - 1} \cdot (x - 1) = \frac{6}{x - 1} \cdot (x - 1)$$

5. If one side of the equation is 0 and the other side can be factored, then we may use the product law* and set each factor equal to 0:

$$\text{Replace} \qquad x(x - 3) = 0$$
$$\text{by} \qquad x = 0 \quad \text{or} \quad x - 3 = 0$$

Whenever it is possible to solve an equation in your head, do so. For example:

The solution of $2x = 8$ is $x = 4$.

The solution of $3x - 15 = 0$ is $x = 5$.

Often, though, some rearrangement is necessary.

E X A M P L E   1   *Solving an Equation*

Solve the equation: $(x + 1)(2x) = (x + 1)(2)$

Solution   We begin by collecting all terms on the left side:

$$(x + 1)(2x) = (x + 1)(2)$$
$$(x + 1)(2x) - (x + 1)(2) = 0$$
$$(x + 1)(2x - 2) = 0 \quad \text{Factor.}$$
$$x + 1 = \quad 0 \quad \text{or} \quad 2x - 2 = 0 \quad \text{Apply the product law.}$$
$$x = -1 \qquad\qquad 2x = 2$$
$$x = 1$$

The solution set is $\{-1, 1\}$.   ■

E X A M P L E   2   *Solving an Equation*

Solve the equation: $\dfrac{3x}{x - 1} + 2 = \dfrac{3}{x - 1}$

*The product law states that if $ab = 0$ then $a = 0$ or $b = 0$ or both equal 0.

Solution First, we note that the domain of the variable is $\{x | x \neq 1\}$. Since the two quotients in the equation have the same denominator, $x - 1$, we can simplify by multiplying both sides by $x - 1$. The resulting equation is equivalent to the original equation, since we are multiplying by $x - 1$, which is not 0 (remember, $x \neq 1$).

$$\frac{3x}{x-1} + 2 = \frac{3}{x-1}$$

$$\left(\frac{3x}{x-1} + 2\right) \cdot (x - 1) = \frac{3}{x-1} \cdot (x - 1) \qquad \text{Multiply both sides by } x - 1; \\ \text{cancel on the right.}$$

$$\frac{3x}{x-1} \cdot (x - 1) + 2 \cdot (x - 1) = 3 \qquad \text{Use the distributive property on} \\ \text{the left side; cancel on the left.}$$

$$3x + (2x - 2) = 3 \qquad \text{Simplify.}$$

$$5x - 2 = 3$$

$$5x = 5 \qquad \text{Add 2 to each side.}$$

$$x = 1 \qquad \text{Divide both sides by 5.}$$

The solution appears to be 1. But recall that $x = 1$ is not in the domain of the variable. Thus, the equation has no solution. ■

■ Now work Problem 19.

Steps for Solving Equations

STEP 1: List any restrictions on the domain of the variable.
STEP 2: Simplify the equation by replacing the original equation by a succession of equivalent equations following the procedures listed earlier.
STEP 3: If the result of Step 2 is a product of factors equal to 0, use the product law and set each factor equal to 0 (procedure 5).
STEP 4: Check your solution(s).

E X A M P L E   3

Solving an Equation

Solve the equation: $x^3 = 25x$

Solution We first rearrange the equation to get 0 on the right side:

$$x^3 = 25x$$

$$x^3 - 25x = 0$$

We notice that x is a factor of each term on the left:

$$x(x^2 - 25) = 0$$

$$x(x + 5)(x - 5) = 0 \qquad \text{Difference of two squares}$$

$$x = 0 \quad \text{or} \quad x + 5 = 0 \qquad \text{or} \quad x - 5 = 0 \qquad \text{Set each factor equal to 0.}$$

$$x = -5 \quad \text{or} \qquad x = 5 \qquad \text{Solve.}$$

The solution set is $\{-5, 0, 5\}$. ■

■ Now work Problem 21.

4

Exercise 4

In Problems 1–36, solve each equation.

1. $6 - x = 2x + 9$

2. $3 - 2x = 2 - x$

3. $2(3 + 2x) = 3(x - 4)$

4. $3(2 - x) = 2x - 1$

5. $8x - (2x + 1) = 3x - 10$

6. $5 - (2x - 1) = 10$

7. $\frac{1}{2}x - 4 = \frac{3}{4}x$

8. $1 - \frac{1}{2}x = 5$

9. $0.9t = 0.4 + 0.1t$

10. $0.9t = 1 + t$

11. $\frac{2}{y} + \frac{4}{y} = 3$

12. $\frac{4}{y} - 5 = \frac{5}{2y}$

13. $(x + 7)(x - 1) = (x + 1)^2$

14. $(x + 2)(x - 3) = (x - 3)^2$

15. $x(2x - 3) = (2x + 1)(x - 4)$

16. $x(1 + 2x) = (2x - 1)(x - 2)$

17. $z(z^2 + 1) = 3 + z^3$

18. $w(4 - w^2) = 8 - w^3$

19. $\frac{x}{x - 3} + 3 = \frac{3}{x - 3}$

20. $\frac{3x}{x + 2} = \frac{-6}{x + 2} - 2$

21. $x^2 = 9x$

22. $x^3 = x^2$

23. $t^3 - 9t^2 = 0$

24. $4z^3 - 8z^2 = 0$

25. $\frac{2x}{x^2 - 4} = \frac{4}{x^2 - 4} - \frac{1}{x + 2}$

26. $\frac{x}{x^2 - 9} + \frac{1}{x + 3} = \frac{3}{x^2 - 9}$

27. $\frac{x}{x + 2} = \frac{1}{2}$

28. $\frac{3x}{x - 1} = 2$

29. $\frac{3}{2x - 3} = \frac{2}{x + 5}$

30. $\frac{-2}{x + 4} = \frac{-3}{x + 1}$

31. $(x + 2)(3x) = (x + 2)(6)$

32. $(x - 5)(2x) = (x - 5)(4)$

33. $\frac{6t + 7}{4t - 1} = \frac{3t + 8}{2t - 4}$

34. $\frac{8w + 5}{10w - 7} = \frac{4w - 3}{5w + 7}$

35. $\frac{2}{x - 2} = \frac{3}{x + 5} + \frac{10}{(x + 5)(x - 2)}$

36. $\frac{1}{2x + 3} + \frac{1}{x - 1} = \frac{1}{(2x + 3)(x - 1)}$

5

Completing the Square

The idea behind the method of **completing the square** is to "adjust" the left side of a second degree polynomial, $ax^2 + bx + c$, so that it becomes a perfect square—the square of a first-degree polynomial. For example, $x^2 + 6x + 9$ and $x^2 - 4x + 4$ are perfect squares* because

$$x^2 + 6x + 9 = (x + 3)^2 \quad \text{and} \quad x^2 - 4x + 4 = (x - 2)^2$$

How do we adjust the second degree polynomial? We do it by adding the appropriate number to create a perfect square. For example, to make $x^2 + 6x$ a perfect square, we add 9.

Let's look at several examples of completing the square when the coefficient of x^2 is 1:

*Perfect squares are discussed in Section 2 of this Appendix.

START	ADD	RESULT
$x^2 + 4x$	4	$x^2 + 4x + 4 = (x + 2)^2$
$x^2 + 12x$	36	$x^2 + 12x + 36 = (x + 6)^2$
$x^2 - 6x$	9	$x^2 - 6x + 9 = (x - 3)^2$
$x^2 + x$	$\frac{1}{4}$	$x^2 + x + \frac{1}{4} = \left(x + \frac{1}{2}\right)^2$

Do you see the pattern? Provided the coefficient of x^2 is 1, we complete the square by adding the square of one-half the coefficient of x:

START	ADD	RESULT
$x^2 + mx$	$\left(\dfrac{m}{2}\right)^2$	$x^2 + mx + \left(\dfrac{m}{2}\right)^2 = \left(x + \dfrac{m}{2}\right)^2$

■   Now work Problem 1.

E X A M P L E   1   *Completing the Square of an Equation Containing Two Variables*

Complete the squares of x and y in the equation

$$x^2 + y^2 - 2x + 4y - 4 = 0$$

Solution   We rearrange the equation, grouping the terms involving the variable x and the variable y.

$$(x^2 - 2x) + (y^2 + 4y) = 4$$

Next, we complete the square of each parenthetical expression. Of course, since we want an equivalent equation, whatever we add to the left side, we also add to the right side.

$$(x^2 - 2x + 1) + (y^2 + 4y + 4) = 4 + 1 + 4$$
$$(x - 1)^2 + (y + 2)^2 = 9$$

The terms involving the variables x and y now appear as perfect squares. ■

■   Now work Problem 7.

The next example illustrates how the procedure of completing the square can be used to solve a quadratic equation.

E X A M P L E   2   *Solving a Quadratic Equation by Completing the Square*

Solve by completing the square: $x^2 + 5x + 4 = 0$

Solution   We always begin this procedure by rearranging the equation so that the constant is on the right side:

$$x^2 + 5x + 4 = 0$$
$$x^2 + 5x = -4$$

Since the coefficient of x^2 is 1, we can complete the square on the left side by adding $\left(\frac{1}{2} \cdot 5\right)^2 = \frac{25}{4}$. Of course, in an equation, whatever we add to the left side must also be added to the right side. Thus, we add $\frac{25}{4}$ to *both* sides:

$$x^2 + 5x + \tfrac{25}{4} = -4 + \tfrac{25}{4}$$
$$\left(x + \tfrac{5}{2}\right)^2 = \tfrac{9}{4}$$
$$x + \tfrac{5}{2} = \pm\sqrt{\tfrac{9}{4}}$$
$$x + \tfrac{5}{2} = \pm\tfrac{3}{2}$$
$$x = -\tfrac{5}{2} \pm \tfrac{3}{2}$$
$$x = -\tfrac{5}{2} + \tfrac{3}{2} = -1 \quad \text{or} \quad x = -\tfrac{5}{2} - \tfrac{3}{2} = -4$$

The solution set is $\{-4, -1\}$. ∎

■   Now work Problem 13.

E X A M P L E   3   *Solving a Quadratic Equation by Completing the Square*

Solve by completing the square: $2x^2 - 8x - 5 = 0$

Solution   First, we rewrite the equation:

$$2x^2 - 8x - 5 = 0$$
$$2x^2 - 8x = 5$$

Next, we divide by 2 so that the coefficient of x^2 is 1. (This enables us to complete the square at the next step.)

$$x^2 - 4x = \tfrac{5}{2}$$

Finally, we complete the square by adding 4 to each side:

$$x^2 - 4x + 4 = \tfrac{5}{2} + 4$$
$$(x - 2)^2 = \tfrac{13}{2}$$
$$x - 2 = \pm\sqrt{\tfrac{13}{2}} = \pm\tfrac{\sqrt{26}}{2}$$
$$x = 2 \pm \tfrac{\sqrt{26}}{2}$$

We choose to leave our answer in this compact form. Thus, the solution set is $\{2 - \sqrt{26}/2, \, 2 + \sqrt{26}/2\}$. ∎

Note: If we wanted an approximation, say, to two decimal places, of these solutions, we would use a calculator to get $\{-0.55, 4.55\}$.

■   Now work Problem 17.

5

Exercise 5

In Problems 1–6, tell what number should be added to complete the square of each expression.

1. $x^2 - 4x$ **2.** $x^2 - 2x$ **3.** $x^2 + \tfrac{1}{2}x$

4. $x^2 - \tfrac{1}{3}x$ **5.** $x^2 - \tfrac{2}{3}x$ **6.** $x^2 - \tfrac{2}{5}x$

In Problems 7–12, complete the squares of x and y in each equation.

7. $x^2 + y^2 - 4x + 4y - 1 = 0$ **8.** $x^2 + y^2 + 4x + 4y - 8 = 0$ **9.** $x^2 + y^2 + 6x - 2y + 1 = 0$

10. $x^2 + y^2 - 8x + 2y + 1 = 0$ **11.** $x^2 + y^2 + x - y - \tfrac{1}{2} = 0$ **12.** $x^2 + y^2 - x + y - \tfrac{3}{2} = 0$

In Problems 13–18, solve each equation by completing the square.

13. $x^2 + 4x - 21 = 0$

14. $x^2 - 6x = 13$

15. $x^2 - \frac{1}{2}x = \frac{3}{16}$

16. $x^2 + \frac{2}{3}x = \frac{1}{3}$

17. $3x^2 + x - \frac{1}{2} = 0$

18. $2x^2 - 3x = 1$

6

Synthetic Division

To find the quotient as well as the remainder when a polynomial function f of degree 1 or higher is divided by $g(x) = x - c$, a shortened version of long division, called **synthetic division,** makes the task simpler.

To see how synthetic division works, we will use long division* to divide the polynomial $f(x) = 2x^3 - x^2 + 3$ by $g(x) = x - 3$.

$$
\begin{array}{r}
2x^2 + 5x + 15 \\
x - 3 \overline{)2x^3 - x^2 \qquad\quad + 3} \\
\underline{2x^3 - 6x^2} \\
5x^2 \\
\underline{5x^2 - 15x} \\
15x + 3 \\
\underline{15x - 45} \\
48
\end{array}
$$

The process of synthetic division arises from rewriting the long division in a more compact form, using simpler notation. For example, in the long division above, the terms in color are not really necessary because they are identical to the terms directly above them. With these terms removed, we have

$$
\begin{array}{r}
2x^2 + 5x + 15 \\
x - 3 \overline{)2x^3 - x^2 \qquad\quad + 3} \\
\underline{- 6x^2} \\
5x^2 \\
\underline{- 15x} \\
15x \\
\underline{- 45} \\
48
\end{array}
$$

Most of the x's that appear in this process can also be removed, provided we are careful about positioning each coefficient. In this regard, we will need to use 0 as the coefficient of x in the dividend, because that power of x is missing. Now we have

$$
\begin{array}{r}
2x^2 + 5x + 15 \\
x - 3 \overline{)2 \quad -1 \qquad 0 \qquad 3} \\
\underline{-6} \\
5 \\
\underline{-15} \\
15 \\
\underline{-45} \\
48
\end{array}
$$

*Long division is discussed in Section 2 of this Appendix.

We can make this display more compact by moving the lines up until the numbers in color align horizontally:

$$
\begin{array}{r}
2x^2 + 5x + 15 \qquad \text{Row 1} \\
\hline
x-3\overline{)2 \quad -1 \quad\ \ 0 \quad\ \ 3} \qquad \text{Row 2} \\
-6 \ -15 \ -45 \qquad \text{Row 3} \\
\hline
\bigcirc \quad\ 5 \quad\ 15 \quad 48 \qquad \text{Row 4}
\end{array}
$$

Now, if we place the leading coefficient of the quotient (2) in the circled position, the first three numbers in Row 4 are precisely the coefficients of the quotient, and the last number in Row 4 is the remainder. Thus, Row 1 is not really needed, so we can compress the process to three rows, where the bottom row contains the coefficients of both the quotient and the remainder:

$$
\begin{array}{r}
x-3\overline{)2 \quad -1 \quad\ \ 0 \quad\ \ 3} \qquad \text{Row 1} \\
-6 \ -15 \ -45 \qquad \text{Row 2 (subtract)} \\
\hline
2 \quad\ 5 \quad\ 15 \quad 48 \qquad \text{Row 3}
\end{array}
$$

Recall that the entries in Row 3 are obtained by subtracting the entries in Row 2 from those in Row 1. Rather than subtracting the entries in Row 2, we can change the sign of each entry and add. With this modification, our display will look like this:

$$
\begin{array}{r}
x-3\overline{)2 \quad -1 \quad\ \ 0 \quad\ \ 3} \qquad \text{Row 1} \\
6 \quad 15 \quad 45 \qquad \text{Row 2 (add)} \\
\hline
2 \quad\ 5 \quad\ 15 \quad 48 \qquad \text{Row 3}
\end{array}
$$

Notice that the entries in Row 2 are three times the prior entries in Row 3. Our last modification to the display replaces the $x - 3$ by 3. The entries in Row 3 give the quotient and the remainder as shown next.

$$
\begin{array}{r}
3\overline{)2 \quad -1 \quad 0 \quad 3} \qquad \text{Row 1} \\
6 \quad 15 \quad 45 \qquad \text{Row 2 (add)} \\
\hline
2 \quad\ 5 \quad 15 \quad 48 \qquad \text{Row 3}
\end{array}
$$

Quotient   Remainder

$$2x^2 + 5x + 15 \qquad R = 48$$

Let's go through another example step by step.

EXAMPLE 1   *Using Synthetic Division to Find the Quotient and Remainder*

Use synthetic division to find the quotient and remainder when

$$f(x) = 3x^4 + 8x^2 - 7x + 4 \quad \text{is divided by} \quad g(x) = x - 1$$

Solution   **STEP 1:**   Write the dividend in descending powers of x. Then copy the coefficients, remembering to insert a 0 for any missing powers of x:

$$3 \quad 0 \quad 8 \quad -7 \quad 4 \qquad \text{Row 1}$$

STEP 2:   Insert the usual division symbol. Since the divisor is $x - 1$, we insert 1 to the left of the division symbol

$$1\overline{)3 \quad 0 \quad 8 \quad -7 \quad 4} \qquad \text{Row 1}$$

STEP 3: Bring the 3 down two rows, and enter it in Row 3:

$$1\overline{)3 \quad 0 \quad 8 \quad -7 \quad 4} \quad \text{Row 1}$$

$$\downarrow \qquad\qquad\qquad\qquad \text{Row 2}$$

$$3 \qquad\qquad\qquad\qquad \text{Row 3}$$

STEP 4: Multiply the latest entry in Row 3 by 1 and place the result in Row 2, but one column over to the right:

$$1\overline{)3 \quad 0 \quad 8 \quad -7 \quad 4} \quad \text{Row 1}$$

$$3 \qquad\qquad\qquad \text{Row 2}$$

$$3 \qquad\qquad\qquad\qquad \text{Row 3}$$

STEP 5: Add the entry in Row 2 to the entry above it in Row 1, and enter the sum in Row 3:

$$1\overline{)3 \quad 0 \quad 8 \quad -7 \quad 4} \quad \text{Row 1}$$

$$3 \qquad\qquad\qquad \text{Row 2}$$

$$3 \quad 3 \qquad\qquad\qquad \text{Row 3}$$

STEP 6: Repeat Steps 4 and 5 until no more entries are available in Row 1:

$$1\overline{)3 \quad 0 \quad 8 \quad -7 \quad 4} \quad \text{Row 1}$$

$$3 \quad 3 \quad 11 \quad 4 \quad \text{Row 2 (add)}$$

$$3 \quad 3 \quad 11 \quad 4 \quad 8 \quad \text{Row 3}$$

STEP 7: The final entry in Row 3, an 8, is the remainder; the other entries in Row 3(3, 3, 11, and 4) are the coefficients (in descending order) of a polynomial whose degree is 1 less than that of the dividend; this is the quotient. Thus,

$$\text{Quotient} = 3x^3 + 3x^2 + 11x + 4 \qquad \text{Remainder} = 8$$

Check: (Divisor)(Quotient) + Remainder

$$= (x - 1)(3x^3 + 3x^2 + 11x + 4) + 8$$
$$= 3x^4 + 3x^3 + 11x^2 + 4x - 3x^3 - 3x^2 - 11x - 4 + 8$$
$$= 3x^4 + 8x^2 - 7x + 4 = \text{Dividend} \qquad \blacksquare$$

Let's do an example in which all seven steps are combined.

E X A M P L E   2   *Using Synthetic Division to Verify a Factor*

Use synthetic division to show that $g(x) = x + 3$ is a factor of

$$f(x) = 2x^5 + 5x^4 - 2x^3 + 2x^2 - 2x + 3$$

Solution The divisor is $x + 3 = x - (-3)$, so the Row 3 entries will be multiplied by -3, entered in Row 2, and added to Row 1:

$$-3\overline{)2 \quad 5 \quad -2 \quad 2 \quad -2 \quad 3} \quad \text{Row 1}$$

$$-6 \quad 3 \quad -3 \quad 3 \quad -3 \quad \text{Row 2}$$

$$2 \quad -1 \quad 1 \quad -1 \quad 1 \quad 0 \quad \text{Row 3}$$

Because the remainder is 0, it follows that $f(-3) = 0$. Hence, by the Factor Theorem, $x - (-3) = x + 3$ is a factor of $f(x)$. $\qquad \blacksquare$

■ Now work Problem 3.

One important use of synthetic division is to find the value of a polynomial.

E X A M P L E   3   *Using Synthetic Division to Find the Value of a Polynomial*

Use synthetic division to find the value of $f(x) = -3x^4 + 2x^3 - x + 1$ at $x = -2$; that is, find $f(-2)$.

Solution   The Remainder Theorem tells us that the value of a polynomial function at c equals the remainder when the polynomial is divided by $x - c$. This remainder is the final entry of the third row in the process of synthetic division. We want $f(-2)$, so we divide by $x - (-2)$:

$$
\begin{array}{r|rrrrr}
-2) & -3 & 2 & 0 & -1 & 1 \\
    &   & 6 & -16 & 32 & -62 \\
\hline
    & -3 & 8 & -16 & 31 & -61 \\
\end{array}
$$

The quotient is $q(x) = -3x^3 + 8x^2 - 16x + 31$; the remainder is $R = -61$. Because the remainder was found to be -61, it follows from the Remainder Theorem that $f(-2) = -61$. ∎

■ Now work Problem 23.

As Example 3 illustrates, we can use the process of synthetic division to find the value of a polynomial function at a number c as an alternative to merely substituting c for x. Compare the work required in Example 3 with the arithmetic involved in substituting:

$$
\begin{aligned}
f(-2) &= -3(-2)^4 + 2(-2)^3 - (-2) + 1 \\
      &= -3(16) + 2(-8) + 2 + 1 \\
      &= -48 - 16 + 2 + 1 = -61
\end{aligned}
$$

As you can see, finding $f(-2)$ may be easier using synthetic division.

6

Exercise 6

In Problems 1–12, use synthetic division to find the quotient $q(x)$ and remainder R when $f(x)$ is divided by $g(x)$.

1. $f(x) = x^3 - x^2 + 2x + 4;$ $g(x) = x - 2$
2. $f(x) = x^3 + 2x^2 - 3x + 1;$ $g(x) = x + 1$
3. $f(x) = 3x^3 + 2x^2 - x + 3;$ $g(x) = x - 3$
4. $f(x) = -4x^3 + 2x^2 - x + 1;$ $g(x) = x + 2$
5. $f(x) = x^5 - 4x^3 + x;$ $g(x) = x + 3$
6. $f(x) = x^4 + x^2 + 2;$ $g(x) = x - 2$
7. $f(x) = 4x^6 - 3x^4 + x^2 + 5;$ $g(x) = x - 1$
8. $f(x) = x^5 + 5x^3 - 10;$ $g(x) = x + 1$
9. $f(x) = 0.1x^3 + 0.2x;$ $g(x) = x + 1.1$
10. $f(x) = 0.1x^2 - 0.2;$ $g(x) = x + 2.1$
11. $f(x) = x^5 - 1;$ $g(x) = x - 1$
12. $f(x) = x^5 + 1;$ $g(x) = x + 1$

In Problems 13–22, use synthetic division to determine whether $x - c$ is a factor of $f(x)$.

13. $f(x) = 4x^3 - 3x^2 - 8x + 4$; $c = 2$

14. $f(x) = -4x^3 + 5x^2 + 8$; $c = -3$

15. $f(x) = 3x^4 - 6x^3 - 5x + 10$; $c = 2$

16. $f(x) = 4x^4 - 15x^2 - 4$; $c = 2$

17. $f(x) = 3x^6 + 82x^3 + 27$; $c = -3$

18. $f(x) = 2x^6 - 18x^4 + x^2 - 9$; $c = -3$

19. $f(x) = 4x^6 - 64x^4 + x^2 - 15$; $c = -4$

20. $f(x) = x^6 - 16x^4 + x^2 - 16$; $c = -4$

21. $f(x) = 2x^4 - x^3 + 2x - 1$; $c = \dfrac{1}{2}$

22. $f(x) = 3x^4 + x^3 - 3x + 1$; $c = -\dfrac{1}{3}$

In Problems 23–28, use synthetic division to find $f(c)$.

23. $f(x) = 5x^4 - 3x^2 + 1$; $c = 2$

24. $f(x) = -2x^3 + 3x^2 + 5$; $c = -2$

25. $f(x) = 4x^5 - 3x^3 + 2x - 1$; $c = -1$

26. $f(x) = -3x^4 + 3x^3 - 2x^2 + 5$; $c = -1$

27. $f(x) = 9x^{17} - 8x^{10} + 9x^8 + 5$; $c = 1$

28. $f(x) = 10x^{15} + 4x^{12} - 2x^5 + x^2$; $c = -1$

ANSWERS

CHAPTER 1 *Exercise 1.1*

1. (a) Quadrant II
  (b) Positive x-axis
  (c) Quadrant III
  (d) Quadrant I
  (e) Negative y-axis
  (f) Quadrant IV

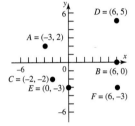

3. The points will be on a vertical line that is 2 units to the right of the y-axis.

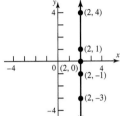

5. $(-1, 4)$ **7.** $(3, 1)$

9. $x\text{min} = -11$
  $x\text{max} = 5$
  $x\text{scl} = 1$
  $y\text{min} = -3$
  $y\text{max} = 6$
  $y\text{scl} = 1$

11. $x\text{min} = -30$
  $x\text{max} = 50$
  $x\text{scl} = 10$
  $y\text{min} = -90$
  $y\text{max} = 50$
  $y\text{scl} = 10$

13. $x\text{min} = -10$
  $x\text{max} = 110$
  $x\text{scl} = 10$
  $y\text{min} = -10$
  $y\text{max} = 160$
  $y\text{scl} = 10$

15. $x\text{min} = -6$
  $x\text{max} = 6$
  $x\text{scl} = 2$
  $y\text{min} = -4$
  $y\text{max} = 4$
  $y\text{scl} = 2$

17. $x\text{min} = -9$
  $x\text{max} = 9$
  $x\text{scl} = 3$
  $y\text{min} = -4$
  $y\text{max} = 4$
  $y\text{scl} = 2$

19. $x\text{min} = -6$
  $x\text{max} = 6$
  $x\text{scl} = 1$
  $y\text{min} = -8$
  $y\text{max} = 8$
  $y\text{scl} = 2$

21. $x\text{min} = -6$
  $x\text{max} = 6$
  $x\text{scl} = 2$
  $y\text{min} = -1$
  $y\text{max} = 3$
  $y\text{scl} = 1$

23. $x\text{min} = 3$
  $x\text{max} = 9$
  $x\text{scl} = 1$
  $y\text{min} = 2$
  $y\text{max} = 10$
  $y\text{scl} = 2$

25. $\sqrt{5}$ **27.** $2\sqrt{2}$ **29.** $2\sqrt{17}$ **31.** $\sqrt{85}$ **33.** $\sqrt{53}$ **35.** 2.625 **37.** $\sqrt{a^2 + b^2}$ **39.** $4\sqrt{10}$ **41.** $2\sqrt{17}$

43. $d(A, B) = \sqrt{13}$
  $d(B, C) = \sqrt{13}$
  $d(A, C) = \sqrt{26}$
  $(\sqrt{13})^2 + (\sqrt{13})^2 = (\sqrt{26})^2$
  Area $= \frac{13}{2}$ square units

45. $d(A, B) = \sqrt{130}$
  $d(B, C) = \sqrt{26}$
  $d(A, C) = \sqrt{104}$
  $(\sqrt{26})^2 + (\sqrt{104})^2 = (\sqrt{130})^2$
  Area $= 26$ square units

47. $d(A, B) = 4$
  $d(A, C) = 5$
  $d(B, C) = \sqrt{41}$
  $4^2 + 5^2 = 16 + 25 = (\sqrt{41})^2$
  Area $= 10$ square units

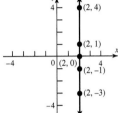
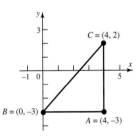

49. $(2, 2); (2, -4)$ **51.** $(0, 0); (8, 0)$ **53.** $(4, -1)$ **55.** $(\frac{3}{2}, 1)$ **57.** $(5, -1)$ **59.** $(1.05, 0.7)$ **61.** $(a/2, b/2)$ **63.** $\sqrt{73}; 2\sqrt{13}; 5$

65. $d(P_1, P_2) = 6; d(P_2, P_3) = 4; d(P_1, P_3) = 2\sqrt{13}$; right triangle

67. $d(P_1, P_2) = \sqrt{68}; d(P_2, P_3) = \sqrt{34}; d(P_1, P_3) = \sqrt{34}$; isosceles right triangle

69. $\dfrac{x - x_1}{x_2 - x_1} = r$ and $\dfrac{y - y_1}{y_2 - y_1} = r$; thus, $x = x_1 + r(x_2 - x_1)$ and $y = y_1 + r(y_2 - y_1)$ **71.** P_2 **73.** $(8, 8)$ **75.** $90\sqrt{2} \approx 127.28$ ft

77. (a) $(90, 0), (90, 90), (0, 90)$ **(b)** 232.4 ft   **(c)** 366.2 ft   **79.** $M_1 = (s/2, s/2); M_2 = (s/2, s/2)$ **81.** $d = 50t$

Exercise 1.2

1.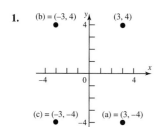

3.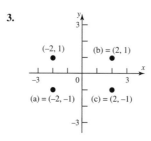

5.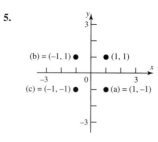

7.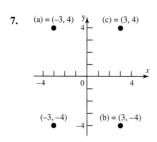

9.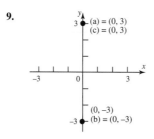

11. (a) 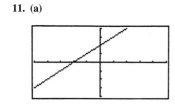   **(b)** 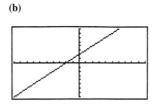   **(c)** 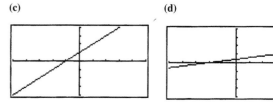   **(d)**

13. (a) 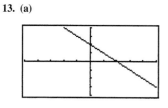   **(b)** 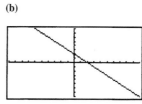   **(c)** 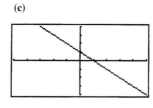   **(d)**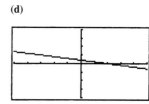

15. (a) 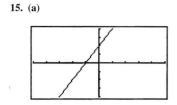   **(b)** 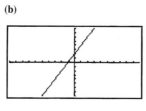   **(c)** 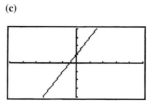   **(d)**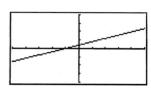

17. (a) 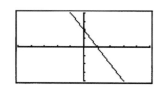   **(b)** 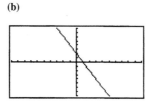   **(c)** 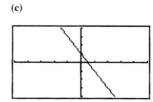   **(d)**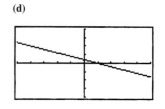

19. (a)   (b)   (c)   (d)

21. (a)   (b)   (c)   (d)

23. (a)   (b)   (c)   (d)

25. (a)   (b)   (c)   (d)

27. (a)   (b)   (c)   (d)

29. (a)   (b)   (c)   (d)

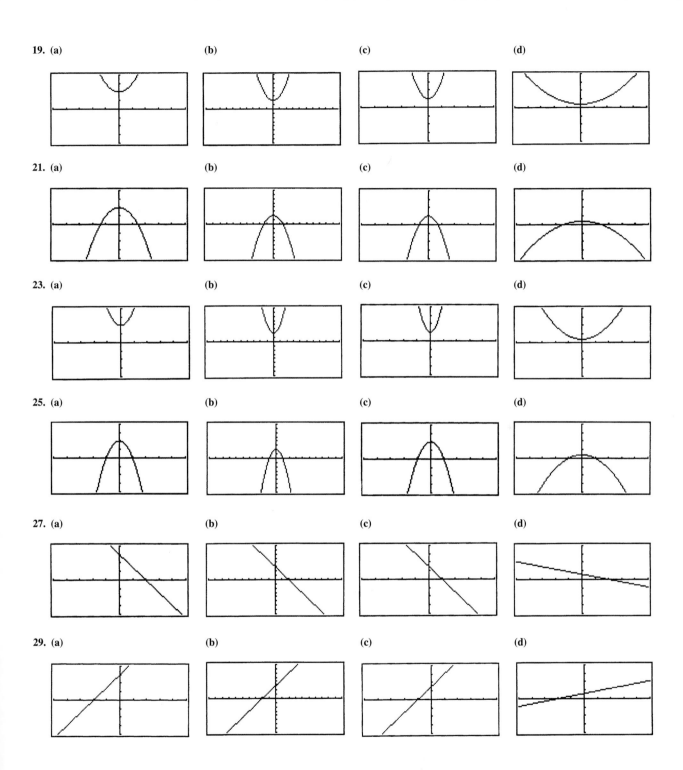

31. (a) $(-1, 0), (1, 0)$ (b) x-axis, y-axis, origin   **33.** (a) $(-\frac{\pi}{2}, 0), (\frac{\pi}{2}, 0), (0, 1)$ (b) y-axis   **35.** (a) $(0, 0)$ (b) x-axis   **37.** (a) $(1, 0)$ (b) none
39. (a) $(-3, 0), (0, 2), (3, 0)$ (b) y-axis   **41.** (a) $(x, 0), 0 \leq x < 2$ (b) none   **43.** (a) $(-1.5, 0), (0, -2), (1.5, 0)$ (b) y-axis
45. (a) none   (b) origin   **47.** $(0, 0)$ is on the graph   **49.** $(0, 2)$ is on the graph   **51.** $(0, 2)$ and $(\sqrt{2}, \sqrt{2})$ are on the graph   **53.** $-\frac{2}{5}$

55. $2a + 3b = 6$

57.

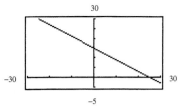

59.

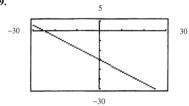

61.

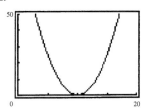

63.

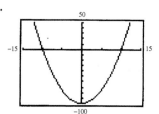

65.

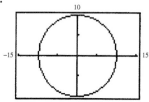

67.

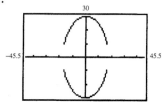

69.

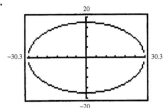

71.

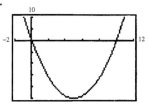

73.

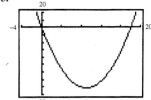

75.

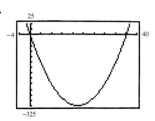

77.

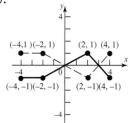

79.

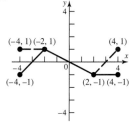

81.

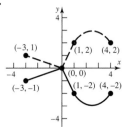

83.

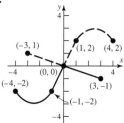

85. $(0, 0)$; symmetric with respect to the y-axis

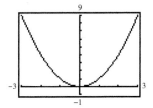

87. (0, 0); symmetric with respect to the origin

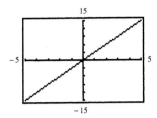

89. (0, 9), (3, 0), (−3, 0); symmetric with respect to the y-axis

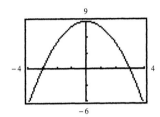

91. (−2, 0), (2, 0), (0, −3), (0, 3); symmetric with respect to the x-axis, y-axis, and origin

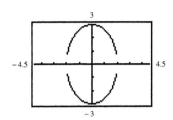

93. (0, −27), (3, 0); no symmetry

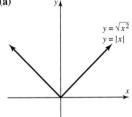

95. (0, −4), (4, 0), (−1, 0); no symmetry

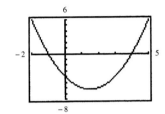

97. (0, 0); symmetric with respect to the origin

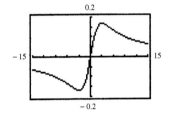

99. (a)

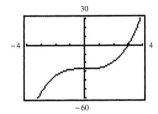

(b) Since $\sqrt{x^2} = |x|$, for all x, the graphs of $y = \sqrt{x^2}$ and $y = |x|$ are the same. **(c)** For $y = (\sqrt{x})^2$, the domain of the variable x is $x \geq 0$; for $y = x$, the domain of the variable x is all real numbers. Thus, $(\sqrt{x})^2 = x$ only for $x \geq 0$. **(d)** For $y = \sqrt{x^2}$, the range of the variable y is $y \geq 0$; for $y = x$, the range of the variable y is all real numbers. Also, $\sqrt{x^2} = |x|$, which equals x only if $x \geq 0$.

Exercise 1.3

1. $\frac{1}{2}$ **3.** −1

5. Slope $= -\frac{3}{2}$

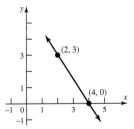

7. Slope $= -\frac{1}{2}$

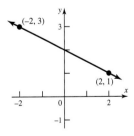

9. Slope $= 0$

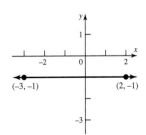

11. Slope undefined

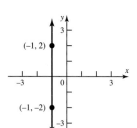

13. Slope $= \dfrac{\sqrt{3}-3}{1-\sqrt{2}} \approx 3.06$

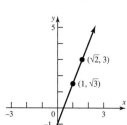

15.

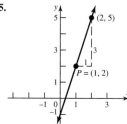

17.

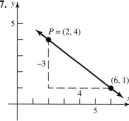

19.

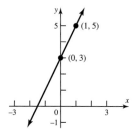

21.

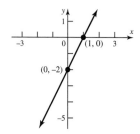

23. $x - 2y = 0$ or $y = \frac{1}{2}x$ **25.** $x + y - 2 = 0$ or $y = -x + 2$ **27.** $3x - y + 9 = 0$ or $y = 3x + 9$
29. $2x + 3y + 1 = 0$ or $y = -\frac{2}{3}x - \frac{1}{3}$ **31.** $x - 2y + 5 = 0$ or $y = \frac{1}{2}x + \frac{5}{2}$ **33.** $3x + y - 3 = 0$ or $y = -3x + 3$
35. $x - 2y - 2 = 0$ or $y = \frac{1}{2}x - 1$ **37.** $x - 2 = 0$; no slope-intercept form
39. Slope $= 2$; y-intercept $= 3$

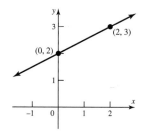

41. $y = 2x - 2$; Slope $= 2$; y-intercept $= -2$

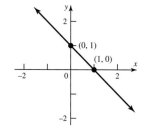

43. Slope $= \frac{1}{2}$; y-intercept $= 2$

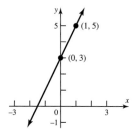

45. $y = -\frac{1}{2}x + 2$; Slope $= -\frac{1}{2}$; y-intercept $= 2$

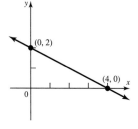

47. $y = \frac{2}{3}x - 2$; Slope $= \frac{2}{3}$; y-intercept $= -2$

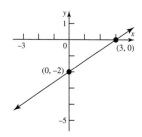

49. $y = -x + 1$; Slope $= -1$; y-intercept $= 1$

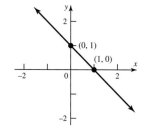

51. Slope undefined; no y-intercept

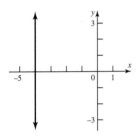

53. Slope $= 0$; y-intercept $= 5$

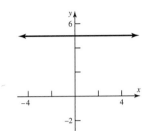

55. $y = x$; Slope $= 1$; y-intercept $= 0$

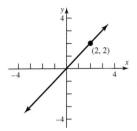

57. $y = \frac{3}{2}x$; Slope $= \frac{3}{2}$; y-intercept $= 0$

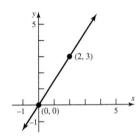

59. $y = 0$ **61.** $°C = \frac{5}{9}(°F - 32)$; approx. $21°C$
63. (a) $P = 0.5x - 100$ **(b)** \$400 **(c)** \$2400

65. $C = 0.10819x + 9.06$; for 300 kWhr, $C = \$41.52$; for 900 kWhr, $C = \$106.43$

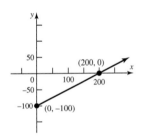

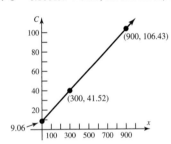

67. (b) **69. (d)** **71.** $x - y + 2 = 0$ or $y = x + 2$ **73.** $x + 3y - 3 = 0$ or $y = -\frac{1}{3}x + 1$ **75.** $2x + 3y = 0$ or $y = -\frac{2}{3}x$
81. Its slope is -1 **83.** Yes, if the y-intercept $= 0$

Exercise 1.4

1. (a) 4 **(b)** $-\frac{1}{4}$ **3. (a)** $-\frac{1}{2}$ **(b)** 2 **5. (a)** $\frac{1}{2}$ **(b)** -2 **7. (a)** $-\frac{3}{5}$ **(b)** $\frac{5}{3}$ **9. (a)** Undefined **(b)** 0 **11.** $2x - y - 3 = 0$ or $y = 2x - 3$
13. $x + 2y - 5 = 0$ or $y = -\frac{1}{2}x + \frac{5}{2}$ **15.** $2x - y + 4 = 0$ or $y = 2x + 4$ **17.** $2x - y = 0$ or $y = 2x$ **19.** $x - 4 = 0$ no slope intercept form
21. $2x + y = 0$ or $y = -2x$ **23.** $x - 2y + 3 = 0$ or $y = \frac{1}{2}x + \frac{3}{2}$ **25.** $y - 4 = 0$ or $y = 4$
27. Center $(2, 1)$; Radius 2; $(x - 2)^2 + (y - 1)^2 = 4$ **29.** Center $(\frac{5}{2}, 2)$; Radius $\frac{3}{2}$; $(x - \frac{5}{2})^2 + (y - 2)^2 = \frac{9}{4}$

31. $(x - 1)^2 + (y + 1)^2 = 1$;
$x^2 + y^2 - 2x + 2y + 1 = 0$

33. $x^2 + (y - 2)^2 = 4$;
$x^2 + y^2 - 4y = 0$

35. $(x - 4)^2 + (y + 3)^2 = 25$;
$x^2 + y^2 - 8x + 6y = 0$

37. $x^2 + y^2 = 4$;
$x^2 + y^2 - 4 = 0$

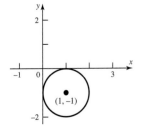

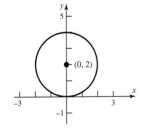

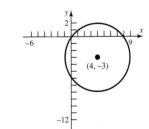

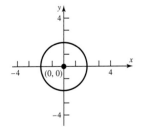

39. $(x - \frac{1}{2})^2 + y^2 = \frac{1}{4}$;
$x^2 + y^2 - x = 0$

41. $r = 2$;
$(h, k) = (0, 0)$

43. $r = 2$;
$(h, k) = (3, 0)$

45. $r = 3$;
$(h, k) = (-2, 2)$

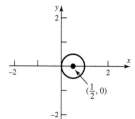

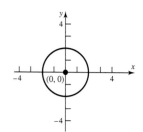

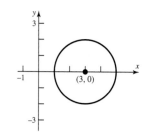

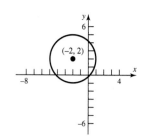

47. $r = \frac{1}{2}$;
$(h, k) = (\frac{1}{2}, -1)$

49. $r = 5$;
$(h, k) = (3, -2)$

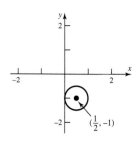

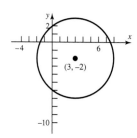

51. $x^2 + y^2 - 13 = 0$ **53.** $x^2 + y^2 - 4x - 6y + 4 = 0$ **55.** $x^2 + y^2 + 2x - 6y + 5 = 0$

57. $P_1 = (-2, 5)$, $P_2 = (1, 3)$, $m_1 = -\frac{2}{3}$; $P_2 = (1, 3)$, $P_3 = (-1, 0)$ $m_2 = \frac{3}{2}$; because $m_1 m_2 = -1$, the lines are perpendicular; the points P_1, P_2, and P_3 thus form a right triangle.

59. $P_1 = (-1, 0)$, $P_2 = (2, 3)$, $m_1 = 1$; $P_3 = (1, -2)$, $P_4 = (4, 1)$, $m_2 = 1$; $P_1 = (-1, 0)$, $P_3 = (1, -2)$, $m_3 = -1$; $P_2 = (2, 3)$, $P_4 = (4, 1)$, $m_4 = -1$; opposite sides are parallel, adjacent sides are perpendicular; the points form a rectangle.

61. (c) **63.** (b) **65.** $(x + 3)^2 + (y - 1)^2 = 16$ **67.** $(x - 2)^2 + (y - 2)^2 = 9$ **69.** 12 square units **71.** 4 square units **73.** 5 square units

75. Refer to Figure 81; $m_1 m_2 = -1$; $d(A, B) = \sqrt{(m_2 - m_1)^2}$; $d(O, A) = \sqrt{1 + m_2^2}$; $d(O, B) = \sqrt{1 + m_1^2}$. Now show that $[d(O, B)]^2 + [d(O, A)]^2 = [d(A, B)]^2$.

77. (a)
$$x^2 + (mx + b)^2 = r^2$$
$$(1 + m^2)x^2 + 2mbx + b^2 - r^2 = 0$$
One solution if and only if discriminant $= 0$
$$(2mb)^2 - 4(1 + m^2)(b^2 - r^2) = 0$$
$$-4b^2 + 4r^2 + 4m^2r^2 = 0$$
$$r^2(1 + m^2) = b^2$$

(b)
$$x = \frac{-2mb}{2(1 + m^2)} = \frac{-2mb}{2b^2/r^2} = \frac{-r^2m}{b}$$
$$y = m\left(\frac{-r^2m}{b}\right) + b = \frac{-r^2m^2}{b} + b$$
$$= \frac{-r^2m^2 + b^2}{b} = \frac{r^2}{b}$$

(c) Slope of tangent line $= m$
Slope of line joining center to point of tangency $= \dfrac{r^2/b}{-r^2m/b} = -\dfrac{1}{m}$

79. $\sqrt{2}x + 4y - 11\sqrt{2} + 12 = 0$ **81.** $x + 5y + 13 = 0$ **83.** All have the same slope, 2; the lines are parallel. **85.** $y = 2$

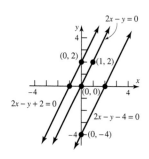

Exercise 1.5

1. Strong linear relation **3.** Strong linear relation **5.** Nonlinear relation
7. (a)   **9. (a)**   **11. (a)**

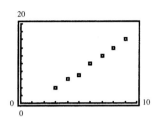

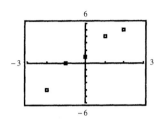

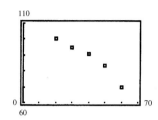

(b) $y = 2.0357x - 2.3571$ **(b)** $y = 2.2x + 1.2$ **(b)** $y = -0.72x + 116.6$
13. (a)   **15. (a)**   **17. (a)**

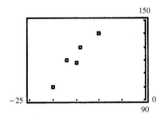

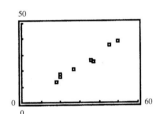

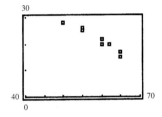

(b) $y = 3.861313869x + 180.2919708$ **(b)** $I = 0.7548890222C + 0.6266346731$ **(b)** $y = -0.8265306122x$
$\qquad\qquad + 70.39030612$

(c) As disposable income increases by \$1,   **(c)** As speed increases by 1 mph,
consumption increases by about \$0.75.   mpg decreases by 0.8265.
(d) \$32,000   **(d)** 20 mpg

19. (a)

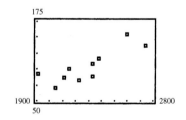

(b) NIBT $= 0.084195641$ sales $- 88.57761368$
(c) As sales increase by \$1, NIBT increases by \$0.08.
(d) \$118.2 billion

Exercise 1.6

1. $y = \frac{1}{5}x$ **3.** $A = \pi x^2$ **5.** $F = \frac{250}{d^2}$ **7.** $z = \frac{1}{5}(x^2 + y^2)$ **9.** $M = \frac{9d^2}{2\sqrt{x}}$ **11.** $T^2 = 8\frac{a^3}{d^2}$ **13.** $V = \frac{4\pi}{3}r^3$ **15.** $A = \frac{1}{2}bh$ **17.** $V = \pi r^2 h$

19. $F = 6.67 \times 10^{-11}\left(\frac{mM}{d^2}\right)$ **21.** 144 ft; 2 sec **23.** 2.25 **25.** 45.45 lb **27.** $\sqrt[3]{6} \approx 1.82$ in. **29.** 900 ft-lb **31.** 384 psi

33. $\frac{720}{49} \approx 14.69$ ohms **35.** $v = \sqrt{gr}$ **37.** 18,001 mph **39.** Approx. 505 mi **41.** $F = \frac{mv^2}{r}$ **43.** By 21% **45.** 9 times

Fill-in-the-Blank Items

1. x-coordinate; y-coordinate **2.** midpoint **3.** y-axis **4.** circle; radius; center **5.** undefined; 0 **6.** $m_1 = m_2$; $m_1 m_2 = -1$ **7.** $kx^2 y^3/\sqrt{t}$

1. F **2.** T **3.** T **4.** F **5.** F **6.** T **7.** T

Review Exercises

1. $2x + y - 5 = 0$ or $y = -2x + 5$ **3.** $x + 3 = 0$; no y-intercept form **5.** $x + 5y + 10 = 0$ or $y = -\frac{1}{5}x - 2$
7. $2x - 3y + 19 = 0$ or $y = \frac{2}{3}x + \frac{19}{3}$ **9.** $-x + y + 7 = 0$ or $y = x - 7$

11.
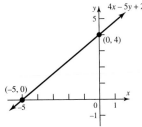

13.
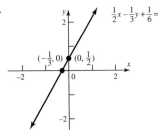

15. $\sqrt{2}x + \sqrt{3}y = \sqrt{6}$
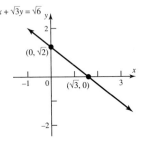

17. Center $(1, -2)$; Radius $= 3$ **19.** Center $(1, -2)$; Radius $= \sqrt{5}$ **21.** Slope $= \frac{1}{5}$; distance $= 2\sqrt{26}$; midpoint $= (2, 3)$
23. Equations in **(c)** **25.** Intercept: $(0, 0)$; symmetric with respect to the x-axis
27. Intercepts: $(0, -2)$, $(0, 2)$, $(4, 0)$, $(-4, 0)$; symmetric with respect to the x-axis, y-axis, and origin
29. Intercept: $(0, 1)$; symmetric with respect to the y-axis **31.** Intercepts: $(0, 0)$, $(0, -2)$, $(-1, 0)$; no symmetry
33. (a)

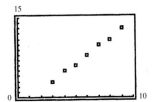

35. (a)

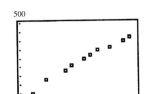

37. (a)

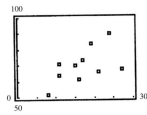

(b) $y = 1.642857143x - 1.857142857$

(b) $y = 2.018273184x + 112.4326176$

(b) Calculus $= 1.018354237$
algebra $+ 52.6898134$
(c) As the algebra score increases
by 1 point, the calculus score
increases by about 1.02 points.
(d) 73

39. $A = ks^2 = \frac{\sqrt{3}}{4}s^2; s = \frac{8}{\sqrt[4]{3}} \approx 6.08$cm **41.** $a \approx 36$ million miles

43. $M = \left(\frac{a}{2}, \frac{b}{2}\right); d(O, M) = \frac{1}{2}\sqrt{a^2 + b^2}; d(A, M) = \frac{1}{2}\sqrt{a^2 + b^2}; d(B, M) = \frac{1}{2}\sqrt{a^2 + b^2}$ **45.** $d(A, B) = \sqrt{13}; d(B, C) = \sqrt{13}$

47. Slope using A and B is -1; slope using B and C is -1 **49.** Center $(1, -2)$; radius $= 4\sqrt{2}; x^2 + y^2 - 2x + 4y - 27 = 0$

CHAPTER 2 *Exercise 2.1*

1. (a) -4 **(b)** -5 **(c)** -9 **(d)** -25 **3. (a)** 0 **(b)** $\frac{1}{2}$ **(c)** $-\frac{1}{2}$ **(d)** $\frac{3}{10}$ **5. (a)** 4 **(b)** 5 **(c)** 5 **(d)** 7
7. (a) $-\frac{1}{5}$ **(b)** $-\frac{3}{2}$ **(c)** $\frac{1}{8}$ **(d)** $\frac{7}{4}$ **9.** $f(0) = 3; f(-6) = -3$ **11.** Positive **13.** $-3, 6,$ and 10 **15.** $\{x| -6 \le x \le 11\}$
17. $(-3, 0), (6, 0) (10, 0)$ **19.** 3 times **21. (a)** No **(b)** -3 **(c)** 14 **(d)** $\{x| x \ne 6\}$ **23. (a)** Yes **(b)** $\frac{8}{17}$ **(c)** $-1, 1$ **(d)** All real numbers
25. Not a function
27. Function **(a)** Domain: $\{x| -\pi \le x \le \pi\}$; Range: $\{y| -1 \le y \le 1\}$ **(b)** Intercepts: $(-\pi/2, 0), (\pi/2, 0), (0, 1)$ **(c)** y-axis
29. Not a function **31.** Function **(a)** Domain: $\{x| x > 0\}$; Range: all real numbers **(b)** Intercept: $(1, 0)$ **(c)** None
33. Function **(a)** Domain: all real numbers; Range: $\{y|y \le 2\}$ **(b)** Intercepts: $(-3, 0), (3, 0), (0, 2)$ **(c)** y-axis

35. Function   **(a)** Domain: $\{x \mid x \neq 2\}$; Range: $\{y \mid y \neq 1\}$ **(b)** Intercept: $(0, 0)$ **(c)** None   **37.** All real numbers   **39.** All real numbers

41. $\{x \mid x \neq -1, x \neq 1\}$ **43.** $\{x \mid x \neq 0\}$ **45.** $\{x \mid x \geq 4\}$ **47.** $(-\infty, -3]$ or $[3, \infty)$ **49.** $(-\infty, 1)$ or $[2, \infty)$ **51.** $A = -\frac{7}{2}$ **53.** $A = -4$

55. $A = 8$; undefined at $x = 3$

57. **(a)**

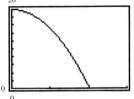

(b) 15.1 m, 14.07 m, 12.94 m, 11.72 m
(c) 1.01 seconds, 1.42 seconds, 1.74 seconds
(d) 2.02 sec

59. $A(x) = \frac{1}{2}x^2$ **61.** $G(x) = 5x$

63. **(a)** $C(x) = 10x + 14\sqrt{x^2 - 10x + 29}$, $0 < x < 5$
(b) $C(1) = \$72.61$
(c) $C(3) = \$69.60$
(d)

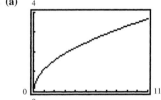

(e) least cost: $x = 2.95$ miles

65. **(a)** $A(x) = (8.5 - 2x)(11 - 2x)$,
(b) $0 \leq x \leq 4.25$, $0 \leq A \leq 93.5$
(c) $A(1) = 58.5$ in.^{2}, $A(1.2) = 52.46$ in.^{2}, $A(1.5) = 44$ in.^{2}
(d)

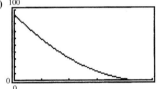

(e) $A(x) = 70$ when $x = 0.64$ in; $A(x) = 50$ when $x = 1.28$ in.

67. **(a)**

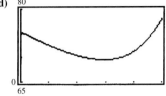

(b) The period T varies from 1.107 seconds to 3.502 seconds.
(c) 81.56 feet

69. Function   **71.** Function   **73.** Not a function   **75.** Function   **77.** Only $h(x) = 2x$

Exercise 2.2

1. C   **3.** E   **5.** B   **7.** F

9. **(a)** Domain: $\{x \mid -3 \leq x \leq 4\}$; Range: $\{y \mid 0 \leq y \leq 3\}$ **(b)** Increasing on $[-3, 0]$ and on $[2, 4]$; decreasing on $[0, 2]$ **(c)** Neither
(d) $(-3, 0)$, $(0, 3)$, $(2, 0)$

11. **(a)** Domain: all real numbers; Range: $\{y \mid 0 < y < \infty\}$ **(b)** Increasing on $(-\infty, \infty)$ **(c)** Neither   **(d)** $(0, 1)$

13. **(a)** Domain: $\{x \mid -\pi \leq x \leq \pi\}$; Range: $\{y \mid -1 \leq y \leq 1\}$
(b) Increasing on $[-\pi/2, \pi/2]$; decreasing on $[-\pi, -\pi/2]$ and on $[\pi/2, \pi]$ **(c)** Odd (symmetric with respect to the origin)
(d) $(-\pi, 0)$, $(0, 0)$, $(\pi, 0)$

15. **(a)** Domain: $\{x \mid x \neq 2\}$; Range: $\{y \mid y \neq 1\}$ **(b)** Decreasing on $(-\infty, 2)$ and on $(2, \infty)$ **(c)** Neither   **(d)** $(0, 0)$

17. **(a)** Domain: $\{x \mid x \neq 0\}$; Range: all real numbers   **(b)** Increasing on $(-\infty, 0)$ and on $(0, \infty)$ **(c)** Odd   **(d)** $(-1, 0)$, $(1, 0)$

19. **(a)** Domain: $\{x \mid x \neq -2, x \neq 2\}$; Range: $\{y \mid -\infty < y \leq 0$ and $1 < y < \infty\}$
(b) Increasing on $(-\infty, -2)$ and on $(-2, 0]$; decreasing on $[0, 2)$ and on $(2, \infty)$ **(c)** Even   **(d)** $(0, 0)$

21. **(a)** Domain: $\{x \mid -4 \leq x \leq 4\}$; Range: $\{y \mid 0 \leq y \leq 2\}$ **(b)** Increasing on $[-2, 0]$ and $[2, 4]$; Decreasing on $[-4, -2]$ and $[0, 2]$ **(c)** Even
(d) $(-2, 0)$, $(0, 2)$, $(2, 0)$

23. **(a)** Domain: $\{x \mid -4 \leq x \leq 4\}$; Range: $\{y \mid 0 < y \leq 4\}$ **(b)** Increasing on $[-4, 0)$; Decreasing on $(0, 4]$ **(c)** Even   **(d)** None

25. **(a)** 2   **(b)** 3   **(c)** -4 **27.** **(a)** 4   **(b)** 2   **(c)** 5

29. **(a)** $-2x + 5$ **(b)** $-2x - 5$ **(c)** $4x + 5$ **(d)** $2x - 1$ **(e)** $\dfrac{2}{x} + 5 = \dfrac{5x + 2}{x}$ **(f)** $\dfrac{1}{2x + 5}$

31. **(a)** $2x^2 - 4$ **(b)** $-2x^2 + 4$ **(c)** $8x^2 - 4$ **(d)** $2x^2 - 12x + 14$ **(e)** $\dfrac{2 - 4x^2}{x^2}$ **(f)** $\dfrac{1}{2x^2 - 4}$

33. **(a)** $-x^3 + 3x$ **(b)** $-x^3 + 3x$ **(c)** $8x^3 - 6x$ **(d)** $x^3 - 9x^2 + 24x - 18$ **(e)** $\dfrac{1}{x^3} - \dfrac{3}{x}$ **(f)** $\dfrac{1}{x^3 - 3x}$

35. **(a)** $-\dfrac{x}{x^2 + 1}$ **(b)** $-\dfrac{x}{x^2 + 1}$ **(c)** $\dfrac{2x}{4x^2 + 1}$ **(d)** $\dfrac{x - 3}{x^2 - 6x + 10}$ **(e)** $\dfrac{x}{x^2 + 1}$ **(f)** $\dfrac{x^2 + 1}{x}$

37. (a) $|x|$ **(b)** $-|x|$ **(c)** $2|x|$ **(d)** $|x - 3|$ **(e)** $\dfrac{1}{|x|}$ **(f)** $\dfrac{1}{|x|}$

39. (a) $1 - \dfrac{1}{x}$ **(b)** $-1 - \dfrac{1}{x}$ **(c)** $1 + \dfrac{1}{2x}$ **(d)** $1 + \dfrac{1}{x - 3}$ **(e)** $1 + x$ **(f)** $\dfrac{x}{x + 1}$

41. 3 **43.** -3 **45.** $3x + 1$ **47.** $x(x + 1)$ **49.** $\dfrac{-1}{x + 1}$ **51.** $\dfrac{1}{\sqrt{x} + 1}$

53. Odd **55.** Even **57.** Odd **59.** Neither **61.** Even **63.** Odd **65.** At most one

67. (a) All real numbers
(b) $(0, -3)$, $(1, 0)$
(d) All real numbers
(c)
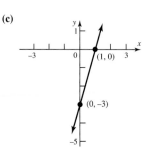

69. (a) All real numbers
(b) $(-2, 0)$, $(2, 0)$, $(0, -4)$
(d) $\{y \mid -4 \le y < \infty\}$
(c)
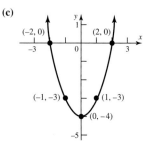

71. (a) All real numbers
(b) $(0, 0)$
(d) $\{y \mid -\infty < y \le 0\}$
(c)
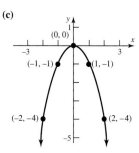

73. (a) $\{x \mid 2 \le x < \infty\}$
(b) $(2, 0)$
(d) $\{y \mid 0 \le y < \infty\}$
(c)
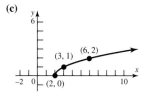

75. (a) $\{x \mid -\infty < x \le 2\}$
(b) $(2, 0)$, $(0, \sqrt{2})$
(d) $\{y \mid 0 \le y < \infty\}$
(c)
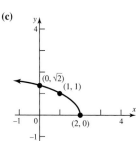

77. (a) All real numbers
(b) $(0, 3)$
(d) $\{y \mid 3 \le y < \infty\}$
(c)
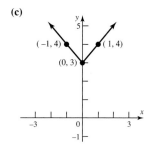

79. (a) All real numbers
(b) $(0, 0)$
(d) $\{y \mid -\infty < y \le 0\}$
(c)
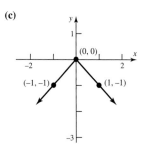

81. (a) All real numbers
(b) $(0, 0)$
(d) All real numbers
(c)
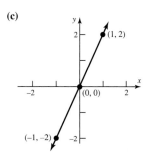

83. (a) All real numbers
(b) $(0, 0), (-1, 0)$
(d) All real numbers
(c)

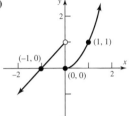

85. (a) $\{x| -2 \le x < \infty\}$
(b) $(0, 1)$
(d) $\{y|0 < y < \infty\}$
(c)

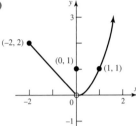

87. (a) All real numbers
(b) $(0, 1)$
(d) $\{-1, 1\}$
(c)

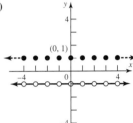

89. (a) All real numbers
(b) $(x, 0)$ for $0 \le x < 1$
(d) Set of even integers
(c)

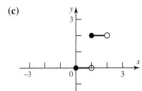

91. (a) All real numbers
(b) $(-2, 0), (0, 4), (2, 0)$
(d) $\{y|y \ge 0\}$
(c)

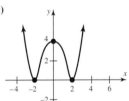

93. 2   **95.** $2x + h + 2$ **97.** $f(x) = \begin{cases} -x & \text{if } -1 \le x \le 0 \\ \frac{1}{2}x & \text{if } 0 < x \le 2 \end{cases}$ (Other answers are possible.)

99. $f(x) = \begin{cases} -x & \text{if } x \le 0 \\ -x + 2 & \text{if } 0 < x \le 2 \end{cases}$ (Other answers are possible.)   **101.** No; $f(-2) = 8$ and $f(2) = 6$

103.

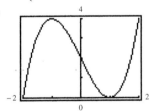

Increasing: $[-2, -1], [1, 2]$
Decreasing: $[-1, 1]$
Local maxima: $(-1, 4)$, local minima: $(1, 0)$

105.

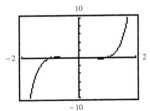

Increasing: $[-2, -0.77], [0.77, 2]$
Decreasing: $[-0.77, 0.77]$
Local maxima: $(-0.77, 0.18)$, local minima: $(0.77, -0.18)$

107. Each graph is that of $y = x^2$, but shifted vertically. If $y = x^2 + k, k > 0$, the shift is up k units; if $y = x^2 + k, k < 0$, the shift is down $|k|$ units.
109. Each graph is that of $y = |x|$, but either compressed or stretched. If $y = k|x|$ and $k > 1$, the graph is stretched; if $y = k|x|, 0 < k < 1$, the graph is compressed.
111. The graph of $y = f(-x)$ is the reflection about the y-axis of the graph of $y = f(x)$.

113.

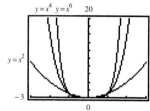

They are all u-shaped and open up. All three go through the point (0, 0). As the exponent increases, the steepness of the curve increases.

115. (a) $30.70
   (b) $203.75

   (c) $C = \begin{cases} 7 + 0.47395x & \text{if } 0 \leq x \leq 90 \\ 15.83 + 0.37583x & \text{if } x > 90 \end{cases}$

   (d)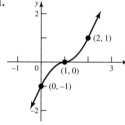

117. (a) $E(-x) = \frac{1}{2}[f(-x) + f(x)] = E(x)$ **(b)** $O(-x) = \frac{1}{2}[f(-x) - f(x)] = -\frac{1}{2}[f(x) - f(-x)] = -O(x)$
   (c) $E(x) + O(x) = \frac{1}{2}[f(x) + f(-x)] + \frac{1}{2}[f(x) - f(-x)] = f(x)$ **(d)** Combine the results of parts **(a)**, **(b)**, and **(c)**.

Exercise 2.3

1. B **3.** H **5.** I **7.** L **9.** F **11.** G **13.** C **15.** B **17.** $y = (x - 4)^3$ **19.** $y = x^3 + 4$ **21.** $y = -x^3$ **23.** $y = 4x^3$

25.

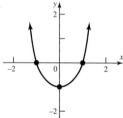

27.

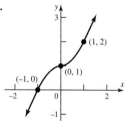

29.

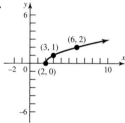

31.

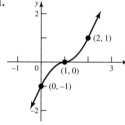

33.

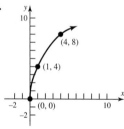

35.

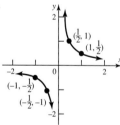

37.

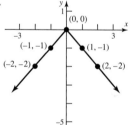

39.

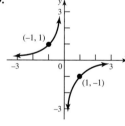

41.

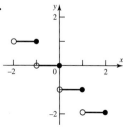

43.

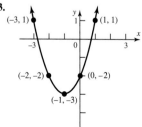

45.

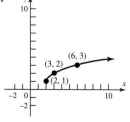

47.

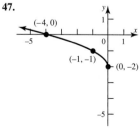

AN14

49.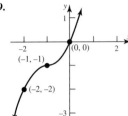

51.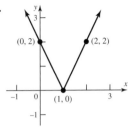

53.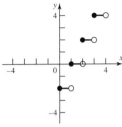

55. (a) $F(x) = f(x) + 3$

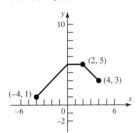

(b) $G(x) = f(x + 2)$

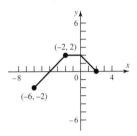

(c) $P(x) = -f(x)$

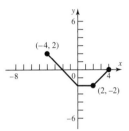

(d) $Q(x) = \frac{1}{2}f(x)$

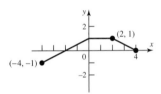

(e) $g(x) = f(-x)$

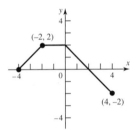

(f) $h(x) = 3f(x)$

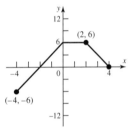

57. (a) $F(x) = f(x) + 3$

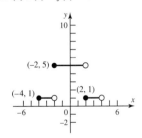

(b) $G(x) = f(x + 2)$

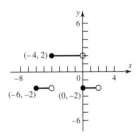

(c) $P(x) = -f(x)$

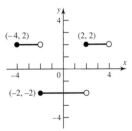

(d) $Q(x) = \frac{1}{2}f(x)$

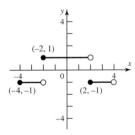

(e) $g(x) = f(-x)$

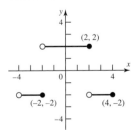

(f) $h(x) = 3f(x)$

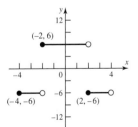

59. (a) $F(x) = f(x) + 3$

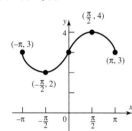

(b) $G(x) = f(x + 2)$

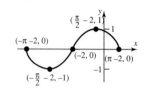

(c) $P(x) = -f(x)$

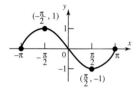

(d) $Q(x) = \frac{1}{2}f(x)$

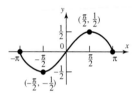

(e) $g(x) = f(-x)$

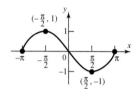

(f) $h(x) = 3f(x)$

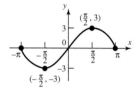

61. (a)

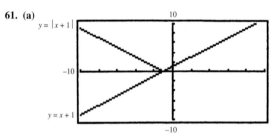

(b)

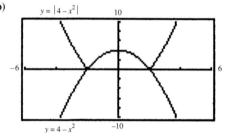

(c)

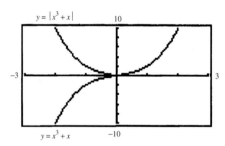

(d) Any part of the graph of $y = f(x)$ that lies below the x-axis is reflected about the x-axis to obtain the graph of $y = |f(x)|$

63. (a)

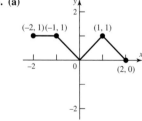

(b)

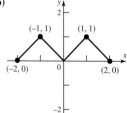

65. $f(x) = (x + 1)^2 - 1$

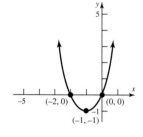

67. $f(x) = (x - 4)^2 - 15$

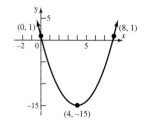

69. $f(x) = (x + \frac{1}{2})^2 + \frac{3}{4}$

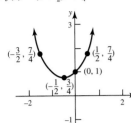

71.

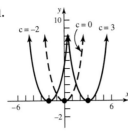

73.

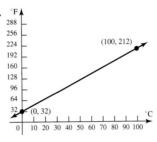

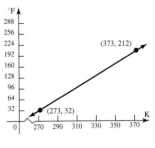

Exercise 2.4

1. (a) $(f + g)(x) = 5x + 1$; all real numbers **(b)** $(f - g)(x) = x + 7$; all real numbers **(c)** $(f \cdot g)(x) = 6x^2 - x - 12$; all real numbers
  (d) $\left(\frac{f}{g}\right)(x) = \frac{3x + 4}{2x - 3}$; all real numbers x except $x = \frac{3}{2}$

3. (a) $(f + g)(x) = 2x^2 + x - 1$; all real numbers **(b)** $(f - g)(x) = -2x^2 + x - 1$; all real numbers **(c)** $(f \cdot g)(x) = 2x^3 - 2x^2$; all real numbers
  (d) $\left(\frac{f}{g}\right)(x) = \frac{x - 1}{2x^2}$; $\{x | x \neq 0\}$

5. (a) $(f + g)(x) = \sqrt{x} + 3x - 5$; $\{x | x \geq 0\}$ **(b)** $(f - g)(x) = \sqrt{x} - 3x + 5$; $\{x | x \geq 0\}$ **(c)** $(f \cdot g)(x) = 3x\sqrt{x} - 5\sqrt{x}$; $\{x | x \geq 0\}$
  (d) $\left(\frac{f}{g}\right)(x) = \frac{\sqrt{x}}{3x - 5}$; $\{x | x \geq 0, x \neq \frac{5}{3}\}$

7. (a) $(f + g)(x) = 1 + \frac{2}{x}$; $\{x | x \neq 0\}$ **(b)** $(f - g)(x) = 1$; $\{x | x \neq 0\}$ **(c)** $(f \cdot g)(x) = \frac{1}{x} + \frac{1}{x^2}$; $\{x | x \neq 0\}$ **(d)** $\left(\frac{f}{g}\right)(x) = x + 1$; $\{x | x \neq 0\}$

9. (a) $(f + g)(x) = \frac{6x + 3}{3x - 2}$; $\{x | x \neq \frac{2}{3}\}$ **(b)** $(f - g)(x) = \frac{-2x + 3}{3x - 2}$; $\{x | x \neq \frac{2}{3}\}$ **(c)** $(f \cdot g)(x) = \frac{8x^2 + 12x}{(3x - 2)^2}$; $\{x | x \neq \frac{2}{3}\}$

  (d) $\left(\frac{f}{g}\right)(x) = \frac{2x + 3}{4x}$; $\{x | x \neq 0, x \neq \frac{2}{3}\}$ **11.** $g(x) = 5 - \frac{7}{2}x$

13. $f + g = x + \frac{1}{x}$

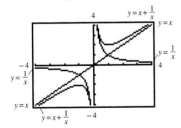

15. $f + g = x^2 + \frac{1}{x}$

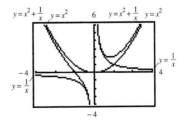

17. $f \cdot g = \frac{x}{x^2 + 1}$

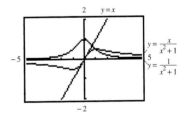

19. $f \circ g = \frac{x^2}{x^2 + 1}$

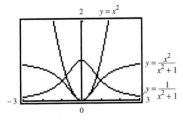

21.

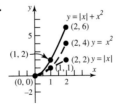

23.

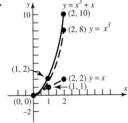

25. (a) 98 **(b)** 49 **(c)** 4 **(d)** 4 **27. (a)** 97 **(b)** $-\frac{163}{2}$ **(c)** 1 **(d)** $-\frac{3}{2}$ **29. (a)** $2\sqrt{2}$ **(b)** $2\sqrt{2}$ **(c)** 1 **(d)** 0
31. (a) $\frac{1}{17}$ **(b)** $\frac{1}{5}$ **(c)** 1 **(d)** $\frac{1}{2}$ **33. (a)** $\frac{3}{5}$ **(b)** $\sqrt{15}/5$ **(c)** $\frac{12}{13}$ **(d)** 0
35. (a) $(f \circ g)(x) = 6x + 3$ **(b)** $(g \circ f)(x) = 6x + 9$ **(c)** $(f \circ f)(x) = 4x + 9$ **(d)** $(g \circ g)(x) = 9x$
37. (a) $(f \circ g)(x) = 3x^2 + 1$ **(b)** $(g \circ f)(x) = 9x^2 + 6x + 1$ **(c)** $(f \circ f)(x) = 9x + 4$ **(d)** $(g \circ g)(x) = x^4$
39. (a) $(f \circ g)(x) = \sqrt{x^2 - 1}$ **(b)** $(g \circ f)(x) = x - 1$ **(c)** $(f \circ f)(x) = \sqrt[4]{x}$ **(d)** $(g \circ g)(x) = x^4 - 2x^2$

41. (a) $(f \circ g)(x) = \dfrac{1-x}{1+x}$ (b) $(g \circ f)(x) = \dfrac{x+1}{x-1}$ (c) $(f \circ f)(x) = -\dfrac{1}{x}$ (d) $(g \circ g)(x) = x$

43. (a) $(f \circ g)(x) = x$ (b) $(g \circ f)(x) = |x|$ (c) $(f \circ f)(x) = x^4$ (d) $(g \circ g)(x) = \sqrt[4]{x}$

45. (a) $(f \circ g)(x) = \dfrac{1}{4x+9}$ (b) $(g \circ f)(x) = \dfrac{2}{2x+3} + 3 = \dfrac{6x+11}{2x+3}$ (c) $(f \circ f)(x) = \dfrac{2x+3}{6x+11}$ (d) $(g \circ g)(x) = 4x+9$

47. (a) $(f \circ g)(x) = acx + ad + b$ (b) $(g \circ f)(x) = acx + bc + d$ (c) $(f \circ f)(x) = a^2x + ab + b$ (d) $(g \circ g)(x) = c^2x + cd + d$

49. $(f \circ g)(x) = f(g(x)) = f(\tfrac{1}{2}x) = 2(\tfrac{1}{2}x) = x; (g \circ f)(x) = g(f(x)) = g(2x) = \tfrac{1}{2}(2x) = x$

51. $(f \circ g)(x) = f(\sqrt[3]{x}) = (\sqrt[3]{x})^3 = x; (g \circ f)(x) = g(x^3) = \sqrt[3]{x^3} = x$

53. $(f \circ g)(x) = f(\tfrac{1}{2}(x+6)) = 2[\tfrac{1}{2}(x+6)] - 6 = x + 6 - 6 = x; (g \circ f)(x) = g(2x-6) = \tfrac{1}{2}(2x - 6 + 6) = x$

55. $(f \circ g)(x) = f\!\left(\dfrac{1}{a}(x - b)\right) = a\!\left[\dfrac{1}{a}(x - b)\right] + b = x; (g \circ f)(x) = g(ax + b) = \dfrac{1}{a}\!\left(ax + b - b\right) = x$

57. $(f \circ g)(x) = 11; (g \circ f)(x) = 2$ **59.** $(f \circ (g \circ h))(x) = 5 - 3x + 4\sqrt{1 - 3x}$ **61.** $((f + g) \circ h)(x) = 9x^2 - 6x + 3 + \sqrt{1 - 3x}$ **63.** $F = f \circ g$

65. $H = h \circ f$ **67.** $q = f \circ h$ **69.** $P = f \circ f$ **71.** $f(x) = x^4; g(x) = 2x + 3$ **73.** $f(x) = \sqrt{x}; g(x) = x^2 + x + 1$ **75.** $f(x) = x^2; g(x) = 1 - \dfrac{1}{x^2}$

77. $f(x) = [[x]]; g(x) = x^2 + 1$ **79.** $-3, 3$ **81.** $S(r(t)) = \tfrac{16}{9}\pi t^6$ **83.** $C(N(t)) = 15000 + 800,000t - 40,000t^2$ **85.** $C = \dfrac{2\sqrt{100 - p}}{25} + 600$

87. Since f and g are odd, $f(-x) = -f(x)$ and $g(-x) = -g(x); (f \circ g)(-x) = f(g(-x)) = f(-g(x)) = -f(g(x)) = -(f \circ g)(x)$; thus, $f \circ g$ is odd.

Exercise 2.5

1. One-to-one
3. Not one-to-one
5. One-to-one

7.

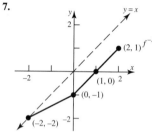

9.

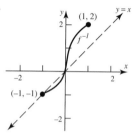

11.

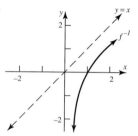

13. $f(g(x)) = f(\tfrac{1}{3}(x - 4)) = 3[\tfrac{1}{3}(x - 4)] + 4 = x; g(f(x)) = g(3x + 4) = \tfrac{1}{3}[(3x + 4) - 4] = x$

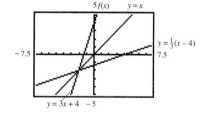

15. $f(g(x)) = 4\!\left[\dfrac{x}{4} + 2\right] - 8 = x; g(f(x)) = \dfrac{4x - 8}{4} + 2 = x$ **17.** $f(g(x)) = (\sqrt[3]{x + 8})^3 - 8 = x; g(f(x)) = \sqrt[3]{(x^3 - 8) + 8} = x$

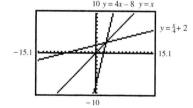

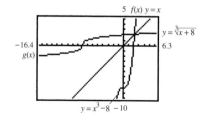

19. $f(g(x)) = \dfrac{1}{1/x} = x$; $g(f(x)) = \dfrac{1}{1/x} = x$

21. $f(g(x)) = \dfrac{2\left(\dfrac{4x-3}{2-x}\right)+3}{\dfrac{4x-3}{2-x}+4} = \dfrac{8x-3x}{-3+8} = x$; $g(f(x)) = \dfrac{4\left(\dfrac{2x+3}{x+4}\right)-3}{2-\dfrac{2x+3}{x+4}} = x$

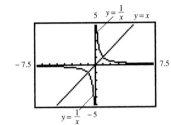

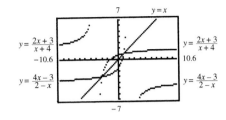

23. $f^{-1}(x) = \frac{1}{3}x$
$f(f^{-1}(x)) = 3(\frac{1}{3}x) = x$
$f^{-1}(f(x)) = \frac{1}{3}(3x) = x$
Domain f = Range $f^{-1} = (-\infty, \infty)$
Range f = Domain $f^{-1} = (-\infty, \infty)$

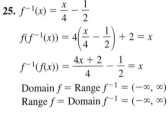

25. $f^{-1}(x) = \dfrac{x}{4} - \dfrac{1}{2}$
$f(f^{-1}(x)) = 4\left(\dfrac{x}{4} - \dfrac{1}{2}\right) + 2 = x$
$f^{-1}(f(x)) = \dfrac{4x+2}{4} - \dfrac{1}{2} = x$
Domain f = Range $f^{-1} = (-\infty, \infty)$
Range f = Domain $f^{-1} = (-\infty, \infty)$

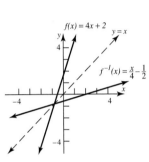

27. $f^{-1}(x) = \sqrt[3]{x+1}$
$f(f^{-1}(x)) = (\sqrt[3]{x+1})^3 - 1 = x$
$f^{-1}(f(x)) = \sqrt[3]{x^3 - 1 + 1} = x$
Domain f = Range $f^{-1} = (-\infty, \infty)$
Range f = Domain $f^{-1} = (-\infty, \infty)$

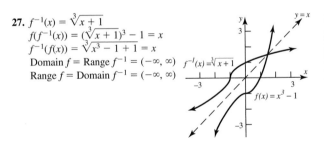

29. $f^{-1}(x) = \sqrt{x-4}$
$f(f^{-1}(x)) = (\sqrt{x-4})^2 + 4 = x$
$f^{-1}(f(x)) = \sqrt{(x^2+4)-4} = \sqrt{x^2} = |x| = x$
Domain f = Range $f^{-1} = [0, \infty)$
Range f = Domain $f^{-1} = [4, \infty)$

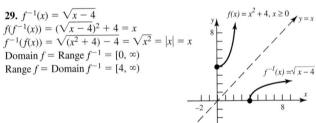

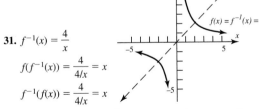

31. $f^{-1}(x) = \dfrac{4}{x}$
$f(f^{-1}(x)) = \dfrac{4}{4/x} = x$
$f^{-1}(f(x)) = \dfrac{4}{4/x} = x$
Domain f = Range f^{-1} = All real numbers except 0
Range f = Domain f^{-1} = All real numbers except 0

33. $f^{-1}(x) = \dfrac{2x+1}{x}$
$f(f^{-1}(x)) = \dfrac{1}{\dfrac{2x+1}{x} - 2} = x$
$f^{-1}(f(x)) = \dfrac{2\left(\dfrac{1}{x-2}\right)+1}{\dfrac{1}{x-2}} = x$
Domain f = Range f^{-1} = All real numbers except 2
Range f = Domain f^{-1} = All real numbers except 0

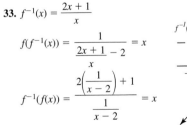

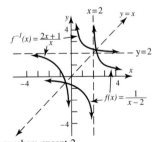

35. $f^{-1}(x) = \dfrac{2 - 3x}{x}$

$$f(f^{-1}(x)) = \dfrac{2}{3 + \dfrac{2 - 3x}{x}} = x$$

$$f^{-1}(f(x)) = \dfrac{2 - 3\left(\dfrac{2}{3 + x}\right)}{\dfrac{2}{3 + x}} = x$$

Domain f = Range f^{-1} = All real numbers except -3
Range f = Domain f^{-1} = All real numbers except 0

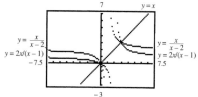

37. $f^{-1}(x) = \sqrt{x} - 2$
$f(f^{-1}(x)) = (\sqrt{x} - 2 + 2)^2 = x$
$f^{-1}(f(x)) = \sqrt{(x + 2)^2} - 2 = |x + 2| - 2 = x$
Domain f = Range f^{-1} = $[-2, \infty)$
Range f = Domain f^{-1} = $[0, \infty)$

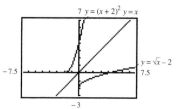

39. $f^{-1}(x) = \dfrac{x}{x - 2}$

$$f(f^{-1}(x)) = \dfrac{2\left(\dfrac{x}{x - 2}\right)}{\dfrac{x}{x - 2} - 1} = x$$

$$f^{-1}(f(x)) = \dfrac{\dfrac{2x}{x - 1}}{\dfrac{2x}{x - 1} - 2} = x$$

Domain f = Range f^{-1} = All real numbers except 1
Range f = Domain f^{-1} = All real numbers except 2

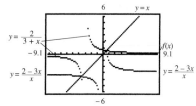

41. $f^{-1}(x) = \dfrac{3x + 4}{2x - 3}$

$$f(f^{-1}(x)) = \dfrac{3\left(\dfrac{3x + 4}{2x - 3}\right) + 4}{2\left(\dfrac{3x + 4}{2x - 3}\right) - 3} = x$$

$$f^{-1}(f(x)) = \dfrac{3\left(\dfrac{3x + 4}{2x - 3}\right) + 4}{2\left(\dfrac{3x + 4}{2x - 3}\right) - 3} = x$$

Domain f = Range f^{-1} = All real numbers except $\frac{3}{2}$
Range f = Domain f^{-1} = All real numbers except $\frac{3}{2}$

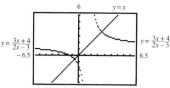

43. $f^{-1}(x) = \dfrac{-2x + 3}{x - 2}$

$$f(f^{-1}(x)) = \dfrac{2\left(\dfrac{-2x + 3}{x - 2}\right) + 3}{\dfrac{-2x + 3}{x - 2} + 2} = x$$

$$f^{-1}(f(x)) = \dfrac{-2\left(\dfrac{2x + 3}{x + 2}\right) + 3}{\dfrac{2x + 3}{x + 2} - 2} = x$$

Domain f = Range f^{-1} = All real numbers except -2
Range f = Domain f^{-1} = All real numbers except 2

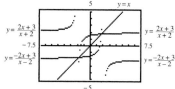

45. $f^{-1}(x) = \dfrac{x^3}{8}$

$$f(f^{-1}(x)) = 2\sqrt[3]{\dfrac{x^3}{8}} = x$$

$$f^{-1}(f(x)) = \dfrac{(2\sqrt[3]{x})^3}{8} = x$$

Domain f = Range f^{-1} = $(-\infty, \infty)$
Range f = Domain f^{-1} = $(-\infty, \infty)$

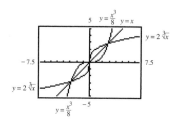

47. $f^{-1}(x) = \dfrac{1}{m}(x - b)$, $m \neq 0$ **49.** No; whenever x and $-x$ are in the domain of f, two equal y values, $f(x)$ and $f(-x)$, are present.

51. Quadrant I   **53.** $f(x) = |x|$, $x \geq 0$ is one-to-one; this may be written as $f(x) = x$; $f^{-1}(x) = x$

55. $f(g(x)) = \frac{9}{5}\left[\frac{5}{9}(x - 32)\right] + 32 = x$; $g(f(x)) = \frac{5}{9}\left[\left(\frac{9}{5}x + 32\right) - 32\right] = x$ **57.** $l(T) = gT^2/4\pi^2$ **59.** $f^{-1}(x) = \dfrac{-dx + b}{cx - a}$; $f = f^{-1}$ if $a = -d$

Exercise 2.6

1. $V(r) = 2\pi r^3$ **3. (a)** $R(x) = -\frac{1}{6}x^2 + 100x$ **(b)**   **(c)** 300; \$15,000   **(d)** \$50

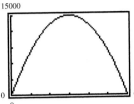

5. (a) $R(x) = -\frac{1}{5}x^2 + 20x$ **(b)**   **(c)** 50; \$500   **(d)** \$10

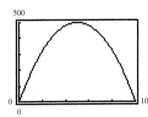

7. (a) $A(x) = -x^2 + 200x$ **(b)** $0 < x < 200$ **(c)** A is largest when $x = 100$ yards   **9. (a)** $C(x) = x$ **(b)** $A(x) = \dfrac{x^2}{4\pi}$ **11.** $A(x) = \frac{1}{2}x^4$

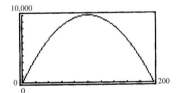

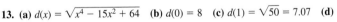

13. (a) $d(x) = \sqrt{x^4 - 15x^2 + 64}$ **(b)** $d(0) = 8$ **(c)** $d(1) = \sqrt{50} = 7.07$ **(d)**   **(e)** 2.73

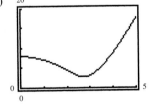

15. (a) $d(x) = \sqrt{x^2 - x + 1}$ **(b)**   **(c)** 0.50   **17.** $d(t) = 50t$

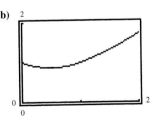

19. (a) $V(x) = x(24 - 2x)^2$ **(b)** The volume is largest when x is 4   **21. (a)** $A(x) = 2x^2 + \dfrac{40}{x}$ **(b)** The area is smallest when x is about 2.15

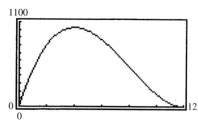

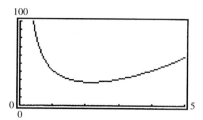

23. (a) $A(x) = x(16 - x^2)$ **(b)** Domain: $\{x \mid 0 < x < 4\}$ **(c)** The area is largest for x about 2.31

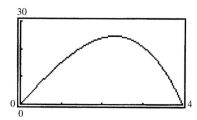

25. (a) $A(x) = 4x(4 - x^2)^{1/2}$ **(b)** $p(x) = 4x + 4(4 - x^2)^{1/2}$
  (c) The area is largest for x about 1.41   **(d)** The perimeter is largest for x about 1.41

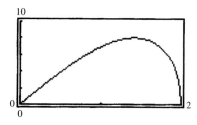

27. (a) $C(r) = 12\pi r^2 + \dfrac{4000}{r}$

  (b) The cost in least for r about 3.75 cm

29. (a) $A(x) = x^2 + \dfrac{25 - 20x + 4x^2}{\pi}$ **(b)** Domain: $\{x \mid 0 < x < 2.5\}$

  (c) The area is smallest for x about 1.40 meters

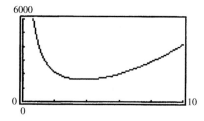

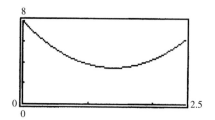

31. (a) $A(r) = 2r^2$ **(b)** $p(r) = 6r$

33. $A(x) = \left(\dfrac{\pi}{3} - \dfrac{\sqrt{3}}{4}\right) x^2$

35. $C = \begin{cases} 95 & \text{if } x = 7 \\ 119 & \text{if } 7 < x \le 8 \\ 143 & \text{if } 8 < x \le 9 \\ 167 & \text{if } 9 < x \le 10 \\ 190 & \text{if } 10 < x \le 14 \end{cases}$

37. $V(h) = \dfrac{\pi}{48} h^3$

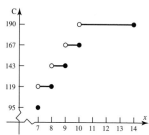

39. (a) $Q = -37.6355P + 1571.7196$ **(b)** $R = -37.6355P^2 + 1571.7196P$ **(c)**

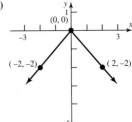

(d) \$20.88   **(e)** 786; \$16,409.39

Fill-in-the-Blank Items

1. independent; dependent   **2.** vertical   **3.** even; odd   **4.** horizontal; right   **5.** $g(f(x)) = (g \circ f)(x)$ **6.** one-to-one   **7.** $y = x$

True/False Items

1. T   **2.** T   **3.** F   **4.** T   **5.** F   **6.** F   **7.** T

Review Exercises

1. $f(x) = -2x + 3$ **3.** $A = 11$ **5. (a)** B, C, D   **(b)** D   **7. (a)** $f(-x) = \dfrac{-3x}{x^2 - 4}$ **(b)** $-f(x) = \dfrac{-3x}{x^2 - 4}$ **(c)** $f(x + 2) = \dfrac{3x + 6}{x^2 + 4x}$

(d) $f(x - 2) = \dfrac{3x - 6}{x^2 - 4x}$ **9. (a)** $f(-x) = \sqrt{x^2 - 4}$ **(b)** $-f(x) = -\sqrt{x^2 - 4}$ **(c)** $f(x + 2) = \sqrt{x^2 + 4x}$ **(d)** $f(x - 2) = \sqrt{x^2 - 4x}$

11. (a) $f(x) = \dfrac{x^2 - 4}{x^2}$ **(b)** $-f(x) = -\dfrac{x^2 - 4}{x^2}$ **(c)** $f(x + 2) = \dfrac{x^2 + 4x}{x^2 + 4x + 4}$ **(d)** $f(x - 2) = \dfrac{x^2 - 4x}{x^2 - 4x + 4}$ **13.** Odd   **15.** Even

17. Neither   **19.** $\{x \mid x \neq -3, x \neq 3\}$ **21.** $(-\infty, 2]$ **23.** $(0, \infty)$ **25.** $\{x \mid x \neq -3, x \neq 1\}$ **27.** $[-1, \infty)$ **29.** $[0, \infty)$

31. (a) All real numbers   **(c)**
(b) $(0, -4), (-4, 0), (4, 0)$
(d) $\{y \mid -4 \leq y < \infty\}$

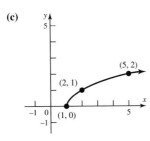

33. (a) All real numbers   **(c)**
(b) $(0, 0)$
(d) $\{y \mid -\infty < y \leq 0\}$

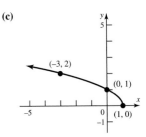

35. (a) $\{x \mid 1 \leq x < \infty\}$ **(c)**
(b) $(1, 0)$
(d) $\{y \mid 0 \leq y < \infty\}$

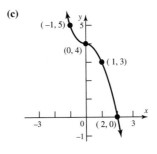

37. (a) $\{x \mid -\infty < x \leq 1\}$ **(c)**
(b) $(1, 0), (0, 1)$
(d) $\{y \mid 0 \leq y < \infty\}$

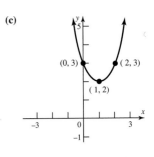

39. (a) All real numbers   **(c)**
(b) $(0, 4), (2, 0)$
(d) All real numbers

41. (a) All real numbers   **(c)**
(b) $(0, 3)$
(d) $\{y \mid 2 \leq y < \infty\}$

43. (a) All real numbers
(b) $(0, 0)$
(d) All real numbers
(c)
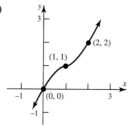

45. (a) $\{x \mid 0 < x < \infty\}$
(b) None
(d) $\{y \mid 0 < y < \infty\}$
(c)
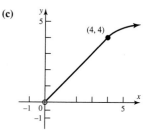

47. (a) $\{x \mid x \neq 1\}$
(b) $(0, 0)$
(d) $\{y \mid y \neq 1\}$
(c)
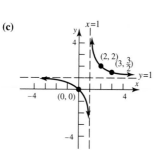

49. (a) All real numbers
(b) $-1 < x \leq 0$ are x-intercepts
0 is the y-intercept
(d) Set of integers
(c)
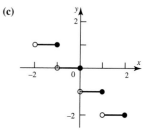

51. $f^{-1}(x) = \dfrac{2x + 3}{5x - 2}$

$$f(f^{-1}(x)) = \frac{2\left(\dfrac{2x+3}{5x-2}\right) + 3}{5\dfrac{2x+3}{5x-2} - 2} = x$$

$$f^{-1}(f(x)) = \frac{2\left(\dfrac{2x+3}{5x-2}\right) + 3}{5\dfrac{2x+3}{5x-2} - 2} = x$$

Domain f = Range f^{-1} = All real numbers except 2/5
Range f = Domain f^{-1} = All real numbers except 2/5

53. $f^{-1}(x) = \dfrac{x + 1}{x}$

$$f(f^{-1}(x)) = \frac{1}{\dfrac{x+1}{x} - 1} = x$$

$$f^{-1}(f(x)) = \frac{\dfrac{1}{x-1} + 1}{\dfrac{1}{x-1}} = x$$

Domain f = Range f^{-1} = All real numbers except 1
Range f = Domain f^{-1} = All real numbers except 0

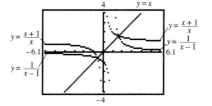

55. $f^{-1}(x) = \dfrac{27}{x^3}$

$$f(f^{-1}(x)) = \frac{3}{(27/x^3)^{1/3}} = x$$

$$f^{-1}(f(x)) = \frac{27}{(3/x^{1/3})^3} = x$$

Domain f = Range f^{-1} = All real numbers except 0
Range f = Domain f^{-1} = All real numbers except 0

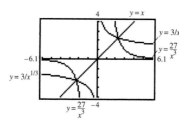

57. (a) -26 **(b)** -241 **(c)** 16 **(d)** -1 **59. (a)** $\sqrt{11}$ **(b)** 1 **(c)** $\sqrt{\sqrt{6} + 2}$ **(d)** 19 **61. (a)** $\frac{1}{20}$ **(b)** $-\frac{13}{8}$ **(c)** $\frac{400}{1601}$ **(d)** -17

63. $(f \circ g)(x) = \dfrac{-3x + 1}{3x + 1}$; $(g \circ f)(x) = \dfrac{6 - 2x}{x}$; $(f \circ f)(x) = \dfrac{3x - 2}{2 - x}$; $(g \circ g)(x) = 9x + 4$

65. $(f \circ g)(x) = 27x^2 + 3|x| + 1$; $(g \circ f)(x) = 3|3x^2 + x + 1|$; $(f \circ f)(x) = 3(3x^2 + x + 1)^2 + 3x^2 + x + 2$; $(g \circ g)(x) = 9|x|$

67. $(f \circ g)(x) = \dfrac{1 + x}{1 - x}$; $(g \circ f)(x) = \dfrac{x - 1}{x + 1}$; $(f \circ f)(x) = x$; $(g \circ g)(x) = x$

69. (a)

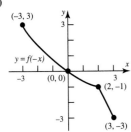

(b)

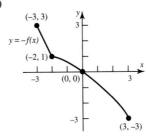

(c)

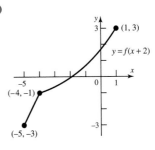

(d)

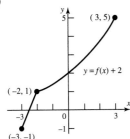

(e)

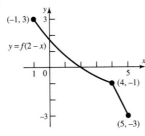

(f)

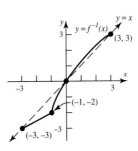

71. $T(h) = -0.0025h + 30$, $0 \le x \le 10{,}000$ **73.** $S(x) = kx(36 - x^2)^{3/2}$; Domain: $\{x \mid 0 < x < 6\}$

CHAPTER 3   *Exercise 3.1*

1. 0.42   **3.** 2.23   **5.** 1.25   **7.** $f(0) = -1, f(1) = 10; 0.21$ **9.** $f(-5) = -58, f(-4) = 2; -4.04$ **11.** $f(1.4) = -0.17536, f(1.5) = 1.40625; 1.41$
13. -3.41 **15.** -1.70 **17.** -0.28 **19.** 3.00   **21.** 4.50   **23.** 0.31, 12.30   **25.** 23.00

Exercise 3.2

1. 7   **3.** -3 **5.** 4   **7.** $\frac{5}{4}$ **9.** 2   **11.** 3   **13.** -1 **15.** $-\frac{4}{3}$ **17.** -18 **19.** -3 **21.** $-\frac{3}{2}$ **23.** -20 **25.** 2   **27.** 0.5   **29.** $\frac{46}{5}$ **31.** 2   **33.** 20
35. $\{0, 9\}$ **37.** $\{-5, 5\}$ **39.** $\{-4, 3\}$ **41.** $\{-\frac{1}{2}, 3\}$ **43.** $\{-4, 4\}$ **45.** $\{3, 4\}$ **47.** $\frac{3}{2}$ **49.** $\{-\frac{2}{3}, \frac{3}{2}\}$ **51.** $\{-\frac{2}{3}, \frac{3}{2}\}$ **53.** $\{-\frac{3}{4}, 2\}$
55. 2   **57.** -1 **59.** 3   **61.** No solution   **63.** $\{0, 4\}$ **65.** $\{0, 9\}$ **67.** 5.91   **69.** 0.41   **71.** $x = \dfrac{b + c}{a}$ **73.** $x = \dfrac{abc}{a + b}$ **75.** $x = a^2$
77. $\{2 - \sqrt{2}, 2 + \sqrt{2}\}$ **79.** $\{2 - \sqrt{5}, 2 + \sqrt{5}\}$ **81.** $\{1, \frac{3}{2}\}$ **83.** No real solution   **85.** $\left\{\dfrac{-1 - \sqrt{5}}{4}, \dfrac{-1 + \sqrt{5}}{4}\right\}$ **87.** $\{0, \frac{9}{4}\}$
89. $\frac{1}{3}$ **91.** $\left\{\dfrac{1 - \sqrt{7}}{3}, \dfrac{1 + \sqrt{7}}{3}\right\}$ **93.** $\left\{\dfrac{1 - \sqrt{33}}{8}, \dfrac{1 + \sqrt{33}}{8}\right\}$ **95.** No real solution   **97.** $\{0.59, 3.41\}$; $\{2 - \sqrt{2}, 2 + \sqrt{2}\}$
99. $\{-2.80, 1.07\}$; $\left\{\dfrac{-\sqrt{3} - \sqrt{15}}{2}, \dfrac{-\sqrt{3} + \sqrt{15}}{2}\right\}$ **101.** $\{-0.85, 1.17\}$; $\left\{\dfrac{1 \pm \sqrt{1 + 4\pi^2}}{2\pi}\right\}$
103. $\{-8.16, -0.22\}$; $\left\{\dfrac{-8\pi \pm \sqrt{64\pi^2 - 12\sqrt{29}}}{6}\right\}$ **105.** $\{-2.64, 2.64\}$; $\{-\sqrt{7}, \sqrt{7}\}$ **107.** $0.25, \frac{1}{4}$ **109.** $\{-0.60, 2.50\}$; $\{-\frac{3}{5}, \frac{5}{2}\}$
111. $\{-0.50, 0.66\}$; $\{-\frac{1}{2}, \frac{2}{3}\}$ **113.** $\{-1.70, 0.29\}$; $\left\{\dfrac{-\sqrt{2} + 2}{2}, \dfrac{-\sqrt{2} - 2}{2}\right\}$ **115.** $\{-2.56, 1.56\}$; $\left\{\dfrac{-1 - \sqrt{17}}{2}, \dfrac{-1 + \sqrt{17}}{2}\right\}$
117. No real solution   **119.** Repeated real solution   **121.** Two unequal real solutions   **123.** $R = \dfrac{R_1 R_2}{R_1 + R_2}$ **125.** $R = \dfrac{mv^2}{F}$ **127.** $r = \dfrac{S - a}{S}$

Exercise 3.3

1. $A = \pi r^2$; $r = $ Radius, $A = $ Area   **3.** $A = s^2$; $A = $ Area, $s = $ Length of a side   **5.** $F = ma$; $F = $ Force, $m = $ Mass, $a = $ Acceleration
7. $W = Fd$; $W = $ Work, $F = $ Force, $d = $ Distance   **9.** $C = 150x$; $C = $ Total cost, $x = $ number of dishwashers
11. \$11,000 will be invested in bonds and \$9,000 in CD's   **13.** Katy will receive \$400,000, Mike \$300,000, and Dan \$200,000
15. The regular hourly rate is \$8.50   **17.** The team got 5 touchdowns   **19.** The length is 19 ft; the width is 11 ft

21. (a) The dimensions are 10 ft by 5 ft **(b)** The area is 50 sq ft **(c)** The dimensions are 7.5 ft by 7.5 ft **(d)** The area is 56.25 sq ft
23. Invest $31,250 in bonds and $18,750 in CD's **25.** $11,600 was loaned out at 8% **27.** The dimensions are 11 ft by 13 ft
29. The dimensions are 5 m by 8 m **31.** The dimensions of the sheet metal should be 4 ft by 4 ft
33. (a) The ball strikes the ground after 6 seconds **(b)** The ball passes the top of the building on its way down after 5 seconds
35. The original price was $147,058.82; purchasing the model saves $22,058.82 **37.** The bookstore paid $44.80
39. Working together, it takes 12 min **41.** Colleen needs a score of 85
43. The defensive back catches up to the tight end at the tight end's 45 yd line **45.** The border will be 2.56 ft wide
47. The dimensions of the reduced candy bar are 11.55 cm by 6.55 cm by 3 cm **49.** The border will be 2.71 ft wide
51. The current is 2.286 mi/hr **53.** Mike passes Dan 1/3 mile from the start, 2 minutes from the time Mike started to race
55. The rescue craft reaches the ship in 2 hrs **57.** Start the auxiliary pump at 9:45 AM **59.** The tub will fill in 1 hr
61. The most you can invest in the CD is $66,667 **63.** The average speed is 49.5 mi/hr
65. Set the original price at $40. At 50% off, there will be no profit at all.

Exercise 3.4

1. $8 + 5i$ **3.** $-7 + 6i$ **5.** $-6 - 11i$ **7.** $6 - 18i$ **9.** $6 + 4i$ **11.** $10 - 5i$ **13.** 37 **15.** $\frac{6}{5} + \frac{8}{5}i$ **17.** $1 - 2i$ **19.** $\frac{5}{2} - \frac{7}{2}i$ **21.** $-\frac{1}{2} + (\sqrt{3}/2)i$
23. $2i$ **25.** $-i$ **27.** i **29.** -6 **31.** $-10i$ **33.** $-2 + 2i$ **35.** 0 **37.** 0 **39.** $2i$ **41.** $5i$ **43.** $5i$ **45.** $\{-2i, 2i\}$ **47.** $\{-4, 4\}$
49. $\{3 - 2i, 3 + 2i\}$ **51.** $\{3 - i, 3 + i\}$ **53.** $\{\frac{1}{4} - \frac{1}{4}i, \frac{1}{4} + \frac{1}{4}i\}$ **55.** $\{-\frac{1}{5} - \frac{2}{5}i, -\frac{1}{5} + \frac{2}{5}i\}$ **57.** $\left\{-\dfrac{1}{2} - \dfrac{\sqrt{3}}{2}i, -\dfrac{1}{2} + \dfrac{\sqrt{3}}{2}i\right\}$
59. $\{2, -1 - \sqrt{3}i, -1 + \sqrt{3}i\}$ **61.** $\{-2, 2, -2i, 2i\}$ **63.** $\{-3i, -2i, 2i, 3i\}$ **65.** Two complex solutions **67.** Two unequal real solutions
69. A repeated real solution **71.** $2 - 3i$ **73.** 6 **75.** 25 **77.** $z + \bar{z} = (a + bi) + (a - bi) = 2a; z - \bar{z} = (a + bi) - (a - bi) = 2bi$
79. $\bar{z} + \bar{w} = \overline{(a + bi) + (c + di)} = \overline{(a + c) + (b + d)i} = (a + c) - (b + d)i = (a - bi) + (c - di) = \bar{z} + \bar{w}$

Exercise 3.5

1. 1 **3.** No real solution **5.** -13 **7.** 3 **9.** 2 **11.** $-\frac{8}{5}$ **13.** 8 **15.** $\{-1, 3\}$ **17.** $\{1, 5\}$ **19.** 1 **21.** 5 **23.** 9 **25.** $\{-4, 4\}$
27. $\{-5, -4\}$ **29.** $-\frac{1}{3}$ **31.** $\{-\frac{3}{2}, 2\}$ **33.** $\{0, 16\}$ **35.** 16 **37.** 1 **39.** $\left\{\left(\dfrac{9 - \sqrt{17}}{8}\right)^4, \left(\dfrac{9 + \sqrt{17}}{8}\right)^4\right\}$ **41.** $\{\sqrt{2}, \sqrt{3}\}$ **43.** $\{-4, 1\}$
45. $\{-2, -\frac{1}{2}\}$ **47.** $\{-\frac{3}{2}, \frac{1}{3}\}$ **49.** $\{-\frac{1}{8}, 27\}$ **51.** $\{-2, -\frac{4}{5}\}$ **53.** $\{-1, 0, 1\}$ **55.** $\{-2, -1, 1, 2\}$ **57.** $\{-1, 1\}$ **59.** $\{-2, 1\}$ **61.** $\{0, 36\}$
63. $\{0, 2\}$ **65.** $\{-5, 0, 4\}$ **67.** $\{-1\}$ **69.** $\{-2, 2, 3\}$ **71.** $\{-1, 1\}$ **73.** $\{0.34, 11.66\}$ **75.** $\{-1.03, 1.03\}$ **77.** $\{-1.85, 0.17\}$
79. $\{\frac{3}{2}, 5\}$ **81. (a)** The depth of the well is 230 ft **(b)**

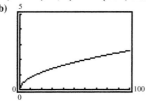

(c)

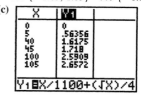

0.112712; 0.0201;
0.013226
As the depth of the
well increases, the
slope decreases.

(d) As the depth of the well increases, the time that elapses before a sound is heard increases at a decreasing rate.

Exercise 3.6

1. $<$ **3.** $>$ **5.** $>$ **7.** $>$ **9. (a)** $0 < 2$ **(b)** $-2 < 0$ **(c)** $9 < 15$ **(d)** $-6 > -10$
11. (a) $2x - 2 < -1$ **(b)** $2x - 4 < -3$ **(c)** $6x + 3 < 6$ **(d)** $-4x - 2 > -4$
13. $\{x | x < 4\}$ or $(-\infty, 4)$ **15.** $\{x | x \geq -1\}$ or $[-1, \infty)$ **17.** $\{x | x > 3\}$ or $(3, \infty)$ **19.** $\{x | x \geq 2\}$ or $[2, \infty)$

21. $\{x | x > -7\}$ or $(-7, \infty)$ **23.** $\{x | x \leq \frac{2}{3}\}$ or $(-\infty, \frac{2}{3}]$ **25.** $\{x | x < -20\}$ or $(-\infty, -20)$ **27.** $\{x | x \geq \frac{4}{3}\}$ or $[\frac{4}{3}, \infty)$

29. $\{x | 3 \leq x \leq 5\}$ or $[3, 5]$ **31.** $\{x | \frac{2}{3} \leq x \leq 3\}$ or $[\frac{2}{3}, 3]$ **33.** $\{x | -\frac{11}{2} < x < \frac{1}{2}\}$ or $(-\frac{11}{2}, \frac{1}{2})$ **35.** $\{x | -6 < x < 0\}$ or $(-6, 0)$

37. $\{x|x < -5\}$ or $(-\infty, -5)$

39. $\{x|x \geq -1\}$ or $[-1, \infty)$

41. $\{x|\frac{1}{2} \leq x < \frac{5}{4}\}$ or $[\frac{1}{2}, \frac{5}{4})$

43. $\{x|x < -\frac{1}{2}\}$ or $(-\infty, -\frac{1}{2})$

45. $\{x|x > 5\}$ or $(5, \infty)$

47. $\{x|x > 3\}$ or $(3, \infty)$

49. $a = 3, b = 5$

51. $a = -12, b = -8$

53. $a = 3, b = 11$
57. $a = 4, b = 16$

55. $a = \frac{1}{4}, b = 1$
59. $\{x|-10 < x < 0\}$

61. $\{x|x \geq -2\}$ or $[-2, \infty)$

63. $a \leq b, c > 0$; $\quad a - b \leq 0$
$(a - b)c \leq 0(c)$
$ac - bc \leq 0$
$ac \leq bc$

65. $\dfrac{a + b}{2} - a = \dfrac{a + b - 2a}{2} = \dfrac{b - a}{2} > 0$; therefore, $a < \dfrac{a + b}{2}$
$b - \dfrac{a + b}{2} = \dfrac{2b - a - b}{2} = \dfrac{b - a}{2} > 0$; therefore, $b > \dfrac{a + b}{2}$

67. $(\sqrt{ab})^2 - a^2 = ab - a^2 = a(b - a) > 0$; thus, $(\sqrt{ab})^2 > a^2$ and $\sqrt{ab} > a$
$b^2 - (\sqrt{ab})^2 = b^2 - ab = b(b - a) > 0$; thus, $b^2 > (\sqrt{ab})^2$ and $b > \sqrt{ab}$

69. $h - a = \dfrac{2ab}{a + b} - a = \dfrac{ab - a^2}{a + b} = \dfrac{a(b - a)}{a + b} > 0$; thus, $h > a$
$b - h = b - \dfrac{2ab}{a + b} = \dfrac{b^2 - ab}{a + b} = \dfrac{b(b - a)}{a + b} > 0$; thus, $h < b$

71. $21 < \text{Age} < 30$ **73. (a)** Male ≥ 73.4 **(b)** Female ≥ 79.7 **(c)** A female can expect to live at least 6.3 years longer
75. The agent's commission ranges from \$45,000 to \$95,000, inclusive. As a percent of selling price, the commission ranges from 5% to 8.6%, inclusive
77. The amount of withholding varies from \$70.62 to \$84.62, inclusive **79.** The usage varies from 675.48 kWhr to 2500.88 kWhr, inclusive
81. The dealer's cost varies from \$7457.63 to \$7857.14, inclusive **83.** Fifth test score ≥ 74
85. The amount of gasoline ranged from 12 to 20 gal, inclusive

Exercise 3.7

1. $\{x|-2 < x < 5\}$ **3.** $\{x|-\infty < x < 0 \text{ or } 4 < x < \infty\}$ **5.** $\{x|-3 < x < 3\}$ **7.** $\{x|-\infty < x < -4 \text{ or } 3 < x < \infty\}$ **9.** $\{x|-\frac{1}{2} < x < 3\}$
11. $\{x|-\infty < x < -1 \text{ or } 8 < x < \infty\}$ **13.** No real solution **15.** $\{x|-\infty < x < -\frac{2}{3} \text{ or } \frac{3}{2} < x < \infty\}$ **17.** $\{x|1 < x < \infty\}$
19. $\{x|-\infty < x < 1 \text{ or } 2 < x < 3\}$ **21.** $\{x|-1 < x < 0 \text{ or } 3 < x < \infty\}$ **23.** $\{x|-\infty < x < -1 \text{ or } 1 < x < \infty\}$ **25.** $\{x|1 < x < \infty\}$
27. $\{x|-\infty < x < -1 \text{ or } 1 < x < \infty\}$ **29.** $\{x|-\infty < x < -1 \text{ or } 1 < x < \infty\}$ **31.** $\{x|-\infty < x < -1 \text{ or } 0 < x < 1\}$
33. $\{x|-\infty < x < -1 \text{ or } 1 < x < \infty\}$ **35.** $\{x|-\infty < x < -\frac{2}{3} \text{ or } 0 < x < \frac{3}{2}\}$ **37.** $\{x|-\infty < x < 2\}$ **39.** $\{x|-2 < x \leq 9\}$
41. $\{x|-\infty < x < 2 \text{ or } 3 < x < 5\}$ **43.** $\{x|-\infty < x < -3 \text{ or } -1 < x < 1 \text{ or } 2 < x < \infty\}$
45. $\{x|-\infty < x < -5 \text{ or } -4 < x < -3 \text{ or } 1 < x < \infty\}$ **47.** $\{x|-1 < x < 8\}$ **49.** $\{x|2.14 \leq x < \infty\}$ **51.** $\{x|-\infty < x < -2 \text{ or } 2 < x < \infty\}$
53. $\{x|-\infty < x \leq -0.5 \text{ or } 1 \leq x < 4\}$ **55.** $\{x|-3.30 < x < -3 \text{ or } 0.30 < x < \infty\}$ **57.** $\{x|-1 \leq x < -0.23 \text{ or } 4.23 \leq x < \infty\}$
59. $\{x|4 < x < \infty\}$ **61.** $\{x|-\infty < x \leq -4 \text{ or } 4 \leq x < \infty\}$ **63.** $\{x|-\infty < x < -4 \text{ or } 2 \leq x < \infty\}$
65. (a) The ball is more than 96 feet above the ground for time t between 2 and 3 seconds, $2 < t < 3$
(b) **(c)** 100 feet **(d)** 2.5 seconds

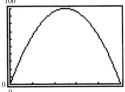

67. (a) For a profit of at least \$50, between 8 and 32 watches must be sold, $8 \leq x \leq 32$
(b) **(c)** \$2000 **(d)** 100 **(e)** **(f)** \$80 **(g)** 20

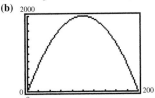
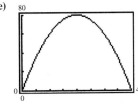

69. $b - a = (\sqrt{b} - \sqrt{a})(\sqrt{b} + \sqrt{a})$; since $a \geq 0$ and $b \geq 0$, then $\sqrt{a} \geq 0$ and $\sqrt{b} \geq 0$ so that $\sqrt{b} + \sqrt{a} \geq 0$; thus, $b - a \geq 0$ is equivalent to $\sqrt{b} - \sqrt{a} \geq 0$ and $a \leq b$ is equivalent to $\sqrt{a} \leq \sqrt{b}$ **71.** $k < 1$

Exercise 3.8

1. $\{-4, 4\}$ **3.** $\{-4, 1\}$ **5.** $\{-1, \frac{3}{2}\}$ **7.** $\{-4, 4\}$ **9.** 2 **11.** $\{-12, 12\}$ **13.** $\{-\frac{36}{5}, \frac{24}{5}\}$ **15.** No real solution **17.** $\{-4, 4\}$ **19.** $\{-1, 3\}$
21. $\{-2, -1, 0, 1\}$ **23.** $(-4, 4)$ **25.** $(-\infty, -4)$ or $(4, \infty)$ **27.** $(1, 3)$ **29.** $[-\frac{2}{3}, 2]$

31. $(-\infty, 1]$ or $[5, \infty)$ **33.** $(-1, \frac{3}{2})$ **35.** $(-\infty, -1)$ or $(2, \infty)$ **37.** $(-\infty, -3)$ or $(-3, \infty)$ **39.** $(-\infty, \infty)$

41. $(0.49, 0.51)$ **43.** $a = 2, b = 8$ **45.** $a = -15, b = -7$ **47.** $a = -1, b = -\frac{1}{15}$ **49.** $\left|\dfrac{a}{b}\right| = \sqrt{\left(\dfrac{a}{b}\right)^2} = \sqrt{\dfrac{a^2}{b^2}} = \dfrac{\sqrt{a^2}}{\sqrt{b^2}} = \dfrac{|a|}{|b|}$

51. $(a + b)^2 = a^2 + 2ab + b^2 \leq |a|^2 + 2|a||b| + |b|^2 = [|a| + |b|]^2$; therefore, $\sqrt{(a + b)^2} \leq \sqrt{(|a| + |b|)^2}$ or $|a + b| \leq |a| + |b|$
53. $|x - 3| < \frac{1}{2}$; $\frac{5}{2} < x < \frac{7}{2}$ **55.** $|x + 3| > 2$; $x < -5$ or $x > -1$ **57.** $|x - 98.6| \geq 1.5$; $x \leq 97.1°F$ or $x \geq 100.1°F$
59. $x^2 - a < 0$; $(x - \sqrt{a})(x + \sqrt{a}) < 0$; therefore, $-\sqrt{a} < x < \sqrt{a}$ **61.** $-1 < x < 1$ **63.** $x \geq 3$ or $x \leq -3$ **65.** $-4 \leq x \leq 4$
67. $x > 2$ or $x < -2$ **69.** $\{-1, 5\}$

Fill-in-the-Blank Items

1. equivalent **2.** identity **3.** add; $\frac{25}{4}$ **4.** discriminant; negative **5.** $-a$ **6.** double; multiplicity 2 **7.** extraneous **8.** negative
9. $-2, 2, -2i, 2i$

True/False Items

1. T **2.** F **3.** F **4.** T **5.** T **6.** (a) T (b) F (c) T **7.** (a) F (b) T (c) T

Review Exercises

1. -12 **3.** 6 **5.** $\frac{1}{5}$ **7.** -5 **9.** No real solution **11.** $\frac{11}{8}$ **13.** $\{-2, \frac{3}{2}\}$ **15.** $\left\{\dfrac{1 - \sqrt{13}}{4}, \dfrac{1 + \sqrt{13}}{4}\right\}$ **17.** $\{-3, 3\}$ **19.** No real solution
21. $\{-2, -1, 1, 2\}$ **23.** 2 **25.** 0 **27.** $\dfrac{\sqrt{5}}{2}$ **29.** $\{-\frac{1}{8}, 1\}$ **31.** $\{-1, \frac{1}{2}\}$ **33.** $-\frac{9}{5}$ **35.** $\left\{\dfrac{m}{1 - n}, \dfrac{m}{1 + n}\right\}$ **37.** $\left\{\dfrac{-9b}{5a}, \dfrac{2b}{a}\right\}$
39. $\{x | 14 \leq x < \infty\}$ **41.** $\{x | -\frac{31}{2} \leq x \leq \frac{33}{2}\}$ **43.** $\{x | -23 < x < -7\}$ **45.** $\{x | -4 < x < \frac{3}{2}\}$ **47.** $\{x | -3 < x \leq 3\}$
49. $\{x | -\infty < x < 1$ or $2 < x < \infty\}$ **51.** $\{x | 1 < x < 2$ or $3 < x < \infty\}$ **53.** $\{x | -\infty < x < -4$ or $2 < x < 4$ or $6 < x < \infty\}$
55. $\{x | -\frac{3}{2} < x < -\frac{7}{6}\}$ **57.** $\{x | -\infty < x \leq -2$ or $7 \leq x < \infty\}$ **59.** $\left\{\dfrac{-1 - \sqrt{3}i}{2}, \dfrac{-1 + \sqrt{3}i}{2}\right\}$ **61.** $\left\{\dfrac{-1 - \sqrt{17}}{4}, \dfrac{-1 + \sqrt{17}}{4}\right\}$
63. $\left\{\dfrac{1 - \sqrt{11}i}{2}, \dfrac{1 + \sqrt{11}i}{2}\right\}$ **65.** $\left\{\dfrac{1 - \sqrt{23}i}{2}, \dfrac{1 + \sqrt{23}i}{2}\right\}$ **67.** $\{-\sqrt{2}, \sqrt{2}, -2i, 2i\}$ **69.** The storm is 3300 ft away

71. The sea plane can go as far as 616 mi **73.** The helicopter will reach the life raft in a little less than 1 hr, 35 min
75. The freight train is 190.67 ft long **77.** 36 seniors went on the trip; each one paid $13.40
79. (a) No (b) Mike wins again (c) Mike wins by $\frac{1}{4}$ meter (d) Mike should line up 5.26316 meters behind the start line (e) Yes

1. D **3.** A **5.** B **7.** E **9.** D **11.** B

13.

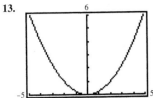

Opens up; Vertex: (0, 0)
Axis of symmetry: $x = 0$
Intercept: (0, 0)

15.

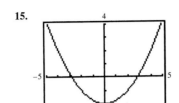

Opens up; Vertex: (0, −2)
Axis of symmetry: $x = 0$
Intercepts: (−2.82, 0), (2.82, 0), (0, −2)

17.

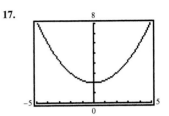

Opens up; Vertex: (0, 2)
Axis of symmetry: $x = 0$
Intercept: (0, 2)

19.

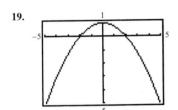

Opens down; Vertex: (0, 1)
Axis of symmetry: $x = 0$
Intercepts: (−2, 0), (2, 0), (0, 1)

21.

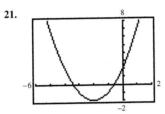

Opens up; Vertex: (−2, −2)
Axis of symmetry: $x = -2$
Intercepts: (−3.41, 0), (−0.58, 0), (0, 2)

23.

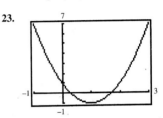

Opens up; Vertex: (1, −1)
Axis of symmetry: $x = 1$
Intercepts: (1.70, 0), (0.29, 0), (0, 1)

25.

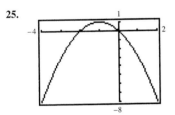

Opens down; Vertex: (−1, 1)
Axis of symmetry: $x = -1$
Intercepts: (−2, 0), (0, 0)

27.

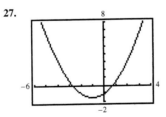

Opens up; Vertex: (−1, −1.5)
Axis of symmetry: $x = -1$
Intercepts: (−2.73, 0), (0.73, 0), (0, −1)

29.

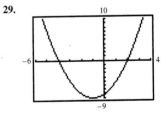

Opens up; Vertex: (−1, −9)
Axis of symmetry: $x = -1$
Intercepts: (−4, 0), (2, 0), (0, −8)

31.

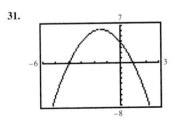

Opens down; Vertex: (−1.5, 6.25)
Axis of symmetry: $x = -1.5$
Intercepts: (−4, 0), (1, 0), (0, 4)

33.

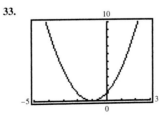

Opens up; Vertex: (−1, 0)
Axis of symmetry: $x = -1$
Intercepts: (−1, 0), (0, 1)

35.

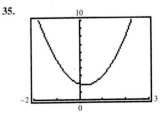

Opens up; Vertex: (0.25, 1.875)
Axis of symmetry: $x = 0.25$
Intercept: (0, 2)

37.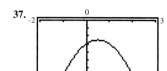

Opens down; Vertex: $(0.5, -2.5)$
Axis of symmetry: $x = 0.5$
Intercept: $(0, -3)$

39.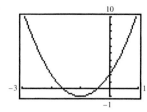

Opens up; Vertex: $(-1, -1)$
Axis of symmetry: $x = -1$
Intercepts: $(-1.57, 0), (-0.42, 0), (0, 2)$

41.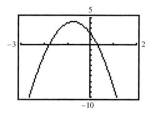

Opens down; Vertex: $(-0.75, 4.25)$
Axis of symmetry: $x = -0.75$
Intercepts: $(-1.78, 0), (0.28, 0), (0, 2)$

43. Minimum value; -21 **45.** Maximum value; 21   **47.** Maximum value; 13   **49.** Opens up;

vertex at $(-1, f(-1))$;
axis of symmetry $x = -1$

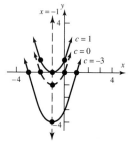

51. Each parabola opens up and passes through $(0, 1)$. Each one has the same shape.
53. Price: $500; maximum revenue: $1,000,000   **55.** 10,000 ft²; 100 ft by 100 ft   **57.** 2,000,000 m²   **59.** 4,166,666.7 m²
61. (a)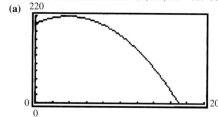

(b) 219.53 ft
(c) 170.02 ft
(d) When the height is 100 ft,
the projectile is 135.69 ft
from the cliff

63. 8 PM   **65.** 18.75 m   **67.** 3 in.   **69.** Width $= \dfrac{40}{\pi + 4} \approx 5.6$ ft; length ≈ 2.8 ft   **71.** $x = \dfrac{16}{6 - \sqrt{3}} \approx 3.75$ ft; other side ≈ 2.38 ft

73. 70 members   **75.** $a = 6, b = 0, c = 2$;   **77.** $\dfrac{a}{2}$ **79.** $\left. \begin{aligned} ah^2 - bh + c &= y_0 \\ c &= y_1 \\ ah^2 + bh + c &= y_2 \end{aligned} \right\}$ $\left. \begin{aligned} y_0 + y_2 &= 2ah^2 + 2c \\ 4y_1 &= 4c \end{aligned} \right\}$ Area $= \dfrac{h}{3}(2ah^2 + 6c) = \dfrac{h}{3}(y_0 + 4y_1 + y_2)$

$f(x) = 6x^2 + 2$

Exercise 4.2

1. Yes; degree 3   **3.** Yes; degree 2   **5.** No; x is raised to the -1 power.   **7.** No; x is raised to the $\frac{3}{2}$ power.   **9.** Yes; degree 4

11. 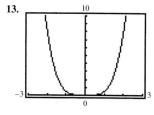   **13.**   **15.**   **17.**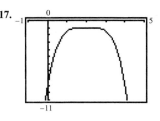

19. 7, multiplicity 1; −3, multiplicity 2; graph touches the x-axis −3 and crosses it at 7   **21.** 2, multiplicity 3; graph crosses the x-axis at 2
23. −½, multiplicity 2; graph touches the x-axis at −½   **25.** 5, multiplicity 3; −4, multiplicity 2; graph touches the x-axis at −4 and crosses it at 5
27. No real zeros; graph neither crosses nor touches the x-axis

29. (a)

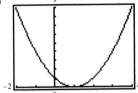

(b) x-intercept: 1; y-intercept: 1
(c) 1: Even
(d) $y = x^2$
(e) 1
(f) Local minima: (1, 0)

31. (a)

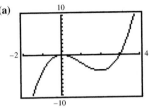

(b) x-intercepts: 0, 3; y-intercept: 0
(c) 0: Even; 3: Odd
(d) $y = x^3$
(e) 2
(f) Local maxima: (0, 0);
Local minima: (2, −4)

33. (a)

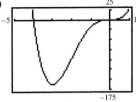

(b) x-intercepts: −4, 0; y-intercept: 0
(c) 0: Odd; −4: Odd
(d) $y = 6x^4$
(e) 3
(f) Local minima: (−3, −162)

35. (a)

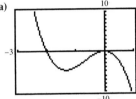

(b) x-intercepts: −2, 0; y-intercept: 0
(c) −2: Odd; 0: Even
(d) $y = -4x^3$
(e) 2
(f) Local minima: (−1.33, −4.74);
Local maxima: (0, 0)

37. (a)

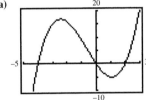

(b) x-intercepts: −4, 0, 2; y-intercept: 0
(c) 0: Odd; 2: Odd; −4: Odd
(d) $y = x^3$
(e) 2
(f) Local maxima: (−2.43, 16.90);
Local minima: (1.09, −5.04)

39. $f(x) = 4x − x^3 = −x(x^2 − 4) = −x(x + 2)(x − 2)$
(a)

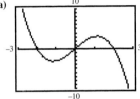

(b) x-intercepts: −2, 0, 2; y-intercept: 0
(c) −2, 0, 2: Odd
(d) $y = -x^3$
(e) 2
(f) Local minima: (−1.15, −3.07);
Local maxima: (1.15, 3.07)

41. (a)

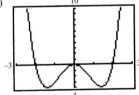

(b) x-intercepts: −2, 0, 2; y-intercept: 0
(c) −2, 2: Odd; 0: Even
(d) $y = x^4$
(e) 3
(f) Local minima: (−1.41, −4), (1.41, −4);
Local maxima: (0, 0)

43. (a)

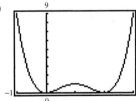

(b) x-intercepts: 0, 2; y-intercept: 0
(c) 0, 2: Even
(d) $y = x^4$
(e) 3
(f) Local minima: (0, 0), (2, 0);
Local maxima: (1, 1)

45. (a)

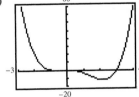

(b) x-intercepts: −1, 0, 3; y-intercept: 0
(c) −1, 3: Odd; 0: Even
(d) $y = x^4$
(e) 3
(f) Local minima: (2.18, −12.39),
(−0.68, −0.54); Local maxima: (0, 0)

47. (a)

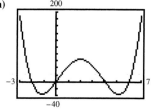

(b) *x*-intercepts: −2, 0, 4, 6; *y*-intercept: 0
(c) −2, 0, 4, 6: Odd
(d) $y = x^4$
(e) 3
(f) Local minima: (−1.16, −36), (5.16, −36);
 Local maxima: (2, 64)

49. (a)

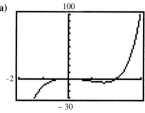

(b) *x*-intercepts: 0, 2; *y*-intercept: 0
(c) 0: Even; 2: Odd
(d) $y = x^5$
(e) 2
(f) Local minima: (1.47, −5.91);
 Local maxima: (0, 0)

51. (a)

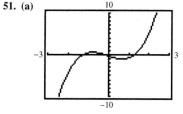

(b) *x*-intercepts: −1.26, −0.20, 1.26;
 y-intercept: −0.31752
(c) −1.26, −0.20, 1.26: Odd
(d) $y = x^3$
(e) 2
(f) Local minima: (0.66, −0.99);
 Local maxima: (−0.79, 0.56)

53. (a)

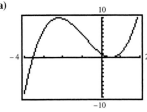

(b) *x*-intercepts: −3.56, 0.50; *y*-intercept: 0.89
(c) −3.56: Odd; 0.50: Even
(d) $y = x^3$
(e) 2
(f) Local minima: (0.50, 0);
 Local maxima: (−2.20, 9.91)

55. (a)

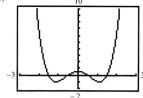

(b) *x*-intercepts: −1.50, −0.50, 0.50, 1.50;
 y-intercepts: 0.5625
(c) −1.50, −0.50, 0.50, 1.50: Odd
(d) $y = x^4$
(e) 3
(f) Local minima: (−1.11, −1), (1.11, −1);
 Local maxima: (0, 0.5625)

57. (a)

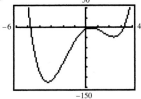

(b) *x*-intercepts: −4.78, 0.45, 3.23;
 y-intercepts: −3.1264785
(c) −4.78, 3.23: Odd; 0.45: Even
(d) $y = x^4$
(e) 3
(f) Local minima: (−3.31, −135.91),
 (2.37, −22.66);
 Local maxima: (0.45, 0)

59. (a)

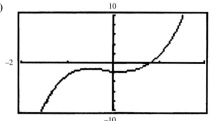

(b) *x*-intercept: 0.83; *y*-intercept: −2
(c) 0.83: odd
(d) $y = \pi x^3$
(e) 2
(f) local maxima: (−0.50, −1.53);
 local minima: (0.20, −2.11)

61. (a)

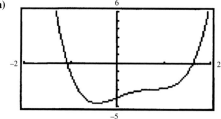

(b) *x*-intercepts: −1.06, 1.61; *y*-intercept: −4
(c) −1.06, 1.61: odd
(d) $y = 2x^4$
(e) 1
(f) local minima: (−0.41, −4.64)

63. (a)

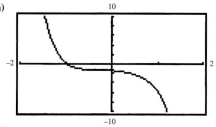

(b) *x*-intercept: −0.97   **65.** c, e, f   **67.** c, e
(c) −0.97: odd
(d) $y = -2x^5$
(e) None
(f) None

1. All real numbers except 3   **3.** All real numbers except 2 and −4   **5.** All real numbers except $-\frac{1}{2}$ and 3   **7.** All real numbers except 2
9. All real numbers   **11.** **(a)** Domain: $\{x \mid x \neq 2\}$; Range: $\{y \mid y \neq 1\}$ **(b)** (0, 0)   **(c)** $y = 1$ **(d)** $x = 2$ **(e)** None
13. **(a)** Domain: $\{x \mid x \neq 0\}$; Range: all real numbers   **(b)** (−1, 0), (1, 0)   **(c)** None   **(d)** None   **(e)** $y = 2x$
15. **(a)** Domain: $\{x \mid x \neq -2, x \neq 2\}$; Range: $\{y \mid -\infty < y \leq 0, 1 < y < \infty\}$ **(b)** (0, 0)   **(c)** $y = 1$ **(d)** $x = -2, x = 2$ **(e)** None
17. **(a)** Domain: $\{x \mid x \neq -1\}$; Range: $\{y \mid y \neq 2\}$ **19.** **(a)** Domain: $\{x \mid x \neq -3, x \neq 3\}$; Range: All real numbers
    (b) (−1.5, 0), (0, 3)   **(b)** (0, 0)
    (c) $y = 2$ **(c)** $y = 0$
    (d) $x = -1$ **(d)** $x = -3; x = 3$
    (e) None   **(e)** None

21. 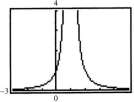   **23.** 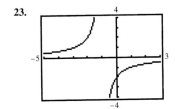   **25.**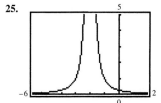

27. 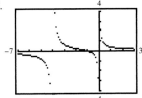   **29.**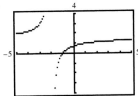

31. Horizontal asymptote: $y = 3$; vertical asymptote: $x = -4$ **33.** No asymptotes
35. Horizontal asymptote: $y = 0$; vertical asymptotes: $x = 1, x = -1$ **37.** Horizontal asymptote: $y = 0$; vertical asymptote: $x = 0$
39. Oblique asymptote: $y = 3x$; vertical asymptote: $x = 0$
41. 1. x-intercept: −1; no y-intercept   **43.** 1. x-intercept: −1; y-intercept: $\frac{3}{4}$
    2. No symmetry   2. No symmetry
    3. Vertical asymptotes: $x = 0, x = -4$ 3. Vertical asymptote: $x = -2$
    4. Horizontal asymptote: $y = 0$, intersected at (−1, 0)   4. Horizontal asymptote: $y = \frac{3}{2}$, not intersected
    5. 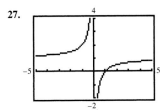   5.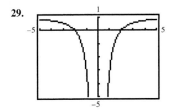

45. 1. No x-intercept; y-intercept; $-\frac{3}{4}$ **47.** 1. No x-intercept; y-intercept: −1
    2. Symmetric with respect to y-axis   2. Symmetric with respect to y-axis
    3. Vertical asymptotes: $x = 2, x = -2$ 3. Vertical asymptotes: $x = -1, x = 1$
    4. Horizontal asymptote: $y = 0$, not intersected   4. No horizontal or oblique asymptotes
    5. 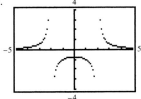   5.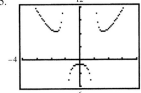

49. 1. x-intercept: 1; y-intercept: $\frac{1}{9}$
2. No symmetry
3. Vertical asymptotes: $x = 3$, $x = -3$
4. Oblique asymptote: $y = x$, intersected at $(\frac{1}{9}, \frac{1}{9})$
5.

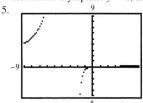

51. 1. Intercept $(0, 0)$
2. No symmetry
3. Vertical asymptotes: $x = 2$, $x = -3$
4. Horizontal asymptote: $y = 1$, intersected at $(6, 1)$
5.

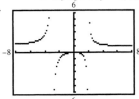

53. 1. Intercept $(0, 0)$
2. Symmetry with respect to origin
3. Vertical asymptotes: $x = -2$, $x = 2$
4. Horizontal asymptote: $y = 0$, intersected at $(0, 0)$
5.

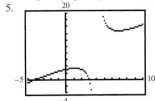

55. 1. No x-intercept; y-intercept: $\frac{3}{4}$
2. No symmetry
3. Vertical asymptotes: $x = -2$, $x = 1$, $x = 2$
4. Horizontal asymptote: $y = 0$, not intersected
5.

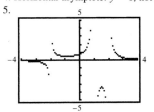

57. 1. x-intercepts: -1, 1; y-intercept: $\frac{1}{4}$
2. Symmetric with respect to y-axis
3. Vertical asymptotes: $x = -2$, $x = 2$
4. Horizontal asymptote: $y = 0$, intersected at $(-1, 0)$ and $(1, 0)$
5.

59. 1. x-intercepts: -1, 4; y-intercept: -2
2. No symmetry
3. Vertical asymptote: $x = -2$
4. Oblique asymptote: $y = x - 5$, not intersected
5.

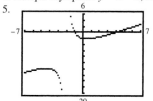

61. 1. x-intercepts: -4, 3; y-intercept: 3
2. No symmetry
3. Vertical asymptote: $x = 4$
4. Oblique asymptote: $y = x + 5$, not intersected
5.

63. 1. x-intercepts: -4, 3; y-intercept: -6
2. No symmetry
3. Vertical asymptote: $x = -2$
4. Oblique asymptote: $y = x - 1$, not intersected
5.

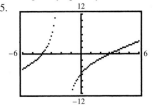

65. 1. x-intercepts: 0, 1; y-intercept: 0
2. No symmetry
3. Vertical asymptote: $x = -3$
4. Horizontal asymptote: $y = 1$, not intersected
5.

67. 1. x-intercepts: -0.84; y-intercept: -2
2. No symmetry
3. Vertical asymptote: $x = 0.74$
4. Horizontal asymptote: $y = 2$, not intersected
5.

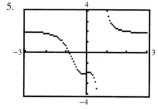

69. 1. x-intercepts: 0.65; y-intercept: -3
2. No symmetry
3. Vertical asymptotes: $x = -4.41$; $x = -0.71$; $x = 0.71$; $x = 4.41$
4. Horizontal asymptote: $y = 0$, intersected at $(0.65, 0)$
5.

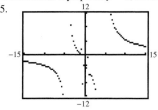

71. 1. x-intercepts: none; y-intercept: 1
2. No symmetry
3. Vertical asymptotes: $x = -5.75$; $x = 0.31$; $x = 5.43$
4. Oblique asymptotes: $y = 5x - 10$, intersected at $(0.35, -8.21)$
5.

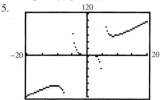

73. 1. x-intercept: -4; y-intercept: 2
2. No symmetry
3. Vertical asymptote: $x = -2$
4. Horizontal asymptote: $y = 1$, not intersected
5.

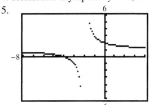

75. 1. x-intercept: $-\frac{1}{3}$; y-intercept: $-\frac{1}{2}$
2. No symmetry
3. Vertical asymptote: $x = 2$
4. Horizontal asymptote: $y = 3$, not intersected
5.

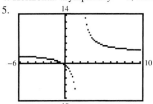

77. 1. x-intercept: $-1, 2$; y-intercept: $\frac{2}{5}$
2. No symmetry
3. Vertical asymptote: $x = -5$; $x = 1$
4. Horizontal asymptote: $y = 1$, intersected at $(\frac{3}{5}, 1)$
5.

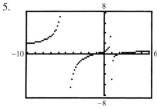

79. 1. No x-intercept; No y-intercept
2. Symmetric about the origin
3. Vertical asymptote: $x = 0$
4. Oblique asymptote: $y = x$, not intersected
5.

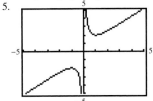

81. 1. x-intercept: -1; no y-intercept
2. No symmetry
3. Vertical asymptote: $x = 0$
4. None
5.

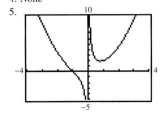

83. 1. No x-intercept; no y-intercept
2. Symmetric about the origin
3. Vertical asymptote: $x = 0$
4. Oblique asymptote: $y = x$, not intersected
5.

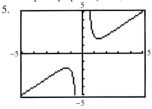

85. 4 must be a zero of the denominator; hence, $x - 4$ must be a factor.   **87.** c, d

Exercise 4.4

1. No; $f(2) = 8$ **3.** Yes; $f(2) = 0$ **5.** Yes; $f(-3) = 0$ **7.** No; $f(-4) = 1$ **9.** Yes; $f(\frac{1}{2}) = 0$ **11.** 7; 3 or 1 positive; 2 or 0 negative
13. 6; 2 or 0 positive; 2 or 0 negative   **15.** 3; 2 or 0 positive; 1 negative   **17.** 4; 2 or 0 positive; 2 or 0 negative   **19.** 5; 0 positive; 3 or 1 negative
21. 6; 1 positive; 1 negative   **23.** $\pm 1, \pm \frac{1}{3}$ **25.** $\pm 1, \pm 3$ **27.** $\pm 1, \pm 2, \pm \frac{1}{4}, \pm \frac{1}{2}$ **29.** $\pm \frac{1}{3}, \pm \frac{2}{3}, \pm 1, \pm 2$ **31.** $\pm \frac{1}{2}, \pm 1, \pm 2, \pm 4$
33. $\pm \frac{1}{6}, \pm \frac{1}{3}, \pm \frac{1}{2}, \pm \frac{2}{3}, \pm 1, \pm 2$ **35.** -1 and 1   **37.** -3 and 7   **39.** -5 and 2   **41.** $-3, -1, 2; f(x) = (x + 3)(x + 1)(x - 2)$
43. $\frac{1}{2}; f(x) = 2(x - \frac{1}{2})(x^2 + 1)$ **45.** $-1, 1; f(x) = (x + 1)(x - 1)(x^2 + 2)$ **47.** $-\frac{1}{2}, \frac{1}{2}; f(x) = 4(x + \frac{1}{2})(x - \frac{1}{2})(x^2 + 2)$
49. $-2, -1, 1, 1; f(x) = (x + 2)(x + 1)(x - 1)^2$ **51.** $-\sqrt{2}/2, \sqrt{2}/2, 2; f(x) = 4(x + \sqrt{2}/2)(x - \sqrt{2}/2)(x - 2)(x^2 + \frac{1}{2})$ **53.** $-5.90, -0.30, 3.00$

55. $-3.80, 4.50$ **57.** $-3.42, 0.31, 12.30$ **59.** $-43.50, 1.00, 23.00$ **61.** $\{-1, 2\}$ **63.** $\{\frac{2}{3}, -1 + \sqrt{2}, -1 - \sqrt{2}\}$ **65.** $\{\frac{1}{3}, \sqrt{5}, -\sqrt{5}\}$
67. $\{-3, -2\}$ **69.** $-\frac{1}{3}$ **71.** $k = 5$ **73.** -7 **75.** If $f(x) - x^n - c^n$, then $f(c) = c^n - c^n = 0$
77. No (use the Rational Zeros Theorem)   **79.** No (use the Rational Zeros Theorem)   **81.** 7 in.
83. All the potential rational zeros are integers. Hence, r either is an integer or is not a rational root (and is therefore irrational).

85.

$$y^3 + by^2 + cy + d = 0$$

$$\left(x - \frac{b}{3}\right)^3 + b\left(x - \frac{b}{3}\right)^2 + c\left(x - \frac{b}{3}\right) + d = 0$$

$$x^3 - \frac{3b}{3}x^2 + 3\left(\frac{b^2}{9}\right)x - \frac{b^3}{27} + b\left(x^2 - \frac{2bx}{3} + \frac{b^2}{9}\right) + cx - \frac{bc}{3} + d = 0$$

$$x^3 - \frac{b^2}{3}x + cx - \frac{b^3}{27} + \frac{b^3}{9} - \frac{bc}{3} + d = 0$$

$$x^3 + \left(c - \frac{b^2}{3}\right)x + \left(\frac{2b^3}{27} - \frac{bc}{3} + d\right) = 0$$

87. $K = \dfrac{-p}{3H}$

$$H^3 + \left(\frac{-p}{3H}\right)^3 = -q$$

$$H^6 + qH^3 - \frac{p^3}{27} = 0$$

$$H^3 = \frac{-q \pm \sqrt{q^2 + \dfrac{4p^3}{27}}}{2} \quad \text{Choose + sign}$$

$$H = \sqrt[3]{\frac{-q}{2} + \sqrt{\frac{q^2}{4} + \frac{p^3}{27}}}$$

89. $x = H + K$; now use results from Problems 87 and 88   **91.** $p = 3, q = -14$; $x = \sqrt[3]{7 + 5\sqrt{2}} + \sqrt[3]{7 - 5\sqrt{2}}$

93. $p = -6, q = 4$; $x = \sqrt[3]{-2 + \sqrt{4 - 8}} + \sqrt[3]{-2 - \sqrt{4 - 8}} = \sqrt[3]{-2 + 2i} + \sqrt[3]{-2 - 2i}$; or $x^3 - 6x + 4 = (x - 2)(x^2 + 2x - 2) = 0$; $x = 2$;

$$x = \frac{-2 \pm \sqrt{4 + 8}}{2} = -1 \pm \sqrt{3}$$

Exercise 4.5

1. $4 + i$ **3.** $-i, 1 - i$ **5.** $-i, -2i$ **7.** $-i$ **9.** $2 - i, -3 + i$
11. Zeros that are complex numbers must occur in conjugate pairs; or a polynomial with real coefficients of odd degree must have at least one real zero.
13. If the remaining zero were a complex number, then its conjugate would also be a zero.
15. $1, -\frac{1}{2} + \frac{\sqrt{3}}{2}i, -\frac{1}{2} - \frac{\sqrt{3}}{2}i$ **17.** $-4 + i$ **19.** $-1 + 5i$ **21.** $-4 + 4i$ **23.** $-18 - 16i$ **25.** $16 - 18i$ **27.** $38 + 31i$
29. $z^3 + (-11 - 2i)z^2 + (40 + 16i)z - 48 - 32i$ **31.** $z^3 - 3z^2 + (3 - i)z - 2 + 2i$
33. $z^4 + (2i - 6)z^3 + (8 - 12i)z^2 + (6 + 18i)z - 9$

Fill-in-the-Blank Items

1. parabola; vertex   **2.** Remainder; Dividend   **3.** $f(c)$ **4.** $f(c) = 0$ **5.** zero   **6.** three; one; two; no   **7.** $\pm 1, \pm\frac{1}{2}$ **8.** $y = 1$
9. $x = -1$ **10.** $3 - 4i$

True/False Items

1. F   **2.** F   **3.** T   **4.** T   **5.** T

Review Exercises

1.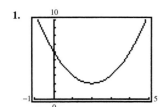

Opens up; Vertex: $(2, 2)$
Axis of symmetry: $x = 2$
Intercept: $(0, 6)$

3.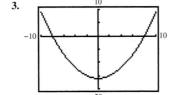

Opens up; Vertex: $(0, -16)$
Axis of symmetry: $x = 0$
Intercept: $(-8, 0), (8, 0), (0, -16)$

5.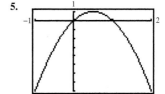

Opens down; Vertex: $(0.50, 1)$
Axis of symmetry: $x = 0.5$
Intercepts: $(0, 0), (1, 0)$

7.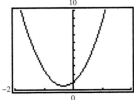

Opens up; Vertex: $(-0.33, 0.5)$
Axis of symmetry: $x = 0.33$
Intercepts: $(0, 1)$

9.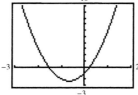

Opens up; Vertex: $(-0.66, -2.33)$
Axis of symmetry: $x = -0.66$
Intercepts: $(-1.54, 0), (0.21, 0), (0, -1)$

11.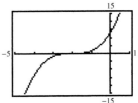

13.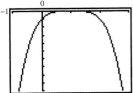

15.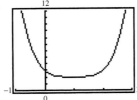

17. Minimum value; 1   **19.** Maximum value; 12   **21.** Maximum value; 16

23. (a)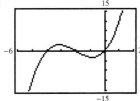

(b) x-intercepts: $-4, -2, 0$; y-intercept: 0
(c) $-4, -2, 0$: odd
(d) $y = x^3$
(e) 2
(f) Local minima: $(-0.84, -3.07)$
Local maxima: $(-3.15, 3.07)$

25. (a)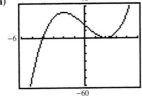

(b) x-intercepts: $-4, 2$; y-intercept: 16
(c) -4: odd; 2 even
(d) $y = x^3$
(e) 2
(f) Local minima: $(2, 0)$
Local maxima: $(-2, 32)$

27. $f(x) = x^3 - 4x^2 = x^2(x - 4)$
(a)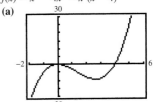

(b) x-intercepts: $0, 4$; y-intercept: 0
(c) 0: even; 4 odd
(d) $y = x^3$
(e) 2
(f) Local minima: $(2.66, -9.48)$
Local maxima: $(0.00, 0.00)$

29. (a)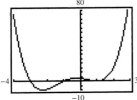

(b) x-intercepts: $-3, -1, 1$; y-intercept: 3
(c) Crosses at $-3, -1$; touches at 1
(d) $y = x^4$
(e) 3
(f) Local minima: $(-2.28, -9.91), (1.00, 0)$
Local maxima: $(-0.21, 3.22)$

31. 1. x-intercept: 3; no y-intercept
2. No symmetry
3. Vertical asymptote: $x = 0$
4. Horizontal asymptote: $y = 2$, not intersected
5.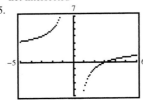

33. 1. x-intercept: -2; no y-intercept
2. No symmetry
3. Vertical asymptotes: $x = 0, x = 2$
4. Horizontal asymptote: $y = 0$, intersected at $(-2, 0)$
5.

35. 1. Intercepts: $(-3, 0)$, $(2, 0)$, $(0, 1)$
2. No symmetry
3. Vertical asymptote: $x = -2$, $x = 3$
4. Horizontal asymptote: $y = 1$, intersected at $(0, 1)$
5.

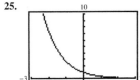

37. 1. Intercept $(0, 0)$
2. Symmetric with respect to the origin
3. Vertical asymptotes: $x = -2$, $x = 2$
4. Oblique asymptote: $y = x$, intersected at $(0, 0)$
5.

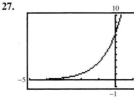

39. 1. Intercept $(0, 0)$
2. No symmetry
3. Vertical asymptote: $x = 1$
4. No oblique or horizontal asymptote
5.

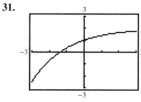

41. 4, 2, or 0 positive; 2 or 0 negative   **43.** $\pm\frac{1}{12}, \pm\frac{1}{6}, \pm\frac{1}{4}, \pm\frac{1}{3}, \pm\frac{1}{2}, \pm\frac{3}{4}, \pm 1, \pm\frac{3}{2}, \pm 3$ **45.** $-2, 1, 4$; $f(x) = (x + 2)(x - 1)(x - 4)$

47. $\frac{1}{2}$, multiplicity 2; -2; $f(x) = 4\left(x - \frac{1}{2}\right)^2(x + 2)$ **49.** 2, multiplicity 2; $f(x) = (x - 2)^2(x^2 + 5)$ **51.** $-2.50, 3.10, 5.32$

53. $-11.30, -.60, 4.00, 9.33$ **55.** $-3.67, 1.33$ **57.** $\{-3, 2\}$ **59.** $\{-3, -1, -\frac{1}{2}, 1\}$ **61.** -2 and 2   **63.** -3 and 5   **65.** 1.52   **67.** 0.93

69. $4 - i$ **71.** $-i, 1 - i$ **73.** $f(z) = z^4 - (5 + i)z^3 + (7 + 5i)z^2 - (3 + 7i)z + 3i$ **75.** $f(z) = z^3 - (6 + i)z^2 + (11 + 5i)z - 6 - 6i$

77. $q(x) = x^2 + 5x + 6$; $R = 0$ **79.** $\{-3, 2\}$ **81.** $\{\frac{1}{3}, 1, -i, i\}$ **83.** 1 is an upper bound; -2 is a lower bound   **85.** $(2, 2)$ **87.** 3.6 ft

91. **(a)** even   **(b)** positive   **(c)** even   **(d)** 0 is a zero of even multiplicity   **(e)** 8

CHAPTER 5   *Exercise 5.1*

1. **(a)** 11.212   **(b)** 11.587   **(c)** 11.664   **(d)** 11.665   **3.** **(a)** 8.815   **(b)** 8.821   **(c)** 8.824   **(d)** 8.825

5. **(a)** 21.217   **(b)** 22.217   **(c)** 22.440   **(d)** 22.459   **7.** 3.320   **9.** 0.427   **11.** B   **13.** D   **15.** A   **17.** E   **19.** A   **21.** E   **23.** B

25.

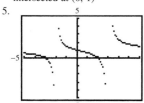

27.

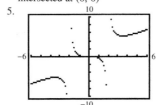

29.

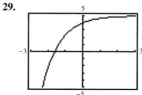

31.

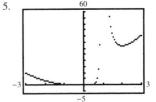

33. $\frac{1}{49}$ **35.** $\frac{1}{4}$ **37.** **(a)** 74%   **(b)** 47%   **39.** **(a)** 44 watts   **(b)** 11.6 watts   **41.** 3.35 milligrams; 0.45 milligrams

43. **(a)** 56%   **(b)** 68%   **(c)** 70%   **(d)** $R = 40\%$ just after 6 days

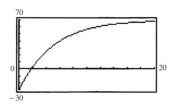

45. **(a)** 5.414 amperes, 7.5854 amperes, 10.38 amperes   **(b)** 12 amperes   **(c)** $I_1(t) = 12(1 - e^{-2t})$ **(d)** 3.343 amperes, 5.309 amperes, 9.443 amperes   **(e)** 24 amperes   **(f)** $I_2(t) = 24(1 - e^{-1/2t})$

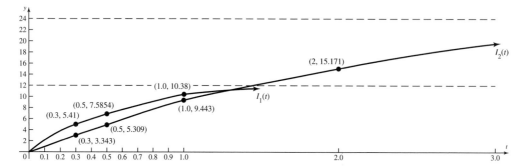

47. (a) 9.23×10^{-3} **(b)** 0.81 **(c)** 5 **(d)** 57.91°, 43.98°, 30.06°

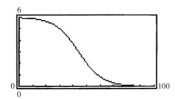

49. $n = 4$: 2.7083; $n = 6$: 2.7181; $n = 8$: 2.7182788; $n = 10$: 2.7182818

51. $\dfrac{f(x+h) - f(x)}{h} = \dfrac{a^{x+h} - a^x}{h} = \dfrac{a^x a^h - a^x}{h} = \dfrac{a^x(a^h - 1)}{h}$ **53.** $f(-x) = a^{-x} = \dfrac{1}{a^x} = \dfrac{1}{f(x)}$

55. (a) $\sinh(-x) = \frac{1}{2}(e^{-x} - e^x)$
$\qquad\qquad = -\frac{1}{2}(e^x - e^{-x})$
$\qquad\qquad = -\sinh x$ **(b)**

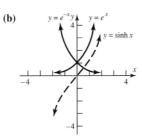

57. $f(1) = 5, f(2) = 17, f(3) = 257, f(4) = 65{,}537;$
$f(5) = 4{,}294{,}967{,}297 = 641 \times 6{,}700{,}417$

Exercise 5.2

1. $2 = \log_3 9$ **3.** $2 = \log_a 1.6$ **5.** $2 = \log_{1.1} M$ **7.** $x = \log_2 7.2$ **9.** $\sqrt{2} = \log_x \pi$ **11.** $x = \ln 8$ **13.** $2^3 = 8$ **15.** $a^6 = 3$ **17.** $3^x = 2$
19. $2^{1.3} = M$ **21.** $(\sqrt{2})^x = \pi$ **23.** $e^x = 4$ **25.** 0   **27.** 2   **29.** -4 **31.** $\frac{1}{2}$ **33.** 4   **35.** $\frac{1}{2}$ **37.** $\{x \mid x < 3\}$ **39.** All real numbers except 0
41. $\{x \mid x < -2 \text{ or } x > 3\}$ **43.** $\{x \mid x > 0, x \neq 1\}$ **45.** $\{x \mid x < -1 \text{ or } x > 0\}$ **47.** 0.511   **49.** 30.099   **51.** $\sqrt{2}$ **53.** B   **55.** D   **57.** A   **59.** E
61. C   **63.** A   **65.** D
67. 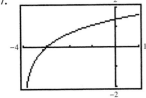   **69.** 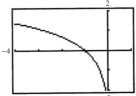   **71.**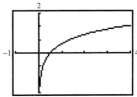

73. 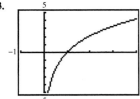   **75.**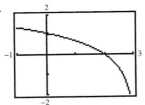

77. (a) $n = 6.93$ so 7 panes are necessary   **(b)** $n = 13.86$ so 14 panes are necessary
79. (a) $d = 127.7$ so it takes about 128 days   **(b)** $d = 575.6$ so it takes about 576 days   **81.** $h = 2.29$ so the time between injections is $2\text{-}2\frac{1}{2}$ hours
83. 0.2695 sec; 0.8959 sec   **85. (a)** $k = 20.07$ **(b)** 91% **(c)** 0.175 **(d)** 0.08   **87.** $y = 20\,e^{0.023t}$; $y = 89.2$ is predicted

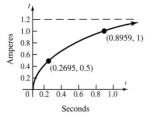

Exercise 5.3

1. $a + b$ **3.** $b - a$ **5.** $a + 1$ **7.** $2a + b$ **9.** $\frac{1}{5}(a + 2b)$ **11.** $\dfrac{b}{a}$ **13.** $2 \ln x + \frac{1}{2} \ln(1 - x)$ **15.** $3 \log_2 x - \log_2(x - 3)$

17. $\log x + \log(x + 2) - 2 \log(x + 3)$ **19.** $\frac{1}{3} \ln(x - 2) + \frac{1}{3} \ln(x + 1) - \frac{2}{3} \ln(x + 4)$ **21.** $\ln 5 + \ln x + \frac{1}{2} \ln(1 - 3x) - 3 \ln(x - 4)$ **23.** $\log_5 u^3 v^4$

25. $-\frac{5}{2} \log_{1/2} x$ **27.** $-2 \ln(x - 1)$ **29.** $\log_2[x(3x - 2)^4]$ **31.** $\log_a\left[\dfrac{25x^6}{(2x + 3)^{1/2}}\right]$ **33.** 2.771   **35.** -3.880 **37.** 5.615   **39.** 0.874

41. 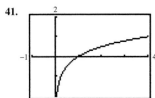   **43.** 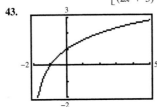   **45.**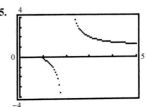

47. $\log_a(x + \sqrt{x^2 - 1}) + \log_a(x - \sqrt{x^2 - 1}) = \log_a[(x + \sqrt{x^2 - 1})(x - \sqrt{x^2 - 1})] = \log_a[x^2 - (x^2 - 1)] = \log_a 1 = 0$

49. $\ln (1 + e^{2x}) = \ln[e^{2x}(e^{-2x} + 1)] = \ln e^{2x} + \ln (e^{-2x} + 1) = 2x + \ln(1 + e^{-2x})$

51. $y = f(x) = \log_a x$; $a^y = x$; $\left(\dfrac{1}{a}\right)^{-y} = x$; $-y = \log_{1/a} x$; $-f(x) = \log_{1/a} x$ **53.** $f(AB) = \log_a AB = \log_a A + \log_a B = f(A) + f(B)$

55. $y = Cx$ **57.** $y = Cx(x + 1)$ **59.** $y = Ce^{3x}$ **61.** $y = Ce^{-4x} + 3$ **63.** $y = \dfrac{\sqrt[3]{C}(2x + 1)^{1/6}}{(x + 4)^{1/9}}$ **65.** 3   **67.** 1

69. If $A = \log_a M$ and $B = \log_a N$, then $a^A = M$ and $a^B = N$. Then $\log_a (M/N) = \log_a (a^A/a^B) = \log_a a^{A-B} = A - B = \log_a M - \log_a N$.

Exercise 5.4

1. $\frac{7}{2}$ **3.** $\{-2\sqrt{2}, 2\sqrt{2}\}$ **5.** 16   **7.** 8   **9.** 3   **11.** 5   **13.** 2   **15.** $\{-2, 4\}$ **17.** 21   **19.** $\frac{1}{2}$ **21.** $\{-\sqrt{2}, 0, \sqrt{2}\}$

23. $\left\{1 - \dfrac{\sqrt{6}}{3}, 1 + \dfrac{\sqrt{6}}{3}\right\}$ **25.** 0   **27.** 2   **29.** 0   **31.** $\frac{3}{2}$ **33.** 3.322   **35.** -0.088 **37.** 0.307   **39.** 1.356   **41.** 0

43. 0.534   **45.** 0.226   **47.** 2.027   **49.** $\frac{9}{2}$ **51.** 2   **53.** -1 **55.** 1   **57.** 16   **59.** 1.92   **61.** 2.78   **63.** -0.56 **65.** -0.70
67. 0.56   **69.** $\{0.39, 1.00\}$ **71.** 1.31   **73.** 1.30

Exercise 5.5

1. $108.29   **3.** $609.50   **5.** $697.09   **7.** $12.46   **9.** $125.23   **11.** $88.72   **13.** $860.72   **15.** $554.09   **17.** $59.71   **19.** $361.93

21. 5.35%   **23.** 26%   **25.** $6\frac{1}{4}$% compounded annually   **27.** 9% compounded monthly   **29.** 104.32 mo; 103.97 mo

31. 61.02 mo; 60.82 mo   **33.** 15.27 yrs or 15 yrs, 4 months   **35.** $104,335   **37.** $12,910.62   **39.** About $30.17 per share or $3017

41. 9.35%   **43.** Not quite. You will have $1057.60. The second bank gives a better deal, since you have $1060.62 after 1 year

45. You have $11,632.73; your friend has $10, 947.89

47. (a) Interest is $30,000   (b) Interest is $38,613.59   (c) Interest is $37,752.73. Simple interest at 12% is best

49. (a) $1364.62   (b) $1353.35   **51.** $4631.93

59. (a) 6.1 years   (b) 18.45 yrs   (c) $mP = P\left(1 + \dfrac{r}{n}\right)^{nt}$

$$m = \left(1 + \dfrac{r}{n}\right)^{nt}$$

$$\ln m = \ln \left(1 + \dfrac{r}{n}\right)^{nt} = nt \ln \left(1 + \dfrac{r}{n}\right)$$

$$t = \dfrac{\ln m}{n \ln \left(1 + \frac{r}{n}\right)}$$

Exercise 5.6

1. (a) 34.7 days; 69.3 days   **(b)** 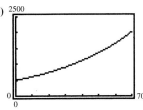   **3.** 28.4 yr   **5. (a)** 94.4 yr   **(b)**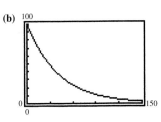

7. 5832; 3.9 days   **9.** 25,198   **11.** 9.797 g   **13. (a)** 9727 years ago   **(b)** 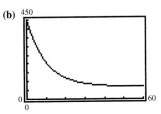   **(c)** 5600 yr

15. (a) 5:18 P.M.   **(b)**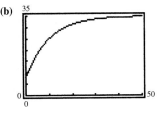

(c) After 14.3 minutes, the pizza will be 160°F.
(d) As time passes, the temperature of the pizza gets closer to 70°F.

17. (a) 18.63°C; 25.1°C   **(b)** 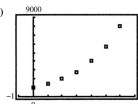   **19.** 7.34 kg; 76.6 hr   **21.** 26.5 days.

Exercise 5.7

1. (a)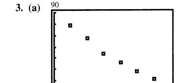

(b) Population = 1000.187781 (1.41414215)^t
(c) $N = 1000.187781e^{0.3465230928t}$
(d) 11,312

3. (a)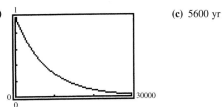

(b) $P = 96.0100692 (0.981490683)^Q$
(c) $Q = 25$

5. (a)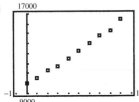

(b) $y = 100.3262508 (0.8768651017)^t$

(c) $N = 100.3262508e^{-0.1314021163t}$

(d) 5.3 weeks

(e) 0.14 grams

7. (a)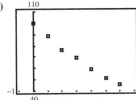

(b) Value $= 10014.4963 (1.056554737)^t$

(c) 5.655%

(d) \$68,682.99

9. (a)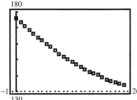

(b) $T = 173.2867183 (0.9889878268)^t$

(c) 41.04

11. (a)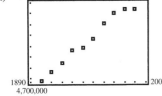

(b) $P = -1163825970 + 154808355.6 \ln t$

(c) 12,469,735 people

13. (a)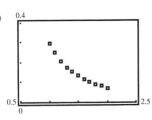

(b) $I = 0.2995056364x^{-2.012000304}$

(c) Close; -2.01 versus -2

(d) 0.056054

15. (a)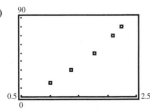

(b) $s = 15.97274236t^{2.001820627}$

(c) $s = \frac{1}{2} \cdot 31.94548472t^{2.001820627}$
$g \approx 31.9455$ feet/sec^2

(d) $t \approx 2.5$ sec

17. (a)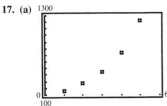

(b) Exponential model:
$y = 99.06645375 (1.652917358)^x$

19. (a)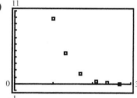

(b) Logarithmic model:
$y = -16.55688464 + 10.31473128 \ln x$

21. (a)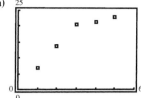

(b) Exponential model:
$y = 346.113717 \, (0.0352268374)^x$

23. (a)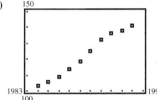

(b) Logarithmic model:
$CPI = -69946.44499 + 9225.465596 \ln t$
(c) 147.7

25. (a)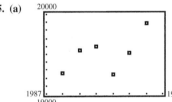

(b) Linear model:
$GDP = 77.45714286t - 134670.9429$
(c) \$19,779

27. (a)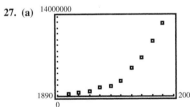

(b) Exponential model:
$P = 2.829551 \times 10^{-25} \, (1.037329901)^t$
(c) 16,066,684

Exercise 5.8

1. 70 decibels **3.** 111.76 decibels **5.** 10 W/m² **7.** 4.0 on the Richter scale
9. 70,794.58 mm; the San Francisco earthquake was 11.22 times as intense as the one in Mexico City.

Fill-in-the-Blank Items

1. (0, 1) and (1, a) **2.** 1 **3.** 4 **4.** sum **5.** 1 **6.** 7 **7.** $x > 0$ **8.** (1, 0) and (a, 1) **9.** 1 **10.** 7

True/False Items

1. T **2.** T **3.** F **4.** F **5.** T **6.** F **7.** F **8.** T

Review Exercises

1. -3 **3.** $\sqrt{2}$ **5.** 0.4 **7.** $\frac{25}{4}\log_4 x$ **9.** $\ln\left[\dfrac{1}{(x+1)^2}\right] = -2\ln(x+1)$ **11.** $\log\left(\dfrac{4x^3}{[(x+3)(x-2)]^{1/2}}\right)$ **13.** $y = Ce^{2x^2}$

15. $y = (Ce^{3x^2})^2$ **17.** $y = \sqrt{e^{x+C} + 9}$ **19.** $y = \ln(x^2+4) - C$

21. 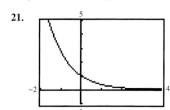 **23.** 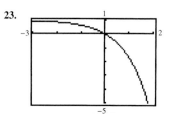 **25.**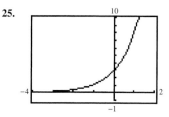

27. 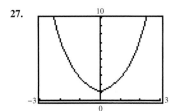 **29.**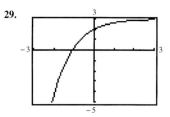

31. $\frac{1}{4}$ **33.** $\left\{\dfrac{-1-\sqrt{3}}{2}, \dfrac{-1+\sqrt{3}}{2}\right\}$ **35.** $\frac{1}{4}$ **37.** 4.301 **39.** $\frac{12}{5}$ **41.** 83 **43.** $\left\{-3, \frac{1}{2}\right\}$ **45.** -1 **47.** -0.609 **49.** -9.327

51. 3229.5 m **53.** 7.6 mm of mercury **55. (a)** 37.3 watts **(b)** 6.9 decibels

57. (a) 71% **(b)** 85.5% **(c)** 90% **(d)** About 1.6 months **(e)** About 4.8 months

59. (a) 9.85 years **(b)** 4.27 years **61.** \$41,669 **63.** 80 decibels **65.** $24,203$ years ago

67. 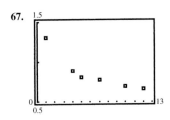 **(b)** Power model:
$$y = 1.298985429x^{-0.2721702954}$$

69. (a) 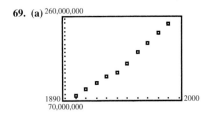 **(b)** Power model:
$$P = 9.545211 \times 10^{-76}t^{25.29069649}$$
(c) $273,930,160$ people

1. B **3.** E **5.** H **7.** C **9.** F **11.** G **13.** D **15.** B

17. $y^2 = 16x$

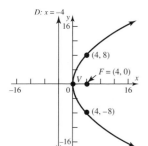

19. $x^2 = -12y$

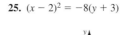
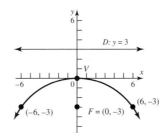

21. $y^2 = -8x$

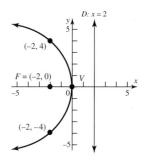

23. $x^2 = 2y$

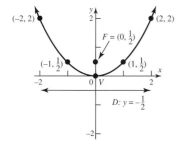

25. $(x - 2)^2 = -8(y + 3)$

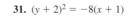
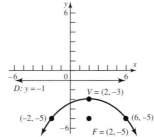

27. $x^2 = \frac{4}{3}y$

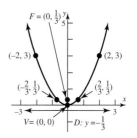

29. $(x + 3)^2 = 4(y - 3)$

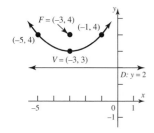

31. $(y + 2)^2 = -8(x + 1)$

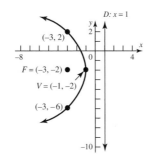

33. Vertex: $(0, 0)$; Focus: $(0, 1)$;
Directrix: $y = -1$

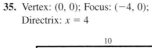

35. Vertex: $(0, 0)$; Focus: $(-4, 0)$;
Directrix: $x = 4$

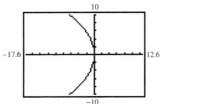

37. Vertex: $(-1, 2)$; Focus: $(1, 2)$;
Directrix: $x = -3$

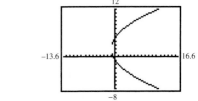

39. Vertex: $(3, -1)$; Focus: $(3, -\frac{5}{4})$;
Directrix: $y = -\frac{3}{4}$

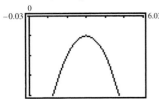

41. Vertex: $(2, -3)$; Focus: $(4, -3)$;
Directrix: $x = 0$

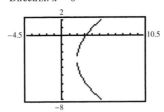

43. Vertex: $(0, 2)$; Focus: $(-1, 2)$;
Directrix: $x = 1$

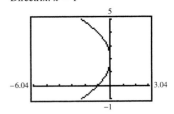

45. Vertex: $(-4, -2)$; Focus: $(-4, -1)$;
Directrix: $y = -3$

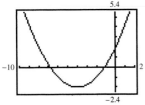

47. Vertex: $(-1, -1)$; Focus: $(-\frac{3}{4}, -1)$;
Directrix: $x = -\frac{5}{4}$

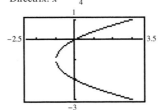

49. Vertex: $(2, -8)$; Focus: $(2, -\frac{31}{4})$;
Directrix: $y = -\frac{33}{4}$

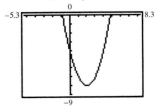

51. $(y - 1)^2 = x$ **53.** $(y - 1)^2 = -(x - 2)$ **55.** $x^2 = 4(y - 1)$ **57.** $y^2 = \frac{1}{2}(x + 2)$
59. 1.5625 ft from the base of the dish, along the axis of symmetry   **61.** 1 in from the vertex   **63.** 20 ft
65. 0.78125 ft   **67.** 4.17 ft from the base along the axis of symmetry   **69.** 24.31 ft, 18.75 ft, 7.64 ft

71. $Ax^2 + Ey = 0$

$$x^2 = -\frac{E}{A}y$$

This is the equation of a parabola with vertex at $(0, 0)$ and axis of symmetry the y-axis. The focus is at $(0, -E/4A)$; the directrix is the line $y = E/4A$. The parabola opens up if $-E/A > 0$ and down if $-E/A < 0$.

73. $Ax^2 + Dx + Ey + F = 0, A \neq 0$

$$Ax^2 + Dx = -Ey - F$$
$$x^2 + \frac{D}{A}x = -\frac{E}{A}y - \frac{F}{A}$$
$$\left(x + \frac{D}{2A}\right)^2 = -\frac{E}{A}y - \frac{F}{A} + \frac{D^2}{4A^2}$$
$$\left(x + \frac{D}{2A}\right)^2 = -\frac{E}{A}y + \frac{D^2 - 4AF}{4A^2}$$

(a) If $E \neq 0$, then the equation may be written as
$$\left(x + \frac{D}{2A}\right)^2 = -\frac{E}{A}\left(y - \frac{D^2 - 4AF}{4AE}\right)$$
This is the equation of a parabola with vertex at $(-D/2A, (D^2 - 4AF)/(4AE))$ and axis of symmetry parallel to the y-axis.

(b)–(d) If $E = 0$, the graph of the equation contains no points if $D^2 - 4AF < 0$, is a single vertical line if $D^2 - 4AF = 0$, and is two vertical lines if $D^2 - 4AF > 0$.

Exercise 6.3

1. C **3.** B **5.** C **7.** D

9. Vertices: $(-5, 0)$, $(5, 0)$
Foci: $(-\sqrt{21}, 0)$, $(\sqrt{21}, 0)$

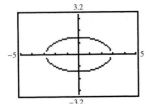

11. Vertices: $(0, -5)$, $(0, 5)$
Foci: $(0, -4)$, $(0, 4)$

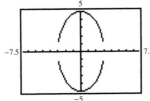

13. Vertices: $(0, -4)$, $(0, 4)$
Foci: $(0, -2\sqrt{3})$, $(0, 2\sqrt{3})$

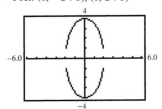

15. Vertices: $(-2\sqrt{2}, 0)$, $(2\sqrt{2}, 0)$; Foci: $(-\sqrt{6}, 0)$, $(\sqrt{6}, 0)$

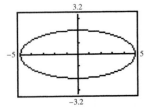

17. Vertices: $(-4, 0)$, $(4, 0)$, $(0, -4)$, $(0,4)$; Focus: $(0, 0)$

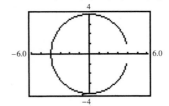

19. $\dfrac{x^2}{25} + \dfrac{y^2}{16} = 1$

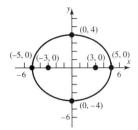

21. $\dfrac{x^2}{9} + \dfrac{y^2}{25} = 1$

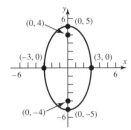

23. $\dfrac{x^2}{9} + \dfrac{y^2}{5} = 1$

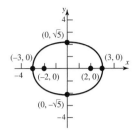

25. $\dfrac{x^2}{4} + \dfrac{y^2}{13} = 1$

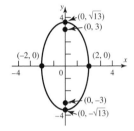

27. $x^2 + \dfrac{y^2}{16} = 1$

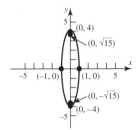

29. $\dfrac{(x+1)^2}{4} + (y-1)^2 = 1$

31. $(x-1)^2 + \dfrac{y^2}{4} = 1$

33. Center: $(3, -1)$; Vertices: $(3, -4)$, $(3, 2)$
Foci: $(3, -1 - \sqrt{5})$, $(3, -1 + \sqrt{5})$

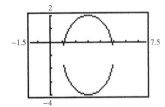

35. Center: $(-5, 4)$; Vertices: $(-9, 4)$, $(-1, 4)$
Foci: $(-5 - 2\sqrt{3}, 4)$, $(-5 + 2\sqrt{3}, 4)$

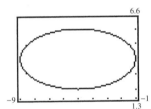

37. Center: $(-2, 1)$; Vertices: $(-4, 1)$, $(0, 1)$
Foci: $(-2 - \sqrt{3}, 1)$, $(-2 + \sqrt{3}, 1)$

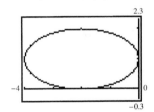

39. Center: $(2, -1)$; Vertices: $(2 - \sqrt{3}, -1)$,
$(2 + \sqrt{3}, -1)$; Foci: $(1, -1)$, $(3, -1)$

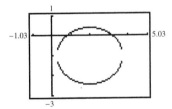

41. Center: $(1, -2)$; Vertices: $(1, -5)$, $(1, 1)$
Foci: $(1, -2 - \sqrt{5})$, $(1, -2 + \sqrt{5})$

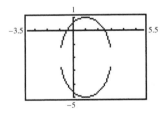

43. Center: $(0, -2)$; Vertices: $(0, -4)$, $(0, 0)$
Foci: $(0, -2 - \sqrt{3})$, $(0, -2 + \sqrt{3})$

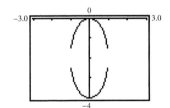

45. $\dfrac{(x-2)^2}{25} + \dfrac{(y+2)^2}{21} = 1$

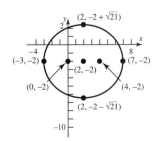

47. $\dfrac{(x-4)^2}{5} + \dfrac{(y-6)^2}{9} = 1$

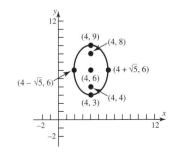

49. $\dfrac{(x-2)^2}{16} + \dfrac{(y-1)^2}{7} = 1$

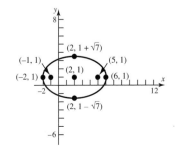

51. $\dfrac{(x-1)^2}{10} + (y-2)^2 = 1$

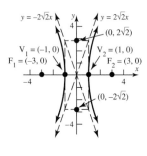

53. $\dfrac{(x-1)^2}{9} + (y-2)^2 = 1$

55.

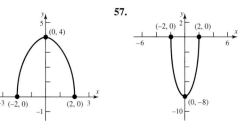

57.

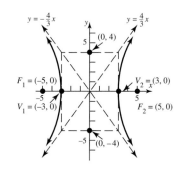

59. $\dfrac{x^2}{100} + \dfrac{y^2}{36} = 1$ **61.** 43.3 ft   **63.** 24.65 ft, 21.65 ft, 13.82 ft   **65.** 0 ft, 12.99 ft, 15 ft, 12.99 ft, 0 ft

67. 91.5 million miles; $\dfrac{x^2}{(93)^2} + \dfrac{y^2}{8646.75} = 1$ **69.** perihelion: 460.6 million miles; mean distance: 483.8 million miles; $\dfrac{x^2}{(483.8)^2} + \dfrac{y^2}{233524} = 1$

71. 30 ft

73. (a) $Ax^2 + Cy^2 + F = 0$
$\qquad Ax^2 + Cy^2 = -F$

If A and C are of the same sign and F is of opposite sign, then the equation takes the form $x^2/(-F/A) + y^2/(-F/C) = 1$, where $-F/A$ and $-F/C$ are positive. This is the equation of an ellipse with center at $(0, 0)$.

(b) If $A = C$, the equation may be written as $x^2 + y^2 = -F/A$. This is the equation of a circle with center at $(0, 0)$ and radius equal to $\sqrt{-F/A}$.

Exercise 6.4

1. B   **3.** A   **5.** B   **7.** C

9. $x^2 - \dfrac{y^2}{8} = 1$

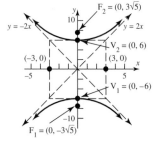

11. $\dfrac{y^2}{16} - \dfrac{x^2}{20} = 1$

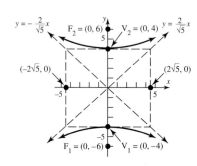

13. $\dfrac{x^2}{9} - \dfrac{y^2}{16} = 1$

15. $\dfrac{y^2}{36} - \dfrac{x^2}{9} = 1$

17. $\dfrac{x^2}{8} - \dfrac{y^2}{8} = 1$

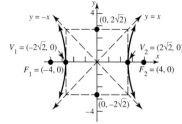

19. Center: $(0, 0)$
Transverse axis: x-axis
Vertices: $(-4, 0)$, $(4, 0)$
Foci: $(-2\sqrt{5}, 0)$, $(2\sqrt{5}, 0)$
Asymptotes: $y = \pm\frac{1}{2}x$

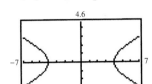

21. Center: $(0, 0)$
Transverse axis: x-axis
Vertices: $(-2, 0)$, $(2, 0)$
Foci: $(-2\sqrt{5}, 0)$, $(2\sqrt{5}, 0)$
Asymptotes: $y = \pm2x$

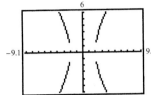

23. Center: $(0, 0)$
Transverse axis: y-axis
Vertices: $(0, -3)$, $(0, 3)$
Foci: $(0, -\sqrt{10})$, $(0, \sqrt{10})$
Asymptotes: $y = \pm3x$

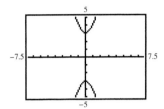

25. Center: $(0, 0)$
Transverse axis: y-axis
Vertices: $(0, -5)$, $(0, 5)$
Foci: $(0, -5\sqrt{2})$, $(0, 5\sqrt{2})$
Asymptotes: $y = \pm x$

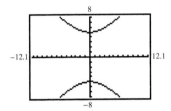

27. $x^2 - y^2 = 1$

29. $\dfrac{y^2}{36} - \dfrac{x^2}{9} = 1$

31. $\dfrac{(x-4)^2}{4} - \dfrac{(y+1)^2}{5} = 1$

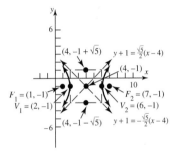

33. $\dfrac{(y+4)^2}{4} - \dfrac{(x+3)^2}{12} = 1$

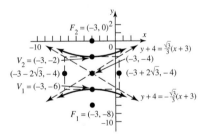

35. $(x-5)^2 - \dfrac{(y-7)^2}{3} = 1$

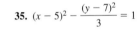

37. $\dfrac{(x-1)^2}{4} - \dfrac{(y+1)^2}{9} = 1$

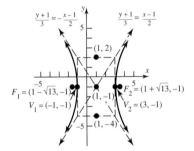

39. Center: $(2, -3)$
Transverse axis: Parallel to x-axis
Vertices: $(0, -3)$, $(4, -3)$
Foci: $(2 - \sqrt{13}, -3)$, $(2 + \sqrt{13}, -3)$
Asymptotes: $y + 3 = \pm\frac{3}{2}(x - 2)$

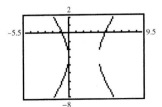

41. Center: $(-2, 2)$
Transverse axis: Parallel to y-axis
Vertices: $(-2, 0)$, $(-2, 4)$
Foci: $(-2, 2 - \sqrt{5})$, $(-2, 2 + \sqrt{5})$
Asymptotes: $y - 2 = \pm2(x + 2)$

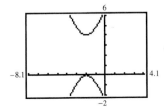

43. Center: $(-1, -2)$
Transverse axis: Parallel to x-axis
Vertices: $(-3, -2)$, $(1, -2)$
Foci: $(-1 - 2\sqrt{2}, -2)$, $(-1 + 2\sqrt{2}, -2)$
Asymptotes: $y + 2 = \pm(x + 1)$

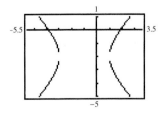

45. Center: $(1, -1)$
Transverse axis: Parallel to x-axis
Vertices: $(0, -1)$, $(2, -1)$
Foci: $(1 - \sqrt{2}, -1)$, $(1 + \sqrt{2}, -1)$
Asymptotes: $y + 1 = \pm(x - 1)$

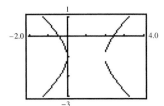

47. Center: $(-1, 2)$
Transverse axis: Parallel to y-axis
Vertices: $(-1, 0)$, $(-1, 4)$
Foci: $(-1, 2 - \sqrt{5})$, $(-1, 2 + \sqrt{5})$
Asymptotes: $y - 2 = \pm 2(x + 1)$

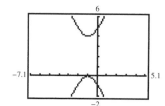

49. Center: $(3, -2)$
Transverse axis: Parallel to x-axis
Vertices: $(1, -2)$, $(5, -2)$
Foci: $(3 - 2\sqrt{5}, -2)$, $(3 + 2\sqrt{5}, -2)$
Asymptotes: $y + 2 = \pm 2(x - 3)$

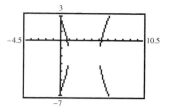

51. Center: $(-2, 1)$
Transverse axis: Parallel to y-axis
Vertices: $(-2, -1)$, $(-2, 3)$
Foci: $(-2, 1 - \sqrt{5})$, $(-2, 1 + \sqrt{5})$
Asymptotes: $y - 1 = \pm 2(x + 2)$

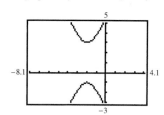

53.

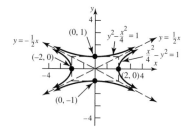

55.

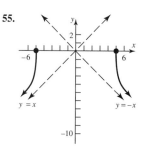

57. (a) The ship will reach shore or a point 64.66 miles from the master station.   **(b)** 0.00086 second   **(c)** (104, 50)
59. (a) 450 ft   **61.** If e is close to 1, narrow hyperbola; if e is very large, wide hyperbola

63. $\dfrac{x^2}{4} - y^2 = 1$; asymptotes $y = \pm \dfrac{1}{2}x$ $y^2 - \dfrac{x^2}{4} = 1$; asymptotes $y = \pm \dfrac{1}{2}x$

65. $Ax^2 + Cy^2 + F = 0$ If A and C are of opposite sign and $F \neq 0$, this equation may be written as $x^2/(-F/A) + y^2/(-F/C) = 1$, where
$Ax^2 + Cy^2 = -F$ $-F/A$ and $-F/C$ are opposite in sign. This is the equation of a hyperbola with center at $(0, 0)$. The transverse axis is
the x-axis if $-F/A > 0$; the transverse axis is the y-axis if $-F/A < 0$.

Exercise 6.5

1.

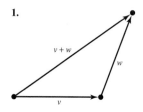

3.

5.

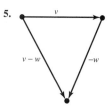

7.

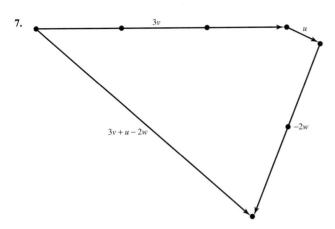

9. $x = A$ **11.** $C = -F + E - D$ **13.** $E = -G - H + D$ **15.** $x = 0$ **17.** 12 **19.** $\mathbf{v} = 3\mathbf{i} + 4\mathbf{j}$ **21.** $\mathbf{v} = 2\mathbf{i} + 4\mathbf{j}$ **23.** $\mathbf{v} = 8\mathbf{i} - \mathbf{j}$

25. $\mathbf{v} = -\mathbf{i} + \mathbf{j}$ **27.** 5 **29.** $\sqrt{2}$ **31.** $\sqrt{13}$ **33.** $-\mathbf{j}$ **35.** $\sqrt{89}$ **37.** $\sqrt{34} - \sqrt{13}$ **39.** $\mathbf{i}$ **41.** $\frac{3}{5}\mathbf{i} - \frac{4}{5}\mathbf{j}$ **43.** $\frac{\sqrt{2}}{2}\mathbf{i} - \frac{\sqrt{2}}{2}\mathbf{j}$

45. $\mathbf{v} = \dfrac{8\sqrt{5}}{5}\mathbf{i} + \dfrac{4\sqrt{5}}{5}\mathbf{j}$ or $\mathbf{v} = -\dfrac{8\sqrt{5}}{5}\mathbf{i} - \dfrac{4\sqrt{5}}{5}\mathbf{j}$ **47.** $\{-2 + \sqrt{21}, -2 - \sqrt{21}\}$ **49.** 460 kph **51.** 218 mph

53.

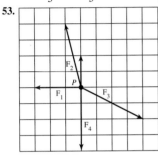

Fill-in-the-Blank Items

1. parabola **2.** ellipse **3.** hyperbola **4.** major; transverse **5.** y-axis **6.** $y/3 = x/2$; $y/3 = -x/2$ **7.** unit

True/False Items

1. T **2.** F **3.** T **4.** T **5.** T **6.** T **7.** T **8.** T

Review Exercises

1. Parabola; vertex $(0, 0)$, focus $(-4, 0)$, directrix $x = 4$
3. Hyperbola; center $(0, 0)$, vertices $(5, 0)$ and $(-5, 0)$, foci $(\sqrt{26}, 0)$ and $(-\sqrt{26}, 0)$, asymptotes $y = \frac{1}{5}x$ and $y = -\frac{1}{5}x$
5. Ellipse; center $(0, 0)$, vertices $(0, 5)$ and $(0, -5)$, foci $(0, 3)$ and $(0, -3)$
7. $x^2 = -4(y - 1)$: Parabola; vertex $(0, 1)$, focus $(0, 0)$, directrix $y = 2$
9. $\dfrac{x^2}{2} - \dfrac{y^2}{8} = 1$: Hyperbola; center $(0, 0)$, vertices $(\sqrt{2}, 0)$ and $(-\sqrt{2}, 0)$, foci $(\sqrt{10}, 0)$ and $(-\sqrt{10}, 0)$, asymptotes $y = 2x$ and $y = -2x$
11. $(x - 2)^2 = 2(y + 2)$: Parabola; vertex $(2, -2)$, focus $(2, -\frac{3}{2})$, directrix $y = -\frac{5}{2}$
13. $\dfrac{(y - 2)^2}{4} - (x - 1)^2 = 1$: Hyperbola; center $(1, 2)$, vertices $(1, 4)$ and $(1, 0)$, foci $(1, 2 + \sqrt{5})$ and $(1, 2 - \sqrt{5})$, asymptotes $y - 2 = \pm\, 2(x - 1)$
15. $\dfrac{(x - 2)^2}{9} + \dfrac{(y - 1)^2}{4} = 1$: Ellipse; center $(2, 1)$, vertices $(5, 1)$ and $(-1, 1)$, foci $(2 + \sqrt{5}, 1)$ and $(2 - \sqrt{5}, 1)$
17. $(x - 2)^2 = -4(y + 1)$: Parabola; vertex $(2, -1)$, focus $(2, -2)$, directrix $y = 0$
19. $\dfrac{(x - 1)^2}{4} + \dfrac{(y + 1)^2}{9} = 1$: Ellipse; center $(1, -1)$, vertices $(1, 2)$ and $(1, -4)$, foci $(1, -1 + \sqrt{5})$ and $(1, -1 - \sqrt{5})$

21. $y^2 = -8x$

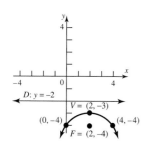

23. $\dfrac{y^2}{4} - \dfrac{x^2}{12} = 1$

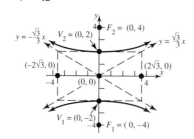

25. $\dfrac{x^2}{16} + \dfrac{y^2}{7} = 1$

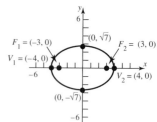

27. $(x - 2)^2 = -4(y + 3)$

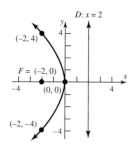

29. $(x + 2)^2 - \dfrac{(y + 3)^2}{3} = 1$

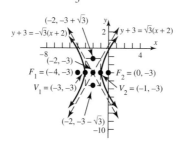

31. $\dfrac{(x + 4)^2}{16} + \dfrac{(y - 5)^2}{25} = 1$

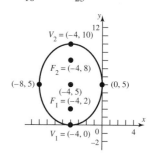

33. $\dfrac{(x + 1)^2}{9} - \dfrac{(y - 2)^2}{7} = 1$

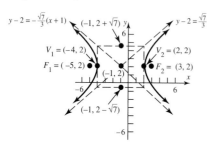

35. $\dfrac{(x - 3)^2}{9} - \dfrac{(y - 1)^2}{4} = 1$

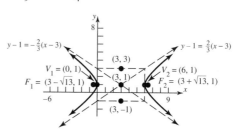

37. $\dfrac{x^2}{5} - \dfrac{y^2}{4} = 1$ **39.** The ellipse $\dfrac{x^2}{16} + \dfrac{y^2}{7} = 1$ **41.** $-20\mathbf{i} + 13\mathbf{j}$ **43.** $\sqrt{5}$ **45.** $\sqrt{5} + 5 \approx 7.24$ **47.** $-\dfrac{2\sqrt{5}}{5}\mathbf{i} + \dfrac{\sqrt{5}}{5}\mathbf{j}$ **49.** $\sqrt{29} \approx 5.39$ mph
51. $\frac{1}{4}$ ft or 3 in   **53.** 19.72 ft, 18.86 ft, 14.91 ft   **55. (a)** 45.24 miles from the Master Station   **(b)** 0.000645 sec   **(c)** (66, 20)

CHAPTER 7   *Exercise 7.1*

1. $2(2) - (-1) = 5$ and $5(2) + 2(-1) = 8$ **3.** $3(2) - 4(\frac{1}{2}) = 4$ and $\frac{1}{2}(2) - 3(\frac{1}{2}) = -\frac{1}{2}$ **5.** $2^2 - 1^2 = 3$ and $(2)(1) = 2$
7. $\dfrac{0}{1 + 0} + 3(2) = 6$ and $0 + 9(2)^2 = 36$ **9.** $3(1) + 3(-1) + 2(2) = 4$, $1 - (-1) - 2 = 0$, and $2(-1) - 3(2) = -8$ **11.** $x = 6, y = 2$
13. $x = 3, y = 2$ **15.** $x = 8, y = -4$ **17.** $x = \frac{1}{3}, y = -\frac{1}{6}$ **19.** Inconsistent   **21.** $x = 1, y = 2$ **23.** $x = 4 - 2y$, y is any real number
25. $x = 1, y = 1$ **27.** $x = \frac{3}{2}, y = 1$ **29.** $x = 4, y = 3$ **31.** $x = \frac{4}{3}, y = \frac{1}{5}$ **33.** $x = 8, y = 2, z = 0$ **35.** $x = 2, y = -1, z = 1$ **37.** Inconsistent
39. $x = 5z - 2$, $y = 4z - 3$ where z is any real number, or $x = \frac{5}{4}y + \frac{7}{4}$, $z = \frac{1}{4}y + \frac{3}{4}$ where y is any real number, or $y = \frac{4}{5}x - \frac{7}{5}$, $z = \frac{1}{5}x + \frac{2}{5}$ where x is any real number
41. Inconsistent   **43.** $x = 1, y = 3, z = -2$ **45.** $x = -3, y = \frac{1}{2}, z = 1$ **47.** $x = \frac{1}{5}, y = \frac{1}{3}$ **49.** $x = 48.15, y = 15.18$
51. $x = -21.47, y = 16.12$ **53.** $x = 0.26, y = 0.06$ **55.** 20 and 61   **57.** 15 ft by 30 ft   **59.** Cheeseburger \$1.55; shake \$0.85
61. 22.5 lb   **63.** Average wind speed 25 mph; average airspeed 175 mph   **65.** 80 \$25 sets and 120 \$45 sets   **67.** \$5.56   **69.** 12, 15, 21
71. $I_1 = \frac{10}{71}, I_2 = \frac{65}{71}, I_3 = \frac{55}{71}$ **73.** 100 orchestra, 210 main, and 190 balcony seats   **79.** $b = -\frac{1}{2}; c = \frac{3}{2}$ **81.** $a = \frac{4}{3}, b = -\frac{5}{3}, c = 1$
83. $x = \dfrac{b_1 - b_2}{m_2 - m_1}$, $y = \dfrac{m_2 b_1 - m_1 b_2}{m_2 - m_1}$ **85.** $y = mx + b$, x is any real number

Exercise 7.2

1. $\begin{bmatrix} 1 & -5 & | & 5 \\ 4 & 3 & | & 6 \end{bmatrix}$ **3.** $\begin{bmatrix} 2 & 3 & | & 6 \\ 4 & -6 & | & -2 \end{bmatrix}$ **5.** $\begin{bmatrix} 0.01 & -0.03 & | & 0.06 \\ 0.13 & 0.10 & | & 0.20 \end{bmatrix}$ **7.** $\begin{bmatrix} 1 & -1 & 1 & | & 10 \\ 3 & 2 & 0 & | & 5 \\ 1 & 1 & 2 & | & 2 \end{bmatrix}$ **9.** $\begin{bmatrix} 1 & 1 & -1 & | & 2 \\ 3 & -2 & 0 & | & 2 \end{bmatrix}$

11.
```
[A]
 [[1  -3  -5   2 ]
  [0   1   6   1 ]
  [0   0  13  16]]
```

13.
```
[A]
 [[1  -3   4   3 ]
  [0   1  -2   0 ]
  [0   0   4  15]]
```

15.
```
[A]
 [[1   -3    2   -6 ]
  [0    1   -1    8 ]
  [0  -15   10  -12]]
```

17.
```
[A]
 [[1  -3   1   -2]
  [0   1   4    2 ]
  [0   0  39   16]]
```

19.
```
[A]
 [[1  -3  -2    3 ]
  [0   1   6   -7 ]
  [0   0  64  -62]]
```

21. $\begin{cases} x = 5 \\ y = -1 \end{cases}$

consistent; $x = 5$, $y = -1$

23. $\begin{cases} x = 1 \\ y = 2 \\ 0 = 3 \end{cases}$

inconsistent

25. $\begin{cases} x + 2z = -1 \\ y - 4z = -2 \\ 0 = 0 \end{cases}$

consistent;
$x = -1 - 2z$,
$y = -2 + 4z$,
z any real number

27. $\begin{cases} x_1 = 1 \\ x_2 + x_4 = 2 \\ x_3 + 2x_4 = 3 \end{cases}$

consistent;
$x_1 = 1$, $x_2 = 4 - x_4$,
$x_3 = 3 - 2x_4$,
x_4 any real number

29. $\begin{cases} x_1 + 4x_4 = 2 \\ x_2 + x_3 + 3x_4 = 3 \\ 0 = 0 \end{cases}$

consistent;
$x_1 = 2 - 4x_4$,
$x_2 = 3 - x_3 - 3x_4$,
x_3, x_4 any real numbers

31. $x = 6$, $y = 2$ **33.** $x = 2$, $y = 3$ **35.** $x = 4$, $y = -2$ **37.** Inconsistent **39.** $x = \frac{1}{2}$, $y = \frac{3}{4}$ **41.** $x = 4 - 2y$, y is any real number

43. $x = \frac{3}{2}$, $y = 1$ **45.** $x = \frac{4}{3}$, $y = \frac{1}{5}$ **47.** $x = 8$, $y = 2$, $z = 0$ **49.** $x = 2$, $y = -1$, $z = 1$ **51.** Inconsistent

53. $x = 5z - 2$, $y = 4z - 3$, where z is any real number, or $x = \frac{5}{4}y + \frac{7}{4}$, $z = \frac{1}{4}y + \frac{3}{4}$, where y is any real number, or $y = \frac{4}{5}x - \frac{7}{5}$, $z = \frac{1}{5}x + \frac{2}{5}$,
where x is any real number

55. Inconsistent **57.** $x = 1$, $y = 3$, $z = -2$ **59.** $x = -3$, $y = \frac{1}{2}$, $z = 1$ **61.** $x = \frac{1}{3}$, $y = \frac{2}{3}$, $z = 1$ **63.** $x = 1$, $y = 2$, $z = 0$, $w = 1$

65. $y = 0$, $z = 1 - x$, x is any real number **67.** $x = 2$, $y = z - 3$, z is any real number **69.** $x = \frac{13}{9}$, $y = \frac{7}{18}$, $z = \frac{19}{18}$

71. $x = \frac{7}{5} - \frac{3}{5}z - \frac{2}{5}w$, $y = -\frac{8}{5} + \frac{7}{5}z + \frac{13}{5}w$, where z and w are any real numbers **73.** $y = -2x^2 + x + 3$ **75.** $f(x) = 3x^3 - 4x^2 + 5$

77. $x =$ liters of 15% H_2SO_4, $y =$ liters of 25% H_2SO_4, $z =$ liters of 50% H_2SO_4: $\begin{cases} x = \frac{5}{2}z - 150 \\ y = 250 - \frac{7}{2}z \end{cases}$

15%	25%	50%	40%
0	40	60	100
10	26	64	100
20	12	68	100

79. If $x =$ Price of hamburgers, $y =$ Price of fries, $z =$ Price of colas, then $x = 2.75 - z$, $y = 0.68 + \frac{1}{3}z$, z any real number.

There is not sufficient information:

x	$2.15	$2.00	$1.85
y	$0.88	$0.93	$0.98
z	$0.60	$0.75	$0.90

81. (a)

Amount Invested At		
7%	9%	11%
0	10,000	10,000
1,000	8,000	11,000
2,000	6,000	12,000
3,000	4,000	13,000
4,000	2,000	14,000
5,000	0	15,000

(b)

Amount Invested At		
7%	9%	11%
12,500	12,500	0
14,500	8,500	2000
16,500	4,500	4000
18,750	0	6250

(c) All the money invested at 7% provides $2100, more than what is required.

83. $I_1 = \frac{4}{15}$, $I_2 = \frac{8}{15}$, $I_3 = \frac{4}{5}$ **85.** $I_1 = 3.5$, $I_2 = 2.5$, $I_3 = 1$

89. If $a_1 \neq 0$,

$$\begin{bmatrix} a_1 & b_1 & | & c_1 \\ a_2 & b_2 & | & c_2 \end{bmatrix} \rightarrow \begin{bmatrix} 1 & \dfrac{b_1}{a_1} & | & \dfrac{c_1}{a_1} \\ a_2 & b_2 & | & c_2 \end{bmatrix} \rightarrow \begin{bmatrix} 1 & \dfrac{b_1}{a_1} & | & \dfrac{c_1}{a_1} \\ 0 & \dfrac{-a_2 b_1}{a_1} + b_2 & | & \dfrac{-a_2 c_1}{a_1} + c_2 \end{bmatrix} \rightarrow \begin{bmatrix} 1 & \dfrac{b_1}{a_1} & | & \dfrac{c_1}{a_1} \\ 0 & \dfrac{-a_2 b_1 + b_2 a_1}{a_1} & | & \dfrac{-a_2 c_1 + c_2 a_1}{a_1} \end{bmatrix}$$

$$\rightarrow \begin{bmatrix} 1 & \dfrac{b_1}{a_1} & | & \dfrac{c_1}{a_1} \\ 0 & 1 & | & \dfrac{-a_2 c_1 + c_2 a_1}{a_1} \cdot \dfrac{a_1}{-a_2 b_1 + b_2 a_1} \end{bmatrix} \rightarrow \begin{bmatrix} 1 & \dfrac{b_1}{a_1} & | & \dfrac{c_1}{a_1} \\ 0 & 1 & | & \dfrac{-a_2 c_1 + c_2 a_1}{-a_2 b_1 + b_2 a_1} \end{bmatrix} \rightarrow \begin{bmatrix} 1 & 0 & | & \dfrac{-b_1 c_2 + b_2 c_1}{-a_2 b_1 + b_2 a_1} \\ 0 & 1 & | & \dfrac{-a_2 c_1 + c_2 a_1}{-a_2 b_1 + b_2 a_1} \end{bmatrix}$$

$$x = \frac{1}{a_1 b_2 - a_2 b_1}(c_1 b_2 - c_2 b_1) = \frac{1}{D}(c_1 b_2 - c_2 b_1), \quad y = \frac{1}{a_1 b_2 - a_2 b_1}(a_1 c_2 - a_2 c_1) = \frac{1}{D}(a_1 c_2 - a_2 c_1)$$

If $a_1 = 0$, then $a_2 \neq 0$, $b_1 \neq 0$, and

$$\begin{bmatrix} 0 & b_1 & | & c_1 \\ a_2 & b_2 & | & c_2 \end{bmatrix} \rightarrow \begin{bmatrix} a_2 & b_2 & | & c_2 \\ 0 & b_1 & | & c_1 \end{bmatrix} \rightarrow \begin{bmatrix} 1 & \dfrac{b_2}{a_2} & | & \dfrac{c_2}{a_2} \\ 0 & b_1 & | & c_1 \end{bmatrix} \rightarrow \begin{bmatrix} 1 & \dfrac{b_2}{a_2} & | & \dfrac{c_2}{a_2} \\ 0 & 1 & | & \dfrac{c_1}{b_1} \end{bmatrix} \rightarrow \begin{bmatrix} 1 & 0 & | & \dfrac{c_2}{a_2} - \dfrac{b_2 c_1}{a_2 b_1} = \dfrac{c_1 b_2 - c_2 b_1}{-a_2 b_1} \\ 0 & 1 & | & \dfrac{c_1}{b_1} = \dfrac{-a_2 c_1}{-a_2 b_1} \end{bmatrix}$$

Exercise 7.3

1. 2 **3.** 22 **5.** -2 **7.** 10 **9.** -26 **11.** $x = 6$, $y = 2$ **13.** $x = 3$, $y = 2$ **15.** $x = 8$, $y = -4$ **17.** $x = 4$, $y = -2$ **19.** Not applicable
21. $x = \frac{1}{2}$, $y = \frac{3}{4}$ **23.** $x = \frac{1}{10}$, $y = \frac{2}{5}$ **25.** $x = \frac{3}{2}$, $y = 1$ **27.** $x = \frac{4}{3}$, $y = \frac{1}{5}$ **29.** $x = 1$, $y = 3$, $z = -2$ **31.** $x = -3$, $y = \frac{1}{2}$, $z = 1$
33. Not applicable **35.** $x = 0$, $y = 0$, $z = 0$ **37.** Not applicable **39.** $x = \frac{1}{5}$, $y = \frac{1}{3}$ **41.** -5 **43.** $\frac{13}{11}$ **45.** 0 or -9 **47.** -4 **49.** 12
51. 8 **53.** 8
55. $(y_1 - y_2)x - (x_1 - x_2)y + (x_1 y_2 - x_2 y_1) = 0$
$(y_1 - y_2)x + (x_2 - x_1)y = x_2 y_1 - x_1 y_2$
$(x_2 - x_1)y - (x_2 - x_1)y_1 = (y_2 - y_1)x + x_2 y_1 - x_1 y_2 - (x_2 - x_1)y_1$
$(x_2 - x_1)(y - y_1) = (y_2 - y_1)x - (y_2 - y_1)x_1$
$y - y_1 = \dfrac{(y_2 - y_1)}{(x_2 - x_1)}(x - x_1)$

57. $\begin{vmatrix} x^2 & x & 1 \\ y^2 & y & 1 \\ z^2 & z & 1 \end{vmatrix} = x^2 \begin{vmatrix} y & 1 \\ z & 1 \end{vmatrix} - x \begin{vmatrix} y^2 & 1 \\ z^2 & 1 \end{vmatrix} + \begin{vmatrix} y^2 & y \\ z^2 & z \end{vmatrix} = x^2(y - z) - x(y^2 - z^2) + yz(y - z)$

$= (y - z)[x^2 - x(y + z) + yz] = (y - z)[(x^2 - xy) - (xz - yz)] = (y - z)[x(x - y) - z(x - y)] = (y - z)(x - y)(x - z)$

59. $\begin{vmatrix} a_{13} & a_{12} & a_{11} \\ a_{23} & a_{22} & a_{21} \\ a_{33} & a_{32} & a_{31} \end{vmatrix} = a_{13}(a_{22}a_{31} - a_{32}a_{21}) - a_{12}(a_{23}a_{31} - a_{33}a_{21}) + a_{11}(a_{23}a_{32} - a_{33}a_{22})$

$= -[a_{11}(a_{22}a_{33} - a_{32}a_{23}) - a_{12}(a_{21}a_{33} - a_{31}a_{23}) + a_{13}(a_{21}a_{32} - a_{31}a_{22})] = -\begin{vmatrix} a_{11} & a_{12} & a_{13} \\ a_{21} & a_{22} & a_{23} \\ a_{31} & a_{32} & a_{33} \end{vmatrix}$

61. $\begin{vmatrix} a_{11} & a_{12} & a_{11} \\ a_{21} & a_{22} & a_{21} \\ a_{31} & a_{32} & a_{31} \end{vmatrix} = a_{11}(a_{22}a_{31} - a_{32}a_{21}) - a_{12}(a_{21}a_{31} - a_{31}a_{21}) + a_{11}(a_{21}a_{32} - a_{31}a_{22})$

$= a_{11}a_{22}a_{31} - a_{11}a_{32}a_{21} - a_{12}(0) + a_{11}a_{21}a_{32} - a_{11}a_{31}a_{22} = 0$

Exercise 7.4

1. $\begin{bmatrix} 4 & 4 & -5 \\ -1 & 5 & 4 \end{bmatrix}$ **3.** $\begin{bmatrix} 0 & 12 & -20 \\ 4 & 8 & 24 \end{bmatrix}$ **5.** $\begin{bmatrix} -8 & 7 & -15 \\ 7 & 0 & 22 \end{bmatrix}$ **7.** $\begin{bmatrix} 28 & -9 \\ 4 & 23 \end{bmatrix}$ **9.** $\begin{bmatrix} 1 & 14 & -14 \\ 2 & 22 & -18 \\ 3 & 0 & 28 \end{bmatrix}$ **11.** $\begin{bmatrix} 15 & 21 & -16 \\ 22 & 34 & -22 \\ -11 & 7 & 22 \end{bmatrix}$

13. $\begin{bmatrix} 25 & -9 \\ 4 & 20 \end{bmatrix}$ **15.** $\begin{bmatrix} -13 & 7 & -12 \\ -18 & 10 & -14 \\ 17 & -7 & 34 \end{bmatrix}$ **17.** $\begin{bmatrix} -2 & 4 & 2 & 8 \\ 2 & 1 & 4 & 6 \end{bmatrix}$ **19.** $\begin{bmatrix} 9 & 2 \\ 34 & 13 \\ 47 & 20 \end{bmatrix}$ **21.** $\begin{bmatrix} 1 & -1 \\ -1 & 2 \end{bmatrix}$ **23.** $\begin{bmatrix} 1 & -\frac{5}{2} \\ -1 & 3 \end{bmatrix}$

25. $\begin{bmatrix} 1 & -1/a \\ -1 & 2/a \end{bmatrix}$ **27.** $\begin{bmatrix} 3 & -3 & 1 \\ -2 & 2 & -1 \\ -4 & 5 & -2 \end{bmatrix}$ **29.** $\begin{bmatrix} -\frac{5}{7} & \frac{1}{7} & \frac{3}{7} \\ \frac{9}{7} & \frac{1}{7} & -\frac{4}{7} \\ \frac{3}{7} & -\frac{2}{7} & \frac{1}{7} \end{bmatrix}$ **31.** $x = 3, y = 2$ **33.** $x = -5, y = 10$ **35.** $x = 2, y = -1$

37. $x = \frac{1}{2}, y = 2$ **39.** $x = -2, y = 1$ **41.** $x = 2/a, y = 3/a$ **43.** $x = -2, y = 3, z = 5$ **45.** $x = \frac{1}{2}, y = -\frac{1}{2}, z = 1$ **47.** $x = -\frac{34}{7}, y = \frac{85}{7}, z = \frac{12}{7}$

49. $x = \frac{1}{3}, y = 1, z = \frac{2}{3}$ **51.** $\begin{bmatrix} 4 & 2 & | & 1 & 0 \\ 2 & 1 & | & 0 & 1 \end{bmatrix} \rightarrow \begin{bmatrix} 1 & \frac{1}{2} & | & \frac{1}{4} & 0 \\ 2 & 1 & | & 0 & 1 \end{bmatrix} \rightarrow \begin{bmatrix} 1 & \frac{1}{2} & | & \frac{1}{4} & 0 \\ 0 & 0 & | & -\frac{1}{2} & 1 \end{bmatrix}$

53. $\begin{bmatrix} 15 & 3 & | & 1 & 0 \\ 10 & 2 & | & 0 & 1 \end{bmatrix} \rightarrow \begin{bmatrix} 1 & \frac{1}{5} & | & \frac{1}{15} & 0 \\ 10 & 2 & | & 0 & 1 \end{bmatrix} \rightarrow \begin{bmatrix} 1 & \frac{1}{5} & | & \frac{1}{15} & 0 \\ 0 & 0 & | & -\frac{2}{3} & 1 \end{bmatrix}$

55. $\begin{bmatrix} -3 & 1 & -1 & | & 1 & 0 & 0 \\ 1 & -4 & -7 & | & 0 & 1 & 0 \\ 1 & 2 & 5 & | & 0 & 0 & 1 \end{bmatrix} \rightarrow \begin{bmatrix} 1 & 2 & 5 & | & 0 & 0 & 1 \\ 1 & -4 & -7 & | & 0 & 1 & 0 \\ -3 & 1 & -1 & | & 1 & 0 & 0 \end{bmatrix} \rightarrow \begin{bmatrix} 1 & 2 & 5 & | & 0 & 0 & 1 \\ 0 & -6 & -12 & | & 0 & 1 & -1 \\ 0 & 7 & 14 & | & 1 & 0 & 3 \end{bmatrix}$

$$\rightarrow \begin{bmatrix} 1 & 2 & 5 & | & 0 & 0 & 1 \\ 0 & 1 & 2 & | & 0 & -\frac{1}{6} & \frac{1}{6} \\ 0 & 1 & 2 & | & \frac{1}{7} & 0 & \frac{3}{7} \end{bmatrix} \rightarrow \begin{bmatrix} 1 & 2 & 5 & | & 0 & 0 & 1 \\ 0 & 1 & 2 & | & 0 & -\frac{1}{6} & \frac{1}{6} \\ 0 & 0 & 0 & | & \frac{1}{7} & \frac{1}{6} & \frac{11}{42} \end{bmatrix}$$

57. $\begin{bmatrix} 0.01 & 0.05 & -0.01 \\ 0.01 & -0.02 & 0.01 \\ -0.02 & 0.01 & 0.03 \end{bmatrix}$ **59.** $\begin{bmatrix} 0.02 & -0.04 & -0.01 & 0.01 \\ -0.02 & 0.05 & 0.03 & -0.03 \\ 0.02 & 0.01 & -0.04 & 0.00 \\ -0.02 & 0.06 & 0.07 & 0.06 \end{bmatrix}$

61. $x = 4.57, y = -6.44, z = -24.07$ **63.** $x = -1.19, y = 2.46, z = 8.27$

65. (a) $\begin{bmatrix} 500 & 350 & 400 \\ 700 & 500 & 850 \end{bmatrix}$; $\begin{bmatrix} 500 & 700 \\ 350 & 500 \\ 400 & 850 \end{bmatrix}$ (b) $\begin{bmatrix} 15 \\ 8 \\ 3 \end{bmatrix}$ (c) $\begin{bmatrix} 11,500 \\ 17,050 \end{bmatrix}$ (d) $[0.10 \quad 0.05]$ (e) $2002.50

67. If $a \ne 0$, $\begin{bmatrix} a & b & | & 1 & 0 \\ c & d & | & 0 & 1 \end{bmatrix} \rightarrow \begin{bmatrix} 1 & \frac{b}{a} & | & \frac{1}{a} & 0 \\ c & d & | & 0 & 1 \end{bmatrix} \rightarrow \begin{bmatrix} 1 & \frac{b}{a} & | & \frac{1}{a} & 0 \\ 0 & -cb + da & | & -\frac{c}{a} & 1 \end{bmatrix}$

$\rightarrow \begin{bmatrix} 1 & \frac{b}{a} & | & \frac{1}{a} & 0 \\ 0 & 1 & | & \frac{-c}{-cb+da} & \frac{a}{-cb+da} \end{bmatrix} \rightarrow \begin{bmatrix} 1 & 0 & | & \frac{d}{ad-bc} & \frac{-b}{ad-bc} \\ 0 & 1 & | & \frac{-c}{ad-bc} & \frac{a}{ad-bc} \end{bmatrix}$. Therefore, $A^{-1} = \frac{1}{D}\begin{bmatrix} d & -b \\ -c & a \end{bmatrix}$.

If $a = 0$, $\begin{bmatrix} 0 & b & | & 1 & 0 \\ c & d & | & 0 & 1 \end{bmatrix} \rightarrow \begin{bmatrix} c & d & | & 0 & 1 \\ 0 & b & | & 1 & 0 \end{bmatrix} \rightarrow \begin{bmatrix} 1 & \frac{d}{c} & | & 0 & \frac{1}{c} \\ 0 & b & | & 1 & 0 \end{bmatrix} \rightarrow \begin{bmatrix} 1 & 0 & | & \frac{-d}{cb} & \frac{1}{c} \\ 0 & 1 & | & \frac{1}{b} & 0 \end{bmatrix}$

$\rightarrow \begin{bmatrix} 1 & 0 & | & \frac{d}{-bc} & \frac{-b}{-bc} \\ 0 & 1 & | & \frac{-c}{-bc} & 0 \end{bmatrix}$. Since $a = 0$, $D = ad - bc = -bc$, so $A^{-1} = \frac{1}{D}\begin{bmatrix} d & -b \\ -c & a \end{bmatrix}$.

Exercise 7.5

1. Proper **3.** Improper; $1 + \dfrac{9}{x^2 - 4}$ **5.** Improper; $5x + \dfrac{22x - 1}{x^2 - 4}$ **7.** Improper; $1 + \dfrac{-2(x - 6)}{(x + 4)(x - 3)}$ **9.** $\dfrac{-4}{x} + \dfrac{4}{x - 1}$ **11.** $\dfrac{1}{x} + \dfrac{-x}{x^2 + 1}$

13. $\dfrac{-1}{x - 1} + \dfrac{2}{x - 2}$ **15.** $\dfrac{\frac{1}{4}}{x + 1} + \dfrac{\frac{3}{4}}{x - 1} + \dfrac{\frac{1}{2}}{(x - 1)^2}$ **17.** $\dfrac{\frac{1}{12}}{x - 2} + \dfrac{-\frac{1}{12}(x + 4)}{x^2 + 2x + 4}$ **19.** $\dfrac{\frac{1}{4}}{(x - 1)} + \dfrac{\frac{1}{4}}{(x - 1)^2} - \dfrac{\frac{1}{4}}{x + 1} + \dfrac{\frac{1}{4}}{(x + 1)^2}$

21. $\dfrac{-5}{x + 2} + \dfrac{5}{x + 1} + \dfrac{-4}{(x + 1)^2}$ **23.** $\dfrac{\frac{1}{4}}{x} + \dfrac{1}{x^2} - \dfrac{\frac{1}{4}(x + 4)}{x^2 + 4}$ **25.** $\dfrac{\frac{2}{3}}{x + 1} + \dfrac{\frac{1}{3}(x + 1)}{x^2 + 2x + 4}$ **27.** $\dfrac{\frac{2}{7}}{3x - 2} + \dfrac{\frac{1}{7}}{2x + 1}$ **29.** $\dfrac{\frac{3}{4}}{x + 3} + \dfrac{\frac{1}{4}}{x - 1}$

31. $\dfrac{1}{x^2 + 4} + \dfrac{2x - 1}{(x^2 + 4)^2}$ **33.** $\dfrac{-1}{x} + \dfrac{2}{x - 3} + \dfrac{-1}{x + 1}$ **35.** $\dfrac{4}{x - 2} + \dfrac{-3}{x - 1} + \dfrac{-1}{(x - 1)^2}$ **37.** $\dfrac{x}{(x^2 + 16)^2} + \dfrac{-16x}{(x^2 + 16)^3}$ **39.** $\dfrac{-\frac{8}{7}}{2x - 3} + \dfrac{\frac{4}{7}}{x + 1}$

41. $\dfrac{-\frac{2}{9}}{x} - \dfrac{\frac{1}{3}}{x^2} + \dfrac{\frac{1}{6}}{x - 3} + \dfrac{\frac{1}{18}}{x + 3}$

Exercise 7.6

1.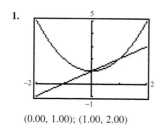

(0.00, 1.00); (1.00, 2.00)

3.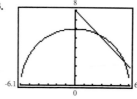

(2.58, 5.41); (5.41, 2.58)

5.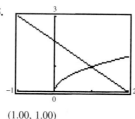

(1.00, 1.00)

7.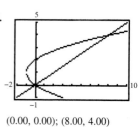

(0.00, 0.00); (8.00, 4.00)

9.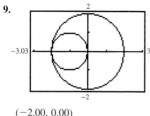

(−2.00, 0.00)

11.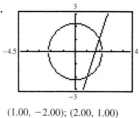

(1.00, −2.00); (2.00, 1.00)

13.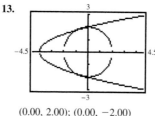

(0.00, 2.00); (0.00, −2.00)

15.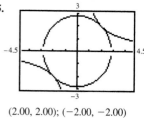

(2.00, 2.00); (−2.00, −2.00)

17.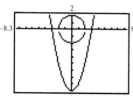

No solution; Inconsistent

19.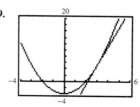

(3.00, 5.00)

21. $x = 1, y = 4; x = -1, y = -4; x = 2\sqrt{2}, y = \sqrt{2}; x = -2\sqrt{2}, y = -\sqrt{2}$ **23.** $x = 0, y = 1; x = -\frac{2}{3}, y = -\frac{1}{3}$

25. $x = 0, y = -1; x = \frac{5}{2}, y = -\frac{7}{2}$ **27.** $x = 2, y = \frac{1}{3}; x = \frac{1}{2}, y = \frac{4}{3}$ **29.** $x = 3, y = 2; x = 3, y = -2; x = -3, y = 2; x = -3, y = -2$

31. $x = \frac{1}{2}, y = \frac{3}{2}; x = \frac{1}{2}, y = -\frac{3}{2}; x = -\frac{1}{2}, y = \frac{3}{2}; x = -\frac{1}{2}, y = -\frac{3}{2}$ **33.** $x = \sqrt{2}, y = 2\sqrt{2}; x = -\sqrt{2}, y = -2\sqrt{2}$

35. No solution; system is inconsistent **37.** $x = \frac{8}{3}, y = 2\sqrt{10}/3; x = -\frac{8}{3}, y = 2\sqrt{10}/3; x = \frac{8}{3}, y = -2\sqrt{10}/3; x = -\frac{8}{3}, y = -2\sqrt{10}/3$

39. $x = 1, y = \frac{1}{2}; x = -1, y = \frac{1}{2}; x = 1, y = -\frac{1}{2}; x = -1, y = -\frac{1}{2}$ **41.** No solution; system is inconsistent

43. $x = 2, y = 1; x = -2, y = -1; x = \sqrt{3}, y = \sqrt{3}; x = -\sqrt{3}, y = -\sqrt{3}$ **45.** $x = 3, y = 2; x = -3, y = -2; x = 2, y = \frac{1}{2}; x = -2, y = -\frac{1}{2}$

47. $x = 3, y = 1; x = -1, y = -3$ **49.** $x = 0, y = -2; x = 0, y = 1; x = 2, y = -1$ **51.** $x = 2, y = 8$ **53.** $x = 0.48, y = 0.61$

55. $x = -1.64, y = -0.89$ **57.** $x = 0.58, y = 1.85; x = 1.81, y = 1.05; x = 0.58, y = -1.85; x = 1.81, y = -1.05$ **59.** $x = 2.34, y = 0.85$

61. 3 and 1; −3 and −1 **63.** 2 and 2; −2 and −2 **65.** $\frac{1}{2}$ and $\frac{1}{3}$ **67.** 5

69. **71.** Solutions: $(0, -\sqrt{3} - 2), (0, \sqrt{3} - 2), (1, 0), (1, -4)$ **73.**

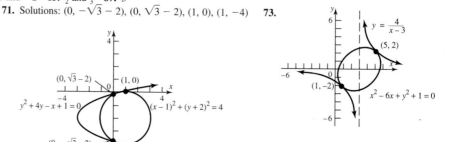

75. 5 in. by 3 in.   **77.** 2 cm and 4 cm   **79.** tortoise: 7 m/hr, hare $7\frac{1}{2}$ m/hr   **81.** 12 cm by 18 cm   **83.** $x = 60$ ft; $y = 30$ ft

85. $l = \dfrac{P + \sqrt{P^2 - 16A}}{4}$; $w = \dfrac{P - \sqrt{P^2 - 16A}}{4}$ **87.** $y = 4x - 4$ **89.** $y = 2x + 1$ **91.** $y = -\frac{1}{3}x + \frac{7}{3}$ **93.** $y = 2x - 3$

95. $r_1 = \dfrac{-b + \sqrt{b^2 - 4ac}}{2a}$

$r_2 = \dfrac{-b - \sqrt{b^2 - 4ac}}{2a}$

Exercise 7.7

1.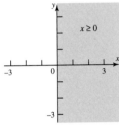

3.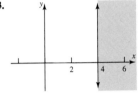

5.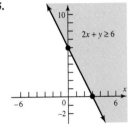

7.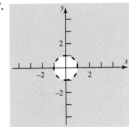

9.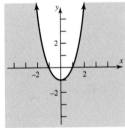

11.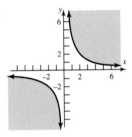

13.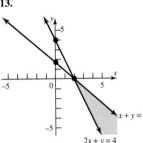

15.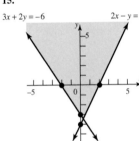

17.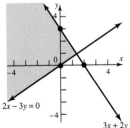

19.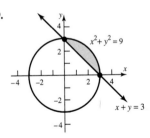

21.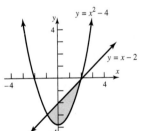

23.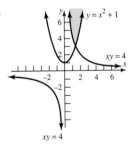

25.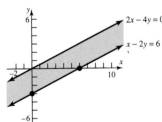

27.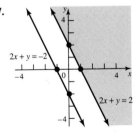

29. No solution
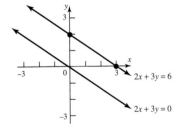

31. Bounded; corner points (0, 0), (3, 0), (2, 2), (0, 3)

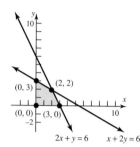

33. Unbounded; corner points (2, 0), (0, 4)

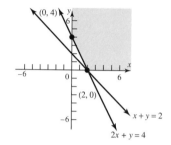

35. Bounded; corner points (2, 0), (4, 0), $(\frac{24}{7}, \frac{12}{7})$, (0, 4), (0, 2)

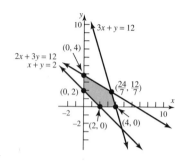

37. Bounded; corner points (2, 0), (5, 0), (2, 6), (0, 8), (0, 2)

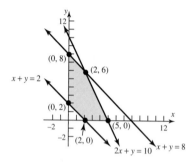

39. Bounded; corner points (10, 0), (0, 5)

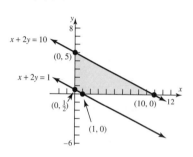

41. $\begin{cases} x \le 4 \\ x + y \le 6 \\ x \ge 0 \\ y \ge 0 \end{cases}$ **43.** $\begin{cases} x \le 20 \\ y \ge 15 \\ x + y \le 50 \\ x \le y \\ x \ge 0 \end{cases}$

45. (a) $\begin{cases} x + y \le 50{,}000 \\ x \ge 35{,}000 \\ y \le 10{,}000 \\ x \ge 0 \\ y \ge 0 \end{cases}$ **(b)**

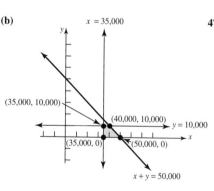

47. (a) $\begin{cases} x \ge 0 \\ y \ge 0 \\ x + 2y \le 300 \\ 3x + 2y \le 480 \end{cases}$ **(b)**

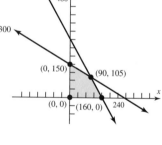

49. (a) $\begin{cases} 30x + 20y \le 1600 \\ 2x + 3y \le 150 \\ x \ge 0 \\ y \ge 0 \end{cases}$ **(b)**

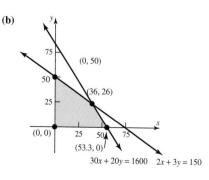

1. Maximum value is 11; minimum value is 3   **3.** Maximum value is 65; minimum value is 4   **5.** Maximum value is 67; minimum value is 20
7. The maximum value of z is 12, and it occurs at the point $(6, 0)$ **9.** The minimum value of z is 4, and it occurs at the point $(2, 0)$
11. The maximum value of z is 20, and it occurs at the point $(0, 4)$ **13.** The minimum value of z is 8, and it occurs at the point $(0, 2)$
15. The maximum value of z is 50, and it occurs at the point $(10, 0)$ **17.** 8 downhill, 24 cross-country; \$1760; \$1920
19. 30 acres of soybeans and 10 acres of corn; maximum profit is \$8900   **21.** $\frac{1}{2}$ hour on machine I; $5\frac{1}{4}$ hours on machine II; \$182.50
23. 100 lbs. of ground beef and 50 lbs of pork; \$97.50   **25.** 10 racing skates, 15 figure skates   **27.** 2 metal samples, 4 plastic samples; \$34
29. (a) 10 first class, 120 coach   **(b)** 15 first class, 120 coach

Fill-in-the-Blank Items

1. inconsistent   **2.** matrix   **3.** determinants   **4.** augmented   **5.** inverse   **6.** square   **7.** identity   **8.** proper   **9.** half-plane
10. objective function   **11.** feasible point

True/False Items

1. F   **2.** T   **3.** F   **4.** F   **5.** F   **6.** F   **7.** F   **8.** T   **9.** T   **10.** T   **11.** T

Review Exercises

1. $x = 2, y = -1$ **3.** $x = 2, y = \frac{1}{2}$ **5.** $x = 2, y = -1$ **7.** $x = \frac{11}{5}, y = -\frac{3}{5}$ **9.** $x = -\frac{8}{5}, y = \frac{12}{5}$ **11.** $x = 6, y = -1$ **13.** $x = -4, y = 3$

15. $x = 2, y = 3$ **17.** Inconsistent   **19.** $x = -1, y = 2, z = -3$ **21.** $\begin{bmatrix} 4 & -4 \\ 3 & 9 \\ 4 & 0 \end{bmatrix}$ **23.** $\begin{bmatrix} 6 & 0 \\ 12 & 24 \\ -6 & 12 \end{bmatrix}$ **25.** $\begin{bmatrix} 4 & -3 & 0 \\ 12 & -2 & -8 \\ -2 & 5 & -4 \end{bmatrix}$

27. $\begin{bmatrix} 8 & -13 & 8 \\ 9 & 2 & -10 \\ 18 & -17 & 4 \end{bmatrix}$ **29.** $\begin{bmatrix} \frac{1}{2} & -1 \\ -\frac{1}{6} & \frac{2}{3} \end{bmatrix}$ **31.** $\begin{bmatrix} -\frac{5}{7} & \frac{9}{7} & \frac{3}{7} \\ \frac{1}{7} & \frac{1}{7} & -\frac{2}{7} \\ \frac{3}{7} & -\frac{4}{7} & \frac{1}{7} \end{bmatrix}$ **33.** Singular   **35.** $x = \frac{2}{5}, y = \frac{1}{10}$ **37.** $x = \frac{1}{2}, y = \frac{2}{3}, z = \frac{1}{6}$

39. $x = -\frac{1}{2}, y = -\frac{2}{3}, z = -\frac{3}{4}$ **41.** $z = -1, x = y + 1, y$ any real number   **43.** $x = 1, y = 2, z = -3, t = 1$ **45.** 5

47. 108   **49.** -100 **51.** $x = 2, y = -1$ **53.** $x = 2, y = 3$ **55.** $x = -1, y = 2, z = -3$ **57.** $\dfrac{-\frac{3}{2}}{x} + \dfrac{\frac{3}{2}}{x - 4}$ **59.** $\dfrac{-3}{x - 1} + \dfrac{3}{x} + \dfrac{4}{x^2}$

61. $\dfrac{-\frac{1}{10}}{x + 1} + \dfrac{\frac{1}{10}x + \frac{9}{10}}{x^2 + 9}$ **63.** $\dfrac{x}{x^2 + 4} - \dfrac{4x}{(x^2 + 4)^2}$ **65.** $\dfrac{\frac{1}{2}}{x^2 + 1} + \dfrac{\frac{1}{4}}{x - 1} - \dfrac{\frac{1}{4}}{x + 1}$ **67.** $x = -\frac{2}{5}, y = -\frac{11}{5}; x = -2, y = 1$

69. $x = 2\sqrt{2}, y = \sqrt{2}; x = -2\sqrt{2}, y = -\sqrt{2}$ **71.** $x = \dfrac{\pm 3(-3 + \sqrt{265})}{2}, y = \dfrac{-3 + \sqrt{265}}{2}$

73. $x = \sqrt{2}, y = -\sqrt{2}; x = -\sqrt{2}, y = \sqrt{2}; x = \frac{4}{3}\sqrt{2}, y = -\frac{2}{3}\sqrt{2}; x = -\frac{4}{3}\sqrt{2}, y = \frac{2}{3}\sqrt{2}$ **75.** $x = 1, y = -1$
77. Unbounded; corner point $(0, 2)$ **79.** Bounded; corner points $(0, 0)$,   **81.** Bounded; corner points $(0, 1)$, $(0, 8)$,
$(0, 2)$, $(3, 0)$ $(4, 0)$, $(2, 0)$

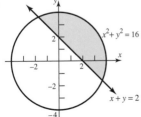

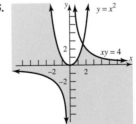

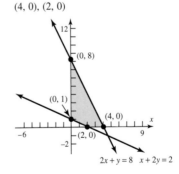

83.

85.

87. The maximum value is 32 when $x = 0$ and $y = 8$ **89.** The minimum value is 3 when $x = 1$ and $y = 0$
91. The maximum value is $\frac{108}{7}$ when $x = \frac{12}{7}$ and $y = \frac{12}{7}$ **93.** 10   **95.** $y = -\frac{1}{3}x^2 - \frac{2}{3}x + 1$ **97.** 70 lbs of \$3 coffee and 30 lbs of \$6 coffee
99. 1 small, 5 medium, 2 large   **101.** 24 ft by 10 ft   **103.** $4 + \sqrt{2}$ in and $4 - \sqrt{2}$ in   **105.** $120\sqrt{10}$ ft
107. Katy gets \$10, Mike gets \$20, Danny gets \$5, Colleen gets \$10   **109.** Katy: 4 hr; Mike: 2 hr; Danny: 8 hr
111. 35 gasoline engines, 15 diesel engines; 15 gasoline engines, 0 diesel engines

CHAPTER 8   *Exercise 8.1*

1. 1, 2, 3, 4, 5   **3.** $\frac{1}{3}, \frac{2}{4} = \frac{1}{2}, \frac{3}{5}, \frac{4}{6} = \frac{2}{3}, \frac{5}{7}$ **5.** 1, -4, 9, -16, 25   **7.** $\frac{1}{2}, \frac{2}{5}, \frac{2}{7}, \frac{8}{41}, \frac{8}{61}$ **9.** $-\frac{1}{6}, \frac{1}{12}, -\frac{1}{20}, \frac{1}{30}, -\frac{1}{42}$ **11.** $1/e$, $2/e^2$, $3/e^3$, $4/e^4$, $5/e^5$
13. $n/(n+1)$ **15.** $1/2^{n-1}$ **17.** $(-1)^{n+1}$ **19.** $(-1)^{n+1}n$ **21.** $a_1 = 2$, $a_2 = 5$, $a_3 = 8$, $a_4 = 11$, $a_5 = 14$
23. $a_1 = -2$, $a_2 = -1$, $a_3 = 1$, $a_4 = 4$, $a_5 = 8$ **25.** $a_1 = 5$, $a_2 = 10$, $a_3 = 20$, $a_4 = 40$, $a_5 = 80$ **27.** $a_1 = 3$, $a_2 = 3$, $a_3 = \frac{3}{2}$, $a_4 = \frac{1}{2}$, $a_5 = \frac{1}{8}$
29. $a_1 = 1$, $a_2 = 2$, $a_3 = 2$, $a_4 = 4$, $a_5 = 8$ **31.** $a_1 = A$, $a_2 = A + d$, $a_3 = A + 2d$, $a_4 = A + 3d$, $a_5 = A + 4d$

33. $a_1 = \sqrt{2}$, $a_2 = \sqrt{2 + \sqrt{2}}$, $a_3 = \sqrt{2 + \sqrt{2 + \sqrt{2}}}$, $a_4 = \sqrt{2 + \sqrt{2 + \sqrt{2 + \sqrt{2}}}}$, $a_5 = \sqrt{2 + \sqrt{2 + \sqrt{2 + \sqrt{2 + \sqrt{2}}}}}$

35. $3 + 4 + \cdots + (n+2)$ **37.** $\frac{1}{2} + 2 + \frac{9}{2} + \cdots + \frac{n^2}{2}$ **39.** $1 + \frac{1}{3} + \frac{1}{9} + \cdots + \frac{1}{3^n}$ **41.** $\frac{1}{3} + \frac{1}{9} + \cdots + \frac{1}{3^n}$

43. $\ln 2 - \ln 3 + \ln 4 - \cdots + (-1)^n \ln n$ **45.** $\sum_{k=1}^{n} k$ **47.** $\sum_{k=1}^{n} \frac{k}{k+1}$ **49.** $\sum_{k=0}^{n} (-1)^k \left(\frac{1}{3^k}\right)$ **51.** $\sum_{k=1}^{n} \frac{3^k}{k}$ **53.** $\sum_{k=0}^{n} (a + kd)$ **55.** 21
59. A Fibonacci sequence

Exercise 8.2

1. $d = 1$; 5, 6, 7, 8   **3.** $d = 2$; $-3, -1, 1, 3$ **5.** $d = -2$; 4, 2, 0, -2 **7.** $d = -\frac{1}{3}; \frac{1}{6}, -\frac{1}{6}, -\frac{1}{2}, -\frac{5}{6}$ **9.** $d = \ln 3$; $\ln 3$, $2 \ln 3$, $3 \ln 3$, $4 \ln 3$
11. $a_5 = 14$; $a_n = 3n - 1$ **13.** $a_5 = -7$; $a_n = 8 - 3n$ **15.** $a_5 = 2$; $a_n = \frac{1}{2}(n-1)$ **17.** $a_5 = 5\sqrt{2}$; $a_n = \sqrt{2}n$ **19.** $a_{12} = 24$ **21.** $a_{10} = -26$
23. $a_8 = a + 7b$ **25.** $a_1 = -13$; $d = 3$; $a_n = -16 + 3n$ **27.** $a_1 = -53$; $d = 6$; $a_n = -59 + 6n$ **29.** $a_1 = 28$; $d = -2$; $a_n = 30 - 2n$
31. $a_1 = 25$; $d = -2$; $a_n = 27 - 2n$ **33.** n^2 **35.** $\frac{n}{2}(9 + 5n)$ **37.** 1260   **39.** 324

41.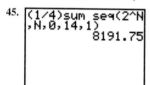
43.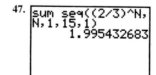
45.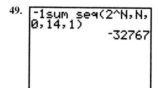

47. $-\frac{3}{2}$ **49.** 1185 seats   **51.** 210 of (1st colors name) and 190 (2nd colors name)

Exercise 8.3

1. $r = 3$; 3, 9, 27, 81   **3.** $r = \frac{1}{2}$; $-\frac{3}{2}, -\frac{3}{4}, -\frac{3}{8}, -\frac{3}{16}$ **5.** $r = 2$; $\frac{1}{4}, \frac{1}{2}, 1, 2$ **7.** $r = 2^{1/3}$; $2^{1/3}, 2^{2/3}, 2, 2^{4/3}$ **9.** $r = \frac{3}{2}$; $\frac{1}{2}, \frac{3}{4}, \frac{9}{8}, \frac{27}{16}$ **11.** Arithmetic; $d = 1$
13. Neither   **15.** Arithmetic; $d = -\frac{2}{3}$ **17.** Neither   **19.** Geometric; $r = \frac{2}{3}$ **21.** Geometric; $r = 2$ **23.** Geometric; $r = 3^{1/2}$
25. $a_5 = 162$; $a_n = 2 \cdot 3^{n-1}$ **27.** $a_5 = 5$; $a_n = (-1)^{n-1}(5)$ **29.** $a_5 = 0$; $a_n = 0$ **31.** $a_5 = 4\sqrt{2}$; $a_n = (\sqrt{2})^n$ **33.** $a_7 = \frac{1}{64}$ **35.** $a_9 = 1$
37. $a_8 = 0.00000004$ **39.** $-\frac{1}{4}(1 - 2^n)$ **41.** $2[1 - (\frac{2}{3})^n]$ **43.** $1 - 2^n$

45.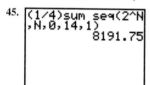
47.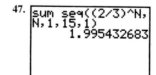
49.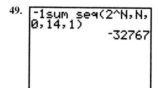

51. $\frac{3}{2}$ **53.** 16   **55.** $\frac{8}{5}$ **57.** $\frac{20}{3}$ **59.** $\frac{18}{5}$ **61.** -4 **63.** (a) 0.775 ft   (b) 8th   (c) 15.88 ft   (d) 20 ft   **65.** \$21,879.11
67. Option 2 results in the most: \$16,038,304; Option 1 results in the least: \$14,700,000   **69.** 1.845×10^{19} **73.** 3, 4, 8, 23, 72
75. A: \$25,250 per year in 5th year, \$112,742 total; B: \$24,761 per year in 5th year, \$116,801 total

1. (I) $n = 1$: $2 \cdot 1 = 2$ and $1(1 + 1) = 2$

(II) If $2 + 4 + 6 + \cdots + 2k = k(k + 1)$, then $2 + 4 + 6 + \cdots + 2k + 2(k + 1) = (2 + 4 + 6 + \cdots + 2k) + 2(k + 1)$
$= k(k + 1) + 2(k + 1) = k^2 + 3k + 2 = (k + 1)(k + 2)$.

3. (I) $n = 1$: $1 + 2 = 3$ and $\frac{1}{2}(1)(1 + 5) = \frac{1}{2}(6) = 3$

(II) If $3 + 4 + 5 + \cdots + (k + 2) = \frac{1}{2}k(k + 5)$, then $3 + 4 + 5 + \cdots + (k + 2) + [(k + 1) + 2] = [3 + 4 + 5 + \cdots + (k + 2)] + (k + 3) =$
$\frac{1}{2}k(k + 5) + k + 3 = \frac{1}{2}(k^2 + 7k + 6) = \frac{1}{2}(k + 1)(k + 6)$.

5. (I) $n = 1$: $3 \cdot 1 - 1 = 2$ and $\frac{1}{2}(1)[3(1) + 1] = \frac{1}{2}(4) = 2$

(II) If $2 + 5 + 8 + \cdots + (3k - 1) = \frac{1}{2}k(3k + 1)$, then $2 + 5 + 8 + \cdots + (3k - 1) + [3(k + 1) - 1]$
$= [2 + 5 + 8 + \cdots + (3k - 1)] + 3k + 2 = \frac{1}{2}k(3k + 1) + (3k + 2) = \frac{1}{2}(3k^2 + 7k + 4) = \frac{1}{2}(k + 1)(3k + 4)$.

7. (I) $n = 1$: $2^{1-1} = 1$ and $2^1 - 1 = 1$

(II) If $1 + 2 + 2^2 + \cdots + 2^{k-1} = 2^k - 1$, then $1 + 2 + 2^2 + \cdots + 2^{k-1} + 2^{(k+1)-1} = (1 + 2 + 2^2 + \cdots + 2^{k-1}) + 2^k = 2^k - 1 + 2^k =$
$2(2^k) - 1 = 2^{k+1} - 1$.

9. (1) $n = 1$: $4^{1-1} = 1$ and $\frac{1}{3}(4^1 - 1) = \frac{1}{3}(3) = 1$

(II) If $1 + 4 + 4^2 + \cdots + 4^{k-1} = \frac{1}{3}(4^k - 1)$, then $1 + 4 + 4^2 + \cdots + 4^{k-1} + 4^{(k+1)-1} = (1 + 4 + 4^2 + \cdots + 4^{k-1}) + 4^k =$
$\frac{1}{3}(4^k - 1) + 4^k = \frac{1}{3}[4^k - 1 + 3(4^k)] = \frac{1}{3}[4(4^k) - 1] = \frac{1}{3}(4^{k+1} - 1)$.

11. (I) $n = 1$: $\dfrac{1}{1 \cdot 2} = \dfrac{1}{2}$ and $\dfrac{1}{1 + 1} = \dfrac{1}{2}$

(II) If $\dfrac{1}{1 \cdot 2} + \dfrac{1}{2 \cdot 3} + \dfrac{1}{3 \cdot 4} + \cdots + \dfrac{1}{k(k + 1)} = \dfrac{k}{k + 1}$, then $\dfrac{1}{1 \cdot 2} + \dfrac{1}{2 \cdot 3} + \dfrac{1}{3 \cdot 4} + \cdots + \dfrac{1}{k(k + 1)} + \dfrac{1}{(k + 1)[(k + 1) + 1]} =$

$\left[\dfrac{1}{1 \cdot 2} + \dfrac{1}{2 \cdot 3} + \dfrac{1}{3 \cdot 4} + \cdots + \dfrac{1}{k(k + 1)}\right] + \dfrac{1}{(k + 1)(k + 2)} = \dfrac{k}{k + 1} + \dfrac{1}{(k + 1)(k + 2)} = \dfrac{k + 1}{k + 2}$.

13. (I) $n = 1$: $1^2 = 1$ and $\frac{1}{6} \cdot 1 \cdot 2 \cdot 3 = 1$

(II) If $1^2 + 2^2 + 3^2 + \cdots + k^2 = \frac{1}{6}k(k + 1)(2k + 1)$, then $1^2 + 2^2 + 3^2 + \cdots + k^2 + (k + 1)^2 =$
$(1^2 + 2^2 + 3^2 + \cdots + k^2) + (k + 1)^2 = \frac{1}{6}k(k + 1)(2k + 1) + (k + 1)^2 = \frac{1}{6}(2k^3 + 9k^2 + 13k + 6) = \frac{1}{6}(k + 1)(k + 2)(2k + 3)$.

15. (I) $n = 1$: $5 - 1 = 4$ and $\frac{1}{2}(9 - 1) = \frac{1}{2} \cdot 8 = 4$

(II) If $4 + 3 + 2 + \cdots + (5 - k) = \frac{1}{2}k(9 - k)$, then $4 + 3 + 2 + \cdots + (5 - k) + 5 - (k + 1) = [4 + 3 + 2 + \cdots + (5 - k)] + 5 - (k + 1)$
$= \frac{1}{2}k(9 - k) + 4 - k = \frac{1}{2}(-k^2 + 7k + 8) = \frac{1}{2}(8 - k)(k + 1) = \frac{1}{2}(k + 1)[9 - (k + 1)]$.

17. (I) $n = 1$: $1 \cdot (1 + 1) = 2$ and $\frac{1}{3} \cdot 1 \cdot 2 \cdot 3 = 2$

(II) If $1 \cdot 2 + 2 \cdot 3 + 3 \cdot 4 + \cdots + k(k + 1) = \frac{1}{3}k(k + 1)(k + 2)$, then
$1 \cdot 2 + 2 \cdot 3 + 3 \cdot 4 + \cdots + k(k + 1) + (k + 1)(k + 2) = [1 \cdot 2 + 2 \cdot 3 + 3 \cdot 4 + \cdots + k(k + 1)] + (k + 1)(k + 2)$
$= \frac{1}{3}k(k + 1)(k + 2) + (k + 1)(k + 2) = \frac{1}{3}(k + 1)(k + 2)(k + 3)$.

19. (I) $n = 1$: $1^2 + 1 = 2$ is divisible by 2.

(II) If $k^2 + k$ is divisible by 2, then $(k + 1)^2 + (k + 1) = k^2 + 2k + 1 + k + 1 = (k^2 + k) + 2k + 2$. Since $k^2 + k$ is divisible by 2 and $2k + 2$
is divisible by 2, therefore, $(k + 1)^2 + k + 1$ is divisible by 2.

21. (I) $n = 1$: $1^2 - 1 + 2 = 2$ is divisible by 2.

(II) If $k^2 - k + 2$ is divisible by 2, then $(k + 1)^2 - (k + 1) + 2 = k^2 + 2k + 1 - k - 1 + 2 = (k^2 - k + 2) + 2k$. Since $k^2 - k + 2$ is divisible
by 2 and $2k$ is divisible by 2, therefore, $(k + 1)^2 - (k + 1) + 2$ is divisible by 2.

23. (I) $n = 1$: If $x > 1$, then $x^1 = x > 1$.

(II) Assume, for any natural number k, that if $x > 1$, then $x^k > 1$. Show that if $x > 1$, then $x^{k+1} > 1$:
$x^{k+1} = x^k \cdot x^1 > 1 \cdot x = x > 1$
$\qquad \uparrow$
$\quad x^k > 1$

25. (I) $n = 1$: $a - b$ is a factor of $a^1 - b^1 = a - b$.

(II) If $a - b$ is a factor of $a^k - b^k$, show that $a - b$ is a factor of $a^{k+1} - b^{k+1}$: $a^{k+1} - b^{k+1} = a(a^k - b^k) + b^k(a - b)$. Since $a - b$ is a factor
of $a^k - b^k$ and $a - b$ is a factor of $a - b$, therefore, $a - b$ is a factor of $a^{k+1} - b^{k+1}$.

27. $n = 1$: $1^2 - 1 + 41 = 41$ is a prime number.
$n = 41$: $41^2 - 41 + 41 = 1681 = 41^2$ is not prime.

29. (I) $n = 1$: $ar^{1-1} = a \cdot 1 = a$ and $a \cdot \dfrac{1 - r^1}{1 - r} = a$, because $r \neq 1$.

(II) If $a + ar + ar^2 + \cdots + ar^{k-1} = a\left(\dfrac{1 - r^k}{1 - r}\right)$, then $a + ar + ar^2 + \cdots + ar^{k-1} + ar^{(k+1)-1} = (a + ar + ar^2 + \cdots + ar^{k-1}) + ar^k =$

$a\left(\dfrac{1 - r^k}{1 - r}\right) + ar^k = \dfrac{a(1 - r^k) + ar^k(1 - r)}{1 - r} = \dfrac{a - ar^k + ar^k - ar^{k+1}}{1 - r} = a\left(\dfrac{1 - r^{k+1}}{1 - r}\right)$.

31. (I) $n = 3$: The sum of the angles of a triangle is $(3 - 2) \cdot 180° = 180°$.

(II) Assume for any k that the sum of the angles of a convex polygon of k sides is $(k - 2) \cdot 180°$. A convex polygon of $k + 1$ sides consists of a convex polygon of k sides plus a triangle (see the illustration). The sum of the angles is $(k - 2) \cdot 180° + 180° = (k - 1) \cdot 180°$. Since Conditions I and II have been met, the result follows.

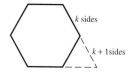

Exercise 8.5

1. 10 **3.** 21 **5.** 50 **7.** 1 **9.** 1.866×10^{15} **11.** 1.483×10^{13} **13.** $x^5 + 5x^4 + 10x^3 + 10x^2 + 5x + 1$
15. $x^6 - 12x^5 + 60x^4 - 160x^3 + 240x^2 - 192x + 64$ **17.** $81x^4 + 108x^3 + 54x^2 + 12x + 1$ **19.** $x^{10} + 5y^2x^8 + 10y^4x^6 + 10y^6x^4 + 5y^8x^2 + y^{10}$
21. $x^3 + 6\sqrt{2}x^{5/2} + 30x^2 + 40\sqrt{2}x^{3/2} + 60x + 24\sqrt{2}x^{1/2} + 8$ **23.** $(ax)^5 + 5by(ax)^4 + 10(by)^2(ax)^3 + 10(by)^3(ax)^2 + 5(by)^4(ax) + (by)^5$
25. 17,010 **27.** $-101,376$ **29.** 41,472 **31.** $2835x^3$ **33.** $314,928x^7$ **35.** 495 **37.** 3360 **39.** 1.00501

41. $\binom{n}{n} = \dfrac{n!}{n!(n - n)!} = \dfrac{n!}{n!0!} = \dfrac{n!}{n!} = 1$ **43.** $2^n = (1 + 1)^n = \binom{n}{0}1^n + \binom{n}{1}(1)(1)^{n-1} + \cdots + \binom{n}{n}1^n = \binom{n}{0} + \binom{n}{1} + \cdots + \binom{n}{n}$ **45.** 1

Exercise 8.6

1. $\{1, 3, 5, 6, 7, 9\}$ **3.** $\{1, 5, 7\}$ **5.** $\{1, 6, 9\}$ **7.** $\{1, 2, 4, 5, 6, 7, 8, 9\}$ **9.** $\{1, 2, 4, 5, 6, 7, 8, 9\}$ **11.** $\{0, 2, 6, 7, 8\}$
13. $\{0, 1, 2, 3, 5, 6, 7, 8, 9\}$ **15.** $\{0, 1, 2, 3, 5, 6, 7, 8, 9\}$ **17.** $\{0, 1, 2, 3, 4, 6, 7, 8\}$ **19.** $\{0\}$
21. $\emptyset, \{a\}, \{b\}, \{c\}, \{d\}, \{a, b\}, \{a, c\}, \{a, d\}, \{b, c\}, \{b, d\}, \{c, d\}, \{a, b, c\}, \{b, c, d\}, \{a, c, d\}, \{a, b, d\}, \{a, b, c, d\}$ **23.** 25 **25.** 40
27. 25 **29.** 37 **31.** 18 **33.** 5 **35.** 175; 125 **37.** (a) 15 (b) 15 (c) 15 (d) 25 (e) 40

Exercise 8.7

1. 30 **3.** 120 **5.** 1 **7.** 336 **9.** 28 **11.** 15 **13.** 1 **15.** 10,400,600
17. $\{abc, abd, abe, acb, acd, ace, adb, adc, ade, aeb, aec, aed$
$\quad bac, bad, bae, bca, bcd, bce, bda, bdc, bde, bea, bec, bed$
$\quad cab, cad, cae, cba, cbd, cbe, cda, cdb, cde, cea, ceb, ced$
$\quad dab, dac, dae, dba, dbc, dbe, dca, dcb, dce, dea, deb, dec$
$\quad eab, eac, ead, eba, ebc, ebd, eca, ecb, ecd, eda, edb, edc\}$; 60
19. $\{123, 124, 132, 134, 142, 143, 213, 214, 231, 234, 241, 243, 312, 314, 321, 324, 341, 342, 412, 413, 421, 423, 431, 432\}$; 24
21. $\{abc, abd, abe, acd, ace, ade, bcd, bce, bde, cde\}$; 10 **23.** $\{123, 234, 124, 134\}$; 4 **25.** 15 **27.** 16 **29.** 8 **31.** 24 **33.** 60 **35.** 120
37. 35 **39.** 1024 **41.** 9000 **43.** $P(5,5) = 5! = 120$ **45.** $C(8,1) \cdot C(15,1) \cdot C(4,1) = 8 \cdot 15 \cdot 4 = 480$ **47.** 336 **49.** 5,209,344
51. 362,880 **53.** 90,720 **55.** 1.156×10^{76} **57.** 15 **59.** (a) 63 (b) 35 (c) 1

Exercise 8.8

1. $S = \{HH, HT, TH, TT\}$; $P(HH) = \frac{1}{4}, P(HT) = \frac{1}{4}, P(TH) = \frac{1}{4}, P(TT) = \frac{1}{4}$
3. $S = \{HH1, HH2, HH3, HH4, HH5, HH6, HT1, HT2, HT3, HT4, HT5, HT6, TH1, TH2, TH3, TH4, TH5, TH6, TT1, TT2, TT3, TT4, TT5, TT6\}$;
each outcome has the probability of $\frac{1}{24}$.
5. $S = \{HHH, HHT, HTH, HTT, THH, THT, TTH, TTT\}$; each outcome has the probability of $\frac{1}{8}$.
7. $S = \{1 \text{ Yellow}, 1 \text{ Red}, 1 \text{ Green}, 2 \text{ Yellow}, 2 \text{ Red}, 2 \text{ Green}, 3 \text{ Yellow}, 3 \text{ Red}, 3 \text{ Green}, 4 \text{ Yellow}, 4 \text{ Red}, 4 \text{ Green}\}$; each outcome has the probability of $\frac{1}{12}$; thus, $P(2 \text{ Red}) + P(4 \text{ Red}) = \frac{1}{12} + \frac{1}{12} = \frac{1}{6}$.
9. $S = \{1 \text{ Yellow Forward}, 1 \text{ Yellow Backward}, 1 \text{ Red Forward}, 1 \text{ Red Backward}, 1 \text{ Green Forward}, 1 \text{ Green Backward}, 2 \text{ Yellow Forward}, 2 \text{ Yellow Backward}, 2 \text{ Red Forward}, 2 \text{ Red Backward}, 2 \text{ Green Forward}, 2 \text{ Green Backward}, 3 \text{ Yellow Forward}, 3 \text{ Yellow Backward}, 3 \text{ Red Forward}, 3 \text{ Red Backward}, 3 \text{ Green Forward}, 3 \text{ Green Backward}, 4 \text{ Yellow Forward}, 4 \text{ Yellow Backward}, 4 \text{ Red Forward}, 4 \text{ Red Backward}, 4 \text{ Green Forward}, 4 \text{ Green Backward}\}$; each outcome has the probability of $\frac{1}{24}$; thus, $P(1 \text{ Red Backward}) + P(1 \text{ Green Backward}) = \frac{1}{24} + \frac{1}{24} = \frac{1}{12}$.
11. $S = \{11 \text{ Red}, 11 \text{ Yellow}, 11 \text{ Green}, 12 \text{ Red}, 12 \text{ Yellow}, 12 \text{ Green}, 13 \text{ Red}, 13 \text{ Yellow}, 13 \text{ Green}, 14 \text{ Red}, 14 \text{ Yellow}, 14 \text{ Green}, 21 \text{ Red}, 21 \text{ Yellow}, 21 \text{ Green}, 22 \text{ Red}, 22 \text{ Yellow}, 22 \text{ Green}, 23 \text{ Red}, 23 \text{ Yellow}, 23 \text{ Green}, 24 \text{ Red}, 24 \text{ Yellow}, 24 \text{ Green}, 31 \text{ Red}, 31 \text{ Yellow}, 31 \text{ Green}, 32 \text{ Red}, 32 \text{ Yellow}, 32 \text{ Green}, 33 \text{ Red}, 33 \text{ Yellow}, 33 \text{ Green}, 34 \text{ Red}, 34 \text{ Yellow}, 34 \text{ Green}, 41 \text{ Red}, 41 \text{ Yellow}, 41 \text{ Green}, 42 \text{ Red}, 42 \text{ Yellow}, 42 \text{ Green}, 43 \text{ Red}, 43 \text{ Yellow}, 43 \text{ Green}, 44 \text{ Red}, 44 \text{ Yellow}, 44 \text{ Green}\}$; each outcome has the probability of $\frac{1}{48}$; thus, $E = \{22 \text{ Red}, 22 \text{ Green}, 24 \text{ Red}, 24 \text{ Green}\}$; $P(E) = n(E)/n(S) = \frac{4}{48} = \frac{1}{12}$.
13. A, B, C, F **15.** B **17.** $\frac{4}{5}; \frac{1}{5}$ **19.** $P(1) = P(3) = P(5) = \frac{2}{9}$; $P(2) = P(4) = P(6) = \frac{1}{9}$ **21.** 0.7 **23.** 0.55 **25.** $\frac{9}{20}$ **27.** $\frac{17}{20}$ **29.** $\frac{5}{18}$ **31.** $\frac{3}{10}$
33. $\frac{2}{5}$ **35.** (a) 0.57 (b) 0.95 (c) 0.83 (d) 0.38 (e) 0.29 (f) 0.05 (g) 0.78 (h) 0.71 **37.** 0.000033068 **39.** (a) $\frac{10}{32}$ (b) $\frac{1}{32}$
41. (a) 0.00463 (b) 0.049 **43.** $\frac{1}{C(30,5)} = 7.02 \times 10^{-6}$; 0.183 **45.** 0.1

Fill-in-the-Blank Items

1. sequence **2.** arithmetic **3.** geometric **4.** Pascal triangle **5.** 15 **6.** union; intersection **7.** 20; 10 **8.** permutation **9.** combination
10. equally likely

True/False Items

1. T **2.** T **3.** T **4.** T **5.** F **6.** F **7.** F **8.** T **9.** F **10.** T **11.** T **12.** F

Review Exercises

1. 120 **3.** 10 **5.** 336 **7.** 56 **9.** $-\frac{4}{3}, \frac{5}{4}, -\frac{6}{5}, \frac{7}{6}, 1, -\frac{8}{7}$ **11.** $2, 1, \frac{8}{9}, 1, \frac{32}{25}$ **13.** $3, 2, \frac{4}{3}, \frac{8}{9}, \frac{16}{27}$ **15.** 2, 0, 2, 0, 2 **17.** Arithmetic; $d = 6; \frac{n}{2}(n + 11)$
19. Neither **21.** Geometric; $r = 8; \frac{8}{7}(8^n - 1)$ **23.** Arithmetic; $d = 4; 2n(n - 1)$ **25.** Geometric; $r = \frac{1}{2}; 6[1 - (\frac{1}{2})^n]$ **27.** Neither **29.** 35
31. $\frac{1}{10^{10}}$ **33.** $9\sqrt{2}$ **35.** $5n - 4$ **37.** $n - 10$ **39.** $\frac{9}{2}$ **41.** $\frac{4}{3}$ **43.** 8
45. (I) $n = 1$: $3 \cdot 1 = 3$ and $\frac{3 \cdot 1}{2}(2) = 3$
    (II) If $3 + 6 + 9 + \cdots + 3k = \frac{3k}{2}(k + 1)$, then $3 + 6 + 9 + \cdots + 3k + 3(k + 1) = (3 + 6 + 9 + \cdots + 3k) + (3k + 3)$
      $= \frac{3k}{2}(k + 1) + (3k + 3) = \frac{3k^2}{2} + \frac{9k}{2} + \frac{6}{2} = \frac{3}{2}(k^2 + 3k + 2) = \frac{3}{2}(k + 1)(k + 2)$.
47. (I) $n = 1$: $2 \cdot 3^{1-1} = 2$ and $3^1 - 1 = 2$
    (II) If $2 + 6 + 18 + \cdots + 2 \cdot 3^{k-1} = 3^k - 1$, then $2 + 6 + 18 + \cdots + 2 \cdot 3^{k-1} + 2 \cdot 3^{(k+1)-1}$
      $= (2 + 6 + 18 + \cdots + 2 \cdot 3^{k-1}) + 2 \cdot 3^k = 3^k - 1 + 2 \cdot 3^k = 3 \cdot 3^k - 1 = 3^{k+1} - 1$.
49. (I) $n = 1$: $1^2 = 1$ and $\frac{1}{2}(6 - 3 - 1) = \frac{1}{2}(2) = 1$
    (II) If $1^2 + 4^2 + 7^2 + \cdots + (3k - 2)^2 = \frac{1}{2}k(6k^2 - 3k - 1)$, then
      $1^2 + 4^2 + 7^2 + \cdots + (3k - 2)^2 + [3(k + 1) - 2]^2 = [1^2 + 4^2 + 7^2 + \cdots + (3k - 2)^2] + (3k + 1)^2 = \frac{1}{2}k(6k^2 - 3k - 1) + (3k + 1)^2$
      $= \frac{1}{2}(6k^3 + 15k^2 + 11k + 2) = \frac{1}{2}(k + 1)(6k^2 + 9k + 2) = \frac{1}{2}(k + 1)[6(k + 1)^2 - 3(k + 1) - 1]$.
51. $x^5 + 10x^4 + 40x^3 + 80x^2 + 80x + 32$ **53.** $32x^5 + 240x^4 + 720x^3 + 1080x^2 + 810x + 243$ **55.** 144 **57.** 84 **59.** (1, 3, 5, 6, 7, 8}
61. {3, 7} **63.** {1, 2, 4, 6, 8, 9} **65.** {1, 2, 4, 5, 6, 9} **67.** 17 **69.** 29 **71.** 7 **73.** 25 **75.** 60 **77.** 128 **79.** 3024 **81.** 70 **83.** 91
85. 1,600,000 **87.** 216,000 **89.** 1260 **91.** (a) 381,024 (b) 1260 **93.** $\frac{3}{20}, \frac{9}{20}$ **95.** $\frac{1}{24}$ **97.** (a) 0.045 (b) 0.318 (c) 0.159
99. (a) 8 (b) 1100 **101.** (a) $(\frac{3}{4})^3 \cdot 20 = \frac{135}{6}$ ft (b) $20 (\frac{3}{4})^n$ ft (c) after the 13th time (d) 140 ft **103.** (a) 0.68 (b) 0.58 (c) 0.32

APPENDIX *Exercise 1*

1. > **3.** > **5.** > **7.** = **9.** < **11.**  **13.** $x > 0$ **15.** $x < 2$ **17.** $x \le 1$ **19.** $2 < x < 5$

21. [0, 4] **23.** [4, 6) **25.** $2 \le x \le 5$ **27.** $x \ge 4$ or $4 \le x < \infty$ **29.** 1 **31.** 5 **33.** 1

35. $\frac{1}{16}$ **37.** $\frac{4}{9}$ **39.** $\frac{1}{9}$ **41.** $\frac{9}{4}$ **43.** 324 **45.** 27 **47.** 16 **49.** $4\sqrt{2}$ **51.** $-\frac{2}{3}$ **53.** y^2 **55.** $\frac{y}{x^2}$ **57.** $\frac{1}{x^3 y}$ **59.** $\frac{25y^2}{16x^2}$ **61.** $\frac{1}{x^3 y^3 z^2}$ **63.** $\frac{-8x^3}{9yz^2}$

65. $\frac{y^2}{x^2 + y^2}$ **67.** $\frac{16x^2}{9y^2}$ **69.** 13 **71.** 26 **73.** 25 **75.** 4 **77.** 24 **79.** Yes; 5 **81.** Not a right triangle **83.** Yes; 25 **85.** 9.4 in.

87. 58.3 ft **89.** 46.7 mi **91.** 3 mi **93.** $a \le b, c > 0; a - b \le 0$ **95.** $\frac{a + b}{2} - a = \frac{a + b - 2a}{2} = \frac{b - a}{2} > 0$; therefore, $a < \frac{a + b}{2}$
                                    $(a - b)c \le 0(c)$
                                      $ac - bc \le 0$ $b - \frac{a + b}{2} = \frac{2b - a - b}{2} = \frac{b - a}{2} > 0$; therefore, $b > \frac{a + b}{2}$
                                        $ac \le bc$

97. No; No **99.** 1 **101.** 3.15 or 3.16
103. The light can be seen on the horizon 23.3 miles distant. Planes flying at 10,000 feet can see it 99 miles away. The ship would need to be 185.8 ft
    tall for the data to be correct.

Exercise 2

1. $10x^5 + 3x^3 - 10x^2 + 6$ **3.** $2ax + a^2$ **5.** $2x^2 + 17x + 8$ **7.** $x^4 - x^2 + 2x - 1$ **9.** $6x^2 + 2$ **11.** $4x^2 - 3x + 1$; remainder 1
13. $4x^2 - 11x + 23$; remainder -45 **15.** $4x^2 + 13x + 53$; remainder 213 **17.** $4x - 3$; remainder $x + 1$

19. $4x - 3$; remainder $-7x + 7$ **21.** $(x - 5)(x + 3)$ **23.** $a(x - 9a)(x + 5a)$ **25.** $(x - 3)(x^2 + 3x + 9)$ **27.** $(3x + 1)(x + 1)$
29. $x^5(x - 1)(x + 1)$ **31.** $3(x - 3)/5x$ **33.** $x(2x - 1)/(x + 4)$ **35.** $5x/[(x - 6)(x - 1)(x + 4)]$
37. $2(x + 4)/[(x - 2)(x + 2)(x + 3)]$ **39.** $(x - 1)(x + 1)/(x^2 + 1)$ **41.** $(x - 1)(-x^2 + 3x + 3)/(x^2 + x - 1)$

Exercise 3

1. $2\sqrt{2}$ **3.** $2x\sqrt[3]{2x}$ **5.** x **7.** $\frac{4}{3}x\sqrt{2}$ **9.** x^3y^2 **11.** x^2y **13.** $6\sqrt{x}$ **15.** $6x\sqrt{x}$ **17.** $15\sqrt[3]{3}$ **19.** $\dfrac{1}{x(2x + 3)}$ **21.** $12\sqrt{3}$ **23.** $2\sqrt{3}$

25. $x - 2\sqrt{x} + 1$ **27.** $\dfrac{\sqrt{2}}{2}$ **29.** $\dfrac{-\sqrt{15}}{5}$ **31.** $\dfrac{\sqrt{3}(5 + \sqrt{2})}{23}$ **33.** $\dfrac{-19 + 8\sqrt{5}}{41}$ **35.** $\dfrac{2x + h - 2\sqrt{x(x + h)}}{h}$ **37.** 4 **39.** -3

41. 64 **43.** $\frac{1}{27}$ **45.** $\dfrac{27\sqrt{2}}{32}$ **47.** $\dfrac{27\sqrt{2}}{32}$ **49.** $x^{17/12}$ **51.** x^2y^4 **53.** $x^{4/3}y^{5/3}$ **55.** $\dfrac{8x^{3/2}}{y^{1/4}}$ **57.** $\dfrac{3x + 2}{(1 + x)^{1/2}}$ **59.** $\dfrac{2 + x}{2(1 + x)^{3/2}}$

61. $\dfrac{4 - x}{(x + 4)^{3/2}}$ **63.** $\frac{1}{2}(5x + 2)(x + 1)^{1/2}$ **65.** $2x^{1/2}(3x - 4)(x + 1)$

Exercise 4

1. -1 **3.** -18 **5.** -3 **7.** -16 **9.** 0.5 **11.** 2 **13.** 2 **15.** -1 **17.** 3 **19.** No real solution **21.** $\{0, 9\}$ **23.** $\{0, 9\}$ **25.** No real solution
27. 2 **29.** 21 **31.** $\{-2, 2\}$ **33.** $-\frac{20}{39}$ **35.** 6

Exercise 5

1. 4 **3.** $\frac{1}{16}$ **5.** $\frac{1}{9}$ **7.** $(x - 2)^2 + (y + 2)^2 = 9$ **9.** $(x + 3)^2 + (y - 1)^2 = 9$ **11.** $(x + \frac{1}{2})^2 + (y + \frac{1}{2})^2 = 1$ **13.** $\{-7, 3\}$ **15.** $\{-\frac{1}{4}, \frac{3}{4}\}$
17. $\left\{\dfrac{-1 - \sqrt{7}}{6}, \dfrac{-1 + \sqrt{7}}{6}\right\}$

Exercise 6

1. $q(x) = x^2 + x + 4$; $R = 12$ **3.** $q(x) = 3x^2 + 11x + 32$; $R = 99$ **5.** $q(x) = x^4 - 3x^3 + 5x^2 - 15x + 46$; $R = -138$
7. $q(x) = 4x^5 + 4x^4 + x^3 + x^2 + 2x + 2$; $R = 7$ **9.** $q(x) = 0.1x^2 - 0.11x + 0.321$; $R = -0.3531$ **11.** $q(x) = x^4 + x^3 + x^2 + x + 1$; $R = 0$
13. No; $f(2) = 8$ **15.** Yes; $f(2) = 0$ **17.** Yes; $f(-3) = 0$ **19.** No; $f(-4) = 1$ **21.** Yes; $f(\frac{1}{2}) = 0$ **23.** 69 **25.** -4 **27.** 15

Index

Abel, Niels, 315, 614
Abscissa (*x*-coordinate), 2
Absolute value, 243–48, 650–51
Absolute value function, 113
Adding *y*-coordinates, 136–37
Addition
  of complex numbers, 207
  of first *n* terms of sequence, 584–85, 590–92,
    595–97
  of matrices, 515–17
  of polynomials, 661–62
  of rational expressions, 669
  of vectors, 455–56, 459, 460
Addition principle of counting, 619
Addition property of inequalities, 222–24
Additive inverse property of vectors, 456
Additive Rule, 632
Advertising, exponential response to, 334–35, 338
Ahmes (Egyptian scribe), 599
Air traffic control problems, 271
Alcohol and driving, function for risk involved in,
  345–47, 350
Algebraic functions, 328. *See also* Polynomial func-
  tions; Rational functions
Algorithm, 302*n*
  division algorithm for polynomials, 302–4, 320
Alternating current in RC circuit problem, 339
Alternating current in RL circuit problem, 338
Altitude problems, 404
Analytic geometry, 408
Ancient tools, estimating age of, 378–79
Anderson, B., xviii, 1959
Annual compounding, 367, 369
Aphelion, 436
Apollonius of Perga, 2
Approximations, 231
Architecture problems, 271, 272
Area
  of circle, 91, 272, 656
  of rectangle, 155, 656
  of triangle, 158–59, 272, 656
Argument of function, 102
Arithmetic mean, 232, 658
Arithmetic progression. *See* Arithmetic sequences
Arithmetic sequences, 588–93, 600
Assignment of probabilities, 630, 631, 638
Associative property
  of matrix addition, 516
  of matrix multiplication, 522
  for vectors, 455
Asymptotes
  of hyperbolas, 443–45
  of rational functions, 288–93
Atmospheric pressure problems, 337, 404

Augmented matrices, 484–85, 487–88
  in echelon form, 490, 492–94
  in reduced echelon form, 491, 493, 494, 495
  row operations on, 487–88, 495–96
Average speed problems, 205
Axis of symmetry (axis)
  of cone, 408
  conjugate, 437
  coordinate, 2
  major axis, 422–31
  minor axis, 422
  of parabola (quadratic function), 257–67, 409
    focus on, equation with, 411
    graphing by hand using, 260–61
    parallel to coordinate axis, equation with, 413–16
  transverse, 437–43, 445–47

Back-substitution, 473, 476, 478, 479
Bacterial growth, 377, 381
Banking problems, 202, 203
Base, 651
  e, 333–35, 339–40
Baseball problems, 13, 628, 629, 644
Basketball problems, 202, 629
Bell, Alexander Graham, 397
Bernoulli, Jacob, 637
Bezout, Etienne, 548
Bill, xi, 1972
Binomial coefficient, 610
Binomials, 660, 663
Binomial Theorem, 578, 607, 610–15
Biology, growth in, 376–77, 390
Bits (binary digits), 619
Blood types, classifying, 620
Boole, George, 637
Border construction problem, 204
Bouncing balls, as geometric series, 601, 646
Bounded graph of system of linear inequalities,
  559–60
Bounds on zeros theorem, 312–15
Box construction problems, 165, 203, 550
BOX function, 25–26
Bracket *x* ([[*x*]]), 113
Braden, T., xi, 1932
Branches of hyperbola, 437
Briggs, Henry, 357
Bürgi, Jobst, 357
Business problems, 49, 202, 203, 204, 243, 253

Calculator, solving linear equation using, 180–81
Calculus, finding components of composite function
  in, 140–41
Calibration of instruments, 453
Carbon dating, 378–79, 381, 405

Cardano, Girolamo, 315, 637
Cartesian coordinate system. *See* Rectangular
    coordinates
Cauchy, Augustin Louis, 231
Cayley, Arthur, 470, 528, 529
CBL Experiment problems, 273, 392–95, 483
Center
    of circle, 55
    of ellipse, 422–31
        at (0, 0), 422–28
        at (*h*, *k*), 428–31
    of hyperbola, 437
        at (0, 0), 437–43
        at (*h*, *k*), 445–47
Challenger disaster problem, 339
Change-of-Base Formula, 355–56
Chemistry problems, 77, 196–97, 202, 233, 273,
    349, 391
Chu Shihchie, 614
Circle(s), 55–60, 408
    area of, 91, 272, 656
    center of, 55
    circumference of, 272, 656
    defined, 55
    general form of equation of, 59–60
    standard form of equation of, 55–59
    unit, 58
Circumference of circle, 272, 656
Closed interval, 649
Code formation problems, 628
Coding matrix, 505
Coefficient(s), 663
    binomial, 610
    correlation, 67
    leading, 318, 654, 660
    of monomial, 660
    of polynomials, 654, 660
        complex polynomial functions, 319–21
Coefficient matrix, 485
Coincident lines, 472
Coin toss, probability in, 630, 636, 640, 645
Colleen, xviii, 1972
Collinear points, 513
Column index of matrix, 484, 513
Column vector, 518–19
Combination locks problem, 628
Combinations, 624–30, 635–36
Combinatorics, 578, 637
Combined inequalities, solving, 228–30
Combined variation, 74–75
Committee formation problems, 625–26, 628, 645
Common difference, 588, 589
Common logarithm (log), 354–55
Common ratio, 594
Commutative property
    of matrix addition, 516
    for vectors, 455
Complement of set, 617
Complete graph, 15, 19–21
Completing the square, 258, 259, 683–86
Complex numbers, 206–11
    matrices and, 528–29
    solving quadratic equations in, 212–13
Complex polynomial functions, 318–21
Complex variable, 318
Components of vector, 458, 459

Composite function, 137–41
Composition, 137
Compound interest, 366–75, 405
    comparing investments using different compound-
        ing periods, 367–68
    computing, 366–67
    continuous compounding, 368–69
    defined, 366
    doubling and tripling time for investment,
        371–72, 375
    formula, 367
    interest rate required to double investment, 371
    present value and, 369–71
Compressions, 126–28, 276–78
Computers, application of sets and counting to, 619
Cone, 163, 408
Congruent triangles, 8
Conics, 407–54. *See also* Ellipses; Hyperbolas;
        Parabolas
    defined, 408
Conjugate, 208–10
Conjugate axis, 437
Conjugate hyperbola, 453
Conjugate Pairs Theorem, 319–20
Connected mode, 114
Consistent systems of equations, 471, 472, 476
Constant, 648
    of proportionality, 71
Constant function, 104, 105, 110–11, 274
Constraints in linear programming, 563
Continuous compounding, 368–69. *See also* Growth
        and decay
Continuous function, 175
Continuous graph, 278
Cooling, Newton's Law of, 359, 379–80, 381–82,
        392–93
Coordinate axes, 2
    major axis parallel to, 428–31
    transverse axis parallel to, 445–47
Coordinate of the point, 2
Coordinates, 648
    rectangular, 2–14
Corner points, 560
Correct to *n* decimal places, 174
Correlation coefficients, 67
Cosine function, hyperbolic, 340
Cost minimization problems, 569
Cost problems, 49, 101, 122, 143, 166, 253
Counting, 618–19
    addition principle of, 619
    general counting principle, 621
    Multiplication Principle of, 622, 626
    without repetition, 622–26
Counting formula, 618
Counting numbers (natural numbers), 648
Cramer, Gabriel, 470
Cramer's Rule, 470, 501, 503–6, 508–10
Craps, probability in game of, 634–35
Credit cards, minimum payments and interest
        charged for, 159–60
Cryptography, 505
Cube function, 90, 111–12
Cube root, 653, 674
Cubes, 656, 663
Cubic equations, 315
Current of stream, finding speed of, 482

Curve fitting
    exponential, 382, 383–85, 392–93
    linear, 64–71, 167
    logarithmic, 382, 385–86, 393
    nonlinear. *See* Nonlinear curve fitting
    power, 382, 386–88, 394–95
    problems, 392–95, 483, 499, 573
Cylinder, volume of, 163
Cylinder construction problems, 160–61, 166, 203,
    550

Daily compounding, 368, 369
Danny, xi, 1970
Dantzig, George, 562*n*
Decay. *See* Growth and decay
Decibels, 397–99
Decimals, 648
Decreasing function, 104–5, 145
Degree
    of monomial, 660, 663
    of polynomial, 654, 660, 663
    of polynomial functions, 274
Dellen, D., xi, 1940
Demand equation, 155–56, 163
Demand problems, 154, 404
Denominator, 667
    partial fraction decomposition and, 532–38
    rationalizing the, 675–76
Dependent equations, 472
Dependent variable, 90–91
Depreciation problems, 602
Depressed equation, 309–10
Descartes, René, 2, 86
    Method of Equal Roots, 551
    Rule of Signs, 306–7, 314
Determinants, 470, 501–13
    properties of, 510–11
    systems of three equations containing three vari-
        ables, 508–10
    3 × 3, 506–8
    2 × 2, 501–6, 510
Diagonal entries, 522
Dice roll, probability in, 631, 640
Dietary requirements problem, 568
Difference. *See also* Subtraction
    common, 588, 589
    of complex numbers, 207
    of two cubes, 663
    of two squares, 662
Difference function, 135, 136
Difference quotient, 103
Diophantus, 599
Directed line segments, 454–55
Direction of vector, 454
Directrix of parabola, 409
Direct variation, 71–73
Dirichlet, Lejeune, 86
Discontinuity, 114
Discount pricing problems, 203
Discriminant of quadratic equations, 187
    negative, 211–13
Distance
    mean, 436
    between P and Q, 651
    speed/distance problems, 205, 547, 550
    using sound to measure, 221

Distance formula, 5–8
Distributive property of matrix multiplication, 522
Dividend, 302, 664
Division
  of polynomials, 663–65
  of rational expressions, 668
  synthetic, 308–12, 320–21, 686–90
Division algorithm for polynomials, 302–4
  with complex coefficients, 320
Divisor, 302, 664
Domain
  of function, 86, 89–90, 96
    inverse function, 146
    logarithmic function, 342–43
    rational function, 286
  of variable, 648
Dot mode, 114, 239
Double root, 184
Doubling time for investment, 371–72, 375
Drug medication problems, 338, 350

e, 333–35, 339–40
Earthquake, magnitude of, 399–401
Eccentricity e, 437, 453
Echelon form, matrix in, 489–94
  reduced echelon form, 491, 493–96
Economics problems, 155–56, 390–91, 395–96, 406
Effective rate of interest, 369
Electrical resistance of wire, 76
Electricity problems, 482–83, 500
Elements of set, 615, 616, 647
Elevation, effect on weight, 101
Elimination, method of, 470, 475–79
  to solve systems of nonlinear equations, 540–43, 546
Ellipses, 408, 422–37
  center of, 422–31
    at (0, 0), 422–28
    at (h, k), 428–31
  defined, 422
  eccentricity e of, 437
  equation of, 422–31
  reflection property of, 431–32
  vertex of, 422
Ellipsis, 648
Empty (null) set, 615
Enclosure problems, 163, 202, 270, 550, 551, 659
Entries of matrix, 484, 513, 522
Environmental concerns problems, 143
Equality
  of complex numbers, 207
  of vectors, 458–59
Equally likely outcomes, probability for, 633–35
Equal matrices, 515
Equal Roots, Descartes' Method of, 551
Equal sets, 616, 647
Equation(s), 680–83
  cubic, 315
  dependent, 472
  depressed, 309–10
  of ellipses, 422–31
  equivalent, 680–81
  exponential, 361–65
  factorable, 219–20
  graphs of, 14–34
    defined, 14
    graphing utility for, 16–21

by hand, 15–16
  intercepts, finding, 21–26, 44
  symmetry, checking for, 26–31
of hyperbolas, 437–47
independent, 472
involving absolute value, 244–45
linear, 35, 43–44, 179–82, 471–72. See also
    Systems of linear equations
logarithmic, 358–61, 364
in one variable, 680
of parabolas, 409–15
quadratic. See Quadratic equations
quadratic in form, 217–19
quartic (fourth-degree), 315
radical, 215–17, 676–77
setting up applied problems, 194–201
sides of, 680
solutions (roots) of, 680
solving, 174–79, 682
of straight line, deriving, 39–47
systems of. See Systems of equations
Equilateral hyperbola, 453
Equilateral triangle, 13, 272
Equivalence property of inequalities, 240
Equivalent equations, 680–81
Equivalent inequalities, 225, 230
Equivalent system of equations, 475–76, 487
Euclid, 191, 600, 614
Euler, Leonhard, 86, 340, 637
Even function, 107–10
Even power function, 275
Events, 632
Expectation, mathematical, 637
Expense computation problems, 204
Explicit form of function, 88
Exponential curve fitting, 382, 383–85, 392–93
Exponential equations, 361–65
Exponential expressions, changing logarithmic expressions to, 341
Exponential functions, 328–40
  base e, 333–35, 339–40
  defined, 329
  graphs of, 329–33
  properties of, 335
Exponential law, 376–82
EXPonential REGression option, 384
Exponents, 651–54
  laws of, 329, 361, 652, 678
  rational, 653–54, 677–79
  relating logarithms to, 341
Extended principle of mathematical induction, 606
Extraneous solutions, 215

Factor(s), 665
  irreducible quadratic, 312, 536–38
  linear, 532–36
  synthetic division to verify, 688
Factorable equations, 219–20
Factorial symbol $n!$, 581–82
Factoring, 182–84, 665–67
Factor Theorem, 304–5, 312, 319, 320–21, 688
Falling objects problems, 76
Family of lines, 63
Farm management problems, 568
Feasible points in linear programming problem,
    563, 564

Federal income tax problems, 167, 233
Fermat, Pierre de, 2, 340, 636, 637
Ferrari, Lodovico, 315
Fibonacci, 599
Fibonacci numbers, 583
Fibonacci sequence, 583, 587
Finance problems, 202, 392
Financial planning problems, 196, 202, 481,
    499–500, 559–60, 561, 562–63, 569
Finite set, 618
First-degree equation. See Linear equations
Focus, foci
  of ellipse, 422
  of parabola, 409–13
Football problems, 202, 204, 629
Force of wind on window problems, 74–75, 76
Forces represented by vectors, 461–62
Fractions
  clearing equation of, 180
  partial, 532
  partial fraction decomposition, 532–38
Fritz, xi, 1974–1987
Frobenius, G., 528
Function(s), 85–172
  absolute value, 113
  algebraic, 328. See also Polynomial functions;
    Rational functions
  argument of, 102
  composite, 137–41
  constant, 104, 105, 110–11, 274
  continuous, 175
  cube, 90, 111–12
  defined, 86–88, 96
  domain of, 86, 89–90, 96
  even and odd, 107–10
  example of, 87–88
  explicit form of, 88
  function notation, 96, 102–3
  graphing techniques, 123–31
  graphs of, 91–92, 96
  greatest-integer, 113–14
  identity, 111
  implicit form of, 88
  increasing and decreasing, 104–5, 145
  independent and dependent variables in, 90–91
  linear, 110, 274
  local maximum/local minimum of, 105–7
  mathematical models involving, 155–67
  objective, 562, 563
  one-to-one, 144–45
    inverse of, 146–52
  operations on, 135–37
  piecewise defined, 115–17
  quadratic. See Quadratic functions
  range of, 86, 87, 90
  reciprocal, 112–13
  as set of ordered pairs, 93–95
  square, 93, 111
  square root, 112
  transcendental, 328. See also Exponential func-
    tions; Logarithmic functions
  values of, finding, 88
  zero, 274
Function keys, 89
Function notation, 96, 102–3
Fundamental Theorem of Algebra, 318–19

Galois, Evariste, 315
Gas laws, 72, 77
Gauss, Karl Friedrich, 318, 470
General counting principle, 621
General form of equation
  of circle, 59–60
  of line, 43–44
Generators of cone, 408
Geometric mean, 232
Geometric progression. *See* Geometric sequences
Geometric sequences, 593–602
Geometric series, 597–99, 601, 602
Geometry, analytic, 408
Geometry formulas, 656
Geometry problems, 7–8, 14, 62, 75–76, 81, 99,
  202, 203, 512–13, 550, 551, 574, 606,
  658–59
Gibbs, Josiah Willard, 463
Grade computation problems, 203
Graph(s), 1–83
  of circles, 55–60
  complete, 15, 19–21
  of equations. *See under* Equation(s)
  of functions, 91–92, 96
    exponential, 329–33
    logarithmic, 343–47, 356
    polynomial, 276–82
    rational, 286–88, 293–98
  of inequality in two variables, 552–55
  linear curve fitting. *See* Linear curve fitting
  of parallel lines, 51–53
  of perpendicular lines, 53–55
  rectangular coordinates, 2–14
    distance between points, 5–8
    graphing utilities for, 3–5
    midpoint formula, 8–9
  of straight line, 35–51
    equations of, deriving, 39–47
    slope of, 35–37
    square screens, 37–39
  of system of inequalities, 555–60
  variation, 71–78
Graphing calculators
  to calculate logarithms with base other than 10
    or *e*, 355–56
  CBL Experiment problems, 273, 392–95, 483
  e^x or $exp(x)$ key on, 333
  to evaluate powers of 2, 328
Graphing utilities, 3–5
  to determine even vs. odd function, 108–10
  to determine where function is increasing and de-
    creasing, 105
  domain of function estimated using, 89–90
  to evaluate 2×2 determinant, 502–3
  finding coordinates of point shown on screen of,
    4–5
  function keys on, 89
  to graph bacterial growth, 377
  to graph circles, 56, 59
  to graph ellipse, 425–26, 427, 430–31
  to graph equations, 16–21
    locating intercepts of equation, 23–26
    logarithmic and exponential, 363–65
    solving equations, 174–79
  to graph hyperbola, 440–42, 446–47
  to graph inequality, 226, 553–54

to graph logarithmic function with base other
    than 10 or *e*, 356
  to graph Newton's Law of Cooling, 379–80
  to graph parabola, 410–12, 415–16
  to graph piecewise defined function on, 115–16
  to graph power functions, 276–78
  to graph radioactive decay, 378–79
  to graph rational function, 287–88, 293–98
  to graph system of linear inequalities, 555–56
  INTERSECT command, 472*n*
  to locate local maxima and minima, 106–7
  min/max settings for square screen, 37–38
  REGression options on, 383, 384, 386, 387
  row operations on, 495–96
  scatter diagrams constructed using, 65
  for sequences, 580, 583, 591, 596–97
  to solve system of equations
    linear equations, 472–73
    nonlinear equations, 539–45
    using Cramer's Rule, 504–6
  TABLE function, 583
Grassmann, Hermann, 463
Gravity, effect of, 99
Greatest-integer function, 113–14
Grouping, factoring by, 666
Growth and decay, 376–82, 383, 391, 392–93

Half-life, 378
Half-open or half-closed intervals, 649
Half-planes, 554–55
Hamilton, William Rowan, 463
Harmonic mean, 232
Harriot, Thomas, 191
Healing of wounds, exponential model of, 338, 350
Heron, 599
Horizontal asymptote, 288, 289, 290–93
Horizontal line, equation of, 41–42
Horizontal-line test, 144–45
Horizontal multiplication of polynomials, 662
Horizontal shifts, 125–26
Horsepower problems, 76
Household voltage problems, 249
Huygens, Christian, 637
Hydrocarbons and carbon monoxide, relating emis-
    sion levels of, 81
Hyperbolas, 408, 409, 437–54
  applications of, 448–50
  asymptotes of, 443–45
  branches of, 437
  center of, 437
    at (0, 0), 437–43
    at (*h, k*), 445–47
  conjugate, 453
  defined, 437
  eccentricity *e* of, 453
  equation of, 437–47
  equilateral, 453
  vertices of, 437
Hyperbolic cosine function, 340
Hyperbolic sine function, 340
Hypotenuse, 655

Identity, 680
Identity function, 111
Identity matrix, 522–23

Identity property, 523
  for vectors, 455–56
Imaginary number, pure, 207
Imaginary unit (*i*), 206
  power of *i*, 210–11
Implicit form of function, 88
Improper rational expression, 532
Improper rational functions, 291, 292, 293
Inconsistent systems of equations, 471, 472, 477
Increasing function, 104–5, 145
Independent equations, 472
Independent variable, 90–91, 102
Index
  of radical, 653, 674
  of sum of first *n* terms of sequence, 584, 585
Individual Retirement Account (IRA), computing
    value of, 369
Induction, mathematical, 578, 603–6
Inequalities, 222–43
  combined, solving, 228–30
  equivalent, 225, 230
  graphing, 649–50
  involving absolute value, 245–48
  linear, 226–31, 552–55
  in one variable, 225
  polynomial, 237–39
  properties of, 222–25, 240
  quadratic, 234–37
  rational, 239–40
  solving, 225–26
  systems of, 552–67
  triangle, 249
  in two variables, 552–55
Inequality signs, 649
Infinite geometric series, 597
Infinite set, 618
Infinity
  notation for, 649–50
  unbounded in negative direction ($H \rightarrow \infty$), 288
  unbounded in positive direction ($H \rightarrow \infty$), 287
Initial point, 454
Integers, 648
  factoring over the, 665
Intensity
  of earthquake, 400–401
  of light, 253
  of sound wave, 397, 398–99, 401
Intercept form of equation of line, 49
Intercepts
  finding, 21–26, 44
  graphing quadratic function by hand using,
    260–61
  slope-intercept form of equation of line,
    44–47
Interest
  charged for credit cards, 159–60
  compound, 366–75, 405
  defined, 366
  problems on, 196–97, 205
  rate of, 196, 366
    effective, 369
  simple, 366
Intermediate Value Theorem, 175, 364
INTERSECT command, 371, 472*n*
Intersecting lines, 472
Intersection of sets, 616–17

Interval notation for inequalities
  involving absolute value, 245–48
  linear inequality solution, 227, 228, 229
  polynomial inequalities, 238
  quadratic inequalities, 234, 236
  rational inequality, 239, 240
Intervals, 649–50
Inverse matrix, 523–28
Inverse of function, 146–52
Inverse variation, 73–74
Investments, 569
  doubling and tripling time for, 371–72, 375
  interest rate required to double, 371
  time to reach goal for, 375
  using different compounding periods, comparing, 367–68
IQ test scores problems, 233
Irrational numbers, 206, 648
Irreducible quadratic factor, 312, 536–38
Isomorphism, 528
Isosceles triangle, 13, 158–59

Jenny, xi, 1977
Joint variation, 74–75
Jonos, R., xi, 1948
Jordan, Camille, 470

Karmarkar, Narendra, 562n
Katy, xi, 1965
Kepler, Johannes, 74, 389n
Keplerian Laws of Planetary Motion, 78, 82, 389n
Kinetic energy problems, 76
Kirchhoff's Rules, 482–83, 500

Latus rectum, 410, 411
Laws of exponents, 329, 361, 652, 678
Leading coefficients, 318, 654, 660
Learning curve function, 350
Least common multiple (LCM), 669–71
Left end point, 649
Legs of right triangle, 655
Leibniz, Gottfried Wilhelm von, 86, 470
Length of major axis, 422
Life expectancy problems, 233
Light, intensity of, 253
Like terms, 660
Limiting magnitude of telescope, 404
Line, straight. See Straight line
Linear algebra, 513
Linear curve fitting, 64–71, 167
Linear equations, 35, 43–44, 179–82, 471–72. See also Systems of linear equations
Linear factors, 532–36
Linear function, 110, 274
Linear inequalities, 226–31, 552–55
  system of, 555–60
Linear programming problem, 231, 470, 562–69
LINear REGression option, 384n
Line segment, 7, 454–55
Lines of best fit, 66–67
Local maximum/local minimum, 105–7, 280
Logarithm(s)
  Change-of-Base Formula, 355–56
  common (log), 354–55
  natural (ln), 354–55, 357
  properties of, 351–54, 356
  relating exponents to, 341

Logarithmic curve fitting, 382, 385–86, 393
Logarithmic equations, 358–61, 364
Logarithmic expressions, changing exponential expressions to, 341
Logarithmic functions, 341–51
  defined, 341
  domain of, 342–43
  graphs of, 343–47, 356
  natural logarithm function (ln), 343–44
  properties of, 342, 347
Logarithmic scales, 397–401
LOng RAnge Navigation system (LORAN), 448–50, 453, 467
Lottery, probability of winning, 639
Loudness of sound, 397–99, 401, 405
Lower bounds to zeros of polynomials, 312–15

Magnitude
  of earthquake, 399–401
  of telescope, limiting, 404
  of vector, 454, 457, 459
Major axis, 422–31
Manufacturing problems, 561
Marginal propensity to consume, 69
Marginal propensity to save, 70
Marketing problems, 390–91
Mathematical expectation, 637
Mathematical induction, 578, 603–6
Mathematical modeling, 194–96
Matrix, matrices, 470
  arranging data in, 514
  augmented, 484–85, 487–88, 490–96
  coding, 505
  coefficient matrix, 485
  column index of, 484, 513
  complex numbers and, 528–29
  defined, 484, 513
  in echelon form, 489–94
    reduced echelon form, 491, 493–96
  entries of, 484, 513
  m by n, 514
  row index of, 484, 513
  row operations on, 486–96
  solving system of linear equations using, 488–95
  square, 514, 521
  zero, 516
Matrix algebra, 513–31
  addition and subtraction of matrices, 515–17
  equal matrices, 515
  history of, 528–29
  identity matrix, 522–23
  inverse matrix, 523–28
  multiplication of matrices, 518–22, 529
Maximizing revenue, 256, 262–63, 270
Maximum area, constructing function defining, 161–62
Maximum linear programming problem, 565–67
Maximum value, 262–63, 264
m by n matrix, 514
Mean
  arithmetic, 232, 658
  geometric, 232
  harmonic, 232
Mean distance, 436
Medians of triangle, 13
Midpoint formula, 8–9

Mike, xi, 1967
Minimum linear programming problem, 565
Minimum payments and interest charged for credit cards, function describing, 159–60
Minimum value, 262–63, 265
Min/max settings
  for complete graph, 20–21
  for square screen, 37–38
Minor axis, 422
Mirrors, problems involving, 467
Mitosis, 376–77
Mixed quotients, 671–72
Mixture problems, 197–98, 205, 481, 494–95, 499, 562, 574
Monomial, 654, 660, 663
Monthly compounding, 368, 369
Motion
  Newton's Second Law of, 257, 456
  of projectile, analyzing, 257, 263–65, 270–71, 272
  uniform motion problems, 198–200, 204, 205, 253
Multiplication
  of complex numbers, 207–8
    by its conjugate, 208–9
  of matrices, 518–22, 529
    of matrix by its inverse, 523–25
  of polynomials, 662–63
  of rational expressions, 668
  of vectors, by numbers, 456–57, 459
Multiplication Principle of counting, 622, 626
Multiplication properties for inequalities, 224–25
Multiplicity of x-intercept, 279–80, 281, 282
Mutually exclusive events, 632

Napier, John, 357
Nappes, 408
Natural logarithm (ln), 354–55, 357
  function, 343–44
Natural numbers (counting numbers), 603–6, 648
Navigation problems, 265, 266, 271
  LORAN system, 448–50, 453, 467
Negative discriminant, 211–13
Negative numbers, square roots of, 212
Negative real numbers, 648
Newton, Isaac, 379n
Newton's Law, 76
Newton's Law of Cooling, 359, 379–80, 381–82, 392–93
Newton's Second Law of Motion, 257, 456
Niccolo of Brescia (Tartaglia), 315
Niklas, P., xi, 1944
Nondistinguishable permutation, 626–27
Nonlinear curve fitting, 382–97
  choosing best model, 388–89, 395–97
  exponential, 382, 383–85, 392–93
  logarithmic, 382, 385–86, 393
  power, 382, 386–88, 394–95
  scatter diagrams of, 382, 383, 384, 386, 387, 389
Nonlinear equations, systems of. See Systems of nonlinear equations
Nonnegative property, 222
Nonsingular matrix, 523
  finding inverse of, 525–26
Null set, 615
Number e, 333–35, 339–40
Numerator, 667
  rationalizing, 676

Objective function, 562, 563
Oblique asymptote, 289, 291–93
Odd function, 107–10
Odd power function, 276
Ohm's law, 230–31
Omar Khayyám, 614
One-to-one function, 144–45
Open interval, 649
Optics problems, 337, 349
Orbit of planet about Sun, problems on, 436
Ordered pair(s), 2
    function as set of, 93–95
Ordinate (*y*-coordinate), 2, 136–37
Origin *0*, 2, 648
    finding distance to point on graph from, 156
    symmetry with respect to, 27–28, 29, 107–10
Outcome(s), 630
    equally likely, 633–35

Parabolas, 275, 408–21
    applications of, 416–17
    defined, 409
    directrix of, 409
    equation of, 409–15
    focus of, 409–13
    graphs of quadratic functions as, 257–67
    reflecting property of, 416–17
    vertex of, 409–16
        at (0, 0), 409–13
        at (*h*, *k*), 413–16
Parabolic arch bridge problems, 326, 421, 423, 467
Paraboloid of revolution, 407, 408, 416, 420, 467
Parallel lines, 51–53, 472
Partial fraction decomposition, 470, 532–38
Partial fractions, 532
Pascal, Blaise, 608, 636, 637
Pascal triangle, 608–10, 612–13, 614, 625, 636
Pat, xi, 1965
Payment period, 366
Peano, Giuseppe, 637
Pendulum
    period of, 101, 134, 154, 387
    simple, 76, 387–88
    swings of, as geometric series, 598–99, 601
Percentage method of withholding for federal
        income tax, 233
Perfect cubes, 663
Perfect roots, 653, 674
Perfect square, 663, 673
Perfect square root, 673
Perihelion, 436
Perimeter of rectangle, 656. *See also* Enclosure
        problems
    fixed, 155, 270
Permutations, 622–24, 626–30, 635–36
Perpendicular lines, 53–55
Physical forces problems, 77–78
Physics problems, 76, 199–200, 202, 203, 204, 221,
        230–31, 242, 243, 253
Piecewise defined function, 115–17
Pixels, 3
Planetary Motion, Keplerian Laws of, 78, 82, 389*n*
Plot of points, 2
Point of tangency, 63
Point-slope form of equation of line, 41–42
Polynomial(s), 654–55, 660–67
    adding and subtracting, 661–62

coefficients of, 654, 660
degree of, 654, 660, 663
dividing, 663–65
factoring, 182–84, 665–67
least common multiple of, 669–71
multiplying, 662–63
prime, 666
standard form of, 655, 661
synthetic division to find value of, 689
terms of, 654, 661
in three variables, 663
in two variables, 663
zero, 654, 661
Polynomial functions, 256, 274–85
    complex, 318–21
    defined, 274
    degree of, 274
    graphing, 276–82
    identifying, 274
    power functions, 275–78, 281
    ratios of. *See* Rational functions
    zeros of, 302–15
        of complex polynomials, 319–21
        Descartes' Rule of Signs to locate, 306–7, 314
        division algorithm for polynomials, 302–4, 320
        Factor Theorem, 304–5, 312, 319, 320–21
        listing all zeros, 321
        number of zeros theorem, 306
        Rational Zeros Theorem, 307–12, 314
        Remainder Theorem, 304, 689
        upper and lower bounds to, 312–15
        using zeros to form polynomial, 321
        zero of multiplicity *m* of *f*, 278–80, 281, 282, 321
Polynomial inequalities, 237–39
Population growth problems, 330, 377, 381, 393,
        397, 405
Position vector, 458
Positive real numbers, 648
Power curve fitting, 382, 386–88, 394–95
Power functions, 275–78, 281
PoWeR REGression option, 387
Powers. *See* Exponents
Powers of *i*, 210–11
Prediction, linear curve fitting and, 68
Present value, 369–71
Price problems, 203, 233, 499
Prime polynomials, 666
Principal, 196, 366, 673
Principal *n*th root of number *a*, 652–53, 674
Principal square root, 673
    of -*N*, 211–12
Probability, 578, 630–40
    applications involving permutations and combina-
        tions, 635–36
    assignment of, 630, 631, 638
    computing, 645
    defined, 630
    for equally likely outcomes, 633–35
    of event *E*, 632
    history of, 636–37
    models, 630–33
Product(s)
    of complex numbers, 207–8
        and its conjugate, 208–9
    of row vector by column vector, 518–19
    scalar, 456, 459, 460
    special, 662

Product design problems, 569
Product function, 135, 136
Production cost problems, 143, 531
Production scheduling problem, 568
Profit computation problems, 49, 531
Profit function problem, 338
Profit maximization problems, 565–67, 569, 575
Programming exercise, 629
Proper rational expression, 532
Proper rational functions, 290
Proper subset, 616
Proportionality, constant of, 71. *See also* Variation
Pure imaginary number, 207
Purity of gold problem, 204
Pythagorean Theorem, 5, 6, 655–56

Quadrants, 3
Quadratic equations, 182–91
    completing the square to solve, 684–85
    defined, 182
    discriminant of, 187, 211–13
    factoring, 182–84
    quadratic formula to solve, 185–90, 212
    in standard form, 182
Quadratic factor, irreducible, 312, 536–38
Quadratic formula, 185–90, 212
Quadratic functions, 256–73, 274
    applications of, 262–67
    defined, 256
    graphing, 257–62. *See also* Parabolas
    x-intercepts of, 260–61
Quadratic inequalities, 234–37
Quadratic type, equation of, 217–19
Quarterly compounding, 368, 369
Quarternions, 463
Quartic (fourth-degree) equations, 315
Quotient(s), 302, 664
    of complex number in standard form, 209
    difference, 103
    mixed, 671–72
    synthetic division to find, 687–88
Quotient function, 135, 136

Radical equation, 215–17, 676–77
Radicals, 653, 673–77
Radical sign, 673
Radicand, 653, 674
Radioactive decay, 378–79, 381, 382, 391
Radius of circle, 55
RANGE, 3, 313–14
Range of function, 86, 87, 90, 146
Rate of interest, 196, 366
    effective, 369
Ratio, common, 594
Rational exponents, 653–54, 677–79
Rational expressions, 667–69
    partial fraction decomposition of, 532–38
    proper vs. improper, 532
    reduced to lowest terms (simplified), 667
    using least common multiple to add, 670–71
Rational functions, 256, 285–302
    asymptotes, 288–93
    defined, 285
    domain of, finding, 286
    graphing, 286–88, 293–98
    improper, 291, 292, 293
    in lowest terms, 286

proper, 290
unbounded in negative direction ($H \to \infty$), 288
unbounded in positive direction ($H \to \infty$), 287
zeros of, 286, 290
Rational inequalities, 239–40
Rationalizing numerators, 676
Rationalizing the denominator, 675–76
Rational numbers, 206, 285, 648
factoring over the, 665
Rational Zeros Theorem, 307–12, 314
Real estate problems, 233
Real number line, 648
Real numbers, 206, 647–51
factoring over the, 665
Reciprocal function, 112–13
Reciprocal of complex number in standard form, 209
Reciprocal property for inequalities, 225
Rectangle
area of, 155, 656
perimeter of, 656
fixed, 155, 270
Rectangular coordinates, 2–14
distance between points, 5–8
graphing utilities for, 3–5
midpoint formula, 8–9
Recursive formula, 582
for arithmetic sequence, 589–90
Recursively defined sequences, 578, 582–84
Reduced echelon form, 491, 493–96
Reflecting property
of ellipses, 431–32
of parabolas, 416–17
Reflecting telescopes, 420
Reflection about x-axis and y-axis, 128–29
graphing exponential functions using, 332–33
graphing logarithmic functions using, 344–45
graphing power functions using, 276–78
Refund, computing, 482
Register, 619
REGression options, 383, 384, 386, 387
Remainder, 302, 664
synthetic division to find, 687–88
Remainder Theorem, 304, 689
Repeated (multiple) zero of f, 279
Repeated solution, 184
Repeating decimals as geometric series, 598
Repetition
counting without, 622–26
permutations with, 626–27
Rescue at sea problems, 205, 253
Resistance due to conductor, 76
Resultant of F_1 and F_2, 461
Revenue maximization, 256, 262–63, 270, 569
Rhind papyrus, 599–600
Richter, C.F., 399n
Richter scale, 399
Right angle, 655
Right circular cone, 408
Right end point, 649
Right triangle, 655–56
Rise, vertical, 35
Root(s)
cube, 653, 674
Descartes' Method of Equal, 551
of multiplicity 2 (double root), 184
perfect, 653, 674

principal nth root of number a, 652–53, 674
square, 212, 653, 673, 674
Roster method, 647
Row index of matrix, 484, 513
Row operations on matrix, 486–95
on graphing utility, 495–96
to solve system of linear equations, 488–95
Row vector, 518–19
Ruffini, P., 315
Rule of Signs, Descartes', 306–7, 314
Rumors, exponential spread of, 338, 350
Run, horizontal, 35
Ryan, xi, 1995

Safe weight problems, 73, 76
Sales commission problems, 233
Sales problems, 470, 476, 482, 483
Salvage value problems, 404
Sample space, 630, 631
Sarah, xi, 1980
Satellite dish problems, 416–17, 420
Satellites in orbit problems, 77
Scalar multiple of matrix, 517
Scalar multiplication, properties of, 518
Scalar product, 456, 459, 460
Scalars, 456–57, 458, 517
Scale, 648
logarithmic, 397–401
Scatter diagrams
finding equation of line from, 66–67
of linear curve fitting, 64–65
of nonlinear curve fitting, 382–84, 386, 387, 389
Schroeder, E., 637
Scrolling, 17
Search and rescue problems, 253
Secant line, 103
Second-degree equation. See Quadratic equations
Seki Kōwa, 470
Semielliptical arch bridge problems, 423, 435, 467
Sequences, 578–602
arithmetic, 588–93, 600
defined, 578
factorial symbol $n!$ and, 581–82
Fibonacci, 583, 587
geometric, 593–602
history of, 599–600
properties of, 585
recursively defined, 578, 582–84
summation notation, 584–85
terms of, 578–81
Series, geometric, 597–99, 601, 602
Set-builder notation, 647
Sets, 615–20, 647
Shanon, xi, 1992
Sharon, xi, 1950
Shifts, graphing functions using, 276–78, 332–33, 344–45
Sidereal year, 388, 389
Sides of equation, 680
Similar triangles, 13n
Simple interest, 196, 366
Simple pendulum, 76, 387–88
Simplex method, 562n
Simpson's Rule, 273
Sine function, hyperbolic, 340

Size reduction problems, 204
Slope
constant of proportionality as, 71
graphing line given point and, 38–39
interpreting, 68
marginal propensity to consume, 69
marginal propensity to save, 70
of parallel lines, 52
of perpendicular lines, 53
of straight line, 35–37
undefined, 35
Slope-intercept form of equation of line, 44–47
Smooth graph, 278
Solution
of equation, 680
to linear programming problem, 564–65
of systems of equations, 471
Solution set, 680
Sound
loudness of, 397–99, 401, 405
measuring distance with, 221
speed of, 253
Sound amplification problems, 404
Sound wave, intensity of, 397, 398–99, 401
Space satellites problems, 337, 349
Special products, 662
Speed. See Velocity
Speed of sound problem, 253
Speed/time/distance problems, 172, 205, 547, 550
Spheres, volume and surface area of, 165
Spring stretching problems, 76
Square(s)
of binomials, 663
difference of two, 662
perfect, 663, 673
Square function, 93, 111
Square matrices, 514, 521
Square root function, 112
Square roots, 212, 653, 673, 674
Square screens, 37–39
Standard deviation, 233
Standard form
of complex number, 206
of equation of circle, 55–59
of polynomials, 655, 661
quadratic equation in, 182
Step function. See Greatest-integer function
Stirling's Formula, 615
Straight line, 35–51
equations of, deriving, 39–47
slope of, 35–37
square screens, 37–39
Strength of beam problem, 172
Stress of materials, measuring, 76
Stretches, 126–28, 276–78
Strict inequalities, 649
Subscript, 654n, 660n
Subset, 616, 647
Substitution, method of, 470, 473–75
back-substitution, 473, 476, 478, 479
to solve systems of nonlinear equations, 539–40, 543–44, 545, 546
Subtraction
of complex numbers, 207
of matrices, 515–17
of polynomials, 661–62
of rational expressions, 669

Sum. *See also* Addition
  of complex numbers, 207
  of infinite geometric series, 597–99
  of two cubes, 663
  of two vectors, 455
Sum function, 135, 136
Summation notation, 584–85
Survey data, analyzing, 618, 620
Survey problems, 639
Suspension bridge problems, 265–67, 271, 420
Sylvester, J.J., 528
Symbol $\left(\dfrac{n}{j}\right)$, 607–10
Symmetry, checking for, 26–31
  with respect to origin, 27–28, 29
    odd function and, 107–10
  with respect to x-axis, 26–29
  with respect to y-axis, 27
    even function and, 107–10
  tests for, 28–31
Synthetic division, 308–12, 320–21, 686–90
Systems of equations, 470–513. *See also* Systems
    of inequalities; Systems of linear equations;
    Systems of nonlinear equations
  consistent vs. inconsistent, 471
  defined, 470
  equivalent, 475–76, 487
  examples of, 470–71
  inverse matrix to solve, 527–28
  solution of, 471
Systems of inequalities, 552–67
  in two variables, 555–60
Systems of linear equations, 472–513
  determinants. *See* Determinants
  matrices. *See* Matrix, matrices
  row operations to solve, 488–95
  three linear equations containing three variables,
    478–80
  two linear equations containing two variables, 472–78
Systems of nonlinear equations, 539–51
  history of, 547–48
  solving, 539–47

TABLE function, 583
Tangency, point of, 63
Tangent line, 63
Tartaglia (Niccolo of Brescia), 315
Telescope
  limiting magnitude of, 404
  reflecting, 420
Temperature conversion problems, 154, 172
Temperature measurement problems, 49, 134, 249
  Newton's Law of Cooling, 359, 379–80, 381–82,
    392–93
Terminal point, 454
Terms
  like, 660
  of polynomial, 654, 661
  of sequences, 578–81
Test number, 235–36
3×3 determinants, 506–8
Time/distance problems, 205
Time value of money. *See* Present value
TRACE function, 23–24, 106, 108–10
Transcendental functions, 328. *See also* Exponential
    functions; Logarithmic functions
Transitive property of inequalities, 222
Transportation cost problems, 166

Transportation problems, 562
Transverse axis, 437–43
  along x-axis, 437–41
  along y-axis, 441–43
  parallel to coordinate axis, 445–47
Tree diagram, 621, 632
Triangle
  area of, 158–59, 272, 656
  congruent, 8
  equilateral, 13, 272
  isosceles, 13, 158–59
  medians of, 13
  Pascal, 608–10, 612–13, 614, 625, 636
  right, 655–56
  similar, $13n$
  Triangle inequality, 249
Trichotomy property, 222
Trinomials, 660, 663
Tripling time for investment, 371–72, 375
Turning points, 280, 281, 282
2×2 determinants, 501–6, 510

Unbounded graph of system of linear inequalities, 560
Uniform motion problems, 198–200, 204, 205, 253
Uninhibited decay, law of, 382
  radioactive decay, 378–79, 381, 382, 391
Uninhibited growth, law of, 376–77, 382, 390
Union of sets, 616–17
Unit, 648
Unit circle, 58
Unit vector, 457, 460–61
Universal set, 617
Upper and Lower Bounds Test, 313–14
Upper bounds to zeros of polynomials, 312–15

Variable(s), 658
  complex, 318
  dependent, 90–91
  domain of, 648
  independent, 90–91, 102
Variation, 71–78
Vectors, 454–64
  adding, 455–56, 459, 460
  applications, 461–63
  components of, 458, 459
  defined, 454
  directed line segments, 454–55
  direction of, 454
  equality of, 458–59
  graphing, 456–57
  history of, 463
  magnitudes of, 454, 457, 459
  multiplying, by numbers, 456–57, 459
  in plane, representing, 457–61
  position, 458
  unit vector, 457, 460–61
Velocity
  relative to air, 461–62, 464
  relative to ground, 461–62, 464
  relative to water, 467
Venn diagrams, 617, 633
Verbal descriptions, translation into mathematical
    expressions, 195
Vertex, vertices
  of cone, 408
  of ellipse, 422
  of graph of system of linear inequalities, 560
  of hyperbola, 437

  of parabola, 409–16
  of quadratic function, 257–62
Vertical asymptote, 288, 289–90
Vertical line, 35, 40–41
Vertical-line test, 92
Vertical multiplication of polynomials, 662
Vertical shifts, 123–26
Vibrating strings problems, 81
Viète, François, 191
Viewing rectangle (window), 3, 37–38
Volume
  of cone, 163
  constructing function expressing, 157–58
  of cube, 656
  of cylinder, 163
Volume formula, 656

Wage computation problems, 196, 202
Weather satellite problems, 63
Weierstrass, Karl, 231
Weight of body problems, 76
Whispering galleries, 431–32, 435, 467
Window (viewing rectangle), 3, 37–38
Wind speed, computing, 482

x-axis, 2
  major axis along, 422–26
  reflection about, 128
  symmetry with respect to, 26–29
  transverse axis along, 437–41
x-coordinate (abscissa), 2
x-intercept, 21
  Intermediate Value Theorem and graphing utility
    to locate, 175
  multiplicity of, 279–80, 281, 282
  of quadratic functions, 260–61
  real zeros of polynomial function as, 278, 279
Xmax, 3
Xmin, 3
Xscl, 3
xy-plane, 2
  quadrants of, 3

Yang Hui, 614
y-axis, 2
  major axis along, 426–28
  reflection about, 129
  symmetry with respect to, 27, 29, 107–10
  transverse axis along, 441–43
y-coordinate (ordinate), 2, 136–37
y-intercept, 21
Ymax, 3
Ymin, 3
Yola, xi, 1966
Yscl, 3

Zero, 648
Zero-coupon bond, computing value of, 370–71,
    374–75
Zero function, 274
Zero-level earthquake, 399
Zero matrix, 516
Zero of multiplicity m of f, 278–80, 281, 282, 321
Zero polynomial, 654, 661
Zeros of polynomial functions. *See under* Polyno-
    mial functions
ZOOM-IN function, 24–25
ZOOM-OUT function, 20–21

Conics

Parabola

$$y^2 = 4ax \qquad\qquad y^2 = -4ax \qquad\qquad x^2 = 4ay \qquad\qquad x^2 = -4ay$$

Ellipse

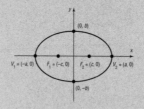 

$$\frac{x^2}{a^2} + \frac{y^2}{b^2} = 1, \quad c^2 = a^2 - b^2 \qquad\qquad \frac{x^2}{b^2} + \frac{y^2}{a^2} = 1, \quad c^2 = a^2 - b^2$$

Hyperbola

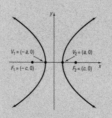

$$\frac{x^2}{a^2} - \frac{y^2}{b^2} = 1, \quad c^2 = a^2 + b^2 \qquad\qquad \frac{y^2}{a^2} - \frac{x^2}{b^2} = 1, \quad c^2 = a^2 + b^2$$

$$\text{Asymptotes:} \quad y = \frac{b}{a}x, \quad y = -\frac{b}{a}x \qquad\qquad \text{Asymptotes:} \quad y = \frac{a}{b}x, \quad y = -\frac{a}{b}x$$